FIFTH EDITION

MODERN CONCEPTS IN BIOCHEMISTRY

Robert C. Bohinski

JOHN CARROLL UNIVERSITY

ALLYN AND BACON, INC.

BOSTON ◇ LONDON ◇ SYDNEY ◇ TORONTO

Composition buyer: Linda Cox
Manufacturing buyer: Ellen Glisker
Cover administrator and designer: Linda K. Dickinson
Editorial-production service: Woodstock Publishers'
    Services
Copy editor: Carol Beal
Text designer: Deborah Schneck

Library of Congress Cataloging-in-Publication Data

Bohinski, Robert C.
  Modern concepts in biochemistry.

  Includes bibliographies and index.
  1. Biological chemistry.   I. Title.
QP514.2.B63   1987   574.19′2   86–14646
ISBN 0–205–08852–X
ISBN (international) 0–205–10555–6

Printed in the United States of America.
10 9 8 7 6 5 4 3 2   91 90 89 88 87

CONTENTS IN BRIEF

CONTENTS

Chapter Five
ENZYMES   173

Chapter Six
NUCLEOTIDES AND NUCLEIC ACIDS   227

Chapter Seven
TRANSCRIPTION: RNA BIOSYNTHESIS   289

Chapter Eight
DNA BIOSYNTHESIS, REPAIR, AND
RECOMBINATION   323

Chapter 9
TRANSLATION: PROTEIN BIOSYNTHESIS   349

PREFACE

Since modern biochemistry is a rapidly evolving field of study, it is critical to employ the most up-to-date teaching materials in general biochemistry courses. The Fifth Edition of *Modern Concepts in Biochemistry* has been painstakingly crafted to reflect contemporary interests and research. Coverage of nucleotides, nucleic acids, genes, and gene activity has been expanded to four chapters. There are now separate chapters on transcription, DNA biosynthesis, and translation, in addition to the chapter on nucleic acids and nucleotides. There is greater emphasis on a number of modern topics, such as biomembranes, membrane receptors, neurotransmitters, growth factors, peptide hormones, non-protein biological catalysts, recombinant DNA, retroviruses, receptor-mediated endocytosis, and regulation of protein biosynthesis.

I have refined the pedagogy that has always distinguished the book. Clinical and medical applications are presented within the context of biochemical principles, not as separate topics. For example, antibodies and interferons are discussed in the protein chapter. Atherosclerosis, cholesterol and LDL receptors are discussed in the chapter on lipid metabolism. Presenting applications and examples in this way enhances and reinforces understanding of biochemical principles and provides immediate motivation for students from a variety of backgrounds.

Instrumental and laboratory techniques are an important part of biochemistry; therefore, a number of methods discussions, separated from the main body of the text, are provided throughout the early chapters. These discussions offer students the opportunity to become acquainted with specific techniques like ion exchange chromatography, electrophoresis, and isoelectric focusing, among others. I hope this added feature will demonstrate the manner in which many important biochemical research questions are addressed.

As always, my goal has been to write a textbook particularly suited to the needs of students who are beginning their study of biochemistry. To this end the more quantitative aspects of the discipline are carefully developed. For example, in the material covering enzyme kinetics, I provide worked-out examples in addition to the derivations of equations. Key terms are highlighted, and many marginal notes facilitate student understanding. References at the end of each chapter have been updated; problem sets have been revised and expanded. A solutions manual containing detailed solutions to all end-of-chapter problems is available.

I welcome comments and suggestions for improvements from students and faculty.

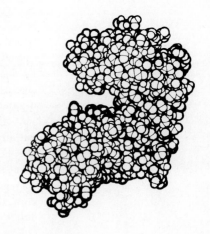

INTRODUCTION

Chemistry—the science dealing with the properties of matter and the changes that matter can undergo—is loosely divided into several overlapping areas or fields: organic chemistry, inorganic chemistry, physical chemistry, analytical chemistry, and others. As its name obviously implies, biochemistry is concerned with the *chemistry of life* in all its forms—animals, plants, bacteria—and the life-infecting viruses. The reason for highlighting the classical divisions of chemistry is to emphasize the broad base of biochemistry, which is part (largely) organic, part inorganic, part physical, and part analytical.

Biochemistry began about a hundred years ago with the realization—initially by only a small number of scientists, then by many, and eventually by all—that life processes involved phenomena that could be explained by the exact sciences of chemistry and physics. The early discoveries were mostly of a very general nature. Owing largely to several pioneering individuals with the creative and imaginative genius to ask the right questions, and even to labor under criticism of colleagues, more particulars of the living state were uncovered. With time (primarily in the past fifty years) the discipline of biochemistry grew and matured and eventually gained respectability and acceptance by both chemists and biologists. (Anyone interested in the history of biochemistry should find the sources listed at the end of the chapter to be very helpful.) Today, biochemistry is a recognized major area of chemistry and is to biology what mathematics is to physics.

The growth of biochemistry has been extensive and expansive, so much so that numerous areas are recognized: bioenergetics, molecular biology, membrane biochemistry, protein biochemistry, plant biochemistry, neurobiochemistry, analytical biochemistry, and numerous others, as well as subspecialties of these divisions. The current annual publication of tens of thousands of research papers in numerous journals, of annual review sources, and of specialized monograph sources reflect the growth and scope of modern biochemistry. Compare the sizes of the three editions of the *Handbook of Biochemistry and Molecular Biology.* The first edition (1968) contained 950 pages; the second edition (1970) grew to 1600 pages; the third edition (1976) exploded into eight volumes having a total of 5000 pages. An updated fourth edition has not been published in ten years, suggesting that the mass of information now available has expanded the feasibility of packaging it into a handbook.

The objective of this book and the course you are taking is not to examine in intricate detail all facts known about all living organisms. The immense biochemical diversity of our biosphere prohibits any such undertaking. Although we will occasionally highlight matters that relate to this diversity, for the most part this book covers concepts, principles, and factual material representative of the remarkable *biochemical unity* exhibited by all forms of life, from bacteria to humans.

I–1 FOCUS ON MOLECULES

All living organisms are composed of chemical substances from both the inorganic and the organic classes. The first tenet of biochemical unity we want to highlight is the following:

All organisms are composed of the same types of substances, in roughly the same proportions, performing the same types of general functions.

In fact, in all organisms one specific inorganic material—**water**—is the most important substance for life.

The major classes of organic substances—*proteins, nucleic acids* (see Note I–1), *carbohydrates,* and *lipids*—are often called biomolecules.

Most are constructed from only six elements, all nonmetals: oxygen, carbon, hydrogen, nitrogen, phosphorus, and sulfur.

The study of these essential organic biomolecules is what biochemistry is all about. How are they synthesized in a living cell? How are they degraded in a cell? How is their degradation linked to the production of useful chemical energy within a cell? How is this energy produced? How are these substances interconverted? How do they enter and leave a cell? How can they be isolated in the laboratory? How can they be synthesized in the laboratory? What is their molecular structure? What specific functions do they perform? What is the explanation for how they function in terms of their molecular structure? How are the genes of chromosomes controlled to regulate the biochemical individuality of an organism? How can the genetic information of an organism be modified by laboratory manipulation? How do cells communicate with each other on a chemical basis? What are the explanations of the abnormal state, that is, the disease state, including mental illnesses? How do antibiotics, antiviral agents, psychotropic drugs, immunosuppressive agents, and other drugs work? What biochemical distinctions exist between normal and cancer cells? What biochemical processes explain the transformation of a normal cell to a cancer cell? How can medical diagnosis be improved by using precise biochemical criteria? How do hormones regulate the activities of an organism, and in what other ways are cells capable of self-regulation? The combination of your instructor and this book should help you to understand some of what we know about these and many other questions regarding the chemical nature of the living state.

Returning to the chemical composition of living cells, your attention is directed to Table I–1. Although the data correspond to the estimated composition of the unicellular bacterium *Escherichia coli,* the values are not greatly different for other organisms, including the human body. However, on a cellular level the composition of an *E. coli* cell is quite different from that of a liver cell, which in turn is different from that of a fat cell of adipose tissue, and so on.

Some other interesting generalizations can be made from the data of Table I–1.

NOTE I–1
There are two types of nucleic acids: DNA, or deoxyribonucleic acid, and RNA, or ribonucleic acid.

1. Cells contain a greater variety of proteins than any other type of material.
2. Approximately 50% of the solid matter of a cell is protein (15% on a wet-weight basis).
3. Cells contain many more protein molecules than DNA molecules.
4. The largest biomolecule in a cell is DNA.

5. *E. coli* and similar bacteria contain only a single unique chromosome, that is, one unique DNA molecule. (The relationship of one chromosome–one DNA means that each human diploid cell contains 46 different DNA molecules in the nucleus.)

6. About 99% of the molecules in a cell are H_2O molecules.

In view of the last item, we can say that the science of biochemistry is concerned with only 1% of the molecules in a living cell. Keep in mind, however, that the specific molecules comprising this 1% and the totality of chemical events in which they participate are unique for each cell.

Although the study of biochemistry emphasizes organic substances, the inorganic materials (frequently termed *minerals*) are also important. Indeed,

TABLE I–1 Approximate chemical composition of a rapidly dividing cell of *E. coli*

MATERIAL	PERCENT OF TOTAL WET WEIGHT[a]	AVERAGE MOLECULAR WEIGHT (g/mole)	APPROX. NUMBER OF MOLECULES PER CELL[b]	DIFFERENT KINDS OF MOLECULES PER CELL[b]
Water	70	18	40 billion	1
Proteins	15	10,000–100,000	2–3 million	2000–3000
Nucleic acids (DNA and RNA total 7%)				
DNA	1	2.5 billion[c]	2 or 4	1
RNA[d]	6			
5S ribosomal RNA		40,000[c]	30,000	1
16S ribosomal RNA		500,000[c]	30,000	1
23S ribosomal RNA		1 million[c]	30,000	1
Transfer RNA		25,000	400,000	40
Messenger RNA		10^5–10^6	100,000	1000
Carbohydrates and metabolites	3	150 (excluding polymers)	200 million	200
Lipids and metabolites	2	750	25 million	50
Inorganic ions (minerals)	1	40	250 million	20
Amino acids and metabolites	0.8	120	30 million	100
Nucleotides and metabolites	0.8	300	12 million	200
Others	0.4	150	15 million	200

This *chromosome* determines the biochemical individuality of ⟶ *E. coli.*

Source: Adapted with permission from James D. Watson, *Molecular Biology of the Gene*, 2nd ed. (Philadelphia: Saunders, 1972).
[a] Approximate values apply to the majority of living organisms existing in a normal state.
[b] Numbers for some materials vary considerably from bacterial cells to cells of higher organisms.
[c] Corresponding substances in cells of higher organism are larger in size.
[d] Other species of RNA are found in animal and plant cells.

several minerals are essential nutrients for all organisms and must be supplied in the diet or natural surroundings. The inorganic elements are present as ionic forms (see Table I–2), existing as free ions or complexed to an organic grouping. At this time we will not itemize the many and varied processes in which they function. Instead, we simply make the following statement, which will become clear later.

> The vital roles of many inorganic ions are associated with their effect on the activity of proteins.

All of the minerals listed in Table I–2 are essential nutrients for humans and, with the exception of iodine, for all other organisms as well. Trace amounts of about a dozen additional elements have been detected in various organisms. However, the importance of many trace elements has yet to be clarified, so research in nutrition is still very active. For example, scientists recently established that nickel (Ni), silicon (Si), and selenium (Se) are essential elements in trace quantities for several organisms, including humans.

TABLE I–2 Partial listing of inorganic substances of biological importance

ELEMENT (ionic form)
Calcium (Ca^{2+})
Chlorine (Cl^-)
Cobalt (Co^{2+})
Copper (Cu^+, Cu^{2+})
Iodine (I^-)[a]
Iron (Fe^{2+}, Fe^{3+})
Magnesium (Mg^{2+})
Manganese (Mn^{2+})
Molybdenum (Mo^{6+})
Phosphorus ($H_2PO_4^-$, HPO_4^{2-})
Potassium (K^+)
Sodium (Na^+)
Sulfur (SO_4^{2-}, S^{2-})
Zinc (Zn^{2+})

[a] Only in vertebrates

I–2 PROTEINS AND NUCLEIC ACIDS: SPECIAL BIOMOLECULES

Although all types of biomolecules are important, nucleic acids (DNA and RNA) and proteins are especially important. They are important because both types of molecules are *informational*, carrying in their chemical structure the instructions that determine what goes on in a cell. The information is stored and replicated in *chromosomes*, which contain *genes* (from the Greek *genos*, meaning "hereditary descent"). A chromosome is a deoxyribonucleic acid (DNA) molecule, and the individual genes are segments of the intact DNA molecule. The number of genes in the 23 pairs of chromosomes (46 DNA molecules) of each human somatic cell is several thousand. When a cell replicates, identical copies of the DNA molecules are produced, and the hereditary line of descent is insured.

Thirty years ago knowledge of DNA was confined primarily to scientists and students in higher education. Today, sixth-grade students learn about it. There is occasional news coverage in newspapers, weekly magazines, and on television. Business, finance, government, investment interests, and environmentalists are also paying attention because of the commercial marketing of products produced by the new methods of genetic engineering.

Generally speaking, *genetic information* refers to any one of several types of instructions carried in the structure of DNA that direct the occurrence of nearly all chemical reactions in the cell.

> Most of this genetic information provides the instructions for the assembly of every individual protein in the cell.

Each protein, unique in its own structure and hence unique in its function, then participates in the processes that characterize the individuality of the cell and the entire organism. The functions performed by proteins are many (see

p. 114), but the most important proteins are those that serve as *enzymes*—catalysts that permit reactions to occur at a fast rate. Virtually every reaction in a living cell requires an enzyme, and most reactions require a specific enzyme.

Protein and nucleic acid molecules are *polymers* (from the Greek *poly,* meaning "several," and *mer,* meaning "unit")—large compounds of high molecular weight. A polymer is composed of smaller low–molecular weight compounds, called *monomers,* successively linked to each other by covalent bonding. Think of a polymer as a chain and the links of the chain as the monomers. In a protein the monomer units are *amino acids;* in a nucleic acid the monomers are *nucleotides.* In later chapters we will examine these biopolymers in greater detail. For now, you need develop only a general understanding of their informational character. So let us consider the following points in conjunction with the displays in Figures I–1 and I–2.

1. Proteins and nucleic acids are *heterogeneous polymers,* meaning that each monomer is not identical [conveyed in Figure I–1(a) by different colors and shading]. In proteins there may be as many as 20 different amino acids, whereas nucleic acids are composed of only 4 different nucleotides. The monomer composition is a characteristic of any polymer. However:

> The specific sequence of the monomers determines the unique individuality of the polymer.

2. From training in other courses you may already know that the expression of genetic information involves the use of the nucleotide sequence in a gene segment of DNA as a set of coded instructions to direct the assembly of a specific amino acid sequence in a protein. A display of that correlation is shown in Figure I–2(a). The complete pattern for the flow of information is commonly represented as $DNA \rightarrow RNA \rightarrow protein$. This symbolism signifies that the nucleotide sequence in a gene of DNA specifies the assembly of a nucleotide sequence in a messenger RNA molecule, which in turn directs the assembly of the amino acid sequence in the protein. The study of the biochemistry of these and related processes is called *molecular biology.*

3. The three-dimensional shape of a protein or a nucleic acid molecule does not correspond to a completely straight line. Rather, the actual shape is due to the way the chain folds, twists, and turns. Chains may also associate with and intertwine with other chains. We will examine all of this structure later. The point here is as follows:

The individual functions performed by these biopolymers are a result

FIGURE I–1 Monomer composition and sequence in biopolymers. Spheres represent monomer units of structure linked to each other by covalent bonding. The value of *n* specifies the size of the polymer, ranging from 50 to 1000 for proteins and 90 to 1 million for nucleic acids.
(a) Heterogeneous polymer, which contains different amounts of different monomers linked in a specific informational sequence.
(b) Homogeneous polymer, which is composed of identical monomer units and thus lacks sequence information.

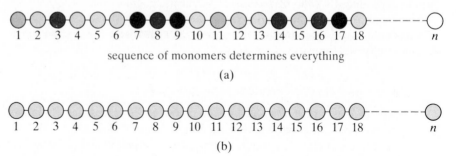

sequence of monomers determines everything

(a)

(b)

of this overall three-dimensional structure [see Figure I–2(b)], which in turn is governed by the sequence of monomeric units.

This is the sense in which the sequence of monomers is informational.

4. Another type of naturally occurring biopolymer is the *homogeneous polymer,* composed of identical monomeric units [Figure I–1(b)]. Substances of this type are represented by some of the polymeric carbohydrates such as cellulose, starch, and glycogen. If you grasp the significance of the principles stated above, you should understand why, in contrast to proteins and nucleic acids, homogeneous molecules are noninformational.

This description of the structure of biopolymers merely scratches the surface and is intended only as an introduction to some important basic concepts and terminology.

I–3   CONCLUSION OF INTRODUCTION— BEGINNING OF LEARNING

Biochemistry is a vast subject. Indeed, it seems overwhelmingly vast even to those of us who carry the label of professional biochemist—and particularly to authors of biochemistry textbooks. Don't let yourself be intimidated by it. You may find it difficult, but you will also discover that it is a most fascinating and exciting scientific discipline. I am confident you will be satisfied that you decided to study biochemistry. Perhaps some of you may even be stimulated to pursue further study.

Here are some study tips and other thoughts. A lot of your study will require memorization. Indeed, you must become as familiar with certain things—such as the names, the structures, and the characteristics of the amino acids—as you are with the alphabet; without hesitation you should be able to rattle off the events of the major metabolic pathways. But do not equate your understanding of the subject with how much and how well you have memorized.

To truly appreciate and understand things, strive to recognize the *logic that permeates the principles of biochemistry.* For example, you will discover that (1) establishing the *relationship of function to structure* (that is, relating the work of a molecule or multimolecular aggregate to its composition and arrangement) and (2) gaining an understanding of the operation of various *regulatory mechanisms* by which a cell or an organism controls its own activities are *two central themes* of modern biochemistry. Neither of these topics can be learned by mere rote memorization. Rather, on the basis of other principles, you should strive to develop an understanding of the rationale for such phenomena in much the same sense as you can appreciate and understand why a basketball is round instead of square, or why each lock has its own key, or why a right hand will fit snugly only in a right-handed glove. There are reasons for things, and in many instances the reasons are fundamentally simple ones.

Do a lot of work with pencil and paper. Solve problems. Design your own problems and exchange them with other students. Talk biochemistry with your associates. Finally, you should anticipate that topics in later chapters are

GGGGGGGGGGGGGGGGTCTTGCTGCGCCTCCGCCTCCTCCTCTG
　　　　　　　　　　　20　　　　　　　　　　　　　　　　　　　　　　　　　　*40*

CTCCGCCACCGGCTTCCTCCTCCTGAGCAGTCAGCCCGCGCG
　　　　　　　　　60　　　　　　　　　　　　　　　　　　　　　　　*80*

　　　　　　　　　　　　　　　met　ala　thr　arg　ser　pro　gly　val　val
CCGGCCGGCTCCGTT ATG GCGACCCGCAGCCCTGGCGTCGTG
　　　　　　　　　100　　　　　　　　　　　　　　　　　　　　　*120*

ile　ser　asp　asp　glu　pro　gly　tyr　asp　leu　asp　leu　phe　cys
ATTAGTGATGATGAACCAGGTTATGACCTTGATTTATTTTGC
　　　　　　　　140　　　　　　　　　　　　　　　　　*160*

ile　pro　asn　his　tyr　ala　glu　asp　leu　glu　arg　val　phe　ile
ATACCTAATCATTATGCTGAGGATTTGGAAAGGGTGTTTATT
　　　　　　　　180　　　　　　　　　　　　　　　*200*

pro　his　gly　leu　ile　met　asp　arg　thr　glu　arg　leu　ala　arg
CCTCATGGACTAATTATGGACAGGACTGAACGTCTTGCTCGA
　　　　　　220　　　　　　　　　　　　　　*240*

asp　val　met　lys　glu　met　gly　gly　his　his　ile　val　ala　leu
GATGTGATGAAGGAGATGGGAGGCCATCACATTGTAGCCCTC
　　　260　　　　　　　　　　　　　*280*

cys　val　leu　lys　gly　gly　tyr　lys　phe　phe　ala　asp　leu　leu
TGTGTGCTCAAGGGGGGCTATAAATTCTTTGCTGACCTGCTG
　　　300　　　　　　　　　　　　*320*

asp　tyr　ile　lys　ala　leu　asn　arg　asn　ser　asp　arg　ser　ile
GATTACATCAAAGCACTGAATAGAAATAGTGATAGATCCATT
　340　　　　　　　　　　　　*360*

pro　met　thr　val　asp　phe　ile　arg　leu　lys　ser　tyr　cys　asn
CCTATGACTGTAGATTTTATCAGACTGAAGAGCTATTGTAAT
　　　　　　　　　　　400　　　　　　　　　　　　　　*420*

asp　gln　ser　thr　gly　asp　ile　lys　val　ile　gly　gly　asp　asp
GACCAGTCAACAGGGGACATAAAAGTAATTGGTGGAGATGAT
　　　　　　　　　440　　　　　　　　　　　　*460*

leu　ser　thr　leu　thr　gly　lys　asn　val　leu　ile　val　glu　asp
CTCTCAACTTTAACTGGAAAGAATGTCTTGATTGTGGAAGAT
　　　　　　　480　　　　　　　　　　　*500*

ile　ile　asp　thr　gly　lys　thr　met　gln　thr　leu　leu　ser　leu
ATAATTGACACTGGCAAAACAATGCAGACTTTGCTTTCCTTG
　　　　　　520　　　　　　　　　　　*540*

val　arg　gln　tyr　asn　pro　lys　met　val　lys　val　ala　ser　leu
GTCAGGCAGTATAATCCAAAGATGGTCAAGGTCGCAAGCTTG
　　　　　　560　　　　　　　　　　　*580*

leu　val　lys　arg　thr　pro　arg　ser　val　gly　tyr　lys　pro　asp
CTGGTGAAAAGGACCCCACGAAGTGTTGGATATAAGCCAGAC
　　　　600　　　　　　　　　　*620*

phe　val　gly　phe　glu　ile　pro　asp　lys　phe　val　val　gly　tyr
TTTGTTGGATTTGAAATTCCAGACAAGTTTGTTGTAGGATAT
　　　640　　　　　　　　　*660*

ala　leu　asp　tyr　asn　glu　tyr　phe　arg　asp　leu　asn　his　val
GCCCTTGACTATAATGAATACTTCAGGGATTTGAATCATGTT
　　　680　　　　　　　*700*

cys　val　ile　ser　glu　thr　gly　lys　ala　lys　tyr　lys　ala　***
TGTGTCATTAGTGAAACTGGAAAAGCAAAATACAAAGCCTAA
　　720　　　　　　　　　*740*　　　*750*

GATGAGAGTTCAAGTTGAGTTTGGAAACATCTGGAGTCCTAT
　　760　　　　　　　　　*780*

(a)

8

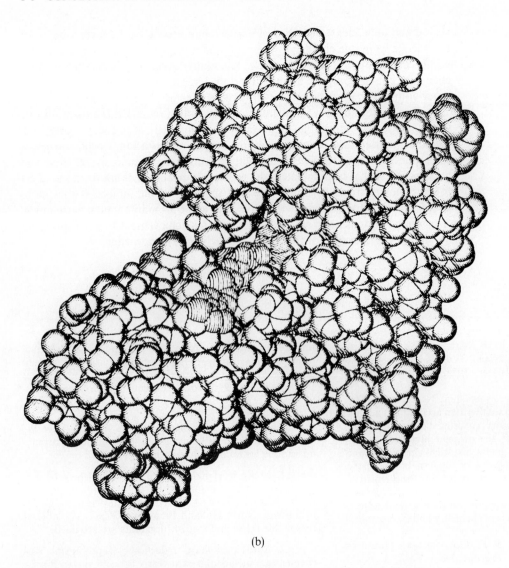

(b)

◀ **FIGURE I–2** Informational biomolecules. (a) Display of **two-dimensional sequence information** in a piece of DNA carrying a *gene coding for a protein*. Nucleotides in DNA are designated as A, G, T, and C and appear in color. Amino acids in the protein are designated with lowercase three-letter abbreviations. The DNA code is read in groups of three nucleotides, each called a triplet, beginning with the start triplet ATG at position 100–102 (underlined) coding for the amino acid methionine (met). The triplet sequence TAA (754–756, shown by three asterisks) signals the end of the coding frame, after specifying 218 amino acid positions. The actual gene depicted codes for the enzyme hypoxanthine phosphoribosyltransferase (HPRT). This particular gene is a candidate for the first attempt to correct a human genetic disease by inserting the gene into an individual suffering from the devastating Lesch-Nyhan disease (see p. 708). (b) Computer-generated graphic of the **three-dimensional structure of a protein molecule** determined by the amino acid sequence, which in turn is determined by the nucleotide sequence in a gene. The protein depicted is not HPRT; it is an enzyme called hexokinase. [*Source:* Sequence display reproduced with permission from D. J. Jolly et al., *Proc. Natl. Acad. Sci., USA,* **80,** 477–481 (1983). Computer graphic provided by T. A. Steitz.]

very much dependent on topics of earlier chapters. Strive to carry over your knowledge from one chapter to the next, and always refer back for review and reinforcement of your understanding. The many topics of an introductory biochemistry course comprise an integrated package, perhaps more so than in any other course you have taken or will take.

The modern history of the human race is full of remarkable and admirable (and some not so admirable) achievements in the areas of science, engineering, medicine, agriculture, communications, space exploration, social progress, economics, and the arts. The future will certainly record many more. However, of all progress that is anticipated, the greatest impact may result from the quest to learn how a living organism operates in terms of the molecules that compose it and then to manipulate the molecules and thus manipulate the organism. A profound statement to say the least, but this achievement is a real prospect for the future. The core of this quest is biochemical knowledge.

LITERATURE

FLORKIN, M., and E. H. STOTZ, eds. *A History of Biochemistry*. Volumes 30 through 33 of *Comprehensive Biochemistry*. New York: Elsevier, 1973. A most thorough treatment.

FRUTON, J. S. *Molecules and Life*. New York: Wiley, 1972. An authoritative and well-written account of scientific developments in biochemistry covering the period 1800–1950. Many excerpts from the original literature are included. A good source to gain a historical perspective and appreciation of the foundations of modern biochemistry.

FRUTON, J. S. "The Emergence of Biochemistry." *Science*, **192**, 327–334 (1976). An interesting discussion of the growth of biochemistry focusing on several aspects of the interplay and conflicts of chemistry and biology since 1800.

GREEN, D. E., and R. F. GOLDBERGER. *Molecular Insights into the Living Process*. New York, London: Academic Press, 1967. This book represents a unified picture of biochemistry from the standpoint of universal principles that apply to all living organisms. Best read after you know some biochemistry.

KARLSON, P. "From Vitalism to Intermediary Metabolism." *Trends Biochem. Sci.*, **1**, N184 (1976). A brief historical article.

The Molecules of Life. An entire issue of *Scientific American* (October 1985) containing outstanding articles on various topics of molecular biology.

1. Many brief items concerning the history of biochemistry and the persons involved are routinely included in the journal *Trends in Biochemical Sciences*, published by Elsevier/North-Holland Biomedical Press. Available at reasonable rates, and student discounts are provided.

2. Autobiographical reflections of prominent biochemists are included in each yearly issue of *Annual Review of Biochemistry*. They provide interesting reading.

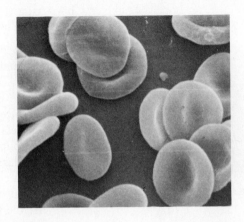

CHAPTER ONE

CELLULAR ORGANIZATION

That our biosphere contains a tremendous variety of cells is common knowledge. For example, in the advanced multicellular (many-celled) organisms of the animal and plant kingdoms, we can identify liver cells, retina cells, nerve cells, brain cells, red blood cells, white blood cells, leaf cells, stem cells, root cells, and several more. Considering just the microbial world, which is composed of both unicellular (single-celled) and multicellular organisms, we see that there are countless individual cell types.

Despite the diversity and individuality, there are among living cells many gross similarities at the level of both cell structure and cell function. Thus we can have a coherent and meaningful discussion of cells according to a general format that highlights elements of cellular unity.

At the functional level the most fundamental similarity is that any cell, sustained by the ingestion and utilization of nutrients and energy from the external environment, is capable of growth and cell division.

> Because the intact cell is the smallest entity capable of these processes, the cell is termed the unit of life. In other words, all living organisms have a cellular structure, and the activity of the whole organism is the result of the individual and collective activities of cells.

The first formal statements of this *cellular theory of life* were not made until 1838 by M. J. Schleiden and T. Schwann, and then again in the late 1850s by R. Virchow. Now in the twentieth century, particularly during the past 30 years, the cellular basis of life has been extrapolated to the level of molecules.

> Thus in molecular terms a cell is what it is and does what it does because it is composed of a characteristic set of protein molecules, RNA molecules, and DNA molecules, with the DNA containing the genes that determine what proteins and RNAs are produced.

Nevertheless, the cell is regarded as the basic unit of life, since it is only at the cellular level that all of the basic characteristics of life are expressed.

Starting at the level of structure, we must realize that a living cell is not an amorphous, indivisible molecular blob. On the contrary, the cell is a highly organized entity consisting of separate and distinguishable parts, each performing important functions in the overall living process. In other words, the cell is like any other machine, with the operation of the complete unit being the result of the combined and integrated operation of its individual parts. In this sense the cell is a complex unit of life. Depending on the type and source of the cell, the degree of complexity is quite variable.

This chapter surveys the types of living cells according to their ultrastructural features. On this basis biologists have segregated living cells as belonging to either of two groups, *prokaryotes* and *eukaryotes,* suggested to have evolved in that order. We will also examine a recent development in research that may radically modify this classification—namely, the possible existence of a third form of life, the archaebacteria.

◇ L A B O R A T O R Y   M E T H O D S ◇

ELECTRON MICROSCOPY

The unaided, normal human eye is capable of observing two or more objects as separate objects as long as the distance between them is about 0.1 mm, or 100 micrometers (100 μm). Below this limit of *resolving power,* the eye is incapable of any magnification.

However, through various types of microscopy we have been able to enormously extend our observations of small objects. A microscope provides both increased resolving power and magnification of the specimen. The effective resolving power varies from one type to another. For example, the best conventional light microscope cannot distinguish between objects that are closer than about one-half the wavelength of illuminating light. Thus with the most perfectly ground lenses and with ordinary white-light illumination, having an average wavelength of 5000 Å (see Note 1–1), the resolution of the light microscope is about 2500 Å, or 0.25 μm. With ultraviolet illumination (of shorter wavelength) and special quartz lenses, a resolution of about 0.17 μm can be achieved. Obviously, both are better than the unaided eye. Indeed, intact living cells with a diameter ranging from about 1 to 20 μm can be seen, the larger ones in considerable detail. However, many still remain invisible, and many details are fuzzy.

◇ **NOTE 1–1**

$$1 \text{ Å} = 10^{-8} \text{ cm} = 10^{-10} \text{ m}$$

(Å is the symbol for angstrom, a unit of length), and

$$1 \text{ } \mu\text{m} = 10^{-6} \text{ m}$$

Therefore

$$1 \text{ Å} = 10^{-4} \text{ } \mu\text{m}$$

or in nanometers

$$1 \text{ Å} = 10^{-1} \text{ nm}$$

Enormous resolving power is provided by the electron microscope. The basic principle of the electron microscope (see Figure 1–1) is the same as that of any microscope—an object is illuminated by radiation and an enlarged image of the interaction is produced. In the electron microscope, however, the radiation does not consist of typical light waves but, rather, is composed of *rays of high-speed electrons. A magnetic condenser lens is used to focus the electron beam on the specimen.* The specimen absorbs some electrons passing through it and scatters others. Electrons that do pass through are focused by magnetic objective lenses on an electron-sensitive photographic film, providing a magnified image of the specimen. The photograph of the image is called an *electron micrograph.*

Transmission Electron Microscope

The preceding description of recording a specimen image by observing the pattern of radiation passing through the specimen is called *transmission microscopy.* Most of the micrographs in this chapter are transmission micrographs. The specimen is prepared as a thin section (a few angstroms thick) cut with an ultramicrotome. The procedure involves first fixing the specimen (dipping it in a solution of glutaraldehyde, for example, to render protein components insoluble), then dehydrating the specimen, and finally embedding the specimen in a plastic block, which is then sliced in the ultramicrotome. For improved contrast the sample may also be stained or shadowed with heavy electron-dense substances such as tungsten salts, osmium salts, or platinum. Consult one of the references for details concerning the various modes of sample preparation. (A special technique, called *freeze etching,* is described in Section 11–5.) Whatever the procedures, by the time the sample is prepared for viewing, it has undergone some rather harsh treatments. Thus there is always the question about how real the image is or whether artifacts produced during preparation are being observed.

As with the ordinary light microscope, the resolving power of the transmission electron microscope is dependent on the wavelength of the incident radiation,

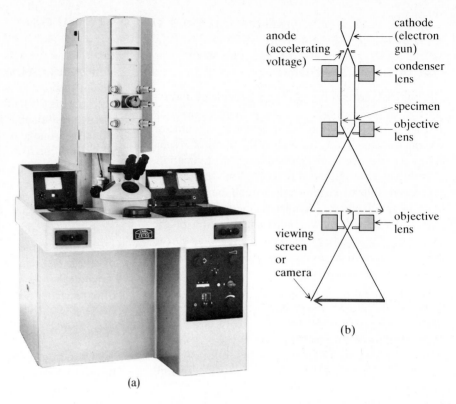

FIGURE 1–1 Electron microscope. (a) Model EM–9S–2 electron microscope manufactured by Carl Zeiss, Inc. (b) Diagrammatic outline of the major components of a transmission electron microscope. The accelerated beam of electrons is focused on the specimen (arrow) by a magnetic field (condenser lens). A series of two objective lenses (each a magnetic coil, also) focuses and magnifies the emerging beam onto an electron-sensitive screen or photographic plate, which depicts the image after development. The entire column assembly is maintained under a very high vacuum. (*Source:* Photograph provided by Carl Zeiss, Inc., New York.)

(a)

(b)

which in this case is determined by the voltage used to accelerate the electrons. For example, if the electron beam is produced under a potential (V) of 50,000 volts, the approximate wavelength (λ) of a ray of high-speed electrons is

$$\lambda \text{ (in Å)} = \frac{12.3}{\sqrt{V}} = 0.05 \text{ Å}$$

and the resolving power (estimated by $\frac{1}{2}\lambda$) is about 0.025 Å. If the instrument were 100% efficient, we would be able to see individual hydrogen atoms. Because of technical difficulties in instrument design, however, this degree of resolution cannot be achieved. The best resolution that can be delivered is about 2 Å, or 0.2 nm. This resolution is still fantastic and is sufficient to show objects in tremendous detail (exemplified by the micrographs displayed in this chapter), including tiny viruses and even some molecules.

Scanning Electron Microscope

Images displaying some *three-dimensional character of surface details* of a specimen can be obtained by scanning electron microscopy (SEM). An SEM image of a red blood cell is shown in Figure 1–2.

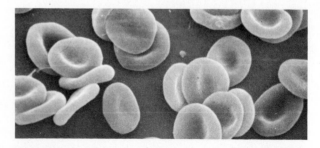

FIGURE 1–2 Image from scanning electron microscopy. The concave, disk shape of human red blood cells (erythrocytes) is shown here. (*Source:* Micrograph provided by Dr. Wray H. Huestis.)

In SEM an electron beam is finely focused to strike a tiny location on the specimen. As electrons strike the surface of the specimen, they cause secondary electrons to be ejected from the surface. A television camera is used to detect the pattern of these secondary electrons as the specimen is moved relative to the finely focused electron beam. The resolving power of a scanning electron microscope is greater than that of a conventional light microscope but less than that of a transmission electron microscope.

The SEM techniques have been recently perfected to obtain three-dimensional images of intracellular structures. Two remarkable examples are displayed later in the chapter (Figures 1–8 and 1–9).

1–1 PROKARYOTES AND EUKARYOTES: TWO LIFE FORMS

Scientific evidence suggests that the first living organisms—small unicellular bacteria—began to evolve about 3.5 billion years ago. (*Note:* The age of the earth is estimated at 4.8 billion years.) Cells in the bacterial kingdom exhibit a simple structural organization: a *cytoplasm* (the intracellular plasma fluid) bounded by a flexible *cell membrane* and a rigid *cell wall*, and sometimes a second outer membrane. This type of simple cell is called a *prokaryote*. It is distinguished from a larger, more complex, and later-appearing (about 1.5 billion years ago) cell type, called a *eukaryote*. The interior of a eukaryotic cell contains several distinct subcellular structures. These structures are entirely membranous or are organized compartments surrounded by a membrane. The most conspicuous (large-size) compartment is the *nucleus,* hence the name *eukaryote,* meaning "true nucleus." The name prokaryote ("before nucleus") is based on the proposed evolutionary development of prokaryotes to eukaryotes. A comparison of prokaryotes and eukaryotes is summarized in Table 1–1, and some descriptive details are given in the following pages.

1–2 PROKARYOTIC CELLS

The organization of a typical prokaryotic cell is depicted in the electron micrographs of the rod-shaped bacterium *Escherichia coli* shown in Figure 1–3. Figure 1–3(a) displays two cells almost completely separated near the end of cell division. The cells are quite small, having a length of about $1-2$ μm and a diameter of about $0.5-1$ μm. The surface of the cell [Figure 1–3(b)] is bounded by a *cell envelope,* composed of three distinct substructures: a *cell wall* sandwiched between an *outer membrane* and an *inner membrane.* The cell wall (CW) is a rigid sheath with a polysaccharide-peptide composition (see Section 10–6) that is coated with an outer membrane (OM) layer of carbohydrate and lipid substances.

In addition to its primary function of providing a rigid protective barrier for the cell, the wall (including its outer layer) may also, according to recent evidence, have important functions in the general physiology of bacteria, such as participation in a molecular communication system between the exterior and

TABLE 1–1 Structural organization of living cells

PROKARYOTIC CELLS
Cytoplasm (no membrane-bounded subcellular structures present)
Cell membrane
Cell wall
Outer coat layer (in some bacteria there is a second membrane)

EUKARYOTIC CELLS
Cytoplasm (with the following subcellular structures):
Nucleus
Mitochondria
Chloroplasts[a]
Golgi body
Endoplasmic reticulum
Vacuoles[a]
Microbodies (lysosomes, peroxisomes, and glyoxosomes[a])
Cytoskeleton network (microtubules, microfilaments, microtrabecular strands)
Cell membrane
Cell wall[b]
Outer coat layer

Note: This list includes only the major subcellular entities; each type of cell, but particularly eukaryotic cells, contain a few other specialized structures; for these and additional information on subjects covered by this chapter, see the text by Dyson cited in the Literature.
[a] In the plant kingdom only.
[b] In plants and bacterial cells, not in animal cells.

the interior of the cell. The cell wall and its outer membrane layer also confer antigenic specificities of the cell, determine the reaction with dyes that are used to stain bacteria for microscopic examination, and provide the binding sites for the attachment of bacterial viruses. The close-up view of the *E. coli* surface in Figure 1–3(b) shows a few attached virus particles (V).

The inner membrane (IM) is also referred to as the *plasma membrane* or the *cytoplasmic membrane*. In addition to providing some protection for the cell, the inner plasma membrane contains several enzymes and is also responsible for determining the *permeability* characteristics of the cell, that is, the passage of substances in and out of the cell.

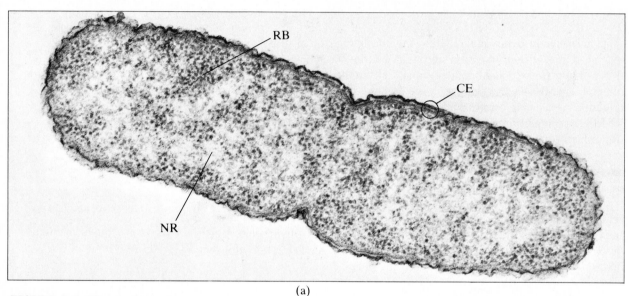

(a)

FIGURE 1–3 Ultrastructure of prokaryotes. (a) Transmission electron micrograph (12,500 ×) of *Escherichia coli* cells near the end of cell division. Visible in each cell are dark electron-dense regions of ribosomes (RB), less electron-dense areas representing the nuclear region (NR), and the cell envelope (CE). (b) Higher magnification (200,000 ×) of the cell surface, showing greater detail of the cell envelope with an outer membrane (OM), the cell wall (CW), and an inner membrane (IM). Faint images of a bacterial virus (V) attached at the surface are also visible. (*Source:* Micrographs prepared and provided by Dr. Robert M. Pfister, Ohio State University.)

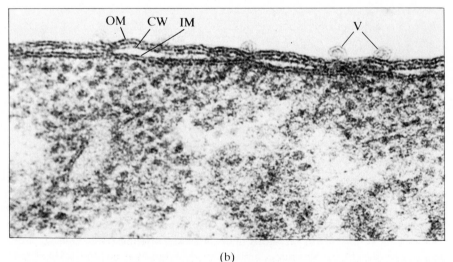

(b)

The interior of a prokaryotic cell is nonspecialized, lacking any distinct membrane-bound compartments. The appearance of light and dark regions within the cell is created by selective fixing and staining methods in the preparation of the sample for viewing. The lighter area corresponds to the *nuclear region* (NR) of the cell, which contains the DNA chromosomal material dissolved in the cytoplasm. In Figure 1–3(b) the faintly visible filamentous network appearing against the light background is interpreted as representing DNA itself. Depending on the growth conditions, an *E. coli* cell may contain one, two, or four DNA molecules. The term *nuclear region* is used here because of the absence of a limiting membrane that would literally separate it from the rest of the cell interior.

The darker, more electron-dense regions represent the presence of *ribosomes* (RB) found throughout the cell interior. Ribosomes are multimolecular aggregates that participate in the assembly of amino acids into protein molecules. An actively metabolizing *E. coli* cell may contain anywhere from 10,000 to 15,000 ribosomes in the cytoplasm.

In addition to containing the dissolved DNA and dispersed ribosomal particles, the cytoplasm contains thousands of other *dissolved materials,* many of which are enzyme proteins. Since dissolved materials are invisible to the electron microscope, this colloidal aqueous solution appears only as a background of low electron density against the remainder of cell structure.

Although not visible in the micrographs of Figure 1–3, several bacteria have protruding appendages at their surface. *Pili* (also called *fimbriae*) are filamentous strands several micrometers long and about 0.01 μm (100 Å) in diameter. Pili are proposed to be channels through which DNA passes between two mating cells during bacterial sexual conjugation. Pili are similar in appearance to, but structurally distinct from, another group of specialized appendages called *flagella,* which are responsible for the motility of bacteria.

1–3   EUKARYOTIC CELLS

Whereas a prokaryotic cell is conspicuous by the absence of distinct membrane-bound particulates within the cytoplasm, just the opposite is true of a eukaryotic cell. This characteristic is shown by the drawing in Figure 1–4, which illustrates an idealized cross-sectional view of a typical eukaryotic cell. Although there is really no such thing as a typical cell of any type, the subcellular organization depicted in Figure 1–4 is common to most eukaryotic cells.

Nucleus

The *nucleus* of a cell (generally, there is only one per cell) is the largest subcellular organelle. It is 10–20 times larger than the *E. coli* prokaryote and is easily seen with a good light microscope and suitable stains. Yet it is only with the electron microscope that the ultrastructure of the nucleus is revealed with good resolution (see Figure 1–5).

The first noteworthy feature, typical of every compartmentalized region of a eukaryotic cell, is the presence of a membrane segregating the nucleus from

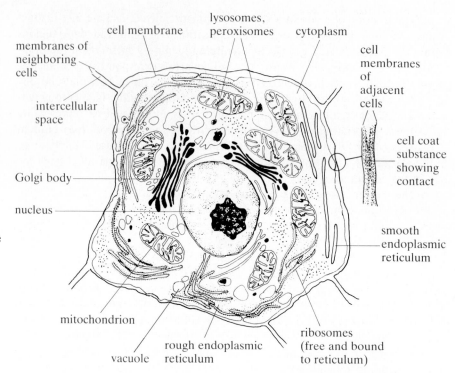

membranes of neighboring cells

intercellular space

cell membrane

lysosomes, peroxisomes

cytoplasm

cell membranes of adjacent cells

cell coat substance showing contact

Golgi body

nucleus

smooth endoplasmic reticulum

mitochondrion

vacuole

rough endoplasmic reticulum

ribosomes (free and bound to reticulum)

FIGURE 1–4 Simplified sketch of a generalized eukaryotic cell. Photosynthetic cells also contain chloroplasts (not shown here). See the text for descriptive information and representative electron micrographs. See p. 27 for a description of the cytoskeletal structure. [*Source:* Adapted, with permission, from R. D. Dyson, *Cell Biology,* 2nd ed. (Boston: Allyn and Bacon, 1978).]

the cytoplasm. The *nuclear membrane* (sometimes called the *nuclear envelope*) is distinct from most other membranes in that it is a *double membrane* separated by an intermembranous space. Another feature is the presence of membrane *pores,* which are interruptions of the double-membrane systems. The nuclear pores provide openings for the exit into the cytoplasm of the large RNA molecules (messenger, transfer, and ribosomal) that are synthesized in the nucleus.

The inside of the nucleus, called the *nucleoplasm,* also contains a degree of internal organization. Particularly evident is the very electron-dense *nucleolus,* a region that is rich in RNA. Current belief is that the nucleolus is the site within the nucleus where ribosomal RNA biosynthesis occurs. Throughout the nucleoplasm, but particularly near the nuclear membrane, there appear regions of less electron density than that found in the nucleolus. These regions are termed *chromatin* and are known to contain the major portion (95% or more) of the total DNA found in the cell associated with histone proteins (see Chapter 6, p. 267). The more or less random distribution of the chromatin is characteristic of the nucleus when the cell is not dividing. At the onset of cell division and during the staged mitotic process, the chromatin regions become highly organized, and the chromosomes then replicate to produce two identical sets.

Note that in describing the structural organization of the nucleus, we also established the primary functions of this specialized organelle. To summarize: the nucleus is the cellular site where genetic information is (1) *stored* as DNA, (2) *transmitted* to the rest of the cell (DNA → RNA → proteins, with protein biosynthesis occurring in the cytoplasm), and (3) *replicated* to ensure perpetuation of the cell line (DNA → DNA).

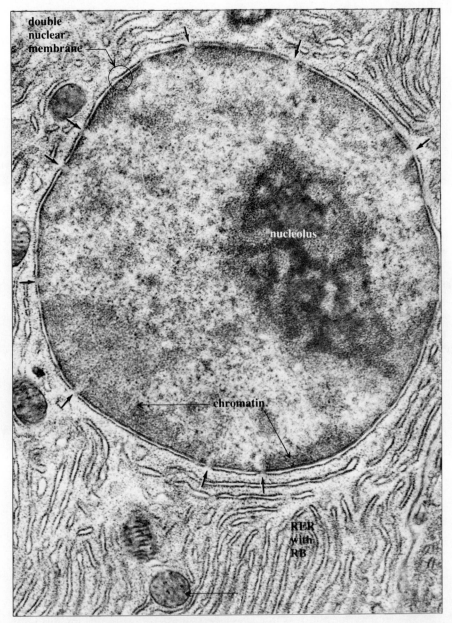

double
nuclear
membrane

nucleolus

chromatin

RER
with
RB

FIGURE 1–5 Electron micrograph of a cross-sectional view of an intact cell nucleus. Clearly visible are the two membranes of the nucleus, with several pores indicated by the arrows around the circumference of the nucleus. Visible within the nucleus and dispersed in the nucleoplasm are the nucleolus and chromatin regions. A large population of rough endoplasmic reticulum (RER) with attached ribosomes (RB) is visible in the cytoplasm surrounding the nucleus. A small number of mitochondria (mito) are also present. This specimen was obtained from the pancreas of a bat; magnification 22,000×. [*Source:* Taken, with permission, from D. W. Fawcett, *An Atlas of Fine Structure: The Cell* (Philadelphia: Saunders, 1966). Photograph provided by D. W. Fawcett.]

Mitochondrion

Although mitochondria are found in virtually all eukaryotic cells, their size, shape, and number are variable from one cell to another. In fact, all three characteristics appear to change in response to shifts in metabolism and as a result of cell aging. In addition, various pathologies are associated with alterations in these characteristics. The significance of such changes to any of these situations is not yet clear.

NOTE 1–2

Some biologists have suggested that even in aerobic cells the number of mitochondria that are actually present in vivo is quite small, and indeed, there may be only one per cell. Proponents of this idea, termed the *unit-mitochondrion hypothesis,* argue that the preparation of samples for observation by electron microscopy destroys these massive mitochondria and that the hundreds to thousands that appear are only artifact fragments of the original whole. While certain yeast cells are so characterized when grown under particular conditions and prepared and examined by special electron microscopic techniques, there is controversy about the general occurrence of this unit-mitochondrial morphology in other eukaryotic cells.

In animal cells the *mitochondrion* is frequently a rod-shaped particle with a length of 1.5–2 μm and a diameter of 0.5–1 μm. In other words, it is approximately a twentieth of the size of the cell nucleus and about equal in size to a prokaryotic bacterial cell such as *E. coli.* Certain generalizations can be made regarding the number of mitochondria per cell. In cells characterized by a high degree of *aerobic* (oxygen-dependent) metabolism, the number per cell might be quite large. For example, each cell of liver tissue is estimated to contain close to a thousand mitochondria. By contrast, cells participating primarily in *anaerobic* (not oxygen-dependent) metabolism, such as the cells of skeletal muscle tissue, contain only a few mitochondria (see Note 1–2).

The origin of the word *mitochondrion* (from the Greek *mitos,* "threadlike," and *chondros,* "grain") comes from the gross structural features of this organelle when it was first observed as a stained body under a light microscope some sixty-five years ago. However, transmission electron microscopy reveals an elaborate internal structure (see Figure 1–6). Like the nucleus, the mitochondrion is bounded by a *double membrane* (DM)—two unit membranes separated by an intermembranous space. However, the mitochondrial membrane system is nonporous. Whereas the outer membrane is smooth and continuous around the mitochondrion, the inner membrane undergoes an extensive and irregular folding within the mitochondrion. These inward folds of the inner membrane are called *cristae.* The fluid-filled interior of the mitochondrion is called the *mitochondrial matrix* (M).

The very dense spots that appear within the matrix are small granules of unknown composition and function. These and other important features of mitochondrial ultrastructure are diagrammatically depicted in the sketch of a mitochondrion in Figure 1–7.

Mitochondria are responsible for the bulk of the *aerobic metabolism* of the cell, which includes the crucial biochemical processes of the *citric acid cycle* and *oxidative phosphorylation.* Together these activities produce nearly all of the energy required for growth and viability of the entire cell. These and other important activities associated with mitochondria will be explored in subsequent chapters. The molecular events responsible for energy production occur within the inner membrane, particularly within the cristae. In Figure 1–7 the mushroom-shaped subunits (respiratory stalks) attached to the matrix-facing side of cristae are associated with this all-important process, the details of which are examined in Chapter 15.

The matrix of the mitochondrion also contains DNA and ribosomes. The amount of DNA present is small, about 2%–4% of the total DNA in the cell. The mitochondrial DNA contains genes coding for *some* of the proteins contained in the mitochondrion. Most mitochondrial proteins are coded for by genes in the nuclear DNA. Since the mitochondrion produces some of its own proteins from its own genetic program, it can be considered a sort of secondary minicell. Indeed, some scientists propose that mitochondria may have evolved from primitive bacteria. Their presence in eukaryotic cells may have occurred via an infection of a developing eukaryotic organism early in that organism's evolution by a parasitic, mitochondrion-like bacterial cell. At this stage the primitive eukaryotic cell may have acquired a natural selective advantage in that the activities of the host cell and the invading bacterial cell proved to be complementary. The two may then have continued to evolve as a unit cell.

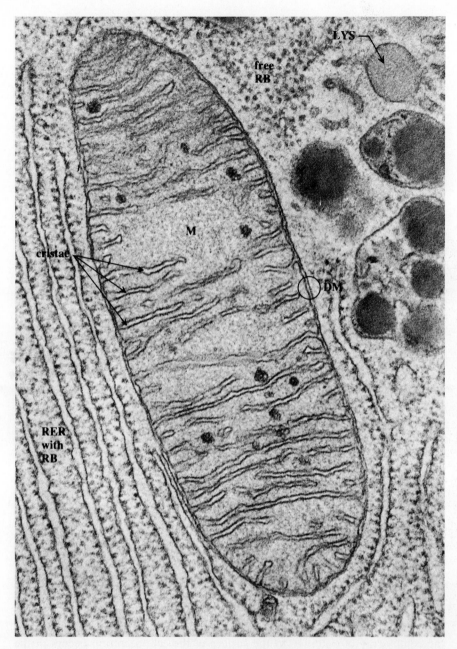

FIGURE 1–6 Electron micrograph of a longitudinal section of a mitochondrion and surrounding cytoplasm from the pancreas of a bat. Note the distinct double membrane (DM) of the mitochondrion and the numerous foldings (cristae) of the inner membrane that project into the matrix (M) of the mitochondrion. The heavily stained small granules in the matrix are of unknown composition and function. They are not found in all mitochondrion. Visible at the left from top to bottom is a region of rough endoplasmic reticulum (RER) with attached ribosomes (RB). Free ribosomes are also present in the upper portion of the micrograph. A lysosome (LYS) can be seen in the upper right corner. Magnification is 95,000×. [*Source:* Taken, with permission, from D. W. Fawcett, *An Atlas of Fine Structure: The Cell* (Philadelphia: Saunders, 1966). Photograph provided by D. W. Fawcett. Original micrograph prepared by Dr. K. R. Porter.]

Endoplasmic Reticulum

First observed in 1953, the *endoplasmic reticulum* is an organelle that occurs in nearly all types of plant and animal eukaryotic cells. It is a netlike system (reticulum) of flattened, membrane-bound regions localized within the cytoplasm (endoplasm) of the cell. Thus unlike the nucleus and the mitochondrion, the endoplasmic reticulum is not a singular, highly ordered entity but is an irregular and interconnected array of membranous vesicles. In many cells it

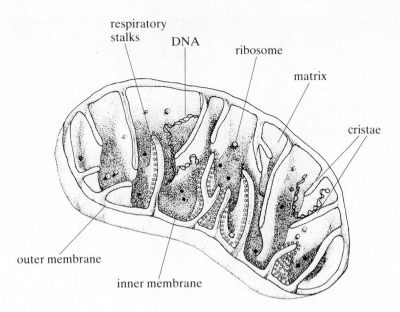

FIGURE 1–7 Mitochondrial ultrastructure. Notice that the cristae are formed from folds in the inner membrane. [*Source:* Adapted, with permission, from R. D. Dyson, *Cell Biology,* 2nd ed. (Boston: Allyn and Bacon, 1978).]

is quite profuse, occupying much of the available intracellular space. Two types are known. One is called *rough endoplasmic reticulum* (RER), due to the presence of ribosome particles attached to the outer surface of the membrane vesicle. The second type is termed *smooth endoplasmic reticulum* (SER) and is characterized by the absence of attached ribosomes.

The transmission electron micrographs in Figures 1–5 and 1–6 show regions of rough endoplasmic reticulum. Figure 1–8(a) shows the RER at greater magnification, and Figure 1–8(b) shows a remarkable three-dimensional image obtained by a recently developed technique for applying scanning electron microscopy to visualizing structures inside a cell.

In addition to containing a variety of enzymes that are localized in the reticular membranes, the reticulum has functions in the biosynthesis of proteins (this function is confined to the rough type that is attached to ribosomes) and in the storage and the transport of proteins that are destined for secretion from the cell. After their synthesis, proteins enter the inner cavity of the reticulum (called the *cisternal space*). By the action of enzymes found in the space compartment, some of the internalized proteins are converted into glycoproteins— proteins with an attached carbohydrate group. Proteins then move through the catacomb array of cisternae to the surface of the cell. Secretion occurs in the form of small vesicles of packaged proteins that are pinched off the SER. Alternatively, protein transport proceeds through the cisternae of the SER to the Golgi body, which packages the protein into the secretory vesicles.

A fact consistent with these roles is that the reticulum, especially the rough variety, is most abundant in those cells known to be specialized sites for the synthesis of various proteins. For example, the partial views in Figures 1–5 and 1–6 (both of which show a high RER content) are of a pancreas cell, which daily produces and secretes rather large amounts of several enzymes that par-

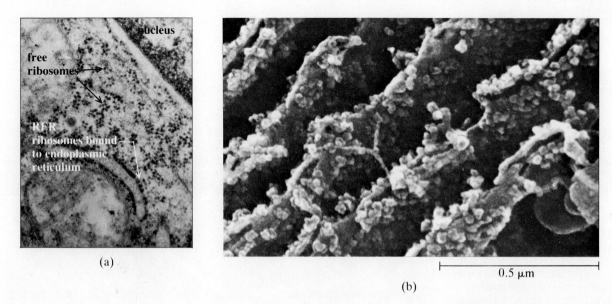

(a)

(b)

0.5 μm

FIGURE 1–8 Micrographs of rough endoplasmic reticulum. (a) High-magnification (134,000 ×) transmission micrograph; many free ribosomes are also visible. (*Source:* Micrograph provided by Donald Mullaly.) (b) Scanning micrograph (91,000 ×) showing a three-dimensional surface view of reticulum membranes with bound ribosomes. (*Source:* Micrograph provided by K. Tanaka and A. Mitsushima.)

ticipate in the digestion of ingested foodstuffs by mammals. However, not all of the protein biosynthesis occurs in association with the RER. The bulk of the proteins produced by the cell for use within the cell itself are assembled at ribosomal clusters found free in the cytoplasm. A region of free ribosomes is seen at the top right of the micrograph shown in Figure 1–6 and also in Figure 1–8.

When whole cells are disrupted by most methods, the reticulum network is randomly cleaved, and smaller fragments are eventually isolated. The reticulum pieces are termed *microsomes*.

Golgi Body

The *Golgi body* is also a network of flattened, membrane-bound vesicles. However, unlike the extensive and pervasive endoplasmic reticulum, the Golgi body is restricted in size. Moreover, the vesicles of the apparatus are frequently stacked in a small cluster and are usually localized near the nucleus or near the apex of specialized secreting cells. Transmission and scanning micrographs of the Golgi body are displayed in Figure 1–9.

As stated earlier, one of the functions of the Golgi apparatus is to accept proteins from the RER–SER system for concentrating and packaging into dense granules, which are then secreted from the cell. Some assembly of carbohydrate groups in glycoproteins also occurs in the Golgi body. Another function is to concentrate certain proteins and package them in membrane-bound bodies (such as lysosomes), which remain in the cell. The Golgi body has also been implicated as the cellular site for the biosynthesis of complex carbohydrate materials, which are ultimately secreted from the cell and then deposited in the exterior coating of the cell. Considerable research is in progress to clarify these

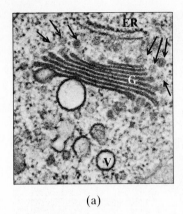

(a)

FIGURE 1–9 Micrographs of the Golgi body. (a) Transmission micrograph of Golgi apparatus (G) showing stacked membranes with small vesicles (arrows) budding or fusing with the edges. Ribosomes on a reticulum surface (ER) are also visible, as are free ribosomes and a variety of other vesicles (V). Magnification is 54,000×. (*Source:* Photograph provided by Andrew Staehlin.) (b) Scanning micrograph (29,000×) showing a three-dimensional surface image of the Golgi body (center of micrograph) surrounded by a network of smooth endoplasmic reticulum. A few sectioned mitochondria are also visible to the right of the center below the Golgi body. (*Source:* Micrograph provided by K. Tanaka and A. Mitsushima.)

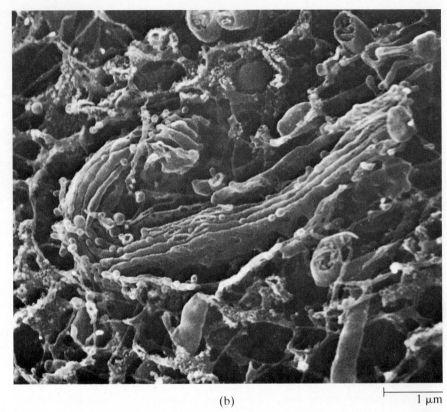

(b)

$\vdash\!\!\dashv$ 1 μm

roles and to investigate others. Since the Golgi apparatus is present in cells that do not specialize in secreting protein as well as in those that do, it may have a general and essential function in the biochemistry and physiology of the cell.

Microbodies: Lysosomes, Peroxisomes, and Glyoxosomes

Nearly all eukaryotic cells contain variable numbers (10–1000 per cell) of small compartments surrounded by a single membrane. These compartments are generally called *microbodies*. These vesicles (about 0.5–1.5 μm in diameter) contain high concentrations of certain proteins and appear as dense regions in electron micrographs. About a hundred different proteins are known to exist in microbodies, but not all are in the same microbody nor in the same cell. These microbodies are formed by pinching off as buds from either the Golgi body or the RER–SER network.

Discovered in 1952, *lysosomes* (see Figure 1–6) contain various types of enzymes, several of which degrade proteins, nucleic acids, and carbohydrates by catalyzing the hydrolysis (bond breaking by water) of bonds in these polymers. As long as such enzymes are confined within the lysosome, they are in a latent state because they are isolated from the substances they degrade. Obviously, if these enzymes were not segregated after their synthesis, the cell would be faced with a built-in, self-destructing capability, since the cell's own nucleic acids and proteins would be the target of the hydrolytic action of the enzymes. However, under certain conditions (which we will not go into here), the lysosome enzymes are utilized to catalyze the intracellular degradation of materials that enter the cell from the outside in unhydrolyzed form. The known occurrence of some thirty mammalian diseases caused by deficiencies of single lysosomal enzymes is strong evidence that the function of lysosomes is rather important to a normal cell.

Discovered in 1966, *peroxisomes* (see Figure 1–10) contain a variety of enzymes that participate in reactions that either produce or degrade peroxides (such as hydrogen peroxide, H_2O_2). Peroxisome size can be as large as that for lysosomes and sometimes smaller (0.1–0.5 μm in diameter). The most prevalent enzyme that occurs is *catalase*, which catalyzes the conversion of toxic H_2O_2 to H_2O and O_2. Other enzymes involved in the metabolism of lipids and carbohydrates have also been detected. In plant cells only, microbodies exist that contain enzymes for converting lipids to carbohydrates through the intermediate production of glyoxylic acid (see p. 639), and hence they are called *glyoxosomes*.

Cell Membrane

Every cell is surrounded by a thin layer (75–95 Å, or 0.0075–0.0095 μm, thick) composed of lipid and protein called the *plasma membrane* or simply the *cell membrane.* It is obviously a distinct part of the cell, but there is evidence (see Figure 1–11) that the cell membrane is not completely separated from other membranes in the cell. From the micrograph we see clearly that the main plasma membrane is *continuous* with the outer membrane of the nucleus. This continuity may apply to all intracellular membranes. That is, all of the individual membranes throughout a eukaryotic cell may be specialized segments of a single, massive membrane system. Part of it is the plasma membrane, another part the nuclear membrane, another part the endoplasmic reticulum, and so on. This continuity does not mean, however, that there is a continuity of chemical composition, molecular ultrastructure, and biochemical function. In each of these characteristics individual segments of this massive membrane (in other words, the cell membrane, the nuclear membrane, the Golgi body, etc.) are unique.

In addition to partitioning one cell from another and providing it with some protection, the cell membrane also plays an important role in the *passage of materials into and out of the cell.* All of the membranes of both eukaryotic and prokaryotic cells and also the membranes of subcellular organelles play such a role. In recent years the list of other functions attributed to the membrane has been growing. These functions vary from species to species, from cell to cell, and from one physiological state to another. Many membrane functions involve the participation of highly specific proteins in the membrane called *receptor proteins,* which operate by binding to a specific substance on one side of the membrane. This event triggers other events in the membrane and then in the cell, thus signaling the cell to do (or not do) something and controlling the extent of what happens. This and other aspects of membrane biochemistry—including the molecular ultrastructure of membranes—are described more fully in later chapters.

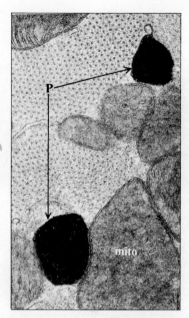

FIGURE 1–10 Peroxisomes (P) as observed in an electron micrograph of a transverse section of mouse myocardial fiber. The peroxisomes appear here as very electron-opaque bodies because the tissue was incubated with a catalase-specific staining mixture prior to preparation for electron microscopy. Obviously, then, these peroxisomes are rich in catalase. The significance of the close association of peroxisomes with mitochondria (mito) is not clear. Magnification is 42,000 ×. [*Source:* Reproduced with permission from *Science,* **185,** 271–273 (1974). Copyright © 1974 by the American Association for the Advancement of Science. Photograph provided by Dr. H. Dariush Fahimi.]

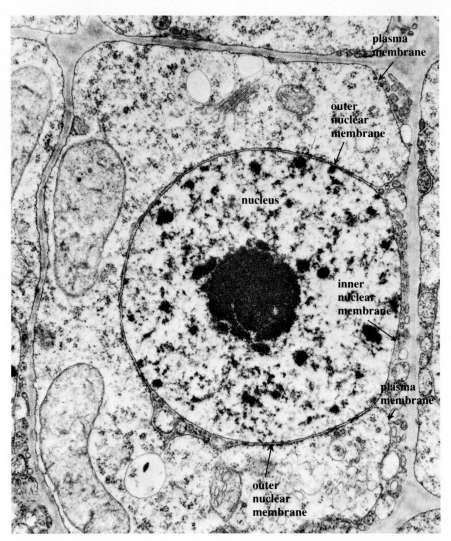

plasma membrane

outer nuclear membrane

nucleus

inner nuclear membrane

plasma membrane

outer nuclear membrane

FIGURE 1–11 Membrane continuity. The continuity of the plasma membrane and the outer nuclear membrane can be seen at two and four o'clock. [*Source:* Reproduced with permission from Z. B. Carothers, "Membrane Continuity Between Plasmalemma and Nuclear Envelope in Spermatogenic Cells of Blasia," *Science,* **175,** 652–654 (1972). Photograph provided by Dr. Z. B. Carothers.]

Cell Wall and Cell Surface Coat

In addition to being protected by the cell membrane, many plant and bacterial eukaryotic cells are protected by a *cell wall* external to the surface membrane. Similar to that found in prokaryotic bacteria, the wall of eukaryotic cells is a rigid, covalent network composed largely of polysaccharide (polymeric carbohydrate) material. In plants the major components are closely packed cellulose fibers, cemented together by other substances. Animal cells generally do *not* have a cell wall.

In most cells (eukaryotes and prokaryotes alike), the cell surface—that is, the cell membrane or cell wall—is frequently coated with a viscous substance called a *cell coat*. The composition of this substance is quite variable from one cell to another, but it is usually known to contain complex polysaccharides in conjunction with other constituents such as proteins and lipids. In addition, some of the membrane proteins protrude into this surface layer. In a multicellular tissue adjacent cells make an irregular contact through their respective coats across the intercellular space (see Figure 1–4). One proposition is that this cellular contact serves as a form of communication among neighboring cells, resulting in a control of their growth and division. The phenomenon has been termed *contact inhibition*. Normal cells exhibit this ability, but tumor cells do not. There is evidence that an alteration in the composition of the cell coat can result in the transformation of normal cells to tumor cells. (See the paper by M. M. Burger listed in the Literature.)

Chloroplast

Although eukaryotic cells of higher animals and plants share most of the structural organization just described, photosynthesizing plant cells contain a unique organelle, the *chloroplast* (from the Greek *chloros,* "green," and *plast,* "formed mass"). Chloroplasts contain the green *chlorophyll* molecules and are the specialized membrane-bound bodies that function in the crucial process of harnessing and converting the energy of the sun into metabolic energy. The energy is then used in the fixation and the conversion of atmospheric CO_2 into carbohydrate material.

The shape and the size of the chloroplast are quite variable. In some cells the chloroplast is similar in both respects to the nucleus—somewhat spherical and very large. In other cells it is cylindrical but two to five times larger than a typical mitochondrion.

As indicated in Figure 1–12, the chloroplast also displays a considerable amount of fine inner structure. The most prominent feature is the presence of several electron-dense stackings. These stackings are called *grana* and represent ordered pilings of flattened membranous systems that presumably originate from the main membrane surrounding the periphery of the chloroplast. The white bodies in the interior of the chloroplast represent large storage vacuoles of the carbohydrate end products of photosynthesis. They are not common to all chloroplasts. The structural and functional characteristics of the chloroplast will be discussed in greater detail in Chapter 16.

Cytoskeleton Elements

The internal cytoplasm matrix of the cell was long assumed to be a structureless gellike medium in which substances are dissolved and subcellular compartments are suspended. This view has changed dramatically in recent years with the discovery that the cytoplasm in eukaryotic cells is permeated by an intricate assembly consisting of *microtubule structures* connected to thinner threads called *microtrabecular strands.* This lattice may also connect with other tubelike structures, called *myofilaments,* associated with the cell membrane (see Figure 1–13). This structure is a fine level, with microtubules having a diameter of 24 nm and a bore of about 15 nm; microtrabecular strands are 2–3 nm thick in the middle and 10 nm where they connect with each other and with the microtubules; and myofilaments have a diameter of about 6 nm (1 nm = 10^{-9} m). All are composed of protein molecules. The detection of the small and delicate micro-

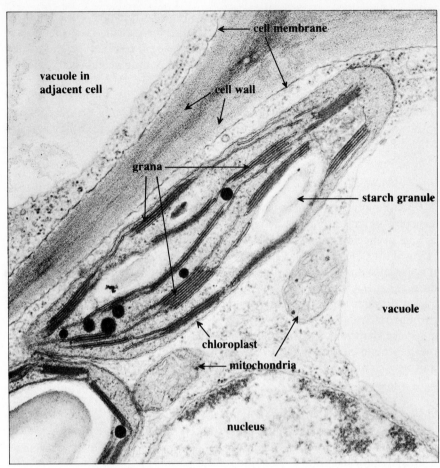

FIGURE 1–12 Electron micrograph showing, in cross-sectional view, pieces of adjacent cells in tomato stem tissue. An intact chloroplast containing several grana can be seen in the lower cell, situated very close to the cell membrane. The thick, cellulose-containing cell wall of each cell is also visible. Other organelles visible in the lower cell are two mitochondria and a portion of the cell nucleus. The large clear region at the right is a portion of a vacuole surrounded by a membrane. Vacuoles are present in most plant cells, particularly in older cells, functioning as storage compartments for dissolved sugars, proteins, oxygen, carbon dioxide, and other substances. See Figures 16–3 and 16–4 for more detail of chloroplast ultrastructure. (*Source:* Photograph provided by Hilton H. Mollenhauer, Charles F. Kettering Research Laboratory, Yellow Springs, Ohio.)

trabecular structure was initially achieved via high-voltage electron microscopy (10^6 V) on whole cells and later by improvements in the sectioning of cells via conventional electron microscopy (10^5 V).

The entire assembly is termed a *cytoskeleton network,* reflecting the thinking that its function is to serve as a molecular scaffold contributing to the shape and the strength of a cell and suspending the various organelles in the cell. Other cell processes proposed to involve this system are cell movement, cell division, the movement of substances from one region of the cell to another and from one organelle to another, and the provision of surfaces on which some chemical reactions may occur.

1–4   ARCHAEBACTERIA: A THIRD LIFE FORM?

Plant and animal cells are eukaryotic; bacterial cells are prokaryotic. This twofold distinction of life forms has been a long-held principle of biology. However, research during the 1970s uncovered a few classes of bacteria having the general structural features of prokaryotes but having some molecular charac-

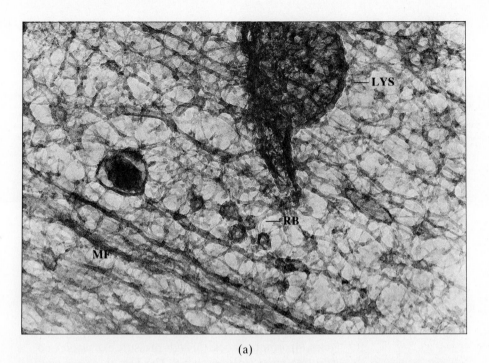

(a)

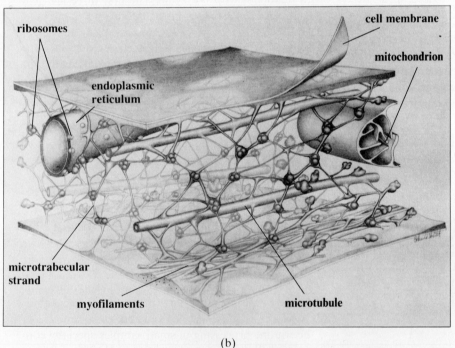

(b)

FIGURE 1–13 Microtrabecular lattice. (a) Electron micrograph of the interlinked filaments of the microtrabecular lattice in a rat cell. The meshwork suspends a lysosome (LYS) and also cross-links a bundle of microfilaments (MF). The small round bodies at several intersections are ribosomes (RB). Magnification is 120,000×. (*Source:* Micrograph prepared by Karen L. Anderson.) (b) Drawing of the microtrabecular lattice illustrating the relationship to other components of the cell cytoplasm. Magnification is 300,000×. (*Source:* Photographs provided by Keith R. Porter.)

teristics inconsistent with the prokaryote kingdom. Extensive evidence argues against the possibility that these bacteria are merely primitive (early) forms of all other prokaryotic bacteria. Rather, biologists propose that these novel bacteria, called *archaebacteria* (so named because they are known to be ancient bacterial cells), may represent a new primary kingdom with a completely different position in the evolutionary history and natural order of life. The revised classification of living cells offers then a threefold distinction of life forms: the prokaryotic archaebacteria, all other prokaryotic bacteria (called the *true bacteria,* or *eubacteria*), and the eukaryotes.

The most common representatives of the archaebacteria are the strict anaerobic bacteria that generate methane (CH_4) by the reduction of carbon dioxide (CO_2). They are called *methanogens* and are found in stagnant water, in sewage treatment plants, and in the intestinal tract of animals. Two other groups are *halophiles* (bacteria that require high concentrations of salt in order to survive) and *thermoacidophiles* (bacteria that require hot temperatures, 80°–90°C, and acidity, pH 2, in order to grow).

Since the molecular distinctions of the archaebacteria prokaryotes from the eubacteria prokaryotes are based on biochemistry to be described throughout this book, we cannot examine them at this time. (See the article by C. R. Woese listed in the Literature.)

◇   **L A B O R A T O R Y**   **M E T H O D S**   ◇

CELLULAR FRACTIONATION

Biochemical research often requires the isolation of a particular subcellular organelle either (1) to study the organelle intact or (2) more commonly, to isolate and then study a specific substance from that organelle. The isolation of any cell part in a pure and undamaged state can be achieved by well-established and, for the most part, relatively routine fractionation procedures. The general approach consists of two phases:

1. Whole cells are disrupted (broken, lysed) to yield a *cell-free system,* called a *homogenate.*

2. *Differential centrifugation* is used to separate the subcellular parts from each other.

If a protein is to be isolated from an organelle, the isolated organelle is disrupted. The isolation of a protein that is originally localized in a membrane can also be accomplished, but often the procedures must be more tedious to avoid damaging the protein during disruption of the membrane.

Cell Disruption

A variety of methods are available to lyse cells. The most common techniques are (1) blending, (2) grinding, (3) sonication, that is, exposure to ultrasonic frequencies, (4) osmotic shock, (5) high-pressure extrusion, and (6) treatment with lysozyme. The first three are generally used in the processing of animal and plant tissue, while the last four (3–6) are employed in the lysis of the smaller bacterial cells. The decision to choose one procedure over another is based on the particular type of cell, the objective of the experimenter, and the mass of material to be processed. In the absence of published guidelines the choice is essentially based on trial and error. What procedure will give the best results? The desired objectives are *maximum disruption of whole cells* and *minimum damage to subcellular compartments,* particularly the organelles to be studied.

Blending is generally accomplished with electric devices of various construction, each offering different advantages and disadvantages. The basis of rupture is quite simple: It is a shearing of cellular tissue by rotating blades.

In grinding methods the cells are merely rubbed against an abrasive and hence against each other. The simplest tools are a mortar and a pestle. Typical abrasives are ground-glass beads, sand, or alumina. For the routine processing of small samples specially constructed glass *homogenizers* are used. These homogenizers consist of a ground-glass barrel fitted with a ground-glass piston providing a clearance of 0.005 in. As the piston moves up and down the barrel containing

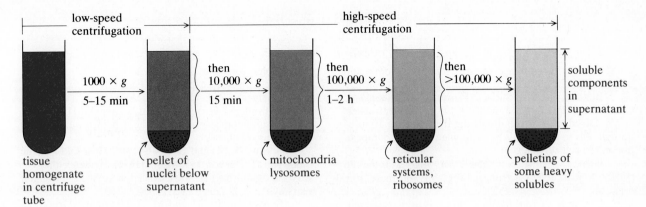

FIGURE 1–14 Outline of cellular fractionation. After the whole cells are lysed, the homogenate is centrifuged in steps. Before one advances from one step to another, the supernatant is separated from the pellet. The number of steps performed depends on the object of the experimenter. Gravitational force is represented here by g.

the sample, the tissue is forced through this small clearance, and cells are disrupted.

Passing sound vibrations of ultrahigh frequency through a cell suspension is a widely used disruption technique. Two reasons for its popularity are that (1) in most cases the subcellular organelles can be recovered in a reasonably intact, undamaged state and (2) the severity of the treatment can be finely controlled.

One of the mildest techniques is that of osmotic shock. Here the cells are first suspended in a solution of high solute concentration, causing the migration of water out of the cell. Then they are transferred to pure water, whereupon the water rushes into the cell and it bursts open.

With the high-pressure extrusion method small cells such as bacteria are efficiently and gently broken by forcing a concentrated suspension through a small opening under several thousand pounds of pressure.

The most delicate procedure is to use lysozyme, but its use is confined to bacterial cells. Lysozyme is an enzyme that breaks up the rigid cell wall structure of bacteria. The destruction of the cell wall leaves a cell protected only by its membrane, called a *protoplast*, which is highly susceptible to being ruptured by osmotic shock.

Regardless of the method utilized, all operations are conducted under carefully controlled conditions and carried out at reduced temperature (2°–4°C) to minimize damage to the particulate organelles and to proteins.

Separation of Organelles

The separation of the soluble cell fluid from the particulate matter, as well as the further fractionation of the latter by differential centrifugation, is based on a simple principle. Since virtually all of the components in the cell-free system have a different mass-to-volume ratio (that is, a different density), heavier bodies will sediment under low speeds and low gravitational forces, while lighter substances will require higher speeds and higher gravitational forces. Particles of intermediate density will obviously require intermediate conditions. Consequently, an efficient fractionation can be achieved by starting on the low side and performing a series of separate and successive centrifugations toward the high side (see Figure 1–14). The steps and conditions specified in the figure approximate only a general and simplified situation. The precise conditions are generally more numerous and complex.

LITERATURE

ALLISON, A. "Lysosomes and Disease." *Sci. Am.,* **217,** 62–72 (1967). Description of the structure of lysosomes and their function in normal and pathological cells.

BURGER, M. M. "Surface Properties of Neoplastic Cells." *Hosp. Pract.,* **8,** 55–62 (July 1973). Excellent article (at the introductory level) discussing the transformation of normal cells to tumor cells by manipulation of the composition of the cell surface coat.

DUSTIN, P. "Microtubules." *Sci. Am.,* **243,** 66–76 (1980). Remarkable electron micrographs of these tiny organelles.

DYSON, R. D. *Cell Biology.* 2nd ed. Boston: Allyn and Bacon, 1978. An excellent textbook. Chapter 1 is devoted to cellular and subcellular ultrastructure and has many electron micrographs. Appendix 1 covers the theory and practice of electron microscopy.

FAWCETT, D. *The Cell.* 2nd ed. Philadelphia: Saunders, 1981. A superb atlas of electron micrographs of cellular and subcellular structures.

HAYAT, M. A. *Basic Techniques for Transmission Electron Microscopy.* Orlando: Academic Press, 1985. A how-to-do-it manual covering methods and procedures for examining prokaryotic and eukaryotic cells.

The Living Cell and *From Cell to Organism.* San Francisco: Freeman, 1965 and 1967. Two books containing a collection of articles from *Scientific American* on cell biology. Dated, but they still provide excellent introductory material on a variety of topics dealing with cellular and subcellular ultrastructure and biological function.

MARGULIS, M. "Symbiosis and Evolution." *Sci. Am.,* **225,** 48–57 (1971). A discussion of the origin and symbiotic evolution of specialized organelles (chloroplasts and mitochondria) of the cells of higher plants and animals.

PORTER, K. R., and J. B. TUCKER. "The Ground Substance of the Living Cell." *Sci. Am.,* **244,** 56–67 (1981).

ROTHMAN, J. E. "The Compartmental Organization of the Golgi Apparatus." *Sci. Am.,* **253,** no. 3, 74–85 (1985). An article with a self-explanatory title.

ROTHMAN, R. J. "The Golgi Apparatus: Two Organelles in Tandem." *Science,* **213,** 1212–1219 (1981). A review of recent developments regarding the structure and the function of this organelle.

SCHOPF, J. W., and D. Z. OEKLER. "How Old Are the Eukaryotes?" *Science,* **193,** 47–49 (1976). Evidence that eukaryotic organisms may have existed as early as 1.5 billion years ago.

SCHWARTZ, R. M., and M. O. DAYHOFF. "Origins of Prokaryotes, Eukaryotes, Mitochondria, and Chloroplasts." *Science,* **199,** 395–402 (1978). A review article describing the tracing of evolutionary history of organisms and organelles by using the amino acid sequence of proteins and the nucleotide sequence of nucleic acids.

TOLBERT, N. E. "Metabolic Pathways in Peroxisomes and Glyoxosomes." *Annu. Rev. Biochem.,* **50,** 133–157 (1981). A review article that includes information on the structure, the distribution, the development, and the isolation of microbodies. The major emphasis is on the metabolic functions of these compartments.

UMBREIT, W. W., R. H. BURRIS, and J. F. STAUFFER. *Manometric and Biochemical Techniques.* 5th ed. Minneapolis: Burgess, 1972. Contains an excellent coverage of the many techniques available for the preparation of animal, plant, and bacterial cell-free extracts.

WHALEY, W. G., M. DAUWALDER, and J. E. KEPHART. "Golgi Apparatus: Influence on Cell Surfaces." *Science,* **175,** 596–599 (1972). A review article devoted to the involvement of the Golgi body in the assembly of proteins destined for use outside the cell.

WOESE, C. R. "Archaebacteria." *Sci. Am.,* **244,** 98–122 (1981). A summary of the biochemical evidence that archaebacteria represent a new primary kingdom of life.

EXERCISES

1–1. Cut a $3\frac{1}{2}$-in. circle out of the center of a sheet of paper. Position the paper over the sketch in Figure 1–4 and identify the parts of the cell. How many different types of membranes and membranous bodies are depicted?

1–2. Assuming that the overall shape of a single bacterial ribosome is approximately spherical, with a diameter of 200 Å, calculate the percentage of the total cell volume occupied by ribosomes in a single *E. coli* cell. Assume further that the shape of an *E. coli* cell is approximated by a cylinder about 2 μm long and about 0.8 μm in diameter

(1 μm = 10^{-6} m). (*Note:* A cylinder of radius r and length or height h has a volume of $\pi r^2 h$; a sphere of radius r has a volume of $\frac{4}{3}\pi r^3$.)

1–3. Using the answer from Exercise 1–2 and the text description of the size of nuclei and mitochondria, express estimates for the volumes of (a) a nucleus and (b) a mitochondrion.

1–4. What percentage of the total cell volume of an *E. coli* cell is occupied by its DNA chromosome? Calculate molecular volume by considering the highly folded DNA to have the following dimensions (of a cylinder) if fully extended: 1.4 mm in length and 20 Å in diameter.

1–5. What percentage of the total mass of an *E. coli* cell (approximately 2×10^{-12} g) is contributed by the DNA chromosome? [Mass-length correlation of DNA: each 30.5 cm of DNA has an average weight of 1 picogram (pg) $(1 \times 10^{-12}$ g).]

1–6. Assume that the nucleus of a human cell has a volume 20 times greater than that of an *E. coli* cell (see answers to Exercises 1–2 and 1–3) and that the 46 DNA molecules in the 46 human chromosomes occupy about 25% of the nucleus volume. If all of the 46 DNA molecules were placed end to end, what would be the overall length? (*Note:* A correlation between molecular volume and the length of DNA can be obtained from the answer and information given in Exercise 1–4.)

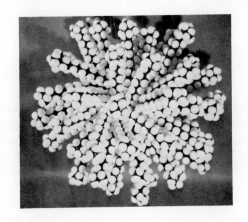

CHAPTER TWO

NONCOVALENT BONDING AND pH BUFFERING

In the study of biochemistry you will encounter many interesting and fascinating things. Indeed, I have never encountered a single student in my classes who found the subject dull in its entirety. The structure and the function of proteins, nucleic acids, and biomembranes are particularly stimulating—even evoking a sense of appreciation for molecular beauty. Then, although it can be bewildering, the integrated maze of reactions responsible for the metabolism of numerous substances also stimulates an appreciation for the complexity and logic of life chemistry. All will be unveiled chapter by chapter, with an early emphasis on proteins and nucleic acids. However, we will start this expedition of study with a focus on purely chemical principles that are fundamental to understanding much of what comes later.

The main topics discussed in this chapter are noncovalent forces of attraction, acid-base ionization, and pH buffers. Although you have had previous exposure to these topics in general and organic chemistry courses, they are reviewed in the following pages with highlights and emphasis of biological significance.

2–1 WATER: THE BIOLOGICAL SOLVENT

Life began in water some 3 billion years ago and continues to be sustained by it. In fact, water is the most abundant material in any living organism, representing approximately two-thirds of the total weight (see Table I–1). That life emerged on our planet is no accident. Water was plentiful, and it had a low freezing point and a high boiling point. Since the earth was neither constantly extremely cold nor hot, the beginning life forms had the opportunity to grow, adapt, and evolve in a liquid system at a moderate temperature. Another important property of water is that at 4°C the density of liquid water is greater than that for ice. Consequently, ice floats, preserving an environment capable of supporting the existence of the countless aquatic organisms.

In the living state the basic role of water is to provide a fluid system in which the physicochemical processes of life can occur. It is the *solvent of the living state*. Moreover, for most photosynthetic cells water is an essential nutrient, being oxidized to support the reductive fixation of CO_2 and providing the major *source of molecular oxygen* O_2 on our planet. Thus water is the single most important substance for life—as we know it.

Polarity of Water

The hydrides of the group VI elements—H_2O, H_2S, H_2Se, and H_2Te—are all planar, trigonal molecules with two pairs of unshared electrons on the central atom. Although these hydrides have similar properties, the magnitudes

NOTE 2–1

Electronegativity refers to the ability of an atom of an element to attract electrons toward its nucleus. Fluorine, the most electronegative element, has an electronegativity value of 4.0. The values of the major elements of which biomolecules are composed are listed in the following table.

ELEMENT	ELECTRO-NEGATIVITY
Fluorine (F)	4.0
Oxygen (O)	3.5
Nitrogen (N)	3.0
Carbon (C)	2.5
Sulfur (S)	2.5
Phosphorus (P)	2.1
Hydrogen (H)	2.1

In a covalent bond the greater the difference in electronegativities of the two atoms, the greater is the polarity of the bond. As described, the O—H bonds in the water molecule are *polar bonds*. In contrast, bonds involving the same atoms, such as C—C, are without electronegativity difference and thus are without polarity. They are *nonpolar*. Bonds involving two atoms of small electronegativity difference, such as C—H, are so weakly polar that they are also considered as being nonpolar.

of the properties for H_2O are far different from the magnitudes for the others. This distinction arises because H_2O is a much more *strongly polar* substance than the others. Polarity is due to the existence of two, noncanceling $\delta^+ - \delta^-$ *dipoles* along each O—H bond axis. Because oxygen has a very high electronegativity relative to that of hydrogen (see Note 2–1), the oxygen atom tends to draw the electrons of each O—H covalent bond to itself. Thus the electrons are shared unequally (see Figure 2–1). Because of the greater electron density near the oxygen atom, it is designated as having a partial negative charge (δ^-). Conversely, each hydrogen atom, now having a reduced electron density, carries a partial positive charge (δ^+). This unequal distribution of electron density always exists along the O—H axis, and hence the dipole is called a *permanent dipole*.

Because "like dissolves like," water is a good solvent for polar solutes and ionic solutes. How good? Well, with a *dielectric constant D* of about 80, water is one of the most polar liquids known (see Table 2–1). In fact, there are few materials that possess D values greater than that for water. The dielectric constant is a measure of the ability of a system to insulate oppositely charged particles (Q^+ and Q^-) from mutual attraction. The force of attraction F between Q^+ and Q^- bodies, separated from each other by a distance r in a medium of dielectric constant D, is given by *Coulomb's law:*

$$F = \frac{Q^+ Q^-}{Dr^2}$$

Thus a polar solvent with a high D value reduces the force of attraction between Q^+ and Q^-, supporting the independent existence of Q^+ and Q^-.

The insulation of oppositely charged ions is due to the engulfment of each ion by a sheath of H_2O molecules. This hydration (or solvation) effect (see Figure 2–2) involves an electrostatic association between the positive ($+$) and negative ($-$) ions and the δ^- and δ^+ regions, respectively, of the permanent dipoles in H_2O molecules. This *permanent-dipole/ion interaction* (δ^+......$\ominus$ or δ^-......$\oplus$) represents just one type of noncovalent force of attraction. We will now examine hydrogen bonding, an immensely important type, and other types are described in the next section.

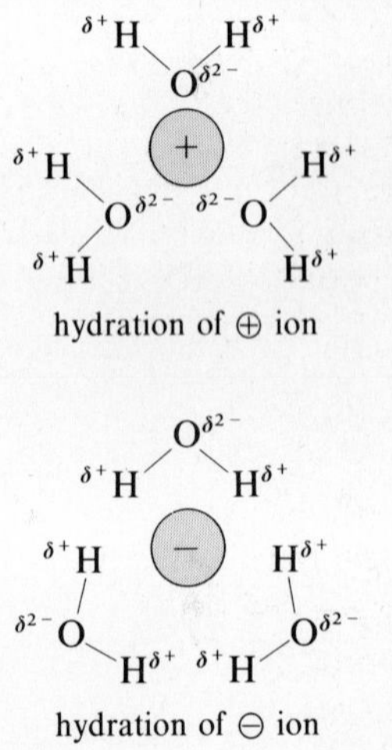

FIGURE 2–1 Polarity of water. The $\leftrightarrow$ indicates unequal sharing of electrons, favoring the more electronegative O and giving rise to partial-charge (δ) centers.

FIGURE 2–2 Hydration effect.

TABLE 2–1 Partial listing of dielectric constants

SUBSTANCE	DIELECTRIC CONSTANT (D)
Water	80.4
Methanol	33.6
Ethanol	24.3
Ammonia	17.3
Acetic acid	6.15
Chloroform	4.81
Ethyl ether	4.34
Benzene	2.28
Carbon tetrachloride	2.24

Hydrogen Bonding

Water has other properties that contribute to its biological importance. Owing to its *high heat of vaporization* (540 cal/g), an organism can dissipate a large quantity of heat through the vaporization of small amounts of water. Because of its *high heat capacity* (1 cal is required to raise the temperature of 1 g of water 1°C), an organism can absorb large amounts of heat without a correspondingly large change in its internal temperature. Both properties contribute to the maintenance of a relatively constant biotemperature.

These properties result from water molecules interacting with each other by a *permanent-dipole/permanent-dipole* force of attraction, called *hydrogen bonding*. A single hydrogen bond is due to the natural electrostatic attraction (see Figure 2–3) that one end of a dipole (the δ^- O) has for the end of another dipole (the δ^+ H) when two dipoles approach each other very closely—about 2.5–3.0 Å. Although a single hydrogen bond is weak, the presence of many can confer considerable bonding energy to a system. For example, 1 mL of liquid water (1 g) contains approximately 3×10^{22} molecules, the majority being involved in small clusters of hydrogen bond associations, with a constant exchange of participating molecules.

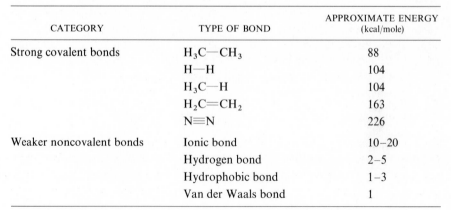

FIGURE 2–3 Electrostatic association between permanent dipoles.

2–2 NONCOVALENT BONDING: STABILIZING AND ORGANIZING FORCES OF NATURE

Covalent bonding is, of course, the major stabilizing factor in organic compounds, also holding atoms together in specific geometric patterns. However, even though bond energies (that is, bond strengths) for noncovalent bonds are less than bond energies for covalent bonds (see Table 2–2), the various noncovalent attractive forces are of immense importance. They are significant for the following reasons, and particularly so for proteins and nucleic acids: (1) They participate in the *organization* and subsequent *stabilization* of the

TABLE 2–2 Summary of some bond energies

CATEGORY	TYPE OF BOND	APPROXIMATE ENERGY (kcal/mole)
Strong covalent bonds	H_3C-CH_3	88
	$H-H$	104
	H_3C-H	104
	$H_2C=CH_2$	163
	$N\equiv N$	226
Weaker noncovalent bonds	Ionic bond	10–20
	Hydrogen bond	2–5
	Hydrophobic bond	1–3
	Van der Waals bond	1

preferred three-dimensional structure of a molecule (see Note 2–2); (2) they contribute to the ability of molecules to undergo *changes in conformation;* and (3) they provide specific patterns of *binding interactions* between substances that combine and react with each other. In other words:

> The structures and the functions of biomolecules—and hence life itself—are in large part determined by noncovalent bonding.

There are seven different types of noncovalent bonding interactions:

1. Ionic bonding:

$$ion \text{''''''''''} ion$$

2. Ion/permanent-dipole bonding:

$$ion \text{''''''''''} dipole_{permanent}$$

3. Hydrogen bonding:

$$dipole_{permanent} \text{''''''''''} dipole_{permanent}$$

4. Ion/induced-dipole bonding:

$$ion \text{''''''''''} dipole_{induced}$$

5. Permanent-dipole/induced-dipole bonding:

$$dipole_{permanent} \text{''''''''''} dipole_{induced}$$

6. Van der Waals bonding:

$$dipole_{induced} \text{''''''''''} dipole_{induced}$$

7. Apolar bonding: nonpolar/nonpolar interactions in a polar medium.

In the preceding description of the properties of water, two of these types were highlighted: ion/dipole$_{permanent}$ interactions and hydrogen bonding. We will now describe the others and also expand the description of hydrogen bonding. Excepting apolar bonding, all other interactions are based on electrostatic attractions between positive and negative charges.

Ionic Bonding

The strongest type of noncovalent bonding involves the attraction between fully charged positive and negative ions, called an *ion pair.* Naturally occurring ion pairs involve any combination of the ion species listed in Table 2–3, such as $-COO^- \text{''''''''''} {}^+H_3N-$. Many of the ion species are part of amino acid structures, described in the next chapter.

NOTE 2–2
The term *conformation* is often used in reference to the three-dimensional structure of a molecule.

TABLE 2–3 Ionic groupings that occur in biomolecules

NEGATIVELY CHARGED IONS		POSITIVELY CHARGED IONS	
$R-COO^-$ carboxylate ion	$R-SO_3^-$ sulfonate ion	$R-NH_3^+$ protonated amino ion	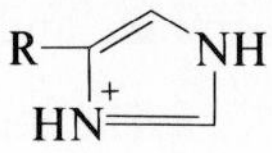 imidazolium ion
$R-S^-$ thiolate ion	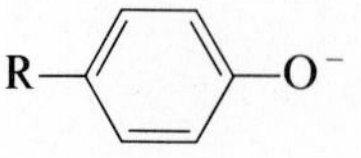 phenolate ion	guanidinium ion	
phosphate dianion	phosphate ion	M^{Z+} (metal ion, such as Mg^{2+}, Fe^{2+}, Zn^{2+})	

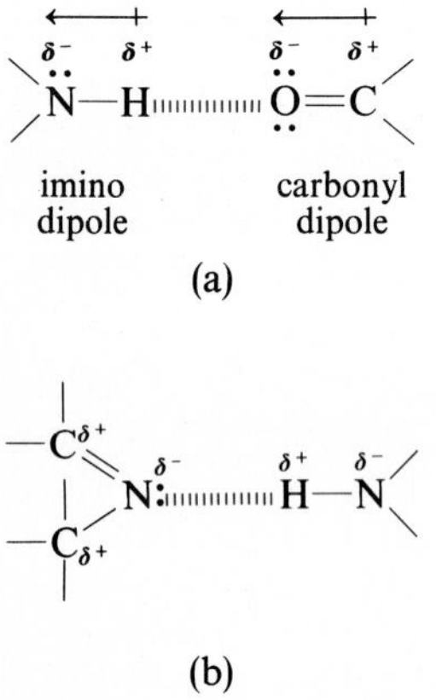

imino   carbonyl
dipole   dipole

(a)

(b)

FIGURE 2–4 Hydrogen bonding in biomolecules. (a) Occurs in both proteins and nucleic acids. (b) Occurs primarily in nucleic acids.

Hydrogen Bonding

Hydrogen bonding is not unique to water but is common to many systems, with the main requirement being the presence of permanent dipoles. It is preferable that one dipole contain a hydrogen atom with a partial positive charge, and the other dipole contain an oxygen or nitrogen atom with a partial negative charge. For example, the large protein and nucleic acid molecules are stabilized in part by hydrogen bonds between dipoles of the types shown in Figure 2–4. *Intermolecular hydrogen bonding* involves dipoles existing in different molecules. *Intramolecular hydrogen bonding* involves dipoles existing in the same molecule.

Although the energy of a single hydrogen bond is small, the presence of several *hydrogen bonds acting cooperatively* can impart considerable stabilization to a structure. One further point: Depending on the spatial orientation of the two dipoles, there are good and less good hydrogen bonds. The best situation occurs when both dipoles are coaxial (see Figure 2–5). The bond energy is less if they are coplanar but not coaxial. It is smallest when the dipoles are neither coplanar nor coaxial.

Van der Waals Bonding

Similar to hydrogen bonding, *van der Waals interactions* also involve attractive forces between dipoles. However, whereas hydrogen bonds involve permanent dipoles of highly polar covalent bonds such as O—H, van der Waals forces involve induced, nonpermanent dipoles of weakly polar covalent bonds such as C—H. Accordingly, the resultant bond energy for van der Waals attractive forces is less than that for hydrogen bonding.

$$>C=O\text{llllllll}H-N<\qquad >C=O\qquad\qquad >C=O$$

(a)   (b)   (c)

A nonpolar bond such as C—H results from the lack of any appreciable difference in the electronegativities of the atoms involved (refer to Note 2–1). Thus a naturally permanent dipole does not exist. However, when molecular orbitals of nonpolar bonds come in close contact, the clouds of negatively charged electrons begin to repel each other, distorting the uniform electron density in a molecular orbital. Thus a dipole is *induced*. The one dipole can induce another, establishing the basis for dipole/dipole interactions to occur (see Figure 2–6). Such interactions would be short-lived, however, because the bond is weak. As the atoms move away, the van der Waals influence is lost, and the dipole is lost.

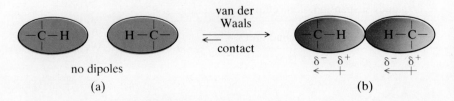

no dipoles
(a)

van der
Waals

contact

$\delta^-\ \ \delta^+$ $\delta^-\ \ \delta^+$

(b)

FIGURE 2–6 Van der Waals bonding. (a) Two nonpolar covalent bonds with a uniform electron density around the atoms. (b) Induced dipoles. At close range a distortion of electron density is induced, and it in turn induces a distortion in the other bond.

Induced dipoles can also form when nonpolar bonds are in van der Waals contact with permanent dipoles and fully charged ions. The respective types of interactions were included in the listing of noncovalent attractions given earlier in the section.

Apolar (Hydrophobic) Bonding

Nonpolar and polar substances are incompatible and cannot coexist as a single phase. However, one can disperse, in a polar solvent, substances that are *amphipathic,* having *both nonpolar and polar character.* The energetic driving force for forming such dispersions is generally called *apolar bonding*—or specifically, *hydrophobic bonding* (water-hating)—if the polar solvent is water.

MICELLES   Consider the example of oleic acid and water. At room temperature oleic acid is an oily liquid and very insoluble in water. We see from the formula in Figure 2–7 that a large part of the oleic acid structure is extremely nonpolar (hence its water insolubility), but owing to the carboxyl group —COOH, there is also some slight polar character.

When water and oleic acid in the —COOH form are shaken together, a white, cloudy suspension results, lasting for a few minutes before separating into the original two phases. The suspension consists of spherical particles called *micelles,* formed under the influence of hydrophobic effects (see Figure 2–8).

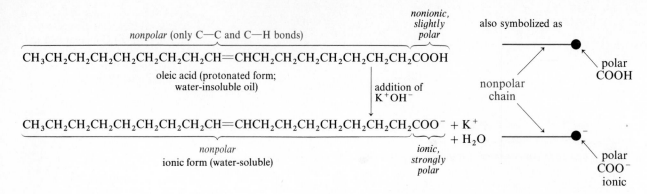

FIGURE 2–7 Oleic acid structure. Two forms are shown, each having a significant difference in polarity at one end of the molecule.

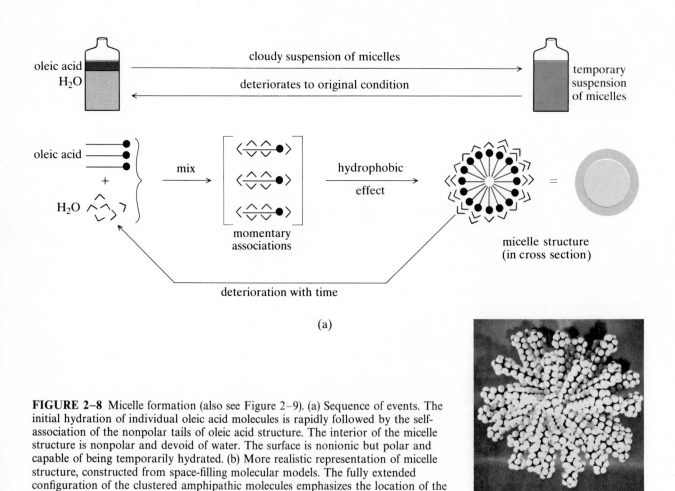

(a)

FIGURE 2–8 Micelle formation (also see Figure 2–9). (a) Sequence of events. The initial hydration of individual oleic acid molecules is rapidly followed by the self-association of the nonpolar tails of oleic acid structure. The interior of the micelle structure is nonpolar and devoid of water. The surface is nonionic but polar and capable of being temporarily hydrated. (b) More realistic representation of micelle structure, constructed from space-filling molecular models. The fully extended configuration of the clustered amphipathic molecules emphasizes the location of the polar groupings on the surface. (*Source:* Photograph provided by Dr. Frederick Menger.)

(b)

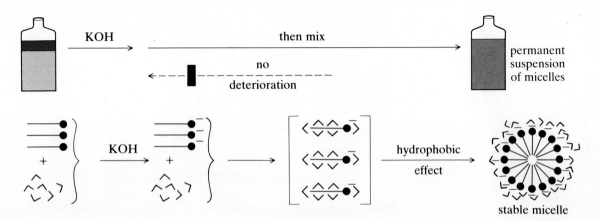

FIGURE 2–9 Micelle formation. Permanent stability can be achieved by converting oleic acid to its ionized $RCOO^-$ form, providing a polar and ionic surface.

When oleic acid molecules are first intermixed with water, a few water molecules tend to orient themselves around each oleic acid molecule, engaging primarily in permanent-dipole/induced-dipole interactions. However, since water is polar and much of the oleic acid structure is nonpolar, the arrangement is incompatible. The *tendency*—that is, the *hydrophobic effect*—is for the oleic acid molecules to attract each other and expel most of the water molecules from their initial orientation. The increased disorder of the water molecules is proposed to provide an entropic driving force for the spontaneous hydrophobic effect.

The mutual attraction of and water expulsion by hundreds of oleic acid molecules produces the micelle aggregate. The core of this aggregate, stabilized by induced-dipole/induced-dipole interactions, consists of the mutually attractive, nonpolar segments of oleic acid molecules with the polar carboxyl groups projecting on the surface. The micelle can be further stabilized as a *hydrated aggregate,* since the polar micelle surface can interact with polar water molecules via permanent-dipole/permanent-dipole interactions. However, this condition exists for only a few minutes because the interaction of —COOH groupings and H_2O is not strong enough (insufficient bond energy) for a permanent engulfment. The micelle aggregates deteriorate, and eventually, two separate layers are re-formed.

A dramatic event occurs if a small amount of strong base (such as KOH) is added before one mixes the oleic acid and water phases (see Figure 2–9). Stable micelles are formed. They form because the surface of the micelle aggregate now has ionic —COO⁻ groups, allowing for stronger hydration forces involving ion/permanent-dipole interactions—strong enough to overcome any tendency for the aggregate to dissociate.

Although the energy of hydrophobic associations is small, they are proposed to be the *major organizing influence* in arranging the structures of proteins, nucleic acids, and biomembranes.

LIPOSOMES   Various types of amphipathic lipid materials can be mixed with water under specific conditions to form *liposomes*. In contrast to the compact clustering of nonpolar structures in the core of a micelle, in a liposome

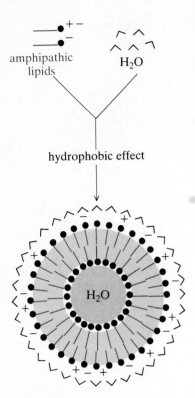

amphipathic lipids

H_2O

hydrophobic effect

H_2O

FIGURE 2–10 Liposome structure. As a result of the hydrophobic effect, certain amphipathic lipids (to be described in Chapter 11) form bilayer aggregates (color shading) with polar inner and outer surfaces. The cross-sectional drawing of a liposome illustrates that the spherical bilayer sheath encapsulates a small volume of water in which a water-soluble drug can be dissolved. After injection into the bloodstream, the drug-carrying liposomes can be internalized by cells and degraded inside the cell to release the drug. Liposome drug delivery prevents the degradation of the drug prior to its entry into cells.

the nonpolar lipids aggregate in a side-by-side fashion, two layers thick, to form concentric bilayer rings completely enclosing a region of water inside (see Figure 2–10). Liposomes have been used in research as artificial model systems simulating the structure and the behavior of biomembranes. More recently, they have been examined for their potential medical use as biodegradable carriers of enzymes and drugs to specific target cells.

2–3 ACID-BASE EQUILIBRIA

Most organic compounds that occur in living organisms exist in a charged state—they are *ions*. Depending on the nature of the charge, they are *anions* (negatively charged), *cations* (positively charged), or *ampholytes* (both negatively and positively charged). In this section we will review some basic principles of acid-base chemistry that explain how these ionic species are formed and that introduce the importance of regulating pH in living organisms.

Ionization

A process that results in the formation of ions is called *ionization*. The phenomenon was first described by Svante Arrhenius in the nineteenth century, and it served as the basis for his classification of acids and bases. An *acid* was defined as a substance that yields hydrogen ions (H^+, protons) upon ionization, and a *base* was defined as a substance that yields hydroxide ions (OH^-) upon ionization. He offered a further distinction in terms of the extent to which ionization occurred. A *strong acid or base* undergoes complete ionization (essentially 100%), and a *weak acid or base* undergoes partial ionization (considerably less than 100%). These statements are summarized by the equations in Figure 2–11.

In these equations a single arrow corresponds to complete dissociation and represents the absence of any undissociated acid or base in solution. A set of arrows corresponds to a dynamic equilibrium between the forward and reverse processes and represents the coexistence of an undissociated acid or base in solution with the corresponding ions. The longer arrow in the set indicates which process is favored and thus specifies which components will be in excess at equilibrium. The (aq) notation designates a water (aqueous) solution.

The majority of acids and bases found in nature are organic compounds and are weak acids or bases. Accordingly, emphasis will be given to these types of compounds. Before proceeding further, however, we will first review an alternative and more useful theory of acids and bases, the *Brönsted classification*, which defines an *acid as a proton-donating* species and a *base as a proton-accepting* species. To illustrate, the ionization of a hypothetical weak acid can be written (and in a more correct way) as

$$HA \ (aq) + H_2O \rightleftharpoons A^- \ (aq) + \ H_3O^+ \ (aq)$$
hydronium ion

General Examples

$$\text{Acids} \begin{cases} \text{HA (strong)} \xrightarrow{\text{H}_2\text{O}} \text{H}^+ \text{ (aq)} + \text{A}^- \text{ (aq)} \\[2em] \text{HA (weak)} \xleftrightarrow{\text{H}_2\text{O}} \text{H}^+ \text{ (aq)} + \text{A}^- \text{ (aq)} \end{cases}$$

$$\text{Bases} \begin{cases} \text{MOH (strong)} \xrightarrow{\text{H}_2\text{O}} \text{M}^+ \text{ (aq)} + \text{OH}^- \text{ (aq)} \\[2em] \text{MOH (weak)} \xleftrightarrow{\text{H}_2\text{O}} \text{M}^+ \text{ (aq)} + \text{OH}^- \text{ (aq)} \end{cases}$$

Specific Examples

$$\text{HCl} \xrightarrow{\text{H}_2\text{O}} \text{H}^+ \text{ (aq)} + \text{Cl}^- \text{ (aq)}$$
(hydrogen chloride)   (chloride ion)

$$\text{CH}_3\text{COOH} \xleftrightarrow{\text{H}_2\text{O}} \text{H}^+ \text{ (aq)} + \text{CH}_3\text{COO}^- \text{ (aq)}$$
(acetic acid)   (acetate ion)

$$\text{NaOH} \xrightarrow{\text{H}_2\text{O}} \text{Na}^+ \text{ (aq)} + \text{OH}^- \text{ (aq)}$$
(sodium hydroxide)   (sodium ion)

$$\text{NH}_4\text{OH} \xleftrightarrow{\text{H}_2\text{O}} \text{NH}_4^+ \text{ (aq)} + \text{OH}^- \text{ (aq)}$$
(ammonium hydroxide)   (ammonium ion)

FIGURE 2–11 Equations for the ionization of acids and bases. Color identifies what species will exist in solution.

Clearly, HA is donating a proton and thus is functioning as a Brönsted acid. The H_2O is accepting the proton from HA and is functioning as a Brönsted base. Similarly, for the reverse process H_3O^+ functions as a Brönsted acid and A^- as a Brönsted base. Because of the relationship between HA and A^-, they are referred to as a *conjugate acid-base pair*. That is, A^- is the conjugate base formed from the dissociation of the original Brönsted acid HA. A similar relationship exists between H_2O and H_3O^+. [*Note:* The hydronium ion, H_3O^+, is the common representation of the hydrated proton, $H^+(H_2O)$.] For reasons of simplification the H_3O^+ notation will not be routinely employed in the discussions to follow, and the presence of water will be assumed. Hence we will merely write

$$\text{conjugate acid} \rightleftharpoons \text{conjugate base} + \text{H}^+$$

Two very common weak-acid groupings that occur in organic compounds are the *carboxyl group* (—COOH) and the *protonated amino group* ($-\text{NH}_3^+$). Their ionization is summarized by the following equations:

$$\text{RCOOH} \rightleftharpoons \text{RCOO}^- + \text{H}^+$$

$$\text{RNH}_3^+ \rightleftharpoons \text{RNH}_2 + \text{H}^+$$

Here arrows of equal length merely specify that the system has the potential to reach an equilibrium condition when the rate of reaction in each direction is the same. The letter R is merely a symbol for the rest of the molecule. Recall that several other ionic species were identified earlier in the chapter (see Table 2–3).

Relative Strengths of Acids

The *relative strength* of a weak acid depends on its tendency to ionize. An acid that is 50% ionized will yield a greater amount of its conjugate base and H^+ than will one that is 5% ionized. At a given temperature the extent of ionization is a constant value that is characteristic of the acid. Since we are dealing with a system that can attain a state of *equilibrium* (that is, a state of balance, with reactants yielding products and being re-formed from products at the same rate), the *law of mass action* can be applied to describe the system. This application defines an *equilibrium constant K* in terms of the amounts of products and reactants that are present.

Since we are dealing in this case with an equilibrium system that results from the ionization of a weak acid, the equilibrium constant is called an *acid ionization constant* and is symbolized as K_a. It is expressed as follows, where [ ] represents molar concentration (moles per liter) of each substance present at equilibrium. For the acid ionization written as

$$\text{conjugate acid} \rightleftharpoons \text{conjugate base} + H^+$$

the equilibrium constant expression is

$$K = \frac{[\text{conjugate base}][H^+]}{[\text{conjugate acid}]} = K_a \qquad \textbf{(2–1)}$$

This expression is *valid for any weak monoprotic acid,* an acid (HA) having only one ionizable hydrogen.

As the formula implies, H_2A and H_3A acids would be classified as *diprotic* and *triprotic acids,* respectively, or *polyprotic acids* generally. In solution polyprotic acids undergo ionization through *separate and sequential steps.* Each step can be considered as the ionization of a monoprotic species; hence for each step a K_a expression can be written. This sequence of steps is illustrated in Table 2–4 for phosphoric acid, H_3PO_4, a weak triprotic acid.

The subscripts attached to the K_a expressions correspond to the successive steps of the ionization, which occur in the order given. The sequence is based on acid strength, with the strongest Brönsted acid dissociating first and the weakest last. Thus the three distinct monoprotic acids are arranged as follows, according to their relative acid strength:

$$H_3PO_4 > H_2PO_4^- > HPO_4^{2-}$$

From Table 2–4 we conclude that relative acid strength can be predicted by merely inspecting K_a values:

The greater the K_a value, the stronger is the acid.

So that the exponential term is avoided, K_a values are commonly expressed as pK_a values, defined as follows:

$$pK_a \equiv \log \frac{1}{K_a} = -\log K_a$$

TABLE 2–4 Ionization of phosphoric acid

| STEP | IONIZATION | IONIZATION CONSTANT | |
		EXPRESSION	VALUE (AT 25°C)
Step 1	$H_3PO_4 \rightleftharpoons H_2PO_4^- + H^+$	$K_{a_1} = \dfrac{[H_2PO_4^-][H^+]}{[H_3PO_4]}$	$= 7.56 \times 10^{-3}$
Step 2	$H_2PO_4^- \rightleftharpoons HPO_4^{2-} + H^+$	$K_{a_2} = \dfrac{[HPO_4^{2-}][H^+]}{[H_2PO_4^-]}$	$= 6.12 \times 10^{-8}$
Step 3	$HPO_4^{2-} \rightleftharpoons PO_4^{3-} + H^+$	$K_{a_3} = \dfrac{[PO_4^{3-}][H^+]}{[HPO_4^{2-}]}$	$= 5.0 \times 10^{-13}$
Overall	$H_3PO_4 \rightleftharpoons PO_4^{3-} + 3H^+$	$K_{a_{net}} = \dfrac{[PO_4^{3-}][H^+]^3}{[H_3PO_4]}$	$= (K_{a_1})(K_{a_2})(K_{a_3})$

For the ionizations of phosphoric acid $pK_{a_1} = 2.12$, $pK_{a_2} = 7.21$, and $pK_{a_3} = 12.3$.

Evaluations of acid strength with pK_a are just the inverse of evaluations with K_a:

The smaller the pK_a value, the stronger is the acid.

Let us return to our generalized equation, (2–1), for any weak monoprotic acid. An important conclusion regarding the nature of this system is apparent if both sides of the equation are multiplied by $1/[H^+]$. This operation gives

$$\frac{[\text{conjugate base}]}{[\text{conjugate acid}]} = \frac{K_a}{[H^+]}$$

and since K_a is a constant, and also since $pH = \log(1/[H^+])$, we have

$$\frac{[\text{conjugate base}]}{[\text{conjugate acid}]} \propto \frac{1}{[H^+]} \propto pH$$

This proportionality states the following:

At any given temperature the ratio of the equilibrium concentrations of the conjugate acid-base pair is solely dependent on the hydrogen ion concentration (pH) of the solution.

A corollary of this statement is as follows:

The pH of a solution consisting of a conjugate acid-base pair is solely dependent on the ratio of their equilibrium concentrations.

Since a living cell is an equilibrium mixture consisting of an abundance of weak acids and their conjugate bases, any change in the physiological pH can cause a significant change in the ionic composition of the cell.

2–4 PRINCIPLES OF pH BUFFERING

The presence of organic biomolecules in ionic forms is determined by the pH of the system. In addition, when two or more ionic forms of the same substance are possible, only one will usually predominate at a given pH. Because of this control by pH of the population of organic ions, living organisms must be capable of preventing excessive changes in the pH of intracellular and extracellular fluids, normally maintained at a pH value of approximately 7—the *physiological pH*. This control is accomplished through the action of *buffer systems*. Without buffers the pH and the ionic environment would be in a constant state of flux, a condition with the potential for serious physiological consequences. For example, a prolonged disturbance of the pH of blood (normally maintained in the narrow range of 7.36–7.42) can be fatal.

The use of buffers in biochemical research is also of immense importance. To undertake the study of whole cells or of subcellular particulates or in many cases to isolate and characterize purified biomolecules—particularly the biopolymers—without attention to pH control is totally impractical.

pH Buffering from Mixtures of Conjugate Acid-Base Pairs

Suppose you add 0.1 mole of solid KH_2PO_4 and 0.1 mole of solid K_2HPO_4 to water and adjust the final volume to 1000 mL (see Figure 2–12). The salts will dissolve and yield 0.1 mole of $H_2PO_4^-$, 0.1 mole of HPO_4^{2-}, and a total of 0.3 mole of K^+, all present in 1 L of solution. The pH of this solution is about 7.2. If now you add 1 mL of NaOH solution containing 0.01 mole of OH^-, the pH of the solution will increase to about 7.3—an increase of only 0.1 pH unit. In contrast, if the same amount of OH^- (0.01 mole) is added to the same volume of pure water with a pH of 7, the pH will change to 12—an

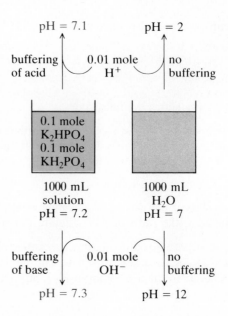

FIGURE 2–12 Results of buffering. Large changes in pH are avoided by the presence of phosphate.

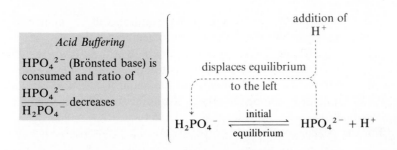

FIGURE 2–13 Acid buffering. As a new equilibrium is achieved, some of the newly formed $H_2PO_4^-$ will ionize to yield a little H^+, so the final system does have a slightly higher H^+ concentration—hence a lower pH—and is slightly more acidic.

increase of 5 pH units. Similar degrees of change apply if 1 mL of an HCl solution containing 0.01 mole of H^+ is added to each system, with the phosphate solution changing from pH 7.2 to 7.1 (decrease of only 0.1 pH unit) and pure water changing from pH 7 to 2 (decrease of 5 pH units). The less drastic pH changes are due to the phenomenon of buffering that occurs in the phosphate solution but not in pure water. This discussion leads us to the following definition:

> A buffer solution is a mixture of a conjugate acid-base pair that is capable of resisting large changes in pH when small amounts of another acid or base are added.

The reason for buffering is that a solution consisting of a conjugate acid-base pair can scavenge the addition of H^+ or OH^-. For H^+ addition the HPO_4^{2-} present acts as an H^+ acceptor to form $H_2PO_4^-$ (see Figure 2–13). Some of the newly formed $H_2PO_4^-$ ionizes to yield $HPO_4^{2-} + H^+$, but the amount of each that is returned is much less than what was consumed. Most of the added H^+ remains as $H_2PO_4^-$. Of course, this chemistry alters the composition of the conjugate acid-base solution: $H_2PO_4^-$ is increased and HPO_4^{2-} is decreased; that is, the base/acid ratio of $HPO_4^{2-}/H_2PO_4^-$ decreases. Such a solution will be capable of buffering further additions of H^+ until the ratio gets lower than $\frac{1}{9}$.

For OH^- addition the $H_2PO_4^-$ present acts as donor of H^+, which consumes the OH^- base (see Figure 2–14). More HPO_4^{2-} is then formed, a small amount of which also reacts with a small amount of H^+ to re-form a small amount of $H_2PO_4^-$. The result is that the amount of free H^+ is slightly less than what was originally present. The change in composition in this case is that $H_2PO_4^-$ is decreased and HPO_4^{2-} is increased; that is, the base/acid

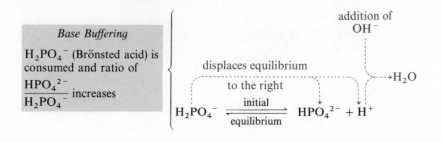

FIGURE 2–14 Base buffering. As a new equilibrium is achieved, some of the newly formed HPO_4^{2-} will also accept a little H^+ to re-form $H_2PO_4^-$, so the final system does have a slightly lower H^+ concentration—hence a higher pH—and is slightly more basic.

ratio of $HPO_4^{2-}/H_2PO_4^-$ increases. Such a solution will be capable of buffering further additions of OH^- until the ratio gets larger than $\frac{9}{1}$.

Within the effective range of $\frac{1}{9}$ to $\frac{9}{1}$ for the base/acid ratio where buffering occurs, the effectiveness of buffering is not constant. Minimum efficiency exists at the stated limits, and *maximum efficiency exists at a base/acid ratio of 1/1*. This 50–50 (equimolar) composition has still another meaning, provided by the following analysis.

For the system

$$H_2PO_4^- \rightleftharpoons HPO_4^{2-} + H^+$$

which is a specific case of

$$\text{conjugate acid} \rightleftharpoons \text{conjugate base} + H^+$$

we have stated that

$$K_a = \frac{[\text{base}][H^+]}{[\text{acid}]} \qquad \text{or} \qquad [H^+] = K_a \frac{[\text{acid}]}{[\text{base}]}$$

Taking the logarithm of both sides of the $[H^+]$ equation, we have

$$\log[H^+] = \log K_a + \log \frac{[\text{acid}]}{[\text{base}]}$$

Multiplying through by a minus sign yields

$$-\log[H^+] = -\log K_a - \log \frac{[\text{acid}]}{[\text{base}]}$$

which can be rewritten as

$$-\log[H^+] = -\log K_a + \log \frac{[\text{base}]}{[\text{acid}]}$$

Now using the substitutions $pH = -\log[H^+]$ and $pK_a = -\log K_a$, we have the following relationship, sometimes called the *Henderson-Hasselbalch equation*:

$$pH = pK_a + \log \frac{[\text{base}]}{[\text{acid}]} \qquad (2\text{--}2)$$

Finally, a meaning of a base/acid ratio of 1/1 is evident:

$$pH = pK_a + \log(1)$$

and since $\log(1) = 0$,

$$pH = pK_a$$

Thus:

> The pH of an equimolar (50–50) solution of a conjugate acid/base pair is equal to its pK_a value. Since such a solution has maximum buffering effectiveness, this value is also the pH at which the solution buffers best.

> The values of $\log(\frac{1}{9})$ and $\log(\frac{9}{1})$ correspond roughly to -1 and $+1$. Hence: The range of pH where buffering is effective corresponds to $pK_a \pm 1$ and is maximum at the pK_a value.

This range applies to any buffer system. For an $HPO_4^{2-}/H_2PO_4^{-}$ solution the range is 6.2–8.2 ($pK_{a_2} = 7.2$).

A graphical summary of our analysis is shown in Figure 2–15. The *titration curve* in this graph depicts two patterns of pH change: (1) from low pH to high pH as strong base is added to a solution of the weak acid (such as $H_2PO_4^{-}$), or looking at it the other way, (2) from high pH to low pH as strong acid is added to a solution of a weak base (such as HPO_4^{2-}). The flatter midsection of the curve corresponds to the buffering region, with the midpoint of that region corresponding to the pK_a value. The shape of this curve applies to *any weak monoprotic acid*. An excess of the base species at pH values on the high side of the pK_a value and an excess of the acid species at pH values on the low side of the pK_a value are also universal interpretations.

The curve in Figure 2–16(a) depicts the process of titrating a weak diprotic acid, and the curve in Figure 2–16(b) shows the process for a weak triprotic acid. Note that each ionization represents a buffering system in different pH regions.

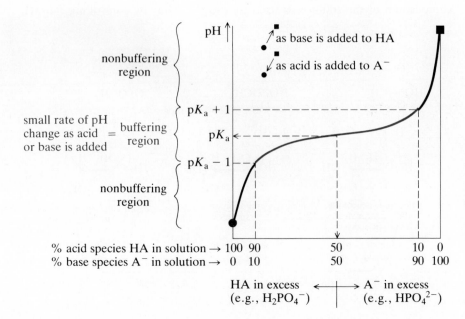

FIGURE 2–15 Titration curve for a weak monoprotic acid. The buffering action is confined to the pH range of $pK_a + 1$ to $pK_a - 1$. This curve is a graphical representation of the Henderson–Hasselbalch equation for the two variables: (1) pH on the vertical axis and (2) composition of HA and A^- on the horizontal axis. As pH changes, the composition changes, and vice versa.

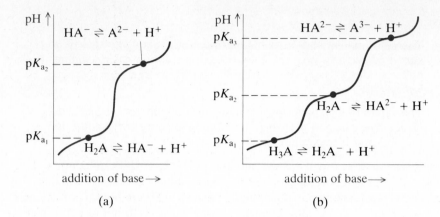

FIGURE 2–16 Generalized titration curves. (a) For a weak diprotic acid. (b) For a weak triprotic acid.

Recapitulation

The expression

$$pH = pK_a + \log \frac{[base]}{[acid]}$$

defines the pH of a solution of a conjugate acid-base pair in terms of the ionization constant of the acid K_a and the relative amounts of the acid and the base. Since K_a is a constant, the pH for a solution of this type must be fixed by the composition of the mixture. Thus by controlling the composition, you can control the pH. We will apply this principle in a later section when describing the preparation of a buffer with specific pH for laboratory use.

There is also an important corollary to this principle—namely, that the composition of the solution is fixed by the pH. Thus by controlling the pH,

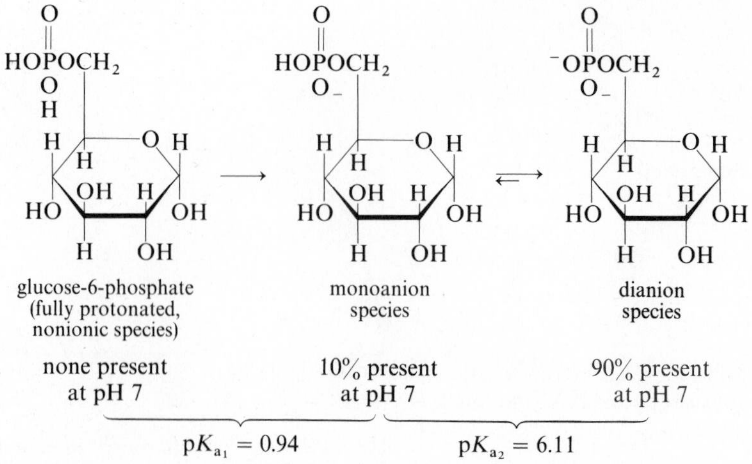

FIGURE 2–17 Glucose-6-phosphate. The structure on the right is the predominant form of glucose-6-phosphate at pH 7.

you can control the composition of the solution. That is, you can control the types of chemical species that are present in solution. It is for this reason that living organisms utilize buffers to control pH. To put it simply:

> The control of physiological pH ensures that the amino acids, nucleotides, proteins, nucleic acids, lipids, and most other biomolecules are maintained in certain ionic forms that are most suited for their structure, function, and even their solubility in water.

As one example, consider the phosphate esters of the simple sugars represented by glucose-6-phosphate (see Figure 2–17). The fully protonated species contains two ionizable hydrogens with $pK_{a_1} = 0.94$ and $pK_{a_2} = 6.11$. In a system (for example, in a living cell) that maintains pH at about 7, which is much greater than 0.94 and about 1 pH unit greater than 6.11, the fully protonated form would not exist at all, about 10% would exist as the monoanion, and about 90% would exist as the dianion.

2–5 BUFFERING OF BLOOD

An interesting illustration of natural buffering in biological systems is provided by what occurs in blood, where the pH is maintained within relatively narrow limits of 7.36–7.40 for venous blood and 7.38–7.42 for arterial blood. Although the inorganic phosphate buffer system of $H_2PO_4^- \rightleftharpoons HPO_4^{2-} + H^+$ is a major natural buffer system, it is of secondary importance in blood because the phosphate concentration in blood is not large enough to accommodate all of the H^+ produced. The bulk of pH buffering in blood is accomplished through the action of three distinct conjugate acid-base pair systems: (1) H_2CO_3 and HCO_3^-, (2) the acid and base species of oxygenated hemoglobin, and (3) the acid and base species of deoxygenated hemoglobin. Before we examine the intricate manner in which these systems operate, a brief description of hemoglobin is needed.

Hemoglobin is an iron-containing protein found in the red blood cells (erythrocytes). It is the most abundant protein in blood, with normal values in the range of 14–16 g/100 mL of whole blood, representing a concentration of about 2 mM (millimolarity). The primary role of all this hemoglobin is to transport oxygen from the lungs to respiring tissues. At the lungs oxygen O_2 molecules enter the erythrocyte and become bound to iron atoms in hemoglobin molecules (see Section 4–5 for details). This *oxygenated hemoglobin* (HbO$_2$) is then carried in the arterial blood to the various tissues of the body, where the oxygen is released. The *deoxygenated hemoglobin* (Hb) then returns to the lungs in the venous blood, ready to participate in the same cycle.

To appreciate what all of this discussion has to do with buffering the pH of blood, consider the following items:

1. Some years ago Bohr determined that the efficiency of O_2 binding was significantly reduced by lowering the pH. Known as the *Bohr effect*, this phenomenon means that the dissociation of oxygenated hemoglobin to deoxygenated hemoglobin would be favored by the presence of H^+. The equation is

$$H^+ + HbO_2 \xrightarrow[\substack{\text{for } O_2 \text{ and thus} \\ \text{dissociation is} \\ \text{favored}}]{\substack{\text{binding of} \\ H^+ \text{ decreases} \\ \text{affinity of Hb}}} HHb + O_2$$

2. When CO_2 enters the erythrocyte, it is rapidly converted to the weak acid H_2CO_3 by action of the important Zn^{2+}-dependent enzyme, *carbonic anhydrase* (see p. 192). How much of the H_2CO_3 ionizes? Since the pH of the cell is about 7.4 and the pK_{a_1} of H_2CO_3 is 6.35, about 90% or more will be ionized to HCO_3^- plus H^+. Thus the input of CO_2 causes an increase in the H^+ concentration, which could cause a potential increase in the acidity of blood if the free H^+ were not eliminated. The equation for the reaction is

$$CO_2 + H_2O \xrightarrow[\substack{\text{anhydrase} \\ (Zn^{2+})}]{\text{carbonic}} H_2CO_3 \underset{\substack{\text{at pH of} \\ \text{blood}}}{\overset{\text{dissociation}}{\rightleftharpoons}} HCO_3^- + H^+$$

3. Like every protein, hemoglobin contains ionizable groupings contributed by some of the amino acids of which it is composed (see Chapter 3). However, a titration curve of a protein is similar in shape to that for a simple compound containing only a single ionizable group, such as the curve shown in Figure 2–15 for $H_2PO_4^-$. The curve would show a single, broad buffering region, and the midpoint would be the pK_a of the protein. Oxygenated hemoglobin ($HHbO_2$) has a pK_a of 6.62, while deoxygenated hemoglobin (HHb) has a pK_a of 8.18. (*Note:* The symbolic abbreviation for each of the two hemoglobins now includes an H, referring to an ionizable hydrogen.) The difference in these pK_a values means that the binding of oxygen has changed a property of the hemoglobin. Specifically, *oxygenated hemoglobin is a stronger acid than deoxygenated hemoglobin*. Moreover, at the pH of blood the equilibrium concentration of each respective conjugate acid-base pair will be quite different. For oxygenated hemoglobin the base form (HbO_2^-) will predominate, whereas for deoxygenated hemoglobin the acid form (HHb) will predominate. The reactions are as follows:

for oxygenated Hb with $pK_a = 6.62$ $HHbO_2 \xrightleftharpoons{\text{at pH} \approx 7.4} HbO_2^- + H^+$
                                                                   (predominates)

for deoxygenated Hb with $pK_a = 8.18$ HHb $\xleftarrow{\text{at pH} \approx 7.4}$ $Hb^- + H^+$
                      (predominates)

All three of these items fit nicely into an explanation of pH control in blood. When CO_2 enters the cell, the H^+ that is generated from H_2CO_3 can react with the predominating base form of oxygenated hemoglobin (HbO_2^-) to form $HHbO_2$. The $HHbO_2$ has a reduced affinity for oxygen (the Bohr effect) and dissociates to yield the acid form of deoxygenated hemoglobin (HHb) and free oxygen. See Figure 2–18. The oxygen diffuses from the erythrocyte and

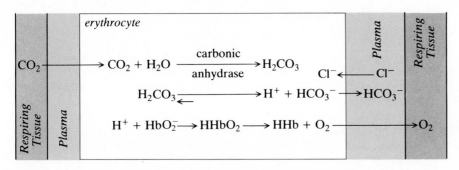

FIGURE 2–18 Events in the erythrocyte when arterial blood is delivered to respiring tissue. An erythrocyte is actually disk-shaped, with a diameter of about 8 μm (see the micrograph in Figure 1–2).

enters the cells of respiring tissues. Because of its high pK_a value, most of the HHb will not ionize at the pH of blood but, rather, will remain as HHb. Thus the increased amount of H^+ caused by the diffusion of CO_2 into the cell has been scavenged by hemoglobin.

The CO_2 is carried in the plasma of the venous blood as HCO_3^-. So that an imbalance of the ionic environment is prevented, this diffusion of HCO_3^- from the erythrocyte is counterbalanced by the movement of Cl^- from the plasma into the erythrocyte, a phenomenon called the *chloride shift.*

When the venous blood reaches the lungs, O_2 and HCO_3^- enter the erythrocyte and Cl^- exits. The O_2 binds to the major hemoglobin species that is present at the pH of blood—namely, HHb—to produce $HHbO_2$. See Figure 2–19. Now, however, the $HHbO_2$ will function as an acid in the presence of HCO_3^- to yield HbO_2^- and H_2CO_3. By the action again of carbonic anhydrase, the H_2CO_3 is converted to H_2O and CO_2, and the latter diffuses into the plasma and ultimately into the lungs.

Although the preceding description oversimplifies things a little, it is a reasonable summary of important events associated with the processes by which O_2 is delivered and CO_2 is eliminated by the circulating blood without any serious alteration in blood pH—despite the production of H^+ from H_2CO_3. At the very most, the pH of venous blood is decreased by only a few hundredths of a pH unit. Thus hemoglobin not only is responsible for the transport of oxygen but also participates as an efficient buffer.

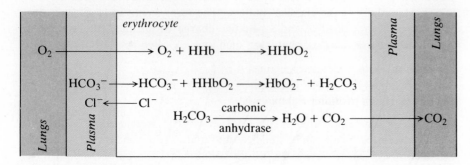

FIGURE 2–19 Events in the erythrocyte when venous blood is delivered to the lungs.

Before leaving this subject, I want to focus again on the change of the acid strength of hemoglobin upon oxygenation.

This example is a specific illustration of how the binding of a small, nonprotein molecule to a much larger protein molecule can cause a significant change in the function of that protein molecule. In view of the structure-function principle, you should appreciate—at least in general terms for now—that the explanation is that the event of binding (of oxygen in this case) modifies the structure of the protein in some fashion and that this modification in turn alters the activity of the protein.

More will be said of this simple but immensely important principle in later chapters.

2–6 LABORATORY USE OF BUFFERS

Requirements for Biological Buffers

The primary naturally occurring buffer systems are the phosphate ($H_2PO_4^-/HPO_4^{2-}$) system, the dissolved proteins, and many weak organic acids. However, as was pointed out earlier, the routine use of buffers in laboratory studies is equally important. Fundamental to both in vivo and in vitro studies is to first establish a physiological pH in the system and then to prevent it from changing appreciably. Moreover, in certain instances the intention may be to determine the effect of pH on the system under study, requiring the preparation of several solutions, each with different pH values.

What considerations apply in selecting a buffer? The most basic requirement is that the pK_a *value be close to the* pH *at which you intend to use the buffer.* Remember that a buffer system buffers best at a pH that equals its pK_a value. For physiological buffering the pK_a should be close to 7 or in a range of about 6–8. Other requirements are *nontoxicity, stability,* and *water solubility.*

Table 2–5 identifies some buffer systems in routine use. In addition to the long-used buffer systems of phosphate and N-tris(hydroxymethyl)aminomethane (abbreviated as TRIS), the list includes two examples of increasingly popular *zwitterionic substances,* which for reasons not yet fully understood often prove superior as biological buffers. (They are called "Good buffers" after N. E. Good who pioneered in the development of these substances to overcome various shortcomings of the classical buffers.) The term *zwitterionic* refers to the presence of both a positive and a negative charge within the same molecular species. Because of this dual character, such ions are often referred to as *dipolar ions.*

Preparation of Buffer Solutions

Buffers can be prepared in either of two ways: (1) both components of the conjugate acid-base pair can be weighed out separately to yield the desired ratio and then dissolved in water; or (2) both components can be obtained from a prescribed amount of only one component, with the second being formed by the

TABLE 2–5 pK_a values for some conjugate acid-base buffer pairs

ACID		BASE	pK_a
Dihydrogen phosphate $H_2PO_4^-$	$\rightleftharpoons$	Monohydrogen phosphate HPO_4^{2-}	7.2
N-Tris(hydroxymethyl)aminomethane (TRIS)			
TRIS·H$^+$ (*protonated form*) $(HOCH_2)_3CNH_3^+$	$\rightleftharpoons$	TRIS (*free amine*) $(HOCH_2)_3CNH_2$	8.3
N-Tris(hydroxymethyl)methyl-2-aminoethane sulfonate (TES)			
TES·$\overset{+}{H}$ (*zwitterionic form*) $(HOCH_2)_3C\overset{+}{N}H_2CH_2CH_2SO_3^-$	$\rightleftharpoons$	TES (*anionic form*) $(HOCH_2)_3CNHCH_2CH_2SO_3^-$	7.55
N-2-Hydroxyethylpiperazine-N′-2-ethane sulfonate (HEPES)			
HEPES·$\overset{+}{H}$ (*zwitterionic form*) $HOCH_2CH_2\overset{+}{N}\langle\rangle NCH_2CH_2SO_3^-$ $\quad H$	$\rightleftharpoons$	HEPES (*anionic form*) $HOCH_2CH_2N\langle\rangle NCH_2CH_2SO_3^-$	7.55

Note: The abbreviations in capital letters for the last three materials are routinely used shorthand designations.

addition of a specified amount of strong acid or strong base. The tools needed to prepare both types are a knowledge of the conjugate acid-base system involved, the pK_a for the system, and the Henderson-Hasselbalch equation. An example of each method will be presented.

TYPE 1 Both components are weighed out separately.

EXAMPLE 2–1 The purpose is to prepare 1 L of a $0.5M$ phosphate buffer at pH 7.5. Assume the availability of H_3PO_4, KH_2PO_4, K_2HPO_4, and K_3PO_4. How should the buffer be prepared?

Solution

Step 1. The first step is always to determine what the principal components of the buffer system will be. This determination is no problem with a monoprotic acid. With a diprotic or polyprotic system, however, the components can vary, depending on the desired pH. In this instance pH 7.5 is specified. The equilibrium system will thus be determined by selecting the ionization having the pK_a value closest to the desired pH. So the desired system is $H_2PO_4^-/HPO_4^{2-}$, with a pK_a value of 7.21. The $H_3PO_4/H_2PO_4^-$ system has a pK_a of 2.12, and the HPO_4^{2-}/PO_4^{3-} system has a pK_a of 12.3. These values are too low and too high, respectively, and neither would be an effective buffer at pH 7.5. Having established the components, we then write the equilibrium equation and identify the conjugate acid-base pair:

$$H_2PO_4^- \rightleftharpoons HPO_4^{2-} + H^+ \qquad pK_{a_2} = 7.21$$
Brönsted acid   Brönsted base

Step 2. Calculate the desired ratio of the acid-base pair from the Henderson-Hasselbalch equation:

$$pH = pK_{a_2} + \log \frac{[HPO_4{}^{2-}]}{[H_2PO_4{}^-]}$$

$$\log \frac{[HPO_4{}^{2-}]}{[H_2PO_4{}^-]} = 7.5 - 7.21 = 0.29$$

$$\frac{[HPO_4{}^{2-}]}{[H_2PO_4{}^-]} = \text{antilog}(0.29) = 1.95$$

Thus the ratio desired is 1.95 parts of $HPO_4{}^{2-}$ to 1 part of $H_2PO_4{}^-$. Since this ratio represents a total of 2.95 parts, we can calculate directly the percentage of each component:

$$\% \ HPO_4{}^{2-} = \frac{1.95}{2.95} \times 100 = 66.2$$

$$\% \ H_2PO_4{}^- = \frac{1.00}{2.95} \times 100 = 33.8$$

As a check on the solution to this point, determine whether the ratio is consistent with the desired pH. The desired pH is on the alkaline side of the pK_a value. Thus there should be a larger concentration of the conjugate base than of the conjugate acid. This result is verified by the ratio calculated.

Step 3. Determine the most feasible means of obtaining the desired components. In this instance the obvious choice is to weigh the desired amount of the potassium salts of the acid-base pair—namely, K_2HPO_4 and KH_2PO_4—which upon dissolution will ionize completely to give both components of the conjugate pair.

NOTE 2–3
Gram formula weights:

$K_2HPO_4 = 174.2$ g/mole
$KH_2PO_4 = 136.1$ g/mole

Step 4. Calculate the amount of each material required. Since the total phosphate concentration was specified as $0.5M$, and since 1 L is desired,

number of moles of K_2HPO_4 required/liter $= (0.662)(0.5) = 0.331$

number of moles of KH_2PO_4 required/liter $= (0.338)(0.5) = 0.169$

Finally, the grams of each required (see Note 2–3) are

$$(0.331 \text{ mole})(174.2 \text{ g/mole}) = 57.7 \text{ g for } K_2HPO_4$$

and $\qquad (0.169 \text{ mole})(136.1 \text{ g/mole}) = 23.0 \text{ g for } KH_2PO_4$

NOTE 2–4
Because K_a is affected by temperature and by the concentration of components in solution, the observed pH at room temperature may not exactly agree with the pH expected. If the buffer is to be used at refrigerated temperature (in the cold room), the final pH check and adjustment must be done in the cold room. Preparing a buffer from a pK_a value that applies at 25°C but using the buffer at 2°–4°C can yield pH values lower than expected by as much as $\frac{1}{2}$ pH unit.

Step 5. Prepare the buffer. Weigh 23.0 g of KH_2PO_4 and 57.7 g of K_2HPO_4, and dissolve these components in about 750 mL of distilled water. At this point check the pH with a pH meter, and adjust if necessary (see Note 2–4). Finally, bring the total volume of the solution to 1 L with distilled water.

TYPE 2. Both components are obtained from the same source.

EXAMPLE 2–2 We need to prepare 1 L of $0.1M$ TRIS buffer of pH 8.3. Assume the availability of crystalline TRIS, $1M$ HCl, and $1M$ NaOH. How should you proceed? (*Note:* In the crystalline state TRIS exists primarily as the free amine—that is, with $-NH_2$.)

Solution

Step 1. The desired equilibrium is

$$(HOCH_2)_3CNH_3{}^+ \rightleftharpoons (HOCH_2)_3CNH_2 + H^+ \qquad pK_a = 8.3$$

Brönsted acid   Brönsted base
protonated   free amine

Step 2. Calculate the base-to-acid ratio by use of the Henderson-Hasselbalch equation:

$$pH = pK_a + \log \frac{[\text{free amine}]}{[\text{protonated amine}]}$$

$$8.3 = 8.3 + \log \frac{[-NH_2]}{[-NH_3{}^+]}$$

$$\frac{[-NH_2]}{[-NH_3{}^+]} = \text{antilog}(0.0) = 1.0$$

Thus the solution should contain 50% free amine and 50% protonated species. Clearly, this step was not really necessary, because a 50–50 acid-base mixture will always exist when the pH of the buffer is equal to the pK_a.

Step 3. Both buffer components will be formed from crystalline TRIS. One-tenth (0.1) mole of crystalline TRIS will be required to yield 1 L of buffer with a total TRIS concentration of $0.1M$. The next problem is to determine the amount of strong acid that will be required to form the desired composition of the acid-base pair. Since the mixture must contain 50% of the protonated species, 0.05 mole of strong acid will be required. This amount will produce 0.05 mole of the $-NH_3{}^+$ form, and 0.05 mole of the $-NH_2$ form will remain; 0.05 mole of H^+ will be provided by 50 mL of $1M$ HCl.

Step 4. Prepare the buffer. Weigh 0.1 mole (12.1 g) of crystalline TRIS and dissolve it in approximately 500 mL of water. Add 50 mL of $1M$ HCl and mix. Check the pH with a pH meter, and adjust if necessary. Bring the total volume to 1 L with water.

2–7 IONIC STRENGTH

A complete description of an ionic solution would consider the types of ions present and the amount of each type. Chemists express this description in terms of the *ionic strength* (symbolized as I), defined as follows:

$$I = \frac{1}{2}\sum_i c_i z_i{}^2 \qquad \textbf{(2–3)}$$

where c_i = molar concentration of the ith ionic species

z_i = electrostatic charge of the ith ionic species

$\sum\limits_i$ = symbol for the summation of all cz^2 terms for each ionic species in solution

Note that the ionic strength is a solution property related to both the concentration of the ions in solution and the ionic nature (that is, the charge) of the ions. Cellular activity is a function of both.

Generally speaking, the ionic strength of solutions used for cellular studies is approximately $0.15M$. This value is commonly referred to as the *optimum physiological ionic strength.* It is considered optimal because, among other things, it maintains a normal water balance in living cells. When whole cells are placed in a solution that does not contain an ionic concentration at least approximately equivalent to the ionic concentration of the intracellular fluid (protoplasm), one of two undesirable things will occur. When the ionic strength of the extracellular fluid is much less than that of the protoplasm, water will enter the cell, causing it to swell and ultimately burst. Alternatively, if the ionic strength of the extracellular fluid is much greater than that of the protoplasm, water will leave the cell, causing it to shrink and collapse. In one case the cell is flooded, and in the other it is dehydrated (see Note 2–5).

The utilization of Equation (2–3) is illustrated next.

NOTE 2–5
The intravenous administration of sterile physiological saline (0.9% NaCl with $I = 0.154M$) to treat dehydration, to prevent postoperative shock, and to replace fluid lost because of hemorrhage is standard medical practice. This solution has no effect on the red blood cells since 0.9% NaCl is *isotonic* with blood, having the same salt concentration and hence the same osmotic pressure as blood. A 5.5% glucose solution can also be used.

EXAMPLE 2–3 Calculate the ionic strength of physiological saline solution that is $0.154M$ NaCl.

Solution

$$I = \frac{1}{2} \sum_i c_i z_i^2$$

$$NaCl \rightarrow Na^+ + Cl^-$$
$$0.154M \quad 0.154M \quad 0.154M$$

Thus

$$I = \tfrac{1}{2}\left[(c_{Na^+} z_{Na^+}^2) + (c_{Cl^-} z_{Cl^-}^2)\right]$$

$$= \tfrac{1}{2}\left[(0.154)(1)^2 + (0.154)(1)^2\right]$$

$$= 0.154M$$

EXAMPLE 2–4 Calculate the ionic strength of the $0.5M$ phosphate buffer (pH 7.5) prepared in Example 2–1.

Solution The buffer contains

$$K_2HPO_4 \rightarrow 2K^+ + HPO_4^{2-} \quad \text{and} \quad KH_2PO_4 \rightarrow K^+ + H_2PO_4^-$$
$$0.331M \quad 2(0.331M) \quad 0.331M \qquad\qquad 0.169M \quad 0.169M \quad 0.169M$$

Therefore

$$I = \tfrac{1}{2}\left[c_{K^+}z^2_{K^+} + c_{H_2PO_4^-}z^2_{H_2PO_4^-} + c_{HPO_4^{2-}}z^2_{HPO_4^{2-}}\right]$$
$$= \tfrac{1}{2}\left[(0.831)(1)^2 + (0.169)(1)^2 + (0.331)(2)^2\right]$$

Note that

$$\text{total } c_{K^+} = 2(0.331) + 0.169 = 0.831$$

So

$$I = 1.16M$$

LITERATURE

BAILAR, J. C., JR. "Some Coordination Compounds in Biochemistry." *Am. Sci.*, **59**, 586 (1971). A review of the metal ions in life processes.

DOUZOU, P., and P. MAUREL. "Ionic Control of Biochemical Reactions." *Trends Biochem. Sci.,* **2**, 14–17 (1977). A short article on the biological significance of ionic strength.

HUGHES, M. N. *The Inorganic Chemistry of Biological Processes.* New York: Wiley, 1975. An introduction to the biochemistry of metal ions. Written for chemists.

MONTGOMERY, R., and C. A. SWENSON. *Quantitative Problems in the Biochemical Sciences.* 2nd ed. San Francisco: Freeman, 1976. Chapters 6, 7, and 8 of this problems book are devoted to ionization and buffers.

PAULING, L. *The Nature of the Chemical Bond.* 3rd ed. Ithaca, N.Y.: Cornell University Press, 1960. A classic description of chemical bonding. Hydrogen bonding is discussed in Chap. 12.

SEGEL, I. H. *Biochemical Calculations.* 2nd ed. New York: Wiley, 1976. A book on how to solve mathematical problems in general biochemistry that can be used in conjunction with standard textbooks. A large number of problems—solved in detail—are accompanied by descriptive background. Approximately one-third of the book is devoted to ionization, titrations, and buffer systems.

SOBER, H. A., ed. *Handbook of Biochemistry: Selected Data for Molecular Biology.* 3rd ed. Cleveland: The Chemical Rubber Company, 1975. A multivolume, in-depth compilation of evaluated data for those engaged in biochemical research. All of the main types of biologically occurring materials are treated.

WILLIAMS, V. R., W. L. MATTICE, and H. B. WILLIAMS. *Basic Physical Chemistry for the Life Sciences.* 3rd ed. San Francisco, London: Freeman, 1978. An excellent text of physicochemical principles designed for students in the life sciences. Acid-base equilibria and buffers are treated in Chap. 4.

EXERCISES

2–1. (a) Disregarding any influence of R groups, classify each of the following functional groups as polar or nonpolar. (b) Which groups do you predict are capable of engaging in hydrogen bonding?

R_3C—S—CR_3	RSO_3H	RCH_3
thioether (sulfide)	sulfonic acid	methyl

$$\overset{O}{\overset{\|}{RC}}-NH_2 \qquad R-\!\!\bigcirc\!\!- \qquad RCH_2OH$$

amide   phenyl   hydroxymethyl

2–2. Do you predict apolar bonding and micelle formation to occur when oleic acid in its RCOOH form is mixed with the liquid solvent carbon disulfide, CS_2? Explain your answer.

2–3. Identify the two conjugate acid-base pairs of the amino acid, glycine, shown here in fully protonated form. For glycine the pK_a value of the ionizable —COOH group is several pH units less than the pK_a value of the ionizable —NH_3^+ group.

glycine $H_3\overset{+}{N}CH_2COOH$

2–4. Carbonic acid, H_2CO_3, has $pK_{a1} = 6.35$ and $pK_{a2} = 10.3$. (a) Draw a titration curve reflecting the pattern of ionization of H_2CO_3 as strong base is added. (b) Identify the regions (pH ranges) where buffering occurs. (c) Identify the predominant chemical species at each of the following pH values: 4, 5, 6, 7, 8, 9, 10, 11, and 12. (d) Identify the pH at which the HCO_3^- species will exist in a 50–50 equilibrium with its conjugate base.

2–5. With an appropriate calculation, verify the statement on p. 54 that more than 90% of the H_2CO_3 formed in erythrocyte cells will ionize to HCO_3^-.

2–6. Most proteins contain some arginine, an amino acid consisting in part of a guanidine ionizable group (see Table 2–3). The pK_a of the monoprotic guanidinium species is about 12. At physiological pH, will the guanidine group of arginine contribute a positive charge, a negative charge, or no charge to the protein?

2–7. In living cells many organic compounds exist as phosphoesters: $ROPO_3H_2$. An example is 3-phosphoglyceraldehyde, shown here with ionizable hydrogens identified in color. (a) Draw the structure of the species of 3-phosphoglyceraldehyde that would predominate under physiological conditions. Explain the basis of your selection. (b) Calculate an estimate of the percentage of this species in solution.

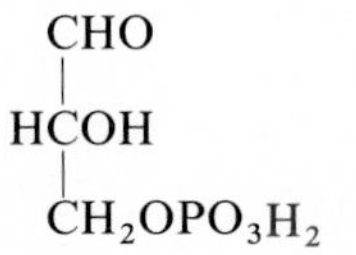

$$
\begin{array}{l}
\text{CHO} \\
| \\
\text{HCOH} \qquad pK_{a_1} = 2.10 \qquad pK_{a_2} = 6.80 \\
| \\
\text{CH}_2\text{OPO}_3\text{H}_2
\end{array}
$$

2–8. Which buffer would be a better choice to protect against pH changes arising from the generation or addition of acid: (a) $0.1M$ phosphate solution at pH 7.71, or (b) $0.1M$ phosphate solution at pH 6.71? Explain your answer.

2–9. Explain why buffer solutions of different concentrations of the same components can be prepared with the same pH. For example:

0.01M phosphate buffer at pH 7.0

0.05M phosphate buffer at pH 7.0

0.10M phosphate buffer at pH 7.0

1.00M phosphate buffer at pH 7.0

2–10. A HEPES buffer solution was prepared by dissolving 0.012 mole of the zwitterionic HEPES species and 0.038 mole of the anionic HEPES species in water and adjusting the final volume of the solution to 1 L. What is the pH of this buffer solution?

2–11. How do you prepare 1 L of a $0.04M$ phosphate buffer at pH 6.6 from crystalline KH_2PO_4? Assume $6M$ NaOH and $6M$ HCl solutions are available.

2–12. How do you prepare the buffer described in Exercise 2–11 from crystalline Na_3PO_4 (gram formula weight = 164) and a solution of $1M$ HCl?

2–13. How do you prepare the buffer described in Exercise 2–11 by using crystalline sources of K_2HPO_4 and KH_2PO_4?

2–14. Crystalline TRIS can be purchased as either the free amine form $[(HOCH_2)_3CNH_2$; molecular weight = 121 g] or as a hydrochloride salt form $[(HOCH_2)_3CNH_3^+Cl^-$; molecular weight = 157.5 g]. If your lab had a supply of the hydrochloride form, how would you prepare a $0.1M$ TRIS buffer of pH 8.5? Assume the availability of $1M$ HCl and $1M$ NaOH solutions.

2–15. Compute the ionic strength of each buffer prepared in Exercises 2–11, 2–12, and 2–13.

2–16. Define the following terms: buffering, ionic strength, physiological pH and physiological ionic strength, acid strength evaluation via pK_a comparison, Bohr effect, carbonic anhydrase, zwitterion, polar and nonpolar bonds, hydrogen bonding, ionic bonding, hydrophobic bonding, van der Waals bonding, liposomes.

CHAPTER THREE

AMINO ACIDS AND PEPTIDES

Because proteins are involved in a greater number and a greater variety of cellular events than any other type of biomolecule, there is a distinct emphasis in the science of biochemistry on proteins. In fact, in one way or another you will be learning about proteins throughout most of this course.

Each protein is unique in terms of its structure and function. We will begin to deal with the reasons for and examples of this individuality in the next chapter. There are also a lot of similarities among proteins. The most common feature is that all proteins are polymers, composed of *amino acids*—the monomers that are covalently linked in sequence to each other by what is called the *peptide bond*. This chainlike structure is called a *polypeptide*, or simply a *peptide*. When a polypeptide is composed of 50 or more amino acid units (an arbitrary minimum), it is placed in the protein category.

Since amino acids are the basic building blocks of all proteins, a knowledge of amino acid biochemistry is needed to understand protein biochemistry. However, the biological importance of the amino acids also includes specific functions involving the metabolism of individual amino acids—a topic covered in Chapter 18. The purposes of this chapter are to acquaint you with the identity of the amino acids, to cover some basic principles regarding their chemistry, to describe two important laboratory methods—electrophoresis and chromatography—and to describe some important naturally occurring peptides, small and large, that function as reducing agents, hormones, antibiotics, growth factors, and neurotransmitters.

3–1 AMINO ACID STRUCTURES

General Features

There are about three hundred different amino acids known to occur in nature. Many are found only in certain life species and some in only one organism. However, all organisms use only twenty of this number in the biosynthesis of proteins—a striking example of biochemical unity in our biosphere.

Most amino acids are *alpha amino acids* having both an amino ($-NH_2$) and a carboxyl ($-COOH$) group attached to the same alpha (α) carbon atom. Because a primary amine function is present, the term primary amino acid is sometimes used. Three equivalent projections of the general structure are shown in Figure 3–1.

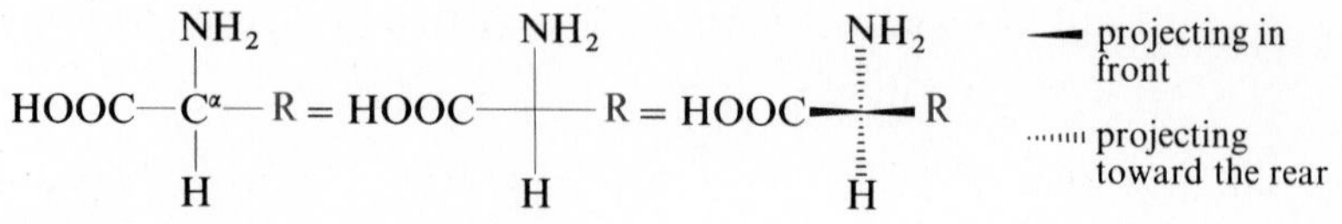

FIGURE 3–1 Equivalent projections of the structure of an alpha amino acid.

$$\text{HOOC}-\underset{\underset{H}{|}}{\overset{\overset{NH_2}{|}}{C^{\alpha}}}-R = \text{HOOC}-\overset{NH_2}{\underset{H}{+}}-R = \text{HOOC}-\overset{NH_2}{\underset{H}{\rightarrow}}-R$$

⟶ projecting in front

········· projecting toward the rear

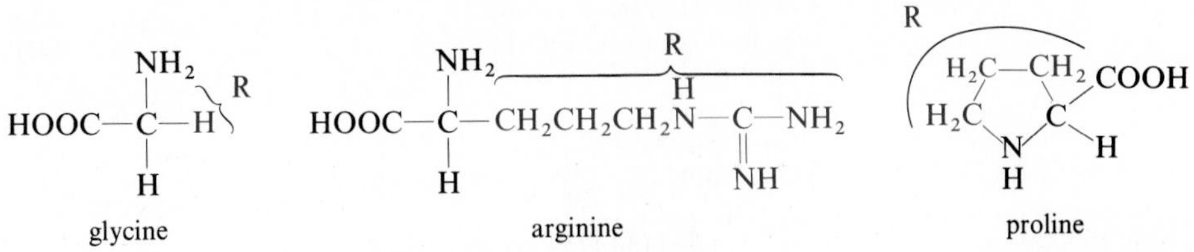

FIGURE 3–2 Some examples of R group structures in amino acids. Also see Table 3–1.

The unique character of each alpha amino acid is conferred by the structure of the R group. In (poly)peptides the amino acid R groups are sometimes called side chains. The R group can vary from a single hydrogen atom in glycine (the simplest amino acid) to a more complex structure such as the guanidine group in arginine (see Figure 3–2). In one case, proline, the R group is part of a cyclic secondary amine structure. Hence proline is referred to as a secondary amino acid and sometimes called an imino acid.

Chemical Composition of Amino Acid R Groups

Table 3–1 gives the structures, common names, shorthand abbreviations, and one-letter symbols of the 20 most important amino acids. The three-letter abbreviations are easy to learn. The one-letter symbols, tabulated together for easy reference, need not be memorized. The seven categories used in Table 3–1 are based on the *chemical composition* of the R group: nonaromatic hydrocarbon (*aliphatic character*), —OH present (*hydroxyl-containing*), S present (*sulfur-containing*), aromatic hydrocarbon (*aromatic character*), additional —COOH function and the corresponding amide derivatives (*acidic/amide-containing*), additional —NH_2 or related function (*basic character*), and a cyclic, secondary amine $>$N—H structure (*imino structure*). Although the categories are not totally exclusive, they do help one in distinguishing the amino acids. Thoroughly familiarize yourself with the structures and the nomenclature used in referring to the R group functions.

Functional Characteristics of Amino Acid R Groups

There are various instances in the book where we will focus on specific features of an individual amino acid. For now we take a broader approach by grouping amino acids according to some important general characteristics. The labels appearing in Table 3–1, such as polar, neutral, and essential, refer to these characteristics.

NEUTRAL/BASIC/ACIDIC   Note that the structures in Table 3–1 depict ionized forms of the amino acids. Later (Section 3–3) we will discuss in some detail the ionization properties of the amino, carboxyl, and R group functions. For now we just introduce some terms. The R group may or may not ionize. If it does not ionize, the R group is *neutral* and carries no charge. When it can ionize, the R group is *basic* if it functions as a Brönsted base to give a positively

TABLE 3–1 Structures, names, abbreviations, and classifications of amino acids

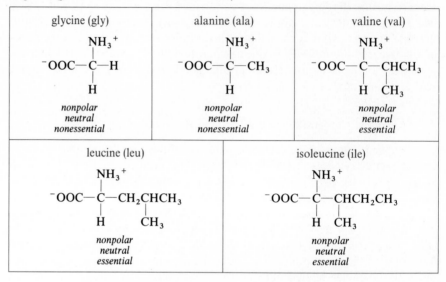

Classification based on the chemical composition of the R group:	
1. Aliphatic 2. Hydroxyl 3. Sulfur 4. Aromatic 5. Acidic (and amides) 6. Basic 7. Imino	$^-OOC—\overset{\overset{+}{NH_3}}{\underset{H}{C}}—R$ *general formula* *showing ionic forms* *existing at about pH 7*

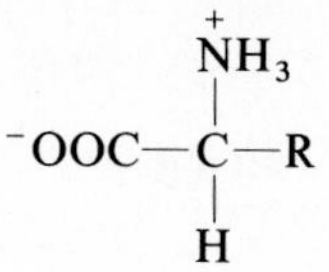

Aliphatic (proline could also be included here)

glycine (gly)

$^-OOC—\overset{NH_3{}^+}{\underset{H}{C}}—H$

nonpolar
neutral
nonessential

alanine (ala)

$^-OOC—\overset{NH_3{}^+}{\underset{H}{C}}—CH_3$

nonpolar
neutral
nonessential

valine (val)

$^-OOC—\overset{NH_3{}^+}{\underset{H}{C}}—CHCH_3$ with CH_3

nonpolar
neutral
essential

leucine (leu)

$^-OOC—\overset{NH_3{}^+}{\underset{H}{C}}—CH_2CHCH_3$ with CH_3

nonpolar
neutral
essential

isoleucine (ile)

$^-OOC—\overset{NH_3{}^+}{\underset{H}{C}}—CHCH_2CH_3$ with CH_3

nonpolar
neutral
essential

Hydroxyl (tyrosine could also be included here)

serine (ser)

$^-OOC—\overset{NH_3{}^+}{\underset{H}{C}}—CH_2OH$

polar
neutral
nonessential

threonine (thr)

$^-OOC—\overset{NH_3{}^+}{\underset{H}{C}}—CHCH_3$ with OH

polar
neutral
essential

Sulfur

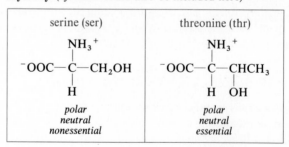

cysteine (cys)

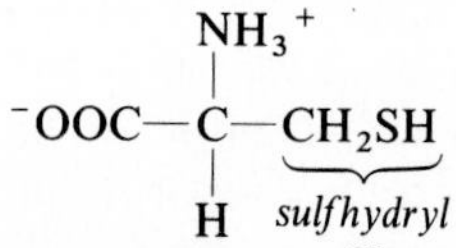

$^-OOC—\overset{NH_3{}^+}{\underset{H}{C}}—CH_2SH$ *sulfhydryl or thiol*

polar
weakly acidic
nonessential

methionine (met)

$^-OOC—\overset{NH_3{}^+}{\underset{H}{C}}—CH_2CH_2—S—CH_3$ *thioether or methyl sulfide*

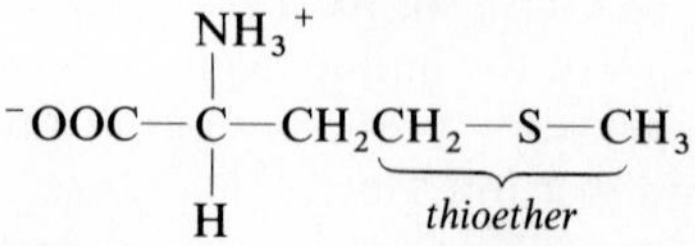

nonpolar
neutral
essential

One-Letter Symbols

A	alanine
C	cysteine
D	aspartic acid
E	glutamic acid
F	phenylalanine
G	glycine
H	histidine
I	isoleucine
K	lysine
L	leucine
M	methionine
N	asparagine
P	proline
Q	glutamine
R	arginine
S	serine
T	threonine
V	valine
W	tryptophan
Y	tyrosine

TABLE 3–1 (*Concluded*)

Aromatic (histidine could be also included here)

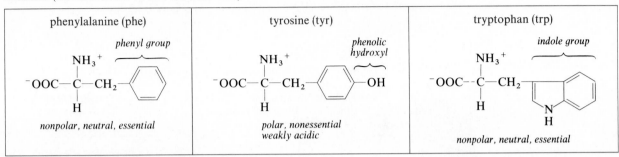

phenylalanine (phe)	tyrosine (tyr)	tryptophan (trp)
phenyl group	*phenolic hydroxyl*	*indole group*
nonpolar, neutral, essential	*polar, nonessential weakly acidic*	*nonpolar, neutral, essential*

Acidic (and corresponding neutral amides)

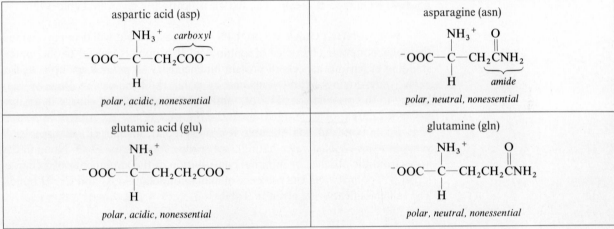

aspartic acid (asp)	asparagine (asn)
carboxyl	*amide*
polar, acidic, nonessential	*polar, neutral, nonessential*
glutamic acid (glu)	glutamine (gln)
polar, acidic, nonessential	*polar, neutral, nonessential*

Basic

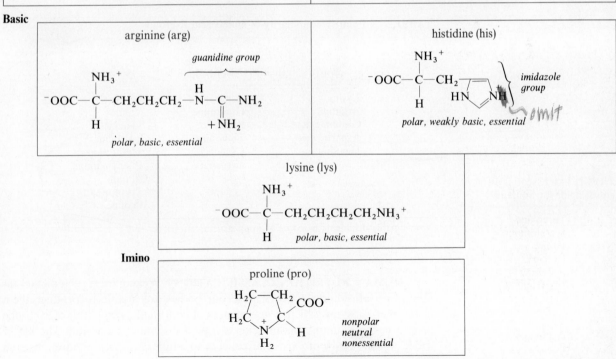

arginine (arg)	histidine (his)
guanidine group	*imidazole group*
polar, basic, essential	*polar, weakly basic, essential* omit

lysine (lys)

polar, basic, essential

Imino

proline (pro)

nonpolar neutral nonessential

TABLE 3–2 Ionization character of amino acid R groups

NEUTRAL (R group does not ionize)			BASIC	ACIDIC (R group does ionize)
Glycine	Serine	Tryptophan	Lysine (strong)	Glutamic acid (strong)
Alanine	Threonine	Glutamine	Arginine (strong)	Aspartic acid (strong)
Valine	Methionine	Asparagine	Histidine (weak)	Cysteine (weak)
Leucine	Proline	Phenylalanine		Tyrosine (weak)
Isoleucine				

charged R group; or it is *acidic* if it functions as a Brönsted acid to give a negatively charged R group. The full classification is given in Table 3–2.

POLAR/NONPOLAR R GROUPS   For reasons that will become obvious, the polar/nonpolar character of amino acid R groups is crucial to the understanding of various aspects of protein biochemistry. A *polar* R group is *hydrophilic* (water-loving), readily solvated by polar H_2O molecules. Polarity may be due to the presence at pH 7 of a full positive or negative charge in an ionizable R group (polar, charged) or to the presence of partial charges in permanent-dipole bonds in nonionizable R groups (polar, uncharged). A *nonpolar* R group is *hydrophobic* (water-hating), not readily solvated by H_2O. Nonpolarity arises from the absence or minimal contribution of the aforementioned characteristics of polarity and the presence of many nonpolar C—C and C—H bonds. The full classification is given in Table 3–3.

TABLE 3–3 Polarity/nonpolarity of amino acid R groups

POLAR			NONPOLAR
charged	(at pH 7)	uncharged	
Lysine		Asparagine	Glycine[a]
Arginine		Glutamine	Valine
Glutamic acid		Serine	Leucine
Aspartic acid		Threonine	Isoleucine
Histidine		Cysteine	Methionine
		Tyrosine	Phenylalanine
			Alanine
			Tryptophan
			Proline

Note: Polarity distinction based on R group structure at pH 7.
[a] Sometimes classified as polar.

TABLE 3–4 Essential dietary amino acids for humans

Threonine
Methionine
Valine
Leucine
Isoleucine
Phenylalanine
Tryptophan
Lysine
Arginine[a]
Histidine[a]

[a] Not after infancy.

HUMAN NUTRITION CLASSIFICATION   We require regular dietary intake of certain amino acids because our bodies are unable to produce them from other substances. The consequences of continual deprivation of even one or two of the essential amino acids can be disastrous—even fatal. The list in Table 3–4 should be of particular concern to adults caring for children, women during pregnancy, vegetarians, and weight watchers.

Modified Forms of the Common Amino Acids Occur in Some Polypeptides

Usually, after amino acids are assembled into a polypeptide chain (sometimes before or during the assembly process), one or more R groups in specific internal positions or the —NH$_2$ or —COOH groups of the amino acids at the ends of the chain may be chemically modified. The modifications occur for various reasons, such as cross-linking two or more polypeptide chains, reversibly changing the activity of the protein, covalently attaching another substance to a polypeptide, and, perhaps, for very specific reasons that are yet unknown. Usually each modification involves the operation of a specific enzyme. Modified forms of most amino acids have been detected, totaling about one hundred fifty. Structures of examples to be described are given in Figure 3–3.

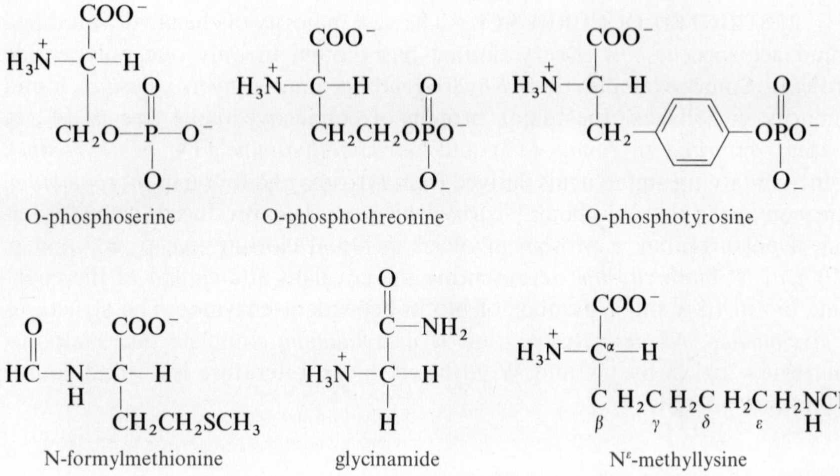

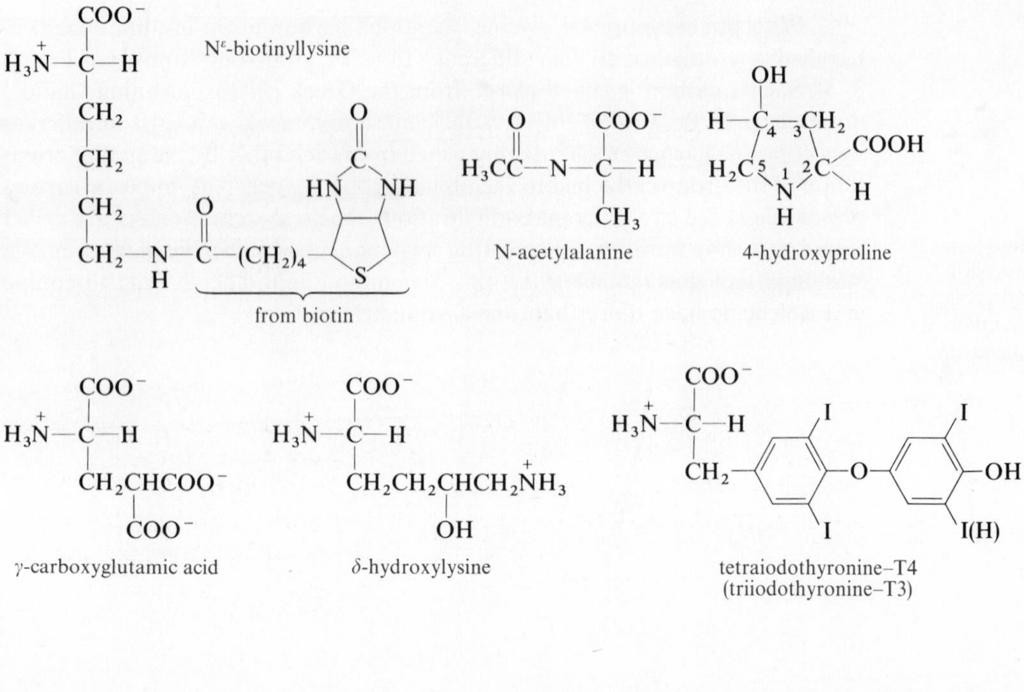

FIGURE 3–3 Some examples of modified amino acids isolated from peptides and proteins. Several of these amino acids will be encountered elsewhere in this book.

GENERAL OCCURRENCE   Some modifications occur in many proteins. The most common example involves the condensation of two sulfhydryl (—SH) functions of two cysteine R groups to yield a *disulfide* (—S—S—) *bond*. The amino acid consisting of two cysteines linked in this way is called *cystine* (see Figure 3–4). The importance of one or more —S—S— bonds in contributing to the folding of a polypeptide chain and in cross-linking two or more polypeptide chains is discussed in the next chapter.

Another event of general importance is *phosphorylation* of either of the three —OH-containing amino acids to yield *O-phosphoserine, O-phosphothreonine,* or *O-phosphotyrosine*. The addition and removal of a phosphate group is an important modification used in nature to regulate the activity of a protein (see Section 5–9). In prokaryotes the biosynthesis of every polypeptide chain begins with *N-formylmethionine* (see Chapter 9).

RESTRICTED OCCURRENCE   The vast majority of chemically modified amino acids occurs on a very limited basis, often in only one polypeptide (protein). Some examples are *4-hydroxyproline* and *δ-hydroxylysine,* found primarily in collagen, the major protein of connective tissue (see p. 142 in Chapter 4); *triiodothyronine* (T3) and *tetraiodothyronine* (T4, or *thyroxine*), iodine-containing amino acids derived from tyrosine and found in *thyroglobulin,* a protein in the thyroid gland; *γ-carboxyglutamic acid,* produced in the formation of prothrombin, a protein involved in blood clotting (see p. 209 and p. 440); and N^{ε}*-biotinyllysine,* representing the covalent attachment of the coenzyme biotin to a small number of biotin-dependent enzymes. The structures of *glycinamide,* N^{ε}*-methyllysine,* and *N-acetylalanine* complete our sampling. The review article by Uy and Wold listed in the Literature is a good source to find more details.

Stereoisomerism of Amino Acids

With the exception of glycine, the alpha carbon atom in amino acids is tetrahedrally attached to four different atoms or groups of atoms (see Figure 3–5). Such a carbon is called *chiral* (from the Greek *cheiros,* meaning "hand") or *asymmetric*. Because of this arrangement, amino acids can exist in different *stereoisomeric configurations,* distinguished from each other by the spatial orientation of the groups attached to the alpha carbon. For each asymmetric carbon present there are two different configurations. The two stereoisomers are called L and D configurations, representing two nonsuperimposable mirror image structures called *enantiomers*. (Of the 20 common amino acids, only threonine and isoleucine have more than one asymmetric carbon.)

FIGURE 3–4 Disulfide bond in cysteine. The interconversion between two sulfhydryl groups and a disulfide linkage involves oxidation/reduction chemistry.

sulfhydryl groups

$$^{-}OOC-\underset{\underset{H}{|}}{\overset{\overset{NH_3^+}{|}}{C}}-CH_2-SH \ + \ HS-CH_2-\underset{\underset{NH_3^+}{|}}{\overset{\overset{H}{|}}{C}}-COO^{-} \quad \overset{oxidation}{\underset{reduction}{\underset{+2H}{\overset{-2H}{\rightleftarrows}}}} \quad ^{-}OOC-\underset{\underset{H}{|}}{\overset{\overset{NH_3^+}{|}}{C}}-CH_2-S-S-CH_2-\underset{\underset{NH_3^+}{|}}{\overset{\overset{H}{|}}{C}}-COO^{-}$$

disulfide bond

cysteine   cysteine   cystine

The L versus D designation for amino acids is based on the relationship of the alpha carbon configuration to the known configuration of the two enantiomers of glyceraldehyde. With the —CHO group oriented up and to the rear ⫶, and the —CH₂OH group oriented down and to the rear ⫶, the L label of glyceraldehyde designates the structure where the —OH group is spatially oriented to the left [see Figure 3–6(a)]. The D form of glyceraldehyde corresponds to the isomer wherein the —OH group is oriented to the right [see Figure 3–6(b)]. In comparison, with the —COOH group directed ⫶ and the —R group directed ⫶, the left and right orientations of the —NH₂ group are used to designate, respectively, the L and D forms of the amino acids.

The biological significance of these configurations is that *only L-amino acids are known to occur in proteins.* Although there is no evidence as yet for the occurrence of D-amino acids in proteins, they do occur in many sources—including humans—both in a free state and as a component of other structures. Examples of the latter include the cell wall material of bacterial cells (see Section 10–6) and many antibiotics (see Section 3–10).

The only physical property distinguishing enantiomers is their ability to rotate a plane of polarized light (of wavelength λ) in equal but opposite directions. The rotation can be observed in an apparatus called a polarimeter, usually by using the D line emission of a sodium lamp as the single-wavelength radiation source. Clockwise rotation is called *dextrorotatory* and given a + sign; counterclockwise rotation is *levorotatory* and given a − sign. Rotation values are expressed as the *specific optical rotation* $[\alpha]$, defined as

$$[\alpha]_\lambda^T = \frac{\alpha_{obs}}{cl}$$

where α_{obs} is the observed rotation measured at temperature T and wavelength λ on a solution with concentration c (in grams per milliliter) contained in a polarimeter tube with length l (in decimeters). For alanine $[\alpha]_D^{25}$ of D-ala = −14° and $[\alpha]_D^{25}$ of L-ala = +14°. In Chapter 4 we will describe the laboratory technique of *circular dichroism*, where the absorption of polarized light is measured as a function of changing wavelength. This technique is a sensitive method

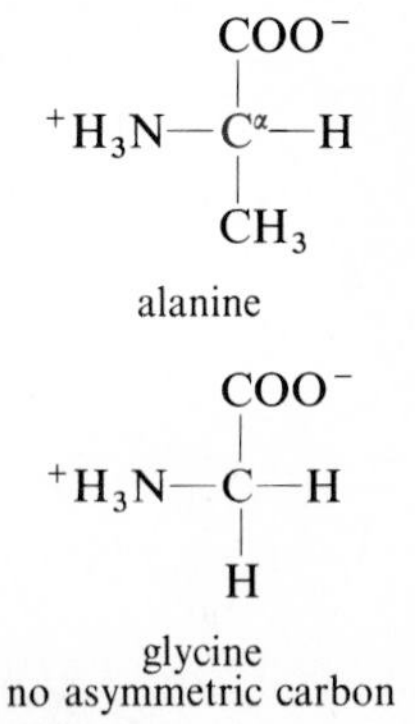

$$^+H_3N—C^\alpha—H$$
$$|\quad CH_3$$

alanine

$$^+H_3N—C—H$$
$$|\quad H$$

glycine
no asymmetric carbon

FIGURE 3–5 Chirality in amino acids. Shown in color, the alpha carbon of alanine is an asymmetric carbon; the same is true of all other amino acids except glycine.

FIGURE 3–6 Stereoisomers. (a) L stereoisomers. (b) D stereoisomers. The L and D stereoisomers are nonsuperimposable mirror images (enantiomers). An L-amino acid is used for protein biosynthesis, while a D-amino acid is not.

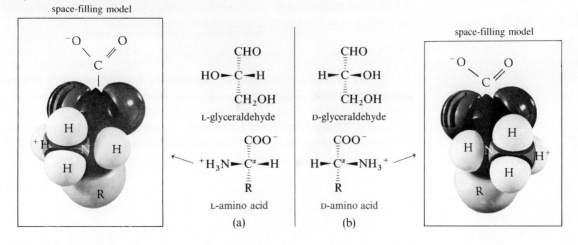

space-filling model

CHO
HO—C—H
CH₂OH
L-glyceraldehyde

CHO
H—C—OH
CH₂OH
D-glyceraldehyde

COO⁻
^+H_3N—C$^\alpha$—H
R
L-amino acid
(a)

COO⁻
H—C$^\alpha$—NH₃$^+$
R
D-amino acid
(b)

space-filling model

for evaluating some features of protein structure and for detecting changes in the structure of polypeptides (proteins) and nucleic acids.

The interconversion of D and L enantiomers is called *racemization*. In living organisms part of the metabolism of amino acids involves D ⇌ L conversions, with the metabolic equilibrium strongly favoring the L forms. When metabolism stops with the death of an organism, D ⇌ L conversions still occur spontaneously at a very slow rate, tending to increase the D/L ratio of each amino acid to a value representative of a nonmetabolic equilibrium. Tens of thousands of years may be required to attain such an equilibrium state. A new *age-dating procedure* based on this phenomenon correlates measurements of the D/L ratio of aspartic acid in fossil bone specimens to the geological age of the specimens. Results of the D/L dating method compare favorably with results obtained by other procedures, such as radiocarbon dating.

Recently, scientists have shown that increases in the D/L ratio of aspartic acid measured in the teeth of *living* mammals also correlates with biological age. Teeth have been examined because the proteins present in teeth (*enamel* and *dentine*) are not regenerated during life. After synthesis early in the animal's development, these proteins remain essentially unaltered throughout the animal's lifetime and thus reflect the nonmetabolic interconversion of D and L forms. Not certain yet is whether amino acid racemization is directly linked to the aging process in living tissue or is simply a secondary consequence of aging.

3–2 (POLY)PEPTIDES

Covalent Bonding Between Amino Acids

In (poly)peptides successive amino acids are covalently bonded to each other through the alpha carboxyl group of one amino acid and the alpha amino group of the next. The resultant amide linkage is called a *peptide bond* (see Figure 3–7). The pattern of one-by-one assembly shown in Figure 3–8 typifies polypeptide synthesis in living cells (biosynthesis) and in the laboratory (chemical synthesis). However, both processes occur in a much more elaborate manner than the direct condensation shown here with loss of water. The nature of the peptide bond is described in more detail in Chapter 4.

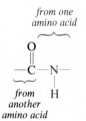

FIGURE 3–7 Peptide bond.

Terminology, Nomenclature, and Shorthand Notations

In the jargon of biochemistry each position in a peptide is termed an amino acid *residue*. If the peptide contains two to ten residues, the substance is referred to as an *oligopeptide*. Frequently, the number of residues in oligopeptides is specified by the use of Greek prefixes, as in *di*peptide, *tri*peptide, *tetra*peptide, and so on. Ordinarily, a peptide with more than ten residues is termed a *polypeptide*. Although cyclic and branched peptides do exist, most consist of a linear, chainlike assembly with two terminal residues. Since the bond between successive residues involves the α-carboxyl and α-amino groups of adjacent amino acids, one terminal residue will possess a free amino group (*N terminus*) and one will have a free carboxyl group (*C terminus*).

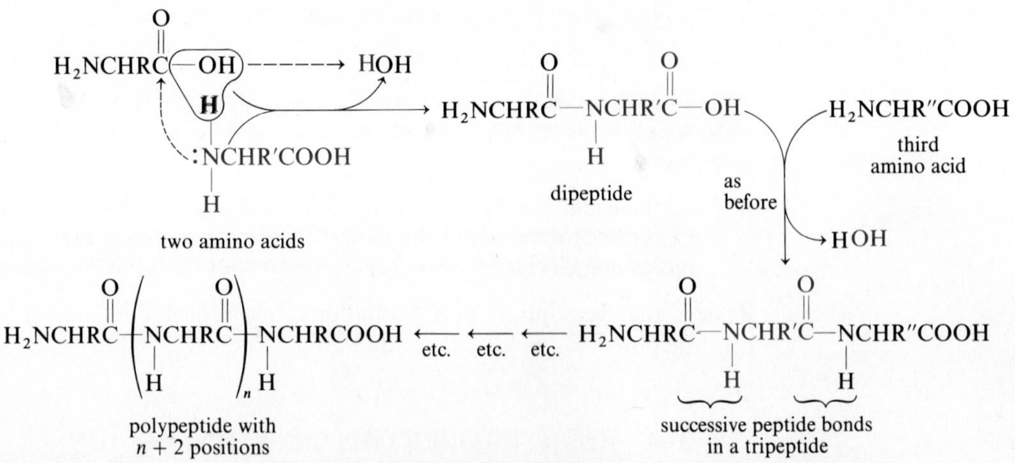

FIGURE 3–8 Formulas of (poly)peptides.

Because the representation of the complete structural formula of a peptide would be cumbersome, shorthand conventions are routinely used. Linear peptides of known residue sequence are named by beginning at the N terminus and designating each residue as an acyl substituent (see Note 3–1) of the α-amino group of the succeeding residue. The peptide bond is designated by a dash [see Figure 3–9(a)]. If the exact sequence of the peptide or any part of it is unknown, the residues are enclosed by parentheses and the dash is replaced by a comma [see Figure 3–9(b)]. An even shorter method for long sequences is to use the abbreviations or code letters (see Table 3–1) to designate each residue. Individual residues are also designated numerically, with the N terminal residue specified as number 1.

NOTE 3–1
Acyl groups are named by dropping the *-ine* or *-ic* ending of the parent name and adding the ending *-yl.*

1 2 3 4 5 6 7 8 ← residue number

arginyl—alanyl—glycyl—arginyl—glutamyl—alanyl—methionyl—lysine

N terminus ——————————————→ (C terminus)
(internal residues)

or arg—ala—gly—arg—glu—ala—met—lys

or R—A—G—R—E—A—M—K

(a)

1 2,3,4 5 6,7 8

arginyl—(alanyl,arginyl,glycine)—glutamyl—(alanyl,methionyl)—lysine

or arg—(ala,arg,gly)—glu—(ala,met)—lys

or R—(A,R,G)—E—(M,A)—K

(b)

FIGURE 3–9 Naming linear peptides. (a) All residue sequences are known. (b) Only a partial sequence is known.

Significance of Amino Acid Sequence

In terms of monomer composition there are three elements of structure to a (poly)peptide: (1) the identity of the amino acids present, (2) the relative amounts of each amino acid, and (3) the sequence in which the residues are connected.

Of these features the *residue sequence is most important*. The sequence of residues determines the overall three-dimensional shape of the molecule, which in turn determines how that molecule will function.

Review the description of informational molecules that was given in the Introduction.

3–3   IONIC PROPERTIES OF AMINO ACIDS AND (POLY)PEPTIDES

Neutral Amino Acids

Depending on pH and in accordance with the principles of acid-base chemistry discussed in Chapter 2, the alpha carboxyl and alpha amino functional groups of an amino acid exist in one of the following interconvertible combinations:

$$\begin{array}{ccccc}
\text{—COOH} & & \text{—COO}^- & & \text{—COO}^- \\
\text{—NH}_3{}^+ & \rightleftharpoons & \text{—NH}_3{}^+ & \rightleftharpoons & \text{—NH}_2
\end{array}$$

| protonated carboxyl | ionized carboxyl | ionized carboxyl |
| protonated amino | protonated amino | free amino |

Recognize that the fully protonated —COOH/—NH$_3{}^+$ species is a *diprotic* Brönsted acid. Thus two ionizations are possible and are described by two separate pK_a values. For most amino acids the pK_{a_1} (for the —COOH group, which is of greater acidity) has a value of about 2, and the pK_{a_2} (for the less acidic —NH$_3{}^+$ group) has a value of 9–10.

The stepwise ionizations for glycine are shown in Figure 3–10. Three

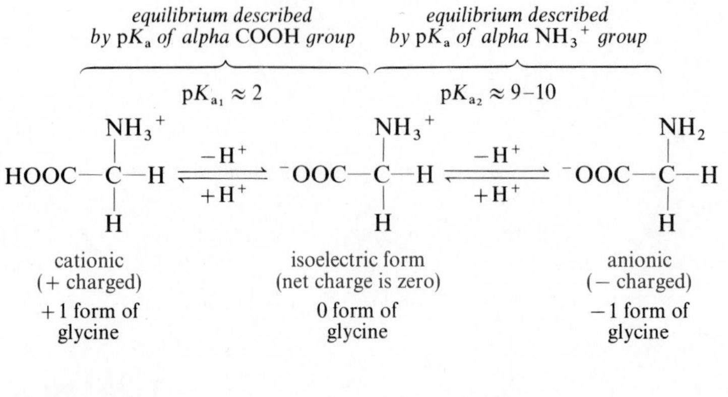

equilibrium described by pK_a of alpha COOH group   *equilibrium described by pK_a of alpha NH$_3{}^+$ group*

p$K_{a_1} \approx 2$ p$K_{a_2} \approx 9$–10

$$\underset{\substack{\text{cationic}\\(+\text{ charged})\\+1\text{ form of}\\\text{glycine}}}{\text{HOOC}-\overset{\overset{\displaystyle \text{NH}_3{}^+}{|}}{\underset{|}{\text{C}}}-\text{H}} \underset{+\text{H}^+}{\overset{-\text{H}^+}{\rightleftharpoons}} \underset{\substack{\text{isoelectric form}\\(\text{net charge is zero})\\0\text{ form of}\\\text{glycine}}}{{}^-\text{OOC}-\overset{\overset{\displaystyle \text{NH}_3{}^+}{|}}{\underset{|}{\text{C}}}-\text{H}} \underset{+\text{H}^+}{\overset{-\text{H}^+}{\rightleftharpoons}} \underset{\substack{\text{anionic}\\(-\text{ charged})\\-1\text{ form of}\\\text{glycine}}}{{}^-\text{OOC}-\overset{\overset{\displaystyle \text{NH}_2}{|}}{\underset{|}{\text{C}}}-\text{H}}$$

FIGURE 3–10 Stepwise ionizations for neutral amino acids. Glycine is the example shown here.

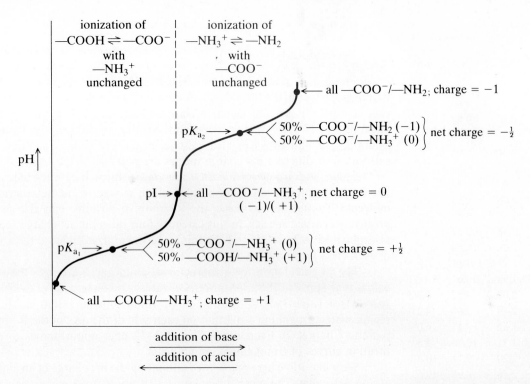

ionization of
$-COOH \rightleftharpoons -COO^-$
with
$-NH_3^+$
unchanged

ionization of
$-NH_3^+ \rightleftharpoons -NH_2$
with
$-COO^-$
unchanged

pH

all $-COO^-/-NH_2$; charge $= -1$

$pK_{a_2} \longrightarrow$

50% $-COO^-/-NH_2$ (−1)
50% $-COO^-/-NH_3^+$ (0)   net charge $= -\frac{1}{2}$

pI $\rightarrow$ all $-COO^-/-NH_3^+$; net charge $= 0$
$(-1)/(+1)$

$pK_{a_1} \longrightarrow$

50% $-COO^-/-NH_3^+$ (0)
50% $-COOH/-NH_3^+$ (+1)   net charge $= +\frac{1}{2}$

all $-COOH/-NH_3^+$; charge $= +1$

addition of base
$\longrightarrow$
addition of acid
$\longleftarrow$

FIGURE 3–11 General titration curve of any neutral amino acid. The pI is on the steep inflection, midway between pK_{a_1} and pK_{a_2}:

$$pI = \frac{pK_{a1} + pK_{a2}}{2}$$

Acidic and basic amino acids show an additional plateau for the ionization of the R group.

different forms can exist, identified by the net charge: $+1, 0,$ and -1. In strongly acidic solutions (pH about 1) only the $-COOH/NH_3^+$ species ($+1$ charge) exists. In strongly basic solutions (pH about 11) only the $-COO^-/-NH_2$ species (-1 charge) exists. Both positions are indicated on the titration curve shown in Figure 3–11. The curve is typical in form for all *neutral* amino acids that do not have an additional ionizable group in the R group (13 of the 20 common amino acids are in this category). The midpoint of the curve represents the pH at which glycine exists in its *isoelectric form* for which the net charge is zero ($-COO^-/-NH_3^+$). The pH at which this condition exists is called the *isoelectric point* and is symbolized as *pI*.

The pI value of any neutral amino acid can be calculated by averaging the pK_a values (see description of Figure 3–11). At any other point on the curve the net charge on the amino acid molecules will be governed by the relative concentrations of the two species in equilibrium with each other. For example, at pH values corresponding to pK_{a_1} and pK_{a_2}, the net charges are $+\frac{1}{2}$ and $-\frac{1}{2}$, respectively. The net charge at any other pH can be evaluated from the Henderson-Hasselbalch equilibrium ratio of the appropriate two species or estimated by a scheme to be described shortly. Two points should be emphasized: (1) The net charge on an amino acid is specified by the pH, and the charge will change as the pH changes; and (2) since no two amino acids have exactly the same ionization tendency—that is, the same pK_a values—no two amino acids will have exactly the same net charge at the same pH. Amino acid pK_a values are given in Table 3–5.

TABLE 3–5 pK_a values

AMINO ACID	α COOH	α NH_3^+	RH or RH^+
gly	2.34	9.60	
ala	2.34	9.69	
val	2.32	9.62	
leu	2.36	9.68	
ile	2.36	9.68	
ser	2.21	9.15	
thr	2.63	10.43	
met	2.28	9.21	
phe	1.83	9.13	
trp	2.38	9.39	
asn	2.02	8.80	
gln	2.17	9.13	
pro	1.99	10.6	
asp	2.09	9.82	3.86^a
glu	2.19	9.67	4.25^a
his	1.82	9.17	6.0^a
cys	1.71	10.78	8.33^a
tyr	2.20	9.11	10.07
lys	2.18	8.95	10.53
arg	2.17	9.04	12.48

^{a} For these amino acids the R group ionization occurs before the α-NH_3^+ ionization.

Ionizable R Groups in Some Amino Acids

The ionization pattern of seven amino acids also includes a third ionizable function in the R group (see Table 3–6). Owing to the presence of the extra —COOH group, *glutamic acid* and *aspartic acid* are termed *acidic amino acids*. At physiological pH 7 the side-chain carboxyl will definitely exist in the ionic state, contributing a full negative charge to the net charge. The —SH and phenolic —OH groups of cysteine and tyrosine respectively, though not as acidic as glu and asp, can also exist as negatively charged species. In this category four different net charge species are possible: $+1, 0, -1, -2$.

Lysine and *arginine* are *basic amino acids*. In each case the side chain at pH 7 will definitely contribute a full positive charge to the net charge of the molecule. Though not as basic, the R group of histidine can also exist as a positively charged species. In this category four different net charge species are also possible: $+2, +1, 0, -1$. Examples of the ionization steps for a member of each class are shown in Figure 3–12.

The pI calculation for a trifunctional amino acid averages the two pK_a values that apply to the appropriate ionizations on either side of the isoelectric species (see Table 3–7). A titration curve would show an additional buffering-region plateau spanning 1 pH unit on each side of the pK_a of the R group (see p. 52 in Chapter 2). Exercises 3–11 and 3–12 offer opportunities to evaluate titration curves of trifunctional amino acids.

We will outline here one way of estimating the net charge of an amino acid at any pH value. The scheme represents a titration curve merely as a straight line labeled with two sliding scales—one for charge and one for pH.

TABLE 3–6 Ionization of R groups in amino acids

AMINO ACID	ADDITIONAL R GROUP IONIZATION	
Glutamic acid (carboxyl group)	$-CH_2CH_2COOH \rightleftharpoons -CH_2CH_2COO^- + H^+$	$pK_a = 3.9$
Aspartic acid (carboxyl group)	$-CH_2COOH \rightleftharpoons -CH_2COO^- + H^+$	$pK_a = 4.3$
Cysteine (sulfhydryl group)	$-CH_2SH \rightleftharpoons -CH_2S^- + H^+$	$pK_a = 8.3$
Tyrosine (phenolic hydroxyl)	$-CH_2-\bigcirc-OH \rightleftharpoons -CH_2-\bigcirc-O^- + H^+$	$pK_a = 10.1$
Histidine (imidazole group)	imidazole $HN{-}NH^+ \rightleftharpoons HN{-}N + H^+$	$pK_a = 6.0$
Lysine (amino group)	$-(CH_2)_4\overset{+}{N}H_3 \rightleftharpoons -(CH_2)_4NH_2 + H^+$	$pK_a = 10.5$
Arginine (guanidine group)	$-(CH_2)_3\overset{H}{N}CNH_2 \underset{\overset{\|}{{}^+NH_2}}{} \rightleftharpoons -(CH_2)_3\overset{H}{N}CNH_2 \underset{\overset{\|}{NH}}{} + H^+$	$pK_a = 12.5$

Note: For the distinctly acidic and basic acids, color identifies the species that would exist at pH 7.

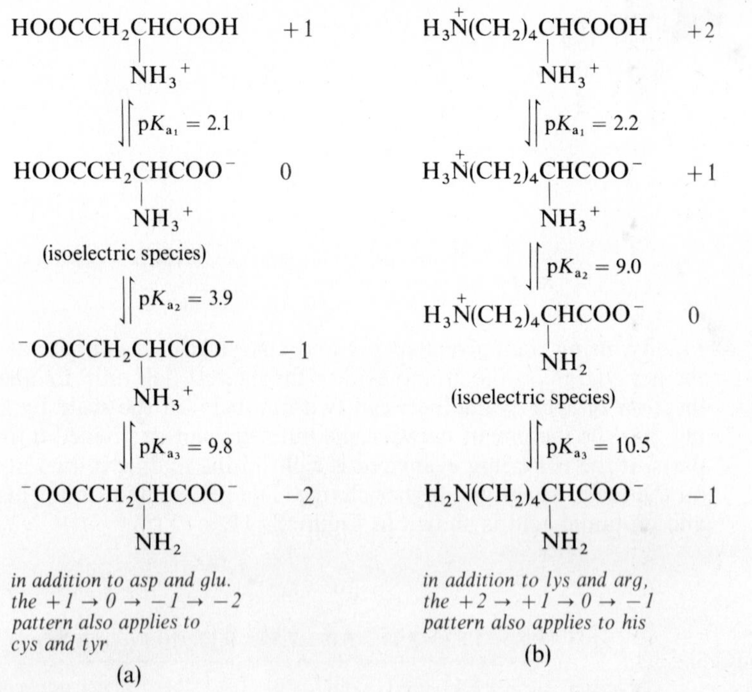

HOOCCH$_2$CHCOOH   +1   H$_3\overset{+}{\text{N}}$(CH$_2$)$_4$CHCOOH   +2
     | |
    NH$_3{}^+$ NH$_3{}^+$

$\Big\Uparrow$ pK$_{a_1}$ = 2.1 $\Big\Uparrow$ pK$_{a_1}$ = 2.2

HOOCCH$_2$CHCOO$^-$ 0   H$_3\overset{+}{\text{N}}$(CH$_2$)$_4$CHCOO$^-$ +1
     | |
    NH$_3{}^+$ NH$_3{}^+$
(isoelectric species)

                                                       $\Big\Uparrow$ pK$_{a_2}$ = 9.0

$\Big\Uparrow$ pK$_{a_2}$ = 3.9   H$_3\overset{+}{\text{N}}$(CH$_2$)$_4$CHCOO$^-$ 0
                                                      |
$^-$OOCCH$_2$CHCOO$^-$ −1   NH$_2$
     | (isoelectric species)
    NH$_3{}^+$

$\Big\Uparrow$ pK$_{a_3}$ = 9.8 $\Big\Uparrow$ pK$_{a_3}$ = 10.5

$^-$OOCCH$_2$CHCOO$^-$ −2   H$_2$N(CH$_2$)$_4$CHCOO$^-$ −1
     | |
    NH$_2$ NH$_2$

in addition to asp and glu.   *in addition to lys and arg,*
the +1 → 0 → −1 → −2   *the +2 → +1 → 0 → −1*
pattern also applies to   *pattern also applies to his*
cys and tyr
          (a)   (b)

FIGURE 3–12 Ionization steps for trifunctional amino acids. (a) Steps for an acidic amino acid (such as aspartic acid). (b) Steps for a basic amino acid (such as lysine).

The procedure is as follows:

1. On a straight line, draw a lower scale of net charge values from +2 to −2 in equal increments of $\frac{1}{2}$ charge unit with a midpoint value of 0. See Figure 3–13.

$$+2 \quad +1\tfrac{1}{2} \quad +1 \quad +\tfrac{1}{2} \quad 0 \quad -\tfrac{1}{2} \quad -1 \quad -1\tfrac{1}{2} \quad -2 \leftarrow \text{net charge}$$

FIGURE 3–13 Scale of net charge values.

2. Display pH values on the top line.

    a. Compute the pI value and assign it to the zero-charge species.
    b. For the amino acid in question, assign pH values to fractionally charged species on the basis of the pK$_a$ values. For example, for an ionization involving the +2 ⇌ +1 species, the pK$_a$ value is the pH where 50% of the molecules are in the +2 state and 50% are in the +1 state. Thus at pH = pK$_a$ the population will behave as if the net charge were +1$\frac{1}{2}$. The three possible patterns for the pK$_a$ assignments are shown in Figure 3–14. Note the relationship of net charge to the pI value. At pH values on the acid side of pI, the net charge is always positive; on the base side of pI, the net charge is always negative.

3. Fill in other pH values for full-charge species.

    a. If the full-charge species is flanked on each side by two pK$_a$ values, the pH at which that species exists is the midpoint of the two pK$_a$ values.
    b. To estimate the pH of full-charge species at the ends of the scale, decrease the first pK$_a$ value by 1 pH unit and increase the last pK$_a$ value by 1 pH unit.

TABLE 3–7 pI calculations

Acidic

$$pI = \frac{pK_{a_1} + pK_{a_2}}{2}$$

$$= \frac{2.09 + 3.86}{2} \quad \text{(aspartic acid)}$$

$$= 2.98$$

Basic

$$pI = \frac{pK_{a_2} + pK_{a_3}}{2}$$

$$= \frac{8.95 + 10.53}{2} \quad \text{(lysine)}$$

$$= 9.74$$

glu, asp, tyr, cys → pK_{a_1} pI pK_{a_2} pK_{a_3}
lys, arg, his → pK_{a_1} pK_{a_2} pI pK_{a_3}
all neutrals → pK_{a_1} pI pK_{a_2}
+2 +1½ +1 +½ 0 −½ −1 −1½ −2
neutral acids
lys, arg, his
glu, asp, tyr, cys

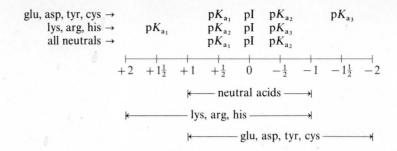

FIGURE 3–14 Possible patterns for pK_a assignments.

4. Finally, using the correlated pH and charge scales, estimate at pH X the net charge on the amino acid to the nearest half unit. Estimate the location of a pH value between two numbers on the scale by assuming that the increment between the numbers can be divided into equal parts. If the pH being evaluated is right in the middle of two numbers on the scale, assign the higher charge. The pH estimation for histidine and glutamic acid is shown in Figure 3–15.

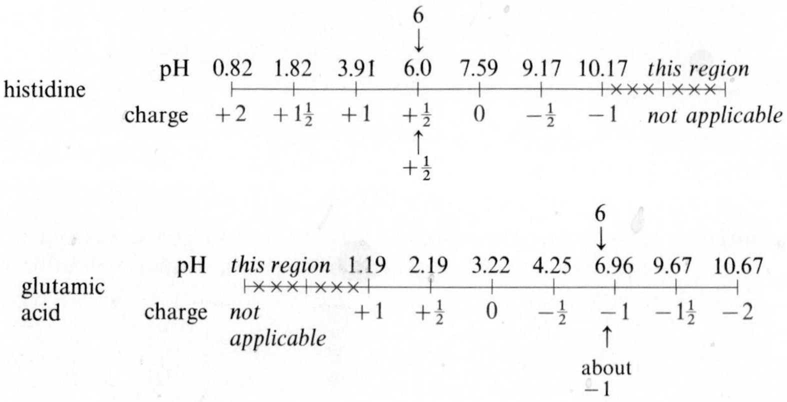

histidine
6
pH 0.82 1.82 3.91 6.0 7.59 9.17 10.17 *this region*
charge +2 +1½ +1 +½ 0 −½ −1 *not applicable*
+½
glutamic acid
6
pH *this region* 1.19 2.19 3.22 4.25 6.96 9.67 10.67
charge *not applicable* +1 +½ 0 −½ −1 −1½ −2
about −1

FIGURE 3–15 Estimation of pH for histidine and for glutamic acid.

(Poly)Peptides

Except for the terminal residues, both the α-carboxyl and α-amino groups of all other residues are involved in peptide bonding. Thus:

> The composite ionic character of a polypeptide depends primarily on the acidic and basic R groups present (see Figure 3–16).

Depending on the prevailing pH, the R groups of glu, asp, tyr, and cys could each contribute negative charge; the R groups of lys, arg, and his could each contribute positive charge. (Obviously, the R groups of the neutral amino acids contribute nothing to the composite charge.)

The analysis of the polyionic character of a heptapeptide is given in Table 3–8 at three different pH values. First, you identify the ionizable functions present, neglecting all that do not ionize. The evaluation of each ionizable function neglects any estimates of fractional charge unless the prevailing pH is within approximately ½ pH unit of the pK_a value for the group being evaluated. Then

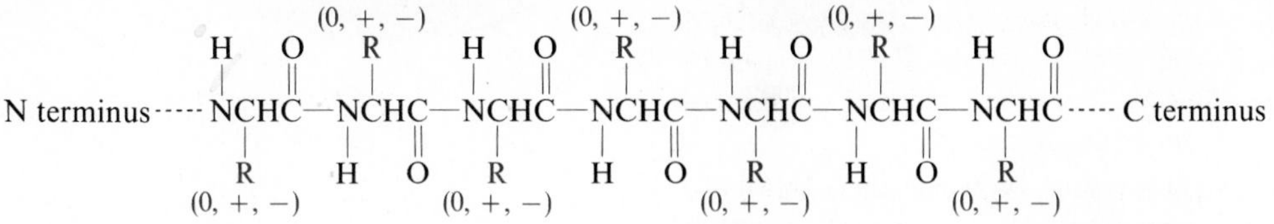

$$N \text{ terminus} \cdots \cdots \text{NCHC—NCHC—NCHC—NCHC—NCHC—NCHC—NCHC} \cdots \cdots C \text{ terminus}$$

FIGURE 3–16 R group side chains. They determine the polyionic character of polypeptides.

$+\frac{1}{2}$ and $-\frac{1}{2}$ values are assigned. At any other pH full charges of $+1$ and -1 or 0 charge are assigned.

The polyionic nature of proteins and biologically active peptides plays a significant role in their overall structure, which in turn controls the biological function. It is also the basis for applying the techniques of ion-exchange chromatography and electrophoresis to the laboratory analysis of amino acids, peptides, and proteins.

◇ **LABORATORY METHODS** ◇

ION-EXCHANGE CHROMATOGRAPHY AND ELECTROPHORESIS

There are many different laboratory methods used in biochemistry. The various techniques of chromatography and electrophoresis are used extensively for such purposes as isolating and separating most substances, measuring the molecular weight of polymers, characterizing different aspects of the structure of substances, and evaluating the purity of isolated substances. We will begin a consideration of those methods based on the ionic character of substances. Other procedures will be described in later chapters.

Principles of Chromatography

First developed in the nineteenth century as a means of separating plant pigments (hence the name *chroma*), the methodology of *chromatography* now includes several techniques. All operate on the same general procedure: *Sample substances, dissolved in a mobile phase, move through a stationary medium, which, by*

TABLE 3–8 Analysis of polyionic character

For the peptide   glu—phe—lys—his—ile—arg—val   (neglect phe and ile)

IONIZABLE FUNCTION	pK_a	R GROUP SPECIES AT PREVAILING pH OF					
		pH 1		pH 7		pH 11.5	
α-Amino of glu	9.6	$-NH_3^+$	$+1$	$-NH_3^+$	$+1$	$-NH_2$	0
α-Carboxyl of val	2.2	$-COOH$	0	$-COO^-$	-1	$-COO^-$	-1
R of glu	4.3	$-(CH_2)_2COOH$	0	$-(CH_2)_2COO^-$	-1	$-(CH_2)_2COO^-$	-1
R of lys	10.5	$-(CH_2)_4NH_3^+$	$+1$	$-(CH_2)_4 NH_3^+$	$+1$	$-(CH_2)_4NH_2$	0
R of his	6.0	$-CH_2$ [imidazole] HN NH	$+1$	$-CH_2$ [imidazole] HN N	0	$-CH_2$ [imidazole] HN N	0
R of arg	12.5	$-(CH_2)_3NHCNH_2$ (NH_2^+)	$+1$	$-(CH_2)_3NHCNH_2$ (NH_2^+)	$+1$	$-(CH_2)_3NHCNH_2$ (NH_2^+)	$+1$
Net charge of peptide			$+4$		$+1$		-1

interacting to varying degrees with individual sub-stances, impedes the flow of these substances to different degrees. The result is that *different substances will migrate at different rates.*

The type of interaction between the stationary phase and the components in the sample is what distinguishes one type of chromatography from another. This interaction is based on one of the following principles: *ion exchange, solubility, adsorption,* or *sieving.* Some procedures utilize a combination of factors.

A common strategy is to pack the stationary-phase material in a column and then allow the moving phase to flow through the column, sometimes with the aid of applied pressure. For an obvious reason such methods are collectively referred to as *column chromatography.*

Ion-Exchange Chromatography

When the solutes to be separated are capable of existing as positively or negatively charged ions, the procedure of *ion-exchange chromatography* is a very effective technique with a high resolving power. The stationary phase consists of a solid substance that actually participates in the process by interacting directly with the components of a mixture. For reasons that will become obvious, these materials are called *ion exchangers.*

The majority of ion exchangers are synthetic materials (called *resins*) fabricated in the form of very small beads. Chemically, each bead is composed of large polymeric chains. The ability of the resin material to function as an ion exchanger is due to the presence of several ionizable chemical groupings attached along the length of the polymeric chain. Thousands of these groupings exist in a single bead (see Figure 3–17).

One of the most widely used resins contains sulfonic acid groupings ($-SO_3H$) that are strongly acidic. Usually, this resin is used in a sodium salt form (such as $-SO_3^-Na^+$) and is called a *cation exchanger.* Once the sample enters the packed column, cation components (X^+) in the sample will reversibly interact with the resin via *electrostatic association.* This attraction results in a cation X^+ component of the sample exchanging with the cation Na^+ at the stationary $-SO_3^-$ sites (see Figure 3–18). Because the resin particles are stationary, the effect of the exchange is to

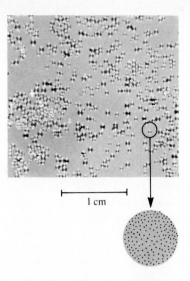

FIGURE 3–17 Spherical ion-exchange resin particles. Each dot in the enlarged sketch represents a functional group (such as $-SO_3^-$) attached to the polymeric structure. These groupings are present on the exterior surface and inside the bead. Interior sites are accessible because the bead is porous. Bead particles with a diameter as small as 10 μm (micrometers) are available, and they give excellent resolution (see Figure 3–19).

retard the movement of ionic solutes through the column.

Throughout the operation of the column, solutes are continually in a cycle of being bound, released, bound, released, and so on. Release into the moving phase is important in order for the solutes to move through the column. The extent of binding will be controlled primarily by the size of the net charge on individual X^+ components in the mixture. *Solutes with a greater net positive charge will interact with the resin particles to a greater degree;* thus they will migrate more slowly through the column.

The liquid emerging from the bottom of the column (the effluent) can be systematically collected in small-volume fractions, and the presence of a solute can be detected by suitable procedures. Detection and quantitative measurement of amino acids and peptides is commonly done by treating the effluent with *ninhydrin* (see Section 3–4). An example of this technique is shown in Figure 3–19, which depicts the analysis of amino acids in human plasma. The profile of distinct, nonoverlapping peaks (from first off the column to last off the column) illustrates the high resolving power of the method.

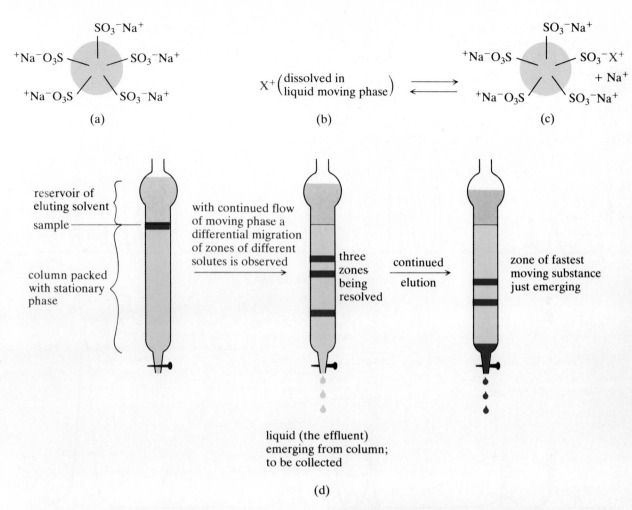

FIGURE 3–18 Reversible events that occur on a cation-exchange column. An exchange in the → direction binds X^+ to the resin. An exchange in the ← direction releases X^+ back to the moving phase. These events occur many times per bead and throughout the entire length of the column, with solutes encountering new beads as they move. (a) Single resin bead with —SO_3^- groups attached. The Na^+ is bound by electrostatic association. (b) Unbound cation. Only in this condition does it move through the column. (c) X^+ now bound to the resin. In this condition the movement of X^+ is retarded. (d) Schematic illustration of the differential migration of solutes through a column. The eluting solvent (on the left) is the liquid moving phase. The column packed with the stationary phase is in contact with the eluting solvent throughout its length. A small volume of the sample is applied to the top of the bed of the solid phase. The effluent (on the right) can be manually or automatically collected and analyzed for the presence of solute.

The late release of lysine, histidine, and arginine reflects the contribution of the positively charged R group of each of these amino acids at the pH of the eluting buffers—starting with a pH 2.9 buffer and finishing with a pH 6.5 buffer. However, note that the (lys, his, arg) order of elution is not consistent with a net charge analysis. On this basis alone you would predict an order of his (with only a partial positive charge at pH 6.5) eluting before lys (with a full charge of +1), followed shortly thereafter by arg (also with a full charge of +1). This and other anomalies in the profile represent the operation of secondary factors that affect the interaction of the resin with substances moving through it. In the case of histidine movement is also impeded because the aromatic imidazole group of his can also bind with similar nonionic, aromatic parts of the resin structure.

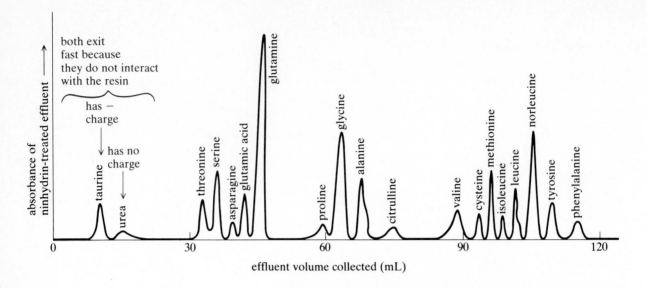

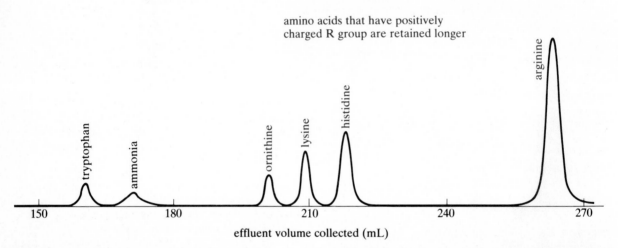

FIGURE 3–19 Cation-exchange column chromatographic analysis of free amino acids in deproteinized (protein-removed) human plasma. The column size was 2.8 × 300 mm ($\frac{1}{8}$ in. × 12 in.); the height of the resin bed was 200 mm; the eluting buffer was an Li$^+$ salt solution; the rate of flow of the moving phase was 7 mL/h; the analysis was completed in about $4\frac{1}{2}$ h. The eluting buffer was changed five times during the operation, covering the pH range of 2.9 to 6.5. The structure of the amino acid ornithine is similar to that of lysine (see p. 678); see the text for an explanation of why histidine moves slower than lysine.

The pumping of eluting buffers from reservoirs, maintaining temperature control of the column, continuous ninhydrin analysis of the eluting solvent, and recorder display of the data can all be done automatically with an instrument called an *amino acid analyzer.* Computer interfacing is also possible.

Anion-exchange resins consist of a polymeric structure containing positively charged functional groupings, usually substituted aminoethyl groups in chloride salt form, such as

$$\text{polymer}\left(\text{CH}_2\text{CH}_2\underset{\underset{\text{H}}{|}}{\overset{\overset{\text{R}}{|}}{\overset{+}{\text{N}}}}\text{R Cl}^-\right)_n$$

where Cl^- would exchange with Y^- components in the sample being analyzed. A popular anion exchanger is (diethylaminoethyl)-cellulose (DEAE-cellulose), where each $R = -CH_2CH_3$. It is particularly efficient in the ion-exchange chromatography of protein mixtures.

High-Performance Liquid Chromatography

A recent advance in chromatographic technology is the preparation of solid-phase column-packing materials (be it an adsorbent, an ion exchanger, or a molecular sieve material) in extremely small bead sizes—as small as $5-10 \, \mu m$ in diameter. Packing a thin column (for example, 250 mm × 4 mm) with microscopic solid-phase particles reduces the dead space between particles and thus maximizes the interaction of substances moving through the column with the solid phase. Since the tight packing reduces the flow rate of the eluting solvent through the column, the application of high pressure (several hundred pounds per square inch) is required. Called *high-performance liquid chromatography* (HPLC), the technique gives outstanding results (superb resolution) in fast time (see Figure 3–20). The HPLC method is applicable for the separation and analysis of nearly all types of substances.

Paper Chromatography

One of the early designs in the development of chromatographic techniques was the use of a sheet of filter paper (cellulose) as an inert stationary support for a liquid. Hence the method is called *paper chromatography*. The paper is moistened by absorbing water vapor; thus the stationary phase is a polar liquid supported on paper. Another solvent (rich in a more nonpolar substance such as *n*-butanol/H_2O; 8/1 v/v) is allowed to migrate up or down the paper via capillary action. When this developing solvent reaches the location (origin) where a portion of the sample was previously spotted on the paper, the material at the origin will be *dissolved to different extents* between the moving nonpolar phase (primarily *n*-butanol) and the stationary polar phase (primarily water). This *partitioning based on solubility* will continue as the developing solvent moves along the length of the paper. Substances more soluble in water will move

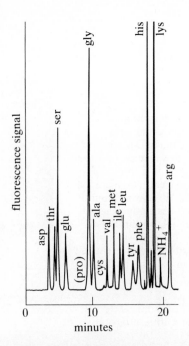

FIGURE 3–20 High-performance liquid chromatography. Compare these results of analyzing this amino acid mixture in less than 30 min by ion-exchange HPLC with those in Figure 3–19 using the conventional procedure. Also, note the sharp and narrow zones in HPLC separations. The fluorescence signal was produced by reacting the effluent of the column with ortho-phthaldehyde (see p. 87).

less than those more soluble in the moving organic phase. The capability of the method is depicted in Figure 3–21.

The extent of migration of a substance is expressed by the R_f value, defined as follows:

$$R_f = \frac{\text{distance traveled by substance}}{\text{total distance traveled by developing solvent}}$$

Both distances are measured from the origin of sample application.

For samples containing many substances the resolution of zones can be increased by drying the paper, then allowing a second solvent to migrate in a direction perpendicular to that of the first solvent. This method is referred to as a *two-dimensional analysis*. The comparison in Figure 3–22 clearly depicts the advantages of the second step.

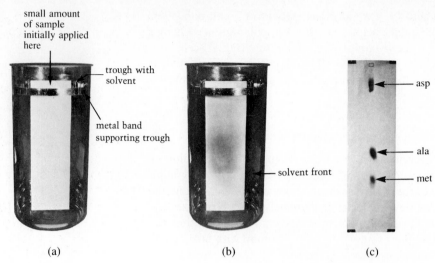

FIGURE 3–21 One-dimensional, descending paper chromatography of an amino acid mixture containing aspartic acid, methionine, and alanine. (a) Paper strip suspended in a trough of developing solvent. No migration has occurred. (b) Solvent front. It has migrated nearly the length of the paper. (c) Appearance of paper strip after removal and treatment with ninhydrin. (Actual size of the jar is 12 in. by 24 in.)

(a)   (b)   (c)

Thin-Layer Chromatography

The use of a solid stationary phase spread as a thin coating on the flat surface of a firm support (glass or plastic) typifies the method called *thin-layer chromatography* (TLC). The solid phase may function as an adsorbent (silica gel is common), an ion exchanger (modified cellulose materials are used), or a sieving agent (a porous gel such as polyacrylamide; see p. 118). The procedure is similar to that of paper chromatography (see Figure 3–23). The popularity of thin-layer chromatography is due to several features: solute resolution and reproducibility of results are good to excellent; most separations are achieved quickly, in 0.5–3 h; it can be used for the analysis of nearly all types of substances; very small amounts (microgram levels) of solutes can be detected; and last but not least, the method is inexpensive and easy to do.

Electrophoresis

Electrophoresis involves the movement of charged particles (ions) in an electric field. The movement occurs in a liquid medium supported by an inert solid substance such as a paper or a gel material. The liquid—a buffer salt solution of known pH and ionic

FIGURE 3–22 Comparison of the resolving power of two types of paper chromatography. (a) One-dimensional paper chromatography. (b) Two-dimensional paper chromatography. A mixture of 12 amino acids was applied to each origin (rectangle); the papers were developed under identical conditions and then sprayed with a ninhydrin solution. In each photograph the ninhydrin-positive spots appear as dark areas against the white background.

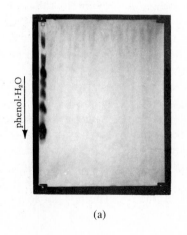

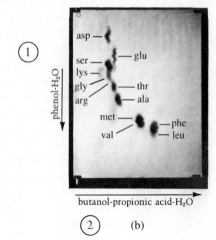

(a)   (b)

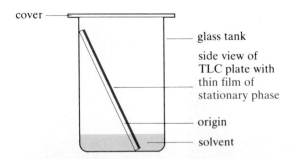

cover — glass tank

side view of TLC plate with thin film of stationary phase

origin

solvent

FIGURE 3–23 Thin-layer chromatography. After sample application the thin-layer plate is placed in a chamber containing the chromatography solvent on the floor of the chamber. The solvent migrates up the thin-layer plate by capillary action. As with paper chromatography, the analysis can be one- or two-dimensional.

strength—serves as a conducting medium for electric current when an external voltage V is applied. The extent of movement of a charged substance (molecule) in an electric field is termed its *electrophoretic mobility* (symbolized by μ). For a spherical molecule not experiencing any strong electrostatic interaction from surrounding ions, the quantity μ is given by

$$\mu = \frac{Q}{6\pi\eta r} V$$

where Q = net charge on molecule

r = radius of molecule (in centimeters)

η = viscosity of liquid medium in which movement occurs

V = applied voltage

Thus in a medium of constant viscosity and at a constant applied voltage, the movement of a charged particle is governed by the *charge-to-size ratio*. That is,

$$\mu \propto \frac{Q}{r} \qquad \frac{\text{charge}}{\text{size}}$$

As the Q/r ratio increases, the mobility increases. If we further assume that the charged components of a mixture do not differ appreciably in molecular size, the

equation is reduced to the direct relationship:

$$\mu \propto Q$$

(for different molecules of the same size in the same medium). As the charge increases, the mobility increases.

PAPER ELECTROPHORESIS Although it is less used than gel methods, we will illustrate the features and capability of electrophoresis with *paper electrophoresis*. A strip or sheet of filter paper, uniformly moistened with a buffer solution, is suspended between two electrode compartments also containing the buffer [see Figure 3–24(a)]. Samples are applied to the paper at the designated origin near either end or at the middle. Under an applied voltage the charged solutes will move as *zones* toward the appropriate pole, making their way through the irregular fibrous matrix of the paper. Using a high voltage gives best results for low–molecular weight components.

The drawing in Figure 3–24(b) illustrates the results of electrophoresing a mixture of four amino acids (glu, leu, his, lys) in a medium buffered at pH 6. The outcome is based on $\mu \propto Q$ (the small differences in size— that is, molecular weight—among these four is neglected) and thus only approximates the actual pattern of migration. To compute relative mobilities on a Q/r basis, divide the quantity pI − pH (a measure of the value of Q) by the molecular weight of the substance.

GEL ELECTROPHORESIS Whereas paper electrophoresis is particularly effective for analysis of low–molecular weight substances, far superior results for high–molecular weight substances (proteins, DNA, and RNA) can be achieved by using *gel electrophoresis*, with a semisolid gel as the supporting medium. *Agarose* gel and *polyacrylamide* gel are the most widely used. These gel media consist of cross-linked polymeric chains conferring a *porous character* to the gel matrix. Agarose gels can provide very large pore sizes; polyacrylamide gels can provide small pore sizes. Moreover, the degree of cross-linking can be controlled to provide uniform pores of any size.

The porosity of the gel matrix means that electrophoretic migration will better reflect differences in the size of the moving ions. Molecular shape will also affect the migration. There are other advantages. The

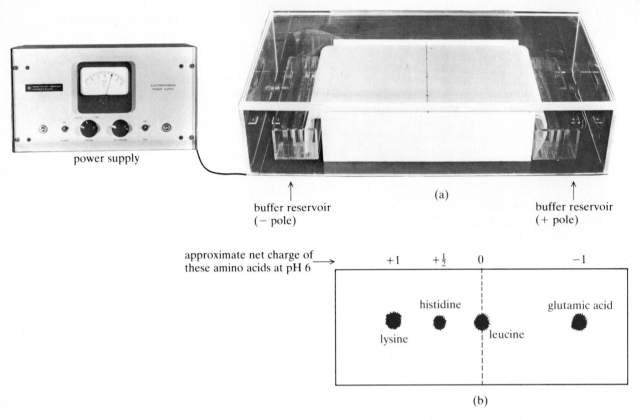

(a)

buffer reservoir
(− pole)

buffer reservoir
(+ pole)

power supply

approximate net charge of
these amino acids at pH 6 →

(b)

FIGURE 3–24 Paper electrophoresis. (a) Actual paper electrophoresis apparatus. (b) Diagrammatic representation of the results for a mixture of amino acids. Initially, a sample of a mixture of all four amino acids was placed at the dashed line (the origin). After electrophoresis and then spraying with ninhydrin, the paper appeared as shown.

gels can be cast as either a rod or as a thin slab; they are firm enough to handle; they provide a clear background for easy viewing of stained materials; after staining they can be prepared for permanent keeping; and because the gel materials are virtually inert, no damage is done to the structure of proteins and nucleic acids as they pass through the gel matrix. Techniques of gel electrophoresis are also described in Chapter 4.

3–4 CHEMICAL REACTIONS OF AMINO ACIDS

Amino acids can participate in many reactions involving either the alpha amino, the alpha carboxyl, or the various R group functions. A survey of all types is beyond the scope of an introductory course. The following material covers a small number of reactions having particular importance.

Detection and Quantitative Measurement

The reaction with *ninhydrin* is a color-producing reaction serving as the basis for the quantitative measurement of all amino acids. The intensity of the color produced with primary amino acids is evaluated by measuring the light absorption at a wavelength of 540 nm. Secondary amino acids like proline give a slightly different product; it is yellow and absorbs maximally at 440 nm. The calculation of concentration is based on the Beer-Lambert relationship (see Appendix I):

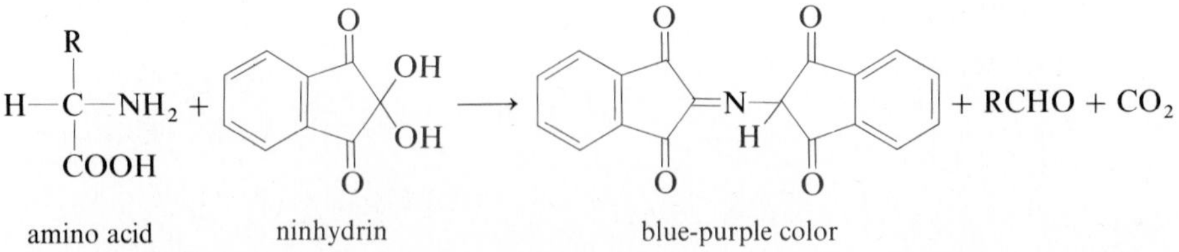

amino acid   ninhydrin   blue-purple color

A new, more sensitive detecting reagent is *ortho-phthaldehyde* (OPA; also called fluoroaldehyde), which can detect levels of amino acids in the picomole (10^{-12}) to femtomole (10^{-15}) range. The high sensitivity is due to the formation of intensely *fluorescent* products. After reaction, exposure to radiation of 360 nm causes a fluorescent emission at 455 nm, the intensity of which correlates to the concentration of the amino acid. The reaction (see Figure 3–25) occurs readily with all of the primary amino acids but not with secondary amino acids. Proline can be detected, however, by first oxidizing it to a primary amine.

Identification of Amino Acids Through Derivatives

Ultraviolet-absorbing *phenylthiohydantoin* (PTH) derivatives, fluorescent *dansyl* derivatives, and yellow *dinitrophenyl* (DNP) derivatives are useful not only for quantitative detection but also for the identification of amino acids (see Figure 3–26). Comparing the paper, thin-layer, or column chromatographic movement of these derivatives prepared from an unknown amino acid with standard derivatives prepared from known amino acids provides the basis of identifying the unknown. Other major applications of these reactions—particularly with *phenylisothiocyanate*, also called Edman's reagent—are in establishing the identity of the N terminal residue in a polypeptide and in determining the amino acid sequence of a polypeptide (see Section 4–2).

Modification of R Groups in Polypeptides

There are many reagents used to chemically modify the R groups of amino acid residues in a protein in order to manipulate the protein for further laboratory study, to determine the presence of an amino acid, or to evaluate whether

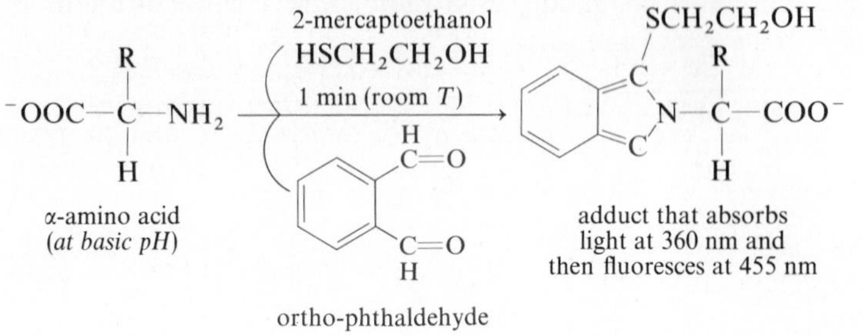

α-amino acid
(*at basic pH*)

ortho-phthaldehyde

adduct that absorbs
light at 360 nm and
then fluoresces at 455 nm

FIGURE 3–25 Ortho-phthaldehyde reaction. All primary amino acids react as shown. The cyclic imino acid proline first requires treatment with sodium hypochlorite (NaOCl) at basic pH to convert the secondary amine to a primary amine function.

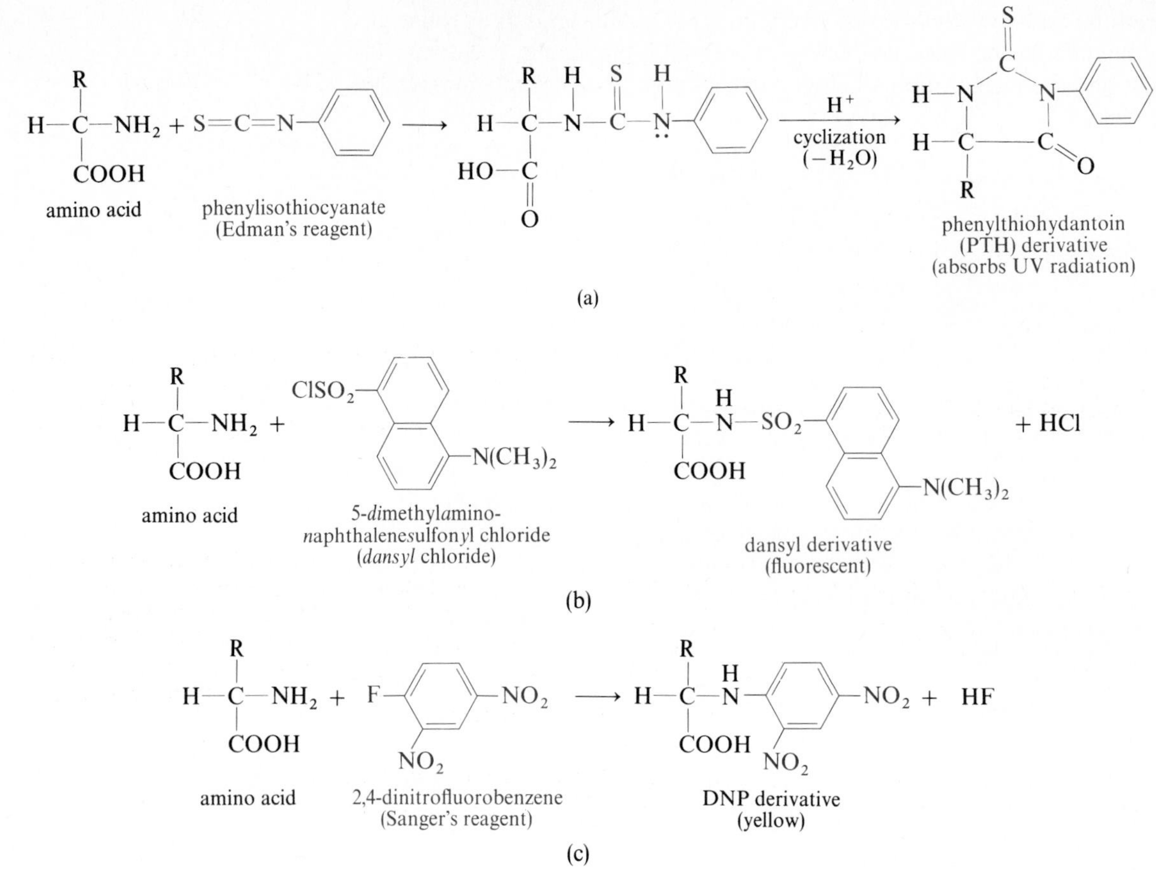

(a)

(b)

(c)

FIGURE 3–26 Additional reactions to detect amino acids. (a) Formation of PTH derivative. (b) Formation of dansyl derivative. (c) Formation of DNP derivative.

an amino acid is essential to the structure and the function of the protein. For example, under mild conditions *tetranitromethane* is highly specific for reacting with the phenolic group of tyrosine (see Figure 3–27). Other examples presented in Chapter 4 include reagents that react with the methionine R group, with the —SH group of cysteine, and with the —S—S— group of cystine.

An important example of R group modification is *radioactive iodination,* a technique of labeling a protein with a radioactive probe. Such a tactic is very useful in assays requiring the detection of extremely small amounts of a specific protein. The phenolic group of tyrosine is particularly susceptible to direct iodination, although the R groups of histidine, cysteine, and methionine can also react. The most commonly used radioactive nuclide of iodine is ^{125}I (a

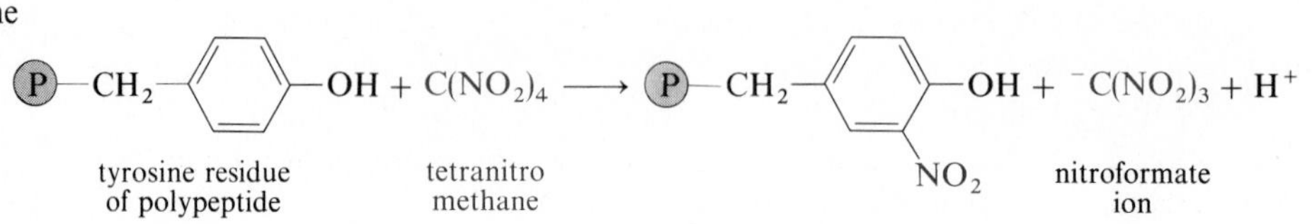

FIGURE 3–27 Tyrosine-specific tetranitromethane reaction. The symbol P represents a polypeptide (protein) containing the tyrosine side chain. The modified polypeptide may be active or inactive, depending on the importance of tyrosine to its structure and function.

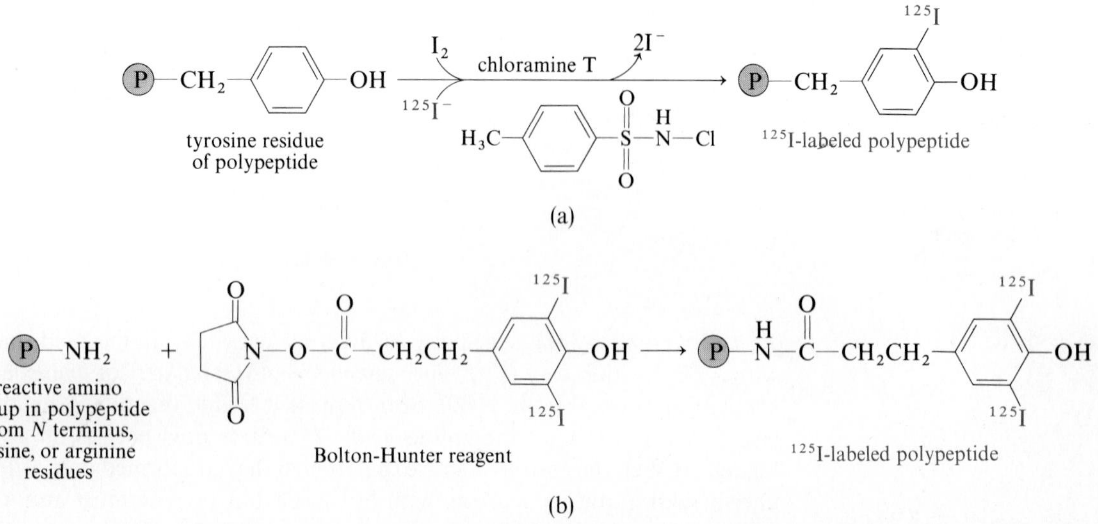

(a)

(b)

FIGURE 3–28 Radioactive iodination of a polypeptide. (a) Direct iodination of a tyrosine side chain. (b) Indirect iodination by attaching an iodinated aryl group to an amino group.

gamma radiation emitter). The labeling is performed by reacting the protein with I_2, $^{125}I^-$, and chloramine T, the latter catalyzing the formation of triiodide ^{125}I—I—I^-, the active iodinating species [see Figure 3–28(a)]. As indicated, the preferred site of iodination is ortho to the phenolic —OH group. If a tyrosine residue is not present in the protein or if the tyrosine R groups are in the interior of the protein structure and not readily accessible for reaction, iodination can be carried out at an amino function in the protein by using an acylating reagent that already carries an iodinated aromatic ring, such as the Bolton-Hunter reagent [see Figure 3–28(b)].

> Radioactively labeled substances are indispensable in much of bio- chemical research. Without ^{14}C, ^3H, ^{35}S, ^{32}P, and other isotopes, many experiments would be difficult and others impossible to per- form. Some information on radioactivity and its measurement are given in Appendix II.

ATP Reacts with Amino Acids in Vivo

Anyone having had a general biology course knows that the substance *adenosine triphosphate* (ATP) has a particular preeminence in the chemistry of all living cells. It is important because ATP is an energy-rich compound used in most cellular reactions and processes that require energy. We will discuss the whole matter of bioenergetics as a separate unit in Chapter 12. Adenosine triphosphate is a purine triphosphonucleotide composed of adenine (Ade) linked to a sugar (ribose) containing a triphosphoanhydride grouping (see Figure 3–29). Most ATP-dependent reactions involve condensation of a nucleophile with the *highly reactive triphosphoanhydride moiety*. Two important reaction types involve (1) attack at the α-phosphorus atom, resulting in the transfer of an Ade-ribose-P (AMP, for adenosine monophosphate) group to the attacking nucleophile, and (2) attack at the γ-phosphorus atom resulting in the transfer of a phosphate (P) group.

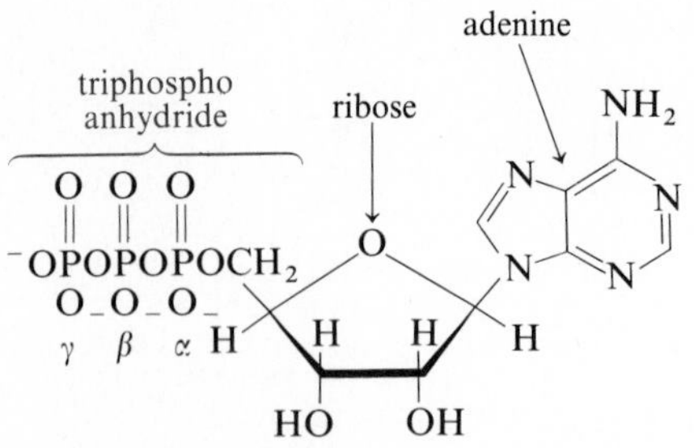

FIGURE 3–29 Adenosine
triphosphate, ATP, or
P-P-P-ribose-Ade. Each P atom
has a partial positive charge (δ^+)
suitable for attack by a
nucleophile $\overset{..}{Y}$ or $\overset{..}{Y}^-$. See
Figure 3–30 for reaction
examples.

Governed by the specificity of different enzymes, amino acids can react
with ATP in either way to produce *aminoacyl-AMP adducts* or *aminoacyl phosphate adducts* (see Figure 3–30). Both represent highly reactive forms of amino
acids because the C of the amino acid $>C=O$, is now part of an anhydride
linkage. In fact, they are so reactive that in vivo they are formed only as transient
intermediates, quickly reacting with other substances. The most important reaction of aminoacyl-AMP adducts is with transfer RNA (tRNA) as the first step
in the ribosome-dependent biosynthesis of polypeptide chains. After aminoacyl-
tRNA adducts bind to the surface of ribosomes, the aminoacyl units condense
with each other to form peptide bonds. There are numerous examples involving
aminoacyl phosphates. Later in this chapter we will encounter one dealing with
the biosynthesis of the tripeptide glutathione.

See p. 354 for
reaction of
this adduct
with transfer RNA

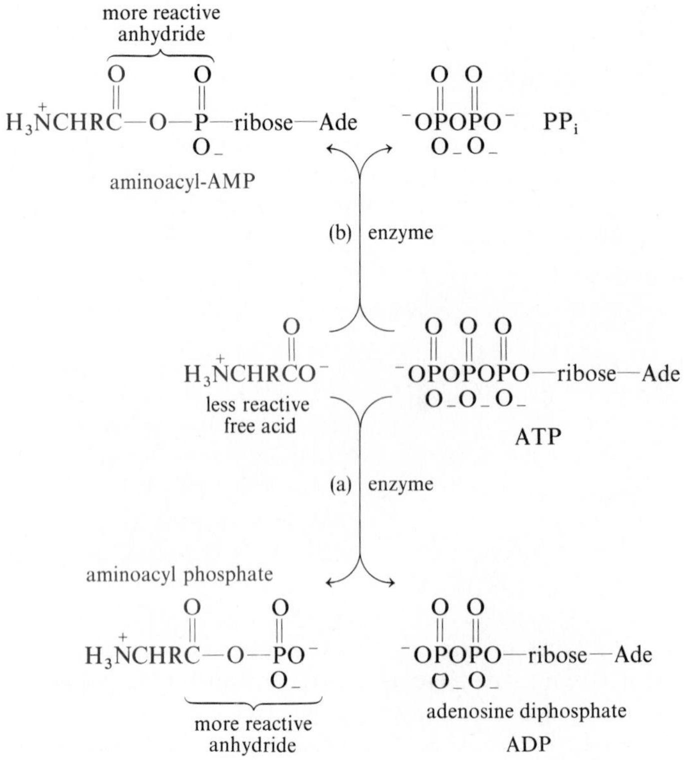

See p. 92 for
an example
involving this
adduct

FIGURE 3–30 ATP reactions
with amino acids. (a) Formation
of an aminoacyl phosphate.
(b) Formation of an aminoacyl-
AMP. In each case the less
reactive free acid is converted to
a more reactive species; PP_i is
inorganic orthophosphate.

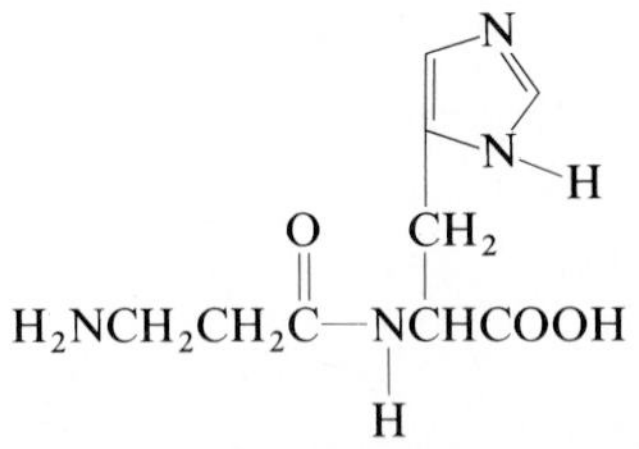

carnosine
(β-alanyl-L-histidine)

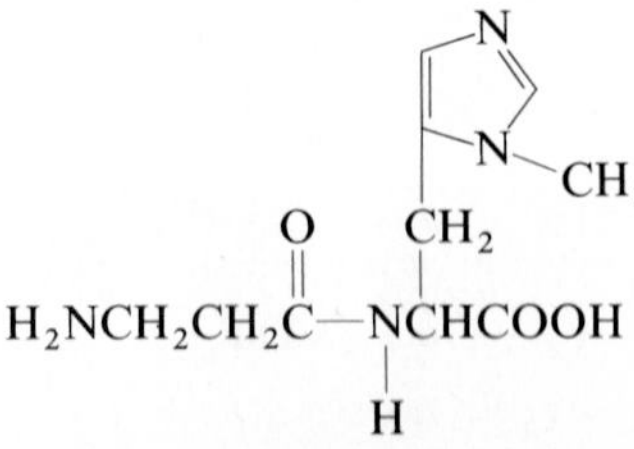

anserine
(β-alanyl-N³-methyl-L-histidine)

$$H_2N\overset{\beta}{C}H_2\overset{\alpha}{C}H_2COOH$$

β-alanine

FIGURE 3–31 Carnosine, anersine, and β-alanine.

3–5 NATURALLY OCCURRING PEPTIDES

Anserine and Carnosine

Obviously, the smallest possible peptide is a dipeptide. Two examples are *carnosine* and *anserine* (see Figure 3–31), both of which are found in muscle tissue of vertebrates, including human muscle. Both contain β-alanine, a structural isomer of α-alanine, in which the amino group is on the beta carbon rather than the alpha carbon. (We will see later that β-alanine is also a component of an important vitamin, pantothenic acid.)

The precise role of these peptides in muscle biochemistry is still unknown—a reminder that even though our knowledge about many things is quite sophisticated, there is a lot we do not know even about simple molecules. Some have suggested, however, that they may function in pH buffering in muscle cells. Both peptides would be efficient buffers of physiological pH owing to the presence in each of the imidazole function of histidine with a pK_a of about 6.

Although it is industrially prepared and not a naturally occurring material, *aspartame* (see Figure 3–32) is another dipeptide that now has considerable importance in the marketplace. Aspartame is the newest FDA-approved artificial sweetener, having a sweetness rating about two hundred times greater than that of table sugar (sucrose). Aspartame is also called Nutrasweet when it is an ingredient in a product and Equal when it is sold as a sugar substitute.

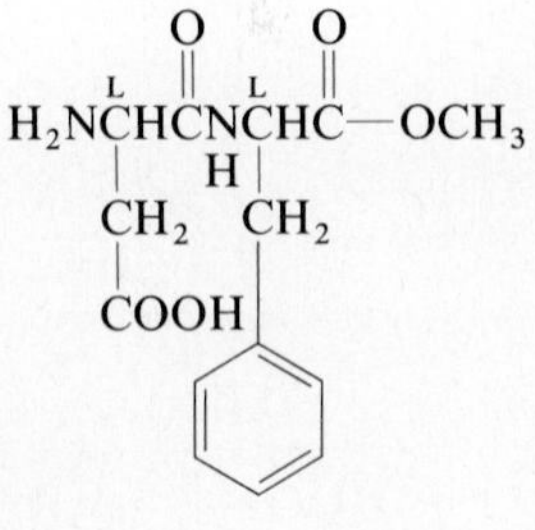

FIGURE 3–32 Aspartame. The peptide name is L-aspartyl-L-phenylalanine (methyl ester).

Glutathione Structure

Gamma-L-glutamyl-L-cysteinyl-glycine, a tripeptide commonly called *glutathione*, is universally distributed in animals, plants, and bacteria and is probably the most abundant simple peptide (see Figure 3–33). The N terminus glutamic acid residue is named gamma glutamyl because the R group γ–COOH rather than the α–COOH participates in peptide bond linkage.

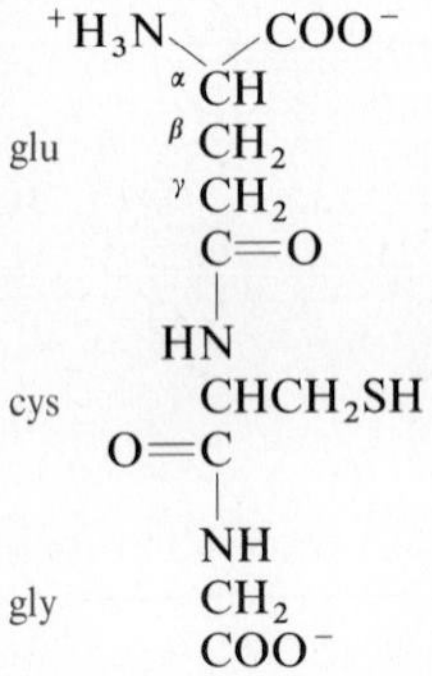

Biosynthesis and Degradation of Glutathione

In living cells glutathione formation is not a ribosome-dependent process. Rather, two specific enzymes (*γ-glutamyl-cysteine synthetase* and *glutathione synthetase*) catalyze two successive, energy-requiring condensation reactions.

FIGURE 3–33 Gamma-L-glutamyl-L-cysteinyl-glycine, or glutathione (glu-cys-gly).

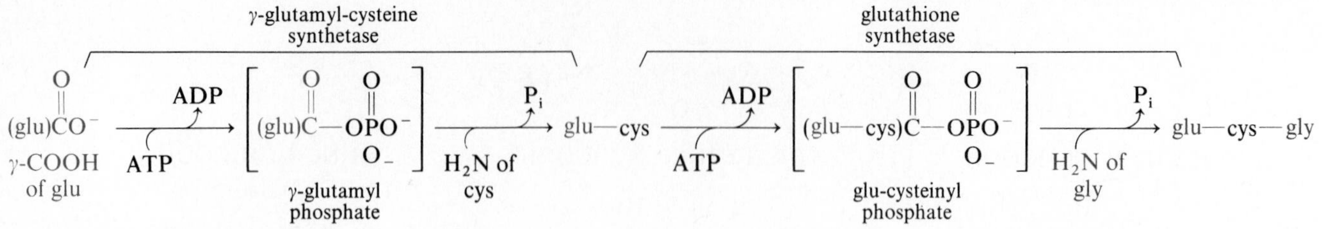

FIGURE 3–34 Glutathione formation.

The energy requirement is fulfilled by the involvement of ATP, producing activated aminoacyl-phosphate intermediates prior to each condensation (see Figure 3–34).

Glutathione degradation occurs in a more indirect manner, involving four enzyme-catalyzed reactions (see Figure 3–35). *γ-Glutamyl transpeptidase* catalyzes transfer of the γ-glutamyl moiety to another amino acid, yielding the cys-gly dipeptide fragment of glutathione and a new γ-glutamyl-amino acid dipeptide. The enzyme *dipeptidase* catalyzes the hydrolysis of cys-gly to free cysteine and glycine.

Free glutamic acid is released in a roundabout way. First, *γ-glutamyl cyclotransferase* catalyzes the intramolecular condensation of the —NH₂ function of γ-glutamyl to the C=O of the peptide bond, cleaving the peptide bond and releasing the free amino acid and 5-oxoproline, a cyclic derivative of the γ-glutamyl residue. Then *5-oxoprolinase* catalyzes the ATP-dependent ring opening of 5-oxoproline to yield glutamic acid.

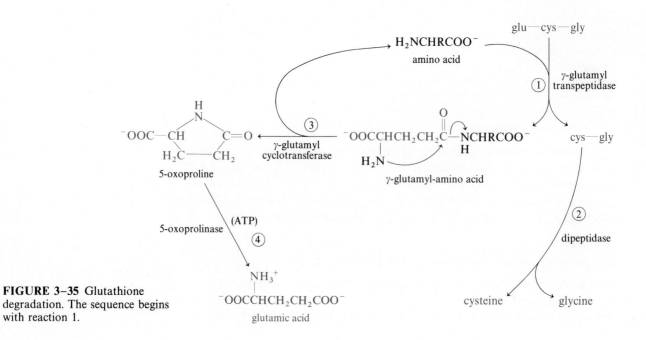

FIGURE 3–35 Glutathione degradation. The sequence begins with reaction 1.

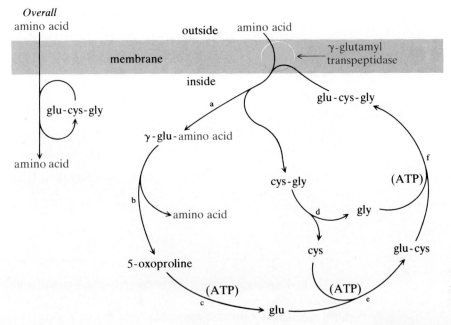

FIGURE 3–36 γ-Glutamyl cycle of glutathione metabolism.

Glutathione-Dependent Membrane Transport of Amino Acids

There are various mechanisms whereby amino acids move across a cell membrane. One scheme, the *γ-glutamyl cycle* (see Figure 3–36), is linked to reactions just described for forming and degrading glutathione. The key feature is that γ-glutamyl transpeptidase is a membrane-localized enzyme. The cycle begins with an amino acid from outside the cell binding to the transpeptidase on the external face of the membrane. In a way not yet clear, the enzyme (with bound amino acid) assumes a position so that the bound amino acid is presented to the interior face of the membrane for reaction (step a) with glutathione. Thereafter, steps b through f involve the other three enzymes of glutathione degradation and the two enzymes of glutathione biosynthesis. All of these enzymes are soluble enzymes, located in the cytoplasm.

Glutathione: An Important Reducing Agent

Approximately 90% of the nonprotein, —SH-containing compounds in mammalian tissue is contributed by glutathione. In other words, glutathione can be considered as a storage form of cysteine. Moreover, it represents a way cells can exploit the —SH function by having glutathione function as a *reducing agent*. When it does, the —SH *reduced form* (GSH) of the tripeptide is converted to an *oxidized form* (GSSG) consisting of two glu-cys-gly chains linked by an *interchain disulfide* (—S—S—) *bond*. As indicated in Figure 3–37, the process is reversible.

The in vivo GSH ⇌ GSSG conversion is linked to two important glutathione functions: One is a *protective role*, and the other is a *regenerative role*.

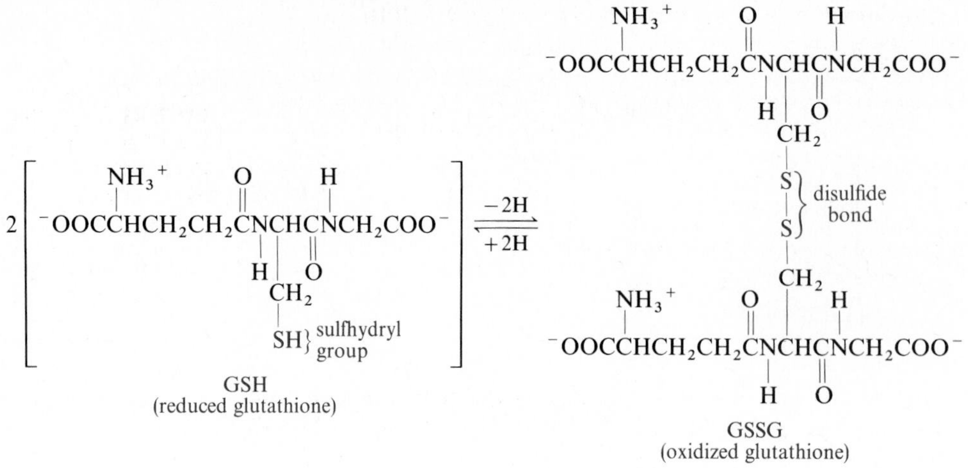

$$
2\left[\begin{array}{c} \overset{NH_3^+}{\underset{|}{}} \quad \overset{O}{\underset{\|}{}} \quad \overset{H}{\underset{|}{}} \\ ^-OOCCHCH_2CH_2CNCHCNCH_2COO^- \\ \overset{|}{H} \quad \overset{\|}{O} \\ \overset{|}{CH_2} \\ SH\} \text{sulfhydryl group} \end{array}\right] \quad \overset{-2H}{\underset{+2H}{\rightleftharpoons}}
$$

GSH
(reduced glutathione)

$$
\begin{array}{c} \overset{NH_3^+}{\underset{|}{}} \quad \overset{O}{\underset{\|}{}} \quad \overset{H}{\underset{|}{}} \\ ^-OOCCHCH_2CH_2CNCHCNCH_2COO^- \\ \overset{|}{H} \quad \overset{\|}{O} \\ \overset{|}{CH_2} \\ \overset{|}{S}\} \text{disulfide bond} \\ \overset{|}{S} \\ \overset{|}{CH_2} \\ \overset{NH_3^+}{} \quad \overset{O}{} \quad \overset{H}{} \\ ^-OOCCHCH_2CH_2CNCHCNCH_2COO^- \\ \overset{|}{H} \quad \overset{\|}{O} \end{array}
$$

GSSG
(oxidized glutathione)

FIGURE 3–37 Interconvertible redox forms of glutathione.

In performing its protective function, reduced glutathione acts as a *scavenger of oxidizing agents,* such as hydrogen peroxide H_2O_2, the hydroxy radical ·OH, and the superoxide anion radical ·O_2^- (see Figure 3–38). Derived from oxygen O_2 (see Section 15–5), these substances can do serious damage to various cellular materials (proteins, nucleic acids, and lipids). However, when they are allowed to react with reduced glutathione, the harmful effects of these oxidizing agents are prevented from occurring. The reaction with reduced glutathione can occur spontaneously (without enzyme involvement), but reaction with H_2O_2 and other organic peroxides can also occur in conjunction with *glutathione peroxidase,* an important enzyme that requires *selenium* for optimal activity. Continued performance of this scavenging role is ensured by the operation of another enzyme, *glutathione reductase,* catalyzing the reduction of oxidized glutathione back to reduced glutathione. The substances NADPH or NADH, either being the reducing agent in this reaction, are discussed in Chapter 12.

The regenerative role is important to some proteins containing —SH func-

FIGURE 3–38 Protective role of glutathione. After reduced glutathione eliminates harmful oxidants (reaction 1), oxidized glutathione is reduced back (reaction 2) to GSH.

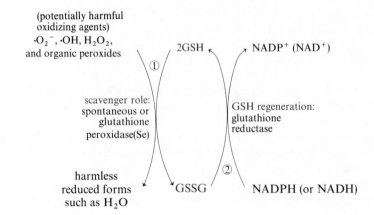

(potentially harmful oxidizing agents)
·O_2^-, ·OH, H_2O_2, and organic peroxides

2GSH   NADP$^+$ (NAD$^+$)

①

scavenger role: spontaneous or glutathione peroxidase(Se)

GSH regeneration: glutathione reductase

②

harmless reduced forms such as H_2O

GSSG   NADPH (or NADH)

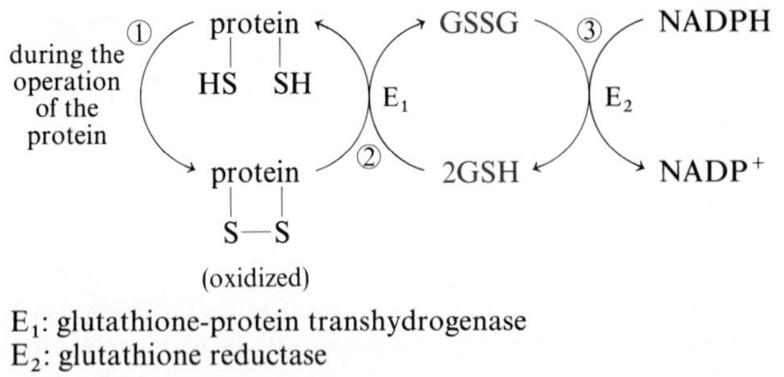

E_1: glutathione-protein transhydrogenase
E_2: glutathione reductase

FIGURE 3–39 Regenerative role of glutathione. Some thiol-containing proteins function by undergoing oxidation (1) and re-formation (2) of the thiol form; the reaction sometimes uses reduced glutathione.

tions that get oxidized to —S—S— linkages during the normal operation of the protein. In such cases an essential event is that the reduced —SH form of the protein be restored, a function relegated to glutathione. As shown in Figure 3–39, the process involves a *glutathione-protein transhydrogenase* enzyme, which is needed to catalyze the —S—S— protein/GSH reaction; and once again the regeneration of reduced GSH from oxidized GSSG is necessary.

Other Glutathione Functions

The glutathione functions just described do not tell the whole story. In fact, there are probably glutathione-dependent processes not yet discovered. Another that is known is the incorporation of glutathione in the biosynthesis of an important class of lipid substances called *leukotrienes*, which we will discuss later (Section 11–4).

3–6 PEPTIDE HORMONES

One of the most fascinating aspects of the living state is that organisms are capable of regulating their chemical processes. There are many facets to this regulation. In the animal kingdom one aspect involves *hormones* (from the Greek *hormaein*, "to rouse or excite"), substances produced in one type of cell and transported to other specific target cells. In mammals (including humans) a large number of different hormones with different regulatory effects are produced by various organs. About fifty are known so far, with more being discovered every year. Nonprotein peptides of small to intermediate size and polypeptide proteins comprise a major group of hormones. We will describe examples of both types in the next several pages. The sampling of protein hormones will focus on a particularly important group, *growth factor hormones*.

The endocrine glands (ductless glands that secrete products directly into the bloodstream, such as adrenals, thyroid, pituitary, thymus, parathyroid, pancreas, gonads, and placenta) are the major hormone-producing organs in mammals—hence the name *endocrinology* for the study of hormones. However, hormone production also occurs in the hypothalamus, the pituitary (also called the hypophysis), the submaxillary gland, and some other specialized cell types.

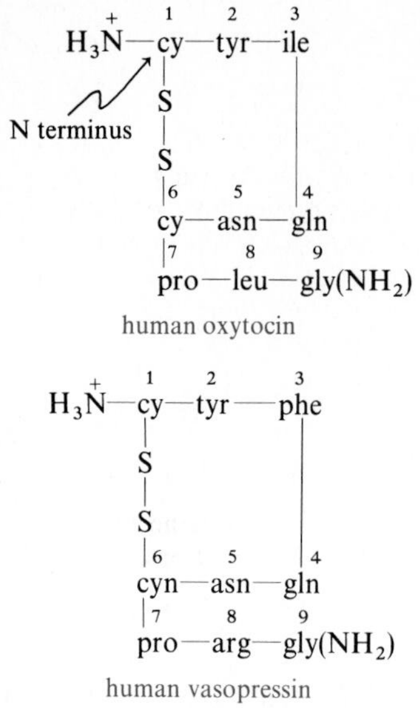

$$\overset{+}{H_3N}-\underset{1}{cy}-\underset{2}{tyr}-\underset{3}{ile}$$

N terminus

$$\underset{6}{cy}-\underset{5}{asn}-\underset{4}{gln}$$
$$\underset{7}{pro}-\underset{8}{leu}-\underset{9}{gly(NH_2)}$$

human oxytocin

$$\overset{+}{H_3N}-\underset{1}{cy}-\underset{2}{tyr}-\underset{3}{phe}$$

$$\underset{6}{cyn}-\underset{5}{asn}-\underset{4}{gln}$$
$$\underset{7}{pro}-\underset{8}{arg}-\underset{9}{gly(NH_2)}$$

human vasopressin

FIGURE 3–40 Human oxytocin and vasopressin. All amino acids are in the L form: gly(NH_2) is glycinamide (see Figure 3–3).

The point is that the scope of modern endocrinology is very broad, concerned with substances produced by many different cell types that regulate nearly every aspect of mammalian biochemistry.

Oxytocin and Vasopressin

The neurohypophysis (the posterior lobe of pituitary gland) releases several hormones, two of which are semicyclic nonapeptides. *Oxytocin* stimulates the contraction of uterine muscle in the pregnant female and the ejection of milk from the mammary glands in lactating females. *Vasopressin* produces potent antidiuretic effects by stimulating the reabsorption of water by the kidney. It also stimulates the contraction of smooth muscle, especially in blood vessels, thus contributing to the control of blood pressure. The similar and yet different effects of these two hormones are consistent with their similar yet different chemical structures. By comparing structures (see Figure 3–40), we can see that residues 3 and 8 are particularly important, since all other residues are identical. The basic arg residue at position 8 is especially important to vasopressin properties. A similar importance for oxytocin properties results from the nonpolar, aliphatic ile residue at position 3. The amide modification of the C terminus, glycinamide, is also necessary for full biological activity.

The overall three-dimensional shape of the oxytocin molecule is depicted in Figure 3–41. As you read through the next several pages, keep in mind that it is this level of structure, conferred by the ordered interactions of all atoms in the specific sequence of amino acids, that determines the structural and functional individuality of any polypeptide.

The biosynthesis of oxytocin and vasopressin begins in the hypothalamus and involves special features that we will discuss later in the chapter. The effects on target cells by these peptides are mediated through the participation of specific *receptor sites* in the cell membrane of the target cells. This feature, too, is discussed later in the chapter.

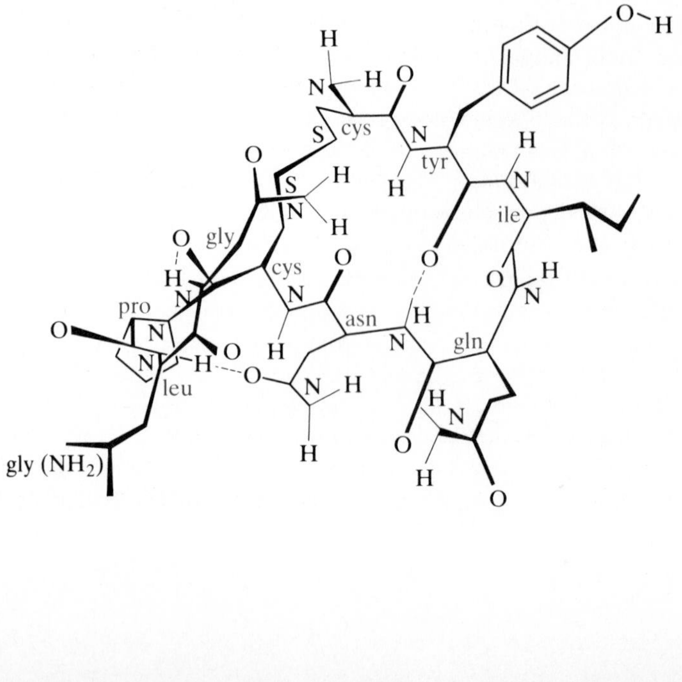

FIGURE 3–41 Three-dimensional skeletal model of oxytocin. [*Source:* Structure reproduced from R. Walter, C. W. Smith, and J. Roy, *Proc. Natl. Acad. Sci., USA,* **73,** 3054 (1976). Drawing provided by Dr. Roderich Walter.]

$$\overset{1}{asp}—arg—val—tyr—\overset{5}{ile}—his—pro—phe—\overset{10}{his}—leu$$

angiotensin I
(weak pressor activity)

$$\overset{1}{asp}—arg—val—tyr—\overset{5}{ile}—his—pro—phe$$

angiotensin II
(potent pressor activity)

FIGURE 3–42 Amino acid sequences of pressor peptides.

Angiotensin

A major participant in the regulation of blood pressure is *angiotensin II*, a linear octapeptide (see Figure 3–42) that elevates blood pressure by stimulating the constriction of blood vessels. Angiotensin II is produced by the removal of a dipeptide fragment from a decapeptide precursor called *angiotensin I*. There is evidence that this conversion occurs as blood passes through the lung, which contains high levels of the converting enzyme. The source of angiotensin I is *angiotensinogen,* a plasma polypeptide protein originally produced in the liver (see Figure 3–43). Angiotensinogen is acted on (in blood) by *renin,* an enzyme that is produced in the kidneys and secreted into blood. Renin cleaves a specific peptide bond in angiotensinogen between residues 10 and 11, releasing the decapeptide fragment, angiotensin I. In addition to its pressor effect, angiotensin II also acts on the central nervous system (brain) to cause thirst and to stimulate the pituitary to increase vasopressin secretion.

An important new class of oral *hypertension drugs* operate by inhibiting the activity of the angiotensin-converting enzyme, thus reducing the formation of angiotensin II and lowering blood pressure. The structure and the mode of action of these drugs, called *captopril* and *enalapril,* are described in Chapter 5.

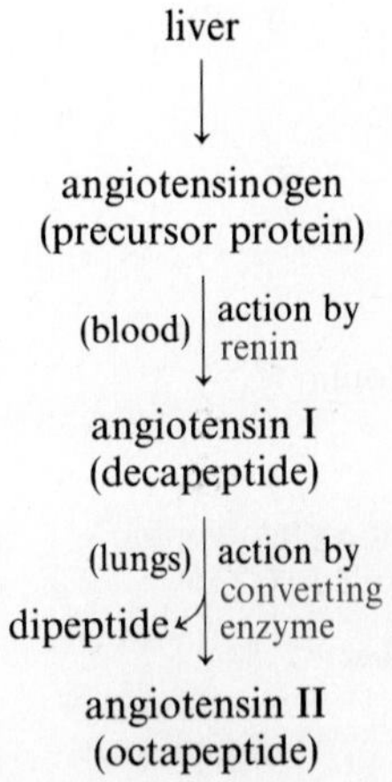

Somatostatin

The hypothalamus produces (at least) three peptide hormones, which in turn control the production of other hormones by the pituitary (see Table 3–9). One of the hypothalamus hormones, *somatostatin,* inhibits the pituitary's production of human growth hormone HGH (this hormone will be discussed later in this chapter). Somatostatin (see Figure 3–44) is also produced by the pancreas and has other regulatory actions in the body. One such action is controlling the release from the pancreas of two other peptide hormones, *insulin* and *glucagon.* Insulin is responsible for lowering levels of blood glucose (sugar); glucagon acts to elevate blood sugar.

Some of the other peptide hormones listed in Table 3–9 are described in the following section.

3–7 GROWTH FACTORS

Substances that stimulate cells to increase their rate of growth and cell division comprise a special class of hormones called *growth factors.* Some are active on a variety of cell types, while others stimulate the proliferation of only

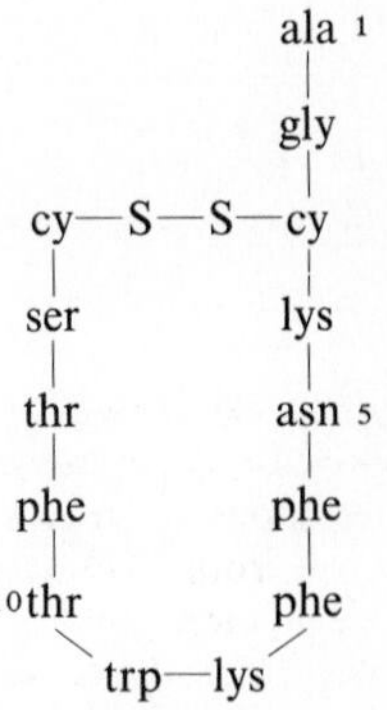

liver
↓
angiotensinogen
(precursor protein)

(blood) | action by renin

angiotensin I
(decapeptide)

(lungs) | action by converting enzyme
dipeptide

angiotensin II
(octapeptide)

FIGURE 3–43 Formation of angiotensin II. Hypertension drugs that inhibit the action of the converting enzyme are described in Section 5–5.

ala 1
|
gly
|
cy—S—S—cy
| |
ser   lys
| |
thr   asn 5
| |
phe   phe
| |
10thr   phe
trp—lys

FIGURE 3–44 Somatostatin.

TABLE 3–9 Peptide and protein hormones in mammals (a partial listing)

SUBSTANCE	NO. OF AMINO ACID RESIDUES (in humans)	PRODUCING SITE	AFFECTED SITE	EFFECTS
Peptides				
Thyrotropin-releasing factor (TRF)	3	Hypothalamus	Pituitary	Each controls release of a specific hormone by the pituitary gland
Luteinizing-releasing factor (LRF)	10	Hypothalamus	Pituitary	
Somatostatin	14	Hypothalamus	Pituitary	
Gastrin	17	Stomach	Stomach	Stimulates HCl secretion
Secretin	27	Stomach	Pancreas	Stimulates the secretion of water and salts
Glucagon	29	Pancreas	Liver	Stimulates breakdown of glycogen to release glucose and the formation of glucose from other substances
Calcitonin	32	Thyroid	Bone, kidney	Inhibits release of calcium from bone and stimulates excretion of calcium and phosphorus; see PTH below
Adrenocorticotrophic hormone (ACTH)	39	Pituitary	Adrenal cortex	Stimulates production of adrenal hormones
Proteins				
Insulin	51 (two chains: 21 and 30)	Pancreas	All cells	Controls carbohydrate, fat, and protein metabolism
Parathyroid hormone (PTH)	84	Parathyroid	Bone, kidney	Stimulates release of calcium from bone and inhibits excretion of calcium; see calcitonin
Human growth hormone (HGH)	191	Pituitary	All tissues	Stimulates the growth of many cell types
Prolactin	198	Pituitary	Mammary gland	Stimulates milk production
Thyroid-stimulating hormone (TSH; thyrotropin)	201 (two chains: 89 and 112)	Pituitary	Thyroid	Stimulates release of thyroxine
Follicle-stimulating hormone (FSH)	202 (two chains: 89 and 113)	Pituitary	Seminiferous tubules (male); ovary (female)	Stimulates production of sperm and maturation of follicle

a single cell type. In either case, the effect of the growth factor may be due to stimulating the synthesis of DNA, RNA, or one or more critical proteins necessary for cell growth and division.

Growth Hormone

The adenohypophysis (the anterior lobe of pituitary gland) releases *growth hormone* (GH; also called *somatotropin*), a polypeptide protein that accelerates the growth, but not the rate of cell division, of many cell types in the body.

The human growth hormone HGH consists of 191 amino acid residues. Its release from the pituitary is also under hormonal control, being stimulated by growth hormone–releasing factor and inhibited by the substance just described, somatostatin.

Hypersecretion of HGH during normal growth stages causes overgrowth of the skeleton, resulting in 7-ft-plus individuals. Hypersecretion after normal growth stages results in *acromegaly,* a disease characterized by enlargement of the bones in the head, of the soft parts of feet and hands, and sometimes of other structures. Hyposecretion of HGH or defects in the action of normal levels of HGH contribute to small but balanced torsos—different from the dwarfism associated with low levels of thyroid activity. Some conditions due to HGH hyposecretion can be successfully treated by intravenous injection of HGH. Severely restricted by a limited supply of HGH, this therapy has been advanced by the recent application of genetic engineering techniques for the design of bacteria capable of producing the human growth hormone in large quantities.

Epidermal Growth Factor and Plasma-Derived Growth Factor

One of the most studied growth factors is *epidermal growth factor* (EGF), synthesized in the submaxillary gland but occurring in many tissues, serum, and urine. The active form (in humans) is a single polypeptide chain of 53 amino acids. The EGF stimulates the growth, proliferation, and differentiation of various types of cells and may play an important role in early prenatal and neonatal development processes. The EGF appears to be virtually identical to another hormone called *urogastrone,* a potent inhibitor of HCl release from the intestinal mucosa. The EGF exerts numerous effects on its different target cells.

The major growth factor present in serum is *plasma-derived growth factor* (PDGF). Discovered in 1975, it is released from blood platelet cells during blood clotting and is a potent stimulator of growth for cells of connective tissue, smooth muscle, and glial cells. There appear to be two forms, PDGF–I and PDGF–II, each composed of about two hundred amino acids. To date, not enough material has been isolated to determine the entire amino acid sequence and perform other structure/function studies.

Skeletal Growth Factor

In 1982 a new human growth factor was discovered that specifically stimulates the growth of bone cells; it is called *skeletal growth factor* (SGF). The SGF protein is very potent: Nearly 1000% increase in growth rate of cultured bone cells treated with an SGF concentration of 0.3 μg/mL has been reported. The SGF is an atypically large growth factor protein of yet uncharacterized structure. Future research to understand structure/function features of SGF offers hope that several bone diseases characterized by the loss of bone tissue can be treated by the administration of SGF. Other important applications may be in hastening growth of broken bones, decreasing the time of recovery, and ensuring the development of a strong overgrowth of new bone. Abnormal hypersecretion of SGF has already been implicated with Paget's disease, a chronic disorder characterized by enlargement and deformity of the skull, spine, and long bones.

TABLE 3–10 Some well-characterized examples of growth factors in vertebrates

Thymosin α_1	Peptide with 29 amino acids; one of several hormone peptides produced in the thymus gland that contribute to the development and differentiation of T cells comprising part of the immune system
Interleukin-2 (IL–2)	Also called T cell growth factor, implying its biological function; a polypeptide with 133 amino acids
Nerve growth factor (NGF)	Active molecule appears to be a dimer of two identical polypeptide chains, each having 118 amino acids; functions in the development and viability maintenance of sympathetic and some sensory neurons

Note: About twenty other growth factors have been detected but are only partially characterized.

Some additional peptide/protein growth factors (*thymosin, interleukin,* and *nerve growth factor*) are described in Table 3–10 for reference.

Cancer and Growth Factors

One of the potentially most beneficial areas of current biochemical research is directed at understanding how normal, growth-controlled cells are transformed into abnormal, growth-uncontrolled cancer cells. Complete understanding will require much study because there are various mechanisms that underlie the normal cell → cancer cell conversion. What is clear, however, is that one mechanism involves some polypeptide substances that mimic some aspect of the biochemical action of naturally occurring growth factors. Substances in this category are proteins representative of *oncogene proteins,* cancer-causing proteins produced in vivo under direction of genes called *oncogenes* that naturally exist in DNA chromosomes but are not normally expressed.

Two oncogenes recently related to growth factors are the *sis*-oncogene and the *erb*-oncogene. The *sis*-oncogene protein is very similar to the plasma-derived growth factor, whereas the *erb*-oncogene protein is very similar to the membrane receptor protein for the epidermal growth factor. Does cancer arise because these oncogene products increase the in vivo occurrence of a growth factor or its receptor, causing an imbalance of growth factor action, or because they are more potent than or sustain their action longer than their natural counterparts? These questions remain unanswered for the time being. We will discuss oncogenes more thoroughly in Section 7–3.

3–8 NEUROTRANSMITTER PEPTIDES (NEUROPEPTIDES)

A *neurotransmitter* is a chemical substance that regulates the transmission of impulses between nerve cells (neurons). In animals various types of neurotransmitters controlling different neurophysiological processes are known. One of the most studied and understood transmitters is *acetylcholine* (see Section 11–7). Several others are produced from the amino acids tyrosine, tryptophan, glutamic acid, glycine, and histidine (see Chapter 18).

tyr—gly—gly—phe—met
    (met)enkephalin

tyr—gly—gly—phe—leu
    (leu)enkephalin

tyr—gly—gly—phe—met—arg—phe
    (met)enkephalin-arg-phe

tyr—gly—gly—phe—met—arg—gly—leu
    (met)enkephalin-arg-gly-leu

tyr—gly—gly—phe—leu—arg—arg—ile—arg—pro—lys—leu—lys
             5   10
    dynorphin

tyr—gly—gly—phe—met—thr—ser—glu—lys—ser—gln—thr—pro—leu—val
                      10   thr
lys—ile—ile—ala—asn—lys—phe—leu
         20
asn—ala—tyr—lys—lys—gly—glu
                  30

     β-endorphin

FIGURE 3-45 Neuropeptides.

In 1975 a new class of neurotransmitters was discovered—*peptides,* or *neuropeptides.* Because they were first detected in extracts of brain tissue, the name *enkephalin* (from a Greek word meaning "in the head") was coined. They are also called *opiate peptides* because their mode of action mimics that of morphine and other opioids. They are present in very small amounts in both vertebrates and invertebrates.

The first two neuropeptides discovered were pentapeptides, differing only in the identity of the C terminus (methionine or leucine). However, this slight difference is responsible for significant differences in neurotransmitter potency: *(met)enkephalin* is about twenty times stronger than (leu)enkephalin in laboratory assays of their effect to provide relief of pain. This analgesic effect appears to be one of the normal functions of opiate peptides. However, they may also be involved in regulating mood and personality. Other known opiate peptides are (met)enkephalin-arg-phe, (met)enkephalin-arg-gly-leu, dynorphin (13 amino acids), and β-endorphin (31 amino acids); see Figure 3–45. Note the sequence of the first five amino acids in the larger peptides.

3-9 MEMBRANE RECEPTORS: AN INTRODUCTION

The preceding descriptions of various bioregulatory peptides and proteins give us an opportunity now to highlight an important aspect of biochemistry common to the operation of them all, namely, the participation of *membrane receptors.* However, their importance is much broader in scope because numerous nonpeptide hormones, nonpeptide neurotransmitters, drugs, toxins, viruses, and various other bioactive materials also operate through membrane receptors. For now, our consideration will deal with general introductory principles. Details regarding some specific receptors are presented in other chapters.

By now, you appreciate that many events in living cells occur in response to the presence of a substance outside the cell. That substance may exert influence (1) while remaining external to the cell or (2) only after it (or some part of it) is translocated into the cell, referred to as *internalization* (see Note 3–2).

> In either situation the first event involved is binding of the substance to a receptor—usually a protein—localized in the membrane of the cell.

NOTE 3–2
For some receptor-bound bioactive peptides the process of internalization by endocytosis, followed by degradation of the peptide via enzymes in lysosomes, provides a way of eliminating the bound peptide and thus eliminating the signal its presence means.

Hereafter, we will refer to the substance being bound as a *ligand*. Establishing the existence of a receptor, measuring the strength and specificity of the binding with its ligand, isolating the receptor, identifying factors that affect the biosynthesis and degradation of the receptor, determining whether the receptor is found in only one or two cell types or widely distributed in many different cells, and identifying how the event of ligand-receptor binding triggers intracellular processes are all matters of biochemical interest.

The isolation of a pure receptor with retention of its normal ligand-binding activity is difficult. To do so requires disrupting the membrane to dislodge the receptor from intimate associations with other protein and lipid components in the membrane. To the extent that this natural membrane geography is essential for maintaining a receptor in an active state, once the geography is destroyed, receptor activity is diminished or even lost, making it difficult to detect. Still, some successes have been achieved.

The most intriguing and most difficult aspect to solve is *how* the ligand-receptor–binding event triggers responses in the cell. It is a difficult aspect because often a complex *cascade of events* is involved, with the initial binding event stimulating a second event, which in turn may stimulate another event, until ultimately a change is effected in such areas as DNA, RNA, or protein biosynthesis, metabolism, or membrane transport. For example, a complex action cascade, totaling six different events with all occurring in the membrane, explains the effect of *cholera toxin*. Another is associated with the regulation of carbohydrate metabolism by the hormone *epinephrine* (also called *adrenalin*). We will examine the many details of the epinephrine process in Chapter 13.

Understanding how the signal of binding is transmitted is of fundamental importance in understanding many biological processes. For example, once the action cascade associated with a cancer-causing growth factor is solved, the biochemical details of the cancerous transformation are then known, making possible the design of a better therapy to stop and even reverse the process.

3–10 PEPTIDE ANTIBIOTICS

Many naturally occurring antibiotics are peptides or contain a small peptide component as part of their overall structure. This group of materials exhibits a tremendous variety of structure, as illustrated in Figure 3–46 for

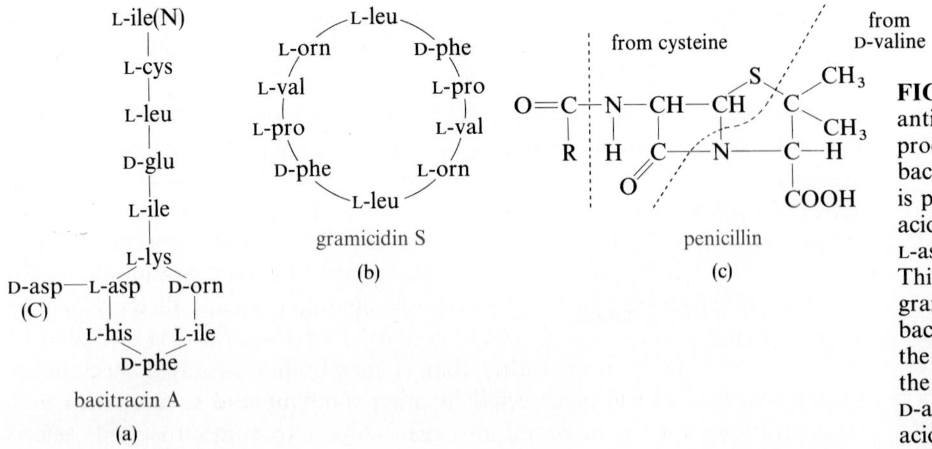

FIGURE 3–46 Peptide antibiotics. (a) Bacitracin A. It is produced by strains of the bacterium *Bacillus licheniformis*. It is partially cyclic, with D and L acids including both D- and L-aspartic acids. (b) Gramicidin S. This antibiotic is one of several gramicidins produced by the bacterium *Bacillus brevis*. Note the totally cyclic structure and the presence of both L- and D-amino acids; L-orn is the amino acid ornithine (see Section 18–4). (c) Penicillin. It is produced by *Penicillium* molds. The most common penicillin has

$$R = \bigcirc\!\!\!\!\!\bigcirc - CH_2 - \text{(benzyl group)}$$

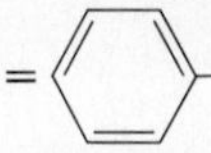

Many others, both naturally occurring and synthetic, are known; see Section 10–6 for the mode of action of penicillin.

benzyl penicillin, bacitracin, and *gramicidin.* The natural biological role of antibiotics is possibly one of self-protection for the microorganisms that produce them. We, of course, use them as chemotherapeutic agents in the fight against disease.

3–11 CYCLOSPORIN A: AN IMMUNOSUPPRESSANT PEPTIDE

In surgical-transplant procedures immunosuppressive agents are used to control rejection of the transplanted tissue. The most successful immunosuppressant to be used in recent years in *cyclosporin A* (see Figure 3–47), one of

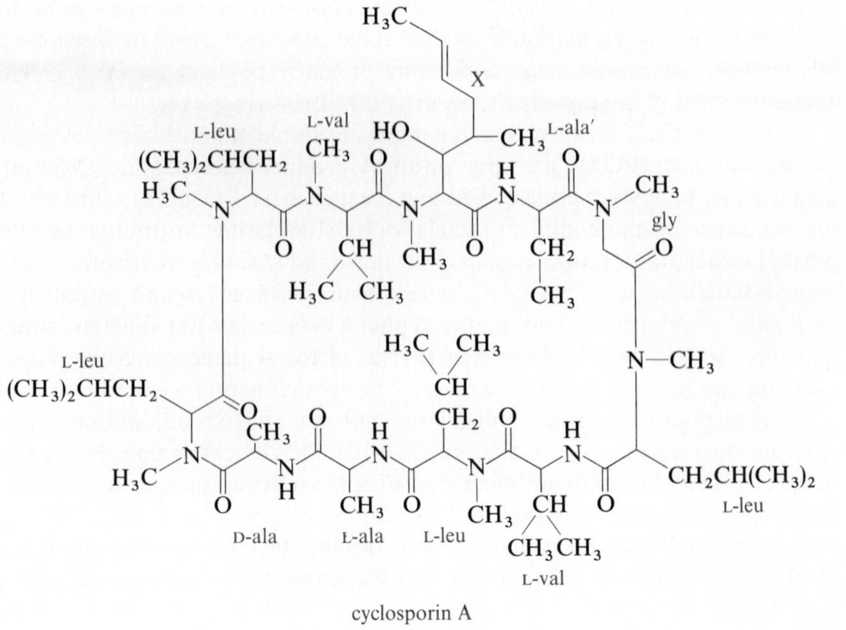

FIGURE 3–47 Cyclosporin A structure. A cyclic undecapeptide consisting of several nonpolar amino acids. Note the presence of a single D-ala residue, of N-methylation on eight of the peptide bonds, of L-homoalanine (L-ala′), and of the unusual amino acid (X) that has not yet been detected in any other naturally occurring polypeptide.

a group of cyclic peptides (11 residues) produced by some fungus organisms. Cyclosporin therapy has dramatically increased the success rate of kidney and liver transplants.

Cyclosporin A appears to act on the immune system by inhibiting initial steps involved in the activation of T cells, thymus-dependent lymphocytes that act directly to incapacitate or destroy tissues bearing foreign antigens or invading microorganisms. Although the precise mode of action of cyclosporin is not yet established, a recent finding is the existence of a highly specific binding protein in the cytoplasm of T lymphocytes that has a very strong affinity for cyclosporin A. The existence of this cytosolic binding protein, called *cyclophilin,* suggests that the immunosuppressant activity of cyclosporin A is mediated by an intracellular mechanism rather than a membrane-associated mechanism. Cyclophilin has also been detected in other nonlymphoid tissues, with high concentrations present in brain, an organ subject to some toxic side effects during cyclosporin A therapy in humans.

3–12 PRECURSOR PROTEINS OF BIOACTIVE PEPTIDES

Some small peptides like anserine, carnosine, glutathione, and antibiotics are assembled in a direct manner, using a specific enzyme to catalyze each successive amino acid condensation. However, other peptides, such as angiotensin, are produced as a fragment from a larger, inactive polypeptide precursor. In Chapter 5 we will again encounter the transformation of

1 inactive protein precursor → 1 active protein fragment

as a mechanism also responsible for the activation of many enzymes (see Section 5–9). What is truly remarkable is that some precursor proteins (*prepro-*) are *multivalent;* that is, two or more different bioactive peptides are released from the same protein. Such proteins are also called *polyproteins.*

A particularly illustrative and complex example of multivalency is *prepro-opiomelanocortin* (POMC), a large pituitary-produced polypeptide (265 amino acid residues in bovine species). Within its sequence the POMC protein contains the sequence of three different melanocyte-stimulating hormones (α-MSH, β-MSH, γ-MSH), adrenocorticotrophic hormone (ACTH), corticotrophinlike intermediate-lobe peptide (CLIP), and six neuropeptides [β- and γ-lipotropins, α-, β-, and γ-endorphins, and (met)enkephalin]—a total of 11 different peptide sequences in one protein. However, several of the sequences are overlapping, meaning that several peptides are released from other peptides (see Figure 3–48).

In response to signals and stimuli not yet understood, specific cutting enzymes operate to carve out all of these peptides by catalyzing the cleavage of specific peptide bonds on each side of a bioactive sequence. A pool of 20 amino acids can yield 400 (that is, 20^2) different dipeptide sequences. However, only certain dipeptide sequences serve as the specific cutting points recognized by the processing enzymes present in cells where the processing occurs (see

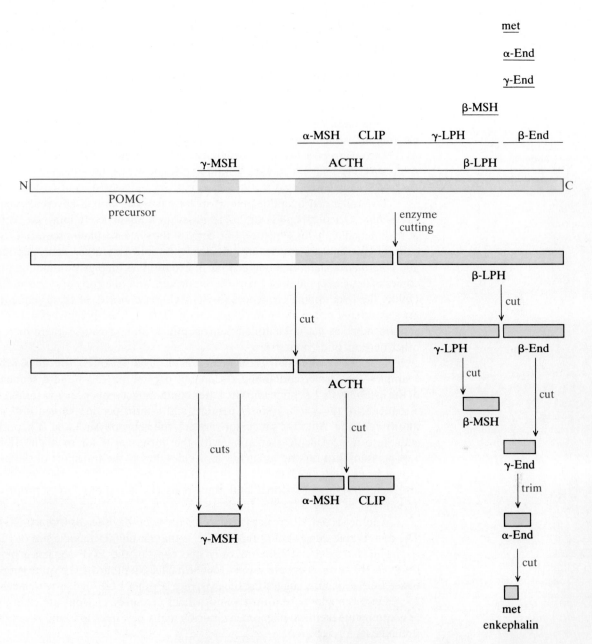

FIGURE 3–48 Proposed steps of prepro-opiomelanocortin (POMC) processing to release various bioactive segments. The solid black horizontal lines at the top indicate the locations of peptide sequences in the shaded areas.

Figure 3–49). These sequences are lys-lys, lys-arg, arg-lys, and arg-arg, all of which are identical in that they consist of two *basic, positively charged* amino acids. Apparently, there is a group of closely related enzymes having this specificity of action.

The enzymes for processing POMC are in the pituitary—some in the anterior lobe and some in the intermediate lobe. The first event in the proposed scenario (see Figure 3–48) is cleavage at the junction of the ACTH/β-lipotropin junction, releasing an ACTH-containing fragment and the intact β-lipotropin

bond cleaved

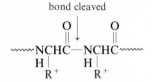

FIGURE 3–49 Common dipeptide sequence recognized by enzymes that catalyze the cutting of polyproteins.

sequence (91 residues). The β-lipotropin (β-LPH) peptide is then cut again into two pieces: γ-lipotropin (58 residues) and β-endorphin (β-End) (31 residues). γ-Lipotropin is then cut to yield β-MSH and a fragment of unknown function, if any. β-Endorphin is cut to release γ-endorphin (17 residues), which is trimmed at the C terminus to yield α-endorphin (16 residues). Any of the various endorphins can be released from the pituitary for transport to target cells or other processing tissues, such as the adrenal gland, where enzymes exist that cut out the amino terminal (met)enkephalin sequence from α-, β-, or γ-endorphins. The ACTH-carrying piece from the original cut is cut again in the anterior lobe to release ACTH and the γ-MSH–containing fragment. In the intermediate lobe the γ-MSH is excised, and the ACTH is cut to yield α-MSH and CLIP.

Two other neuropeptide precursors have recently been discovered: *prepro-enkephalin A,* a polypeptide of 267 residues (in humans) containing six copies of the (met)enkephalin sequence, one copy of the (leu)enkephalin sequence, one copy of the (met)enkephalin-arg-phe sequence, and two other neuropeptides; and *prepro-enkephalin B,* a polypeptide of 256 residues (in pigs) containing three copies of (leu)enkephalin, one copy of dynorphin, and one copy of β-endorphin. Unlike the overlapping sequences in POMC, the majority of these sequences are separately located along the entire chain. Both of these precursors may also contain other as yet unidentified neuropeptides, and, of course, there may be other neuropeptide precursors.

Prepro-oxytocin and *prepro-vasopressin* (166 residues) are two other examples, the former containing a sequence copy of oxytocin and a sequence of the *neurophysin I* peptide and the latter containing a sequence of vasopressin, a sequence of the *neurophysin II* peptide, and a third peptide sequence of unknown function. After the precursors are cut, the neurophysin I and II peptides enter into noncovalent association with the hormones (I for oxytocin, II for vasopressin), thus serving as carrier molecules during the transport of the hormones from the hypothalamus to the pituitary. At the pituitary the hormones and neurophysins are stored and then released in response to appropriate stimuli.

Another recent discovery is *prepro-epidermal growth factor* (prepro-EGF). The remarkable thing about prepro-EGF is that despite its massive size of 1217 amino acid residues, it contains only one copy of the EGF sequence of 53 residues. However, there are seven other sequences in prepro-EGF quite similar to the EGF sequence, suggesting the existence of other EGF-like growth factors.

Once thought to be a rarity, multivalent precursor proteins are emerging as common biochemical phenomena, raising many new questions and providing fertile fields for research.

3–13 LABORATORY CHEMICAL SYNTHESIS OF POLYPEPTIDES

The strategy and techniques for a multistep laboratory synthesis of sequence-defined peptides were first developed in the 1950s. The first success was achieved with the synthesis of the eight residues of oxytocin. In 1963 R. B.

Merrifield devised a revolutionary method, *solid-phase synthesis*, making possible the assembly of even long polypeptides in less time and better overall yield than before. The novel approach of Merrifield was to attach the first amino acid (AA_1) residue to an insoluble, virtually inert, solid polymeric (pol) material. The pol-AA_1 adduct was then reacted with the next amino acid, AA_2, to yield pol-AA_1-AA_2. Another step of condensation yields pol-AA_1-AA_2-AA_3, and so on.

> Because the condensation product formed in each step is attached to an insoluble, solid polymer, it is relatively easy to separate and wash the pol-(AA_n) product of one step before attempting the next step of condensation to yield pol-(AA_{n+1}).

The current state of the art has improved conditions for polymer attachment and for each cycle of condensation and separation, and everything can be done with an automated computer-programmed instrument, a *peptide synthesizer*. Small peptides (5–10 residues) can be synthesized in a day or two (see also Note 3–3). Two notable successes are the syntheses of the ribonuclease polypeptide (an enzyme) with 124 residues and of the human growth hormone with 191 residues.

Various types of polymer materials, fabricated in small-bead form, can be used. One type, pol-$C_6H_4CH_2Cl$, contains benzyl chloride —$C_6H_4CH_2Cl$ functions that react with the α–COOH group of the first amino acid to form the anchor bond (see Figure 3–50). However, the amino acid must first be chemically *modified to block* the α–NH_2 function (and sometimes also an R group function), preventing it from reacting with the polymer or with the α–COOH of another molecule of the amino acid. Thus blocked amino acids have only one potentially reactive group, namely, the α–COOH.

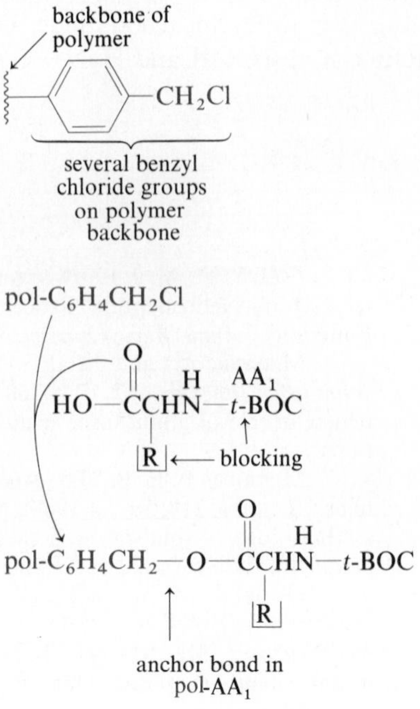

NOTE 3–3
Solid-phase methods of chemical synthesis for polynucleotide chains, used in the laboratory synthesis of DNA, have also been recently developed (see Section 6–9).

FIGURE 3–50 Solid-phase chemical synthesis. In the initial step an amino acid is covalently attached via its carboxyl group to the functional groups of an otherwise inert solid polymer.

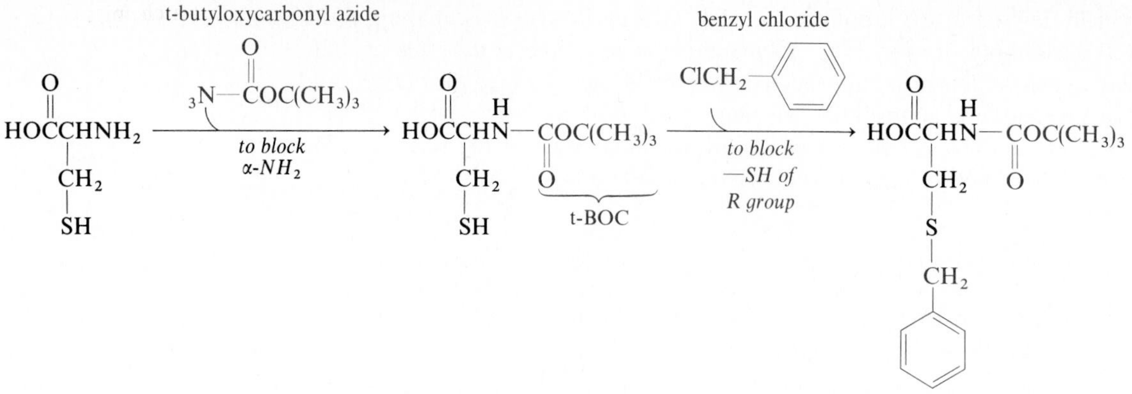

FIGURE 3–51 Blocking reactions of amino acids. The first reaction can be used to block the α–NH_2 group of any amino acid. The second reaction illustrates R group blocking for cysteine.

Note that blocked amino acids are used throughout all cycles of the entire synthesis. One way to block α-amino groups is treatment with *t-butyloxycarbonyl* (*t*-BOC) *azide* to form *t*-BOC derivatives (see Figure 3–51). There are a variety of reagents used to block the different R group functions in cys, thr, ser, glu, asp, arg, lys, and tyr. For example, the —SH of cysteine can be blocked with benzyl chloride. The important criteria of blocking groups are as follows: Blocking groups for R side-chain functions must remain attached throughout all cycles of the operation; blocking groups for the α–NH_2 functions must be selectively removed after each cycle of condensation; and after the complete polypeptide is assembled, all R-blocking groups need to be removed under conditions that will not cleave any of the peptide bonds in the product.

The scenario of solid-phase polypeptide synthesis is outlined in Figure 3–52. Note that the step of peptide bond formation in each cycle occurs in the presence of a condensing agent (dicyclocarbodiimide, DCC, is a popular one) to promote the interaction of —COOH and H_2N— functions.

LITERATURE

General

BADA, J. L., and S. E. BROWN. "Amino Acid Racemization in Living Mammals: Biochronological Application." *Trends Biochem. Sci.,* **5**, III–V (1980). A brief review article.

BLOOM, F. E. "Neuropeptides." *Sci. Am.,* **245**, 148–168 ((1981). A good survey of this rapidly developing field.

BUMPUS, F. M. "Angiotensin Antagonists in Relation to Hypertension." *Hosp. Pract.,* **9**, 80–92 (1974). A discussion of the molecular basis for the biological actions of angiotensin II in regard to its amino acid sequence and three-dimensional structure.

JAMES, R., and R. A. BRADSHAW. "Polypeptide Growth Factors." *Annu. Rev. Biochem.,* **53**, 259–292 (1984). A thorough review article.

LOH, Y. P., M. J. BROWNSTEIN, and H. GAINER. "Proteolysis in Neuropeptide Processing and Other Neural Functions." *Annu. Rev. Neurosci.,* **7** (1984). A review article.

MEISTER, A., and M. E. ANDERSON. "Glutathione." *Annu. Rev. Biochem.,* **52**, 711–760 (1983). A comprehensive review article of glutathione metabolism and cellular functions.

MERRIFIELD, R. B. "The Automatic Synthesis of Proteins." *Sci. Am.,* **218**, 56–74(1968). A synopsis of the chemical methodology of solid-phase peptide synthesis, with specific details on the synthesis of insulin.

ONDETTI, M. A., and D. W. CUSHMAN. "Enzymes of the Renin-Angiotensin System and Their Inhibitors." *Annu. Rev. Biochem.,* **51**, 283–308 (1983). A review article of renin and the angiotensin-converting enzyme.

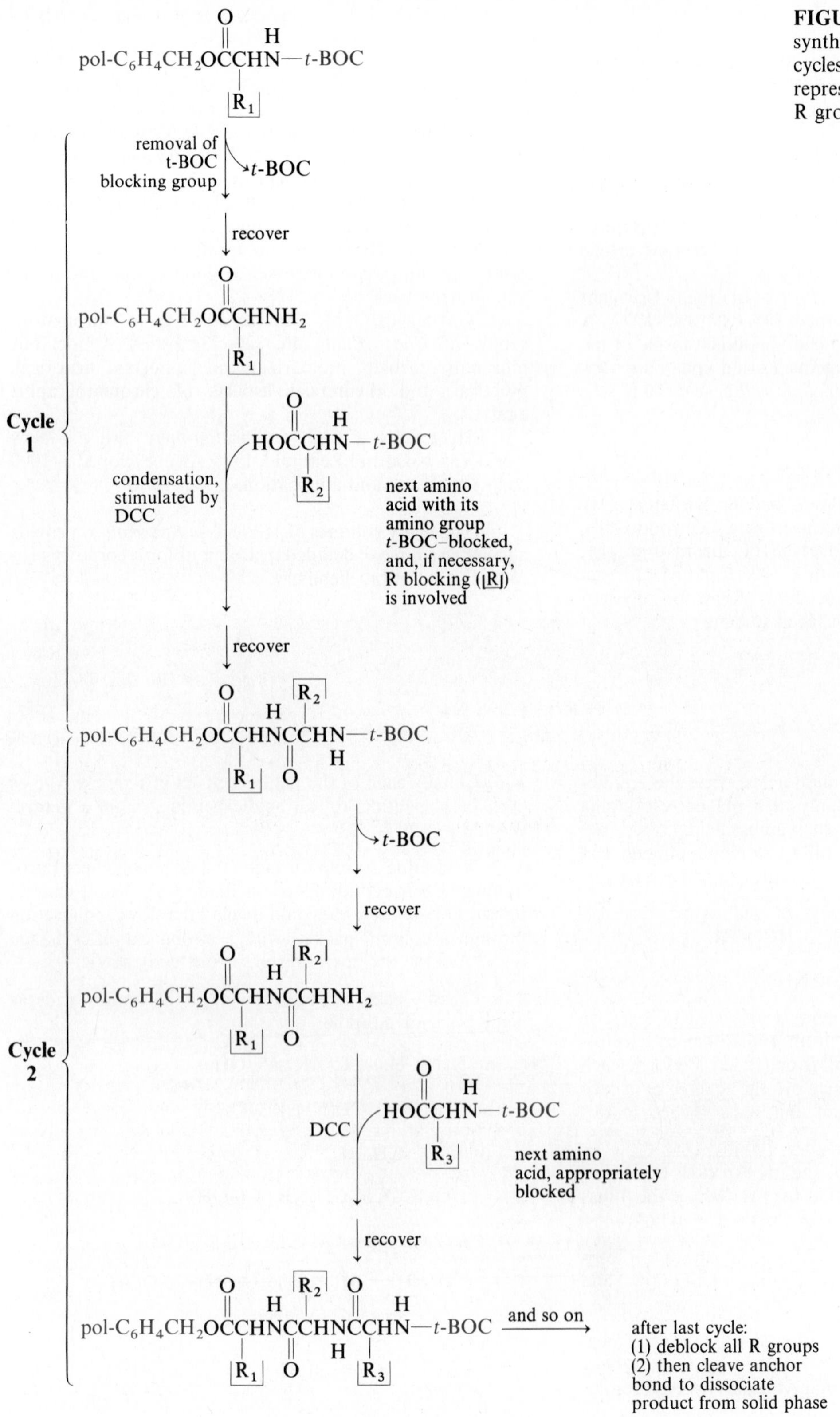

FIGURE 3–52 Solid-phase synthesis of polypeptides. Two cycles are summarized; ⌊R⌋ represents a chemically blocked R group.

$$\text{pol-C}_6\text{H}_4\text{CH}_2\text{OCCHN}-t\text{-BOC}$$

with O (double bond) above, H above N, and ⌊R₁⌋ below

Cycle 1

removal of t-BOC blocking group → t-BOC

↓ recover

$$\text{pol-C}_6\text{H}_4\text{CH}_2\text{OCCHNH}_2$$

with O above, ⌊R₁⌋ below

$$\text{HOCCHN}-t\text{-BOC}$$

with O above, H above N, ⌊R₂⌋ below

next amino acid with its amino group t-BOC–blocked, and, if necessary, R blocking (⌊R⌋) is involved

condensation, stimulated by DCC

↓ recover

$$\text{pol-C}_6\text{H}_4\text{CH}_2\text{OCCHNCCHN}-t\text{-BOC}$$

with O, ⌊R₂⌋, H above; ⌊R₁⌋, O, H below

→ t-BOC

↓ recover

$$\text{pol-C}_6\text{H}_4\text{CH}_2\text{OCCHNCCHNH}_2$$

with O, ⌊R₂⌋, H above; ⌊R₁⌋, O below

Cycle 2

DCC

$$\text{HOCCHN}-t\text{-BOC}$$

with O above, H above N, ⌊R₃⌋ below

next amino acid, appropriately blocked

↓ recover

$$\text{pol-C}_6\text{H}_4\text{CH}_2\text{OCCHNCCHNCCHN}-t\text{-BOC}$$

with O, ⌊R₂⌋, O, H above; ⌊R₁⌋, O, H, ⌊R₃⌋ below

→ and so on

after last cycle:
(1) deblock all R groups
(2) then cleave anchor bond to dissociate product from solid phase

SNYDER, S. H. "Opiate Receptors and Internal Opiates." *Sci. Am.*, **237**, 44–54 (1977). Enkephalins and endorphins are described.

STEWART, J. M., and J. D. YOUNG. *Solid-Phase Peptide Synthesis*. Pierc Chemical, 1984. Detailed laboratory manual for performing this technique.

TAGER, H. S., and D. F. STEINER. "Peptide Hormones." *Annu. Rev. Biochem.*, **43**, 509–538 (1974). A review article dealing with structure and biosynthesis.

UY, R., and F. WOLD. "Posttranslational Covalent Modification of Proteins." *Science*, **198**, 890–896 (1977). A review article dealing with the many modified forms of the common 20 acids found in proteins. For an updated review see the article by F. Wold in *Annu. Rev. Biochem.*, **50** (1981).

Biochemical Methods

BREWER, J. M., A. J. PESCE, and R. B. ASHWORTH. *Experimental Techniques in Biochemistry*. Englewood Cliffs, N.J.: Prentice-Hall, 1974. Coverage of chromatography, electrophoresis, ultracentrifugation, spectroscopy, radioactivity, and immunological procedures. Rigorous physicochemical treatment for the advanced student.

CLARK, J. M., and R. L. SWITZER. *Experimental Biochemistry*. 2nd ed. San Francisco: Freeman, 1977. One of the best laboratory manuals available. Contains excellent introductory-level discussions of the theory and applications of several methods of biochemistry.

FREIFELDER, D. *Physical Biochemistry: Applications to Biochemistry and Molecular Biology*. 2nd ed. San Francisco: Freeman, 1982. The theory and application of several laboratory techniques are discussed. A good source, and available in paperback.

GAUCHERµ G. M. "An Introduction to Chromatography." *J. Chem. Educ.*, **46**, 729–733 (1969). A brief but informative article summarizing the important historical, practical, and theoretical aspects of chromatographic analysis.

HEFTMANN, E., ed. *Chromatography*. 3rd ed. New York: Van Nostrand Reinhold, 1975. An encyclopedia (1000 pages) of theory and application.

The many volumes of *Methods in Enzymology* provide a valuable source of detailed instructions for laboratory procedures used in biochemistry.

EXERCISES

3–1. For each of the given amino acids, write the equilibrium reactions that would apply at a pH corresponding to each of the pK_a values for each amino acid: (a) alanine; (b) glutamic acid; (c) arginine; (d) α_1, ε-diaminopimelic acid ($pK_{a_1} = 1.8$, $pK_{a_2} = 2.2$, $pK_{a_3} = 8.8$, $pK_{a_4} = 9.9$), whose formula is

$$\text{HOOCCH(CH}_2)_3\text{CHCOOH}$$
$$\text{NH}_2 \qquad \text{NH}_2$$

3–2. Would the first three amino acids listed in Exercise 3–1 be completely separated from each other by electrophoresis at a pH of (a) 2, (b) 7, or (c) 12? Perform each evaluation by estimating, for each pH, the net charge of each amino acid to the nearest half unit, as described in this chapter.

3–3. The difference between pI and pH expresses the degree of net charge on an amino acid at the pH. Why is the difference computed by pI − pH rather than pH − pI?

3–4. Using three-letter abbreviations, illustrate all possible tripeptides composed of alanine, methionine, and asparagine.

3–5. Draw the complete structures of the following peptides (show all ionizable groups in the protonated state): (a) methionyl-glutamine; (b) glutamyl-aspartyl-penylalanine; (c) phe-arg-trp-ile; (d) VIP.

3–6. Classify each of the peptides given in Exercise 3–5 as (a) a basic peptide, (b) an acidic peptide, or (c) a neutral peptide.

3–7. Neglecting secondary factors that might affect partitioning, which peptide listed in Exercise 3–5 and characterized in Exercise 3–6 would display the slowest migration through a column packed with a strong cation-exchange resin? Assume the operating pH is distinctly acidic.

3–8. Classify each of the following alpha amino acids as being polar or nonpolar.

(a) $H_2N(CH_2)_3CH(NH_2)COOH$

(b) $CH_3(CH_2)_5CH(NH_2)COOH$

(c)
$$\overset{\text{CHCH}_3}{\underset{}{\|}}$$
$$\text{HOOCCCH}_2\text{CH(NH}_2)\text{COOH}$$

(d) $CH_3CH(NH_2)CH_2CH(NH_2)COOH$

(e) $CH_3CH{=}CHCH_2CH_2CH(NH_2)COOH$

(f)
$$\overset{\text{CH}_2}{\text{H}_2\text{C}{-}{-}\text{C}{-}\text{COOH}}$$
$$\text{NH}_2$$

3–9. A polypeptide that binds strongly to a column packed with a strong cation-exchange resin at pH 3.5 can be washed off the column by passing a buffer of pH 8 through the column. Explain why.

3–10. Of those amino acids containing an ionizable R group, which amino acid has the R group of weakest acidity? Which has an R group that would be interconverting between its conjugate pair forms at about pH 6.5?

3–11. The titration curve in Figure 3–53 profiles the ionization of glutamic acid. Identify the following on the curve: (a) the three pK_a values; (b) the pH at which a 50–50 mixture of the -1 and -2 species of glutamate will exist; (c) the pH range in which glutamic acid will always be carrying a net positive charge; (d) a pH range in which the conjugate acid-base pair of 0 glu and -1 glu species will act as a buffer.

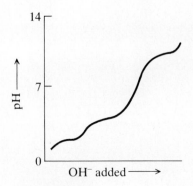

FIGURE 3–53

3–12. The general shape of the titration curve for arginine is shown in Figure 3–54. Identify points on the curve for (a) the three pK_a values; (b) the pH range in which arginine will always carry a net negative charge; (c) the pH at which the $+1$ species of arginine will exist almost exclusively; (d) the pH at which a 10/90 equilibrium mixture of $+2$ arg/$+1$ arg will exist.

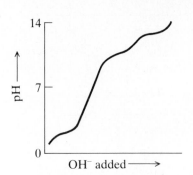

FIGURE 3–54

3–13. A small sample of a pure material was subjected to thin-layer chromatography. After drying, the plate was sprayed with sulfuric acid and heated. The appearance of the charred plate is represented in Figure 3–55. What is the R_f value of this material in the solvent system that was used?

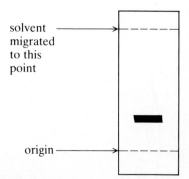

solvent migrated to this point

origin

FIGURE 3–55

3–14. From the information in Table 3–11, determine which combination of developing solvents (Ph + BuAc or Ph + BuP) will provide optimum resolution by two-dimensional paper chromatography of a mixture containing all seven of the amino acids listed. Proceed by making a sketch of the paper sheet as it would appear after ninhydrin treatment. (Represent the paper sheet as a 5-in. square and the ninhydrin zones as spheres with a diameter of $\frac{3}{8}$ in. You will find it convenient to use graph paper with ten squares to the inch.)

3–15. In the design of a solid-phase chemical synthesis of angiotensin I, what amino acid will be anchored to the solid phase in the first step?

3–16. Define these terms: disulfide bond, peptide bond, chirality, N terminus, pI, column chromatography, ninhydrin, HPLC, electrophoretic mobility, PTH derivative, hormone, solid-phase chemical synthesis.

TABLE 3–11 R_f values of amino acids on Whatman no. 1 paper

AMINO ACID	DEVELOPING SOLVENT		
	Ph[a]	BuAc[b]	BuP[c]
Glutamic acid	0.33	0.28	0.20
Lysine	0.42	0.12	0.13
Glycine	0.42	0.23	0.29
Alanine	0.58	0.30	0.37
Valine	0.78	0.51	0.48
Serine	0.35	0.22	0.33
Methionine	0.80	0.50	0.58

[a] Phenol saturated with water.
[b] *n*-Butanol-glacial acetic acid-water (12:3:5).
[c] *n*-Butanol-pyridine-water (1:1:1).

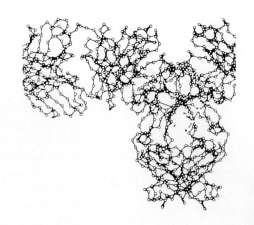

CHAPTER FOUR

PROTEINS

In view of the many previous references to the biological importance of proteins, little else need be said as introduction to this chapter. Let us now examine the proteins more closely, with primary focus on their molecular structure. We will begin with a consideration of protein classification. In addition to contributing some organization to the subject, classification provides an opportunity to obtain an overview of the extensive involvement of proteins in biological processes and an overview of much of the terminology used in reference to the biochemistry of proteins.

The chapter also covers some other laboratory methods in biochemistry in addition to those covered in the previous chapter. They are intrusive to the flow of the chapter design, but they can help you to learn descriptive material. Although placed in this protein chapter, the laboratory techniques are more general in application, used also in the analysis of other types of substances, including nucleic acids. The chapter concludes with a description of protein binding and an introduction to the principle of allosterism (cooperative effects), to be described further in the next chapter.

4–1 CLASSIFICATIONS

Different Functions

As indicated by the following listing, proteins perform many and varied roles in nature. Although the list of proteins is not a ranking of importance (because they are all important), those proteins functioning as enzymes are properly positioned at the top of the list.

1. *Catalytic proteins* are *enzymes* that increase the rate at which chemical reactions occur in cells. Many also have a regulatory characteristic, controlling when a reaction occurs. Most cells contain several hundred different enzymes, with some capable of producing as many as fifteen hundred. Chapter 5 is devoted to enzymes in general, and individual enzymes are described throughout all chapters.

2. *Structural proteins* lack a true dynamic function (that is, they are not involved in chemical reactions). These proteins confer structural *support*. Examples are *collagen,* found in the connective tissues of vertebrates, and various proteins in the membranes of cells, such as *spectrin,* found in the membrane of red blood cells.

3. *Contractile proteins* can reversibly tighten and relax their molecular shape. Examples are *myosin* and *actin* in muscle tissue and similar proteins in many cell membranes and in the microtrabecular lattice described in Chapter 1.

4. *Natural-defense proteins* provide protection against foreign substances, cells, and viruses. Examples are the *antibodies* in the gamma globulin fraction of blood and proteins called *interferons,* which have been linked to an antiviral defense.

5. *Digestive proteins* are the various enzymes present in gastrointestinal secretions. These proteins catalyze the degradation of dietary foodstuffs to smaller substances. Examples are *trypsin* and *chymotrypsin.*

6. *Transport proteins* function in conveying a substance from one place to another. The best-known example is *hemoglobin,* transporting oxygen in the blood; various *membrane transport proteins* are found in membranes and function in the movement of substances across the membrane.

7. *Blood proteins* are the numerous proteins (many are enzymes) involved in various blood processes such as blood clotting, the dissolution of blood clots, and the transport of various substances such as vitamin B_{12} and cholesterol.

8. *Hormone proteins* are produced by one type of cell, and they regulate the actions of other types of cells. An example is *insulin.*

9. *Growth factor proteins* stimulate the rate of cell growth and cell division. Some examples (*growth hormone, PDGF and EGF*) were described in Chapter 3.

10. *Electron transfer proteins* function in the flow of electrons between an initial electron donor and a terminal electron acceptor. An example is the *cytochrome* group of proteins, participating in the transfer of electrons to suitable acceptors, such as oxygen in aerobic organisms. They can also be classified as enzymes.

11. *DNA-binding proteins* reversibly bind to DNA. Depending on the protein, they may regulate the expression of genes (repressor proteins), promote the unwinding of the duplex structure of DNA, stabilize the unwound structure of DNA, or promote the binding of other proteins to DNA.

12. *Chromosomal proteins* are related to DNA-binding proteins. They occur in association with DNA in the nucleus of eukaryotic cells. The major group is the *histone* proteins, which may function in the regulation of genes in the chromosome.

13. *Membrane receptor proteins* are specific proteins, located in the membranes of cells and subcellular organelles, that bind to a specific substance on the exterior side of the cell, generating a signal that triggers other events to occur in the membrane and ultimately in the cell. Most neurotransmitters have membrane receptors; most hormones have membrane receptors; many drugs operate through membrane receptors.

14. *Ribosomal proteins* are specific proteins in association with specific RNA molecules to form *ribosomes,* the multimolecular aggregates participating in protein biosynthesis.

15. *Storage proteins* are used as a nutritional energy and amino acid pool, particularly in plant seeds.

16. *Toxin proteins* are proteins in the venom of poisonous reptiles responsible for the toxicity to the mammalian nervous system and disease-causing proteins, such as *cholera toxin,* produced by pathogenic bacteria.

17. *Vision proteins* are light-sensitive proteins, *opsins* and *rhodopsins,* participating in the molecular events associated with sight.

Different Compositions

There are two composition categories based on what the naturally occurring protein is composed of.

1. *Conjugated proteins* consist of a polypeptide component in association with a nonpeptide component (organic or inorganic), called a *prosthetic group*. Subclasses are identified by the type of prosthetic group: *glycoprotein* (with a carbohydrate prosthetic group), *metalloprotein* (a metal ion), *hemoprotein* (a heme group), *flavoprotein* (a flavin group), *phosphoprotein* (a phosphate group), *lipoprotein* (a lipid), and *nucleoprotein* (a nucleic acid). Occasionally, two or more identical or different prosthetic groups may be present.

2. *Nonconjugated proteins* are composed only of polypeptide material. No prosthetic group is attached.

Different Shapes and Solubilities

There are two categories based on shapes and solubilities.

1. *Fibrous proteins* have very elongated molecular shapes [see Figure 4–1(a)]. Usually, several elongated polypeptide chains are bunched or wrapped together, producing multimolecular threads or filaments. Fibrous proteins are generally insoluble in water and, depending on the protein, are characterized by high or low degrees of tensile strength, elasticity, resiliency, and brittleness and by various textures.

2. *Globular proteins* have a much more compact structure than fibrous proteins, owing to a highly contorted pattern of folding, bending, and twisting along the polypeptide chain [see Figure 4–1(b)]. The overall shape ranges from nearly spherical structures to varying degrees of elliptical structures. Globular proteins are generally more soluble than fibrous proteins. In addition, globular proteins are usually much *more delicate* structures than fibrous proteins. A greater variety and number of proteins are in the globular class. Indeed, excepting the contractile and structural proteins, all other proteins listed in the previous functional classification are globular proteins.

Whether a protein is fibrous or globular, we customarily refer to four levels (aspects) of structure for a protein molecule.

• The *primary level* refers to the *identity, relative amount,* and, ultimately, the exact *linkage sequence of the amino acids* present in the polypeptide chain(s). If any prosthetic group is present, its identity and amount may also be included at this level.

• The *secondary level* refers to the ability of the backbone of the polypeptide chain to assume *ordered orientations stabilized by cooperative hydrogen bonding*.

• The *tertiary level* refers to the complete, *three-dimensional architecture* of the entire protein molecule, including the orientation of any associated prosthetic group.

• The *quaternary level* applies to protein molecules composed of *two or more polypeptide chains* (dimers, trimers, tetramers, and so forth), with the added stipulation that the aggregate be *held together solely by various noncovalent forces of interaction* between individual chains.

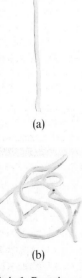

(a)

(b)

FIGURE 4–1 Protein models. (a) Simple wire model to illustrate the extended, threadlike structure of a fibrous protein. (b) Folded, compact structure of a globular protein.

The first three levels of structure apply to all proteins, and the fourth applies to many, but not all, proteins.

Before discussing each of the features of structure, we will first consider some laboratory procedures that are very useful in the isolation and the characterization of proteins.

◇ **LABORATORY METHODS** ◇

MOLECULAR WEIGHT MEASUREMENT AND PURITY EVALUATIONS

Various procedures are available for determining the molecular weights of biopolymers. Three methods are described here. All of these procedures are also utilized in the evaluation of sample purity, and the chromatographic and electrophoretic techniques can be used for the isolation and separation of biopolymers.

Gel-Permeation Column Chromatography

Gel-permeation column chromatography (also called *molecular sieve* chromatography) uses a solid stationary phase consisting of tiny, spherical gel particles that operate as molecular sieves. The particles are made from substances having a cross-linked chemical structure analogous to that of a screen. Two different types of materials are cross-linked *agarose* or *dextran* (both carbohydrate materials) and cross-linked *polyacrylamide*. When placed in a water medium, the tiny gel particles absorb water and swell into larger granules having a *porous* network, with the size of the pores determined by the degree of cross-linking in the structure. A slurry of the granules is used to pack a vertical column, a small volume of the sample to be analyzed is placed on top of the column bed, and a buffer solution is continuously run through the column. On entry into the column bed, substances in the sample begin to reversibly interact with the gel granules by *entering and leaving* a granule through its pores. Entry into the stationary granules will retard the movement of a substance down the column, small molecules being retarded more than larger ones because they enter more easily and permeate more of the granule interior. Molecules with sizes greater than the dimensions of the pores will not penetrate at all, and thus, dissolved in the moving phase, they will move right through the column. The essential events are diagramed in Figure 4–2.

Substances present in the effluent of the column can be detected by measuring absorption of ultraviolet (UV) light (at 280 nm for proteins; at 260 nm for nucleic acids) or, if the sample contained some ^{14}C or ^3H to begin with, by measuring the level of radioactivity. Peaks in the chromatographic profile represent individual zones of proteins (or whatever) that have been separated from each other on the basis of *differences in their molecular size*. Each component in the pattern is identified by its *elution volume* (V_e), the volume of effluent collected that corresponds to the apex of a peak. A pattern showing only a single peak indicates the purity of the original sample applied to the column.

The unknown molecular weight of a protein can be measured as follows:

1. Calibrate a gel-permeation column by determining the elution volumes required to displace several pure proteins of known molecular weight from the column.

2. Determine, on the same column run under the same conditions, the elution volume of the protein under study.

3. Compare the data via the linear empirical relationship between the logarithm of molecular weight and V_e (see Figure 4–3).

Note: Molecular weight (MW) is commonly expressed with dimensions of grams per mole. Thus the mass of a molecule can be calculated by dividing MW by the Avogadro number (6.023×10^{23} molecules per mole). However, biochemists and biologists sometimes use the *dalton* as a unit of mass in describing the size of ribosomes, viruses, and various other multimolecular structures for which the terms molecule and mo-

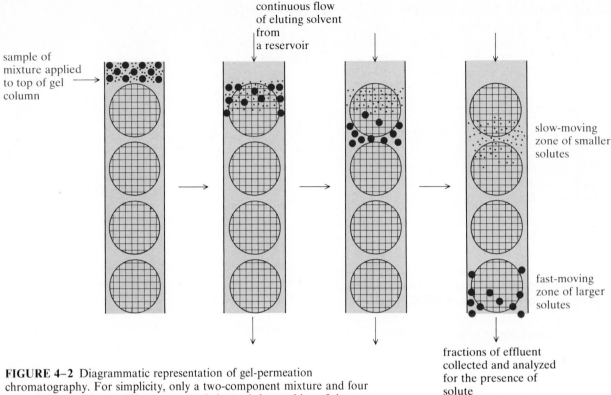

sample of
mixture applied
to top of gel
column

continuous flow
of eluting solvent
from
a reservoir

slow-moving
zone of smaller
solutes

fast-moving
zone of larger
solutes

fractions of effluent
collected and analyzed
for the presence of
solute

FIGURE 4–2 Diagrammatic representation of gel-permeation
chromatography. For simplicity, only a two-component mixture and four
gel granules are shown. (The exaggerated size and the stacking of the
granules are not realistic. Actually, in any one plane of the column
there would be several granules, and in the the whole column, a
countless number.) The grid network represents pores of the granule.
After elution begins, the smaller solute molecules will enter and penetrate
the gel granules to a greater extent than the larger molecules. Because
of this greater interaction with the stationary phase, the small
molecules have a slower rate of migration through the column.

lecular weight are inappropriate. The dalton is defined
as $\frac{1}{12}$ the mass of one atom of the carbon isotope ^{12}C
and equals 1.661×10^{-24} g, identical with the offi-
cially defined atomic mass unit (amu). For example, the
mass of a single *E. coli* ribosome, an aggregate composed of about sixty different molecules, is logically
expressed as 2.6×10^6 daltons.

Although there is a subtle difference in expressing
a molecular weight as X g/mole or as X daltons, the
two are sometimes used interchangeably. For example,
a protein having a molecular weight of 25,000 g/mole
can be described as a protein molecule having a mass
of 25,000 daltons (or 25 kilodaltons, 25 kd). Either
expression yields 4.15×10^{-20} g for the mass of the
molecule.

Polyacrylamide Gel Electrophoresis

Polyacrylamide gel electrophoresis (PAGE) is a powerful tool of the biochemist for separation, isolation,
purity evaluation, and molecular weight measurements. A uniform polyacrylamide gel matrix (not as
beads) can be cast either as a thin slab sandwiched
between glass plates or as a rod in a tube. The cross-
linked chemical structure of polyacrylamide provides
a porous network through which charged particles
must move according to their respective Q/r values
(refer to pp. 84–86). Ease and rapidity of operation,
high sensitivity of detection, the lack of damage done
to delicate substances such as proteins and nucleic
acids, and resolution of complex mixtures into the

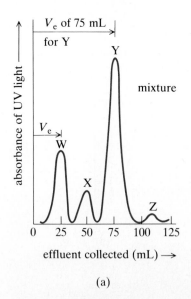

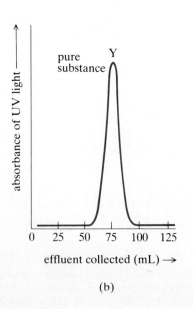

(a)   (b)

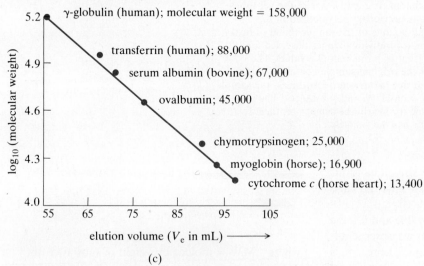

(c)

FIGURE 4–3 Gel-permeation chromatography. (a) Chromatographic profile of a mixture of substances (W, X, Y, and Z) illustrating the potential to separate individual components. (b) Profile of a pure substance—only a single peak. (c) Correlation of molecular weight and elution volume for proteins from a gel-permeation column. [*Source:* Data taken from the *Handbook of Chromatography,* vol. 1, (Boca Raton, Fla.: CRC Press, 1972).]

individual components are all advantages of PAGE. The stages of a gel tube analysis are diagramed in Figure 4–4. Some photographs of actual gels appear in Section 4–9.

A highly useful variation of the standard PAGE procedure for protein analysis is to first treat the protein sample (be it pure or a mixture) with *sodium dodecyl sulfate* (SDS) prior to application for electrophoresis. The method is referred to as SDS–PAGE. Sodium dodecyl sulfate is an ionic detergent (negatively charged) that unravels the folded structure of a protein:

$$CH_3CH_2CH_2CH_2CH_2CH_2CH_2CH_2CH_2CH_2CH_2CH_2OSO_3{}^-Na^+$$

Sodium dodecyl sulfate
(ionic detergent)

Then it is adsorbed along the elongated surface of the polypeptide chain. The adsorption of many negatively charged SDS molecules on the polypeptide chain confers a net negative charge to that polypeptide. Moreover, *polypeptides of different sizes will adsorb an*

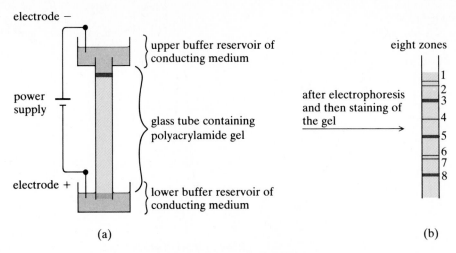

FIGURE 4–4 Gel tube electrophoreses. (a) Initial setup. However, the actual apparatus can accommodate several tubes simultaneously. As implied, the sample in this case consists of materials that are negatively charged, and migration occurs toward the reservoir of opposite charge. The actual size of the gel tube is $\frac{1}{4}$ in. by 3 in. The polyacrylamide gel was previously polymerized in the presence of a conducting medium. A small volume of the sample (band of heavy color) is applied to the top of the column. (b) Individual zones observed for each substance with a different mobility (that is, a different Q/r value). The column shows eight zones, implying that the original sample contained eight different substances, three major ones and five minor ones. Obviously, the sample was not pure. Substance 8 has the greatest mobility, and substance 1 has the smallest. If all substances were of the same (or nearly the same) size, then 8 has the greatest net negative charge and 1 has the smallest.

amount of SDS in proportion to their sizes. The result is that polypeptides of different sizes will have the same (or nearly so) Q/r value.

If mobility is governed by the Q/r value and if there is no difference in Q/r values, why do we observe different rates of electrophoretic migration? The reason is that the sieving capability of the polyacrylamide gel matrix allows smaller polypeptides to move faster than larger polypeptides, even though they have the same Q/r value. On the basis of a calibration procedure and an empirical relationship similar to that of gel-permeation chromatography, SDS–PAGE provides another approach for molecular weight measurements (see Figure 4–5).

Analytical Ultracentrifugation

In an ultraspeed centrifuge (20,000–70,000 rpm with gravitational forces of 50,000–500,000 × g) the rate of sedimentation dx/dt for a substance is related to the size (MW) of the substance in the following way:

$$MW = \frac{RT(dx/dt)}{\omega^2 x D(1 - \bar{V}\rho)}$$

where   MW = molecular weight of substance (in grams per mole)

   R = molar gas constant (8.314×10^7 g·cm^2/s^2/degree/mole)

   T = temperature (in degrees Kelvin)

   D = diffusion constant of substance, a value reflecting shape of sedimenting molecules and must be measured independently (in square centimeters per second)

   $\bar{V}$ = partial specific volume of substance, also measured independently (most proteins have a value close to 0.74 mL/g)

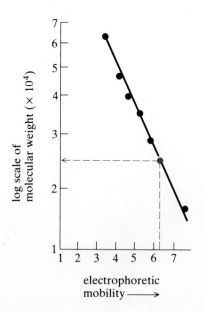

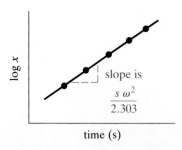

FIGURE 4–6 Evaluation of sedimentation coefficient *s* from data obtained in an analytical ultracentrifuge. Refer also to Figure 4–7.

FIGURE 4–5 Estimation of molecular weight by SDS–PAGE. The line constructed through the black points corresponds to the mobilities of proteins of known molecular weight. Plotting the mobility of a protein of unknown size (color point) yields a molecular weight of 25,000.

ρ = density (in milliliters per gram) of liquid medium in which sedimentation is occurring

ω = speed of centrifuge (in radians per second; see Note 4–1)

x = distance (in centimeters) moved by substance at any time *t*, *measured from center of rotation*

The *sedimentation coefficient s* is defined in terms of dx/dt, ω, and x as follows:

$$s = \frac{dx/dt}{\omega^2 x}$$

It yields, after substitution into the previous equation,

$$\text{MW} = \frac{RTs}{D(1 - \bar{V}\rho)}$$

In sedimentation analysis the distance migrated by the substance is observed at various times after sedimentation starts. The *s* value is then calculated from the slope of a plot of $\log(x)$ versus time *t* (see Figure 4–6). From *s* and independently measured values for

D, $\bar{V}$, and ρ, the molecular weight is calculated from the preceding equation. In comparison with the other two methods, this method is more time-consuming, requires a special instrument, and is not used as routinely. However, it does give accurate results, and there are other applications of sedimentation analysis beyond the scope of this book.

Sedimentation is observed by using an analytical ultracentrifuge, a centrifuge equipped with a special optical system and photographic hardware to detect the boundary that results during sedimentation (see Figure 4–7). The boundary exists between the pure solvent and the zone of solution containing the sedimenting material. If the original sample is pure, only one boundary will be present—detected as a single symmetrical peak in the recorded image, as shown in Figure 4–8.

The sedimentation of most biopolymers is on the order of 10^{-13} s. So that the exponent is eliminated, *s* values are usually reported in *svedbergs* (S), with 1 svedberg = 10^{-13} s. (T. Svedberg pioneered in the design of ultracentrifuges; he was awarded the Nobel Prize in 1926.) Thus a substance with a sedimentation coefficient of 2.5×10^{-13} s has a 2.5S value. Most proteins, nucleic acids, and multimolecular particles composed of proteins and nucleic acids have values in the range of 1 to 200S. As molecular size increases, the S value increases.

Note 4–1 Revolutions per minute (rpm) can be converted to radians per second by the relationship

$$\omega = \frac{r}{60}(2\pi)$$

where *r* is the known revolutions per minute.

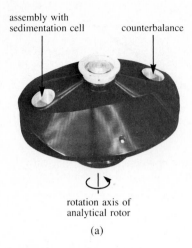

rotation axis of
analytical rotor

(a)

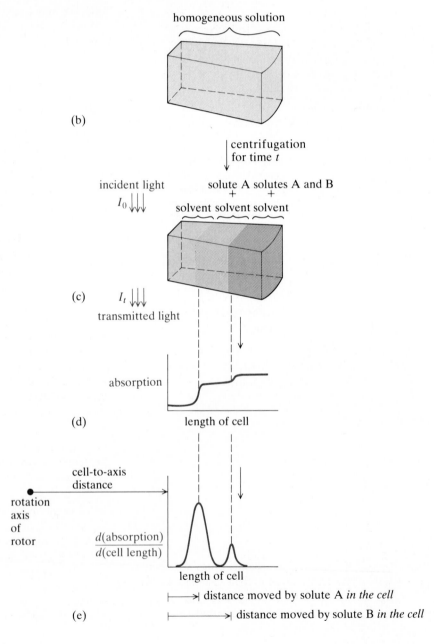

FIGURE 4–7 Events of sedimentation in an analytical ultracentrifuge. The diagram illustrates the formation and the detection of boundaries during sedimentation. The example depicts a minor, faster-moving (B) protein component in the presence of a major, slower-moving protein (A). (a) Rotor used in the analytical ultracentrifuge. A small volume of sample (1.0 mL) is placed in a sedimentation cell sandwiched between optically clear glass surfaces in an assembly that is placed in the rotor. (b) Enlarged view of the centrifuge cell placed in the rotor. The compartment is initially filled with solution containing two solutes, A and B, dissolved in a solvent. (c) Three distinct absorbing regions (two boundaries) that result during centrifugation. While the rotor spins, light passes through the cell, interacting with the solutes. Light absorption is shown here. (d) Absorption pattern. Within a region the absorption is constant, but it increases sharply at the interface of the two regions. (e) Change of absorption along the cell length. The optic system converts the pattern shown in part (d) to the pattern shown here. Each peak in the final pattern corresponds to a boundary between distinct regions.

4–2 PRIMARY STRUCTURE

The largest single sequence determined for a protein composed of only one polypeptide chain is 1021 amino acid residues for the enzyme β-galactosidase (molecular weight MW = 116,000). The largest total sequence determined for a protein composed of two or more different chains is 1320 residues for the four chains of an antibody molecule (MW = 150,000; antibodies are discussed later in this chapter).

There are two methods that can be used to evaluate the sequence of a polypeptide. One method involves procedures performed directly on the polypeptide to be sequenced. A direct method described in this section utilizes the Edman reagent (phenylisothiocyanate) to identify amino acids as they are removed one at a time from the N terminus. The second method is indirect, not even requiring that a sample of the polypeptide be available. The indirect method, mentioned in Chapter 6 (Figure 6–38), deduces the sequence of amino acids from the previously determined sequence of nucleotides in the gene of DNA that codes for the biosynthesis of the polypeptide.

Before examining the strategy of the direct Edman method, let us first consider some other aspects of primary structure.

Preliminary Studies

Prior to sequencing, some important preliminary determinations are usually performed. All operations require a *pure sample* of the protein. The isolation of pure proteins is discussed later in the chapter. The steps in the preliminary determinations are as follows:

1. The molecular weight is measured. This measurement can be done in various ways, with the three most widely used methods being *molecular sieve chromatography, SDS-polyacrylamide gel electrophoresis,* and *analytical ultracentrifugation.* All were described in the preceding Laboratory Methods section. Because each method is affected by the shape and solution properties of protein molecules, an approximate molecular weight value is obtained. Thus a value obtained by any one method is usually corroborated with that obtained by another method. The true molecular weight can be obtained by summing the weight contribution of each amino acid residue known to be present in the protein.

2. If the sample is a conjugated protein, the type and the amount of the prosthetic group are determined. Before sequencing, the prosthetic group is preferably removed from the polypeptide component (see item 3).

3. Proteins having a quaternary level of structure are treated in order to dissociate the intact aggregate into the individual polypeptide chains, which are then separated and sequenced separately. Simple but effective ways to dissociate oligomeric proteins involve changing the pH of the protein solution, treating it with a detergent, mild heating, altering the ionic strength of the protein solution, or freezing and thawing it in repeated cycles. Understanding why these conditions result in dissociation will come after study of Sections 4–5 and 4–8. Such treatments are also effective in dissociating prosthetic groups from a polypeptide.

sedimentation

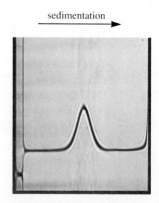

FIGURE 4–8 Photograph of a single symmetrical peak for sedimentation of a pure sample of the protein bovine serum albumin.

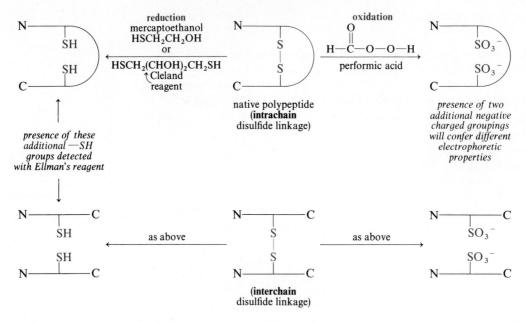

FIGURE 4–9 Detection of —S—S— bonds. Whether by oxidation or reduction, two chains will be produced. Since the individual chains are smaller than the original protein, they can be detected by molecular sieve chromatography or SDS-polyacrylamide gel electrophoresis.

4. The presence of any *interchain* or *intrachain disulfide bond(s)* (—S—S—) is determined (see Figure 4–9). Detection is based on selective chemical modification techniques that cleave the disulfide bond. One approach is to cleave the bond oxidatively; *performic acid* is a frequently used oxidizing agent. If the product has an increased electrophoretic mobility due to an increased negative charge of the sulfonate (—SO_3^-) functions produced, one or more disulfide bonds must have been present. Gel-permeation chromatography and/or gel electrophoresis are also useful in demonstrating whether the disulfide bond was interchain or intrachain and, if interchain, whether the connected chains are identical or different.

An alternative tactic is to cleave the —S—S— linkage reductively. *Mercaptoethanol* ($HSCH_2CH_2OH$) or *dithiothreitol* (Cleland's reagent) will do the job, with electrophoresis and/or chromatographic techniques used to monitor the results.

Without going into details, suffice it to say that the oxidative and reductive methods can be quantified to determine the exact number of disulfide linkages. If disulfide bonds are present, they are usually cleaved prior to sequencing. When different chains are connected, —S—S— cleavage is followed by separation of the chains, which are sequenced separately.

5. The evaluation of disulfide bonds in the native protein is usually done in conjunction with determining the presence of cysteine sulfhydryl (—SH) groups not involved in disulfide bonds. One quantitative method is to treat the protein with *5,5'-dithiobis-2-nitrobenzoic acid* (Ellman's reagent), which reacts selectively with free —SH groups to give a modified protein and the thionitrobenzoate anion (see Figure 4–10). The latter is strongly colored ($\lambda_{max} = 412$ nm) and serves as the basis for a quantitative spectrophotometric assay. (One-mole unit of the anion is produced for each —SH group that reacts.)

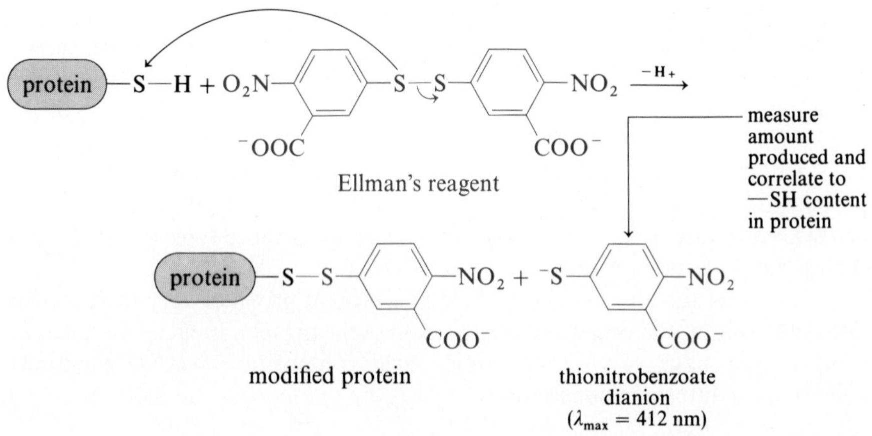

FIGURE 4–10 Detection of free —SH function.

Amino Acid Composition

The identity and the amount of each amino acid present in a polypeptide is determined in the following way:

$$polypeptide \xrightarrow[\substack{H_2O}]{\substack{Step\ 1 \\ \text{hydrolysis of all} \\ \text{peptide bonds}}} \substack{\text{mixture of} \\ \text{free amino acids} \\ \text{called a} \\ \text{polypeptide} \\ \text{hydrolyzate}} \xrightarrow[]{\substack{Step\ 2 \\ \text{chromatographic} \\ \text{analysis}}} \substack{\text{amino acid} \\ \text{composition} \\ \text{solved}}$$

The standard method for degradation is *acid-catalyzed hydrolysis*, usually with $6N$ HCl for 12–36 h at 100°–110°C under N_2. These conditions will cleave every peptide bond without racemization (no L → D or D → L changes). However, a serious problem with HCl hydrolysis is that the indole structure of tryptophan, when it is present, is completely destroyed. Although they have other disadvantages, treatments that are nondestructive for tryptophan are acid hydrolysis with methanesulfonic acid in place of HCl and base-catalyzed hydrolysis with NaOH. Acid hydrolysis also releases NH_3 from the R groups of glutamine and asparagine, with these residues then detected as glutamic and aspartic acids. Eventually, the exact content of glutamine and asparagine can be resolved from information obtained in the sequencing procedures.

The hydrolyzate is analyzed by use of an automated amino acid analyzer operating via ion-exchange column chromatography and ninhydrin or ortho-phthaldehyde detection (see Figure 3–25). By measurement of the amount of each amino acid obtained from the hydrolysis of a known amount of polypeptide of known molecular weight, the frequency of occurrence of each amino acid in the polypeptide can be expressed. For example, if 0.13 millimole of leucine (17.1 mg) was recovered from 0.01 millimole of polypeptide (150 mg of material having MW = 15,000), the polypeptide chain contains 13 residues (0.13/0.01) of leucine.

TABLE 4–1 Amino acid composition of cytochrome *c* (human)

(Class) AMINO ACID	NUMBER OF RESIDUES (% of Total)	
(Aliphatic)		
• Glycine	13	
• Alanine	6	
• Valine	3	(35%)
• Leucine	6	
• Isoleucine	8	
(Hydroxyl)		
Serine	2	(9%)
Threonine	7	
(Sulfur)		
Cysteine	2	(5%)
• Methionine	3	
(Aromatic)		
• Phenylalanine	3	
Tyrosine	5	(9%)
• Tryptophan	1	
(Acidic)		
Aspartic acid	3	(11%)
Glutamic acid	8	
(Amides)		
Asparagine	5	(7%)
Glutamine	2	
(Basic)		
Lysine	18	
Histidine	3	(22%)
Arginine	2	
(Imino)		
• Proline	4	(4%)
Total residues	104	

Note: The • identifies amino acids with nonpolar, hydrophobic R groups.

A tabulation of amino acid content for the polypeptide component of the hemoprotein cytochrome *c* appears in Table 4–1. As the table shows, cytochrome *c* contains at least one residue of all 20 amino acids—a feature true of many proteins but not all. Another quasi-typical feature is that 45% of the residues are hydrophobic and nonpolar, the range for most proteins being 35%–50%. Still another typical feature is the low content of tryptophan. In several proteins tryptophan is not even present. Cytochrome *c* is atypical in that there are twice as many basic residues as acidic residues. A more typical distribution slightly favors acidic residues.

Knowing the amino acid content of a protein provides very little insight into details about the shape and the function of the protein. In fact, two different proteins may have very similar amino acid compositions yet have entirely different structures and functions.

Sequence Determination (Terminal Residues)

The sequencing of a polypeptide chain usually begins with the determination of the two terminal residues.

C TERMINUS   *Hydrazine* will react at the carbonyl grouping of each peptide bond, cleaving the peptide bond and producing acyl hydrazine derivatives of each residue—except the C terminus residue, which is released as a free amino acid (see Figure 4–11). The C terminus is not modified because its alpha carboxyl group is not involved in peptide linkage.

Another method exploits the specificity of a naturally occurring exopeptidase enzyme, called carboxypeptidase (see Figure 4–12). *Peptidases* (also called *proteases* or *proteolytic enzymes*) are enzymes that catalyze the hydrolytic cleavage of peptide bonds; *exopeptidases* act specifically only on peptide bonds at the end of a polypeptide chain; *carboxypeptidases* act specifically only on the C terminal peptide bond, releasing the C terminal residue as a free amino acid. (Available commercially, carboxypeptidase is one of a group of enzymes in mammals that function in the digestion of proteins.) Difficulties arise, however, because the enzyme does not stop after removing the initial C terminal residue but continues to attack new C terminal bonds every time the chain is shortened. Therefore monitoring the rate of release of amino acids is necessary.

N TERMINUS   The enzyme *aminopeptidase,* another naturally occurring exopeptidase but having the opposite specificity of carboxypeptidase, can be used to determine the N terminus. However, it also presents the disadvantage of continuous action. Chemical methods are much better (also see Note 4–2).

Either of three substances can be used to identify the N terminus (see Figure 4–13): *2,4-dinitrofluorobenzene* (Sanger's reagent), *dansyl chloride,* or *phenylisothiocyanate* (Edman's reagent). Each reagent can react with the —NH$_2$ function of the N terminus, producing a modified polypeptide. Acid hydrolysis is then used to release the N terminal residue in derivative form, which can be extracted and identified by chromatographic comparison against pure derivatives of known amino acids. Detection is based on characteristic light-absorbing properties of each derivative: DNP derivatives absorb visible light and are in-

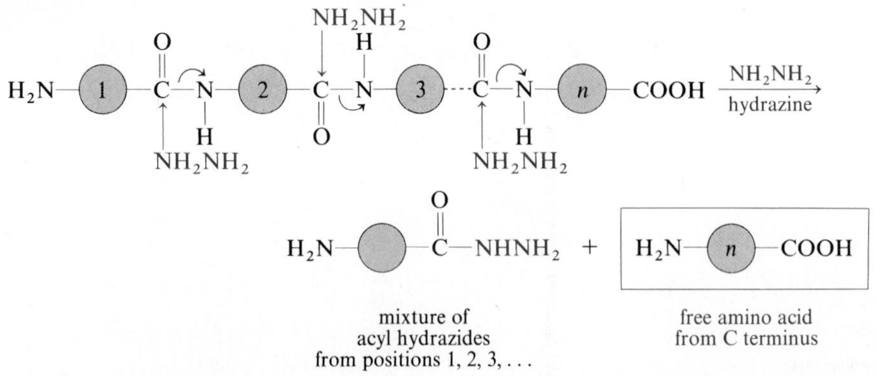

FIGURE 4–11 C terminus identification with hydrazine. The only free amino acid formed is the one corresponding to the C terminus (extract and identify by chromatography against standards). In this figure and others we will use

$$H_2N\!-\!\!\bigcirc\!\!-COOH$$

as a symbolic representation of an amino acid.

tensely yellow; PTH derivatives absorb UV light; dansyl derivatives are fluorescent—after excitation by absorbing UV light, they emit intense light at characteristic wavelengths. Fluorescent-based assays are valuable because extremely small amounts of material can be detected.

Sequence Determination (Internal Residues)

In the aforementioned N terminus identification procedures there is one notable distinction:

The Edman procedure is the only one that results in the release of the modified N terminal residue without cleaving any other peptide bond.

Thus, having extracted and identified the first PTH-amino acid, the remaining polypeptide, now shortened by one residue, can be treated again with phenylisothiocyanate to identify the new N terminus which corresponds to residue 2 in the original polypeptide. Successive cycles of this *subtractive Edman degradation* would continue to identify residues 3, 4, 5, . . . *n*.

NOTE 4–2
These procedures are based on the pioneering work in the early 1950s by F. Sanger, the first to develop a successful strategy for sequencing polypeptides. Sanger received the Nobel Prize (1958) for this work. He later received a second Nobel Prize (1980) for developing methods to establish the sequence of monomeric units in DNA

FIGURE 4–12 N and C terminus identifications by using exopeptidases.

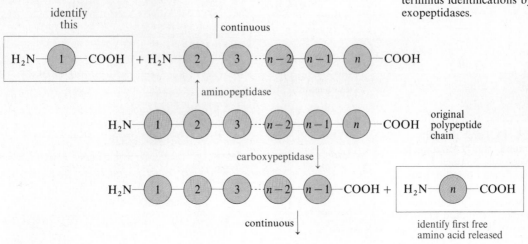

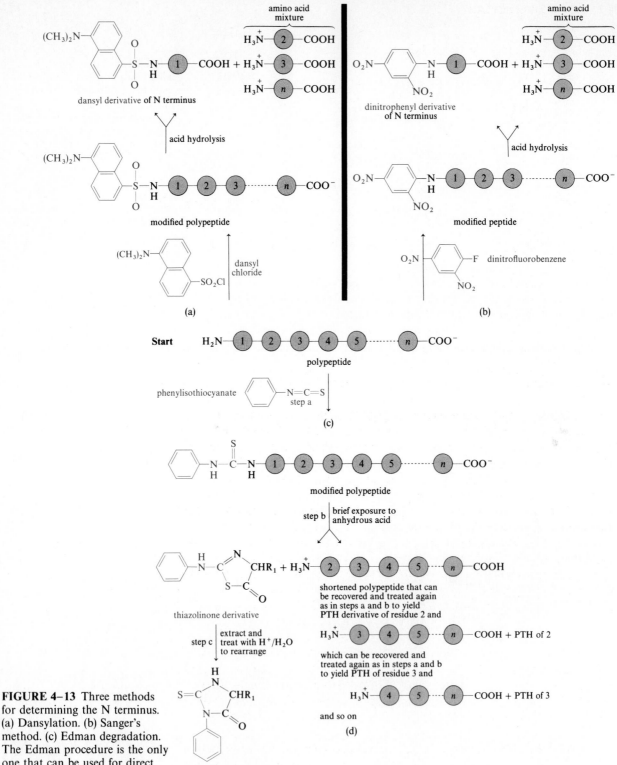

FIGURE 4–13 Three methods for determining the N terminus. (a) Dansylation. (b) Sanger's method. (c) Edman degradation. The Edman procedure is the only one that can be used for direct sequencing. (d) Procedure steps for subtractive degradation.

TABLE 4–2 Site-specific cleavage of internal peptide bonds

$$\left(\begin{array}{c} N \\ \text{terminus} \end{array}\right)\text{--------}CH\text{---}C\overset{\overset{O}{\|}}{\underset{\underset{\underset{H_2O}{\text{enzyme or}}}{H}}{\text{---}N}}\text{---}CHR\text{--------}\left(\begin{array}{c} C \\ \text{terminus} \end{array}\right)$$

enzyme or
chemical reagent

ENZYMES	R_x PREFERRED	CHEMICAL REAGENT
Trypsin	Lys, arg	
Chymotrypsin	Phe, tyr, trp	
Sa protease	Asp, glu	
Fm protease	Pro	
	Met	$Br-C\equiv N$

With the use of automated peptide-sequencing instruments to perform all operations, maximum efficiency of sequential Edman degradation is usually limited to about 20–40 cycles. For larger polypeptide chains an obvious strategy is used prior to any sequencing procedure, namely, degrading the large polypeptide into smaller fragments suitable in size for Edman sequencing, with any fragment that may still be too large being fragmented further. However, the fragmentation pattern must occur with some degree of specificity in order to produce a small number of fragments. Specificity is also of some help in determining the position occupied by the fragments in the intact polypeptide.

ENZYME CLEAVAGE Cleavage specificity of internal peptide bonds via enzymes is achieved by using *endopeptidases*, having a strong preference to operate at only certain peptide bonds. Two long-used enzymes are *trypsin* and *chymotrypsin,* both digestive proteolytic enzymes isolated from mammals. Two recent additions are *Sa protease* (isolated from the bacterium *Stapylococcus aureus*) and *Fm protease* (isolated from the bacterium *Flavobacterium meningosepticum*). Their preferred actions are summarized in Table 4–2.

CHEMICAL CLEAVAGE The use of enzymes is complemented by a small number of chemical reagents that act preferentially with certain R groups in such a way that a cleavage of the peptide bond in that position also occurs. The best example is *cyanogen bromide* (BrCN in H_2O), which is *methionine-specific*. The reaction involves some interesting chemistry, as outlined in Figure 4–14.

OVERALL STRATEGY The isolation and sequencing of one set of fragments obtained from only one treatment (trypsinolysis, for example) does not completely solve the problem. The fragments still need to be aligned in a proper order. Thus the same procedure is repeated—with chymotrypsin, with

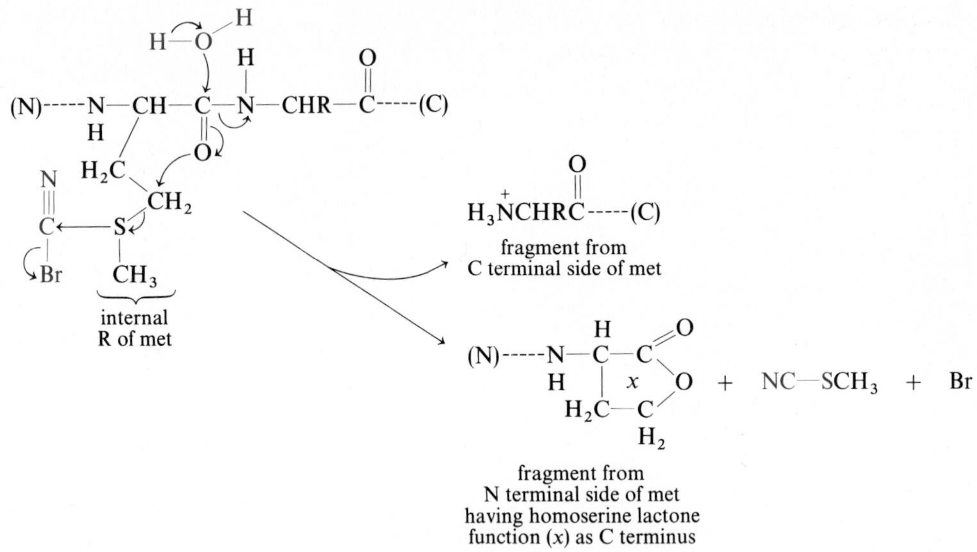

FIGURE 4–14 Specific cleavage of a polypeptide chain at internal methionine residues with cyanogen bromide.

cyanogen bromide, with a combination of both enzymes, or with either enzyme under different conditions—in order to obtain a *collection* of different peptide fragments formed by the cleavage of different peptide bonds in the same polypeptide. Fragments from different treatments are aligned by identifying all of their *overlapping sequences,* revealing the sequence of the parent polypeptide. The strategy is diagramed in Figure 4–15. *Fingerprinting,* a quick way of sequencing the same polypeptide chain obtained from different sources, is described later in the chapter.

Refinements in the techniques of HPLC and gel electrophoresis have greatly improved the process of purifying very small amounts of proteins and peptides of scarce biological origin. Additional refinements in the detection of amino acids with ortho-phthaldehyde and dansyl chloride and in the sequencing chemistry via Edman degradation now permit amino acid analysis and amino acid sequencing to be done with as little as 5–50 picomole of sample. Such a small sample represents an increase in sensitivity of more than a thousandfold over the past ten years. The Hunkapiller, Strickler, and Wilson paper cited in the Literature provides an excellent review of these methods.

Some Principles Regarding Amino Acid Sequence

First, we present a dogma:

There is overwhelming evidence that the amino acid sequence is the principle determinant of the overall structure of the protein and, hence, determines what function the protein performs. For proteins not dependent on a prosthetic group, amino acid sequence is the only determinant.

Now we give some generalizations based on comparing sequences, a task that is currently done by computer analysis:

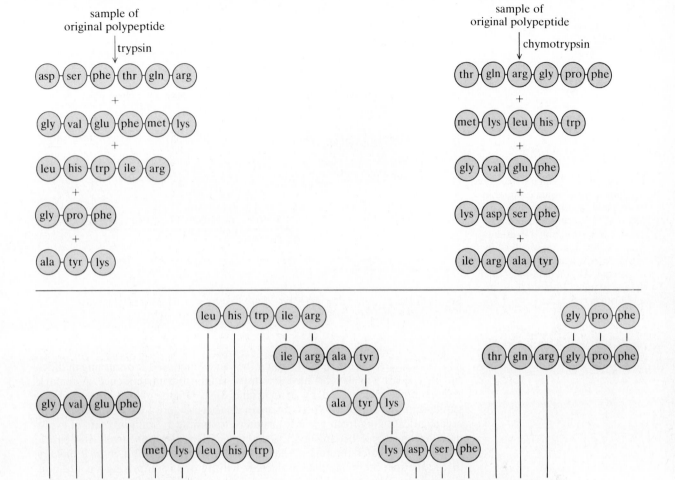

FIGURE 4–15 Mapping of an amino acid sequence. (1) Each treatment will yield a mixture of smaller peptides according to the presence of susceptible bonds in the original polypeptide. (2) Individual peptides in each mixture are isolated via chromatography and/or electrophoresis. (3) The composition and the sequence of each peptide are determined. (4) Peptide fragments are aligned on the basis of overlapping sequences. For simplicity only a small polypeptide is shown here.

• There is no single partial sequence or group of partial sequences common to all polypeptides.

• Every possible combination of two successive amino acids has been detected.

• Proteins having different functions usually have different sequences, although there are some interesting exceptions.

• Proteins of similar function have varying degrees of similar sequences—usually slight, but sometimes extensive. (Sequence similarity is called *homology*.) Sequence comparisons for such proteins can help determine what part (or parts) of the polypeptide is (are) crucial to their participation in certain types of reactions. Sequence comparisons of this sort also help in tracing the evolution of genes (see the next item).

$$
\begin{array}{l}
\underset{}{\underline{gly}}—asp—val—glu—\overset{5}{lys}—\underline{gly}—lys—lys—ile—\overset{10}{\underline{phe}}—ile—met—lys—\overset{15}{\underline{cys}}—ser—\\[1.2em]
gln—\underline{cys—his}—thr—\overset{20}{val}—glu—lys—gly—gly—lys—his—\overset{25}{\underline{lys}}—thr—\underline{gly}—pro—\\[1.2em]
asn—\underline{leu}—his—\underline{gly}—\overset{35}{leu}—phe—gly—\underline{arg}—lys—\overset{40}{thr}—\underline{gly}—gln—ala—pro—\underline{gly}—\\[1.2em]
tyr—ser—\underline{tyr}—thr—\overset{50}{ala}—\underline{ala—asn}—lys—\overset{55}{asn}—lys—gly—ile—ile—trp—\overset{60}{\underline{gly}}—\\[1.2em]
lys—asp—thr—leu—\overset{65}{met}—glu—\underline{tyr—leu}—glu—\overset{70}{\underline{tyr—pro—lys—lys—tyr}}—\overset{75}{\underline{ile}}—\\[1.2em]
\underline{pro—gly—thr—lys}—\overset{80}{\underline{met}}—ile—\underline{phe}—val—\underline{gly}—\overset{85}{ile}—lys—\underline{lys—lys}—glu—glu—\\[1.2em]
\underline{arg}—ala—asp—leu—\overset{95}{ile}—ala—tyr—leu—lys—\overset{100}{lys}—ala—thr—tyr—\overset{104}{glu}
\end{array}
$$

FIGURE 4–16 Primary structure of human cytochrome *c*. Lines in color underneath the symbols identify 35 residues, including a consecutive stretch of 11 residues from 70–80, that are identical in cytochrome *c* from 38 different organisms. Although not identified here, functional homology occurs at 23 other residues.

• The same proteins performing the same function but occurring in different life species will usually have extensive similarities in sequence. A remarkable example is provided by cytochrome *c* (see Figure 4–16), for which the sequence of the single polypeptide chain has been determined from 38 different organisms, ranging from yeast to primates, spanning 1.2 billion years of evolution. Among these organisms, the chain length varies only from 104 to 112 residues. Of greater interest is that 35 of the residues exhibit *total homology;* that is, they are identical in all molecules, with 11 of 35 comprising a continuous segment. Another 23 sites exhibit *functional homology,* meaning that the residues are virtually identical because the R groups are of the same type: lys or arg; glu or asp; ser or thr; ala, val, leu, or ile; and others. Radical changes occur in only six positions. Evidently, the extent of sequence homology is closely related with evolutionary relationships, with less variation common to a protein in organisms of the same phylogenetic class. For example, humans and chimpanzees (both primates) have totally identical cytochrome *c* molecules.

• The same protein performing the same function in different members of the same species will almost always have the same sequence. When a variation does occur, the overall function of the protein may or may not be affected. (See the following discussion for a description of hemoglobin variants in humans.)

Specific Examples of the Significance of Amino Acid Sequence

A particularly convincing proof that the individuality of a protein is determined by amino acid sequence was supplied by the chemical synthesis of the enzyme ribonuclease. One by one, 124 residues were linked in a sequence identical to that of the native polypeptide. The result: The synthetic product exhibited the same biological activity as the native enzyme. There are many other similar results with other synthetic polypeptides.

Biological evidence is unfortunately available from several diseases. A notable example is *sickle-cell anemia,* a hereditary disease affecting red blood cells.

The malfunction is attributable to the presence of *abnormal hemoglobin molecules* (HbS) in the sickle cells, so named because of the characteristic abnormal shape they assume. Normal red blood cells (see Figure 4–17) are disk-shaped; sickle cells are often elongated, and many are crescent-shaped.

An intact human hemoglobin molecule ($\alpha_2\beta_2$) consists of two copies of the α polypeptide chain and two copies of another slightly different β chain (see Section 4–5). The hemoglobin defect is traced to a point mutation in the gene coding for the β chain of hemoglobin, resulting in a change of a single residue during assembly of the β chain. The β chain of normal adult hemoglobin (HbA) contains glutamic acid in position 6, whereas the β chain of HbS contains valine in position 6. All other β chain residues are identical, and the α chains are completely identical in the two hemoglobins. Despite this small difference (involving only two residues out of 574), the HbS molecule does not work properly.

The consequence of the β^6(glu) $\rightarrow$ β^6(val) substitution is that HbS molecules tend to stick to each other, forming large, elongated multimolecular aggregates and causing the red blood cell to change shape. The aggregation also diminishes the oxygen-binding capacity of HbS. Still other consequences are that the deformed cells can get trapped in narrow blood vessels, further retarding the delivery of any oxygen that is bound; and the sickle cells rupture easily—hence the anemia. All of these events happen because a neutral and nonpolar R group (from valine) replaces a polar, negatively charged R group (from glutamic acid) in position 6 of the two β chains. About three hundred hemoglobin variants have been discovered. Most involve amino acid substitutions, but a few are due to amino acid deletions. Some examples are listed in Table 4–3.

To conclude at this point that every amino acid residue in every position is indispensable to the normal structure and function of a protein would be incorrect. For example, many of the 300 hemoglobin sequence variants still behave normally. The amino acid residues involved in these cases would be considered dispensable in the sense that substitutions can occur without any effect on the molecule's activity. The distinction between *indispensable* (essential) and *dispensable* (nonessential) residues will be amplified further in our discussion of enzymes in the next chapter.

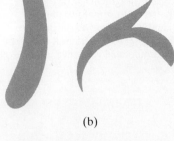

(a)

(b)

FIGURE 4–17 Human red blood cells. (a) Normal disk shape as viewed by SEM. (*Source:* Photograph provided by J. Dirando.) (b) Sketches of sickle cells.

TABLE 4–3 Examples of some hemoglobin sequence variants

Hb	RESIDUE CHANGE
Substitutions	
C	Glu $\rightarrow$ lys (β^6)
E	Glu $\rightarrow$ lys (β^{26})
Sydney	Val $\rightarrow$ ala (β^{67})
D$_{Punjab}$	Glu $\rightarrow$ gln (β^{125})
G	Asn $\rightarrow$ lys (α^{68})
Q	Asp $\rightarrow$ his (α^{75})
M$_{Iwate}$	His $\rightarrow$ tyr (α^{87})
Deletions	
Tours	Thr (β^{87}) deleted
Lyon	Lys-val (β^{17-18}) deleted

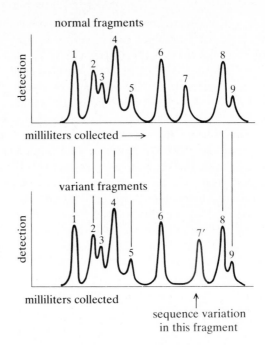

FIGURE 4–18 Column chromatographic fingerprinting. Comparing the elution pattern of variant fragments with the pattern of normal fragments identifies fragments (in this case only one) that contain a variation in amino acid sequence.

In searching for sequence variants, one need not perform a total sequencing procedure every time the protein is isolated from a new member of the species. To find and identify the residue quickly, one may use the technique of *fingerprinting*. In this procedure samples of the normal chain and the chain to be tested for variance are separately fragmented under identical conditions. The two fragment mixtures are then separately analyzed under identical conditions by high-resolution column chromatography or by a combined procedure of chromatography and electrophoresis. If the two chains are identical, the profile of the separated fragments is identical. If they are not identical, slightly different fragment profiles are observed (see Figure 4–18). For example, if a single substitution occurred without changing the pattern in which peptide bonds were cleaved, the normal- and variant-fragment mixtures would differ in only one fragment. All one needs to do then is isolate and sequence the variant fragment and compare it with the sequence of the normal fragment.

4–3   SECONDARY STRUCTURE

Polypeptide Backbone and the Peptide Bond

The *backbone* of a polypeptide chain consists of three atoms from each residue in the chain: a nitrogen involved in one peptide bond, an alpha carbon (C^α), and a carbon involved in another peptide bond, with the pattern of $N—C^\alpha—C$ repeating itself along the entire chain. A fully extended representation of this pattern is shown in Figure 4–19. Solid, regular bonds (—) are all in the plane of the paper; ▬ represents bonds directed out of the plane

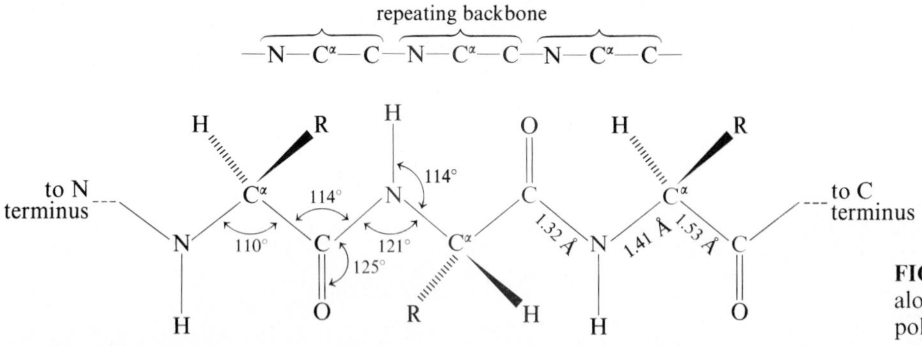

repeating backbone

$$-\text{N}-\text{C}^\alpha-\text{C}-\text{N}-\text{C}^\alpha-\text{C}-\text{N}-\text{C}^\alpha-\text{C}-$$

to N terminus

to C terminus

FIGURE 4–19 Bonding pattern along the backbone of a polypeptide chain.

toward the viewer; and ⸺ represents bonds directed behind the plane away from the viewer.

The specified bond angles and bond lengths in Figure 4–19 reveal several important features of the peptide bond and the backbone:

1. The four atoms of the peptide bond and the two attached alpha carbon atoms are all in the same spatial plane. The H and R groups on the alpha carbons are projected out of the plane (see Figure 4–20).

2. The C=O and N—H bond dipoles are in opposite directions relative to the C—N bond axis. This arrangement is called a *trans alignment of dipoles*. In addition, the two alpha carbon atoms also have a trans alignment.

3. As a consequence of the spatial orientation in item 2, and given the L configuration of each residue, the R groups on each of the alpha carbon atoms are arranged in a repeating *trans* fashion.

4. The C—N peptide bond distance of 1.32 Å is intermediate in length between that of a double covalent bond (1.21 Å) and that of a single covalent bond (1.47 Å). This distinction suggests that the C—N peptide bond linkage has *some double-bond* (pi bond) *character*. This feature can be explained in terms of either a resonating or a tautomeric structure:

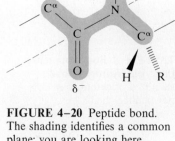

FIGURE 4–20 Peptide bond. The shading identifies a common plane; you are looking here directly at the top of this plane. The long dashed line identifies the axis of the C—N bond serving as the reference for the trans nomenclature.

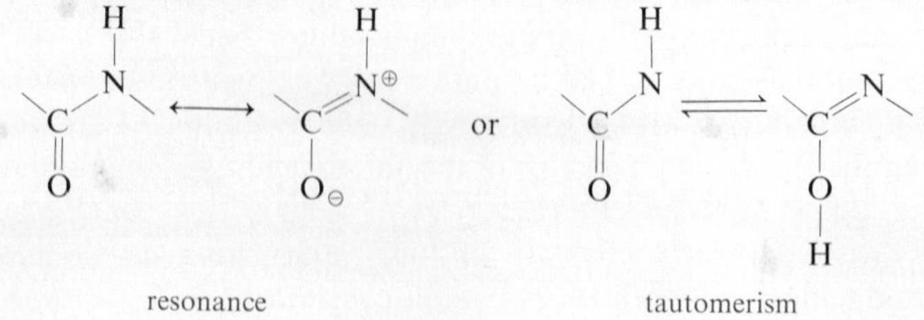

resonance   or   tautomerism

An important significance of this characteristic is described in the next section.

Rotational Flexibility of the Polypeptide Backbone

Naturally occurring polypeptide chains are not rigidly maintained in the fully extended structure represented above. Rather, because of free rotation around the majority of bonds along the backbone, the chain can easily change

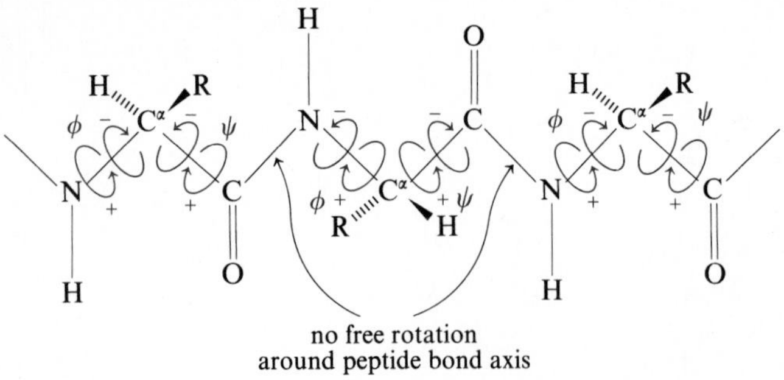

no free rotation
around peptide bond axis

FIGURE 4–21 Flexibility of the polypeptide backbone. Limited only by steric hindrance, free rotation occurs around the N—C$^\alpha$ and C$^\alpha$—C axes but not at the peptide bond.

its orientation. Of the three types of bonds along the backbone, only two exhibit free rotation: the C$^\alpha$—C and N—C$^\alpha$ bonds.

Rotation around the C—N peptide bond axis is severely restricted because of its double-bond character.

The notations ϕ and ψ are used to identify the specific rotational values for each bond. In the fully extended orientation the ϕ and ψ values are both considered to be 0°. Then when the chain is viewed from the N terminus, any clockwise rotation through 180° is assigned a positive value, and any counterclockwise rotation through 180° is assigned a negative value. See Figure 4–21.

Backbone Orientations in Naturally Occurring Polypeptide Chains

The particular type of orientation assumed by a polypeptide chain is the result of the pattern of rotation around the ϕ and ψ bonds. Three major types of orientation are found in nature: *helical, sheet,* and *random.* The helical and sheet forms (first discovered by Linus Pauling and R. Corey in 1951 through X-ray diffraction studies on synthetic polypeptides) are ordered geometric arrangements due to a repetition of the same ϕ value for successive C$^\alpha$—N bonds and of the same ψ value for successive C$^\alpha$—C bonds.

TABLE 4–4 Bond rotations that yield hydrogen bond–stabilized orientations of polypeptide chains

ORIENTATION	APPROXIMATE DEGREES OF ROTATION	
	C$^\alpha$—N BONDS ϕ	C$^\alpha$—C BONDS ψ
α-Helix, 3.6$_{13}$ (right-handed)	−60	−57
Sheet (antiparallel)	−140	+135
Sheet (parallel)	−119	+113

HELIX STRUCTURE In polypeptide chains composed of L amino acids, the most common helical orientation is the *right-handed alpha-helix,* where each $\phi = -60°$ and each $\psi = -57°$ (approximate values). See Table 4–4. This specific (ϕ, ψ) combination is special because a very stable arrangement results. It is stable for two reasons. First, there is little or no crowding of atoms, particularly among the R side-chain groups of the amino acids. Second, and of greater importance, the C=O and N—H dipoles of neighboring peptide bonds are optimally oriented (nearly coaxial) for maximum dipole-dipole interaction, thereby producing a network of *intrachain cooperative hydrogen bonding* (see Figure 4–22).

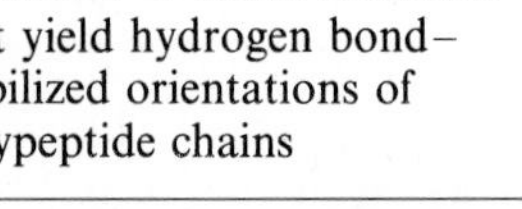

Figure 4–23 displays the α-helix geometry. Note that the intrachain hydrogen bonding is a direct consequence of the trans orientation in the coplanar peptide bond. Characteristic specifications of the α-helix geometry are as follows: The helix pitch (distance spanned by one turn) is 5.4 Å; there are 3.6 amino acid residues in each complete turn; and the N—H dipole of residue n is hydrogen-bonded with the C=O dipole of residue $n + 4$. The α-helix is also designated as a 3.6$_{13}$ helix, with the subscript designating the number of

carbonyl dipole   imino dipole

$$\overset{\delta^+}{C}=\overset{\delta^-}{O} \cdots\cdots \overset{\delta^+}{H}—\overset{\delta^-}{N}$$

FIGURE 4–22 Hydrogen bonding between peptide bond dipoles.

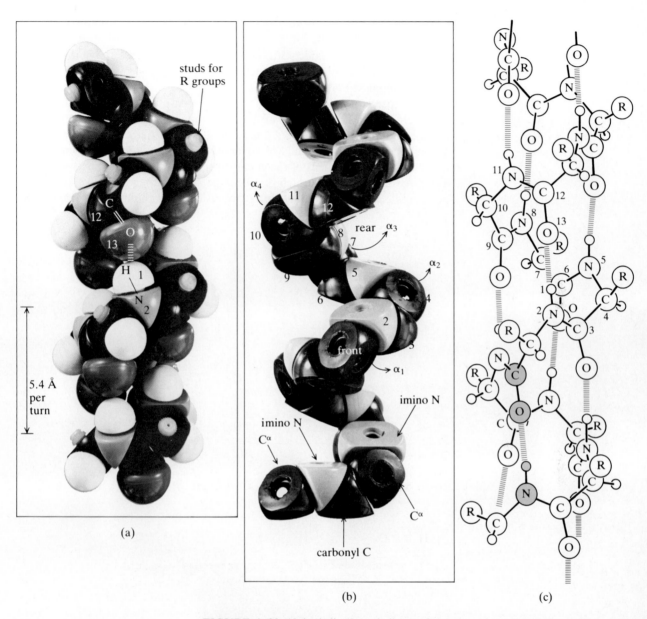

FIGURE 4–23 Alpha-helix (3.6_{13} helix) conformation of a polypeptide chain.
(a) Space-filling model, with all C=O, N—H, and C^α—H linkages inserted
and corresponding to the segment in part (c). The R groups of the alpha carbons
are shown as studs so that they do not obstruct the view of the hydrogen
bonds that exist between successive turns of the helix. (b) Backbone of the
α-helix. (c) Stick model. The numbers 1 through 13 identify the 13 consecutive
atoms of a closed loop in the α geometry. (*Source:* Drawing of the stick model
reprinted from Linus Pauling, *The Nature of the Chemical Bond*. Copyright
1939 and 1940 by Cornell University. Third edition © 1960 Cornell University.
Used by permission of Cornell University Press.)

consecutive atoms present in a closed loop between the interacting O and H atoms (refer to Figure 4–23). The only other widespread helix structure found in proteins is a right-handed 3_{10} helix, which is slightly less stable than the α-helix. Another important helix structure is found in only one protein, namely, collagen (discussed later in this chapter).

The percentage of α-helix content in globular proteins varies from protein to protein, ranging from none at all (0%) to as much as 80%–90%. (A polypeptide chain could never involve 100% α-helix content and still be a globular protein. It would be a fibrous protein.) Moreover, when α-helix content is present, it may occur in just one segment of the chain or in two or more separated segments. The largest consecutive stretch of α-helix geometry in a segment of a globular polypeptide chain spans about thirty-five residues, or about ten complete turns

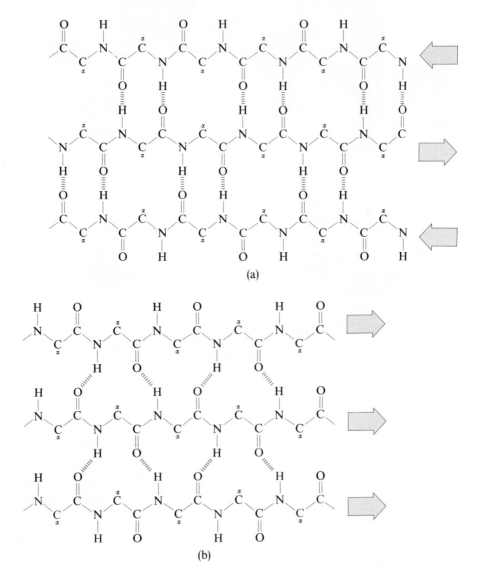

(a)

(b)

FIGURE 4–24 Two β sheet structures. (a) Antiparallel β sheet. (b) Parallel β sheet. For each structure only the backbone is shown. Each alpha carbon would have H and R groups projecting in front of and behind the plane of the paper. A space-filling model of stacked-sheet structures is shown in Figure 4–28.

of the helix. Obviously, a minimum stretch of α-helix geometry requires four residues for one turn.

Two factors that definitely *interrupt the α-helix orientation* are (1) the presence of *proline,* the cyclic structure of which naturally puts a bend in the polypeptide backbone, and (2) the presence of *electrostatic repulsion,* due either to a cluster of positively charged R groups from lys and arg or a cluster of negatively charged R groups from glu and asp.

SHEET STRUCTURES   There are two (ϕ, ψ) combinations (see Table 4–4) giving rise to ordered structures that are not helical. They are called *β-sheet structures,* and though different from the α-helix, they are also stabilized by cooperative hydrogen bonding between the same C=O and N—H dipoles. However, whereas helical structures arise from hydrogen bonding within a continuous segment of the polypeptide backbone, sheet structures do not.

Hydrogen bonding in sheet structures can occur among two or more different segments of the same chain (*intrachain sheet*) or among two or more segments of different chains (*interchain sheet*). Intrachain sheet structures occur primarily in globular proteins; interchain structures occur in fibrous proteins. In either case, two patterns of sheet structure are possible, depending on the alignment of the different chains or segments (see Figure 4–24). If they are aligned in the same direction from one terminus to the other, the arrangement is termed *parallel β sheet.* If they are aligned in opposite directions, the arrangement is *antiparallel β sheet.* Although both occur, the antiparallel alignment is somewhat more stable because the C=O and N—H dipoles are better oriented for optimum interaction.

RANDOM STRUCTURES   When a segment of a polypeptide chain has no repetitive (ϕ, ψ) combination, the resulting structure has *no repeating* geometric pattern; it is *random.* However, in the native conformation of an intricately folded globular chain, random segments are still properly designated as being ordered. Although this terminology may seem confusing, the point is that because of the ordered interactions involving R groups, constraints exerted by the possible presence of one or more disulfide bonds, and perhaps interactions with a cofactor or prosthetic group, a *particular random arrangement is preferred.* Thus in the context of the highly ordered native conformation of the globular protein, any nonhelical or nonsheet segment of the chain could be referred to as being in a highly *ordered random* orientation. A summary is presented in Table 4–5.

TABLE 4–5 Summary of secondary structure in proteins

TYPE OF PROTEIN	ORIENTATION OF POLYPEPTIDE CHAIN(S)
Fibrous	Exclusively helical
	Exclusively sheet
Globular	Part helix, part sheet, and part random
	Part helix and part random
	Part sheet and part random
	Entirely random

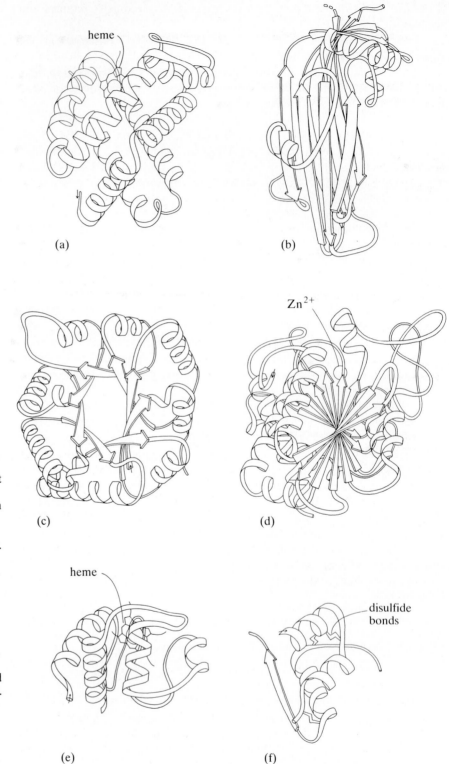

FIGURE 4–25 Globular protein structures. (a) β chain of hemoglobin. It is a bundle of eight helical segments, with no β sheet (see also Section 4–5). (b) Southern bean mosaic virus protein. This protein has layers of antiparallel β sheets, with some helix geometry. (c) Triose phosphate isomerase. This protein has an open, twisted core of parallel β sheets, with a perimeter of α-helix segments. (d) Carboxypeptidase. It has a tight, twisted region of parallel and antiparallel β sheets, with some helix geometry. (e) Cytochrome c. It has about 35% helix content, which is a result of the amino acid sequence displayed in Figure 4–16. (f) Insulin. It consists of two chains connected by two of three —S—S— bonds. (*Source:* Drawings provided by J. Richardson.)

Examples from the globular protein group are illustrated in Figure 4–25. Spiral ribbons represent segments of an α-helix; arrows represent segments of a β sheet, with ⟹ for parallel and ⇌ for antiparallel sheets; a meandering ribbon represents segments of an ordered random structure. The schematic drawings are the creation of Jane S. Richardson, who is pioneering in the attempt to correlate the molecular anatomy of proteins with molecular function and to trace the evolutionary history of proteins. Her brilliant monograph cited in the Literature is required reading for those interested in protein structure.

TIGHT TURNS   The most conspicuous features of globular protein anatomy are the many changes in directions of the polypeptide backbone. Those segments responsible for an abrupt change in direction of almost 180° are called *tight turns,* usually involving about four successive amino acid residues. Glycine, proline, and hydrophilic (water-loving) residues are preferred: glycine, because its small R group (H) does not contribute much steric interference; proline, because it is cyclic and naturally contributes to changing the chain direction; and hydrophilic residues, because tight turns are usually located near the surface of the molecule where they can interact with solvent water molecules. One example of a tight turn is shown in Figure 4–26; it is a type stabilized in part by a single hydrogen bond.

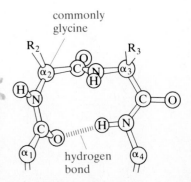

FIGURE 4–26 Example of a tight turn in a globular polypeptide.

4–4   TERTIARY STRUCTURE

General Shape

In the previous Laboratory Methods section we considered how protein size (that is, molecular weight) can be measured. Recall that the sedimentation method requires independently measured values of the diffusion constant D and partial specific volume $\bar{V}$ of the protein. Knowing the values of $\bar{V}$ and D (measured in a medium having a viscosity η) allows one to also evaluate the overall shape of a protein of size MW. This evaluation is done by calculating the *dissymmetry ratio* f/f_0 from the following relationship:

$$\frac{f}{f_0} = \left[\frac{R^3 T^3}{162\pi^2 (\text{MW}) \bar{V} N^2 D^3 \eta^3} \right]^{1/3} \qquad (N \text{ is Avogadro's number})$$

The symbols f and f_0 refer to frictional coefficients reflecting the drag on protein molecules as they flow through a liquid medium. Drag forces will obviously be affected by the shape of the molecules. Symbol f is the frictional coefficient for the actual shape of a molecule, and f_0 is the frictional coefficient for the molecule if it were a perfect sphere.

Thus the f/f_0 ratio is an *index of deviation* from a spherical molecular structure. If $f/f_0 = 1$, there is no deviation, and the actual shape is that of a perfect sphere. For actual shapes that are more elliptical and even elongated (see Figure 4–27), $f/f_0 > 1$. Some examples are given in Table 4–6.

TABLE 4–6 Dissymmetry ratios of proteins

PROTEIN	f/f_0	MOLECULAR WEIGHT[a]
Enolase	1.00	63,300
Insulin	1.07	12,650
Myoglobin (horse heart)	1.10	17,000
Lactate dehydrogenase (beef heart)	1.13	133,000
Hemoglobin (human)	1.16	66,000
Cytochrome c (beef heart)	1.19	13,400
Albumin (human serum)	1.29	68,500
Gamma globulin (human)	1.51	153,000
Fibrinogen[b] (human)	2.34	340,000
Myosin[b] (cod)	3.63	500,000

[a] Note the lack of any correlation between size and shape.
[b] Both of these proteins have an extensive amount of fibril character.

insulin

albumin

hemoglobin

gamma globulin

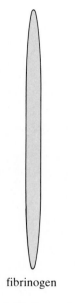

fibrinogen

FIGURE 4–27 Approximate shapes of some protein molecules.

Fibrous Proteins

The three-dimensional structure of fibrous proteins has been described in the previous section.

Fibrous polypeptide chains are entirely in a sheet or a helical conformation.

However, naturally occurring fibrous chains do not usually exist as free, non-associated chains. Rather, two, three, or several chains form *polychain aggregates*. For example, the fibroin proteins in silk are composed of several antiparallel β sheet chains stacked side by side as well as above and below each other (see Figure 4–28). In contrast, the protein fibers of wool, hair, feather, horns, and nails are composed of coiled helical chains to give a larger, multichain structure, which in turn is bundled or coiled together to give an even larger superstructure. The various levels of structure are analogous to those found in a rope.

One of the most understood fibrous proteins is *collagen,* the principal constituent of connective tissue and the major structural protein in all animal life. About 30% of total body protein and 6% of total body weight is collagen. One estimate of its abundance on earth is 2 trillion pounds.

As shown in Figure 4–29, each macrocollagen fiber of connective tissue is an aggregate of collagen molecules. Each collagen molecule is about 3000 Å in length and has an approximate molecular weight of about 300,000. The collagen molecule, also called *tropocollagen,* has a coiled-coil structure, specifically, a *triple helix.* Each of the three chains has a left-handed helical geometry, but they are coiled together to give a right-handed triple helix.

The helical orientation of each chain is more extended than the α-helix: a pitch of 9 Å for collagen versus 5.4 Å for the α-helix. Of course, the α-helix is ruled out anyway, because of the presence of proline and *hydroxyproline* (hyp) (see Figure 4–30), their sum accounting for about 25% of the residues present. Glycine residues occupy about 33% of the chain, and the rest of the chain is

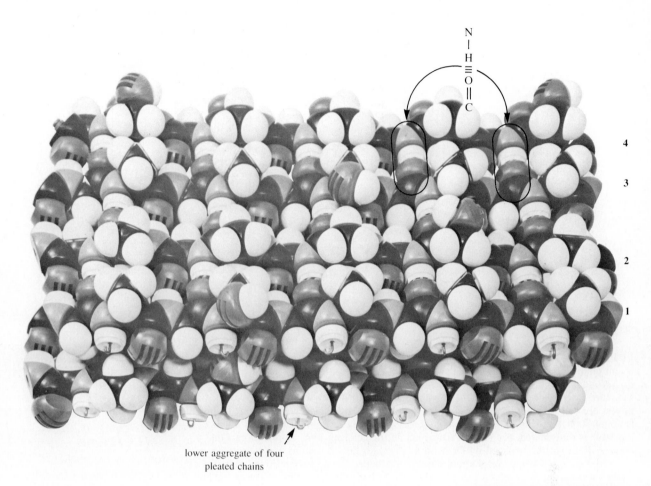

lower aggregate of four
pleated chains

FIGURE 4–28 Structure of silk protein. It has an upper layer of four polypeptide chains, each in the sheet orientation, on top of a lower layer of four polypeptide chains, also in the sheet orientation. Additional layers would be packed similarly. Tight packing of successive layers is possible when R groups are small, as in glycine, alanine, and serine.

made up of small amounts of most of the other amino acids, including some *hydroxylysine* (see Figure 4–31). The sequence is somewhat repetitive, with glycine found in almost every third position and tripeptide segments of gly—X—pro, gly—X—hyp, and gly—pro—hyp being very common. The pattern

$$— —gly— —gly— —gly$$
$$1 \quad 2 \quad 3 \quad 1 \quad 2 \quad 3 \quad 1 \quad 2 \quad 3$$

appears to be an absolute requirement for forming the triple helix.

The coiled-coil structure of the triple helix is stabilized by extensive interchain hydrogen bonding and some covalent cross-links between chains involving lysine and hydroxylysine. Thus the structure is *rigid* and *inflexible*. The packing of collagen molecules into collagen fibers, which also involves some covalent cross-links between tropocollagen molecules, confers additional rigidity and inflexibility to the fiber. Collagen structure—all the way down to its amino acid composition and sequence—is well suited for a biological role as a tough *structural* protein. For instance, a 1 mm thread of collagen will support about 20 lb (10 kg).

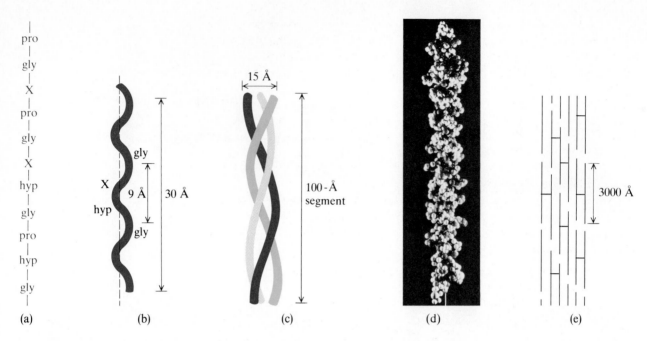

FIGURE 4–29 Levels of collagen structure. (a) Primary structure, an amino acid sequence of a unit chain. Every third residue is glycine; X represents any one of the usual amino acids. (b) Secondary structure, a left-handed helical conformation of an individual chain. (c) Tertiary structure, a triple-helix arrangement of collagen (also called tropocollagen). Three left-handed helices are coiled around each other with a right-handed twist. The 100-Å segment shown is about 3000 Å in length. (d) Space-filling model, showing tertiary structure. A 3000-Å molecule is represented as a single vertical line in the collagen fibril shown in part (e). (*Source:* Photograph provided by Dr. Karl A. Piez.) (e) Polymolecular association, an ordered packing of tropocollagen molecules in a collagen fibril of connective tissue. In some fibrils the adjacent molecules are covalently cross-linked. This structure is a very rigid macrostructure.

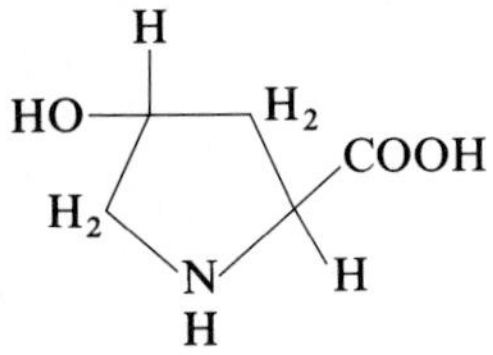

FIGURE 4–30 Hydroxyproline (hyp).

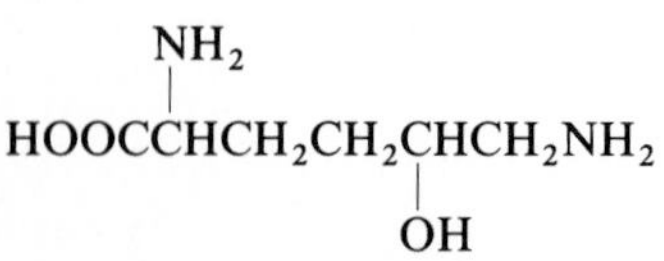

FIGURE 4–31 Hydroxylysine (hyl).

Collagen biosynthesis begins with the assembly of the individual polypeptide chains. Two different chains, designated as $\alpha 1$ and $\alpha 2$, are assembled, and the composition of the triple helix is $(\alpha 1)_2(\alpha 2)$. Prior to the forming of the triple helix, the *hydroxylation* of specific R groups occurs. This modification involves the operation of two similar but separate enzymes, one for hydroxylating proline residues and the other for lysine residues. Both enzymes use O_2 as the source of the —OH group, and there is evidence that both require *ascorbic acid* (vitamin C) for optimal activity. Some researchers propose that the presence of hydroxyproline contributes to additional hydrogen bonding for stabilizing the triple helix. Some of the —OH groups are also used to attach carbohydrate structures to collagen. Hereditary disorders of the hydroxylation steps result in serious disturbances in which mechanical properties of tissues such as skin and ligaments are impaired.

Globular Proteins

In 1960 J. C. Kendrew, M. F. Perutz, and co-workers reported success—the first of its kind—in determining the complete three-dimensional structure of a protein molecule. This remarkable feat has since been accomplished with about a hundred different proteins, nearly all of which are globular proteins. All of these structures were solved by using the powerful method of *X-ray diffraction analysis,* which is in fact the only method to solve three-dimensional molecular structure with atom-by-atom detail. A resolving power of 1.5 Å can be achieved, meaning that most of the atoms (except for the small H atoms) in a molecule can be unambiguously located. The method is technically sophisticated but conceptually simple; it can be thought of as a form of molecular microscopy.

In X-ray diffraction analysis a solid sample of a substance is exposed to a beam of X-radiation (high-energy rays of very short wavelength). Crystals give sharper results than fibers. As X rays pass through the crystalline lattice, they strike atoms and are then scattered or reflected at various angles. In addition, the intensity of the scattered radiation is governed by the electron cloud around the scattering atoms. In other words:

> *Both* the pattern of the scattering and the intensity of the scattering reflect the atomic structure of the sample.

The scattered radiation is detected by photographic film [see Figure 4–32(a)]. Thousands of diffraction patterns are collected from the same sample at different angles of exposure to the X rays; then a computer is used to calculate, for each pattern, the intensity of diffraction at each point and the spacings between each point. By methods we will not consider here, these data are then used to construct a three-dimensional model of the molecular structure.

From Figures 4–32(b) and 4–32(c) we see that the molecular conformation of a globular protein is much more compact than the threadlike structure of a fibrous protein. As indicated in the f/f_0 discussion, globular conformations can be nearly spherical or have elliptical shapes. Irregularities on the molecular surface, such as the cavity shown in Figure 4–32(c), are often crucial topographical features involved in the function of the protein.

However you view this discussion—be it with fascination, excitement, or perhaps even an appreciation of what could be termed the beauty of macromolecules—keep the following principle in mind:

> The preferred native conformation of a protein represents the ultimate structural individuality of the protein, which in turn is responsible for the functional individuality of the protein.

The native conformation of a protein is preferred because among countless arrangements it represents a conformation of *minimum energy,* or putting it another way, a conformation of *maximum stability.* However, do not conclude that every molecule of a given protein has exactly the same conformation into which it is frozen forevermore. It is more correct to think of a protein as follows:

> A protein is capable of existing in a small number of closely related and easily interconvertible conformations of minimum energy.

Indeed, the principle of *interconvertible protein conformations* is one of the hallmarks of modern biochemistry.

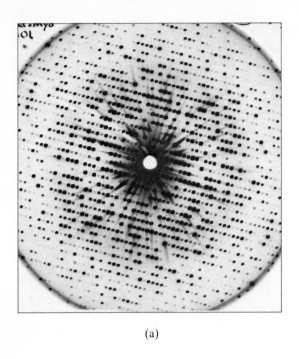

(a)

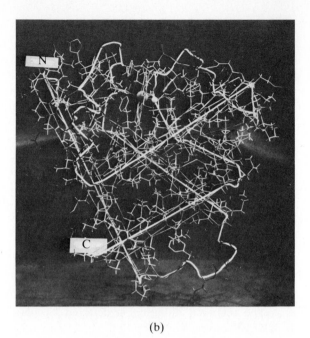

(b)

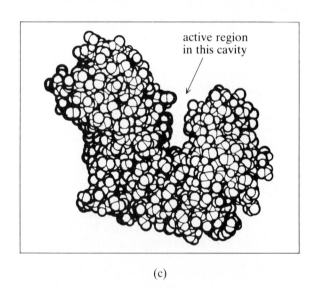

active region
in this cavity

(c)

FIGURE 4–32 X-ray diffraction analysis.
(a) Photograph of a single X-ray diffraction pattern
of a myoglobin crystal. (b) Skeletal model of the
three-dimensional arrangement of atoms in the
myoglobin molecule. There are seven α-helix
segments and no sheet structures in myoglobin.
(c) Computer-generated, space-filling representation
of the enzyme hexokinase from yeast. This model
illustrates more realistically the true tightly packed
structure and irregular surface topography of a
globular protein. [*Source:* Parts (a) and (b) taken with
permission from J. C. Kendrew, "Myoglobin and the
Structure of Proteins," *Science*, **139**, 1259–1266
(1963). Photographs provided by J. C. Kendrew.
Photograph in (c) provided by T. A. Steitz.]

Stabilization of Tertiary Structure in Globular Proteins

The maintenance of the intricate conformation of a globular protein is
achieved by various types of stabilizing forces (see Figure 4–33). When present,
one or more disulfide linkages have an obvious influence on conformation.
However, stabilization is achieved primarily because of *noncovalent interactions,*
which we examined in Section 2–2. In terms of the interactions in a polypeptide

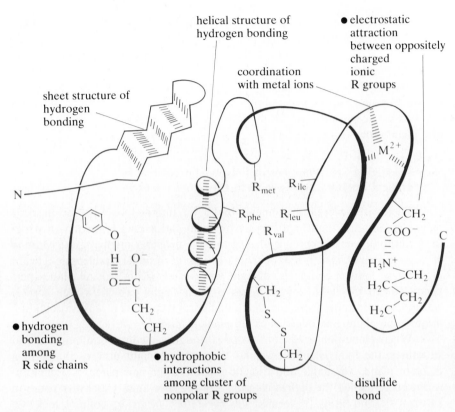

FIGURE 4–33 Diagrammatic representation of stabilizing forces in globular proteins. Only one of each type of bond is shown. The frequency of occurrence of each type varies from protein to protein. The sheet and helical structures, the disulfide bonds, and the metal ion coordination can all be absent or can all be present, or any combination can be present. The other types of interactions (identified by a •) are found in all globular proteins.

these noncovalent bonding patterns are as follows:

• *Ion-ion forces of attraction* occurring between side chains having oppositely charged ionic groups, such as those between lys(+)/glu(−) or arg (+)/asp(−).

• *Ion-dipole forces of attraction* occurring between charged R groups and various dipole functions in the molecule, such as (1) glu(−)/δ^+ of the HO function of thr and (2) lys(+)/δ^- of a C=O function of a peptide bond.

• *Peptide bond hydrogen bonding* occurring in helical and sheet regions.

• *Nonpeptide bond hydrogen bonding* involving various dipole functions on R groups, such as ser(OH)/his(imidazole).

• *Prosthetic group interactions,* when present, such as those between a positively charged metal ion and negatively charged R groups of asp/glu.

•*Hydrophobic interactions* occurring among the nonpolar side chains of leu, val, ile, ala, phe, and other nonpolar amino acids, mostly in the *interior* core of the molecule.

Hydrophobic interactions are proposed to be especially important.

In addition to stabilizing a folded protein, the pattern of clustering hydrophobic residues away from water (the interior core is virtually devoid of water) results in the polar, hydrophilic R groups being directed to the exterior surface of the molecule. This arrangement provides the basis for the protein to interact with polar, solvent molecules of water and other dissolved materials. Although this molecular micelle arrangement (refer to p. 42) is common to most globular proteins, there are some important exceptions. Particularly important are proteins that extensively penetrate or completely span a membrane. The surface of a membrane-localized protein often has many exposed hydrophobic residues, a feature consistent with the nonpolar, hydrophobic environment in the interior of a membrane.

Evaluating Conformations of Polymers in Solution by Optical Rotatory Dispersion and Circular Dichroism

There are several phenomena based on interactions between electromagnetic radiation (light) and matter (molecules). Lacking symmetry in structure, chiral molecules exhibit *optical activity,* a property involving the *rotation of plane-polarized light.* A monochromatic (that is, of single wavelength λ) beam of ordinary light is composed of light waves that oscillate in all planes perpendicular to the light source. When ordinary light passes through certain substances called polarizers, the emerging light is composed of waves oscillating in only one plane. Such light is called *plane-polarized.*

When plane-polarized light passes through a medium containing chiral structures, the light is *refracted,* and the emerging light now oscillates in a different plane. The angle and direction defining the new plane of oscillation is a property called the *optical rotation* α of the substance. Clockwise rotation between $0°$ and $+90°$ is termed *dextrorotatory behavior;* counterclockwise rotation between $0°$ and $-90°$ is called *levorotatory behavior* (see Figure 4–34). When measurements of α are made over a range of different wavelengths, the change in α as a function of wavelength is called an *optical rotatory dipersion* (ORD) *spectrum.*

In the related technique of *circular dichroism* (CD) the incident plane of polarized light is first passed through a substance that converts the plane of light into two circular components (see Figure 4–35). Each circular component moves through space as a wave that rotates around the axis of propagation. One component has a clockwise (right-handed) rotation; the other has a counterclockwise (left-handed) rotation. Whereas ORD analysis depends on the refraction of light, circular dichroism is based on *different degrees of absorption* of these two circular waves of light as they pass through an asymmetrical substance. For reasons we will not describe, this difference is expressed in terms of a property called *ellipticity,* symbolized as θ. Similar to an α value, a θ value also ranges from $-90°$ to $+90°$ and also represents an angle describing the beam of emerging light, altered by the differential absorption of the two circular components. Like ORD spectra, CD spectra are obtained by measuring θ at different wavelengths of incident light.

Although neither individual α or θ values of a substance at any given wavelength nor ORD or CD spectra are unique for any given substance, they are characteristic and reproducible properties. There is one other important point: Mirror image structures of the same substance—that is, enantiomers with opposite chirality or handedness—will have equal but opposite α and θ values at the same wavelength and also have inverted ORD and CD spectra.

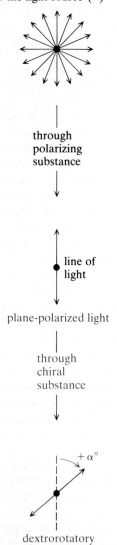

end-on view of beam of ordinary light, with waves oscillating in all planes perpendicular to the light source (●)

through polarizing substance

line of light

plane-polarized light

through chiral substance

$+\alpha°$

dextrorotatory effect

FIGURE 4–34 Plane-polarized light and optical activity.

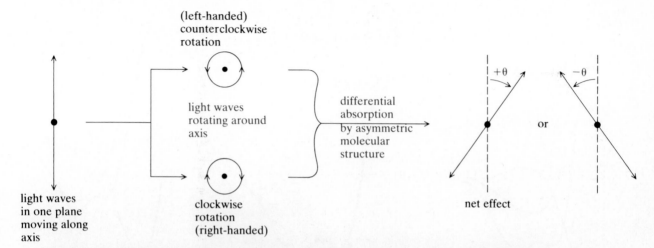

FIGURE 4–35 Circular dichroism. The net effect can be thought of as having changed the angle of the incident plane.

Applications (see Figure 4–36) of ORD and CD analysis are to estimate the content of helical and sheet geometries in protein structures and to observe changes and interconversions in the total conformation of proteins and nucleic acids in response to changing conditions. Because CD spectra often exhibit more distinct features than ORD spectra, the former tend to be more useful. For example, after one obtains a CD spectrum for a polypeptide, a computer analysis is made to determine the percentage combination of helical, sheet, and random geometries that will satisfy the observed CD spectrum. This curve-fitting analysis uses θ values obtained from control polypeptides known to exist exclusively in either helix, sheet, or random structures. Although there are many reports of close agreement between the methods of ORD/CD analysis and X-ray diffraction in evaluating helical and sheet content of a polypeptide, the ORD/CD techniques are not absolutely accurate in every instance. Thus X-ray diffraction is still the ultimate procedure for this purpose.

Any transitory changes occurring in solution of the total conformation of a complex protein or nucleic acid structure can be evaluated by comparing CD spectra under different conditions, such as varying pH, varying temperature, or in the presence of varying amounts of a substance that interacts with the protein or nucleic acid. Similar spectra suggest that few changes occur; slightly different spectra reflect subtle changes in conformation; grossly different spectra reflect large conformational transitions. The application of CD spectra in the discovery of a new conformation of DNA is described in Chapter 6.

4–5   QUATERNARY STRUCTURE OF GLOBULAR PROTEINS

Oligomers

The hemoprotein myoglobin (see Figure 4–37) occurs in muscle tissue, binding and storing O_2 when the cellular level of oxygen is high, then releasing the bound O_2 when cells need it. Myoglobin is composed of a single polypeptide

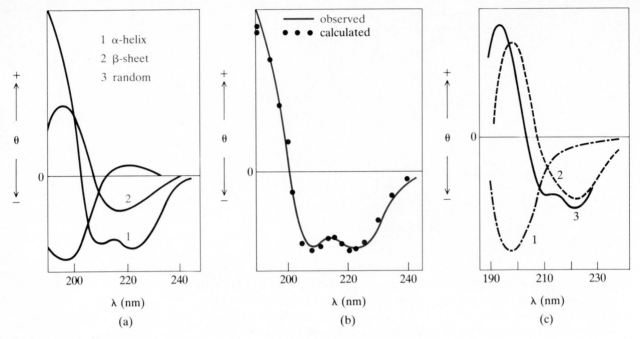

FIGURE 4–36 Circular dichroism (CD) spectra and their use. (a) Reference CD spectra showing characteristic profiles for a synthetic polypeptide examined under conditions that yield different geometries. (b) Observed and calculated spectra. The observed CD spectrum for myoglobin was fitted to additive θ values from the three reference CD spectra in part (a) to yield 69.3% helix, 4.7% sheet, and 27.0% random coil, in good agreement with X-ray diffraction data. (c) Observed changes in the conformation of an opiate peptide (β-endorphin) in response to the presence of a membrane lipid. Curve 1 represents endorphin with no lipid (mostly random coil); curve 2 represents endorphin with lipid (some helix content); curve 3 represents endorphin with three times more lipid than in case 2 (substantial helix character).

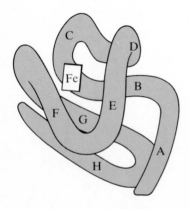

FIGURE 4–37 Myoglobin, which is nonoligomeric.

chain (153 residues; MW = 17,500) in association with one heme prosthetic group. The heme group is described later in this section. The globular chain consists of eight helical segments and seven direction-changing turns, six of which are tight turns. The helix content is about 80%. There are no intrachain disulfide bonds and no β sheet structure.

Hemoglobin (see Figure 4–38), another hemoprotein, occurs in red blood cells, where it binds to O_2, transporting it throughout the body and releasing it to oxygen-dependent tissues. Hemoglobin (MW = 64,500) is composed of four polypeptide chains, each in association with a heme group. In normal adult hemoglobin (HbA) there are two copies of an α chain (141 residues) and two copies of a β chain (146 residues), to yield a tetramer $\alpha_2\beta_2$ (see Note 4–3). Although they are distinctly different, the α and β chains are structurally very similar to each other and also very similar to the myoglobin chain. Allowing for differences in chain length, there is about 40% sequence homology between the α and β chains and about 15% sequence homology among all three chains. Each is coded for by a different gene located on different chromosomes. There is convincing evidence that these separate genes evolved from the same ancestral gene.

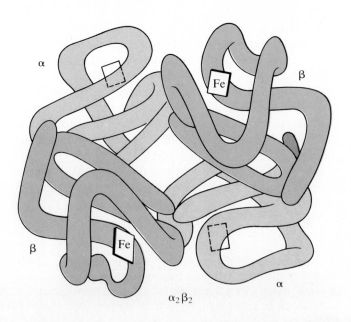

FIGURE 4–38 Hemoglobin ($\alpha_2\beta_2$), which is oligomeric.

Hemoglobin provides an example of the quaternary level of protein structure:

> The quaternary level is an association of two or more polypeptide chains linked only by noncovalent forces of attraction among R groups at the surface of the chains. No covalent bonds such as —S—S— link the chains.

Proteins in this group are called *oligomers* (dimers, trimers, tetramers, and so on), and individual chains are often called *subunits*. If the subunit chains are identical, the protein is a *homogeneous oligomer;* if they are different (as in hemoglobin), they are called *heterogeneous oligomers.*

Of what advantage to cells is the oligomeric structure of proteins? In the context of the structure-function principle, the packing of more structural information into a molecule should mean that additional elements of function result. One function exhibited by some oligomers is *regulation of activity* by interconversion between *associated and dissociated states.* An important specific example is described in the next chapter for the activation/deactivation of an enzyme called protein kinase (see Figure 5–37). Another very important regulatory aspect of oligomeric structures is called *cooperativity;* it is described later in this chapter, using hemoglobin as a representative case, and is further described, in terms of its significance to enzymes, in the next chapter (Section 5–9). Still another consequence of oligomeric structure is the formation of *hybrid forms* of a protein.

Hybrid Oligomers

When chains of different polypeptides associate to yield an oligomer, sometimes they *associate in different proportions to yield hybrid oligomeric structures* that perform the same function. The phenomenon is primarily en-

NOTE 4–3
A common practice of biochemists is to use lowercase Greek letters and numerical subscripts as symbols for individual chains and their number in an oligomer. Thus a heterogeneous tetramer composed of two copies of each of two different chains is symbolized as $\alpha_2\beta_2$. Unfortunately, another system is sometimes used, capital letters (M, B, H, L, or any other). The letter signifies such features as size (H = heavy, L = light), tissue (M = muscle, H = heart, B = brain), or something else.

countered with enzymes, and the hybrid forms are called *isoenzymes,* or *isozymes.* For example, *lactate dehydrogenase* (LDH), a widespread enzyme functioning in carbohydrate metabolism, is a tetramer that can occur in either of five hybrid forms: H_4, H_3L, H_2L_2, HL_3, and L_4. In some cells all five forms may be present. The H and L subunits are so designated because one predominates in heart muscle (H) and the other predominates in liver (L). Since the polypeptide subunits have a different amino acid composition and hence a different ionic character, each hybrid oligomer has a different composite ionic character. Thus, electrophoresis is commonly used to separate them.

There is no single reason why hybrid proteins exist. We know that different cell types contain a set of hybrid proteins in a ratio characteristic of that cell type. Generally speaking, however, the significance of hybrid proteins is explained as follows:

> Hybrid proteins confer a spectrum of protein activity to a cell—not a spectrum of different activities but a spectrum of different degrees of the same activity or a spectrum of different responses to inhibitors, activators, or other signals (such as temperature) that regulate the activity.

In other words, a set of hybrid proteins is one way a specific type of cell efficiently adjusts to a changing environment and/or to the changing needs of that cell. Sometimes, different isoenzyme forms predominate in different compartments of a cell, with the response of the isoenzymes fine-tuned to different events occurring in the different compartments (for example, mitochondrion versus cytoplasm).

There is an important medical application of isoenzymes, namely, diagnosing disease, particularly myocardial infarction. Significantly elevated blood levels of certain enzymes (H_4–LDH, to a lesser extent H_3L–LDH, and MB–CPK) can be expected shortly after heart cells die and lyse. The CPK enzyme is *creatine phosphokinase,* a dimeric protein that exists in three hybrid forms: MM, MB, and BB. The MM form predominates in skeletal muscle, and BB predominates in brain; most other cell types have characteristic ratios of two or three forms. However, the only tissue with a sizable amount of MB–CPK is heart, accounting for about 15% of the total CPK activity in heart cells. A disadvantage of the CPK diagnosis is that elevated MB levels remain elevated for only about 24 h after the infarct, whereas the H_4–LDH elevation can persist for several days. A disadvantage of the LDH diagnosis is that other disease states can also give rise to elevated blood levels of H_4–LDH and H_3L–LDH. In contrast, the MB–CPK elevation gives a much more specific indication of cardiac cell death.

Other protein markers indicating the existence, severity, and/or duration of a particular disease exist, and new ones will certainly be discovered in the future. In addition, advancements are constantly being made in the development of methodology and instrumentation to perform fast, accurate, and precise quantitative measurements of protein activity.

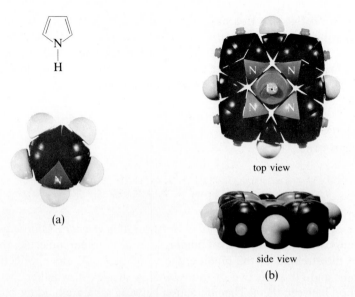

top view

side view

(b)

(a)

(c)

Figure 4–39 Heme group. (a) Pyrrole. (b) Space-filling model of a heme. (c) Protoporphyrin IX heme group in oxymyoglobin and oxyhemoglobin.

Heme Group of Hemoproteins

In a previous subsection the structures of myoglobin and hemoglobin were contrasted. In the next section we will compare their oxygen-binding characteristics. Before doing so, however, we first examine the *heme prosthetic group*. When present in enzymes, it is considered as a coenzyme.

A heme group is an *iron-coordinated, cyclic tetrapyrrole*. The cyclic tetrapyrrole ring system (also called a *porphyrin*) is composed of four pyrrole units connected in ring fashion by one-carbon methenyl ($=CH—$) bridges. All atoms in the ring lie in the same plane; the structure is flat. In both myoglobin and hemoglobin the iron is maintained in the reduced ferrous state (Fe^{2+}) and displays a coordination number of 6. In addition to coordination bonding with the N of each of the four pyrrole units, Fe is also bonded to the polypeptide chain (the globin moiety) through two specific histidine residues occupying the fifth and sixth coordination sites perpendicular to each face of the tetrapyrrole plane. When the proteins are oxygenated, the *O_2 binds to Fe,* displacing one of the histidine groups. Heme groups differ in terms of the eight side chains attached to the four pyrrole units. The groups shown in Figure 4–39 are characteristic of the *protoporphyrin IX* structure in the heme of both myoglobin and hemoglobin.

4–6 PROTEIN BINDING AND THE COOPERATIVE EFFECT

General Model of Protein Binding

For most globular proteins (P) the *initial action event is to bind with another substance,* which we will generally refer to as a *ligand* (L). The P and L interaction almost always occurs at a precise location on the surface of the protein, called the *binding site*. In a protein molecule there may be one binding site (as in myoglobin, with one heme group that binds one O_2 molecule) or *n* binding sites for the same ligand (as in hemoglobin, with four heme sites for $4O_2$). The oligomeric structure of proteins makes the existence of multiple sites particularly feasible. The process can be represented simply as

$$P + nL \rightleftharpoons PL_n \qquad \text{where} \qquad n = 1, 2, 3, \ldots$$

for which we can write

$$K = \frac{[PL_n]}{[P][L]^n}$$

where *K* is the equilibrium *binding constant,* or *association constant.* Of interest to the biochemist are (1) the number of binding sites per protein molecule, the value of *n*; (2) the strength of the binding, the value of *K*; (3) any specificity of P for ligands; and (4) when two or more binding sites exist, whether there is any interaction between them. The interaction between separate binding sites is called the *cooperative effect*.

Laboratory Measurements of Binding

Various methods can be used for precise quantitative measurements of protein binding. Although they differ in approach, each method involves measuring the *saturation function* (symbolized as *v*) of the protein at various concentrations of ligand. The saturation function *v* is defined as the ratio of concentration of bound ligand $[L]_b$ to concentration of total protein $[P]_t$:

$$v = \frac{[L]_b}{[P]_t}$$

The limits of *v* are 0 and *n*. When $v = 0$, no ligand is bound. When $v = n$, the maximum amount of ligand is bound, and the protein is referred to as being *fully* (100%) *saturated*. When $v = \frac{1}{2}n$ the condition is 50% saturated; and so forth. The relationship of *v* to ligand concentration [L] is satisfied by the equation

$$v = \frac{nK[L]^c}{1 + K[L]^c}$$

where exponent *c*, which is related to *n*, is a measure of the interaction between sites. This characteristic is discussed shortly.

In the technique of *membrane ultrafiltration* a known amount of free P and free L (*free* meaning unbound) are mixed in a compartment constructed of a porous-membrane material. The pore size of the membrane is of molecular dimensions but small enough to prevent passage of the high–molecular weight protein and, of course, any ligand that becomes bound to the protein. Only free ligand will filter through.

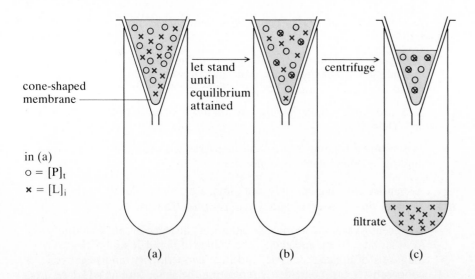

FIGURE 4–40 Membrane ultrafiltration. (a) Cone-shaped membrane supported in a centrifuge tube. A solution of free protein P and free ligand L is placed in it at the start. Both concentrations $[P]_t$ and $[L]_i$ are known. (b) Equilibrium state. The initial mixture is left to stand until an equilibrium of

$$P + L \rightleftharpoons PL$$

$$\bigcirc + \times \rightleftharpoons \otimes$$

is attained. (c) Separation of unbound ligand from the equilibrium mixture. Centrifugal force will displace the free ligand through the small pores of the membrane filter. Protein that contains bound ligand and free protein remain behind. Measure the amount of the free ligand $[L]_f$ in the filtrate. By calculation the amount of bound ligand is $[L]_b = [L]_i - [L]_f$.

Figure 4–40(a) depicts a cone-shaped membrane compartment supported in a centrifuge tube. After one lets the P and L mixture stand [Figure 4–40(b)] until equilibrium is reached ($P + L \rightleftharpoons PL$), the system can be centrifuged [Figure 4–40(c)] to force the passage of any unbound ligand through the membrane. The filtrate is recovered, and the amount of free ligand $[L]_f$ present is measured. Precise and accurate measurements on extremely small amounts of material can be made by using radioactively labeled ligand. Since one knows the initial concentration of ligand $[L]_i$, the value of $[L]_b$ is merely $[L]_i - [L]_f$. The procedure is then repeated for several other concentrations of ligand with the same amount of protein.

Cooperative and Noncooperative Patterns of Binding

A *direct plot* of v versus $[L]_i$ will yield either a *hyperbola* or a *sigmoid* (S-shaped) curve [see Figures 4–41(a) and 4–41(b)]. Each curve represents a different type of binding. A hyperbolic plot signifies one of the following:

1. A protein molecule having a single binding site for L.

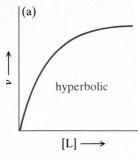

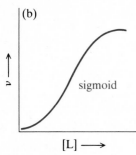

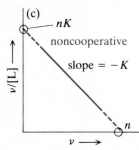

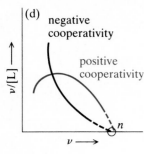

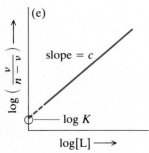

2. A protein molecule having two or more binding sites that are *non-cooperative*. That is:

The event of binding at one site has no influence on the same event of binding at other sites in the molecule. In other words, the binding events at noncooperative sites are wholly independent of each other.

3. A protein molecule having two or more sites that are *negatively cooperative*. That is:

The event of binding at one site interferes with the same event of binding at other sites in the molecule.

A sigmoid plot signifies only one thing: a protein molecule having two or more binding sites that are *positively cooperative*. That is:

The event of binding at one site enhances the same event of binding at other sites in the molecule. The values of n and K and a clearer diagnosis of noncooperativity, positive cooperativity, and negative cooperativity can be made by replotting the same data as $v/[L]_i$ (y axis) versus v (x axis). This plot is called a *Scatchard plot* [see Figures 4–41(c) and 4–41(d)]. It is based on rearranging the binding equation given earlier to the form shown below for a linear relationship. (The c exponent is overlooked for the moment.)

$$\frac{v}{[L]} = nK - Kv$$

$$y = b - mx$$

When $v/[L]_i = 0$, the equation predicts that $v = n$. Thus extrapolating a Scatchard plot through the v axis identifies the number of binding sites present. If $n = 1$ or $n > 1$ with noncooperativity, the Scatchard plot is linear with a slope of $-K$. Nonlinear Scatchard plots indicate $n > 1$ with cooperativity. If $n > 1$ with negative cooperativity, a plain convex curve is observed. If $n > 1$ with positive cooperativity, a curve passing through a maximum is observed.

To evaluate c, one can rearrange the original binding equation (without overlooking c) still another way, called the *Hill equation* (after A. V. Hill), to yield the linear relationship

$$\log\left(\frac{v}{n - v}\right) = c \log[L] + \log K$$

$$y = mx + b$$

In a *Hill plot* [Figure 4–41(e)] the y and x coordinates are $\log(v/n - v)$ and $\log[L]$, respectively, with c as the slope and $\log K$ as the y axis intercept. A

FIGURE 4–41 Graphical evaluations of protein-ligand binding data.
(a) Hyperbolic direct plot. (b) Sigmoid direct plot. (c) Scatchard plot showing noncooperativity. (d) Scatchard plot showing negative and positive cooperativity. (c) Hill plot.

value of $c = 1$ corresponds to noncooperative sites; a value of $c > 1$ indicates positive cooperativity; and a value of $c < 1$ indicates negative cooperativity.

Cooperative effects are but part of a broader scope applying to the action of multiple-site proteins, because many proteins have multiple sites for different ligands. All aspects of multiple-site proteins are referred to as *allosteric effects,* which are developed further in the next chapter on enzymes. There we will reemphasize the significance of such effects to molecular regulation and also describe some models that provide a molecular explanation of how cooperative effects (that is, allosteric effects) arise. For now only the following general explanation of cooperative effects is emphasized:

> Communication between interacting sites is achieved by changes in protein conformation. A binding event at the first site results in a change of protein conformation at other sites, enhancing or impeding the same binding event at those sites.

Before ending the discussion, let us illustrate the cooperative effect with hemoglobin.

Oxygen Binding to Hemoglobin: Strong Positive Cooperativity

A comparison of O_2 binding between myoglobin and hemoglobin is presented in the direct plots shown in Figure 4–42. Myoglobin yields a classical, noncooperative binding curve ($c = 1$). With only one O_2-binding heme group, cooperativity is not even a possibility with myoglobin. The sigmoid curve for hemoglobin reflects a strong degree of positive cooperativity ($c = 2.8$). Scientists have established that each O_2-heme interaction significantly enhances the next, achieving an enhancement factor of about 400 for the last O_2 bound:

$$\text{Hb} \xrightarrow{\text{O}_2} \text{HbO}_2 \xrightarrow[\text{enhanced}]{\text{O}_2} \text{Hb(O}_2)_2 \xrightarrow[\substack{\text{enhanced} \\ \text{more}}]{\text{O}_2} \text{Hb(O}_2)_3 \xrightarrow[\substack{\approx 400 \text{ times easier} \\ \text{than the first}}]{\text{O}_2} \text{Hb(O}_2)_4$$

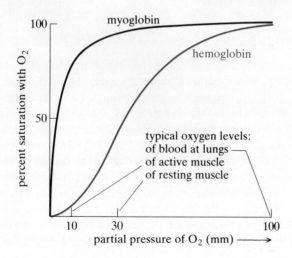

FIGURE 4–42 O_2-binding patterns of myoglobin and hemoglobin.

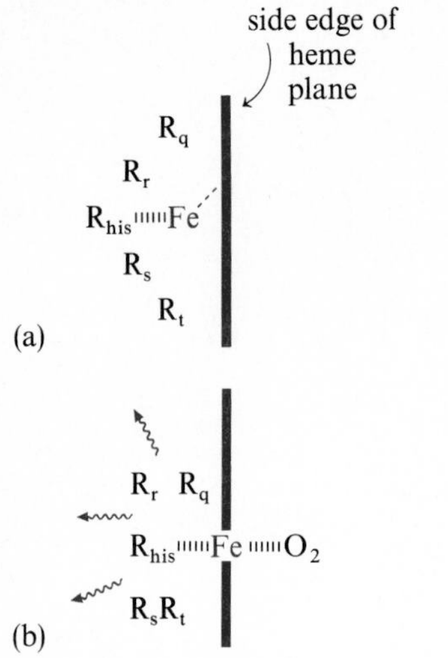

FIGURE 4–43 Heme conformations. (a) In deoxyhemoglobin. The Fe^{+2} is slightly out of the heme plane. (b) In oxyhemoglobin. The Fe^{+2} is pulled into the heme plane, pulling in R_{his} and thus triggering additional local R group realignments, causing still other ($\rightsquigarrow$) realignments.

2,3-bisphosphoglycerate
BPG

symbolized as

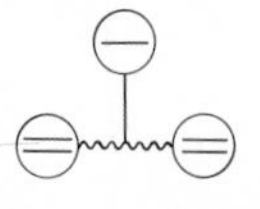

$$Hb(O_2)_4 + BPG \underset{\substack{\text{at lungs } O_2 \\ \text{displaces} \\ \text{BPG}}}{\overset{\substack{\text{at respiring} \\ \text{tissues BPG} \\ \text{displaces } O_2}}{\rightleftharpoons}} Hb(BPG) + 4O_2$$

FIGURE 4–44 Hemoglobin regulator, BPG. In Figure 4–45 BPG is symbolized as shown here.

In addition to showing the noncooperative/cooperative comparison, the patterns in Figure 4–42 also reveal why the two proteins are well suited for their respective biological roles. Myoglobin is a good storage protein because it is easily saturated and then releases oxygen only when tissue levels are low. Hemoglobin requires much greater O_2 levels to get saturated—a condition that is satisfied in the lungs. It is a good transport protein because oxygen is released at oxygen levels where myoglobin remains saturated.

M. F. Perutz (Medical Research Council Laboratory of Molecular Biology in Cambridge, England) has studied hemoglobin structure since 1937. The first success came about twenty years later with the solution of a crude outline of hemoglobin structure. By the mid 1970s detailed structures of both deoxyhemoglobin and oxyhemoglobin were solved, revealing subtle but significant differences between the two (see Figure 4–43). Further studies established the following feature:

The trigger of conformational changes in hemoglobin appears to be an initial movement of Fe^{2+} in the heme group when O_2 binds.

In deoxyhemoglobin the heme iron lies slightly outside the common plane of the heme group by about 0.6 Å. This out-of-plane alignment is due to the coordination of the Fe with N of an imidazole group of histidine. Apparently, so that local crowding of the histidine group with other R groups is overcome, the imidazole/Fe coordination displaces the position of the Fe. When O_2 binds to Fe from the other side of the heme plane, the Fe is pulled back into the heme plane—a movement that affects the imidazole/Fe interaction on the other side, pulling the histidine R group closer to the heme plane. Owing to the close contact of atoms, the local movement of the histidine R group sets in motion other R group realignments, generating still others throughout the entire subunit. The new conformation of the subunit can affect its interaction with other subunits, causing subtle shifts in the overall three-dimensional alignment of subunits in the intact oligomer. Compared with the position of the two $\alpha\beta$ dimers in deoxyhemoglobin, one $\alpha\beta$ pair in oxyhemoglobin has rotated about 15° in relation to the other. Many additional details are presented in the articles by Perutz cited in the Literature.

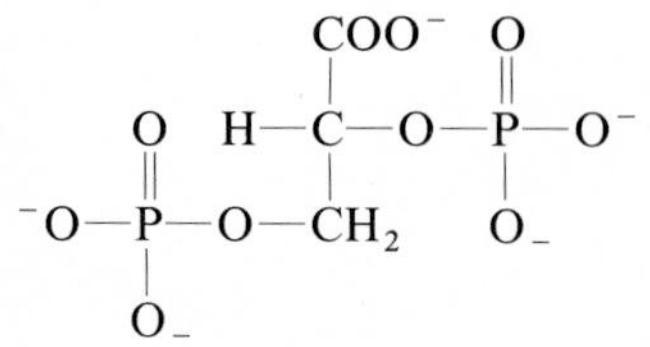

Additional Regulatory Features of Hemoglobin

The cooperative effect of O_2 in hemoglobin is supplemented by other regulatory devices. Recall that in Chapter 2 we mentioned the Bohr effect, where the binding of H^+ promotes the dissociation of O_2 from oxyhemoglobin. Another important hemoglobin regulator is *2,3-bisphosphoglycerate* (BPG) (see Figure 4–44), present in red blood cells in about the same concentration as hemoglobin. Like the H^+ effect, the binding of BPG to oxyhemoglobin *promotes the dissociation of bound* O_2 for uptake by respiring tissues.

An interesting feature of BPG binding is that there is only one binding site in the intact molecule, occurring at the region of contact between the two β chains. Consistent with the highly negatively charged structure of BPG, the binding site for BPG is composed of six positively charged residues contributed by the two β subunits. The BPG decreases the affinity of hemoglobin

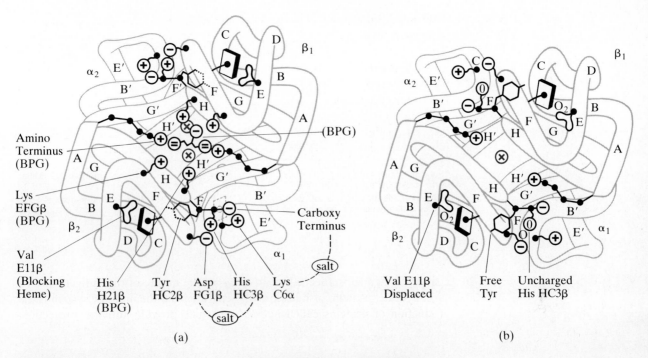

(a)   (b)

FIGURE 4–45 Structural changes in hemoglobin associated with the binding of 2,3-bisphosphoglycerate (BPG). (a) Deoxyhemoglobin with bound BPG. The BPG binding site in the $\beta\beta$ contact region is composed of three positive residues from each of the β chains. (b) Oxyhemoglobin. The valine residues in the vicinity of the heme are directed toward heme when BPG binds, weakening the O_2........Fe bond, and O_2 is released. The overall difference in alignment of subunits is indicated by the $\otimes$ markers in deoxyhemoglobin, which get superimposed in the oxyhemoglobin structure as R groups and subunits shift positions. [*Source:* Reproduced with permission from "X-ray Studies of Protein Mechanisms." *Annu. Rev. Biochem.,* **41,** 815 (1972). Drawings provided by R. E. Dickerson.]

for O_2 by cross-linking the β subunits, pushing them away from each other and causing conformational changes favoring the release of bound oxygen. Inspect the structures shown in Figure 4–45.

4–7 ANTIBODIES AND INTERFERONS: NATURAL-DEFENSE PROTEINS

All types of organisms have self-defense mechanisms providing some protection against the invasion of foreign substances, be they chemicals, viruses, or even other cells. The complex and intricate operation of the *immune system* is particularly important in vertebrate systems (see Note 4–4). A major component of the immune system involves the production of *antibodies* by plasma cells, mature forms of B lymphocytes (B cells) that arise from the differentiation of stem cells in bone marrow. An adult human may be able to produce as many as 10^6–10^7 different B cells, each capable of producing a single antibody against

NOTE 4–4
In addition to involving antibody-producing B cells, the total operation of the immune system involves *macrophage cells,* or *T-cells,* which are lymphocytes derived from the thymus, a protein called *complement,* and other factors. A good introductory source book is the text by Golub listed in the Literature.

TABLE 4–7 Comparison of human immunoglobulins

TYPE[a]	STRUCTURE[b]	MOLECULAR WEIGHT (g/mole)	RELATIVE ABUNDANCE IN SERUM (%)	CARBOHYDRATE CONTENT (%)
IgG	$\kappa_2\gamma_2$	150,000	71	3
IgA	$(\kappa_2\alpha_2)_n$	180,000–500,000	22	8
IgM	$(\kappa_2\mu_2)_5$	950,000	7	12
IgD	$\kappa_2\delta_2$	186,000	very small	?
IgE	$\kappa_2\varepsilon_2$	200,000	very small	11

[a] Subclasses and sub-subclasses of each type are proposed to exist, differing, among other things, in the size of the heavy chain.
[b] Symbol κ designates one of two classes of light chains; the other is designated λ. Each type of antibody also occurs with λ in place of κ.

a single foreign substance called an *antigen*. The nature of the antigen may be carbohydrate, protein, or nucleic acid. Separate from the immune system is the production of proteins called *interferons,* which provide antiviral protection.

Antibodies

The *gamma globulin fraction* of blood serum contains the antibodies, also called *immunoglobulins.* There are five different types: IgG (the most prevalent), IgA, IgM, IgD, and IgE (see Table 4–7).

The IgG molecule consists of two copies of each of two different polypeptide chains, all interconnected by interchain —S—S— bonds. There are also many intrachain disulfide bonds. The IgG structure is designated as $\kappa_2\gamma_2$, with κ representing a light chain (MW $\approx$ 25,000) and γ a heavy chain (MW $\approx$ 50,000). A structure of $\lambda_2\gamma_2$ also occurs with another class of light chain λ.

Amino acid sequences of κ and γ (done during the 1960s) revealed a striking feature of structure: different IgG antibodies display, in each chain, both highly *constant* (C) and highly *variable* (V) sequence regions. The V region spans approximately the first 110 residues in each chain. In IgG the V regions of both a light and a heavy chain contribute to the formation of an *antigen-binding domain*. There are two such domains in the molecule (see Figure 4–46), and IgG is said to be *bivalent.* The remarkable specificity in recognizing antigens is due to the diversity of structure generated in the V regions.

In addition to having a natural biological role, antibodies are extremely useful as laboratory tools. They are utilized (1) in performing quantitative measurements of minute amounts of a material in complex mixtures, a procedure called *radioimmunoassay* (RIA); (2) in detecting specific proteins after an electrophoretic analysis, a procedure called *immunoelectrophoresis;* and (3) in isolating a specific protein from a complex mixture by *immunoaffinity chromatography.* These procedures all exploit the *binding specificity* of antibodies (RIA is discussed shortly, and affinity chromatography is discussed later in the chapter).

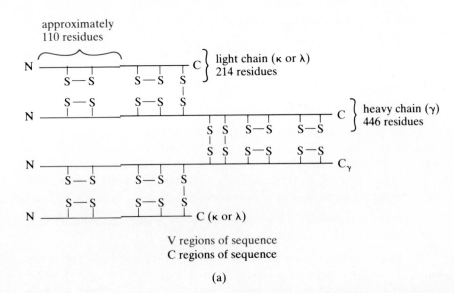

approximately
110 residues

N ——|—|——|—|—|— C } light chain (κ or λ) 214 residues
  S—S   S—S   S

  S—S   S—S   S
N ——|—|——|—|—|————————— C } heavy chain (γ) 446 residues
          S S S—S S—S

          S S S—S S—S
N ——————|—|——|—|—|— C_γ
  S—S   S—S   S

  S—S   S—S   S
N ——|—|——|—|— C (κ or λ)

V regions of sequence
C regions of sequence

(a)

FIGURE 4-46 Schematic illustrations of the three-dimensional structures of an IgG antibody molecule. (a) Primary structure. (b) Tertiary structure in chain form. (c) Tertiary structure showing V and C regions.

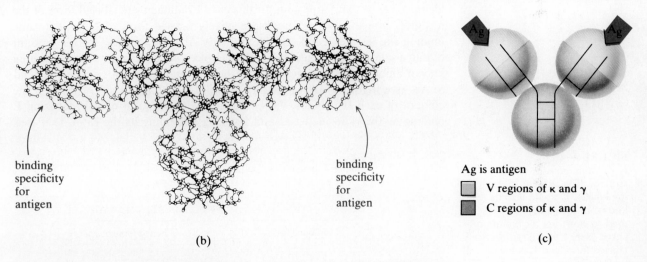

binding
specificity
for
antigen

binding
specificity
for
antigen

(b)

Ag is antigen

▢ V regions of κ and γ
■ C regions of κ and γ

(c)

These antibody-based procedures work best with purified antibodies, which until recently were difficult to obtain in large amounts. However, in 1975 C. Milstein and G. Kohler reported success in fusing a normal B cell, producing only a single antibody, to an abnormal myeloma tumor cell. While B cells cannot be grown in laboratory culture, the procedure is routine for myeloma cells. Among the hybrid cells formed in the fusion (called *hybridomas*) they observed that one had retained both the ability to produce the same B cell antibody and the immortal character of the myeloma cell. Thus the antibody specificity of the original B cell was cloned, and the supply became unlimited. Potentially, the same thing can be done with any antibody-producing B cell.

The pure antibodies produced by hybridoma cells are called *monoclonal antibodies,* and many have been prepared in the past few years. In addition to improving the quality of antibody-based laboratory procedures, monoclonal antibodies have potentially important medical applications in the areas of diagnosis, prevention, and cure of disease. A good review of applications can be found in the articles by Yelton and Scharff and by Marx listed in the Literature.

Radioimmunoassay (RIA)

Specific, ultrasensitive, and accurate quantitative procedures are essential to research and clinical laboratories. An RIA procedure satisfies all these requirements. Two things are needed to perform RIA: a radioactive form of the substance being assayed and an antibody preparation (ideally monoclonal) recognizing the substance as a specific antigen.

Sensitivity is provided by the use of radioactivity (see Appendix II) and specificity by the antibody.

The following description refers to antibody as Ab and to its antigen as ligand L. The ligand might be a steroid, a drug, a peptide, or a polypeptide. The RIA strategy exploits the principle of *competitive binding* between virtually identical ligands and a protein. In this case a radioactively labeled form of the ligand **L** competes with a nonradioactively labeled form of L for the same binding site on the ligand-specific antibody Ab_L. In an assay for the amount of L in a sample, a small aliquot of the sample is added to a solution containing known amounts of Ab_L and **L**. With both L and **L** present, either can bind to Ab_L (see Figure 4–47). Sometimes, **L** will bind; sometimes, L will bind. The amount of each that binds will be governed by the ratio of L/**L**, favoring L binding as the ratio increases and favoring **L** binding as the ratio decreases.

FIGURE 4–47
Radioimmunoassay (RIA) strategy. The amount of nonlabeled L would also be known in the control systems used to construct a calibration curve. Here **L** is the radioactively labeled form, and L is the nonlabeled form.

$$\underbrace{\mathbf{L} \quad + \quad Ab_L \quad + \quad L}_{\text{unbound forms}} \quad \rightleftharpoons \quad \underbrace{\mathbf{L} \sim Ab_L + L \sim Ab}_{\text{bound forms}}$$

known amount → **L**
known amount → Ab_L
unknown amount → L

After equilibration the antibody-bound ligand species are separated from the unbound forms. By measuring the amount of $\mathbf{L}-AB_L$ formed and comparing it against control values using known concentrations of both L and **L**, one can evaluate the amount of L in the original sample (see Figure 4–48). Bound **L** can be evaluated directly by measuring the amount of radioactivity in $\mathbf{L}-Ab_L$ or indirectly by measuring the amount of radioactivity associated with free **L** at equilibrium and subtracting it from the amount of radioactivity present initially.

Interferons

After contracting an initial viral infection, animals have the ability to resist future infections from the same or different virus. In 1957 A. Isaacs and J. Lindemann proposed that this protection was due to the production by the animal of a specific protein that interfered with the process of viral multiplication. Hence the protein was called *interferon*. After years of debate the existence and the protein nature of interferon were confirmed. Isolation was difficult because interferon occurs in very small amounts—in the range of nanograms (10^{-9}) to picograms (10^{-12}).

Knowledge of interferon biochemistry has progressed rapidly in recent years. For example, the structure consists of a single polypeptide chain with 155 residues; the amino acid sequence is solved; crystals of pure interferon have been prepared to start X-ray diffraction studies; some progress has been made in understanding how interferon is produced and how it works in conferring an antiviral state (see p. 370 in Chapter 9), and a gene for interferon has been isolated and cloned in bacteria.

Interferon has some remarkable properties. It is an extremely active substance having some 10^6–10^9 units of antiviral activity per milligram of protein. This activity level approaches the correlation of *one cell being protected by one molecule of interferon*. In addition, activity is not greatly affected by very acid pH (even as low as pH 2), by heating to 60°C, or by exposure to sodium dodecyl sulfate. Most proteins are severely damaged by such treatments (see the next section).

There is an intense interest in the use of interferons for treating human diseases, including cancer. Although there have been some promising results in clinical trials, more study is needed, because interferon biochemistry is more complicated than once thought. Indeed, interferon is now known to be associated with a bewildering array of other biological effects besides its antiviral role. Moreover, there are at least 14 different interferons produced by at least three different cells—leukocytes, fibroblasts, and thymus-derived lymphocytes (T cells). Thus a unified theory of interferon biochemistry grows more elusive. The book by Friedman cited in the Literature is an excellent source for further introductory information.

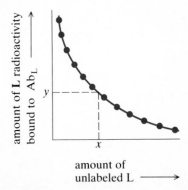

FIGURE 4–48 Idealized standard curve for a radioimmunoassay procedure. The standard curve is prepared with varying amounts of L but always with the same amount of L and Ab_L. For the sample being assayed, y = radioactivity level for the sample containing L, and x = concentration of L in the sample.

4–8   DENATURATION OF PROTEINS

Being a very delicate state, the native conformation of a globular protein is subject to alteration by various chemical and/or physical agents (see Table 4–8) without any change in the amino acid sequence. This loss of native conformation is called *denaturation*. Depending on whether or not the protein is oligomeric, whether or not the loss of conformation is partial or complete, and whether or not biological activity is lost, the term has a variety of meanings (see Figure 4–49). Note that with some proteins the process is reversible. The restoration of both conformation and biological activity provides further evidence that the secondary, tertiary, and quaternary levels of protein structure are predetermined by primary structure.

The biochemist must make every effort to prevent denaturation during isolation of a protein, during storage of the isolated protein, and during an assay with it for activity. Accordingly, one must follow these steps: (1) Most of the handling is done at reduced temperature to avoid thermal denaturation; (2) proteins are stored in a deep freeze at −20° to −80°C; (3) buffers are

TABLE 4–8 Protein-Denaturing Agents

High temperatures
Low pH and high pH
Mercaptoethanol[a]
Guanidine hydrochloride (6*M*)
Urea (6–8*M*)
Vigorous stirring or shaking
Detergents (sodium dodecyl sulfate, SDS)

[a] For proteins containing —SH groups, this agent has a protective effect.

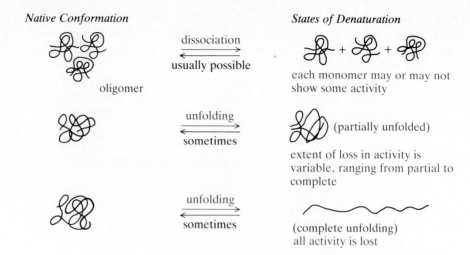

Native Conformation   *States of Denaturation*

dissociation
usually possible

oligomer

each monomer may or may not
show some activity

unfolding
sometimes

(partially unfolded)

extent of loss in activity is
variable, ranging from partial to
complete

unfolding
sometimes

(complete unfolding)
all activity is lost

FIGURE 4–49 Native
conformation and states of
denaturation.

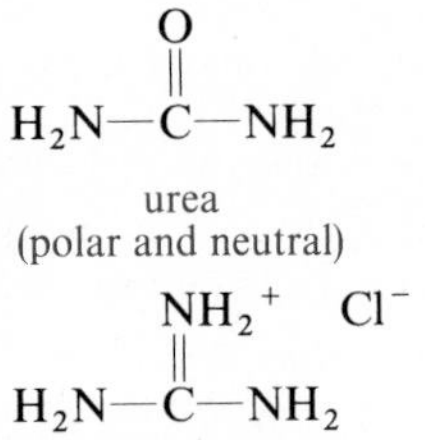

$$O$$
$$\|$$
$$H_2N—C—NH_2$$

urea
(polar and neutral)

$$NH_2^+ \quad Cl^-$$
$$\|$$
$$H_2N—C—NH_2$$

guanidine hydrochloride
(polar and ionic)

(a)

$$HSCH_2CH_2OH$$

mercaptoethanol
(reducing agent)

(b)

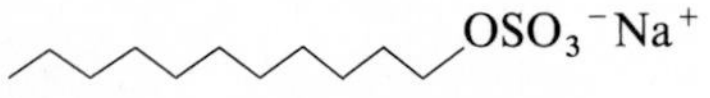

$$OSO_3^-Na^+$$

SDS (sodium dodecyl sulfate)
(nonpolar and ionic)

(c)

FIGURE 4–50 Denaturants
and their effects. (a) Urea and
guanidine hydrochloride. Both
interfere with the native
patterns of ion-ion
associations and hydrogen
bonding. (b) Mercaptoethanol.
This agent reduces
—S—S—covalent bonds.
(c) SDS. This agent disrupts
ion-ion associations and the
nonpolar/nonpolar
associations.

employed to maintain the natural polyionic character of the protein; and
(4) physical trauma such as shaking is avoided or kept to a minimum.

Sometimes, however, the experimental objective requires a deliberate
denaturation of a protein. For instance, you may wish to establish oligomeric
structure, to isolate one of the subunits, to unravel a polypeptide chain exposing
peptide bonds for maximum fragmentation by trypsin, or to do a number of
other things. Deliberate denaturation involves *disrupting* one or more types of
stabilizing forces. Extremes of pH disrupt ion-pair interactions, such as
—COO$^-$........$^+$H$_3$N— by converting, at low pH, —COO$^-$ to —COOH and
by converting, at high pH, —NH$_3^+$ to —NH$_2$. The effects of high temperature
and vigorous shaking are obvious. The effects of other denaturants are given
in Figure 4–50.

4–9 ISOLATION OF PURE PROTEINS

To study a protein, one must *isolate it in pure form*. This technique
is easier said than done. Several different steps are generally required, each
designed to eliminate a greater portion of unwanted material from the sample
of the previous step. Since there is no universal isolation procedure for every
protein, the choice of what to do in each step is based on trial and error. A good
procedure recovers the maximum amount of desired protein in a pure, nondenatured state with as few steps as possible, usually 4–6 steps. Fractions
obtained in each step must be assayed for content of both protein and activity.
Dividing the total activity of a fraction by its protein content yields the *specific
activity* of that fraction (see Note 4–5). During purification the specific activity
increases, representing the continual enrichment of the total protein recovered
with respect to the protein being isolated.

A typical isolation scheme is illustrated by the data given in Table 4–9
and the photographs in Figure 4–51. Note the use of *percentage yield* and

TABLE 4–9 Isolation of galactose-binding protein from *Escherichia coli*

STEPS OF ISOLATION	TOTAL PROTEIN[a] (mg)	TOTAL BINDING ACTIVITY[b] (units)	SPECIFIC ACTIVITY (units/mg)	PERCENTAGE YIELD[c]	PURIFICATION[d]
1. Cell-free extract (osmotic shock)	2600	940	0.39	100	1
2. Protamine precipitation	2352	798	0.34	85	1
3. Ammonium sulfate precipitation	1664	728	0.43	75	1
4. DEAE-cellulose chromatography	432	488	1.15	52	3
5. Hydroxyapatite chromatography	46	322	7.0	34	18
6. DEAE-Sephadex chromatography	18	208	12.0	22	31

Source: Reproduced with permission from H. Anraku, "Transport of Sugars and Amino Acids in Bacteria," *J. Biol. Chem.,* **243**, 3316–3122; 3123–3127 (1968).
[a] Protein is measured by the Lowry method (see Appendix I).
[b] Activity of the protein was assayed by placing a protein fraction in a semipermeable tubing and dialyzing it for several hours against a solution containing a known amount of ^{14}C-galactose. The amount of radioactive galactose bound to the protein was calculated by subtracting the radioactivity of the dialyzate from that in the tubing. One unit of binding activity represents the binding of 1×10^{-9} mole of galactose.
[c] (Total activity of any step/total activity of initial extract) × 100.
[d] (Specific activity of any step/specific activity of initial extract).

purification to summarize the progress of the isolation (see footnotes *c* and *d* in Table 4–9). The protein in the illustration participates in the movement of simple sugars such as galactose and glucose across the bacterial cell membrane. The functions of the protein are to bind with the sugar and then, in conjunction with other membrane components, to transport the sugar to the inside of the cell.

The first step involves the rupture of the intact cells to yield a cell-free extract. The treatment with protamine did not eliminate much protein (2600 mg to 2352 mg). The reason for using protamine—a protein with a high arginine

NOTE 4–5
The term *specific activity* is used extensively in biochemistry. It expresses the amount of activity of a specific substance per unit weight of a sample containing the substance.

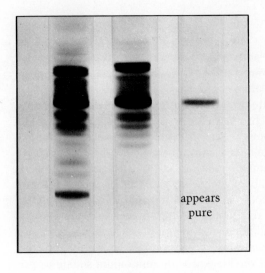

appears pure

(a)

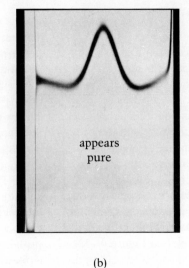

appears pure

(b)

FIGURE 4–51 Progress of the protein isolation summarized in Table 4–9. (a) Polyacrylamide gel rods stained after electrophoresis to detect proteins. Lane A is the original cell-free extract; B shows what remains after ammonium sulfate precipitation; C shows what remains after DEAE-Sephadex chromatography. (b) Photograph of the sedimentation of protein from DEAE-Sephadex (lane C). The presence of a single symmetrical peak indicates purity. (*Source:* Photographs provided by Y. Anraku.)

content, giving it many positively charged R groups—is to eliminate nucleic acids (RNA and DNA). Nucleic acids have many negatively charged groups and thus associate with protamine, yielding an insoluble complex. The third step consists of *ammonium sulfate fractionation,* a tactic used extensively in protein isolation. Addition of ammonium sulfate to protein solutions decreases the solubility of proteins. The effect, generally called the *salting-out* phenomenon, is dependent on the concentration of ammonium sulfate added and the nature of the protein. The last three steps are column-chromatographic methods based on anion exchange (DEAE-cellulose), adsorption (hydroxyapatite), and anion exchange plus molecular sieve (DEAE-Sephadex).

The progress of the protein isolation is readily apparent from the stained polyacrylamide gels in Figure 4–51. Note the gradual elimination of protein until one band is observed in the last fraction, suggesting that the sample is pure. A standard practice is to perform two or more checks of purity. In this case analytical ultracentrifugation was used.

Some other traditional tactics used in protein isolation include precipitation with organic solvents such as acetone, mild heating to change protein solubilities, and isoelectric precipitation, which is based on the fact that a protein has minimum solubility at a pH corresponding to its isoelectric point (pI). *Affinity chromatography* and *isoelectric focusing* (both described in the following Laboratory Methods section) are newer and particularly efficient procedures.

Reading note *b* in Table 4–9, you see that the activity assay of the galactose-binding protein used ^{14}C-labeled galactose. This assay illustrates an important point made previously (Section 3–4): Radioactively labeled substances are indispensable in much of biochemical research.

◇ **L A B O R A T O R Y   M E T H O D S** ◇

AFFINITY CHROMATOGRAPHY AND OTHER TECHNIQUES

In 1968 C. B. Anfinsen, P. Cuatrecasas, and co-workers developed a new technique for protein isolation that takes advantage of a characteristic virtually unique to the protein being isolated—the *specific binding* that often occurs between a protein and another substance L (for ligand). Usually, L is the substance that the protein naturally binds with in vivo, an analogous material synthesized in the laboratory, or an antibody prepared against L as an antigen. This technique, *affinity chromatography,* is a column-chromatographic method using a column filled with a solid polymeric stationary phase to which the substance L has been previously attached by firm covalent bonding (see Figure 4–52). As a mixture of proteins moves through the column,

the only protein that will have its movement retarded is the one that binds to L. All other proteins lacking this binding ability move right on through the column.

A state-of-the-art example of protein isolation via affinity chromatography is illustrated in Table 4–10. The protein is leukocyte interferon, one of the human interferons. However, the protein was not isolated from human leukocyte cells but from *E. coli* bacteria that had been genetically engineered through recombinant DNA techniques to produce the protein by splicing the human interferon gene into the DNA of the bacteria. One kilogram of bacteria was lysed; the cell-free extract, having perhaps 1000–1500 different proteins, was treated with ammonium sulfate to precipitate some unwanted protein. The supernatant from

protein has minimum solubility at pH at pI.

$$P + L \rightleftharpoons PL$$

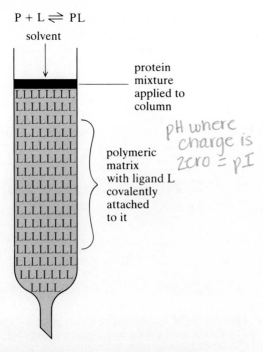

FIGURE 4–52 Exploitation of binding specificity in affinity chromatography. As the protein mixture enters the column, the P specific for binding to L will be adsorbed, and other proteins will move through and exit. The polymeric matrix is polyacrylamide, agarose, dextran, or glass (silica).

the salt fractionation step was applied to an affinity column containing bound antibody, specifically, a monoclonal antibody for interferon. The interferon was washed off the column in 95% yield with a purification factor of 1150. Further studies established that this interferon preparation contained only very minor protein contaminants. Finally, the bacteria-produced interferon proved to be structurally and functionally comparable to the natural human interferon.

Isoelectric Focusing

An electrophoresis technique, *isoelectric focusing* (IF), *exploits differences in the* pI *values* of proteins. (Recall that pI is the isoelectric point, the pH at which the net molecular charge is zero.) The method is capable of exploiting very small differences in pI—as little as 0.01–0.0025 pH unit. A polyacrylamide or agarose gel, cast in tube or slab form, is used as a supporting medium.

Initially, the gel is cast in a solution containing a mixture of low–molecular weight substances called *ampholytes.* Under applied voltage the different ampholyte materials migrate in such a manner as to set up a linear *gradient of* pH along the length of the gel medium. (We will not examine the basis for forming the gradient.) Then the protein sample is applied on top of the gel, and the voltage is reapplied.

As proteins migrate in the gel, each encounters a medium of changing pH, and a titration takes place changing the net charge of each protein—changing continuously as the protein moves to a different position in the gel having a different pH. This titration occurs until each protein reaches a location of pH at which the protein has been titrated to its isoelectric state. At this point, because it now has no net positive or negative charge ($Q = 0$), the protein stops moving. The result is a concentration of each protein in the original mixture into an extremely sharp band—much sharper than the band obtained with conventional electrophoresis. Compare the much finer stained bands of protein in the gel pattern in Figure 4–53 with that shown in Figure 4–51(a).

TABLE 4–10 Purification of recombinant human interferon by affinity chromatography using a monoclonal interferon-antibody

STEP	VOLUME (mL)	TOTAL PROTEIN[a] (mg)	TOTAL ACTIVITY[b] (units)	SPECIFIC ACTIVITY (units/mg)	PURIFICATION	YIELD (%)
1. Ammonium sulfate fraction	700	37,100	7.4×10^9	2.0×10^5	1.0	100
2. Interferon-containing fractions from an affinity column	27	30	7.0×10^9	2.3×10^8	1150	95

Source: T. Staehlin, D. S. Hobbs, H. Kung, C. Y. Lai, and S. Pestka, *Proc. Natl. Acad. Sci., USA,* **256,** 9750–9754 (1981).
[a] Protein measured by the Lowry method (see Appendix I).
[b] Activity assay based on the inhibition of growth of a virus on cultured bovine kidney cells.

FIGURE 4–53 Protein bands from isoelectric focusing.

FIGURE 4–54 Separation of *E. coli* proteins. The *E. coli* cells were grown in a medium containing ^{14}C-labeled amino acids to incorporate a radioactive label into cellular proteins. The cells were lysed by sonication, and a 25-μL portion of the lysate (containing 10 mg of protein) was examined by two-dimensional electrophoresis. Afterward, the gel slab was exposed to a piece of X-ray film to detect the location of ^{14}C-labeled proteins. Developing the film reveals (dark spots) where the film was exposed to the radiation of ^{14}C. The film image is called a *radioautograph* (or autoradiography). (*Source:* Photograph provided by Patrick H. O'Farrell.)

Two-Dimensional Electrophoresis

Apply a protein sample to a polyacrylamide gel rod and perform isoelectric focusing. After this step, lay the gel along the edge of an SDS-polyacrylamide gel slab, and carry out electrophoresis again. Under applied voltage proteins will move from the isoelectric-focusing gel to the SDS-polyacrylamide gel. This combination of two separate electrophoretic techniques, called *two-dimensional electrophoresis,* gives outstanding resolution of complex protein mixtures (see Figure 4–54). Each spot represents a different protein—about 1100 total. (Anyone want to confirm this?) P. O'Farrell, who developed the technique, estimates that it should be possible to resolve as many as 5000 proteins.

Such patterns represent a type of molecular (genetic) fingerprint characteristic of the source from which the protein mixture was obtained—anatomy at the molecular level. Research is in progress to catalog these fingerprints from various human tissues (normal and diseased) with the hope that such data will have applications in diagnosing a specific disease condition or a stage of a disease.

LITERATURE

ANFINSEN, C. B. "Principles That Govern the Folding of Protein Chains." *Science,* **181,** 223–230 (1973). A review article on the cooperative interactions of amino acid side chains resulting in the formation of the native conformation of proteins.

BORNSTEIN, P. "The Biosynthesis of Collagen." *Annu. Rev. Biochem.,* **43,** 567–604 (1974). A review article.

CERAMI, A., and C. M. PETERSON. "Cyanate and Sickle-Cell Disease." *Sci. Am.,* **232,** 45–50 (1975). A description of sickle-cell anemia and the abnormal sickle-cell hemoglobin.

CLARK, J. M., and R. L. SWITZER. *Experimental Biochemistry.* 2nd ed. San Francisco: Freeman, 1977. A laboratory manual containing coverage on the theory and application of the major laboratory methods in biochemistry.

CUATRECASAS, P., M. WILCHEK, and C. B. ANFINSEN. "Selective Enzyme Purification by Affinity Chromatography." *Proc. Natl. Acad. Sci., U.S.A.,* **61,** 636–643 (1968). Original research paper describing the development of this method.

DAYHOFF, M. O., ed. *Atlas of Protein Sequence and Structure.* Vol. 5. Washington, D.C.: National Biomedical Research Foundation, 1978. A collection of known amino acid sequences of peptides and proteins with a comparison of sequences of the same molecule isolated from different sources. Some coverage of nucleic acid sequences.

DICKERSON, R. E. "The Structure and History of an Ancient Protein." *Sci. Am.,* **226,** 58–72 (1972). Discussion of the structure of cytochrome *c* and its evolution as a molecule over 1.2 billion years ago. The amino acid sequence from 38 different species is examined. Interesting discussion on some specifics regarding the relationship of function to amino acid sequence and three-dimensional structure.

FREIFELDER. D. *Physical Biochemistry.* 2nd ed. San Francisco: Freeman, 1982. Introductory-level treatment of numerous laboratory techniques for the separation, isolation, and characterization of macromolecules. Available in paperback.

FRIEDMAN, R. M. *Interferons: A Primer.* New York: Academic Press, 1981. An excellent introductory source. 150 pages.

GOLUB, E. S. *The Cellular Basis of the Immune Response.* 2nd ed. Sunderland, Mass.: Sinauer Associates, 1981. A superb introduction. Available in paperback.

HIRS, C. H. W., and S. N. TIMASHEFF, eds. *Enzyme Structure.* Vols. 11, 25, 27, 47, 48, and 49 of *Methods in Enzymology.* New York: Academic Press, 1967, 1972, 1973, 1978, 1978, and 1978. Part of a multivolume work devoted to practical aspects of biochemical studies, particularly those dealing with the isolation and assay of enzymes. Volumes 11 and 47 contain much information on the techniques available for the study of the primary level of protein structure, such as determination of amino acid composition, end group analysis, separation of polypeptide subunits, cleavage of disulfide bonds, separation of peptides, and sequence determination.

HUNKAPILLER, M. W., J. E. STRICKLER, and K. J. WILSON. "Contemporary Methodology for Protein Structure Determination." *Science,* **226,** 304–311 (1984). A review article describing laboratory techniques for isolation, amino acid analysis, and amino acid sequencing of picomole quantities of proteins and peptides.

KENDREW, J. C. "Myoglobin and the Structure of Proteins." *Science,* **139,** 1259–1266 (1963). A paper adapted from the author's address on accepting the Nobel Prize in chemistry in 1962. Emphasis is given to the use of X-ray crystallography in deciphering protein structures. The treatment is nonmathematical and suitable for beginning students.

MARX, J. L. "Monoclonal Antibodies in Cancer." *Science,* **216,** 283–285 (1982). A research news article that surveys current and future applications.

MOSBACH, K. "Enzymes Bound to Artificial Matrixes." *Sci. Am.,* **224,** 26–33 (1971). Discussion of the binding of proteins to inert polymers and the significance to affinity chromatography and to uses in industry and medicine.

NEURATH, H., and R. L. HILL, eds. *The Proteins.* 3rd ed. New York: Academic Press, 1975. A planned eight-volume reference work dealing with the isolation, composition, structure, and function of proteins.

PAULING, L., R. B. COREY, and H. R. BRANSON. "The Structure of Proteins: Two Hydrogen-Bonded Helical Configurations of the Polypeptide Chain." *Proc. Natl. Acad. Sci., U.S.A.,* **37,** 205–211 (1951). The original paper describing the nature of the alpha helix.

PERUTZ, M. F. "The Hemoglobin Molecule." *Sci. Am.,* **211,** 64–76 (1964). A description of the three-dimensional structure of the hemoglobin molecule by the primary investigator.

———. "Hemoglobin Structure and Respiratory Transport." *Sci. Am.,* **239,** 92–125 (1978). Superb article on the cooperative effect of oxygen binding. Details of the molecular changes are described.

RICHARDSON, J. "The Anatomy and Taxonomy of Protein Structure." *Adv. Protein Chem.,* **34,** 167–339 (1981). Proteins are classified according to distinctions and similarities in secondary and tertiary levels of structure.

TANFORD, C. "The Hydrophobic Effect and the Organization of Living Matter." *Science,* **200,** 1012–1018 (1978). A review article on the importance of hydrophobic interactions in the assembly and organization of membranes and proteins.

WALSH, K. A., et al. "Advances in Protein Sequencing." *Annu. Rev. Biochem.,* **50,** 261–284 (1981). A thorough review article of experimental methods.

YELTON, D. E., and M. D. SCHARFF. "Monoclonal Antibodies." *Am. Sci.,* **68,** 510–516 (1980). A short review article.

ZUCKERHANDL, E. "The Evolution of Hemoglobin." *Sci. Am.,* **212,** 110–118 (1965). An article comparing the amino acid sequences of the alpha and beta chains of hemoglobin molecules from different species and showing how this comparison provides a basis for establishing evolutionary relationships among organisms on a chemical level, in terms of the evolution of a molecule common to these organisms.

EXERCISES

4–1. Verify that about 45% of the amino acid residues in cytochrome *c* (human) are nonpolar.

4–2. A protein is found to have a molecular weight of 85,400. Using both expressions for the dimensions of MW (see p. 118), calculate the mass of a single molecule of this protein.

4–3. Which treatment (trypsin, chymotrypsin, Fm protease, Sa protease, or CNBr) with yield a mixture of an octapeptide, a nonapeptide, and a tridecapeptide from the large polypeptide chain of insulin? Which treatment will yield no peptide fragments?

4–4. Four pure proteins were used as standards to construct a standard curve for a molecular weight analysis via SDS-gel electrophoresis. Protein 1, with a molecular weight of 15,000, was the smallest protein. Protein 2 (MW = 35,000) moved only 39% as far as protein 1. Protein 3 (MW = 25,000) moved only 63% as far as protein 1. Protein 4 (MW = 20,000) moved only 81% as far as protein 1. Construct the standard curve, and then determine the molecular weight of an unknown protein that had a mobility (under the same conditions) midway between that of proteins 2 and 3.

4–5. Which is the largest substance in Figure 4–3(a): W, X, Y, or Z?

4–6. If a solution of the same unknown protein identified in Exercise 4–4 were examined by gel-permeation column chromatography under conditions applicable to the standard curve shown in Figure 4–3(c), what would be the elution volume for the unknown protein?

4–7. The mobility of a pure protein at pH 8.2 in regular PAGE is represented in Figure 4–55(a). After treatment with sodium dodecyl sulfate and subsequent analysis of the treated protein via SDS–PAGE, the result in Figure 4–55(b) was obtained. What conclusion can you make regarding the structure of the original pure protein?

4–8. Is the pI value of the pure protein described in Exercise 4–7 less than or greater than pH 8.2? Explain your answer.

4–9. Separate samples of a protein-containing solution are analyzed by SDS–PAGE and cation-exchange column chromatography, giving the results shown in Figure 4–56. What conclusions can you make concerning the composition of the original protein solution?

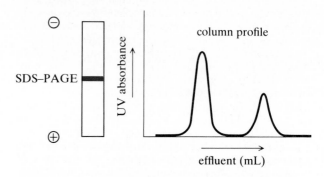

FIGURE 4–56 Results for Exercise 4–9.

4–10. A mixture of five peptides (P1, P2, P3, P4, P5) is subjected to paper electrophoresis at pH 8.5. After electrophoresis, staining of the paper revealed the pattern of migration shown in Figure 4–57. Given the following pI values for each peptide—9.0 for P1, 5.5 for P2, 10.2 for P3, 8.2 for P4, and 7.2 for P5—identify which zone corresponds to each peptide. Assume that each peptide has a molecular weight close to 1200. If another peptide P6 (with pI = 10.2 and a molecular weight about 600) is added to the original mixture, where will you predict it to move?

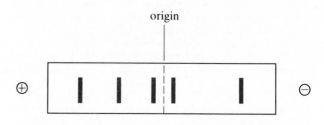

FIGURE 4–57 Staining pattern for Exercise 4–10.

4–11. Consider the following facts: On a gel-permeation column protein P migrates as a single zone corresponding to a molecular weight of 66,000. After treatment with Cleland's reagent, SDS–PAGE analysis of the treated protein shows two bands, and the slower-moving band corresponds to a molecular weight of 38,000. What can you conclude about the structure of this protein?

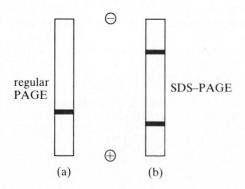

FIGURE 4–55 Results for Exercise 4–7. (a) Regular PAGE. (b) SDS–PAGE.

4–12. Consider the following facts: (1) A 15-mg sample of a nondenatured protein P (MW = 15,000) quantitatively reacts with 3 μmole of Ellman's reagent; (2) after complete denaturation of the same amount of protein P and then reaction of the denatured protein with an excess amount of Cleland's reagent, the treated protein quantitatively reacts with 6 μmole of Ellman's reagent. What can you conclude about the original structure of this protein?

4–13. A pentapeptide obtained from treatment of a protein with trypsin was shown to contain arginine, aspartic acid, leucine, serine, and tyrosine. For determination of the amino acid sequence, the peptide was cycled through the Edman degradation procedure three times. The composition of the peptide remaining after each cycle was as follows:

After cycle 1: arginine, aspartic acid, leucine, serine.

After cycle 2: arginine, aspartic acid, serine.

After cycle 3: arginine, serine.

What is the sequence of the pentapeptide?

4–14. A pure protein is treated with performic acid. You have proved that a reaction has occurred by showing that the untreated protein is eluted faster than the treated protein from the same molecular sieve column. However, polyacrylamide electrophoresis of the product mixture after performic acid treatment reveals only one very distinct and sharp zone. What can you conclude from this information regarding the structure of the protein?

4–15. Which hemoglobin variant in Table 4–11 will you predict may be less functional than normal adult hemoglobin HbA? Explain your answer.

TABLE 4–11 Data for Exercise 4–15

HEMOGLOBIN VARIANT	DESCRIPTION OF DIFFERENCE RELATIVE TO HbA
1	Ala rather than val in position 67 of β chain
2	Asp rather than his in position 50 of α chain

4–16. If a mixture containing proteins A, B, and C is analyzed by gel-permeation column chromatography, which of the elution profiles in Figure 4–58 best represents the differential movement of A, B, and C through the column? (Given: The molecular weights of A, B, and C are 150,000, 75,000, and 65,000, respectively. The swollen gel granules have an exclusion limit of approximately 100,000.)

4–17. Separate treatments of polypeptide P with trypsin and chymotrypsin gave the peptide fragments listed in Table 4–12. Amino acid residues are identified by the use of one-letter symbols (see Table 3–1). Dansylation of P followed by acid hydrolysis yielded dansyl-serine. Aspartic acid was the only free amino acid detected after treatment of P with hydrazine. (a) Display the sequence of polypeptide P. (b) To what extent will Ellman's reagent react with this polypeptide?

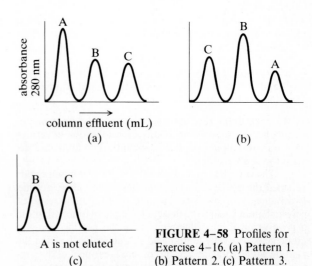

FIGURE 4–58 Profiles for Exercise 4–16. (a) Pattern 1. (b) Pattern 2. (c) Pattern 3.

TABLE 4–12 Peptide fragments for Exercise 4–17

TRYPSIN	CHYMOTRYPSIN
IR	KVEGDTKPELELTLKYF
FSR	NKAAVTMPSSKLKVAF
PGLR	DMLTRIRNGQAW
YFQGK	ANVLKEEGFIEDF
IYKLQD	QGKVVAEISQRF
VEGDTK	SRPGLRIY
VAFANVLK	SMQDPIF
PELELTLK	KLQD
VVAEISQR	
AAVTMPSSK	
EEGFIEDFK	
SMQDPIFDMLTR	
NGQAWNKAAVTMPSSKLK	

4–18. Explain why the carbonyl oxygen and the imino hydrogen atoms contributed by the same peptide bond do not enter into hydrogen bond formation with each other.

4–19. Table 4–13 summarizes the sedimentation of a protein ($D = 6.4 \times 10^{-7}$ cm^2/s, $\bar{V} = 0.74$ mL/g) in an analytical ultracentrifuge operating at 58,000 rpm and 20°C. The density of the solvent was 0.998 g/mL. Calculate the molecular weight of the protein.

TABLE 4–13 Sedimentation data for Exercise 4–19

t (s)	x (cm)
500	6.22
1000	6.34
1500	6.47
2000	6.61
2500	6.76

4–20. Which of the following statements is most correct in regard to a fibrous polypeptide chain? (a) All of the ϕ angles of rotation are identical. (b) All of the ψ angles of rotation are different. (c) All of the ϕ and ψ angles of rotation are different. (d) All of the ϕ and ψ angles of rotation are identical, and $\phi = \psi$. (e) Only some of the ϕ and ψ angles of rotation are different.

4–21. Prior to the development of sequential Edman degradation, peptide sequencing required the collection of various data on overlapping peptide fragments. The following description is an example.

The complete hydrolysis of an unknown nonapeptide revealed the presence of glutamic acid, two valine, glycine, two lysine, tyrosine, threonine, and phenylalanine residues. The first amino acid to be detected as a phenylthiohydantoin derivative on Edman degradation of the peptide was glutamic acid. The only amino acid detected after treating the peptide with hydrazine was threonine. Treatment of the peptide with trypsin and chymotrypsin gave three fragments in each case: T1, T2, T3 and C1, C2, C3, respectively. None of the trypsin fragments were identical to the chymotrypsin fragments; C2 and T2 proved to be dipeptides; C1 and T1 were tripeptides; and C3 and T3 were tetrapeptides. Hydrolysis of C3 followed by paper chromatography revealed only three ninhydrin-sensitive spots. The N terminal residue of T3 was shown to be phenylalanine, and the C terminus was threonine. The N terminus of C1 was glycine, and the C terminus was the same as in T3. Fragment C2 was shown to contain tyrosine and glutamic acid. Fragment T1 was composed of lysine, tyrosine, and glutamic acid. The N terminus of T2 was valine, and the N terminus of C3 was lysine. At basic pH the C3 fragment migrated with a net charge of $+2$.

Use all of this information to construct as much of the sequence for the original nonapeptide as the data permit. (Assume that the specificity of trypsin and chymotrypsin is limited to the text description.)

4–22. A research biochemist wanted to determine the best way to store a protein, having already designed an efficient procedure for its isolation. She took a portion of a solution containing 50 mg of the pure protein obtained from the last step and placed it in a refrigerator for two weeks at $4°C$. When analyzed after two weeks, this protein solution exhibited 360 units of activity. An equal portion of the same protein solution was lyophilized (that is, freeze-dried), and the powdered protein was stored at $-20°C$. When the powder was analyzed two weeks later (after redissolving), a total of 310 units of activity were measured. A third portion of the solution containing 60 mg of the pure protein was placed directly in a freezer operating at $-20°C$. When it was analyzed two weeks later (after thawing), a total of 390 units of activity were measured. What storage condition did the biochemist probably decide to use (at least after a two-week storage period)? Why do you suppose the question is qualified with the phrase in parentheses?

4–23. An enzyme was purified to constant specific activity; various other tests also supported the isolation of a pure protein. Polyacrylamide gel electrophoresis of a sample from

TABLE 4–14 Binding data for Exercise 4–25

INITIAL CONCENTRATION OF LIGAND (mM)	CONCENTRATION OF FREE LIGAND AFTER EQUILIBRATION WITH PROTEIN (mM)
0.570	0.466
0.970	0.800
1.51	1.28
2.27	1.99
3.43	3.09
5.88	4.98
8.90	8.45

the last step of the isolation procedure gave a stained gel slab with one major zone and two minor zones. Furthermore, the protein associated with each of the three zones catalyzed the same reaction. What can you conclude?

4–24. Before attempting to answer this question, study footnotes c and d in Table 4–9. Now verify the 1150-fold purification in 95% yield of human interferon via affinity chromatography as summarized in Table 4–10.

4–25. The results of a binding analysis of protein P for ligand L are summarized in Table 4–14. In each assay the protein concentration was 0.57 mM. Evaluate the values of n and K, and determine whether the protein exhibits cooperativity. If there is cooperativity, establish the type—that is, positive or negative—by using a Hill plot to evaluate the Hill coefficient c for a comparison to the value of n.

4–26. The results of a binding analysis of protein P for ligand L are summarized in Table 4–15. In each assay the protein concentration was 0.40 mM. Evaluate the values of n and K and whether the protein exhibits cooperativity. If there is cooperativity, establish the type—that is, positive or negative—by using a Hill plot to evaluate the Hill coefficient c for a comparison to the value of n.

4–27. Define the following terms: prosthetic group, protease, trans peptide bond, oligomeric protein, denaturation, native conformation, salting-out, affinity chromatography, S value, heme, exopeptidase, essential residues.

TABLE 4–15 Binding data for Exercise 4–26

INITIAL CONCENTRATION OF LIGAND (mM)	CONCENTRATION OF BOUND LIGAND AFTER EQUILIBRATION WITH PROTEIN (mM)
0.750	0.040
1.00	0.120
1.25	0.240
1.67	0.360
2.31	0.480
3.75	0.600
6.08	0.680
10.00	0.720

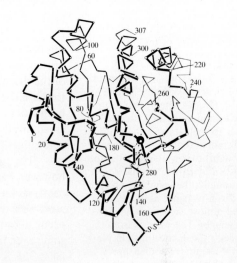

CHAPTER FIVE

ENZYMES

After nearly nine years of effort, in 1926 J. B. Sumner reported having isolated, from jack beans, an enzyme named *urease*, which he claimed was a protein substance. His findings were initially labeled by some as preposterous. Continued research by Sumner and independently by J. Northrop and W. M. Stanley resulted in the isolation of a few more enzymes, and finally in 1935 the protein nature of enzymes was universally accepted (see Note 5–1). The impact of the discovery is reflected by the selection of Sumner, Northrop, and Stanley to share the Nobel Prize in 1946. In the past 50 years about 2500 enzymes have been isolated from all types of organisms. Several of these enzymes have been crystallized, and about 50 have had their complete three-dimensional structure determined by X-ray diffraction.

The *function* of an enzyme is to *increase the rate* of a reaction. Enzymes, however, have three unequalled characteristics. First, they are the *most efficient catalysts* known, with very small (micromolar) quantities of an enzyme able to accelerate a reaction to an extremely fast rate. In fact, most cellular reactions occur about a million times faster than they would in the absence of enzymes, some even faster. Second, the majority of enzymes are distinguished by a *specificity of action* in that virtually every conversion of a reactant (called a *substrate*) to a product is catalyzed by a preferred enzyme. In fact, several enzymes exhibit absolute specificity, meaning that they act on only one substrate to yield only one product. The third and perhaps most remarkable characteristic is that the actions of many enzymes are *regulated;* that is, they are capable of changing back and forth from a state of low activity to one of high activity. The regulated actions of enzymes comprise an elaborate system by which organisms can control all of their activities. Gradually, you will appreciate the following feature:

> The individuality of a living cell is due in large part to the unique set of enzymes that it is genetically programmed to produce.

If even one is missing or defective, the results can be disastrous.

5–1   ENZYME NOMENCLATURE

Some enzymes have seemingly nondescript names such as trypsin, pepsin, renin, and lysozyme. For the most part, however, enzymes are named with an *-ase* ending, on the basis of the type of reaction they catalyze and the identity of the substrates involved. For example, the enzyme catalyzing the decarboxylation of histidine is named *histidine decarboxylase:*

Another catalyzing the removal of two hydrogen atoms from ethyl alcohol to yield acetaldehyde is named *alcohol dehydrogenase:*

$$\underset{\substack{\text{ethyl}\\\text{alcohol}}}{CH_3\overset{\displaystyle H}{\underset{\displaystyle H}{C}}-OH} \xrightarrow[\quad 2H\quad]{\substack{\text{alcohol}\\\text{dehydrogenase}}} \underset{\text{acetaldehyde}}{CH_3\overset{\displaystyle H}{C}=O}$$

And the hydrolysis of urea to ammonia and carbon dioxide involves the enzyme *urease:*

$$\underset{\text{urea}}{H_2N-\overset{\displaystyle O}{\overset{\displaystyle \|}{C}}-NH_2} \xrightarrow[\quad H_2O\quad]{\text{urease}} 2NH_3 + CO_2$$

As more enzymes of various functions were discovered, other names were formed: oxidases, oxygenases, kinases, thiokinases, mutases, transaldolases, transketolases, phosphorylases, phosphatases, polymerases, topoisomerases, and several others. Do not be alarmed. Gradually, you will become familiar with most types.

A more systematic scheme of nomenclature was first suggested in 1965 and revised in 1972 (see Note 5–2). This system categorizes all enzymes into *six main classes* (divisions) on the basis of the general type of chemical transformation they catalyze (see Table 5–1). The system is designed to zero in on the specific identity of each enzyme by dividing each main class into subclasses and sub-subclasses. The use of a numbering system throughout the scheme allows each enzyme to be assigned a numerical code, such as 2.1.3.4, where the first number specifies the main class, the second and third numbers correspond

NOTE 5–2
The detailed recommendations of the Nomenclature Committee of the International Union of Biochemistry are documented in *Enzyme Nomenclature* (Academic Press, 1984). This source comprises a catalog of all known enzymes. A current computer listing is regularly maintained as new enzymes are discovered.

TABLE 5–1 Main enzyme classes according to the International Enzyme Commission

MAIN CLASS	TYPE OF REACTION CATALYZED
1. Oxidoreductases	Oxidation-reduction reactions of all types
2. Transferases	Transfer of an intact group of atoms from a donor to an acceptor molecule
3. Hydrolases	Hydrolytic (H_2O participates) cleavage of bonds
4. Lyases	Cleavage of C—C, C—O, C—N, and other bonds by means other than hydrolysis or oxidation; includes reactions that eliminate water to leave double bonds or add water to a double bond
5. Isomerases	Interconversion of various isomers, such as cis $\rightleftharpoons$ trans, L $\rightleftharpoons$ D, aldehyde $\rightleftharpoons$ ketone
6. Ligases	Bond formation due to the condensation of two different substances, with energy provided by ATP

NOTE 5–3
A partial breakdown of class 4
is given here to illustrate
the indexing of histidine
decarboxylase.

4. Lyases
    4.1. Carbon-carbon lyases
        (cleavage of C—C bond)
        4.1.1. Carboxy-lyases
            (cleavage of
            C—COO⁻ bond)
            4.1.1.22. histidine carboxy-
                  lyase
                  (cleavage of
                  C—COO⁻ bond in
                  histidine)
        4.1.2. Aldehyde-lyases
    4.2. Carbon-oxygen lyases
        (cleavage of C—O bond)
    4.3. Carbon-nitrogen lyases
        (cleavage of C—N bonds)
    4.4. Carbon-sulfur lyases
        (cleavage of C—S bonds)
    4.5. Carbon-halogen lyases
    4.6. Phosphorus-oxygen lyases

to specific subclasses and sub-subclasses, and the final number represents the serial listing of the enzyme in its sub-subclass (see Note 5–3). For example, histidine decarboxylase (the traditional name) is identified as histidine carboxy-lyase, 4.1.1.22; alcohol dehydrogenase as alcohol:NAD oxidoreductase, 1.1.1.1; urease as urea amidohydrolase, 3.5.1.5.

The use of the systematic classification/nomenclature is required in most of the professional research journals. However, the older system is still widely used, mostly in monographs and textbooks, including this one.

5–2 COFACTOR- (COENZYME-) DEPENDENT ENZYMES

All enzymes are globular proteins, with each enzyme having a specific function because of its specific protein structure. However, the optimum activity of many (but not all) enzymes depends on the *cooperation of nonprotein substances* called *cofactors*. The molecular partnership of protein and cofactor is termed a *holoenzyme* (see Figure 5–1) and exhibits maximal catalytic activity. The protein component, stripped of its cofactor, is termed an *apoenzyme;* it exhibits very low activity—frequently none at all.

There are two categories: the *inorganic cofactors,* which include several inorganic ions such as Zn^{2+}, Mg^{2+}, Mn^{2+}, Fe^{2+}, Cu^{2+}, K^+, and Na^+, and the *organic cofactors,* which consist of about a dozen substances of diverse structure. The organic cofactors are usually called *coenzymes*. The cofactor participation of inorganic ions represents (in part) the reason these materials are essential nutrients for every organism.

> Coenzymes (the organic cofactors) have a special significance in mammalian nutrition because most are produced from some of the water-soluble vitamins or are one and the same with a vitamin.

For example, the vitamin *riboflavin* (B_2) is ingested and converted to either of two cofactors, flavin adenine dinucleotide (FAD) or flavin mononucleotide (FMN).

Table 5–2 gives the names and vitamin relationships of several major coenzymes and a brief statement of their function. Note that this description is

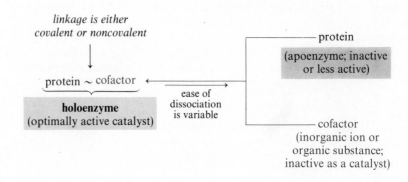

FIGURE 5–1 Cofactors. Some enzymes require two or three different cofactors, and usually one of them is an inorganic ion.

TABLE 5–2 Coenzymes: Name, function, and vitamin relationship

COENZYME	TYPE OF REACTION	GROUP TRANSFERRED	VITAMIN PRECURSOR	STRUCTURE DISPLAYED
Nicotinamide adenine dinucleotide (NAD$^+$)	Oxidation-reduction	H (electrons)	Niacin	Figure 12–18
Nicotinamide adenine dinucleotide phosphate (NADP$^+$)	Oxidation-reduction	H (electrons)	Niacin	Figure 12–18
Flavin adenine dinucleotide (FAD); flavin mononucleotide (FMN)	Oxidation-reduction	H (electrons)	Riboflavin (B$_2$)	Figure 12–22
Cytochrome heme groups (iron-containing)	Oxidation-reduction	Electrons		Figure 15–10
Coenzyme A	Activation and transfer of acyl groups	$R-\overset{\overset{\displaystyle O}{\|}}{C}-$	Pantothenic acid	Figure 12–13
Lipoic acid	Acyl group transfer	$R-\overset{\overset{\displaystyle O}{\|}}{C}-$		Figure 14–15
Thiamine pyrophosphate	Acyl group transfer	$R-\overset{\overset{\displaystyle O}{\|}}{C}-$	Thiamine (B$_1$)	Figure 13–48
Biotin	CO$_2$ fixation	CO$_2$	Biotin (H)	Figure 17–13
Pyridoxal phosphate	Transamination of amino acids and other reactions	$-NH_2$	Pyridoxal (B$_6$)	Figure 18–6
Tetrahydrofolic acid	Metabolism of one-carbon fragments	$-CH_3$; $-CH_2-$; or $-CHO$	Folic acid	Figure 18–25
Cobamide coenzymes	Specialized (see p. 696)		B$_{12}$	Figure 18–28

given in terms of the general or specific type of reaction in which the coenzymes participate. Once again, do not be alarmed. You do not need to understand all of these reactions at this time. Throughout succeeding chapters we will describe each coenzyme as the need arises.

The type of association between the cofactor and the apoenzyme protein component varies. In some cases they exist separately and become bound to each other only during the course of the reaction. In other cases they are always bound together, sometimes very firmly, by covalent bonding.

Generally speaking, the role of a cofactor is either (1) to alter the three-dimensional structure of the protein and/or the bound substrate to maximize the interaction between the substrate and enzyme, or (2) to actually participate in the overall reaction as another substrate. The organic coenzymes operate primarily according to role (2). The chemistry of this participation is usually described in terms of the coenzyme acting as a donor or acceptor of a particular chemical grouping relative to the other substrate(s). The grouping may be CO$_2$, a methyl ($-CH_3$) group, an amino ($-NH_2$) group, or electrons, to name just a few. Accordingly, the coenzymes are sometimes called *group transfer agents*. The examples that follow should help reinforce the meaning of group transfer.

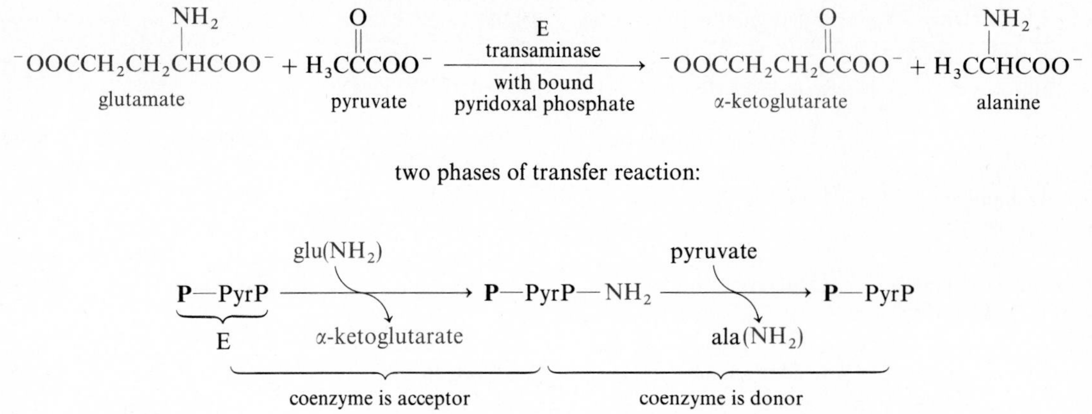

$$^-OOCCH_2CH_2\overset{\overset{\displaystyle NH_2}{|}}{C}HCOO^- + H_3C\overset{\overset{\displaystyle O}{\|}}{C}COO^- \xrightarrow[\substack{\text{with bound} \\ \text{pyridoxal phosphate}}]{\substack{E \\ \text{transaminase}}} {}^-OOCCH_2CH_2\overset{\overset{\displaystyle O}{\|}}{C}COO^- + H_3C\overset{\overset{\displaystyle NH_2}{|}}{C}HCOO^-$$

glutamate   pyruvate   α-ketoglutarate   alanine

two phases of transfer reaction:

$$\mathbf{P}-PyrP \xrightarrow[\alpha\text{-ketoglutarate}]{glu(NH_2)} \mathbf{P}-PyrP-NH_2 \xrightarrow[ala(NH_2)]{pyruvate} \mathbf{P}-PyrP$$

E

coenzyme is acceptor   coenzyme is donor

FIGURE 5–2 Transfer of an amino (—NH₂) group by pyridoxal phosphate (PyrP)-dependent transaminases. The transfer reaction occurs in the two phases shown in the lower part of the diagram. Symbol **P** is an apoenzyme protein to which the coenzyme PyrP is covalently attached.

The first example summarizes the transfer of an amino (—NH$_2$) group occurring during a reaction catalyzed by a single enzyme (see Figure 5–2). The pyridoxal phosphate (PyrP)–dependent transaminase enzyme catalyzes the overall reaction in two successive stages. First, the amino group of glutamate (the substrate donor of the amino group) is transferred to the coenzyme portion of the apoenzyme. Second, the amino group is transferred from the apoenzyme to pyruvate (the substrate acceptor of the amino group), completing the overall reaction. Complete details of the chemical events involved are presented elsewhere (see Section 18–3).

The second example illustrates a transfer of electrons occurring as a result of two separate oxidation-reduction reactions catalyzed by two separate enzymes (see Figure 5–3). The two enzymes utilize different forms of the same coenzyme, nicotinamide adenine dinucleotide (NAD). In the first reaction (E$_1$ catalyzed), the oxidized form of the coenzyme (NAD$^+$) acts as an oxidizing agent (electron acceptor) in the oxidation of an HO(H)C$<$ carbon to an O=C$<$ carbon. (The formation of the active-protein–NAD$^+$ holoenzyme

FIGURE 5–3 Electron transfer catalyzed by NAD-dependent dehydrogenases. Electron transfer in dehydrogenation reactions involves the net transfer of two hydrogen atoms (2H), usually as a hydride ion ($\colon H^{-1}$ with two electrons) and H$^+$.

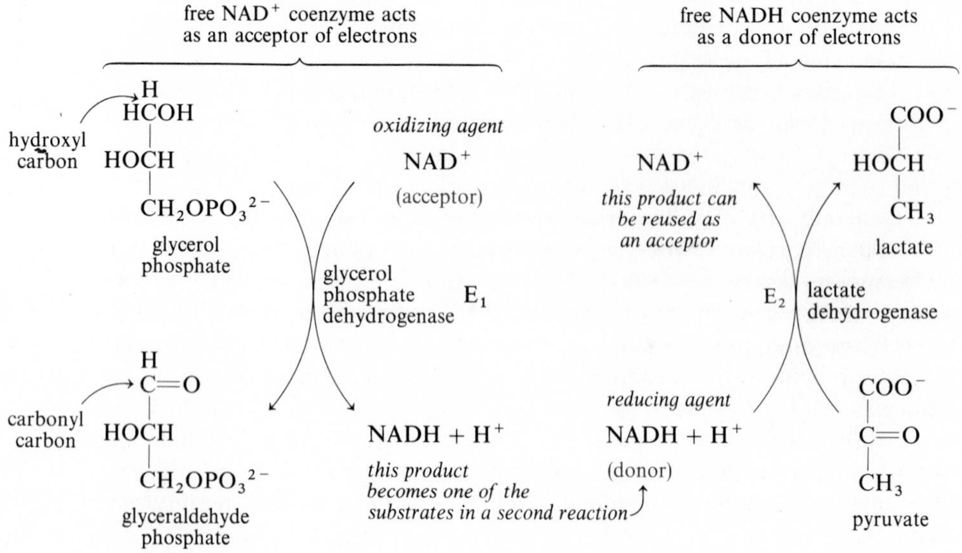

complex occurs during the reaction.) The coenzyme itself is reduced in this process and dissociates from the protein component in its reduced form (NADH). In the second reaction (E_2 catalyzed), the NADH acts as a reducing agent (electron donor) in the reduction of $O=C<$ to $HO(H)C<$. The two separate reactions are thus linked to each other by a common participant, serving as a product in one reaction and a reactant in the other. Such reactions are referred to as *coenzyme-coupled reactions,* a common scenario in the maze of chemical reactions in living cells.

5–3 FUNDAMENTAL PRINCIPLES OF CATALYSIS

In energetic terms the progress of a chemical reaction does not proceed directly from reactants to products. Rather, according to *transition state theory,* products are formed only after reactant species have (1) *collided in an optimum spatial orientation* that will lead to a reaction between them and (2) acquired sufficient *energy to attain a transition state* wherein the chemistry of bond breaking and/or bond forming is imagined to be in some stage of development. Obviously, the more efficiently the transition state is formed, the greater the speed (velocity, rate) of the reaction is.

The basic essentials of this statement are diagramed in Figure 5–4(a). The key feature is the difference in energy levels between the initial reactant state and the excited transition state—a difference called the *energy of activation,* E_{act}. Without exception, every chemical reaction has a value of E_{act} characteristic of the chemistry of the reaction and the conditions under which the reaction occurs. You may recall from general chemistry and physical chemistry courses that the actual value of E_{act} can be evaluated experimentally by measuring the rate of the reaction at different temperatures.

In terms of transition state theory, the action of a catalyst increases the rate of a chemical reaction by providing a *different path* by which the reaction occurs, a path characterized by a *lower* energy of activation. Since less energy must be acquired to achieve the transition state, the transition state can be attained with greater frequency, and the reaction rate is increased. Although the E_{act} is reduced in the presence of a catalyst, the E_{net} *remains unchanged* [see Figure 5–4(b)].

Providing a reaction path with a reduced energy of activation is not the only characteristic of enzyme action, although it usually is the major one. Another important feature is the ability of the enzyme to bind and spatially orient the reacting molecules with each other and with the enzyme in such a way as to maximize the occurrence of a productive reaction [see item (1) given previously]. The binding and subsequent reaction occur at a specific location on the surface of an enzyme, called the *active site,* which we will discuss more thoroughly in later sections.

Those interested in a more detailed and mathematical description of the physicochemical features of enzyme catalysis according to transition state theory should consult a physical chemistry textbook. However, you will not find any definitive theory explaining how and why an enzyme lowers the E_{act}. Pre-

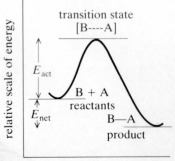

some parameter representative of the progress of reaction $\longrightarrow$
(a)

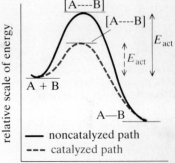

some parameter representative of the progress of reaction $\longrightarrow$
(b)

FIGURE 5–4 Energy profile diagrams. (a) Hypothetical reaction B + A → B—A. The finite difference between the energy levels of the ground state of reactants and the excited (transition) state of reactants is the *energy of activation,* E_{act}. The illustration depicts an energy-yielding reaction, with the net output of energy E_{net} corresponding to the difference between the ground states of A + B and B—A. (b) Smaller energy of activation in the presence of a catalyst. The catalyst has *no effect* on E_{net}.

cise knowledge of this sort remains elusive for various reasons. For one thing, not all enzymes operate in exactly the same manner; second, the details of binding and bond forming/bond cleavage at the active site are complex and often numerous.

5–4   PRINCIPLES OF ENZYME KINETICS

Kinetics is the study of reaction rates, and it provides the basis for understanding how a reaction occurs. How many steps are involved? What is the chemistry occurring in each step? Which is the slowest-occurring step and thus the step that will limit the rate of the overall reaction? A description of a reaction in these terms is called the *mechanism* of the reaction. Before considering the mechanism of enzyme-catalyzed reactions, we must cover one other topic—*rate equations*.

Although reaction rates are measured experimentally, the relationship of rate to the concentration of reactants can be expressed in simple equation form. In the writing of these equations *rate constants* (symbolized as *k*) are used. At constant temperature the *k* value is a constant, characteristic of how rapidly the reaction occurs.

To illustrate, consider the hypothetical reversible reaction of $A \rightleftharpoons B + C$. Each of the two chemical events (forward and reverse reactions) has a *k* value:

$$A \underset{k_r}{\overset{k_f}{\rightleftharpoons}} B + C$$

A rate equation for each event can be written according to the following format:

rate (or velocity *v*) = (rate constant) × (product of all reactant concentrations)

Thus for the forward reaction

$$A \xrightarrow{k_f} B + C$$

we write

$$v = k_f[A]$$

And for the reverse reaction

$$A \xleftarrow{k_r} B + C$$

we write

$$v = k_r[B][C]$$

Note 5–4
The brackets [ ] symbolize concentration of a substance, usually as molarity M, millimolarity mM, or micromolarity μM.

Writing the equation does not make it valid. A valid rate equation conforms to laboratory data, which establish the degree to which *v* is dependent on the concentration of each reactant. This concentration is reflected in the exponent of each [ ] term (see Note 5–4), which in the previous equations is implied as being 1. An exponent value of 1 means that the reaction rate increases linearly with reactant concentration—doubling for each twofold increase in reactant concentration. This situation is called *first-order* kinetics. A value of 0 means that the reaction rate is independent of reactant concentration—not changing as the concentration of the reactant changes. This situation is called a *zero-order* kinetic relationship.

Enzyme-Substrate Complex

Although the detailed mechanism of action is unique for each enzyme, all enzymes operate in the same general way. The first insight into enzyme behavior was provided by V. Henri (1903) and later by L. Michaelis and Maud L. Menten (1913). Henri and the team of Michaelis and Menten proposed basically the same model, but Michaelis and Menten based theirs on data collected from carefully designed and controlled laboratory experiments. The Henri-Michaelis-Menten model of enzyme action is still the *foundation of enzyme kinetics*.

The experimental work of Michaelis and Menten was done with an extract of yeast rich in invertase, an enzyme that catalyzes the hydrolysis of the carbohydrate sucrose (see Figure 5–5). Data were collected on changes in the *initial velocity* v_0 in two separate experiments. The experiments are described next.

1. When the substrate concentration [S] was held constant while the amount of enzyme [E] was varied, a linear increase in velocity was observed with increasing concentration of enzyme present [see Figure 5–6(a)].

2. In experiments of the reverse type, when the enzyme concentration was held constant while the amount of substrate was varied, a nonlinear hyperbolic relationship between velocity and substrate concentration was observed [Figure 5–6(b)]. Note that the hyperbolic plot could be described as a transition from first-order kinetics (with $v_0 \propto [S]^1$) to zero-order kinetics (with $v_0 \propto [S]^0$).

A simple model was formulated to explain such behavior. It was proposed that the enzyme E could reversibly combine with the substrate S to form an *intermediate complex* of enzyme and substrate, ES, which then decomposes to

sucrose
(composed of
glucose and fructose)

H_2O | hydrolysis
catalyzed by
the enzyme
invertase

free glucose
+
free fructose

FIGURE 5–5 Reaction catalyzed by invertase. In this reaction the substrate sucrose, which is dextrorotatory, is converted to products that give a levorotatory solution. Recording this change in optical activity over time is a convenient way of measuring the rate of this reaction.

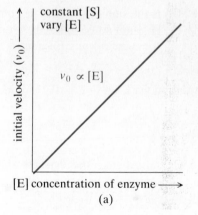

(a)

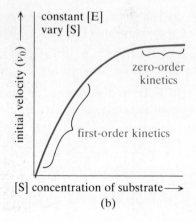

(b)

FIGURE 5–6 Kinetic features of an enzyme-catalyzed reaction. (a) Linear relationship of reaction rate to amount of enzyme present. (b) Hyperbolic relationship of reaction rate to amount of substrate present. In zero-order kinetics rate is independent of substrate concentration. In first-order kinetics rate is dependent on substrate concentration, with $n = 1$.

yield product(s) P and the free enzyme in its original form:

$$E + S \rightleftharpoons ES \rightarrow P + E$$

Implied here is that the transformation of S to P occurs at the level of the intermediate ES complex.

The validity of the model was established by deriving a mathematical equation consistent with the experimental data relating v_0 to [E] and v_0 to [S] as collected in the kinetic study of invertase action (refer to Figure 5–6). Various derivations were suggested by different individuals using different assumptions concerning the ES model. The conditions in the following derivation are attributable in part to the thinking of Michaelis and Menten (1913) and in part to Briggs and Haldane (1925). The derivation requires that the ES model be restated with an identification of rate constants for the three events: k_1, k_2, and k_3. The style used here assigns odd-numbered rate constants to forward reactions and even-numbered constants to the reverse reactions. (The Enzyme Commission recommends a different style: positive subscripts for a forward reaction and a negative subscript for the corresponding reverse reaction, such as k_1 and k_{-1}.) Thus the overall reaction can be written as

$$E + S \underset{k_2}{\overset{k_1}{\rightleftharpoons}} ES \overset{k_3}{\longrightarrow} P + E$$

Here are the steps in the derivation.

1. For the ES intermediate a *steady-state equilibrium* is attained very rapidly (Briggs and Haldane). A steady-state for ES means that the rate of formation of ES is the same as the rate of disappearance of ES. In other words, there is no change in the concentration of ES with time. The rate equations are

$$\text{rate of ES formation} = k_1[E][S]$$

$$\text{rate of ES disappearance} = k_3[ES] + k_2[ES]$$

(there are two events for ES decomposition)

Thus $\qquad\qquad\qquad k_1[E][S] = k_3[ES] + k_2[ES] \qquad\qquad\qquad$ **(5–1)**

2. The concentration of *total enzyme* $[E_t]$ is the sum of the enzyme combined with substrate [ES] and the free enzyme $[E_f]$ not so complexed:

$$[E_t] = [ES] + [E_f] \qquad\qquad\qquad\qquad \textbf{(5–2)}$$

3. In terms of the rate of product formation via $ES \overset{k_3}{\longrightarrow} E + P$, the initial velocity of the reaction is given by

$$v_0 = k_3[ES] \qquad\qquad\qquad\qquad \textbf{(5–3)}$$

4. A *maximum initial velocity* V_{max} will be attained when the concentration of ES reaches a maximum. This maximum concentration will occur when all of the available enzyme is complexed with substrate—that is, when $[E_f] = 0$.

This condition is termed *saturation* of the enzyme with substrate. When $[E_f] = 0$, however, $[ES]_{max} = [E_t]$; thus the maximum velocity will be directly proportional to the total enzyme concentration:

$$\text{maximum } v_0 = V_{max} = k_3[ES]_{max} = k_3[E_t] \qquad (5\text{–}4)$$

The information in Equations (5–1), (5–2), (5–3), and (5–4) can be used to develop a rate equation relating v_0 to $[S]$. Using Equation (5–1), where $[E]$ is $[E_f]$, we obtain an expression for $[ES]$:

$$[ES] = \frac{k_1}{k_2 + k_3}[E_f][S] \qquad (5\text{–}5)$$

The composite–rate constant term can be defined as one constant:

$$K_m = \frac{k_2 + k_3}{k_1}$$

The m subscript refers to Michaelis-Menten, and K_m is called the Michaelis-Menten constant as a tribute to their pioneering study.

Substitution with K_m into Equation (5–5) yields

$$[ES] = \frac{[E_f][S]}{K_m} \qquad (5\text{–}6)$$

From item 3 we can express $[ES] = v_0/k_3$; then we set this expression equal to Equation (5–6):

$$\frac{v_0}{k_3} = \frac{[E_f][S]}{K_m} \quad \text{or} \quad v_0 = \frac{k_3[E_f][S]}{K_m} \qquad (5\text{–}7)$$

From item 2 we can express $[E_f] = [E_t] - [ES]$, which can substitute for $[E_f]$ in Equation (5–7) to yield

$$v_0 = \frac{k_3[E_t][S] - k_3[ES][S]}{K_m} \qquad (5\text{–}8)$$

Substitute, in Equation (5–8), the relationships $k_3[E_t] = V_{max}$ (from item 4) and $k_3[ES] = v_0$ (from item 3), to yield

$$v_0 = \frac{V_{max}[S] - v_0[S]}{K_m}$$

Finally, collect terms and solve for v_0, to yield

$$v_0 = \frac{V_{max}[S]}{K_m + [S]}$$

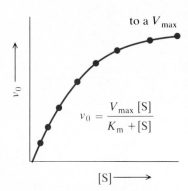

FIGURE 5–7 Graphical
representation of the Michaelis-
Menten equation. See also Figure
5–9.

This last statement is the form of the classical *Michaelis-Menten kinetic equation* corresponding to that of a rectangular hyperbola [Figure 5–6(b)], with v_0 and [S] as coordinates and with the constant V_{max} as the asymptote (see Note 5–5) of the graph (see Figure 5–7). This relationship, the ES model from which it was derived, and the laboratory procedure of measuring v_0 while varying [S] for a constant [E] are the foundations of enzyme kinetics.

The Michaelis-Menten equation must also be consistent with the observation [Figure 5–6(a)] that the velocity is linearly dependent on the amount of enzyme present. This agreement can be seen by substituting for V_{max} from Equation (5–4):

$$v_0 = \frac{k_3[E_t][S]}{K_m + [S]}$$

Since k_3 and K_m are constant and [S] is held constant, we have

$$v_0 = \frac{(\text{constant})[E_t](\text{constant})}{\text{constant} + \text{constant}}$$

And by combining all constants, we obtain

$$v_0 = (\text{constant})[E_t] \qquad \text{or} \qquad v_0 \propto [E_t]$$

Reversibility of Many Enzymes

The original ES model with ES → E + P implies that an enzyme does not operate in the reverse direction; that is, the action is *irreversible*. However, many enzymes do exhibit *reversible catalysis*, capable of operating in either the forward or reverse reaction. Enzymes in this category do not require a new model to explain their action. Only a slight modification is required, namely, that ES ← E + P is also possible:

$$E + S \underset{k_2}{\overset{k_1}{\rightleftharpoons}} ES \underset{k_4}{\overset{k_3}{\rightleftharpoons}} E + P$$

However, the added dimension of the k_4 step can usually be safely neglected, and the Michaelis-Menten rate equation will still apply. It can be neglected since initial rates of reaction are measured. Thus the occurrence of ES ← E + P is low because very little P will be available in the early stages of the reaction during which time the v_0 is recorded.

The irreversible or reversible aspect of enzyme operation will be encountered at various times in our future discussions of cellular metabolism.

Active-Site Events: E + S → ES and ES → E + P (Preliminary Consideration)

The enzyme-substrate complex ES (see Figure 5–8) is a *real* chemical species; a few have actually been isolated. The formation of the ES complex is the result of the binding of S to E, and as such, this event is a perfect example

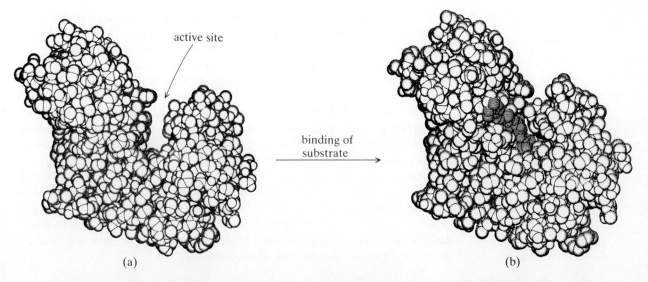

active site

binding of
substrate →

(a)   (b)

FIGURE 5–8 Computer-
generated images of a substrate
binding to an active site. The
enzyme depicted is hexokinase.
Another display is given in Figure
5–28. (a) Free enzyme with a
vacant active site. (b) ES complex,
with atoms of bond substrate
identified in color. (*Source*:
Photographs provided by T.
Steitz.)

of a basic concept referred to earlier:

> The initial event of any protein associated with a dynamic function
> is one of binding with another substance.

Without binding, nothing happens. In fact, the binding event is often the sole
basis for the specificity of enzyme action. The next event of ES → E + P represents the chemistry of breaking and forming bonds to yield the products of the
reaction:

$$E + S \xrightarrow{\text{binding}} ES \xrightarrow{\text{transformation}} E + P$$

> Although catalysis is primarily dependent on events occurring in the
> second phase, the initial event of binding does contribute to the
> events of catalysis by specifically positioning S and rendering it more
> susceptible to the action of the catalytic residues.

> The surface of each enzyme contains at least one specific location, called
> the *active site,* where the events of binding and chemical conversion occur.
> For most enzymes the active-site region comprises only about 5% of the total
> molecular surface of the protein.

> The active site consists of a spatially ordered cluster of a few amino
> acid R groups (and possibly a cofactor), some of which participate
> in the binding of substrate and some of which participate in the
> chemistry of product formation. (*Note:* Some R groups may serve
> a dual function.)

> The R groups comprising the active site do not occupy adjacent positions
> along the polypeptide chain. Rather, most of the residues occupy distant posi-
> tions along the chain and are brought close together by virtue of the many
> bends, folds, and twists of the polypeptide backbone (see Figures 5–26 and
> 5–28 later in the chapter). These *binding residues* and *catalytic residues* are
> obviously *essential* to the activity of the enzyme.

Each enzyme contains other equally essential amino acid residues, although they are not directly involved in active-site events. These residues are the crucial *structural residues,* which through their ordered interactions contribute to the formation and the stabilization of the ordered conformation of the entire molecule and thus are responsible for forming the active site. A cofactor may also help in this regard. A severe change in the identity of any of these essential residues (binding, catalytic, or structural) can greatly affect enzyme activity; in fact, a total loss in activity can result. Those residues not involved in any of these events are called *nonessential residues.* The number of essential versus nonessential residues varies from enzyme to enzyme.

Of course, the optimum conformation of the active site—and hence maximum activity—is also dependent on the maintenance of physiological conditions that do not denature the protein: A temperature of 35°–37°C, a pH of 6.5–7.5, and an ionic strength of about 0.15 apply to most enzymes. Some exceptions are those enzymes that operate under extreme conditions, such as pepsin in the acidic gastric juice of the stomach.

Multifunctional Enzymes

A long-standing axiom of biochemistry has been that each catalytic polypeptide unit (or subunit in an oligomer) of an enzyme molecule contains only a single active site. In oligomeric enzyme molecules, depending on the number of subunits present, 2, 3, . . . , *n* active sites can exist. Furthermore, although the typical active site participates in only one type of chemical reaction, it is not uncommon for an active site to catalyze a small number of different reactions. The obvious examples are enzymes that can act reversibly where the S → P and S ← P reactions occur at the same active site.

In recent years a new dimension of enzyme catalysis has been discovered that modifies these descriptions of a typical enzyme. The new discovery is that some enzyme molecules are *multifunctional;* that is:

> Separate and distinct active-site regions, each catalyzing a different chemical reaction, can occur in the same enzyme molecule.

One remarkable example is *fatty acid synthetase,* a molecule composed of a single polypeptide chain that has seven different active site domains. Fatty acid synthetase is described in greater detail in Section 17–2.

Two Substrate Reactions

The $E + S \rightleftharpoons ES \rightarrow E + P$ model of enzyme kinetics actually applies to a *unimolecular reaction* wherein one reacting substrate is converted to product(s): S → P. If only one product is formed, the overall reaction is labeled as being a Uni-Uni process. Although there are some examples of this process, most enzyme-catalyzed reactions involve more than one reacting substrate, with *bimolecular* (two substrates) *processes* being particularly common. The overall bimolecular process may involve different results, such as two substrates yielding one product,

$$\text{Bi-Uni:} \qquad S_1 + S_2 \rightarrow P$$

or two substrates yielding two products,

$$\text{Bi-Bi:} \qquad S_1 + S_2 \rightarrow P_1 + P_2$$

The obvious question is whether the unimolecular model of $E + S \rightleftharpoons ES \rightarrow E + P$ still applies to bimolecular reactions. The answer is that it does as long as the initial concentration of one of the two reacting substrates is in large excess so that its concentration never becomes limiting. The biochemical term is a *saturating concentration*. Thus for a two-substrate reaction one can separately measure (1) v_0 as a function of varying the concentration of S_1 in the presence of a saturating concentration of S_2 and then (2) v_0 as a function of varying the concentration of S_2 in the presence of a saturating concentration of S_1. From one experiment we would obtain K_m and V_{max} values for S_1, and from the second we would obtain K_m and V_{max} values for S_2. The values are not necessarily identical.

Our consideration of bimolecular reactions gives us the opportunity to discuss an additional aspect of the Michaelis-Menten model, namely, its oversimplification. A more realistic representation is as follows, indicating that several intermediates may actually be involved and that the product is in association with the enzyme prior to the event of its dissociation to yield free P:

$$E + S \rightleftharpoons \underbrace{ES \rightleftharpoons ES' \rightleftharpoons EP' \rightleftharpoons EP}_{\substack{\text{ES of Michaelis-Menten} \\ \text{theory refers to these reactions}}} \rightleftharpoons E + P$$

In this representation, after the initial binding of substrate there may be stages of substrate modification and product development.

For Bi-Bi reactions ($S_1 + S_2 \rightarrow P_1 + P_2$) there are three possible reaction pathways:

1. *Random sequential:* Either S_1 or S_2 can bind first, but both must be bound before any products are formed. Product release can be in either order. The course of a totally reversible random sequential Bi-Bi path is represented as follows (rate constants for each step are omitted for clarity):

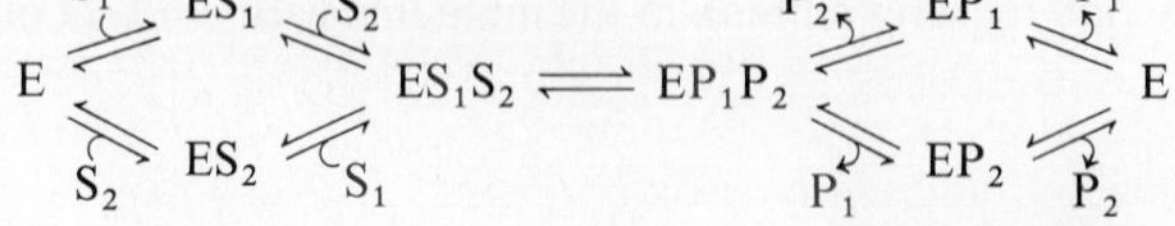

2. *Ordered sequential:* S_1 and S_2 bind in a specific order but both must be bound before any products are formed, with product release also occurring in a specific order. The representation is

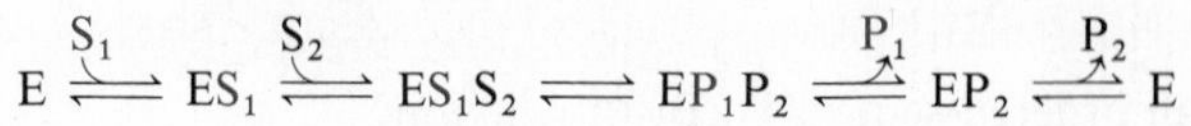

3. *Ordered Ping-Pong:* S_1 and S_2 bind in a specific order and products are formed in a specific order, but product formation does not require that both

S_1 and S_2 be bound simultaneously. The overall path consists of two Uni-Uni steps: $S_1 \rightarrow P_1$ and $S_2 \rightarrow P_2$. The representation is

$$E \xrightleftharpoons{S_1} ES_1 \rightleftharpoons E'P_1 \xrightleftharpoons{P_1} E' \xrightleftharpoons{S_2} E'S_2 \rightleftharpoons EP_2 \xrightleftharpoons{P_2} E$$

Note that the Ping-Pong mechanism implies that the enzyme is altered during the reaction so that the second substrate S_2 binds with a chemically modified form of the enzyme E', whereas S_1 binds with the original E. An example of the Ping-Pong pathway is the transaminase enzyme illustrated earlier in the chapter (Figure 5–2) showing the modified form of the enzyme to be E—PyrP—NH$_2$ after the amino group of glutamate is transferred to the pyridoxal phosphate coenzyme component.

We will not discuss how one can experimentally evaluate which of the three paths applies to a particular enzyme-catalyzed Bi-Bi reaction. Those interested can refer to the monograph by Segel cited in the Literature.

Cleland Notation

The outline of an enzyme-catalyzed reaction can also be shown in a more simplified representation called the *Cleland notation* (after W. W. Cleland, one of the early pioneers in the field of enzyme kinetics). A horizontal line is used to represent the progress of the reaction. Above the line arrows pointing at the line indicate the entry of reacting substrates, and arrows pointing away from the line indicate the release of products. Letters below the line indicate the form of the enzyme (E, ES_1, EP_2, E', and so on) present at various points throughout the reaction, with interconversions of enzyme forms shown in parentheses. The following list gives some examples.

- For a Uni-Uni reaction:

- For a random sequential Bi-Bi reaction (the possibility of alternate paths during the reaction requires a break in the main line with a subset of two parallel lines):

- For an ordered sequential Bi-Bi reaction:

- For a Ping-Pong, Bi-Bi reaction:

$$
\begin{array}{c}
\quad\;\; S_1 \qquad\quad\; P_1 \; S_2 \qquad\qquad P_2 \\
\quad\;\; \downarrow \qquad\qquad \uparrow\;\; \downarrow \qquad\qquad \uparrow \\
\hline
E \quad (ES_1 \rightleftharpoons E'P_1) \quad E' \quad (E'S_2 \rightleftharpoons EP_2) \quad E
\end{array}
$$

Measurement of K_m and V_{max}

The values of K_m and V_{max} are kinetic constants reflecting the action of an enzyme, so they provide a basis for understanding how the enzyme operates. Noting changes in K_m and/or V_{max}—changes due to different reaction conditions, to the presence of another substance, or to subjecting of the enzyme to some treatment that modifies its protein structure—provides even more insight. These values are obtained by analyzing kinetic data collected as described earlier: Initial velocities are separately measured at different concentrations of substrate, using the same amount of enzyme in each assay. Remember that if the reaction involves two substrates ($S_1 + S_2 \rightarrow P$), separate experiments are performed on each substrate, using an excess, saturating concentration of the substrate whose concentration is not being varied.

A *direct plot* of v_0 versus [S] (see Figure 5–9 and Table 5–3) permits a visual *estimation* of V_{max} and from V_{max} an estimation of K_m. Obviously, the V_{max} estimate is done from the upper plateau region of the hyperbolic curve. The K_m estimate is easy, arising from the relationship (to be discussed later) that $K_m = $ [S] when $v_0 = \frac{1}{2}V_{max}$.

TABLE 5–3 Data for Figure 5–9

[S] Millimolar (mM)	v_0 (ΔA_{405}/min)
0.50	0.075
0.75	0.090
2.00	0.152
4.00	0.196
6.00	0.210
8.20	0.214
10.0	0.230

Note: The data were collected on the enzyme acid phosphatase by the students in the biochemistry laboratory at John Carroll University.

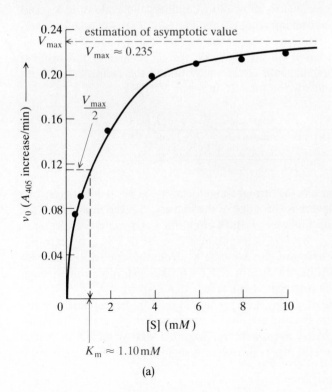

(a)

p-nitrophenyl phosphate (S_1)

(S_2) H_2O ⟶ acid phosphatase

O_2N—⟨ ⟩—OH + $HOPO_3^{2-}$

p-nitrophenol P_1 phosphate P_2

(b)

FIGURE 5–9 Direct Michaelis-Menten plot. (a) Plot of the data given in Table 5–3. Initial rate v_0 is expressed as the increase in absorbance A at 405 nm/min, reflecting the appearance of the product being formed in the reaction. The inherent uncertainty in the graphical estimation of V_{max} means that the value of K_m is approximate. (b) Complete reaction. After treatment with OH^-, product P_1 absorbs light at 405 nm.

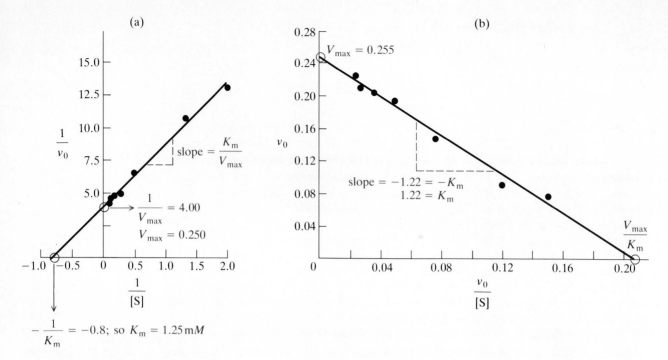

FIGURE 5–10 Alternative plots of the Michaelis-Menten equation. (a) Lineweaver-Burk plot. (b) Eadie-Hofstee plot. The same data for acid phosphatase were used as in the direct hyperbolic plot in Figure 5–9.

The uncertainty in visually evaluating the V_{max} asymptote of the direct plot can be overcome by resorting to computer curve analysis of the direct plot or by using a different graphical format. The rationale of an alternative graphical format is to obtain a *linear plot* to allow more confidence in evaluating K_m and V_{max} values from the slope and intercept points. The coordinates of a *Lineweaver-Burk plot* [see Figure 5–10(a)] are $1/v_0$ (for the y axis) versus $1/[S]$ (for the x axis). These coordinates are based on an algebraic rearrangement of the original Michaelis-Menten hyperbolic rate equation by taking the reciprocal of both sides of the equation and solving for $1/v_0$ in terms of $1/[S]$:

$$v_0 = \frac{V_{max}[S]}{K_m + [S]} \quad \xrightarrow[\text{rearrangement}]{\text{algebraic}} \quad \frac{1}{v_0} = \frac{K_m}{V_{max}}\frac{1}{[S]} + \frac{1}{V_{max}}$$

$$y = mx + b$$

The new equation has the straight-line form of $y = mx + b$ between two variables (y and x), where m is the slope of the line and b is the intercept of the line of the y axis. In the Lineweaver-Burk equation $y = 1/v_0$, $x = 1/[S]$, $m = K_m/V_{max}$, and $b = 1/V_{max}$. By first determining the value of $1/V_{max}$ by extrapolation through the $1/v_0$ axis, one can evaluate K_m from the slope of the line. Alternatively, K_m can be directly determined by further extrapolation through the $1/[S]$ axis where the intercept is equal to $-1/K_m$.

The jamming of several $1/v_0$, $1/[S]$ points in the Lineweaver-Burk format can bias the placement of the line, leading to erroneously high or low values for both K_m and V_{max}. A better approach [see Figure 5–10(b)], called an *Eadie-Hofstee plot* and based on the algebraic rearrangement indicated below, is to plot

v_0 (for the vertical y axis) versus $v_0/[S]$ (for the horizontal x axis):

$$v_0 = \frac{V_{\text{max}}[S]}{K_{\text{m}} + [S]} \quad \xrightarrow[\text{rearrangement}]{\text{algebraic}} \quad v_0 = -K_{\text{m}}\frac{v_0}{[S]} + V_{\text{max}}$$

$$y = mx + b$$

As this equation shows, K_{m} and V_{max} are easily evaluated: $-K_{\text{m}}$ equals the slope of the line, V_{max} equals the intercept value on the v_0 axis, and $V_{\text{max}}/K_{\text{m}}$ equals the intercept value on the $v_0/[S]$ axis. Note also that the data points are more evenly spaced than in the Lineweaver-Burk plot.

A quite different method of plotting, called a *direct linear plot,* has been described by R. Eisenthal and A. Cornish-Bowden. The following rearrangement,

$$v_0 = \frac{V_{\text{max}}[S]}{K_{\text{m}} + [S]} \quad \xrightarrow[\text{rearrangement}]{\text{algebraic}} \quad V_{\text{max}} = \frac{v_0}{[S]}K_{\text{m}} + v_0$$

$$y = mx + b$$

treats V_{max} and K_{m} as variables and treats $[S]$ and v_0 as constants. This may seem like a strange maneuver, but, in fact, once a v_0 is experimentally measured for a given $[S]$, any combination of K_{m} and V_{max} can satisfy that one (v_0, $[S]$) combination. In that sense K_{m} and V_{max} can be considered as variables.

Data are handled in the following way (see Figure 5–11): (1) The y axis is labeled V_{max} (it is a velocity coordinate), and the x axis is labeled K_{m} (it is a substrate concentration coordinate); (2) for each (v_0, $[S]$) combination the v_0 value is marked on the vertical velocity axis and the $-[S]$ value is marked on the horizontal substrate concentration axis (*Note:* when $V_{\text{max}} = 0$, $K_{\text{m}} = -[S]$); (3) a line is then drawn connecting these two points and extending into the right quadrant of the graph; (4) steps (2) and (3) are repeated for every other (v_0, $[S]$) combination. Now one evaluates the region in space where all of the lines most closely crowd each other—a graphical point that identifies the one K_{m} and V_{max} combination that most closely satisfies all of the (v_0, $[S]$) combinations. The plot is easy, and there is no need to compute $1/v_0$, $1/[S]$, or $v_0/[S]$ values before plotting.

Significance of V_{max}

As the term implies, the maximum velocity is an expression of the upper-limit efficiency of operation for a given amount of an enzyme. To compare different enzymes, however, one must first express V_{max} in terms of the *same molar amount* of each enzyme. This conversion of V_{max} yields a value called the *molecular activity* or *turnover number* of the enzyme, representing the moles of substrate reacted per mole of enzyme per unit time (usually, 1 min).

Consider the example of carbonic anhydrase, an important Zn^{2+}-containing enzyme in blood that catalyzes the reaction

$$CO_2 + H_2O \xrightleftharpoons[]{\text{carbonic} \atop \text{anhydrase}} H_2CO_3$$

FIGURE 5–11 Direct linear plot (Eisenthal-Cornish-Bowden) of the Michaelis-Menten equation. The same data were used as in Figures 5–9 and 5–10.

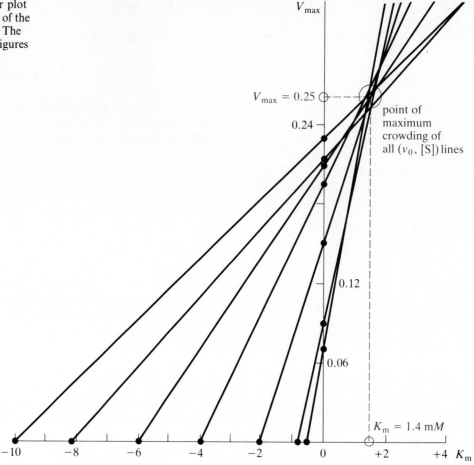

A Michaelis-Menten kinetic assay under optimum conditions (pH about 7, T of $35°$–$37°$) shows that 1 μg of enzyme exhibits a V_{max} of 1.2×10^{-3} mole of CO_2 reacted per minute. The molecular weight of carbonic anhydrase is 30,000 g/mole; hence 1 μg of enzyme represents 0.000001/30,000 mole of enzyme, that is, 3.33×10^{-11} mole. Thus

$$\text{molecular activity} = \frac{V_{max}}{\text{moles of E present}} = \frac{1.2 \times 10^{-3} \text{ mole } CO_2 \text{ reacted/minute}}{3.33 \times 10^{-11} \text{ mole enzyme}}$$

$$= 36 \times 10^6 \text{ mole } CO_2 \text{ reacted/minute/mole enzyme}$$

In other words, one molecule of carbonic anhydrase will catalyze the reaction of 36 million molecules of CO_2 in 1 min—that's fast. Carbonic anhydrase represents one of the fastest-working enzymes. The molecular activity of most enzymes is in the range of 1000–10,000. Even these values represent fast catalysis when one considers that most organic reactions in the absence of a catalyst require several minutes or several hours, even when heated to higher temperatures.

Two other commonly used expressions for the amount of enzyme activity are the *enzyme unit* and the *specific activity*. Traditionally, the term *unit* designates the amount of enzyme that can convert 1 μmole of substrate into product in 1 min under standard conditions. Depending on the enzyme, however, other dimensions may be more convenient, dimensions such as 1 unit = 1 mmole substrate reacted per minute or 1 μmole/s. The specific activity is merely the number of enzyme units per unit weight of protein.

Significance of K_m

When $v_0 = V_{max}$, all active sites are occupied and there are no free E molecules. This condition is called 100% saturation. At 50% saturation, where $v_0 = \frac{1}{2}V_{max}$, the Michaelis-Menten equation states that

$$\frac{V_{max}}{2} = \frac{V_{max}[S]}{K_m + [S]}$$

which reduces to

$$K_m + [S] = 2[S]$$

or
$$K_m = [S] \qquad (\text{only when } v_0 = \tfrac{1}{2}V_{max})$$

Thus K_m (having units of concentration) represents the amount of substrate required to bind with half of the available enzyme and producing half of the maximal velocity. As a rough approximation, the value of K_m can be considered as representing the concentration of the substrate in a living cell.

The K_m values can also be used to evaluate the *specificity of action* of a given enzyme toward similar substrates. The general rule is as follows: The lower the K_m value, the better (more preferred) is the substrate. To illustrate, we will use some data (Table 5-4) collected from assays on hexokinase, an enzyme that catalyzes the conversion of simple sugars to phosphoesters. In three separate experiments, performed under identical conditions with the same amount of enzyme, the K_m values for three different sugar substrates were measured.

A comparison of K_m values reveals that the amount of glucose required for 50% saturation is 1000 times less than that required for 50% saturation with allose. The interpretation is that the enzyme has a more efficient path with glucose than with allose. This efficiency may be due to a more efficient binding of the sugar to yield ES or a more efficient conversion of ES to products—or both may be more efficient. In other words, the enzyme has a preference for (is more specific in its action toward) glucose. Inspection of the two structures suggests that the spatial orientation of merely one hydroxyl (OH) group on one carbon atom (number 3) is critical. But at the same time, the sugar mannose shows a K_m value very close to the glucose value, suggesting that the spatial orientation of the OH group at carbon atom 2 is not as critical to enzyme action. (*Note:* The correlation of lower K_m values with a preferred substrate is

TABLE 5–4 K_m values of hexokinase (from brain)

The K_m values are for the sugar substrate. The concentrations of the ATP substrate and the Mg^{2+} cofactor were held constant. The reaction is

$$\text{sugar} + \text{ATP} \xrightarrow[\text{Mg}^{2+}]{\text{hexokinase}} \text{sugar phosphate} + \text{ADP}$$

With the sugar glucose,	With the sugar allose,	With the sugar mannose,
^{1}CHO	^{1}CHO	^{1}CHO
H—^{2}C—OH	H—^{2}C—OH	HO—^{2}C—H
HO—^{3}C—H	H—^{3}C—OH	HO—^{3}C—H
H—^{4}C—OH	H—^{4}C—OH	H—^{4}C—OH
H—^{5}C—OH	H—^{5}C—OH	H—^{5}C—OH
^{6}CH$_2$OH	^{6}CH$_2$OH	^{6}CH$_2$OH
as substrate, $K_m = 8 \times 10^{-6} M$	as substrate, $K_m = 8 \times 10^{-3} M$ $= 8000 \times 10^{-6} M$ (1000 times greater than K_m for glucose)	as substrate, $K_m = 5 \times 10^{-6} M$ (same order of magnitude as K_m for glucose)

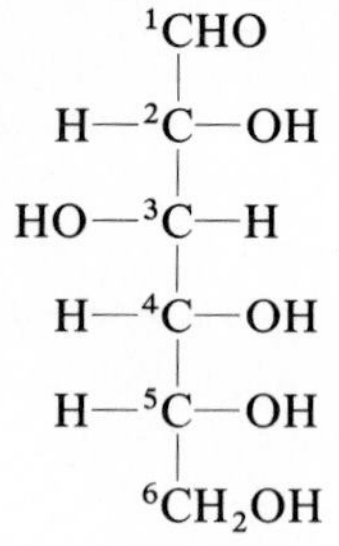

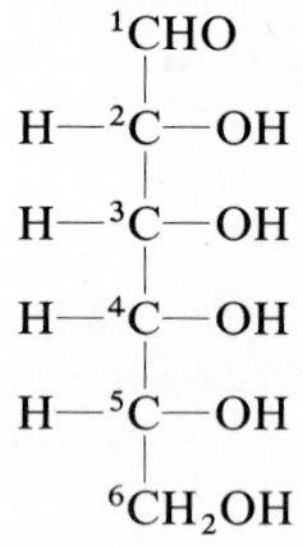

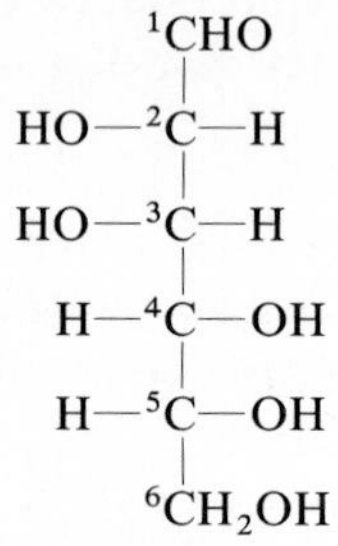

not as valid when different enzymes are compared, particularly when the type of substrates and the chemical transformations involved are not similar.)

This same principle can be illustrated another way based on the definition of $K_m = (k_2 + k_3)/k_1$. If we assume that k_2 is much greater than k_3—that is, the conversion of ES to products is much slower than the dissociation of ES back to E and S—the value of K_m can then be approximated by k_2/k_1. This assumption provides a basis for us to focus only on the events represented by E + S $\rightleftharpoons$ ES. As shown in Figure 5–12, the equilibrium affinity constant of the E + S $\rightleftharpoons$ ES binding is equal to k_1/k_2. If we keep the same rate constant for the two events but focus on the dissociation of ES $\rightleftharpoons$ E + S, the equilibrium dissociation constant is equal to k_2/k_1. Thus when the assumption $K_m = k_2/k_1$ is valid, K_m can

FIGURE 5–12 K_m value. For many enzymes the K_m value provides an inverse measure of the strength of binding between E and S. This interpretation applies when $k_3 \ll k_2$, which is a condition common to many enzymes.

At Equilibrium for ES Formation

$$E + S \underset{k_2}{\overset{k_1}{\rightleftharpoons}} ES$$

$$\text{rate}_{\text{forward}} = k_1[E][S]$$
$$\text{rate}_{\text{reverse}} = k_2[ES]$$

Hence

$$K_{eq} = K_{\text{affinity}} = \frac{[ES]}{[E][S]} = \frac{k_1}{k_2}$$

At Equilibrium for ES Dissociation

$$ES \underset{k_1}{\overset{k_2}{\rightleftharpoons}} E + S$$

$$\text{rate}_{\text{forward}} = k_2[ES]$$
$$\text{rate}_{\text{reverse}} = k_1[E][S]$$

Hence

$$K_{eq} = K_{\text{dissociation}} = \frac{[E][S]}{[ES]} = \frac{k_2}{k_1}$$

$$K_m = \frac{k_2 + k_3}{k_1} \approx \frac{k_2}{k_1} = \frac{1}{K_{\text{affinity}}} = K_{\text{dissociation}}$$

be expressed as a *dissociation constant* of ES or the reciprocal of the affinity constant of E for S. A low K_m, then, reflects a low tendency for ES to dissociate to E + S or, conversely, a high affinity of E for S.

5–5 ENZYME INHIBITION

Antibiotics, insecticides, herbicides, poisons and various drugs that combat pain, inflammation, viral infections, and cancers are all substances with which we are familiar. The effect of many of these substances is due to their ability to *interfere with (inhibit)* the operation of cellular proteins, usually a specific enzyme. Enzyme inhibition also occurs naturally, contributing to the normal patterns of bioregulation. In the research laboratory the use of enzyme inhibitors assists in elucidating the mechanism of action of enzymes, which, among other things, can assist in the development of more efficient drugs.

In general terms, the action of an *inhibitor* involves the binding of the inhibitor with some form of the enzyme, resulting in a total or partial loss in the activity of the enzyme to transform substrate(s) into product(s). There are different types of inhibitors, differing in their mode of action. The following discussion focuses on competitive inhibition, noncompetitive inhibition, irreversible inhibition, and uncompetitive inhibition.

Competitive Inhibition

A *competitive inhibitor* I is a substance that reversibly binds with the free form of an enzyme E to produce a binary EI complex incapable of binding S. Thus when E, S, and I are present, E can bind with S to yield ES, or E can bind with I to yield EI (see Figure 5–13). However, E cannot bind I and S simultaneously to yield a ternary EIS complex:

The binding of I and S is mutually exclusive.

The interaction of enzyme and inhibitor is accounted for by identifying an equilibrium *dissociation constant* K_I for EI $\rightleftharpoons$ E + I. In addition, the influence of the inhibitor's presence on the S $\xrightarrow{\text{E}}$ P conversion is accounted for by designating *apparent values* for the Michaelis-Menten constant and the maximum velocity: K'_m and V'_{max}.

Since the formation of EI reduces the population of E available for interaction with substrate S, the velocity decreases. However, because a competitive inhibitor combines reversibly with the enzyme, this type of inhibition can be *overcome by merely increasing the concentration of the substrate,* with the greater population of substrate molecules favoring the formation of a larger percentage of the normal ES complex.

There are various explanations for the mode of action of a competitive inhibitor, the most classical of which is that I and S have similar chemical structures and they *compete with each other to bind at the same active site.* If I

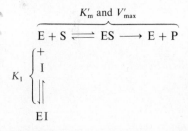

FIGURE 5–13 Competitive inhibitor. Here K_I is the inhibitor dissociation constant for EI $\rightleftharpoons$ E + I.

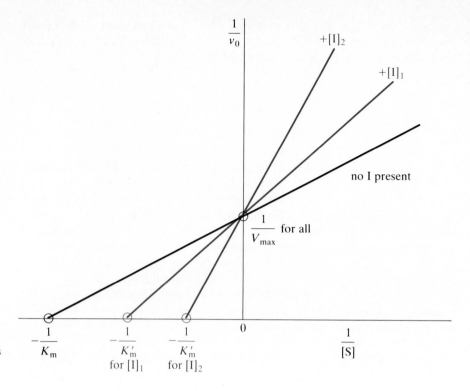

FIGURE 5–14 Diagnosis of pure competitive inhibition via Lineweaver-Burk plotting. The results for two inhibited assays are shown in color, where $[I]_2$ is greater than $[I]_1$.

and S bind at different locations on the surface of the enzyme, the mutual exclusion of I and S binding can be explained by the binding of I at the inhibitor-binding site, blocking the binding of S at the active site (a total or partial obstruction) or causing a conformational change to occur in the enzyme that ultimately results in an alteration of the active-site conformation so that S cannot bind at the active site. The conformation effect is a special case of *allosteric inhibition* that we will discuss later in the chapter (Section 5–9).

The type of enzyme inhibition is diagnosed by measuring v_0 versus $[S]$ in the absence and presence of I and then evaluating the data to determine the effect on the K_m and V_{max} values. These effects can be conveniently evaluated by displaying the results of both the uninhibited and inhibited assays on a Lineweaver-Burk plot (see Figure 5–14). In the case of competitive inhibition (see Note 5–6), V_{max} is *unaffected* ($V'_{max} = V_{max}$), but K_m is *increased* ($K'_m > K_m$). Because of the competition between I and S for the same site, a greater amount of substrate should be required for half saturation—hence the increase in K_m. However, when the system is saturated with enough S to overcome the presence of I, the maximum velocity should be unaffected, as though no inhibitor were present.

The K'_m (the apparent K_m in the presence of an inhibitor) is greater than K_m by a factor of $(1 + [I]/K_I)$, where $[I]$ is the initial concentration of inhibitor present. (*Note:* After calculating K_m from the uninhibited assay and K'_m from the inhibited assay for a known value of $[I]$, you have enough information to then calculate K_I.)

NOTE 5–6
For competitive inhibition:

$$-\frac{1}{K'_m} = -\frac{1}{K_m(1 + [I]/K_I)}$$

$$\frac{1}{V'_{max}} = \frac{1}{V_{max}}$$

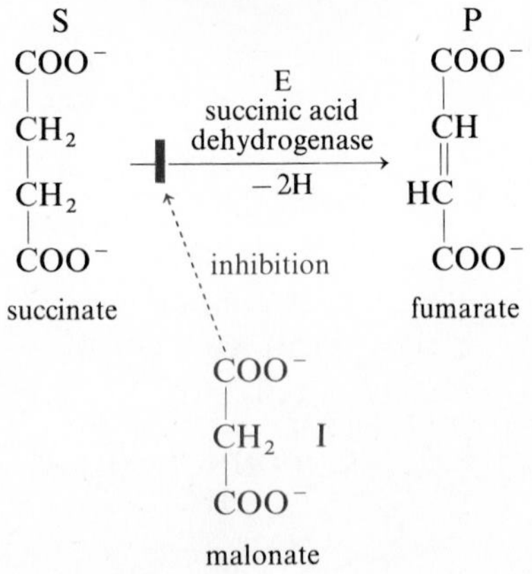

FIGURE 5–15 Competitive inhibition of succinic acid dehydrogenase. In this case if the ratio of [I]/[S] is only $\frac{1}{50}$, there is still 50% inhibition.

A classical example of competitive inhibition is shown by succinic acid dehydrogenase (see Figure 5–15 and Note 5–7). Malonic acid with one less —CH$_2$— group than succinic acid, the natural substrate, is a potent competitive inhibitor of this enzyme.

Many chemotherapeutic drugs function as competitive inhibitors. For example, several of the *sulfa drugs* used to combat microbial infections in mammals are structurally related to para-aminobenzoic acid (PABA). The PABA is a vital precursor in the microbial synthesis of folic acid (see Figure 5–16), which in turn is converted to tetrahydrofolic acid (see Figure 18–25), an extremely important coenzyme for several enzymes involved in the biosynthesis of the nucleic acids RNA and DNA. When the sulfa drug is administered, the immediate effect is the inhibition of the enzyme that catalyzes the PABA-incorporating step in the production of folic acid. A reduced level of folic acid will then reduce the production of nucleic acids, causing the death of the infectious organism.

The hypertension drugs *Captopril* and *Enalopril* are synthetic compounds that competitively inhibit the action of angiotensin-converting enzyme (see Figure 3–43 and Figure 5–17). Both drugs are proline derivatives, resembling the proline-containing region adjacent to the phe-his peptide bond in angiotensin I that is cleaved by the angiotensin-converting enzyme. Recent clinical studies

NOTE 5–7
The symbolism

$$S \xrightarrow[X]{} P$$

is used throughout this book to represent the inhibition by substance X of the enzyme catalyzing the S → P reaction.

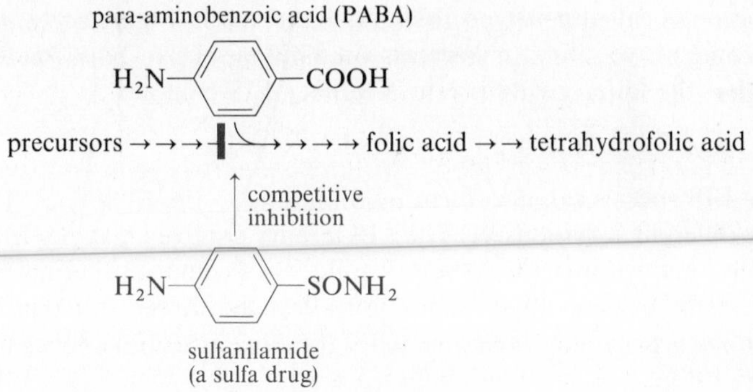

FIGURE 5–16 Competitive inhibition by sulfa drugs. The arrows represent several enzymatic steps involved with one enzyme catalyzing the incorporation of *p*-aminobenzoic acid. Tetrahydrofolic acid is an essential coenzyme in the biosynthesis of purines and pyrimidines which in turn are used for RNA and DNA biosynthesis.

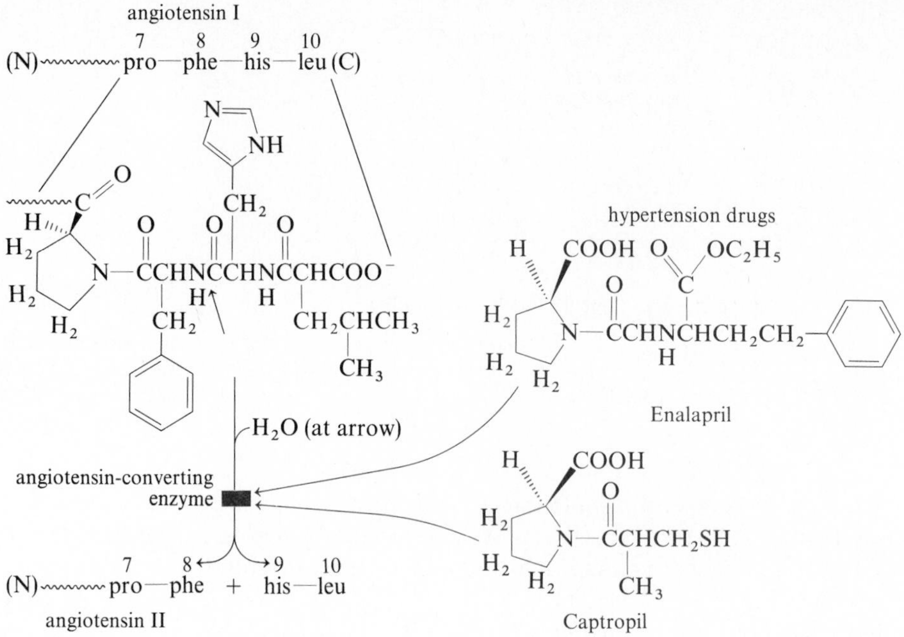

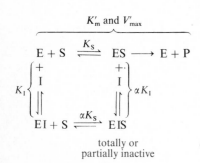

FIGURE 5–17 Competitive inhibition of hypertension drugs.

suggest that low doses of these drugs are very effective in lowering blood pressure without serious side effects impairing normal body functions.

Among many, two other examples of drugs acting via competitive inhibition are *allopurinol,* used to control gout (see Figure 18–44), and *fluorouracil,* used to treat cancer (see Figure 18–40).

In some instances competitive inhibition also explains *product inhibition,* wherein the product P of the S → P conversion inhibits the enzyme E responsible for forming the product in the first place. In later chapters we will encounter some examples of how product inhibition contributes to the normal patterns of bioregulation in living cells.

Noncompetitive Inhibition

When the effect of I is not overcome by increasing the concentration of S, the inhibition is called *noncompetitive.* As indicated in Figure 5–18, the action of a noncompetitive inhibitor involves the binding of I to both free E and the ES complex, the latter giving rise to a ternary EIS complex.

So I and S bindings are not mutually exclusive.

The EIS species can also form by S binding to the EI species. The model may have different variations: (1) The EIS ternary complex may be either totally or partially inactive in terms of catalyzing S → P, and (2) the initial binding of S or I to the enzyme may or may not affect the subsequent binding of the other. *Pure noncompetitive inhibition* (with total inactivation) applies to the case where $\alpha = 1$; that is, I and S bindings do not influence each other. When I and

FIGURE 5–18 Noncompetitive inhibition. Here K_S is the substrate dissociation constant for ES $\rightleftharpoons$ E + S or ESI $\rightleftharpoons$ EI + S, and K_I is the inhibitor dissociation constant for EI $\rightleftharpoons$ E + I or ESI $\rightleftharpoons$ ES + I.

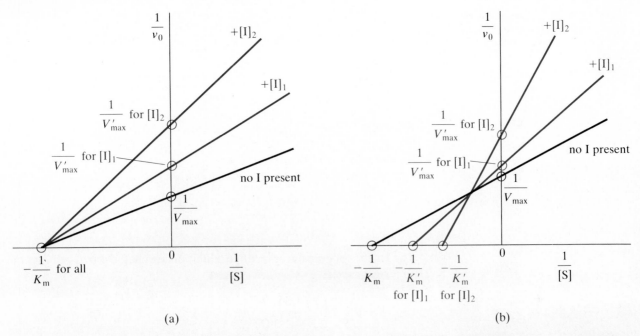

(a)   (b)

S bindings are interdependent ($\alpha \neq 1$), the effect is termed *mixed noncompetitive inhibition*, or merely *mixed inhibition* (see Note 5–8).

The kinetics of pure noncompetitive inhibition are characterized [see Figure 5–19(a)] by a decrease in the maximum velocity (V'_{max} is less than V_{max}) but no change in the K_m value ($K'_m = K_m$). The relationship of V'_{max} to V_{max} (see Note 5–8) involves the same factor identified in the previous discussion of competitive inhibition. In mixed inhibition both V_{max} and K_m are affected (see Note 5–8). In Figure 5–19(b) the characteristics for $\alpha > 1.0$ are shown: The maximum velocity is reduced, and the Michaelis-Menten constant is increased (K'_m is greater than K_m).

Uncompetitive Inhibition

When a nonproductive, ternary EIS complex forms only as a result of I binding to the ES species, the inhibition is called *uncompetitive* (see Figure 5–20). In this case increasing the concentration of S actually enhances the effect of the inhibitor, because more of the I-binding ES species is made available.

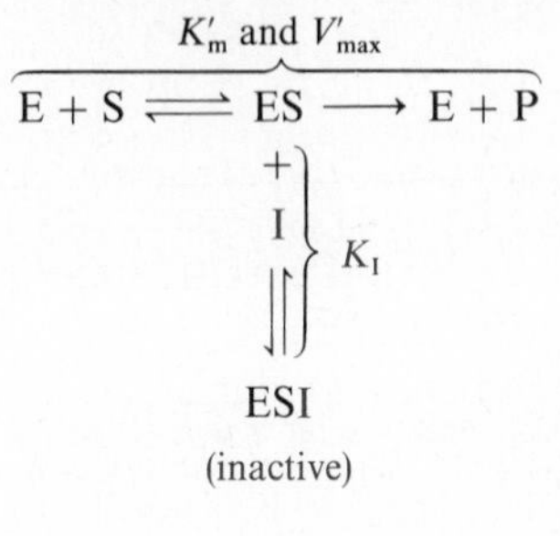

$$\overbrace{\text{E} + \text{S} \rightleftharpoons \text{ES}}^{K'_m \text{ and } V'_{max}} \longrightarrow \text{E} + \text{P}$$

$$\left.\begin{array}{c} + \\ \text{I} \\ \Big\Updownarrow \end{array}\right\} K_I$$

ESI
(inactive)

FIGURE 5–19 Lineweaver-Burk plots. (a) Pure noncompetitive inhibition. (b) Mixed inhibition. In each case the results for two inhibited assays are shown, where $[I]_2$ is greater than $[I]_1$.

NOTE 5–8
For pure noncompetitive inhibition:

$$\frac{1}{V'_{max}} = \left(1 + \frac{[I]}{K_I}\right)\frac{1}{V_{max}}$$

For mixed inhibition:

$$\frac{1}{V'_{max}} = \left(1 + \frac{[I]}{K_I}\right)\frac{1}{V_{max}}$$

and

$$-\frac{1}{K'_m} = -\frac{1}{K_m(1 + [I]/K_I)}$$

FIGURE 5–20 Uncompetitive inhibition.

FIGURE 5–21 Lineweaver-Burk plots of uncompetitive inhibition. The results for two inhibited assays are shown, where $[I]_2$ is greater than $[I]_1$.

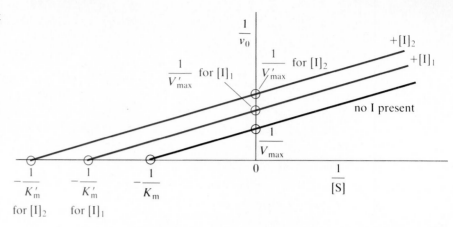

The kinetics of pure uncompetitive inhibition are characterized (see Figure 5–21) by changes in both V and K values:

$$\frac{1}{V'_{max}} = \left(1 + \frac{[I]}{K_I}\right)\frac{1}{V_{max}} \qquad -\frac{1}{K'_m} = -\frac{(1 + [I]/K_I)}{K_m}$$

V'_{max} is less than V_{max} and K'_m is less than K_m. The decrese in K_m may seem to be contrary to the action of an inhibitor. In fact, a decrease in K_m is commonly associated with an activator of enzyme action. The explanation is that by I binding to and consuming ES, the $E + S \rightleftharpoons ES$ equilibrium is shifted to favor ES formation, and so less S is necessary to achieve 50% saturation.

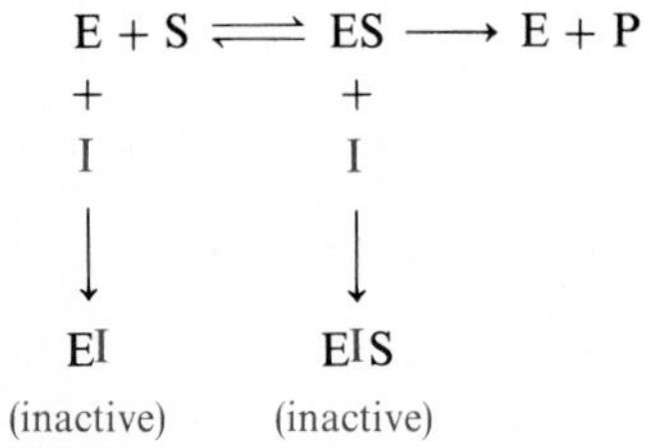

Irreversible Inhibition

Many inhibitors bind *irreversibly* (see Figure 5–22) to E and/or ES. And because V_{max} is reduced, the inhibition is sometimes labeled as noncompetitive. Inhibitors of this type often become *covalently attached* to E and/or ES. The poisonous character of Hg^{2+}, Pb^{2+}, and arsenic compounds is based on this effect.

Inhibitors in this category can also help to identify essential amino acid R groups involved in structure and/or operation of the enzyme. For example,

$$E + S \rightleftharpoons ES \longrightarrow E + P$$
$$+ \qquad\qquad +$$
$$I \qquad\qquad\quad I$$
$$\downarrow \qquad\qquad\quad \downarrow$$
$$EI \qquad\qquad EIS$$
(inactive)   (inactive)

FIGURE 5–22 Irreversible inhibition. The inhibitor may react with free E or ES.

FIGURE 5–23 Identifying amino acid R groups essential to enzyme action.

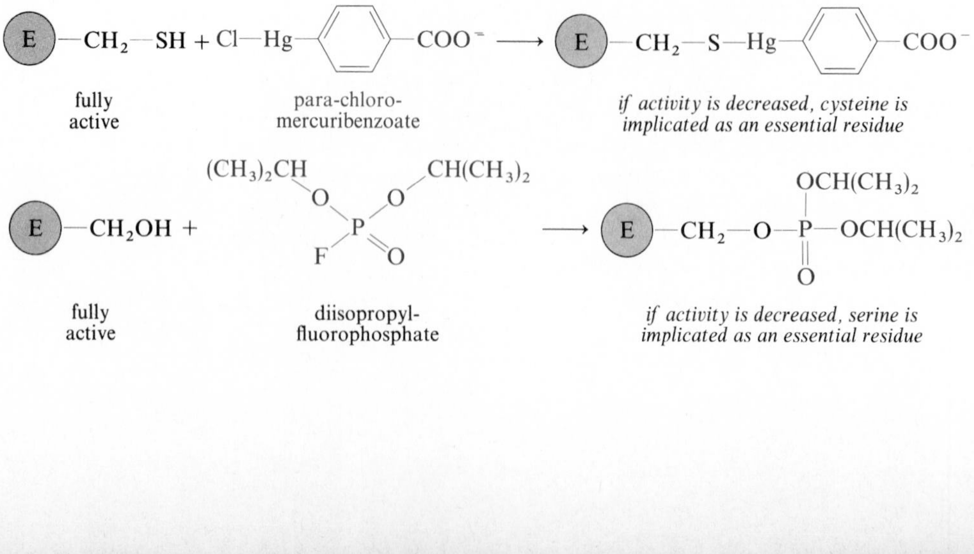

(E)—CH_2—SH + Cl—Hg⟨⟩—COO^- ⟶ (E)—CH_2—S—Hg⟨⟩—COO^-

fully active   para-chloro-mercuribenzoate   *if activity is decreased, cysteine is implicated as an essential residue*

(E)—CH_2OH + diisopropylfluorophosphate ⟶ (E)—CH_2—O—P($OCH(CH_3)_2$)($OCH(CH_3)_2$)=O

fully active   diisopropylfluorophosphate   *if activity is decreased, serine is implicated as an essential residue*

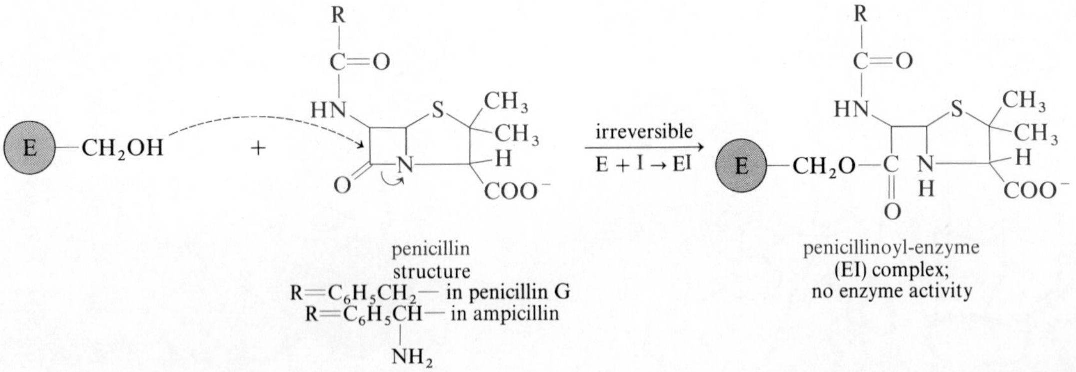

FIGURE 5–24 Penicillin as an irreversible inhibitor. Here E is one of the enzymes involved in the biosynthesis of the cell wall in bacteria; a serine —CH_2OH group is at the active site of the enzyme.

p-chloromercuribenzoate and *diisopropylfluorophosphate* covalently attach to cysteine sulfhydryl groups and serine hydroxyl groups, respectively. Thus if inhibition is detected with either material, we have strong experimental evidence for the essential participation of the R groups of cysteine or serine (see Figure 5–23).

A noteworthy example of a drug operating as an irreversible inhibitor is *penicillin*, the most widely used antibiotic. Penicillin acts by inhibiting one of the enzymes involved in the assembly of the bacterial cell wall (see Section 10–40. Cells lacking a cell wall (they are called *protoplasts*) lyse very easily. The target enzyme of penicillin therapy is converted to an inactive EI form as a result of the chemistry shown in Figure 5–24.

The excessive and indiscriminate use of antibiotics has resulted in the rapid evolution of bacterial strains resistant to antibiotics. One remedy to this problem is to discover or synthesize antibiotics against which bacteria are not yet resistant. Such a development occurred in 1981 with the discovery of a new class of antibiotics called the *monobactams* (see Figure 5–25), which are structurally related to the penicillins and probably act in the same way. Another example of a drug inhibiting an enzyme by covalent modification is *aspirin* (see Figure 11–19).

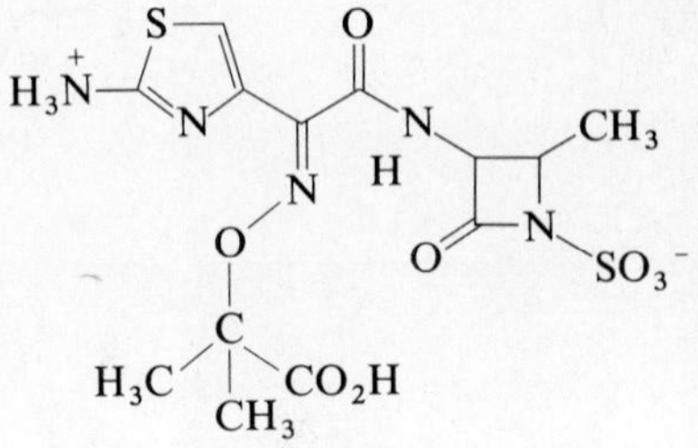

FIGURE 5–25 Monobactam antibiotic. Early testing suggests that the monobactams may be as efficient and as safe as the penicillins.

5–6 ACTIVE-SITE EVENTS: A CLOSER LOOK

The $E + S \rightleftharpoons ES \rightarrow E + P$ processes do not occur in the same way for all enzymes, though similar enzymes often exhibit similar mechanisms. There are, however, some unifying features that explain the chemistry of catalysis. For one thing, the only amino acids able to function in $ES \rightarrow E + P$ as catalytic residues are those with chemically reactive side chains; cysteine (with sulfhydryl), serine (with hydroxyl), threonine (with hydroxyl), glutamic acid (with carboxyl), aspartic acid (with carboxyl), lysine (with amino), arginine (with guanidine), tyrosine (with aromatic hydroxyl), and histidine (with imidazole). All of these amino acids, as well as exposed hydrophobic residues, can also serve as binding residues.

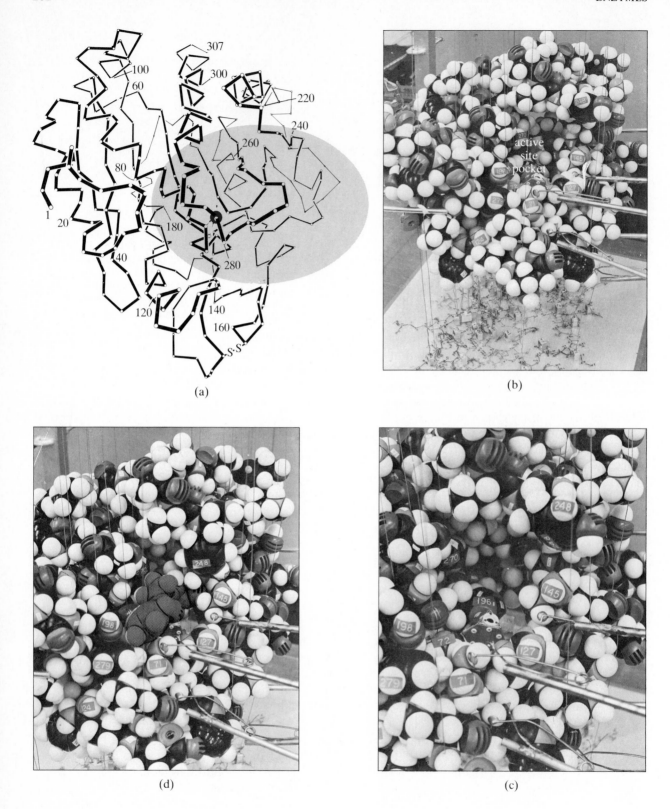

(a)

(b)

(d)

(c)

Another feature is that the catalytic residues generally operate via a combination of two or all three of the following chemical explanations:

1. *Bond strain catalysis,* involving interactions among the cluster of active-site R groups and S that distort bond angles or stretch bond lengths in S. Sometimes, this strain may occur as a result of the covalent attachment of S to E.

2. *Acid-base catalysis,* involving active-site R groups that participate as donors or acceptors of protons (that is, as Brönsted acids or bases) or as acceptors or donors of electrons (that is, as Lewis acids or bases).

3. *Orientation catalysis,* referring to the role of E in positioning the substrates at the same location in such a way that they are in an optimum spatial alignment to react with each other.

These principles are illustrated in the descriptions for the solved mechanisms of the enzymes *carboxypeptidase* and *chymotrypsin.* The carboxypeptidase mechanism illustrates a *concerted set* of events occurring virtually at the same time; the chymotrypsin example illustrates a *sequential mechanism,* with separate events occurring in a precise order.

Carboxypeptidase A is a zinc-containing enzyme consisting of 307 amino acid residues in a single polypeptide chain with one disulfide bond and a molecular weight of 34,000. A drawing of the chain folding is shown in Figure 5–26(a). Years of study by many researchers using several methods, including

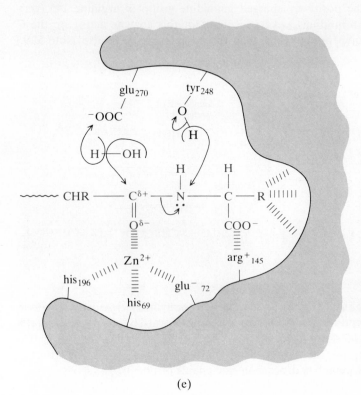

(e)

FIGURE 5–26 Study of carboxypeptidase A structure and function. (a) Polypeptide chain conformation of carboxypeptidase A. The sphere is Zn^{2+}. (b) Space-filling model of the region corresponding to the shaded area in part (a). This region includes the active site. (c) Close-up view of the active-site location, consisting of a deep pocket, a coordinated Zn^{2+}, and other crucial residues. (d) Same region as shown in part (c), with a model substrate in the binding pocket. (e) Coordinated acid-base catalysis of peptide bond cleavage by water and the binding of the R group deep in the pocket. [*Source:* Part (a) reproduced with permission from W. N. Lipscomb, "Structure and Mechanism in the Enzymatic Activity of Carboxypeptidase A and Relations to Chemical Sequence," *Chem. Res.,* **3,** 81–89 (1970). Drawing provided by Dr. Lipscomb. Photographs for parts (b), (c), and (d) provided by Dr. John Sebastian.]

the solution of the three-dimensional structure by X-ray diffraction, have provided the following major facts:

1. The zinc ion (Zn^{2+}) is an essential cofactor for catalytic activity.

2. Two separate histidine residues (69 and 196) and a glutamic acid residue (72) are responsible for holding the Zn^{2+} in its position to interact optimally with the C terminal peptide bond of the substrate polypeptide.

3. An arginine residue (145) engages in an important binding interaction with the substrate.

4. A tyrosine residue (248), in conjunction with the Zn^{2+}, interacts with the substrate to sensitize the peptide bond in the substrate, which is cleaved by water.

5. A deep pocket is at the surface to help anchor the substrate by binding the R group of the C terminal residue of the polypeptide substrate.

The drawings and models in Figure 5–26 summarize one interpretation of these interactions. The key chemical events in the breaking of the C—N peptide bond (see Figure 5–27) are the increased polarity of the carbonyl O═C bond due to the electron-attracting influence of the Zn^{2+} (it functions as a Lewis acid) and the proton-donating influence of tyrosine 248 (it functions as a Brönsted acid). These events are accompanied by the attack of water under the influence of glutamic acid 270 to complete the hydrolysis of the C terminal peptide bond. The electrostatic binding of the negative —COO^- group of the substrate with the positively charged guanidine group of arginine 145 (very near the R group binding pocket) accounts for this enzyme acting on the C terminus of a polypeptide (rather than the N terminus). (See also Note 5–9.)

NOTE 5–9
Carboxypeptidase A (CPA) is active toward all polypeptides that do *not* have arginine, lysine, or proline as the C terminus. Animals also produce carboxypeptidase B (CPB), which acts only if the C terminus is arginine or lysine.

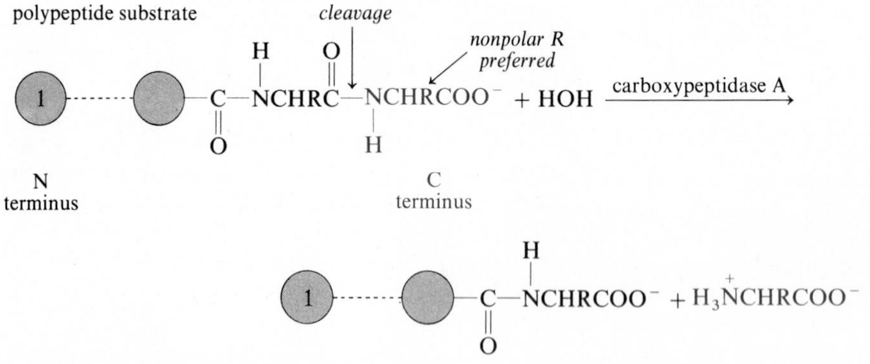

FIGURE 5–27 Overall exopeptidase action of carboxypeptidase A. Active-site events are shown in Figure 5–26.

Chymotrypsin (see Figure 5–28) catalyzes the hydrolysis of internal peptide bonds in a polypeptide substrate with a preference toward bonds where aromatic amino acids donate their C═O group. The specificity implies that the active site has a binding pocket composed of R group residues compatible with the neutral/nonpolar/planar hydrocarbon side chains of phe, trp, and tyr.

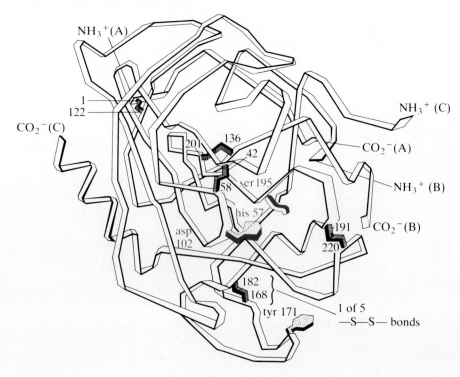

FIGURE 5–28 Chymotrypsin. This intact protein is composed of three polypeptide chains (A, B, C) connected by disulfide bonds. (*Source:* Drawing provided by Dr. D. M. Blow.)

The proposed reaction mechanism (see Figure 5–29) consists of two steps comprising an overall Ping-Pong pathway. The first begins with the covalent attachment of the polypeptide substrate to E via an acylation of the ser 195 residue. The reactivity of the serine 195 is considerably enhanced because of the influence of H^+ abstraction by the N atom of the imidazole group of his 57, an influence that increases the nucleophilic character of the oxygen of the serine —OH group. Then with some influence from a carboxylate side chain of asp 102, H^+ from his 57 is donated to the covalent ES adduct to promote the cleavage of the C—N bond, releasing the C terminal part of the original substrate. Release of the other product occurs in the second step (a deacylation) and involves participation of the second substrate, H_2O. The his 57 again abstracts H^+, but this time for water, promoting the formation of a new C—O bond involving the acyl group carbon. Finally, H^+ from his 57 is donated to O of ser 195, with the breaking of the O—C bond and the release of the product.

5–7 ENZYME SPECIFICITY

The specificity of enzyme action has been mentioned several times, and examples have been given. However, there are still some important aspects to consider.

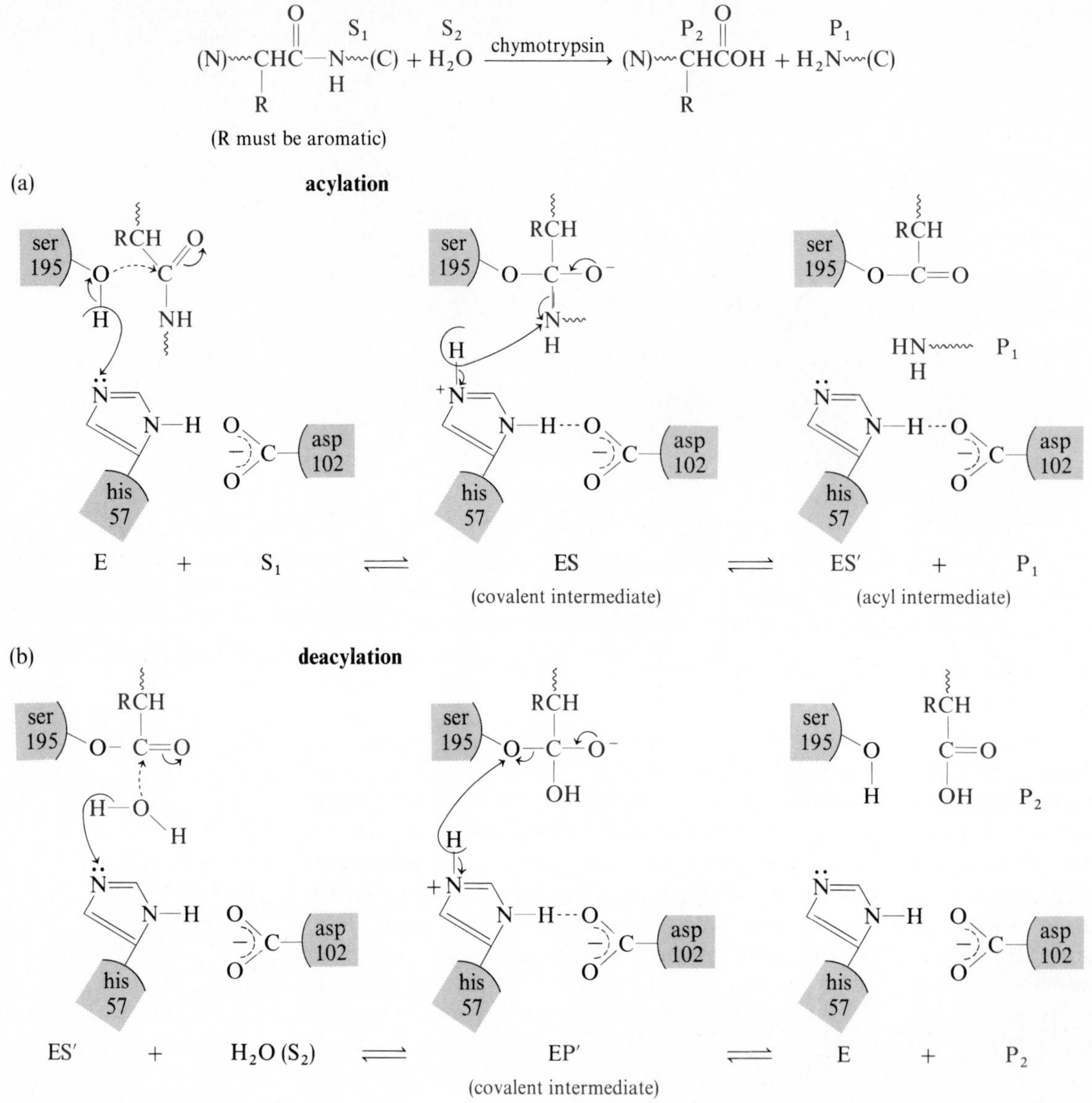

FIGURE 5–29 Chymotrypsin mechanism of action, involving acid-base catalysis and a covalent intermediate. (a) Acylation. (b) Deacylation. The crucial binding interactions for the aromatic R group are not shown.

Some enzymes exhibit *absolute specificity;* each enzyme catalyzes only one S → P reaction. This behavior implies that the molecular topography of the active site is not only highly ordered but is also rigid. There is only one S with a complementary topography that can be properly "fitted" *and* acted upon at the active site. (In 1894, long before the protein nature of enzymes was even established, Emil Fischer first proposed this type of thinking in his lock-and-key hypothesis for biological catalysts.)

A particularly interesting aspect of absolute specificity is the *stereospecific action* exhibited by many enzymes. A remarkable type of stereospecific catalysis occurs when the enzyme distinguishes identical chemical groupings in a sub-

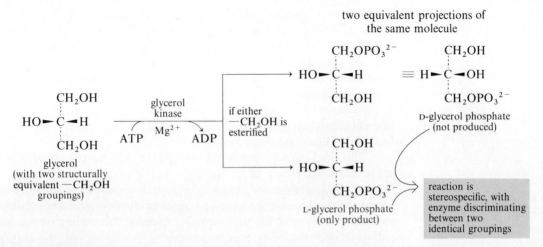

two equivalent projections of the same molecule

D-glycerol phosphate (not produced)

L-glycerol phosphate (only product)

reaction is stereospecific, with enzyme discriminating between two identical groupings

glycerol (with two structurally equivalent —CH₂OH groupings)

FIGURE 5–30 An example of stereospecific catalysis by enzymes.

strate as being different. Consider the example of glycerol kinase, an enzyme that catalyzes the conversion of glycerol to glycerol phosphate, using the substance ATP as the phosphate donor. As shown in Figure 5–30, if the phosphate group is transferred to the OH of carbon 1, the D stereoisomer is produced. If the transfer is to the OH of carbon 3, the L isomer is produced. Although we may reasonably expect that there is a 50–50 chance of either happening because both OH groups are structurally identical (in the same sense that all four hydrogen atoms in CH_4 methane are identical), the enzyme always *produces only the* L *isomer*. In the language of stereochemistry, a substance with a plane of symmetry is asymmetrically converted into only one of two or more possible stereoisomers.

The explanation of this riddle is provided by the *asymmetry of the active-site locus* on the enzyme. A. G. Ogston displayed keen insight when he proposed this theory in 1948 before much was known about active sites. Ogston reasoned that if a symmetrical substrate binds by a one-of-a-kind, three-point attachment to an asymmetric active site, only one of two identical groups in S is properly positioned near the catalytic site—and always the same one (see Figure 5–31). In other words:

> Identical chemical groups in a substrate become different after binding at the microenvironment of the active site of the enzyme.

Several other examples will be encountered in later chapters.

FIGURE 5–31 Asymmetry of the active site. For the sake of simplicity a dual function of binding and catalysis is shown for one residue (R_z); the catalytic area (dashed line) may contain other residues. In the symmetrical substrate the two blocks might represent, for example, the two —CH₂OH groups in glycerol. The ES product displays a one-of-a-kind fit followed by a selective modification of one group in the substrate.

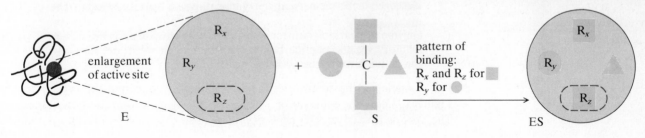

enlargement of active site

E

pattern of binding:
R_x and R_z for ⬛
R_y for ⬤

S

ES

Absolute specificity, however, is not characteristic of all enzymes; most exhibit *relative specificity.* They can catalyze the same type of reaction with any one of a few, several, or many structurally similar substrates. Rigid structures are obviously inconsistent with this behavior. Conformational flexibility, however subtle, is required. The protein conformation and the active-site orientation in particular must be capable of undergoing subtle but significant changes in order to bind and/or catalyze the transformation of similar but different substrates. The example of hexokinase used earlier in the chapter (see Table 5–4) illustrated the point.

This thinking was first proposed by D. Koshland about thirty years ago in his *induced-fit theory* of enzyme action. The premise is remarkably simple: As a substrate binds at the active-site region, R groups at the site shift orientation slightly, assuming the necessary orientation to engage the substrate in a potentially productive enzyme-substrate complex. Depending on the degree of conformational flexibility that is possible at the active site, the enzyme may accept a few to many different substrates.

> The ability of enzymes—and proteins in general—to assume different conformations resulting in a change in the behavior of the protein is of tremendous importance.

Indeed, it is a theme that permeates much of the material in this book.

5–8 ZYMOGENS

zymogen
(inactive precursor protein)

| modification
| by protease

release of
active protein

FIGURE 5–32 Zymogen precursors. For each zymogen a specific modification in structure is involved.

In living cells several proteins are originally synthesized in an inactive form called a *zymogen* (or *preprotein, proenzyme,* or *preenzyme*). Later, usually in response to some other biochemical signal, the inactive zymogen precursor is converted into the bioactive protein (see Figure 5–32).

Consider the examples of trypsin and chymotrypsin, two intestinal enzymes participating in the breakdown of ingested protein to the constituent amino acids, which are then absorbed into the bloodstream. The inactive precursors, *trypsinogen* and *chymotrypsinogen,* are originally synthesized in the pancreas. This inactivity is crucial to the well-being of the pancreas cells that synthesize these digestive proteins. If the enzymes were produced in an active form within the cell, this situation would be potentially self-destructing, since any of the other proteins in the cell would be the immediate targets of their proteolytic action. In fact, the disease *pancreatitis* (which is sometimes fatal) is caused by the early release of trypsin and chymotrypsin in the pancreas.

> Zymogen activation is an enzyme-catalyzed process wherein the operating enzyme (protease, proteinase, peptidase, or proteolytic enzyme are various names) catalyzes the hydrolytic cleavage of one or more specific peptide bonds in the zymogen substrate, releasing the active form of the protein.

For example, in the small intestine trypsinogen is converted to trypsin (see Figure 5–33) by the removal of a hexapeptide fragment from the N terminus. The process is catalyzed by *enteropeptidase* (also called *enterokinase*), another

$$\text{trypsinogen} \xrightarrow{\text{enteropeptidase}} \text{trypsin} + \text{val}—(\text{asp})_4—\text{lys}$$

(first crucial activation)

$$\text{trypsinogen} \xrightarrow{\text{trypsin}} \text{trypsin} + \text{val}—(\text{asp})_4—\text{lys}$$

(*229 residues*)   (*223 residues*)

$$\text{chymotrypsinogen} \xrightarrow{\text{trypsin}} \pi\text{-chymotrypsin}$$

(*245 residues;*
single chain with
five —S—S— bonds)   (*245 residues;*
two chains with
five —S—S—bonds)

$$\pi\text{-chymotrypsin} \xrightarrow{\pi\text{-chymotrypsin}} \alpha\text{-chymotrypsin} + \text{ser}—\text{arg} + \text{thr}—\text{asn}$$

(*241 residues;*
three chains
with two of the five
—S—S— bonds
linking the chains)

FIGURE 5–33 Processing of trypsinogen and chymotrypsinogen zymogens.

enzyme secreted in small amounts by the intestinal mucosa. Then trypsin itself performs the same operation on more trypsinogen. Trypsin also catalyzes the first of two stages in the formation of a stable and active chymotrypsin. The action of trypsin specifically cleaves the peptide bond between residues 15 and 16 in chymotrypsinogen, an inactive protein composed of a single polypeptide chain with 245 residues and five intrachain disulfide bonds. The product of the trypsin action is π-chymotrypsin which is an active proteolytic enzyme. In fact π-chymotrypsin can act on other molecules of π-chymotrypsin as substrates, specifically cleaving three other peptide bonds in π-chymotrypsin between residues 13 and 14, between residues 146 and 147, and between residues 148 and 149. These cuts excise two dipeptide fragments and yield a more stable form of chymotrypsin (the α form), now composed of three polypeptide chains held together by disulfide bonds.

Zymogens are inactive because they lack an active site. The *binding and catalytic residues* are present, but they *are not properly aligned*. The cleavage of one or more peptide bonds in the zymogen triggers new R group interactions throughout the molecule to produce a new conformation wherein the active-site residues assume new positions optimal for catalysis. In other words, a modification in the primary level of structure in a protein results in a modification of the tertiary level of structure in the protein.

Zymogen activation also operates in the process of blood clotting. The last step in blood clotting is the conversion of the soluble blood protein *fibrinogen* to protein *fibrin*. The fibrin molecules then undergo end-to-end and side-to-side aggregation, yielding insoluble strands of fibrin, which form the clot. The fibrinogen → fibrin conversion (see Figure 5–34) is catalyzed by the action of *thrombin,* another enzyme in blood that cleaves four specific peptide bonds in fibrinogen. Thrombin in turn is produced from the inactive zymogen *prothrombin*. The prothrombin activation requires Ca^{2+} and other blood proteins, called *clotting factors*, that participate in an elaborate cascade (not shown here) of activation steps. One of these proteins, designated *factor VIII*, is called the

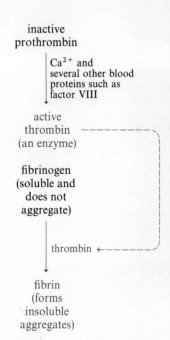

inactive
prothrombin

Ca^{2+} and
several other blood
proteins such as
factor VIII

active
thrombin
(an enzyme)

fibrinogen
(soluble and
does not
aggregate)

thrombin

fibrin
(forms
insoluble
aggregates)

FIGURE 5–34 Conversion of fibrinogen to fibrin.

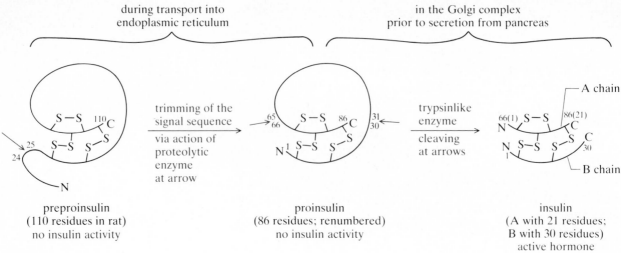

during transport into
endoplasmic reticulum

in the Golgi complex
prior to secretion from pancreas

trimming of the
signal sequence
via action of
proteolytic
enzyme
at arrow

trypsinlike
enzyme
cleaving
at arrows

A chain

B chain

preproinsulin
(110 residues in rat)
no insulin activity

proinsulin
(86 residues; renumbered)
no insulin activity

insulin
(A with 21 residues;
B with 30 residues)
active hormone

FIGURE 5–35 Conversion of preproinsulin to insulin.

antihemophilic factor because its absence in blood is the cause of the common type of hemophilia.

The operation of protein-cutting enzymes applies to the production of other types of bioactive proteins and smaller peptides. Some we have already discussed, namely, various hormone peptides and neurotransmitter peptides (see Section 3–8). Another example is the active form of *insulin,* which is produced from *preproinsulin,* a single polypeptide chain of 110 amino acid residues (in the rat), as a result of cleaving three specific peptide bonds in the sequence shown in Figure 5–35.

First, as part of the membrane transport process of getting the preproinsulin precursor into the cisternae compartment of the endoplasmic reticulum for transport to the Golgi body, the ala 24 —phe 25 peptide bond is cleaved (see Note 5–10). Then before secretion, a trypsinlike enzyme in the islet cells of the pancreas cleaves the 30–31 and 65–66 peptide bonds in proinsulin to excise a 35-residue fragment and form the final insulin molecule. Thus 59 of the 110 residues in the original preproinsulin protein are removed.

In diabetes the low levels of blood insulin can have various explanations: a diseased pancreas, an abnormally small pancreas, a failure of the pancreas to release insulin, or mutations in the insulin gene, giving rise to the production of defective insulin molecules. For example, a phe → leu replacement at position B24 has recently been associated with a defective insulin molecule. Another mutation that has been detected causes a defective proinsulin molecule that cannot be trimmed to insulin.

NOTE 5–10
The removal from a newly synthesized polypeptide of an amino terminal fragment of 15–30 amino acid residues—called a *signal sequence*—often occurs when a protein is transported across a membrane. This aspect of protein processing is discussed more fully in Section 9–5.

5–9 NATURAL REGULATION OF ENZYMES (PROTEINS)

General Principles of Active-Inactive Conversions of Proteins

The ability of a living organism to adjust its processes in response to external and internal influences is called *bioregulation.* This biological control is achieved by *molecular control,* mainly by controlling the *interconversion of active and inactive forms of proteins,* most notably those proteins that operate as enzymes:

$$\text{protein in active state} \xrightleftharpoons[\text{signal B}]{\text{signal A}} \begin{array}{l}\text{protein in}\\\text{inactive or}\\\text{less active state}\end{array}$$

molecular regulation

After proteins are synthesized, their activity can be regulated by (1) *covalent modification,* (2) *noncovalent binding of ligands,* (3) *classical competitive and noncompetitive inhibition,* and (4) *allosteric effects.* (The listing is not a ranking of importance.) In these cases the active-inactive transition occurs rapidly. Regulation by the process of (5) *gene repression,* which occurs more slowly, involves control of the biosynthesis of a protein—an obvious way of regulating a protein-dependent process, since there can be no activity if there is no protein:

$$\text{amino acids} \xrightarrow[\text{biosynthesis}]{\text{gene-directed}} \text{protein}$$

gene regulation

We have already considered the effects of classical inhibition, and gene control will be discussed in Chapter 7.

Covalent Modification

The cleavage of peptide bonds in zymogen activation is a regulatory modification, but it is an irreversible process: Zymogens are activated, but the active products are not deactivated by reconversion to the zymogen state. We now focus on events of chemical modification involving a *reversible,* chemical change of side-chain R groups on the surface of a protein.

PHOSPHORYLATION   The most common and probably most important occurrence involves the *phosphorylation/dephosphorylation* of —OH functions of serine (primarily), threonine, and tyrosine R groups. [Tyrosine phosphorylation (discussed shortly) was detected only recently (1980).] The reversible modification is enzyme-catalyzed but with a different type of enzyme operating in each direction. *Protein kinases* catalyze the transfer of a phosphoryl group ($-PO_3^{2-}$) from ATP to the R group, and *protein phosphatases* catalyze the removal of phosphate by hydrolyzing the phosphorylated R group (see Figure 5–36). Often, the enzymes involved operate only on specific protein substrates and almost always on a specific —OH group in the protein. Usually, the occurrence of phosphorylation activates the protein, but important examples of the reverse situation are known.

Conformational changes are the result of a new set of R group interactions triggered by the conversion of the *polar, uncharged* —OH group to the *more polar, −2-charged* $-OPO_3^{2-}$ group, or vice versa. This $-OH/-OPO_3^{2-}$ change represents a significant *difference in structural information.* The change will first trigger subtle reorientations in other R groups in the immediate vicinity, which in turn may then trigger a wave of R group reorientations through-

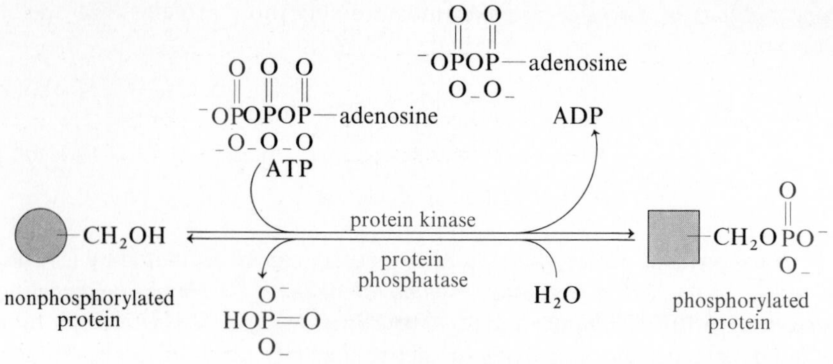

FIGURE 5–36 Phosphorylation and dephosphorylation of proteins. The ● and ■ symbolize the same protein in different three-dimensional conformations. The ○ ⇌ □ symbolism implies that the covalent modification causes a change in the conformation of the intact protein.

out other regions of the molecule, even throughout the entire molecule. The molecule is not denatured but is sufficiently changed in conformation to have its *functional domains altered* and thus have greater or less activity.

Although tyrosine phosphorylation may account for only about 0.05% of all protein phosphorylations, evidence is accumulating that the specific phosphorylation of tyrosine residues is associated with important processes. Most noteworthy is evidence that *tyrosine-specific protein kinases* represent one of the types of proteins coded for by genes, called *oncogenes,* responsible for triggering the transformation of normal cells to tumor cancer cells. (Oncogenes are discussed more fully in Section 7–3.) Other tyrosine-specific protein kinases are associated with the action of at least two polypeptide growth factors, *epidermal growth factor* (EGF) and *platelet-derived growth factor* (PDGF), mitogenic substances that bind to membrane receptors in a cell and stimulate cell division. The association of tyrosine-specific protein kinases with tumorigenic and mitogenic substances strongly suggests that at least one pathway of cellular-growth control involves tyrosine phosphorylation. There is also some recent evidence suggesting that the mode of action of insulin is involved with the activation of a cellular tyrosine-specific protein kinase.

SULFHYDRYL/DISULFIDE REDOX   There are some examples of the control of structural and functional changes in a protein due to the formation/removal of —S—S— bonds involving cysteine side chains in the protein. Clearly, the presence or absence of a disulfide bridge in a polypeptide can affect protein conformation.

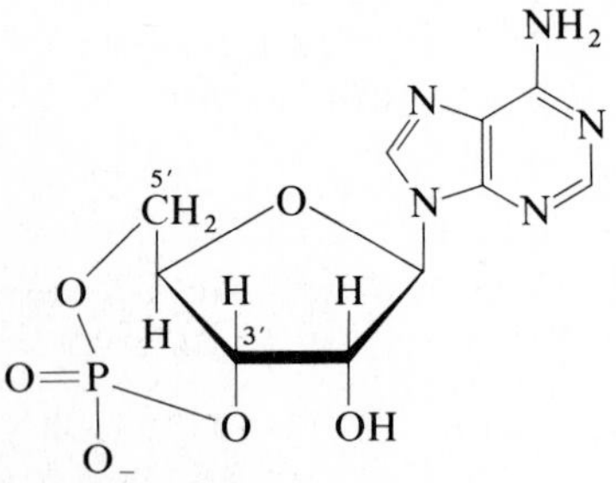

FIGURE 5–37 Structure of 3′,5′-cyclic AMP, a cyclic nucleotide. See Figure 6–20 for details of its formation from ATP and its degradation to noncyclic AMP. The presence of the internal phosphodiester linkage between the 3′ and 5′ carbons is the basis for the cyclic designation.

Noncovalent Ligand Binding

The mere binding/debinding of a substance (generally, we will call it a ligand) is sufficient to affect the transition of a protein between different structural and functional forms. Let us consider a specific example of immense importance, namely, the control by *cyclic AMP* (cAMP)(see Figure 5–37) of a protein kinase (for now, knowing *which* protein kinase is involved is not important; we will discuss its involvement in regulating metabolism in later chapters).

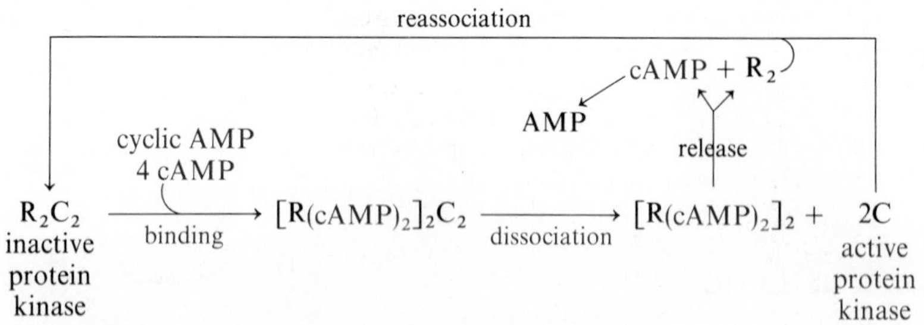

$$R_2C_2 \xrightarrow[\text{binding}]{\text{cyclic AMP}} [R(cAMP)_2]_2C_2 \xrightarrow[\text{dissociation}]{} [R(cAMP)_2]_2 + 2C$$

reassociation

cAMP + R_2

AMP

release

R_2C_2
inactive
protein
kinase

$[R(cAMP)_2]_2C_2$

$[R(cAMP)_2]_2$ + 2C
active
protein
kinase

FIGURE 5–38 Control of a protein kinase by cAMP. The specific protein kinase affected in this way is called cyclic AMP–dependent protein kinase, or cAMP-dPK.

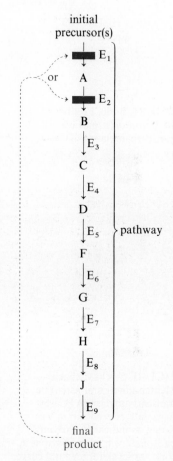

FIGURE 5–39 Feedback inhibition. The product of the sequence can inhibit the activity of an enzyme in one of the early reactions, blocking that reaction and hence blocking all subsequent reactions. The pathway is a sequence of successive enzyme-catalyzed reactions through several intermediates. The final product will exert feedback control when the amount being produced exceeds the needs of the cell.

As indicated in Figure 5–38, the inactive form of the protein kinase is a heterogeneous tetramer R_2C_2 composed of two copies of two different polypeptides—R (for regulatory subunit) and C (for catalytic subunit). Each R subunit has two binding sites for cAMP, and the consequence of cAMP binding is the dissociation of the tetramer to release the free C subunit, which is the active protein kinase. The dissociation of cAMP and its subsequent conversion to noncyclic AMP releases R_2, which can then reassociate with C to restore the inactive R_2C_2. Increases in the cellular concentration of cAMP restart the activation.

Two other important ligands that control the function of other proteins are Ca^{2+} and *guanosine triphosphate* (GTP). Examples will be encountered in subsequent chapters. In fact, throughout the entire book there will be many illustrations of the principles of protein regulation.

Allosterism: Secondary-Site Effects

EXPERIMENTAL BACKGROUND The term *feedback inhibition* (see Figure 5–39) refers to the inhibition exerted by the final product of a multi-step reaction sequence on one of the enzymes operating in an early step of the sequence. Since all steps are linked—the product of one becomes the substrate of the next—any control at an early stage will obviously control the entire sequence. Such control is a major strategy of bioregulation in all living cells.

The first molecular explanation of feedback inhibition was provided in 1962 from studies with the enzyme aspartate transcarbamylase (ATCase), which catalyzes the first step in the biosynthesis of the substances UTP and CTP, both pyrimidine nucleotides. Substances UTP and CTP are used further in the biosynthesis of the nucleic acids RNA and DNA. As indicated in Figure 5–40, CTP (but not UTP) exerts feedback inhibition on ATCase activity.

The experimental clue to ATCase behavior was provided by a standard Michaelis-Menten study: measuring v_0 at different concentrations of aspartate with carbamyl phosphate in excess. The plot of v_0 versus [aspartate] [see Figure 5–41(a)] gave a *sigmoid* (S-shaped) curve rather than the classical hyperbola. Such behavior is referred to as *cooperative kinetics* and has the same general meaning as stated previously (see Section 4–6). For ATCase the aspartate substrate is acting cooperatively, with low levels of aspartate improving the catalytic action of ATCase for higher levels of aspartate.

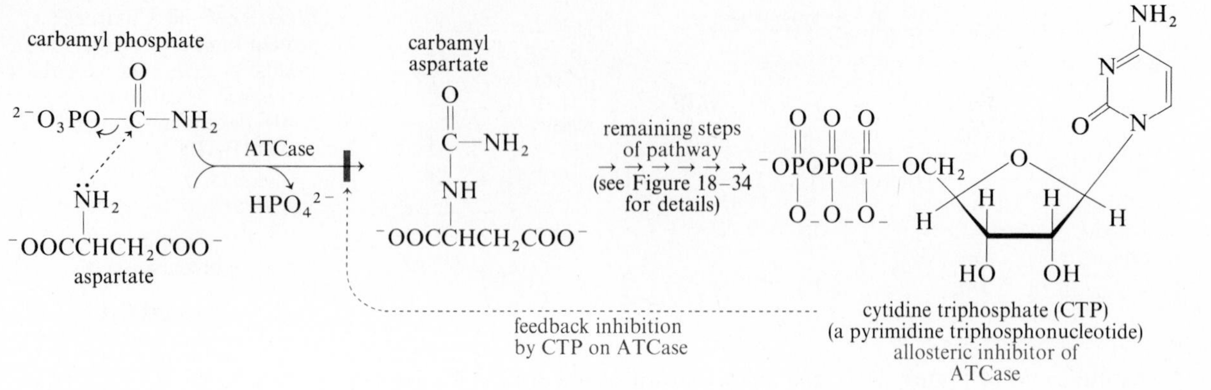

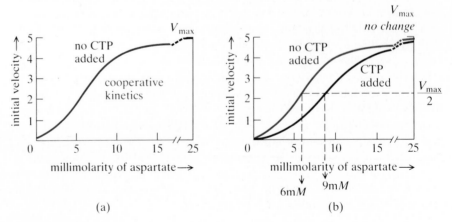

carbamyl phosphate

ATCase

HPO₄²⁻

carbamyl aspartate

remaining steps of pathway → → → → → → (see Figure 18–34 for details)

cytidine triphosphate (CTP) (a pyrimidine triphosphonucleotide) allosteric inhibitor of ATCase

aspartate

feedback inhibition by CTP on ATCase

FIGURE 5–40 Feedback inhibition on aspartate transcarbamylase (ATCase).

When v_0 measurements were repeated in the presence of CTP, the inhibition was distinguished by an accentuated sigmoid curve [see Figure 5–41(b)], indicating that the effect of CTP was to *increase the degree of the cooperative effect of aspartate*. That is, although the presence of CTP did not affect the V_{max} value, a greater amount of substrate was required to achieve the same value for any v_0 less than V_{max}—suggesting competitive inhibition. However, because CTP and aspartate are not structurally similar, the possibility of I and S competition for the same site seems unlikely.

If I and S do not compete for the same site, researchers suggested then that *separate* and *different sites* are involved: one domain having the true active site for S binding and P formation and another domain having a site for binding I. Furthermore, these separate sites necessarily have the ability to *influence each other*. In the case of ATCase the binding of CTP to its site inhibits the binding of aspartate to its site. Convincing proof of this reasoning was obtained by subjecting ATCase to various protein-denaturing treatments, some of which resulted in the selective destruction of the CTP-binding site without damage to the substrate-binding site. In other words, the treated enzyme was still capable of forming carbamyl aspartate at the active site, but it was incapable of binding CTP and thus was not inhibited by CTP.

FIGURE 5–41 Michaelis-Menten kinetics of aspartate transcarbamylase (ATCase). Aspartate concentration was varied as indicated; carbamyl phosphate was maintained at a constant level. (a) Sigmoid curve, reflecting the cooperative effect. (b) Comparing the effect of added CTP, the inhibitor. The black curve is for the CTP effect; the color curve is the same as the curve in part (a). Note that in the presence of CTP, 50% more aspartate is required (9mM versus 6mM) to achieve the indicated v_0. [*Source:* Data taken with permission from J. C. Gerhart and A. B. Pardee, "The Enzymology of Control by Feedback Inhibition," *J. Biol. Chem.*, **237**, 891–896 (1962).]

(a)

(b)

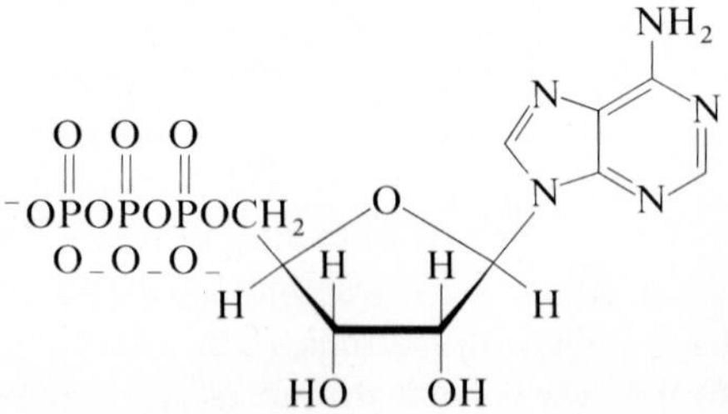

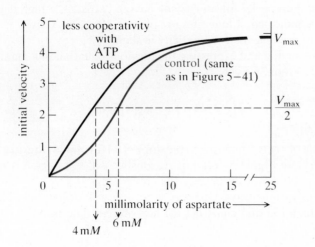

FIGURE 5–42 Adenosine triphosphate (ATP), a purine nucleotide and an activator of ATCase.

Substance ATCase exhibits still another regulatory property, namely, an activation by the substance ATP (see Figure 5–42), a purine triphosphonucleotide. The ATP activation (see Figure 5–43) is reflected by a diminished sigmoid character in the rate curve. Thus ATP *decreases the need for cooperative binding of the substrate.* Indeed, in the presence of enough ATP the cooperative effect of aspartate virtually disappears. The ATP activation was also established to be caused by ATP binding to a separate secondary site. In fact, the ATP-binding site is the same as the CTP-binding site. In view of the close structural similarity between ATP and CTP, it is easy to appreciate this result. What is remarkable, however, is that the separate binding of each one transmits an opposite effect on the active site.

SUMMARY OF GENERAL PRINCIPLES   The ATCase study gave a new insight into the regulatory properties of enzymes in particular and proteins in general. To recapitulate:

> Events of binding at one location on the protein surface can cause a change in the conformation and the activity at another location. This phenomenon is referred to as allosterism (*allo-*, meaning "other or several" and *-sterism,* meaning "spatial conformation").

A protein displaying such behavior is called an *allosteric protein,* and substances causing the effect are called *allosteric effectors*–be they inhibitors, activators, or the substrates themselves.

FIGURE 5–43 Comparison of aspartate transcarbamylase kinetics in the presence and absence of adenosine triphosphate (ATP). The greater velocities at low levels of substrate and the diminished need of the substrate cooperative effect (loss of sigmoid character) typify activation. In the presence of ATP, 33% less aspartate is required to achieve the indicated v_0 (4mM versus 6mM). [*Source:* Data taken with permission from J. C. Gerhart and A. B. Pardee, "The Enzymology of Control by Feedback Inhibition," *J. Biol. Chem.,* **237,** 891–896 (1962).]

When the interactions occur between separate but identical sites responding to the same effector, the effects are called *homotropic*. Homotropic effects always occur when the substrate acts cooperatively in the absence of an inhibitor or activator. The binding of aspartate to ATCase in the absence of CTP or ATP and the binding of O_2 to hemoglobin are examples. When the interactions involve separate and different sites responding to the binding of different effectors, the effects are called *heterotropic*. The inhibiting effect of CTP on aspartate's binding to ATCase is heterotropic, as is the activating effect of ATP on aspartate binding.

Kinetics of Allosteric Enzymes

A simple rate equation can be derived satisfying sigmoid kinetics of an enzyme with n equivalent substrate-binding sites. It is

$$v_0 = \frac{V_{max}[S]^c}{K + [S]^c} \qquad \text{or} \qquad \frac{v_0}{V_{max}} = \frac{[S]^c}{K + [S]^c}$$

The ratio of initial velocity at any saturation (v_0) to that at 100% saturation (V_{max}) is analogous to the fractional-saturation function discussed in the previous chapter. In fact, if you refer to page 154, you will find the v_0/V_{max} equation above to be identical to that for cooperative binding, and with some thought you should be able to appreciate why.

The exponent c is called the *cooperativity coefficient* (sometimes the Hill coefficient). The value of c, which can only be determined experimentally, indicates the type and degree of cooperativity. For an enzyme with n active sites that show no cooperativity, $c = 1.0$ and the above rate equation reduces to the classical Michaelis-Menten rate equation; if the n sites show positive cooperativity, $c > 1.0$; and if the n sites show negative cooperativity, $c < 1.0$. The difference between n and c is a measure of the extent of cooperativity. (The actual integer value of n needs to be evaluated in some other way—for example, by equilibrium-binding assays or by solving the structure of the protein.) The less the difference between the two numbers, the greater is the degree of cooperativity. For example, for $c = 1.7$ for an enzyme having two sites, a high degree of cooperativity is indicated; if four sites are present, the cooperativity is moderate; if six sites are present, the cooperativity is slight.

By rearranging and expressing the previous equation in a manner similar to the Hill equation, we get

$$\log\left(\frac{v_0}{V_{max} - v_0}\right) = c \log[S] - \log K_m$$

with the value of c corresponding to the slope of a Hill plot for $\log[v_0/(V_{max} - v_0)]$ versus $\log[S]$ (see Figure 5–44). For reasons we will not pursue, a Hill plot of actual data is often not linear over the entire range of [S]. This problem can be handled in various ways. One way is to overlook deviations from linearity at low and high [S] and construct the best straight line through intermediate points.

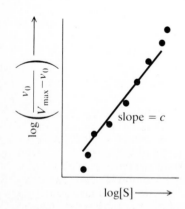

FIGURE 5–44 Hill plot. The V_{max} value can be evaluated by a Lineweaver-Burk or an Eadie-Hofstee plot.

Molecular Explanations of Allosteric Effects

There are two widely accepted descriptions of allosteric behavior: (1) the *indirect concerted model* proposed in 1965 by Monod, Wyman, and Changeux and (2) the *direct sequential model* proposed in 1966 by Koshland. The following descriptions are given in terms of an oligomeric protein in order to account for both homotropic and heterotropic effects. The homotropic effect requires an oligomeric structure. For the sake of simplicity the descriptions use a dimeric protein composed of two subunits. We will not pursue how one can evaluate which of the two models may apply to the behavior of a particular protein.

INDIRECT CONCERTED MODEL The indirect concerted model is based on the mass action principle of chemical equilibrium. The key feature is as follows:

> Initially, different molecules of the same protein exist in two different conformations that are in equilibrium with each other.

One conformation (labeled R, for relaxed) is composed of subunits, each having an optimal site for binding substrate S and, if there is one, an activator A (see Figure 5–45). In other words, the *R conformation is the active state*. The other conformation (labeled T for tight) is also composed of subunits, each lacking an optimal site for S binding but having an optimal site for an inhibitor, if there is one. The *T conformation is the inactive state*. Initially, there are more molecules in the T state than in the R state.

A homotropic cooperative effect is explained as follows [see Figure 5–46(a)]: When the first S molecules bind to yield RS and RS_2, the original $T \rightleftharpoons R$ equilibrium will respond to the loss of R by some conversion of $T \rightarrow R$ to establish a *new equilibrium*. This conversion will give a *greater population of optimal active sites* than initially existed, and additional S can now be more readily bound. Thus a small amount of S binding causes some molecules in the T conformation to change into the R conformation. The effect of S on T is indirect, since S causes T to change to R without binding to T. Because the model depends on an integrated set of conversions, the influence is said to be *concerted*. With increasing levels of S, a $T \rightarrow R$ conversion will continue until there are no molecules left in the T conformation.

Heterotropic effects [see Figures 5–46(b) and 5–46(c)] are explained by similar reasoning on the existence of optimal activator A sites in the active R

FIGURE 5–45 Preexisting equilibrium of conformational states in the indirect concerted model. In the inactive conformation each subunit has an occluded site for substrate binding; S cannot bind. If there is an inhibitor I, it will preferably bind to a T site. In the active conformation the sites for binding substrate S are optimally oriented in each subunit; S can bind. If there is an activator A, it will preferably bind to an R site.

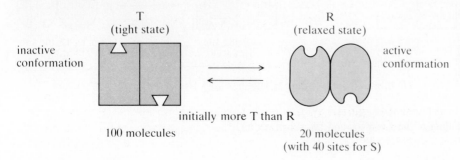

T
(tight state)

R
(relaxed state)

inactive
conformation

active
conformation

initially more T than R

100 molecules

20 molecules
(with 40 sites for S)

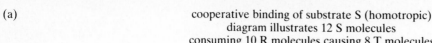

(a)

cooperative binding of substrate S (homotropic)
diagram illustrates 12 S molecules
consuming 10 R molecules, causing 8 T molecules
to convert to 8 R molecules (hypothetical)
result: 44 sites for S now exist

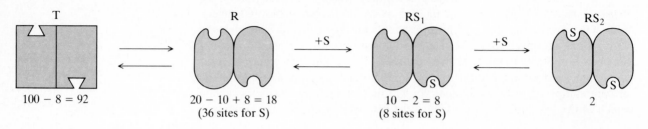

T

$100 - 8 = 92$

R

$20 - 10 + 8 = 18$
(36 sites for S)

RS_1

$10 - 2 = 8$
(8 sites for S)

RS_2

2

$\longrightarrow$ equilibrium shift yielding more R sites for S

(b)

activator A binding (itself cooperative) diminishes need for
cooperativity of S binding; optimal sites for
binding A exist only in R state; A binding
causes $T \rightarrow R$ conversion

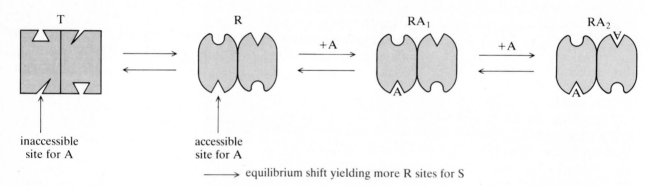

T

inaccessible
site for A

R

accessible
site for A

RA_1

RA_2

$\longrightarrow$ equilibrium shift yielding more R sites for S

(c)

inhibitor I binding (itself cooperative) increases need for
cooperativity of S binding; optimal sites for
binding I exist only in T state; I binding causes
$T \leftarrow R$ conversion

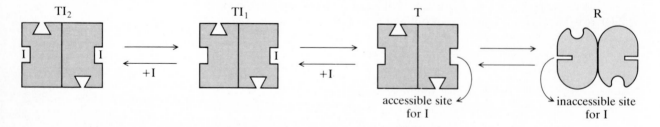

TI_2

TI_1

T

accessible site
for I

R

inaccessible site
for I

$\longleftarrow$ equilibrium shift yielding less R sites for S

FIGURE 5–46 Indirect, two-state model of allosteric effects. (a) Homotropic binding
of the substrate. (b) Heterotropic binding of the activator and the substrate. (c)
Heterotropic binding of the inhibitor and the substrate.

218

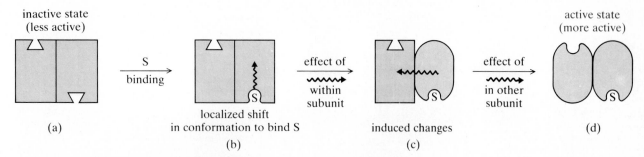

FIGURE 5–47 Direct, sequential model of positive cooperativity of substrate binding. (a) Initial inactive state of dimeric enzyme, with two active sites that are partially accessible to substrate S. (b) Localized shift in conformation. It optimizes the binding of S, which in turn triggers other R group interactions throughout the entire subunit. (c) Induced changes. As indicated, because subunits are associated by noncovalent R group interactions, any conformational change in one will induce a change in the other. (d) Active state. The second site is now more accessible to bind additional S.

conformation and of optimal inhibitor I sites in the inactive T conformation. The presence of A will yield RA and RA_2, causing a $T \rightarrow R$ conversion, increasing the population of optimal active sites for S, and requiring a smaller degree of cooperative S binding. The sigmoid character of the rate curve will diminish. And if sufficient A is present to convert all T conformers into R, RA, and RA_2, there will be virtually no cooperative binding of S (see Figure 5–43). The presence of I will yield TI and TI_2, causing a $T \leftarrow R$ conversion, diminishing the population of optimal active sites for S, and requiring a greater degree of cooperative S binding to overcome the I influence. The sigmoid curve will be accentuated [see Figure 5–41(b)].

DIRECT SEQUENTIAL MODEL The direct sequential model is based on the *flexibility* of protein structure inherent in the induced-fit principle. There is no preexisting equilibrium between active and inactive states. The effector (S, A, I) alters the activity of the same molecule to which it binds—the effect is direct.

Consider the homotropic, positive cooperativity of substrate binding (see Figure 5–47). The key feature is as follows:

> The initial binding of an S molecule to an active site in one sub-
> unit induces conformational changes within that subunit, which in
> turn induce further conformational changes in the other subunit.

The result is that the second active site assumes a more favorable orientation for binding the second S molecule. The initial occurrence and subsequent transmission throughout the molecule of the conformational changes involves the domino type of R group interactions previously described in this chapter.

Figure 5–48 explains the allosteric effect of an inhibitor according to the direct model.

Although both models can account for positive cooperativity, only the induced-fit model is capable of explaining *negative cooperativity,* where the initial

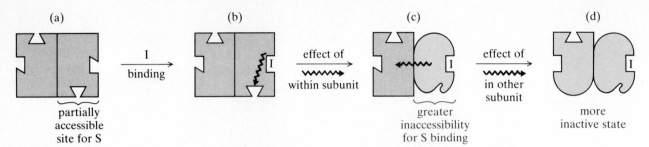

FIGURE 5–48 Direct, sequential model of allosteric inhibition. (a) Dimeric enzyme molecule with partially accessible sites for inhibitor I and substrate S in each subunit. (b) Binding of I. It causes a conformation change within the same subunit that, as shown in **(c)**, makes the binding of S even less likely. (d) More inactive state. The effect shown here that is transmitted to the other subunit illustrates positive cooperativity for I toward more I. After the second I binds, a conformation change at the second S site similar to that shown in (c) will occur. A variation on this diagram could also account for negative cooperativity for any effector. For example, the first I binding could result in the second I site being even less accessible than it was initially.

binding of an effector diminishes further binding of the same effector. This type of behavior has been observed for some allosteric proteins.

5–10 RESEARCH, INDUSTRIAL, AND MEDICAL USES OF ENZYMES

Of the more than two thousand different enzymes described to date, about two hundred are commercially available in small quantities from various biochemical suppliers, and about twenty are available in larger bulk amounts. (In 1983 worldwide sales totaled about $400 million, up nearly 25% from 1979.) The enzymes-for-sale era reflects their use as highly specialized molecular "tools" in various applications exploiting their specificity and rate enhancement effects.

Research Uses

The previously described use of trypsin, chymotrpysin, carboxypeptidase, and other enzymes in the evaluation of a polypeptide's amino acid sequence illustrates one particular research application. In Chapter 6 you will learn how other enzymes are used in sequencing methods applied to the nucleic acids DNA and RNA. Currently, a major application is in the area of recombinant DNA research (Section 6–9), where enzymes are used to map the locations of genes on DNA, to isolate gene segments of DNA, to assist in the chemical synthesis of DNA fragments, and to insert new genes into DNA. Other types of applications include the design of fast, specific, and sensitive quantitative laboratory procedures, the lysis of bacterial cells, and the use of enzymes as catalysts in one or more reactions (particularly stereospecific ones) involved in the chemical synthesis of various types of biological and nonbiological materials.

Their use by synthetic chemists is an area in its infancy, with great potential to improve on many difficult aspects affecting reaction selectivity and yield in the synthesis of structurally complicated compounds. Reporting some promising initial successes, some investigators are even attempting to redesign better enzymes (more selective in their action) by changing the structure of the enzyme and/or the reaction conditions and to design wholly synthetic, nonprotein catalysts that mimic the operation of enzymes.

Industrial Uses

Large-scale industrial processes, such as product synthesis or waste treatment, demand maximum yield at minimum cost. Recently, interest has grown in the use of enzymes to achieve these objectives. Since reaction and recovery conditions are often harsh, industrial processes are designed to use an *immobilized enzyme*—that is, an enzyme attached to an inert support in the reaction vessel as a way of protecting the fragile protein. Another attractive feature of immobilization is the possibility of recovering the enzyme after one use and reusing it, thus lowering cost. However, there are many difficulties, and to date, successful on-line processes are few in number. One particularly efficient process uses glucose isomerase to convert glucose-rich corn syrup into fructose-rich corn syrup used in the manufacture of soft drinks and many other packaged products. Continued research and development will undoubtedly revise some industrial processes and give rise to new ones, spawning a new technology of producing the required enzyme in large amounts.

Methods using a single pure enzyme are supplemented by those using whole microbial cells to detoxify or eliminate waste material and to produce high-protein biomass as an animal feed additive and, in all likelihood, eventually as a human diet additive. In 1981 Imperial Chemical Industries (England) put into operation a 50,000-ton-per-year production capacity for growing a bacterium (*Methylophilus methylotropus*) on methanol. After the cells are recovered and dried, a powder having a crude protein content of 72% is obtained (soybean meal is 45% protein). Moreover, the protein is rich in several of the amino acids essential for animal nutrition, and the powder is also rich in vitamins.

Medical Uses

Some blood diseases, such as hemophilia, can be treated by the administration of blood from nonhemophiliacs. Given such exceptions, the ultimate application of enzymes in medicine—treating those diseases caused by a missing or defective enzyme by replacing the enzyme—is not yet a reality. There is, however, a recently developed medical application of enzymes having great significance: the treatment of coronary heart disease by using enzymes to dissolve blood clots in coronary arteries, estimated to be the cause of 90% of all heart attacks. Before describing the treatment, though, we must briefly describe another aspect of blood biochemistry.

Fibrin coagulation (clot forming) is normally followed within days by *fibrinolysis* (see Figure 5–49), a process involving the proteolytic action of a

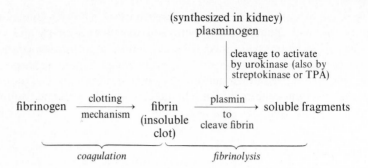

FIGURE 5–49 Formation and dissolution of blood clots (TPA is tissue plasminogen activator; see text).

blood enzyme called *plasmin,* which cleaves the insoluble, coagulated fibrin into smaller peptide fragments that are soluble in blood—hence the clot dissolves. Plasmin itself is produced from an inactive zymogen, *plasminogen,* in a conversion that is physiologically activated by *urokinase,* another proteolytic enzyme.

Since urokinase is a natural clot-dissolving activator, an obvious application is to use it as a clinical treatment for just that in attending to coronary heart disease. Such a treatment was developed about ten years ago. Then a better source of another plasmin-releasing enzyme was discovered (urokinase is isolated from human urine). The source is a *Streptococcus* bacterium that excretes a proteolytic enzyme that not only mimics the action of urokinase but is also more potent in releasing plasmin from plasminogen. It is called *streptokinase.* Either or both enzymes are administered by intravenous injection or preferably by injection directly into a clogged artery through a catheter. The treatment has had good but not outstanding results, and there is a serious complication, namely, the destruction of the normal clotting mechanism, causing major bleeding in patients.

Recently, however, a new plasmin-releasing enzyme was discovered, and it may prove to be more effective and much less troublesome than urokinase and streptokinase. The enzyme, *tissue plasminogen activator* (TPA), is a natural human protein present in only small amounts in the bloodstream. The excitement over TPA is caused by its apparent capability of activating plasmin formation to dissolve the fibrin clot without *deactivating* other blood proteins participating in the complex and delicately balanced process of clotting in response to injury. Clinical studies are currently underway. Anticipating favorable results and eventual FDA approval, a genetic engineering firm has already developed the recombinant DNA techniques to produce TPA in large quantities.

5–11   NEW TYPE OF BIOLOGICAL CATALYST

All enzymes are proteins—a dogma of biochemistry, or so we have thought. Well, now a modification is necessary because of the discovery of the first *nonprotein* biological catalyst, an *RNA* (ribonucleic acid) *molecule.* In the mid

1970s S. Altman was the first to suggest a catalytic function for RNA in combination with protein, but the proposal met with indifference. Then in 1982 T. Cech reported that an RNA substance alone (no association with protein was necessary) catalyzed a chemical reaction. Cech's RNA catalyst, however, did not represent a true catalyst since it did not satisfy one criterion—emerging from the reaction in its original form and available to catalyze the same reaction again. In other words, the RNA did not exhibit turnover.

Then in the period 1983–1984 Altman and N. Pace obtained convincing evidence for a true RNA catalyst exhibiting rate enhancement, specificity, and turnover. The RNA was obtained from *ribonuclease P,* an enzyme that participates in vivo in the conversion of precursor transfer RNA molecules to mature transfer RNA by removing a segment of the precursor (see Section 7–2). When isolated from cells, ribonuclease P is obtained as an RNA-protein complex consisting primarily of RNA. They established that after complete dissociation of the complex and separation of the RNA from protein, the free protein was catalytically inactive but the free RNA still exhibited catalytic activity. The activity of the RNA was less than the activity for the intact RNA-protein complex, suggesting that the protein acts as a cofactor for the RNA.

This discovery now raises the obvious question about whether there are other naturally occurring RNA molecules acting as catalysts or whether the RNA of ribonuclease P is an oddity. Whatever, after your study of Chapter 6, you will appreciate that the RNA three-dimensional structure is intricate and flexible enough for an RNA molecule to function as a catalyst. One interesting implication of this finding is the support it gives to the suggestion that on primitive earth protein catalysts may have evolved after nucleic acids, rather than vice versa.

LITERATURE

General

ALPER, J. "Better Weapons for Antibiotic Warfare." *High Technol.,* 61–65 (December 1983). General article describing the development of new antibiotics.

BISHOP, J. M. "Oncogenes." *Sci. Am.,* **246,** 80–94 (1982). A description of cancer-causing viruses in terms of the modification of host proteins by a phosphorylating kinase protein.

BLOW, D. M. "Structure and Mechanism of Chymotrypsin." *Accounts Chem. Res.,* **9,** 145–152 (1976). A review article.

BLOW, D. M., and T. A. STEITZ. "X-Ray Diffraction Studies of Enzymes." *Annu. Rev. Biochem.,* **39,** 63–100 (1970). A good review of structural details of seven enzymes in terms of the mechanism and the specificity of their catalytic action.

BOYER, P. D., H. LARDY, and K. MYRBACK, eds. *The Enzymes.* 2nd and 3rd eds. New York: Academic Press, 1960 and 1970. A valuable multivolume work covering most aspects of biocatalysis with emphasis on reaction types. Volume 1 (second edition, 1960) is devoted to fundamentals of enzyme catalysis. Volumes 1 and 2 (third edition, 1970)

contain information on structure, control, kinetics, and mechanism.

CHAN, S. J., and D. F. Steiner. "Preproinsulin, a New Precursor in Insulin Biosynthesis." *Trends Biochem. Sci.,* **2,** 250–252 (1977). A short review article.

COLOWICK, S. P., and N. O. KAPLAN, eds. *Methods in Enzymology.* New York: Academic Press. Articles in this encyclopedic work provide detailed instructions of laboratory procedures in biochemistry. One or more volumes are published every year. As of 1985, about 90 volumes have been published.

CORNISH-BOWDEN, A. *Fundamentals of Enzyme Kinetics.* Boston and London: Butterworth, 1979. An excellent compact treatment (211 pages) of the essential principles of enzyme kinetics. The level of mathematics is elementary, and practical aspects are included.

GUERRIER-TAKADA, C., and S. ALTMAN. "Catalytic Activity of an RNA Molecule Prepared by Transcription in Vitro." *Science,* **223,** 285–286 (1984). Proof that the RNA component of ribonuclease P can act as catalyst. In the same issue (p. 266), also see the Research News article, "First True RNA Catalyst Found," by R. Lewin.

HUNTER, T. "Phosphotyrosine—A New Protein Modification." *Trends Biochem. Sci.,* 246–249 (July 1982). A short review article.

KOSHLAND, D. E. "Correlation of Structure and Function in Enzyme Action." *Science,* **142,** 1533–1541 (1963). A review article defining the nature of the catalytic process from the standpoint of protein structure. Included is a definitive summary of the induced-fit theory.

LIPSCOMB, W. N. "Structure and Mechanism in the Enzymatic Activity of Carboxypeptidase A and Relations to Chemical Sequence." *Accounts Chem. Res.,* **3,** 81–89 (1970). A review article.

MAUGH, T. H. "A Renewed Interest in Immobilized Enzymes." *Science,* **223,** 474–476 (1984).

———. "Need a Catalyst? Design an Enzyme." *Science,* **223,** 269–271 (1984).

———. "Semisynthetic Enzymes Are New Catalysts." *Science,* **223,** 154–156 (1984). Brief but informative research news articles.

NEURATH, H., and K. A. WALSH. "Role of Proteolytic Enzymes in Biological Regulation." *Proc. Natl. Acad. Sci., USA,* **73,** 3825–3832 (1976). An excellent short review article of zymogen activations.

NORD, F. F., ed. *Advances in Enzymology.* New York: Wiley. A publication composed of annual volumes (since 1942) devoted to reviewing progress in enzymology. The articles, written by authorities, deal with general and specific subjects. This publication is an extremely useful reference work for researchers, teachers, students, and writers of biochemistry textbooks.

ROSEN, O. M., and E. G. KREBS, eds. *Protein Phosphorylation.* New York: Cold Spring Harbor Laboratories, 1981. Two volumes of research papers on many aspects of phospho $\rightleftharpoons$ dephospho conversion of protein.

Scientific American. The entire September 1981 issue contains articles dealing with the industrial use of enzymes.

SEGAL, H. L. "Enzymatic Interconversion of Active and Inactive Forms of Enzymes." *Science,* **180,** 25–32 (1973). A review article.

———. *Enzyme Kinetics.* New York: Wiley, 1975. A thorough treatment of many subjects.

STROUD, R. M. "A Family of Protein-Cutting Proteins." *Sci. Am.,* **231,** 74–88 (1974). An excellent article discussing the structure and function of serine-containing proteolytic enzymes, including chymotrypsin.

Aspartate Transcarbamylase

GERHART, J. C., and A. B. PARDEE. "The Enzymology of Control by Feedback Inhibition." *J. Biol. Chem.,* **237,** 891–896 (1962). The first study providing evidence for the existence of secondary sites in enzymes and their role in the modulation of catalytic activity.

KANTROWITZ, E. R., S. C. PASTRA-LANDIS, and W. N. LIPSCOMB. "*E. coli* Aspartate Transcarbamylase." *Trends Biochem. Sci.,* **5,** 124–128 and 150–153 (1980). Two short review articles on the kinetic (I) and structural (II) features of ATCase.

Allosterism

KOSHLAND, D. E. "Protein Shape and Biological Control." *Sci. Am.,* **230,** 52–64 (1973). Excellent article discussing allosterism, with an explanation according to the principles of induced-fit theory.

MONOD, J., J. P. CHANGEUX, and F. JACOB. "Allosteric Proteins and Cellular Control Systems," *J. Mol. Biol.,* **6,** 306–329 (1963). The original article describing the theory of allosterism.

MONOD, J., and J. P. CHANGEUX. "On the Nature of Allosteric Interactions: A Plausible Model." *J. Mol. Biol.,* **12,** 88–118 (1965). Original article explaining the formulation of the two-state equilibrium model for enzyme regulation.

PERUTZ, M. F. "Hemoglobin Structure and Respiratory Transport." *Sci. Am.,* **239,** 92–125 (1978). A superb article describing the allosteric properties of hemoglobin in precise molecular detail according to the T $\rightleftharpoons$ R model.

E X E R C I S E S

5–1. For each of the following reactions, identify the enzyme as an oxidoreductase, transferase, hydrolase, lyase, isomerase, or ligase, and give its main-class numerical designation (CoE represents coenzyme).

(a)
$$CH_3\overset{O}{\overset{\|}{C}}COO^- \xrightarrow[\text{(CoE)}]{E} CH_3\overset{O}{\overset{\|}{C}}H + CO_2$$

(b)
$$CH_3\overset{O}{\overset{\|}{C}}H + NADH \xrightarrow{E} CH_3CH_2OH + NAD^+$$

(c) D-amino acid $\xrightarrow{E}$ L-amino acid

(d)
$$CH_3\overset{O}{\overset{\|}{C}}COO^- + CO_2 + ATP \xrightarrow[\text{(CoE)}]{E}$$
$$^-OOCCH_2\overset{O}{\overset{\|}{C}}COO^- + ADP + P_i$$

(e) adenine-ribose + H_2O $\xrightarrow{E}$ adenine + ribose

(f) $^-OOCCH{=}CHCOO^- + H_2O \xrightarrow{E} {^-}OOCCH_2\overset{OH}{\overset{|}{C}}HCOO^-$

5–2. Describe the interpretation of the ES model, and derive the Michaelis-Menten rate equation—without referring to the text.

5–3. The enzyme glutamic acid dehydrogenase catalyzes the reaction

$$\overset{\overset{\displaystyle NH_3^+}{|}}{^-OOCCH_2CH_2CHCOO^-} + NAD^+ \longrightarrow$$

L-glutamate

$$\overset{\overset{\displaystyle O}{||}}{^-OOCCH_2CH_2CCOO^-} + NH_3 + NADH + H^+$$

α-ketoglutarate

The dependence of initial velocity on the concentration of L-glutamate is given in Table 5–5. The initial concentration of NAD^+ was held constant in each case. Calculate the K_m and V_{max} of the enzyme by the direct Michaelis-Menten plot and by the Lineweaver-Burk method. (*Note:* The reaction velocity is expressed in terms of the rate of change in the absorbance at 360 nm, reflecting the formation of NADH.)

TABLE 5–5 Data for Exercise 5–3

L-GLUTAMATE CONCENTRATION (mM)	INITIAL VELOCITY (CHANGE IN ABSORBANCE AT 360 nm/min)
1.68	0.172
3.33	0.250
5.00	0.286
6.67	0.303
10.0	0.334
20.0	0.384

5–4. For the system described in Exercise 5–3, what is the value of the dissociation constant for the dehydrogenase and L-glutamate? What is the value for the affinity constant?

5–5. Evaluate K_m and V_{max} from the data given in Exercise 5–3 (a) by the Eadie-Hofstee format and (b) by the direct, linear plot according to Eisenthal and Cornish-Bowden.

5–6. Kinetic data for an enzyme-catalyzed reaction assayed in the absence and presence of an inhibitor I ($4.5 \times 10^{-2}\ M$) are given in Table 5–6.

TABLE 5–6 Kinetic data for Exercise 5–6

SUBSTRATE (M)	$v_0 (-I)$	$v_0 (+I)$
	(mmole/min)	
0.5×10^{-4}	0.71	0.43
1.0×10^{-4}	1.07	0.71
2.0×10^{-4}	1.50	1.05
3.5×10^{-4}	1.80	1.41
5.0×10^{-4}	1.88	1.60

are given in Table 5–6. Determine whether the inhibition is competitive, noncompetitive, or uncompetitive. What is the value of K_I for the enzyme-inhibitor complex?

5–7. Pyruvate kinase, a dimer with two active sites, catalyzes the reaction

$$\underset{\text{(PEP)}}{\text{phosphoenolpyruvate}} + ADP \longrightarrow \underset{\text{(PYR)}}{\text{pyruvate}} + ATP$$

The velocities of the pyruvate kinase reaction, obtained at several concentrations of PEP and in the presence of an excess amount of ADP, are given in Table 5–7. Two sets of v_0 are given: one in the absence of fructose-1,6-bisphosphate (FBP) and the second in the presence of FBP. From direct Michaelis-Menten plots, evaluate (a) whether pyruvate kinase displays positive cooperativity for PEP and (b) the type of allosteric effect, if any, for FBP.

TABLE 5–7 Kinetic data for Exercise 5–7

[PEP] (mM)	$v_0 (-FBP)$	$v_0 (+FBP)$
	(mmole/min)	
0.020	0.006	0.090
0.030	0.010	0.104
0.055	0.031	0.115
0.085	0.065	0.118
0.150	0.110	0.124
0.200	0.125	0.130

5–8. Construct Hill plots for the two sets of kinetic data given in Exercise 5–7, and evaluate the Hill coefficient for each. Do the values of the Hill coefficients, when compared with a value of 2 active sites present in pyruvate kinase, corroborate your conclusions made in Exercise 5–7? Explain.

5–9. In the presence of a saturating amount of substrate, 1.4 mg of an enzyme results in a velocity of 6.2 mmole of product formed per minute. The molecular weight of the enzyme is 52,500 g/mole. What is the molecular activity of the enzyme?

5–10. After a preincubation with *p*-chloromercuribenzoate, the binding of an enzyme with its substrate was no different from that of the untreated enzyme, but the catalytic activity of the treated enzyme was found to be 40% less. What conclusion can be drawn from this type of observation?

5–11. Knowing how the K_m and V_{max} values of an enzyme are affected by a competitive inhibitor, construct, on the same graph, a set of three plots that illustrate how competitive inhibition is diagnosed by the Eadie-Hofstee format. Use one plot for the control with no I present, a second plot for $[I]_1$ present, and a third plot for $[I]_2$, where $[I]_2$ is greater than $[I]_1$.

5–12. As an exercise in algebra (for a change of pace), convert the Michaelis-Menten equation to (a) the Lineweaver-Burk equation and (b) the Eadie-Hofstee equation.

5–13. Given the preferred specificity of carboxypeptidase B (see Note 5–10) and the proposed mechanism of action for carboxypeptidase A [see Figure 5–26(e)], propose a likely mechanism of action for carboxypeptidase B.

5–14. An enzyme is competitively inhibited to different degrees by the same concentration of three separate inhibitors, I_1, I_2, and I_3. The inhibitor dissociation constants are $K_{I_1} = 0.1 mM$, $K_{I_2} = 0.01 mM$, and $K_{I_3} = 1.0 mM$. Which inhibitor will probably produce the greatest inhibition? The least inhibition?

5–15. What type of Michaelis-Menten plot will be obtained if the concentration of CTP is increased above that specified in the labels and caption of Figure 5–41(b)? Explain the reason for this result in terms of the two-state model for allosteric effects.

5–16. According to the principles of the induced-fit model, draw a series of diagrams, as shown in Figure 5–48, explaining the following: A dimeric allosteric protein has separate substrate, inhibitor, and activator sites; the inhibitor displays negative cooperativity toward itself and inhibits both substrate and activator binding.

5–17. Define these terms: ES, cofactor, coenzyme, holoenzyme, apoenzyme, kinase, dehydrogenase, oxidoreductase, stereo-specificity, active site, 100% saturation, induced-fit theory, protein kinase, allosterism, homotropic, heterotropic, R and T.

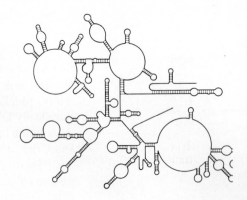

CHAPTER SIX

NUCLEOTIDES AND NUCLEIC ACIDS

In the 1860s F. Miescher isolated an acidic substance from cell nuclei that he termed nuclein and later *nucleic acid*. The biological function of this material was not discovered until nearly a century later when (in the 1940s) Avery, MacLeod, and McCarty established that nucleic acid material—and specifically DNA—was responsible for carrying hereditary information. When the solution for the molecular structure of DNA was reported in 1953 by Watson and Crick, a new era in biochemistry and biology began.

There are two classes of nucleic acids found in every living organism: *ribonucleic acid* (RNA) and *deoxyribonucleic acid* (DNA). Viruses, on the other hand, contain only one type, either RNA or DNA. The biological functions of nucleic acids include the storage, replication, recombination, and transmission of genetic information. In short, they are the molecules that determine what a living cell is and what a living cell does. The classes and occurrence of nucleic acids are summarized in Table 6–1.

All nucleic acids are polymeric substances covering a wide spectrum of molecular size. Transfer RNA is the smallest nucleic acid, having a molecular weight of approximately 25,000. At the other extreme, representing some of the largest substances yet known, individual molecules of DNA range in molecular weight from 1,000,000 to 1,000,000,000.

With each passing year several new findings are reported concerning the biochemistry of nucleic acids. Merely trying to keep pace with new developments is a major chore. In Chapters 6, 7, and 8 we will examine many, but not all, aspects of this area. The primary consideration in this chapter will be nucleic acid structure. We will also describe recombinant DNA techniques with which it is possible to isolate individual genes, to clone isolated genes, and to design novel life forms by inserting foreign genes into a host cell.

TABLE 6–1 Classes and occurrence of nucleic acids

NUCLEIC ACID	OCCURRENCE
DNA	
Nuclear DNA	Nucleus of eukaryotes
Cellular DNA	Prokaryotes
Plasmid DNA	Prokaryotes
Mitochondrial DNA	Mitochondrion of eukaryotes
Chloroplast DNA	Chloroplasts
Viral DNA	Animal, plant, and bacterial viruses
RNA	
Messenger RNA	Prokaryotes and eukaryotes
Ribosomal RNA	Prokaryotes and eukaryotes
Transfer RNA	Prokaryotes and eukaryotes
Small nuclear RNA	Eukaryotes
Viral RNA	Animal, plant, and bacterial viruses
Subviral RNA	Free molecules of RNA

6–1 NUCLEOTIDES: THE MONOMERIC COMPONENTS OF NUCLEIC ACIDS

Nucleotide Composition

Both RNA and DNA are polymers composed of monomeric units called *nucleotides;* hence a nucleic acid is also called a *polynucleotide.* A single nucleotide consists of three chemical parts: inorganic **phosphate** P, a simple **sugar** S, and either a **purine** or a **pyrimidine.** The purines and pyrimidines are often called *nitrogen bases,* or simply *bases* B. In a nucleotide structure the three parts are attached to each other in the order phosphate—sugar—base, or P—S—B. In a polynucleotide ester bonds link the sugar and phosphate components of adjacent nucleotide monomers (see Figure 6–1). Since the sugar and the phosphate within a nucleotide monomer are also linked via an ester bond, the S—P—S linkage along the backbone of a polynucleotide chain is called a *phosphodiester bond.* The nitrogen bases are not involved in any covalent linkages other than their attachment to the sugar moieties along the backbone.

It is the sequence of nitrogen bases along the invariant sugar—phosphate backbone that determines the unique structure of DNA and RNA molecules.

The terms *nucleotide sequence* and *base sequence* are used interchangeably.

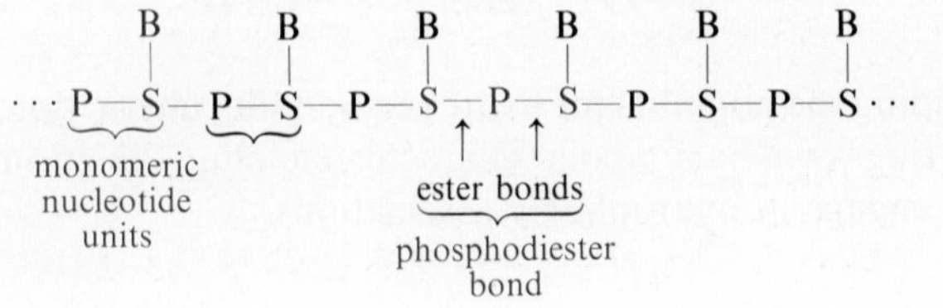

FIGURE 6–1 Symbolic representation of a polynucleotide chain. (See Figure 6–26 for a complete formula.)

Sugars: Ribose in RNA and Deoxyribose in DNA

A complete description of carbohydrate structures is given in Chapter 10. The objective here is merely to show the structures of the two simple sugars characteristic of RNA or DNA.

The nucleotides of RNA contain *β-D-ribose,* and the nucleotides of DNA contain *β-D-2-deoxyribose.* Both are *pentose sugars* (five carbon atoms), with the most stable form of each being a *furanose ring system.* As shown in Figure 6–2, the two differ in structure only at carbon 2 (C2), with deoxyribose having

FIGURE 6–2 Pentose sugars. (a) *β-D-ribose* of RNA. (b) *β-D-2-deoxyribose* of DNA.

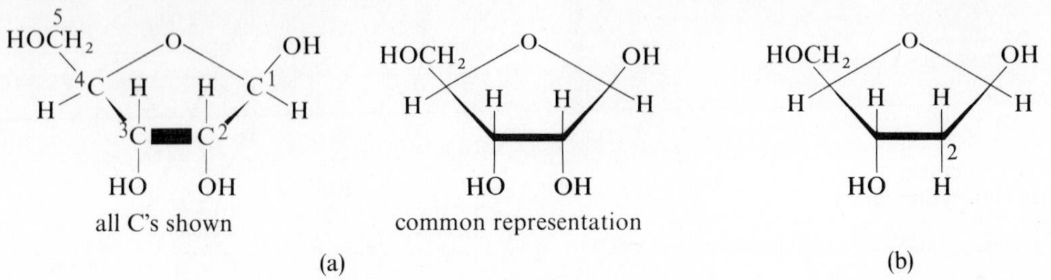

all C's shown · common representation

(a)

(b)

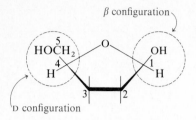

β configuration

D configuration

FIGURE 6–3 Basis for β and D designations in furanose ring system.

—H in place of the —OH group. In other words, deoxyribose is a reduced form of ribose. The β and D symbols (see Figure 6–3) refer to the particular configurations at the C1 and C4 positions in the furanose ring. For now, understanding why is not essential. This and other points are dealt with in Chapter 10. Shortly, we will identify how the nitrogen bases are linked at the C1 position and how the C3—OH and C5—OH groups participate in the phosphodiester bonds.

Purines and Pyrimidines

The bases found most frequently in RNA and DNA are the purines, *adenine* and *guanine*, and the pyrimidines, *cytosine, thymine,* and *uracil* (see Table 6–2). Their structures and systematic names are given in Figure 6–4, and their occurrence in nucleic acids is listed below:

In DNA: A, G, C, T

In RNA: A, G, C, U

The space-filling models in Figure 6–5 clearly illustrate the coplanar bonding pattern in these *aromatic, heterocyclic ring systems.* Although A, G, C, T, and U contain

$$\overset{\delta^+}{C}=\overset{\delta^-}{O} \quad \text{and} \quad \overset{\delta^-}{N}—\overset{\delta^+}{H}$$

dipoles, providing the potential for hydrogen bonding interactions, the aromatic nature of the rings confers nonpolar character, enabling the purines and pyrimidines to also engage in hydrophobic associations.

TABLE 6–2 Abbreviations and symbols for bases and nucleosides

Bases

Adenine	Ade
Guanine	Gua
Cytosine	Cyt
Uracil	Ura
Thymine	Thy
Purine	Pur
Pyrimidine	Pyr
Base	Base or B

Nucleosides

Adenosine	Ado or A
Guanosine	Guo or G
Cytidine	Cyd or C
Uridine	Urd or U
Thymidine	Tho or T
Purine nucleoside	Puo or R
Pyrimidine nucleoside	Pyd or Y
Any nucleoside	Nuc or N

Note: Sometimes, the A, G, C, U, and T symbols are also used for the bases.

adenine
6-aminopurine

guanine
2-amino-6-oxypurine

uracil
2,4-dioxypyrimidine

cytosine
2-oxy-4-amino
pyrimidine

thymine
5-methyl-2,4-
dioxypyrimidine

FIGURE 6–4 Structures of major purines and pyrimidines in nucleic acids.

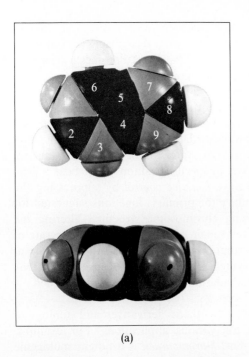

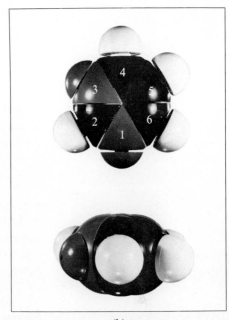

(a)   (b)

FIGURE 6–5 Space-filling models of bases in RNA and DNA. (a) Purine structure. (b) Pyrimidine structure. Both purines and pyrimidines are flat molecules.

NATURALLY OCCURRING MODIFIED BASES   Examples of some nitrogen bases occurring less frequently in nature are shown in Figure 6–6. *Dihydrouracil* and *4-thiouracil* are just two of several minor components of RNA found specifically in transfer RNA. Various methylated bases are found in the DNA of prokaryotes and eukaryotes and in RNAs (see Note 6–1). In the DNA of eukaryotes *5-methylcytosine* (m^5Cyt or m^5C) is particularly important. In various higher plants and vertebrates the m^5C content may range from 0.5 to 7.0 mole % (that is, moles of m^5C per 100 mole bases). In prokaryotes and some lower eukaryotes the DNA also contains small amounts of a methylated purine, *6-methyladenine*. Some transfer RNA molecules also contain small amounts of dimethylated bases.

In the DNA of prokaryotes the natural existence of methylated bases is known to be responsible for at least one specific function, namely, providing protection to the cellular DNA from the action of restriction enzymes present in the cell. In bacterial cells restriction enzymes are responsible for degrading

NOTE 6–1
Specific enzymes participate in adding the methyl group to bases after the polynucleotide is assembled. The source of the methyl group is an activated methyl sulfonium compound,

$$-\overset{+}{\underset{|}{S}}-CH_3,$$ produced from the amino acid methionine (see Figure 18–27).

FIGURE 6–6 Some modified bases occurring in RNA and DNA.

5-methylcytosine
(DNA)

6-N-methyladenine
(DNA)

4-thiouracil
(in some transfer RNAs)

dihydrouracil
(in some transfer RNAs)

the foreign DNA of a bacterial virus that enters the cell. Methylated bases in the cellular DNA prevent the restriction enzyme from recognizing the cell DNA as a substrate. Lacking methylated bases, the viral DNA becomes a target for restriction enzyme binding and subsequent degradation.

Given the widespread occurrence of 5-methylcytosine in specific base sequences of eukaryotic DNA, its presence is likely to be of major importance in the functioning of DNA. However, what function(s) of DNA is(are) determined by methylation is not yet definitely established. Some researchers propose that it has a major part in regulating the in vivo expression of genetic information. Specifically, the existence of m^{5}C bases in specific positions in or near a gene on the chromosome is proposed to have *negative control* by suppressing the expression of the gene. Other secondary (or primary) functions suggested for m^{5}C in eukaryotes involve stabilizing chromosome structure, altering the conformation of DNA, assisting in the repair of DNA, enhancing the rate of mutation at certain DNA sites, contributing to the transformation of normal cells to tumor cells, and, as in prokaryotes, providing some protection to DNA against certain types of enzymatic degradation. Further research is needed.

SYNTHETIC MODIFIED BASES   Synthetic derivatives and analogs of purines and pyrimidines are used for various applications. For example, *6-mercaptopurine* is an antitumor agent; *5-bromouracil* is a potent mutagenic agent; *5-fluorouracil* is a drug used in treating cancer; and *9[(2-hydroxyethoxy)-methyl]guanine,* is also called *acyclovir,* is a potent antiviral agent (see Figure 6–7).

Acyclovir is the first antiviral drug to gain approval as an ointment in the treatment of some viral infections. It is particulary effective in the treatment of genital herpes, a serious disease of epidemic proportions affecting about 20 million individuals in the United States. The acyclovir treatment decreases the healing time of painful herpes sores and the duration of the painful state of an active infection. Acyclovir does not, however, prevent the virus from spreading through the body and taking up residence in a dormant state in sensory neurons. The active virus can erupt at any time, causing excessive discomfort. Rates of recurrence of one-to-two times a month are not unusual. However, recent clinical studies have established that taken orally on a regular basis, acyclovir can significantly reduce the frequency of these recurrences. Already in use in the United Kingdom, the oral drug is currently under FDA review for licensing in the United States.

Acyclovir represents a new generation of antiviral drugs that prevent the multiplication of a virus with minimal harmful effects on other cells in the body.

FIGURE 6–7 Synthetic modified bases. Their uses are described in the text.

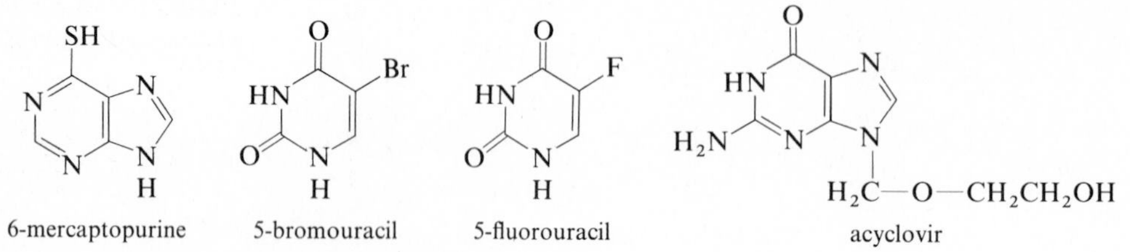

6-mercaptopurine   5-bromouracil   5-fluorouracil   acyclovir

The antiviral action of acyclovir is initiated by an enzyme, thymidine kinase, present only in cells infected by the herpes virus. The enzyme produces phosphorylated acyclovir, which in turn is converted by two other natural enzymes of the cell to a product that inhibits the replication of the virus.

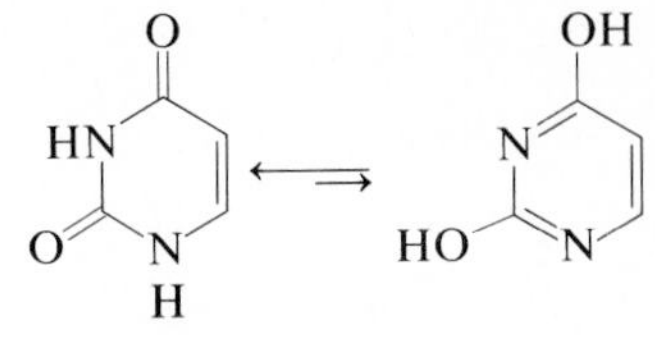

TAUTOMERISM   Except for adenine, the other four major bases exhibit *tautomerism* and thus are capable of existing in either of two isomeric forms: the *keto form,* with $>C=O$ functions, and the *enol form,* with $>C(H)OH$ functions (see Figure 6–8). The keto structures are more stable and thus predominate under physiological conditions.

UV ABSORPTION   Because of their aromatic nature, purines and pyrimidines *absorb ultraviolet* (UV) *radiation.* Consequently, nucleotides and nucleic acids also absorb UV radiation. This property is the basis for several applications, such as (1) laboratory methods of detection and quantitative measurement of nucleotides and nucleic acids, (2) the viewing of biological specimens via ultraviolet microscopy, (3) the mutagenic effect of UV radiation (see Section 8–2), and (4) sterilization by UV radiation.

Although each purine and pyrimidine has a unique absorption spectrum (see Figure 6–9), each shows maximum absorption at or close to 260 nm. Proteins also absorb UV radiation, owing primarily to the presence of the aromatic amino acids (trp and tyr). Protein absorption is maximal at 280 nm (see Appendix I).

keto
(more stable)   enol

FIGURE 6–8 Tautomerism in uracil.

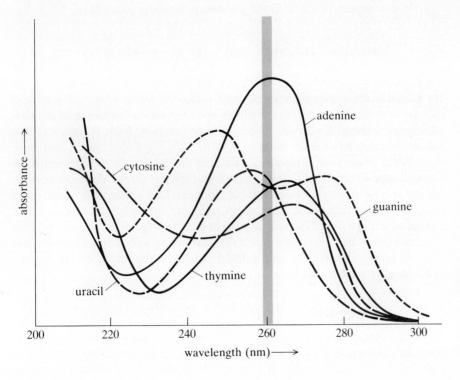

FIGURE 6–9 Ultraviolet absorption spectra of the major purines and pyrimidines. All spectra were measured on solutions having the same molar concentration and the same pH of 7. As indicated, adenine is the strongest absorber, and cytosine is the weakest.

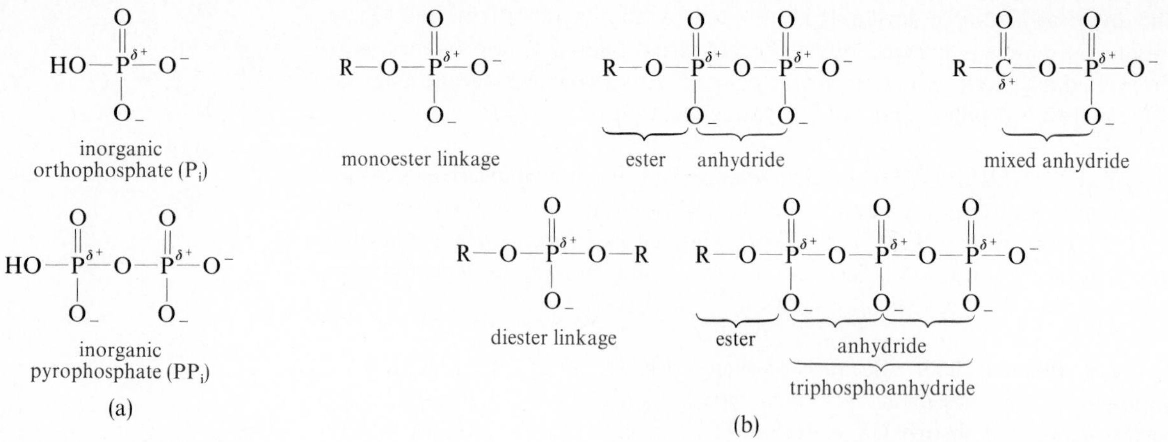

FIGURE 6–10 Phosphate and organophosphorus structures. (a) Inorganic forms. (b) Organic forms.

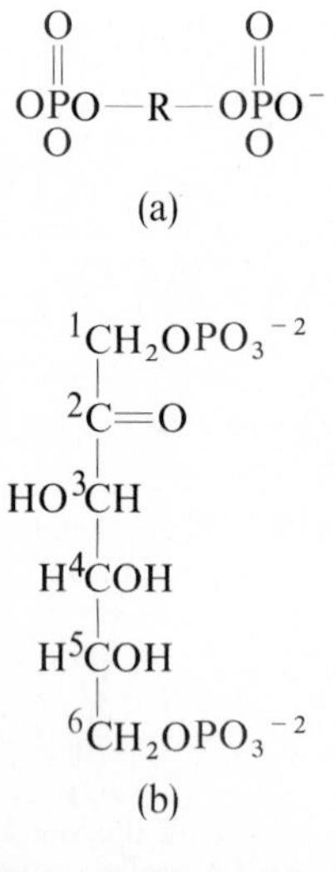

FIGURE 6–11 Bisphosphate nomenclature. (a) Pattern of phosphate bonding. (b) Specific example, fructose-1,6-bisphosphate.

Phosphate

In the vast majority of cases when phosphate occurs in an organic biomolecule, it is present in one of three ways: (1) as part of an ester linkage, (2) as part of a monoanhydride or dianhydride linkage, or (3) as part of combined ester and anhydride linkages. We will encounter each type in the next several pages. The display of structures in Figure 6–10 highlights the electrophilic character (δ^+) of the P atom, which explains the reactions of these species with nucleophilic substances.

In nucleotides (see p. 237) the phosphate bonding patterns of

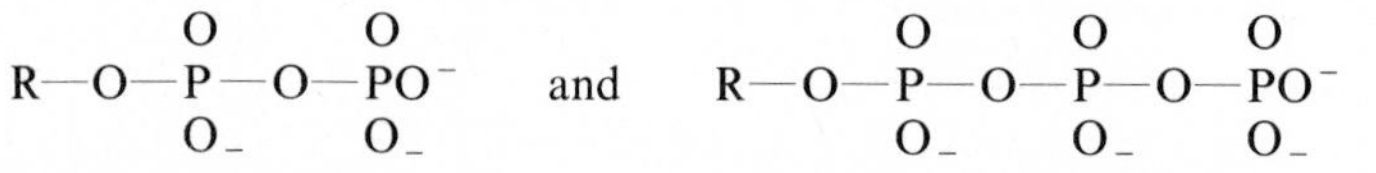

$$R—O—P—O—PO^- \quad \text{and} \quad R—O—P—O—P—O—PO^-$$

are named as nucleoside **di**phosphate and nucleoside **tri**phosphate, respectively, with diphosphate representing the (di)phosphoanhydride function and triphosphate representing the (tri)phosphoanhydride function. Each is linked to the parent structure by an ester bond.

When phosphate groups are separately linked at two or three different locations (see Figure 6–11), a **bis** (for two) and **tris** (for three) scheme of nomenclature is now preferred. For example, fructose-1,6-**bis**phosphate identifies a phosphorylated species of fructose having two phosphates—one in ester linkage at carbon 1 and the second also in ester linkage at carbon 6. An older nomenclature scheme does not make this distinction in bonding patterns, and fructose-1,6-bisphosphate is also called fructose-1,6-diphosphate. The bis/tris convention is used in this book.

Nucleoside Structure

The sugar—base portion of a nucleotide is called a *nucleoside*. The covalent linkage usually involves the C1 atom of the sugar and the N1 atom of a pyrimidine or the N9 atom of a purine. According to carbohydrate nomen-

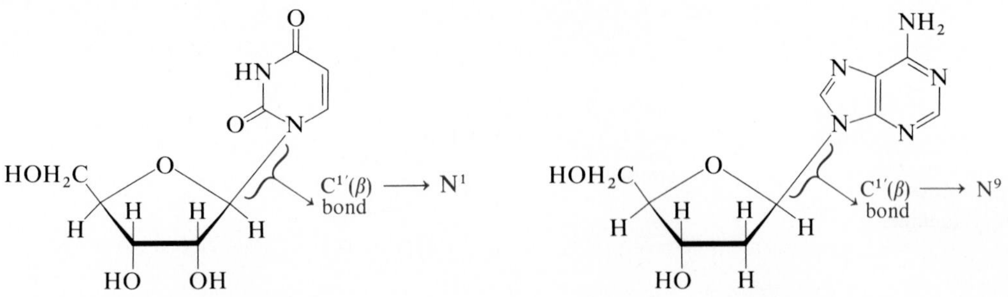

uridine
1-β-D-ribofuranosyluracil

(a)

deoxyadenosine
9-β-D-2'-deoxyribofuranosyladenine

(b)

FIGURE 6–12 Nucleosides.
(a) Pyrimidine ribonucleoside.
(b) Purine deoxyribonucleoside.

clature, the linkage is called a *glycosidic bond.* So that there is no confusion in numbering, the atoms of the carbohydrate unit are differentiated by a prime superscript. The accepted trivial names of the common ribonucleosides are *adenosine, guanosine, uridine,* and *cytidine* (see Table 6–2 for abbreviations). The common deoxyribonucleosides are *deoxyadenosine, deoxyguanosine, deoxycytidine,* and *deoxythymidine* (or simply thymidine). (The only instance where a thymine ribonucleoside occurs is in transfer RNA; it is called ribosylthymine.) On a more systematic basis the nucleosides are named as β-D-ribosyl or β-D-2'-deoxyribosyl derivatives of the purine or pyrimidine, as indicated in Figure 6–12.

Although we have described nucleosides as part of the nucleotide structure, many free nucleosides also occur naturally. Some have valuable clinical uses. For example, a *streptomyces* organism secretes *puromycin* (Figure 6–13), a potent antibiotic that operates by functioning as an inhibitor of protein biosynthesis. Various microorganisms produce *arabinosyl adenine* (ara-A) (Figure 6–14) and *arabinosyl cytosine* (ara-C) (Figure 6–15), which contain the sugar β-D-arabinose (Figure 6–16) rather than β-D-ribose. Both substances have proven successful as potent antiviral and anticancer agents without having extensive adverse effects on the host cell. Substance ara-A is another antiherpes drug, effective primarily against herpes type I infections of the eye. A major drawback of ara-A is that it is rapidly inactivated by a normal enzyme of human cells. This drawback has been overcome by the design of *carbocyclic ara-A,* an analog of ara-A resistant to degradation by the human enzyme.

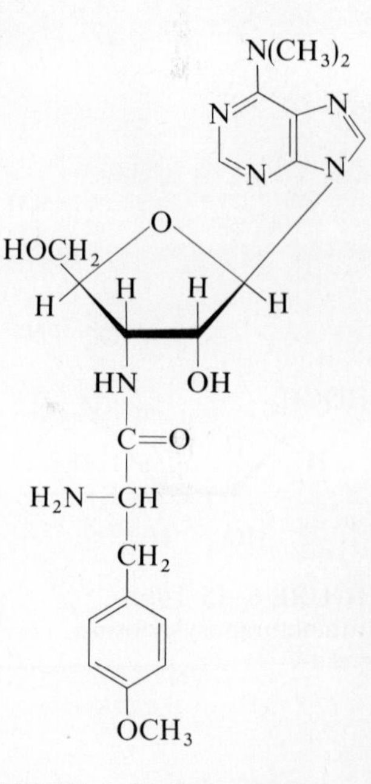

FIGURE 6–13 Puromycin.

Nucleotide Structure

A nucleotide (see Note 6–2) consists of a phosphate group in ester linkage to the sugar unit of a nucleoside. Usually the linkage involves the 5' position on the pentose. Depending on the identity of the pentose, all nucleotides can be categorized as either ribonucleotides or deoxyribonucleotides. Then according to the number of phosphate residues present, there are monophosphonucleotides, diphosphonucleotides, or triphosphonucleotides. All types occur in living cells.

NOTE 6–2
Occasionally, we will use the following symbolism for nucleotides:

Ade—rib—P

Ade—rib—P—P

Ade—rib—P—P—P

Other shorthand conventions are described in Section 6–3.

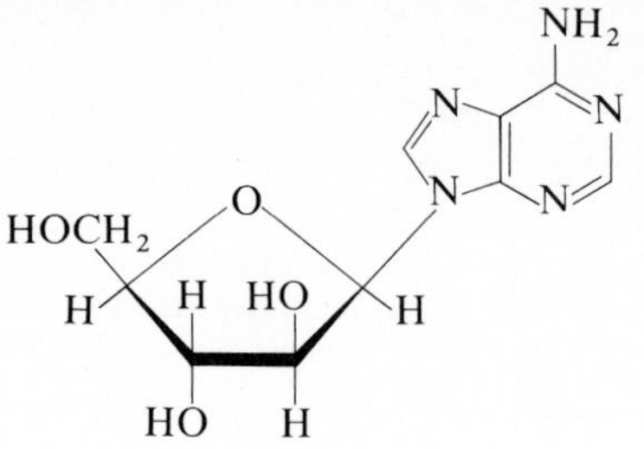

FIGURE 6–14 9-β-D-Arabinofuranosyladenine, or ara-A.

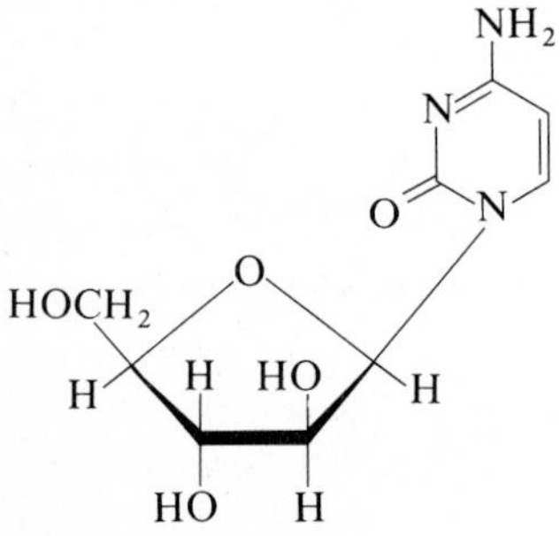

FIGURE 6–15 1-β-D-Arabinofuranosylcytosine, or ara-C.

FIGURE 6–16 β-D-Arabinose.

Specific nucleotides are named as nucleoside esters of phosphoric acid: nucleoside monophosphate, nucleoside diphosphate, and nucleoside triphosphate. For example, the adenosine family (see Figure 6–17) consists of *adenosine-5'-monophosphate* (5'-AMP), *adenosine-5'-diphosphate* (5'-ADP), and *adenosine-5'-triphosphate* (5'-ATP), with the symbolism in parentheses being shorthand notation. The deoxy counterparts are named deoxyadenosine-5'-monophosphate (5'-dAMP), and so on. Monophosphonucleotides (that is, nucleoside monophosphates) are also named as acyl acids of the parent nucleoside, because of the presence of the acidic phosphate group. Some examples are *adenylic acid, deoxyadenylic acid, uridylic acid,* and *thymidylic acid.*

Cyclic Nucleotides

Adenosine-3',5'-cyclic monophosphate (cyclic AMP, or cAMP) (see Figure 6–18) is a universally occurring nucleotide of immense biological importance. It was first isolated in 1959 by E. Sutherland (Nobel Prize, 1971) and co-workers as part of investigations into the mechanism of action of certain hormones, such as adrenalin, in regulating carbohydrate metabolism. They established that the immediate action of adrenalin is to activate the enzyme responsible for the production of cAMP. In turn, the cAMP then acts to control the activity of other cellular enzymes, frequently by allosteric activation.

The term *secondary messenger* is used to highlight the intermediate role of cAMP in this relationship, in which adrenalin is an example of a primary regulator (primary messenger). (Details are described in Chapter 13.) Since the original work with cAMP and adrenalin, the secondary-messenger strategy has been established as operating in many diverse processes in all types of organisms and cells. The secondary messenger is frequently cAMP, but there are other substances that perform a similar role. Some of these substances will be described in later chapters.

Two other cyclic nucleotides have also been discovered: *cyclic GMP* (see Figure 6–19) and *cyclic CMP*. Little is known about cCMP. However, cGMP has been found in many sources and is proposed to act in the same manner as cAMP, but usually in reverse—that is, as an inhibitor of enzymes. In some cases cAMP and cGMP exert their opposite effects on the same enzyme.

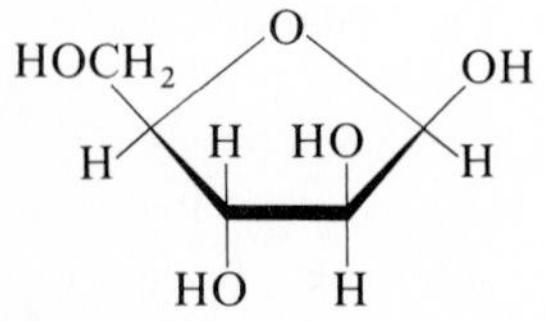

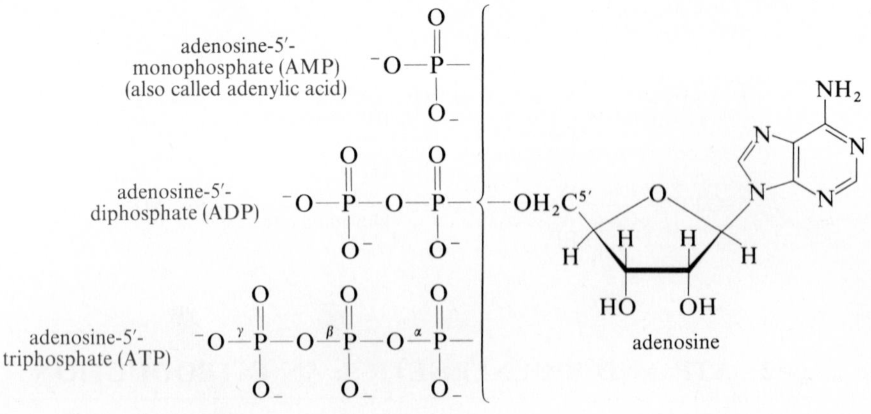

adenosine-5′-
monophosphate (AMP)
(also called adenylic acid)

adenosine-5′-
diphosphate (ADP)

adenosine-5′-
triphosphate (ATP)

adenosine

(a)

◀ **FIGURE 6–17** Mono-, di, and triphosphonucleotides (5′) of adenosine. (a) Structures. (b) Photo of space-filling model.

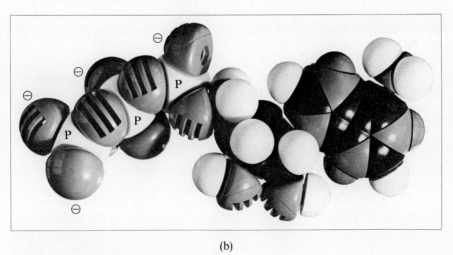

(b)

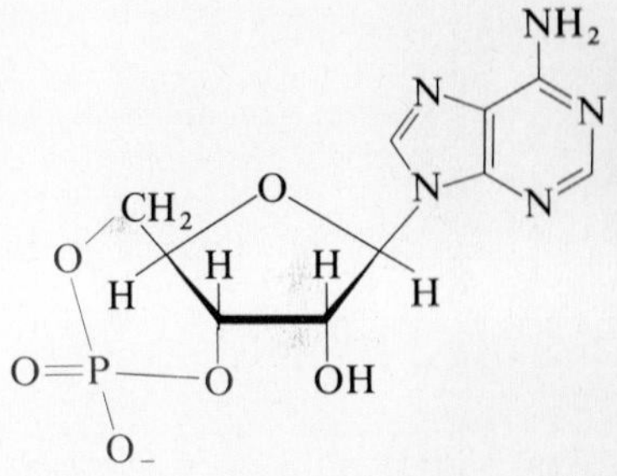

FIGURE 6–18 Adenosine-3′,5′-cyclic monophosphate, or 3′,5′-cyclic AMP. Cyclic refers to an intramolecular 3′,5′-phosphodiester linkage.

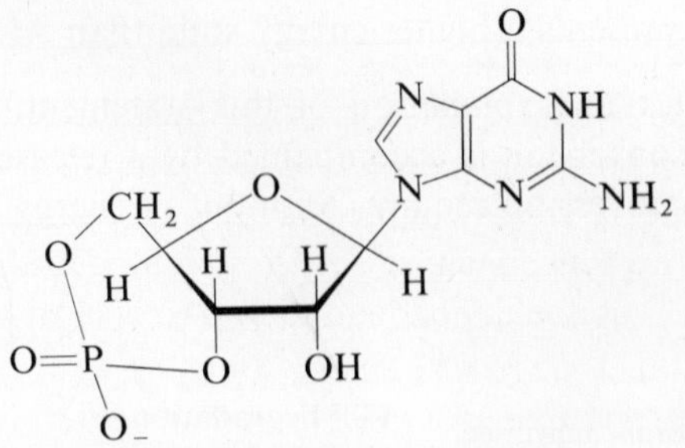

◀ **FIGURE 6–19** Guanosine-3′,5′-cyclic monophosphate, or 3′,5′-cyclic GMP.

The cyclic nucleotides are formed in vivo from the corresponding triphosphonucleotides by the enzymes *adenylate cyclase* and *guanylate cyclase* (see Figure 6–20). Metabolic turnover also involves *cyclic phosphodiesterase,* an enzyme catalyzing the degradation of the cyclic diesters on the 3′ side, yielding the corresponding noncyclic 5′-monophosphonucleotide.

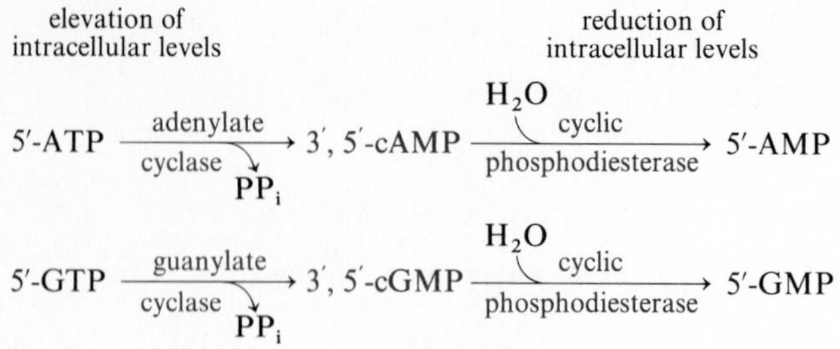

FIGURE 6–20 Formation and degradation of cyclic nucleotides. Primary regulators (such as certain hormones) control the activity of the cyclase enzymes. Sometimes, the activity of phosphodiesterase is also controlled by primary regulators.

$$\underset{\text{elevation of}}{\underset{\text{intracellular levels}}{}}$$

$$5'\text{-ATP} \xrightarrow[\text{cyclase}]{\text{adenylate}} 3', 5'\text{-cAMP} \xrightarrow[\text{phosphodiesterase}]{\overset{H_2O}{\text{cyclic}}} 5'\text{-AMP}$$

with PP_i released at the adenylate cyclase step, and at the right the reduction of intracellular levels.

$$5'\text{-GTP} \xrightarrow[\text{cyclase}]{\text{guanylate}} 3', 5'\text{-cGMP} \xrightarrow[\text{phosphodiesterase}]{\overset{H_2O}{\text{cyclic}}} 5'\text{-GMP}$$

with PP_i released at the guanylate cyclase step.

6–2 ATP AND BIOENERGETICS: AN INTRODUCTION

Triphosphonucleotides participate in many enzyme-catalyzed reactions involved in the metabolism of all types of compounds. For example, CTP participates in phospholipid biosynthesis (see Section 17–3), UTP functions in the biosynthesis and interconversions of various carbohydrates (see Section 13–5), and they are all utilized in the biosynthesis of RNA and DNA (see Chapters 7 and 8).

While all of these functions are important to normal cellular processes, there is one central role of ATP that is of extreme importance:

> In all living cells ATP is the central molecule in the flow of chemical energy, being formed to store energy and being degraded to transfer energy.

A preliminary, compact treatment of this function is summarized in Figure 6–21, showing the interconversion of ATP and ADP.

The ATP → ADP degradation involves loss of the terminal γ phosphate grouping in ATP; the ADP → ATP synthesis replaces the terminal phosphate. This interconversion, however, is only half of the description. The rest involves the energetics of the processes. In this regard the simple fact is that ATP is a much less stable molecule compared with ADP and P_i.

> Hence ATP represents a higher energy state than ADP and P_i.

(See Section 12–2 for an explanation of this statement.) Consequently, the ATP → ADP + P_i conversion is accompanied by a release of energy, and the ADP + P_i → ATP conversion requires an input of energy.

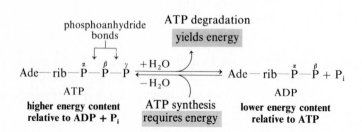

FIGURE 6–21 Energy relationship of ATP and ADP.

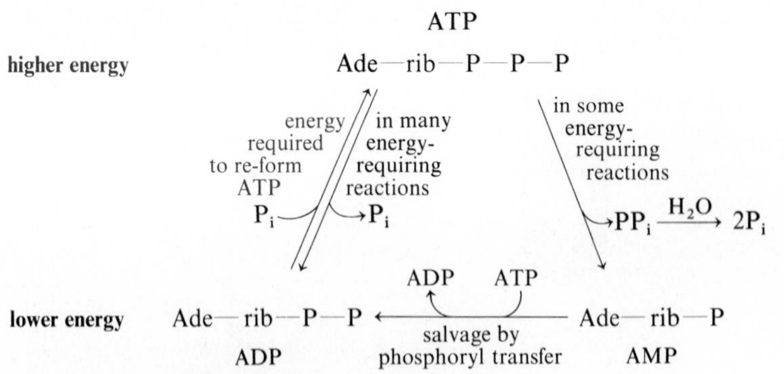

ATP

higher energy   Ade—rib—P—P—P

energy required to re-form ATP — P_i

in many energy-requiring reactions → P_i

in some energy-requiring reactions

→ PP_i $\xrightarrow{H_2O}$ $2P_i$

ADP   ATP

lower energy   Ade—rib—P—P ← ——————— Ade—rib—P

salvage by phosphoryl transfer

ADP   AMP

FIGURE 6–22 Energy cycle in living cells.

A similar description also applies to the hydrolysis of the other phosphoanhydride bond in ATP to yield AMP and PP_i (pyrophosphate). The ATP is re-formed from AMP by the route of AMP → ADP → ATP, and the inorganic pyrophosphate is hydrolyzed to inorganic orthophosphate P_i (see Figure 6–22).

To summarize, the ATP $\rightleftharpoons$ ADP + P_i, the ATP → AMP + PP_i, and the AMP → ADP conversions comprise the molecular basis for the flow of chemical energy within all living cells. When a cell degrades high-energy foodstuffs such as carbohydrates, the energy available is used to form ATP. Then when a cell uses energy (for a process such as biosynthesis, muscle contraction, or transport of substances across membranes), ATP is degraded to provide the energy. The ATP–ADP–AMP system and other aspects of bioenergetics are examined in more detail in Chapter 12.

6–3 NUCLEIC ACIDS: SOME GENERAL ASPECTS

Extraction and Isolation of Nucleic Acids

From cells, subcellular fractions, or viruses, RNA and DNA can be extracted by exploiting some of the properties listed in Table 6–3. The different solubilities in salt solutions can also be used to separate RNA from DNA.

TABLE 6–3 Properties of DNA and RNA

DNA PROPERTIES	RNA PROPERTIES
Insoluble in dilute solutions of NaCl	Soluble in dilute solutions of NaCl
Soluble in concentrated solutions of NaCl	Insoluble in alcohol
Insoluble in alcohol	Can be dissociated from protein by treatment with a detergent or a phenol
Can be dissociated from protein by treatment with a detergent or a phenol	

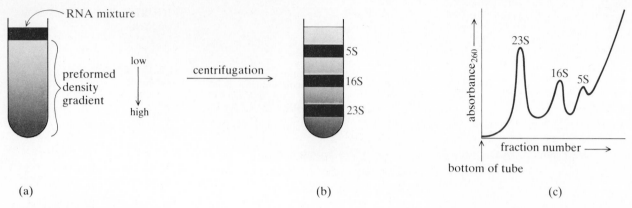

FIGURE 6–23 Rate zonal density gradient ultracentrifugation. (a) Beginning setup. (b) After centrifugation. Zones of nucleic acid are detected by piercing the bottom of the tube and continuously monitoring the UV absorbance (at 260 nm) of the fluid seeping out of the tube. (c) Absorbance profile corresponding to the sedimentation pattern in (a).

To isolate individual components of an RNA or DNA mixture, one can use various procedures of *column chromatography* (ion-exchange, adsorption, molecular sieve, or affinity chromatography) and *gel electrophoresis* (either acrylamide or agarose). Examples of electrophoresis separations are given later in this chapter.

Another widely used method is *density gradient ultracentrifugation,* of which there are two types. In the *rate zonal* technique (see Figure 6–23) a sample is layered on top of a liquid medium having a preformed linear gradient of density. The density gradient is usually prepared with sucrose solutions. The rates of sedimentation for polymers of different sizes in a medium of nonuniform density are often sufficiently different to achieve an acceptable level of separation. As indicated in Figure 6–23, nucleic acids of different size are commonly designated by their Svedberg (S) values.

A variation that is particularly effective for DNA separations is the *isopycnic technique,* where the density gradient is formed during centrifugation (see Figure 6–24). Cesium chloride (CsCl) or cesium trifluoroacetate solutions are used. Initially, the centrifuge tube is filled with the salt solution containing the dissolved DNA, and a uniform density exists along the length of the tube. As the rotor spins, a concentration gradient of cesium chloride forms. Each DNA migrates to and forms a band at a position in the tube where the density of CsCl is identical to the density of DNA; this density is called the *buoyant density* (ρ_B) of the DNA. The individual zones can be collected as described in Figure 6–23. Another purpose for measuring ρ_B is discussed later in the chapter.

A powerful method to isolate a specific nucleic acid is the technique of *nucleic acid hybridization.* We will defer discussion of this method until later in the chapter (see Section 6–8).

Hyperchromic Effect Shown by Most Nucleic Acids

Like proteins, most nucleic acids exist in ordered conformations, with helical twists being particularly prevalent. We will get to this business shortly. However, extending your understanding of protein denaturation to nucleic acids, you might expect that certain treatments will also disrupt ordered structures of DNA and RNA. The use of heat is particularly effective. As temperature increases, the gradual disruption of the nucleic acid conformation can be monitored by measuring the increase in the UV absorption of the solution. This procedure is called the *hyperchromic effect*. The UV absorption will increase because as the ordered conformation is disrupted, the purine and pyrimidine bases of the polynucleotide chain will become more exposed and hence more accessible to the incident UV radiation. When denaturation is complete, no further increase in absorption will occur.

The term *melting* is used to refer to the thermal denaturation of nucleic acids, and a plot of absorbance at 260 nm versus temperature is called a *melting curve* (see Figure 6–25). The midpoint of the hyperchromic rise in the curve is termed the *melting temperature* T_m. Under a specific set of conditions, individual nucleic acids have melting curves characterized by the size of the hyperchromic effect, the steepness of the hyperchromic rise, and the T_m value. For DNAs the T_m value is very high, in the range 85°–100°C. The use of T_m values for estimating the nucleotide composition of a DNA molecule is described elsewhere in the chapter (see Section 6–4).

Covalent Backbone of Polynucleotide Chains

Polypeptides are composed of amino acids linked by peptide bonds. Polynucleotides are composed of nucleotides linked by phosphodiester bonds. Since the phosphate group includes two ester bonds involving specifically the 3′ and 5′ sugar carbons of successive nucleotides, the linkage is called a $3' \rightarrow 5'$ *phosphodiester bond*. The linkage is

$$-\overset{3'}{\underset{|}{C}}-O-\overset{\overset{O}{\|}}{\underset{\underset{O_-}{|}}{P}}-O-\overset{5'}{\underset{|}{C}}-$$

A complete formula representation of a pentanucleotide is shown in Figure 6–26. From left to right the structure progresses from the 5′ terminus (with a phosphate group) to the 3′ terminus (with a free —OH group on the 3′ carbon). The much larger chains of RNA and DNA are termed *nucleic acids* because of the presence of several mildly acidic phosphate groups along the chain. At pH 7 the phosphate group (with a pK_a of about 6) is almost completely ionized, and so the nucleic acid backbone is *polyanionic* (many negative charges).

For obvious reasons shorthand designations of polynucleotide chains are routinely used (see Figure 6–26). One convention uses a vertical line and diagonal slashes to represent the sugar unit, with phosphodiester bonds represented by the letter P between the 3′ and 5′ slashes. An even simpler representation uses pN to symbolize a 5′-monophosphonucleotide and Np a 3′-nucleotide. A polynucleotide is then shown as a succession of pN notations, reading from left to right from the 5′ terminus to the 3′ terminus. The simplest representation uses just the single-letter code, with a p to identify only the phosphorylated 5′ terminus.

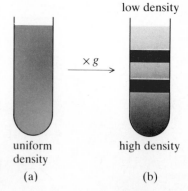

FIGURE 6–24 Isopycnic density gradient ultracentrifugation. (a) Homogeneous solution containing sample. (b) After application of gravitational force This force produces a linear density gradient, and substances in the sample migrate to a position equal to their density.

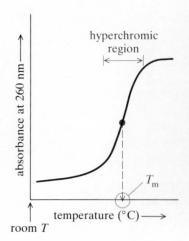

FIGURE 6–25 Nucleic acid melting curve depicting the hyperchromic effect observed with heating.

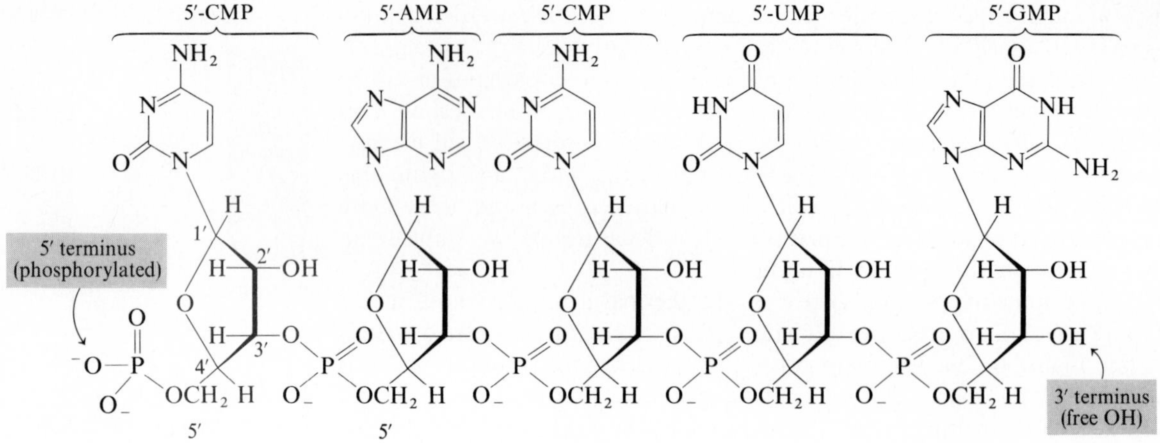

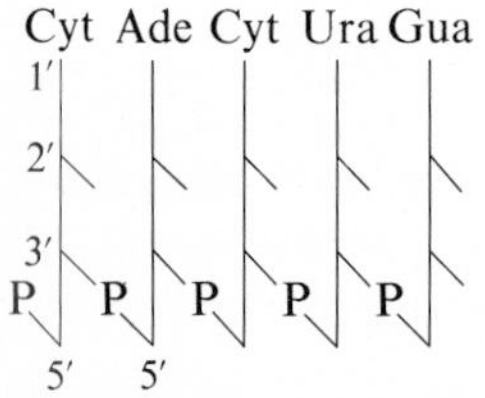

FIGURE 6–26 Pentanucleotide structure illustrating $3' \rightarrow 5'$ phosphodiester bonds between adjacent nucleotides. As indicated, each phosphate group is ionized at about pH 7.

shorthand representations of the pentanucleotide

pCpApCpUpG

or   pC—A—C—U—G
       5'   3'

or   pCACUG

$2' \rightarrow 5'$ Phosphodiester Bonding in a Unique Oligonucleotide

Interferons are a class of proteins produced in extremely small amounts by animal cells after infection by a virus (refer to Section 4–7), providing protection from subsequent viral infections. In the past few years much has been learned about the genetics and structure of interferons and about how an interferon may confer an antiviral state to cells. One discovery is that interferon-treated cells produce several new proteins. One of these proteins is *oligonucleotide polymerase,* an enzyme that forms small amounts of short *oligoadenylate* chains from ATP. The trimer $2-5A_3$ is shown in Figure 6–27. The most interesting structural feature is that the nucleotides are linked by $2' \rightarrow 5'$ phosphodiester bonds, never previously detected in any naturally occurring substance.

Recent studies suggest that $2-5A_3$ acts as a selective inhibitor of the biosynthesis of proteins required for viral multiplication. How a virus stimulates the production of interferons, how interferons stimulate the production of the oligonucleotide polymerase, what roles are played by the other new proteins, explanations for other interferon effects on cell functions, and clinical testing of interferon therapy as an antiviral agent and as an anticancer agent are issues still under intensive study.

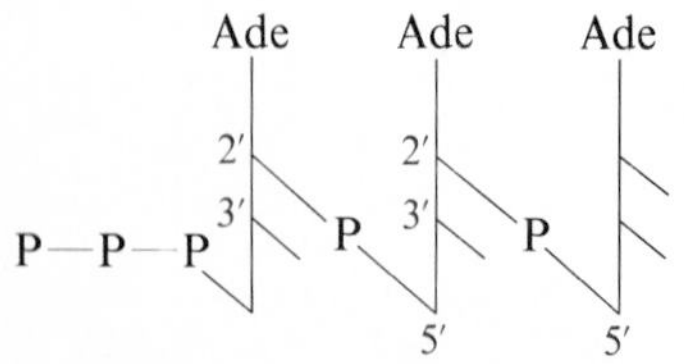

Shortly, we will describe another recent discovery for $2' \rightarrow 5'$ phosphodiester bonding, namely, its involvement in producing branched intermediates during the biosynthesis of messenger RNA in eukaryotes.

FIGURE 6–27 2–5A trimer produced in response to interferon.

6–4 BASE COMPOSITION AND BASE SEQUENCE OF NUCLEIC ACIDS

Base Composition of RNA

The percentage of A, G, C, and U of an RNA can be determined by routine procedures involving two steps: (1) complete hydrolytic degradation of the RNA into a mixture of its constituent nucleotides and (2) an analysis of the mixture by chromatography (usually, an ion-exchange column method). Complete hydrolysis of RNA can be achieved by heating the RNA with NaOH or by using enzymes called *ribonucleases*. Catalysis with a base (see Figure 6–28) involves the $2'$—OH of a ribonucleotide, which loses a proton to OH^-, yielding $2'$—O^-, which then attacks the P of the diester bond. This step cleaves the diester bond and forms intermediate cyclic-$2',3'$-monophosphonucleotides, which are then hydrolyzed on either side to yield a $2'$-NMP or a $3'$-NMP. (We will discuss nuclease enzymes later.) The detection of nucleotides containing modified bases requires special precautions, which we will not consider.

Unlike the base composition for DNA described next, except for certain viral RNAs, base compositions for RNA show no common patterns other than that all four bases are always present. Each RNA has a different base composition both in general terms (percentage of purines and percentage of pyrimidines) and in specific terms (percentage of each specific base).

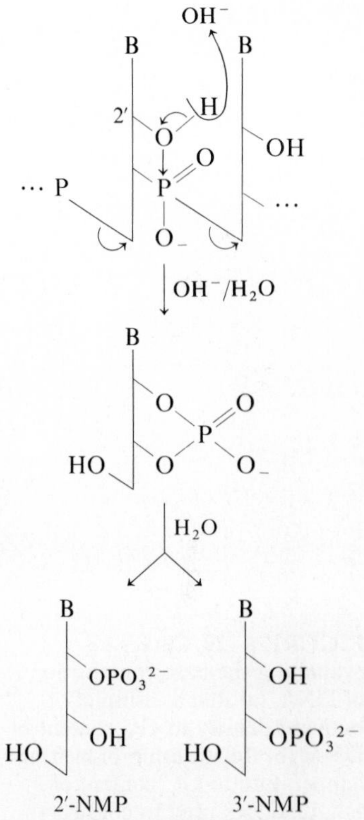

Base Composition of DNA

The percentage of A, G, C, and T of a DNA can also be determined by using nucleases to degrade the DNA and subsequent chromatographic analysis of the nucleotide mixture. However, base-catalyzed hydrolysis with NaOH does not work because of the absence of the $2'$—OH group on the deoxyribose moieties.

A DNA base composition can be more easily evaluated from T_m (melting temperature) and ρ_B (buoyant density) measurements. After the collection of much data, the empirical linear relationships in Figure 6–29 allow one to calculate the sum of G and C present in a DNA. For reasons explained shortly, the sum of G and C content automatically identifies the content of all four bases.

There are some important generalizations regarding the patterns of base composition in DNA, regardless of source (except for some viral DNAs). These generalizations have come to be known as *Chargaff's rules,* after E. Chargaff, who first recognized them about thirty five years ago. Represented by the data in Table 6–4, these generalizations are as follows:

1. The number of purine bases (A + G) is balanced by the number of pyrimidine bases (T + C); that is, the ratio of purines to pyrimidines is approximately one (pur/pyr = 1.0).

2. The number of adenine residues is balanced by the number of thymine residues; that is, the ratio of adenine to thymine is approximately one (A/T = 1.0).

FIGURE 6–28 Base-catalyzed hydrolysis of phosphodiester bonds in RNA. The symbol B is for base.

$$\rho_B = 1.660 + \frac{0.098\,(\%G + \%C)}{100}$$

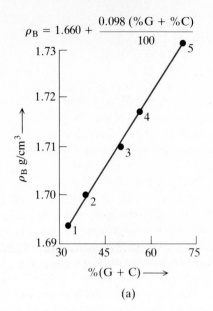

$$T_m = 69.3 + 0.41\,(\%G + \%C)$$
when conducted in a solution containing $0.2M$ Na$^+$

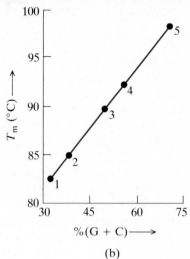

(a)   (b)

FIGURE 6–29 Curves for evaluating the base composition of DNA. (a) Relationship of buoyant density to GC content of DNA. (b) Relationship of melting temperature to GC content of DNA. Values used to construct the graphs are for the following DNAs: (1) *Staphylococcus aureus,* (2) calf thymus, (3) *Escherichia coli,* (4) *Brucella abortus,* and (5) *Streptomyces griseus.* [*Source:* Data taken from the *Handbook of Biochemistry and Molecular Biology,* 3rd ed. (Cleveland: CRC Press, 1976).]

3. The number of guanine residues is balanced by the number of cytosine residues; that is, the ratio of guanine to cytosine is approximately one (G/C = 1.0).

Nuclease Enzymes

Before starting the subject of base sequence, let us briefly consider some of the characteristics of *nucleases* (also called *phosphodiesterases*), which are enzymes that catalyze the hydrolytic cleavage of phosphodiester bonds in nucleic acids. In living cells nucleases perform important functions in many phases of DNA and RNA biochemistry, including the recombination of genes. In the re-

TABLE 6–4 Representative data of purine and pyrimidine distribution in DNA

	SOURCE			
ITEM	HUMAN LIVER	CARROT	E. coli	λ VIRUS
%A	30.3	26.7	23.8	26.0
%G	19.5	23.1	26.0	23.8
%C[a]	19.9	23.2	26.4	24.3
%T	30.3	26.9	23.8	25.8
A/T	1.00	0.99	1.00	1.01
G/C	0.98	1.00	0.98	0.98
pur/pyr	0.99	0.99	0.99	0.99

[a] Includes any contribution from 5-methylcytosine.

search laboratory purified nucleases have proven to be immensely valuable and even indispensable tools.

A broad distinction exists between *endonucleases,* acting only on internal bonds, and *exonucleases,* acting only on terminal diester bonds. However, there are several other elements of nuclease specificity, such as the ability (1) to act either on RNA, DNA, or both; (2) to act on single-stranded (ss) and/or double-stranded (ds) nucleic acids; (3) to produce an *a*-type or *b*-type cleavage (see Figure 6–30); (4) to recognize either the 3′ or 5′ terminus, in the case of exonucleases; (5) to recognize either purine or pyrimidine sites (this feature is called *base specificity*); and (6) to recognize only certain base sequences (this feature is called *sequence specificity*). Many nucleases have been isolated from numerous sources; a few examples are given in Table 6–5.

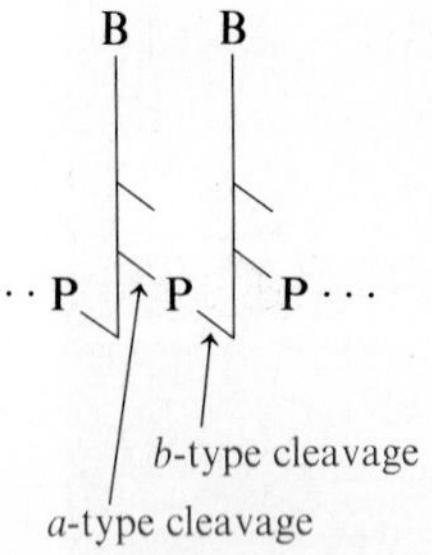

Endonucleases acting at specific sequences in DNA—and only at a specific bond in that sequence—were first discovered in *E. coli* and characterized by H. O. Smith and W. Arber in 1970 (Nobel Prize, 1978). Occurring only in prokaryotes, these enzymes have been termed *restriction endonucleases,* or simply *restriction enzymes.* Their site-specific action is used in the laboratory

FIGURE 6–30 Two types of cleavage at a phosphodiester bond.

TABLE 6–5 Functional characterization of some nucleases

ENZYME	SUBSTRATE	CHARACTERISTICS OF ACTION	CLEAVAGE
1. Ribonuclease A (bovine pancreas)	ssRNA endo(*b*)	Pyrimidine specific	··· Pyr p↓Np ···
2. Ribonuclease T1 (*Aspergillus oryzae*)	ssRNA endo(*b*)	G specific	··· Gp↓Np ···
3. Ribonuclease U$_2$ (*Ustilago sphaerogena*)	ssRNA endo(*b*)	A specific	··· Ap↓Np ···
4. Ribonuclease PhyM (*Physarum polycephalum*)	ssRNA endo(*b*)	A + U specific	··· Ap↓Np ··· NpUp↓N ···
5. Ribonuclease CL3 (chicken liver)	ssRNA endo(*b*)	C specific	··· Cp↓Np ···
6. Nuclease S1 (*Aspergillus oryzae*)	ssDNA, endo(*a*) ssRNA	Random locations	··· N↓pNp ···
7. Deoxyribonuclease I (bovine pancreas)	ssDNA, endo(*a*) dsDNA	Random locations	··· N↓pNp ···
8. Exonuclease VII (*Escherichia coli*)	ssDNA exo(*a*)	Processive cleavage from either the 3′ or 5′ terminus	pN↓^{1}pN↓^{2}p ···
9. Exonuclease III (*Escherichia coli*)	dsDNA exo(*a*)	Processive cleavage only from 3′-/OH terminus	··· pN↓^{2}pN↓^{1}pN ···
10. EcoR I[a] (*Escherichia coli*)	dsDNA endo(*a*)		5′ ··· pG↓pApApTpTpC ··· 3′
11. Hae III[a] (*Haemophilus aegyptius*)	dsDNA endo(*a*)	See Figure 6–52	5′ ··· pGpG↓pCpC ··· 3′
12. Pst I[a] (*Providencia stuartii*)	dsDNA endo(*a*)		5′ ··· pCpTpGpCpA↓pG ··· 3′

[a] Restriction enzymes cleaving a specific bond in a specific sequence of DNA.

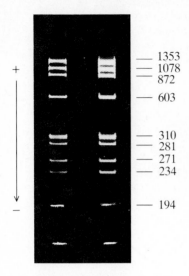

+

1353
1078
872

603

310
281
271
234

194

−

FIGURE 6–31 Cleavage pattern of φX174 DNA with the Hae III restriction enzyme. The side-by-side duplicate patterns of two separate treatments illustrate the reproducible specificity of enzyme action. The enzyme-treated DNA samples are applied to the top of a polyacrylamide gel slab; after electrophoresis the gel is soaked in a solution of *ethidium bromide* (EtBr), a DNA stain. When the stained gel is examined under UV light, intensely fluorescent bands appear, revealing the presence of EtBr–DNA complexes. The numbers at the right specify the length of each restriction fragment as the number of base pairs. (*Source:* Photograph provided by Bio-Rad Laboratories, Richmond, California.)

to cleave a DNA molecule at very precise locations to yield a specific set of DNA fragments, depending on the number of times the restriction site occurs along the DNA molecule. Figure 6–31 displays the characteristic cleavage pattern of φX174 DNA (a small bacterial virus DNA) with HaeIII: ten DNA fragments are obtained. The caption describes the detection of DNA with *ethidium bromide,* a sensitive fluorescent stain capable of detecting nanogram quantities of DNA.

About two hundred restriction enzymes, each with a different sequence specificity, have been discovered, and at present about sixty of these are commercially available. They are used in laboratory procedures for sequencing DNA, for mapping the locations of genes along a chromosome, and for cutting out intact genes from one DNA chromosome and then splicing them into another DNA chromosome. Later in this chapter we will consider the reason for their remarkable specificity and also illustrate their use in recombinant DNA procedures.

The natural biological function of restriction enzymes in prokaryotes is to provide protection against viral infection. These enzymes preferentially degrade viral DNA after its entry into the host cell. Hence the expression of the viral chromosome and viral multiplication are restricted. The host cell protects its own chromosome from the degradative action of its own restriction enzymes in the following way: restriction sites in the host chromosome are modified by methylation (refer to Note 6–1), preventing their recognition by the restriction enzyme.

Ligase Enzymes

In addition to the cleaving nuclease enzymes, living cells also contain enzymes called *ligases* that catalyze the *joining of two ends* of a polynucleotide chain (see Figure 6–32). The biological importance of ligase enzymes in the replication and recombination of DNA will be illustrated in Chapter 7. They too are valuable tools in recombinant DNA strategies for connecting different pieces of DNA.

Primary Structure: Base Sequences

Base-sequencing procedures for RNA were first designed by R. Holley (Nobel Prize, 1968) in the 1960s. Using a strategy similar to sequencing amino acids in a polypeptide chain—cutting RNA into smaller fragments with endonucleases, sequencing individual fragments by one-at-a-time removal of nucleotides by using exonucleases, and aligning fragments according to overlapping sequences—Holley established the sequence of 77 bases in a transfer RNA mole-

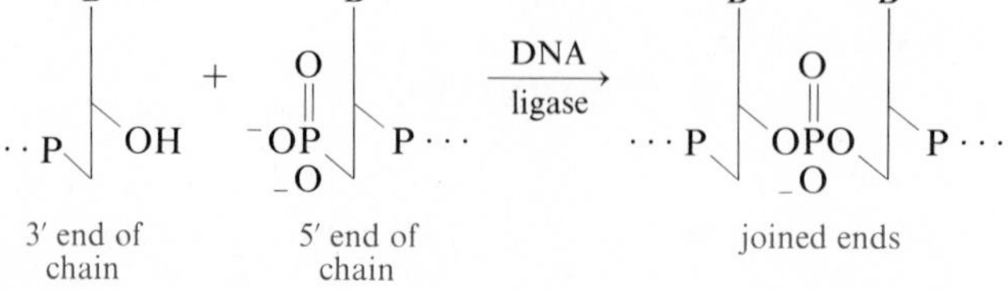

FIGURE 6–32 Action of a DNA ligase enzyme. ATP and Mg^{2+} are required for optimal activity.

cule. Since then, the same strategy has been used to sequence many different transfer RNAs, other small-size RNAs, and even some large RNAs such as the 3569 bases of the RNA chromosome of a bacterial virus. Later in this section we will describe an improved method for sequencing RNA.

Until 1977, the only way to sequence bases in DNA was to use an enzyme to synthesize a complementary RNA copy of the DNA (see Note 6–3), sequence the RNA copy, and transcribe the RNA base sequence into the DNA base sequence. Then in 1977, working independently, F. Sanger and co-workers and the team of W. Gilbert and A. Maxam designed two different procedures to sequence DNA directly (Nobel Prize to Sanger and Gilbert in 1980). The Gilbert-Maxam method has also been adapted to sequence RNA molecules. These methods are fast and reliable, and require only small amounts of DNA. Three spectacular achievements so far are the entire sequence of (1) the 5375 bases in the DNA chromosome of the ϕX174 bacterial virus, (2) the 16,569 bases in the human mitochondrial DNA chromosome, and (3) the 48,502 bases in the DNA chromosome of the lambda (λ) bacterial virus. Work is also nearing completion on the DNA of the Epstein-Barr animal virus with 170,000 bases.

Before going any further, we should point out that neither method allows the continual and progressive identification of thousands of bases in an intact DNA molecule. Rather, restriction enzymes are used to generate different sets of fragments having sizes within the sequencing capability of each method. The maximum capability of each method is about two hundred consecutive bases. As shown in Figure 6–33, different overlapping sets of fragments are obtained from the same DNA target by using different restriction enzymes, each cleaving at a specific base sequence, as shown in the examples in Table 6–5.

These revolutionary sequencing procedures are now in routine use to do what was once thought of as nearly impossible. At present, DNA sequences are being reported at the annual rate of about 400,000 bases per year. This work has had an enormous impact on advancing the understanding of how information is packaged into genes, how genes are regulated, and many other questions regarding molecular genetics. For assistance in the sorting and the analysis of all this sequence information, several computer data bases have been and are being established both in private and public sectors. Software is also available to search for homologies in DNA base sequences from various sources, to inspect sequences for regulatory locations, to inspect sequences for the location of any restriction sites, and to do other tasks.

Gilbert-Maxam Method of Sequencing Bases in DNA

Both the Sanger and the Gilbert-Maxam methods of base sequencing have the same general design: to produce from the target DNA a *nested set of oligonucleotides* differing in length from each other by only one nucleotide. For

NOTE 6–3
The meaning of complementary base sequences is fully described later in the chapter. An example in DNA is the sequence

$$\cdots A A C G T C T A G G \cdots$$

which corresponds to the sequence

$$\cdots U U G C A G A U C C \cdots$$

in RNA.

FIGURE 6–33 Generating DNA fragments with restriction enzymes. Hypothetical cleavage sites for two different restriction enzymes are identified by the coded arrowheads; ten fragments are from treatment with ○—→; seven fragments are from treatment with ●—→. The size of the DNA shown here (1200 residues) does not represent a native DNA molecule, which may be 5–10,000 times larger. The 1200-residue DNA may itself be a restriction fragment obtained from the larger parent DNA.

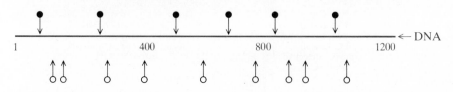

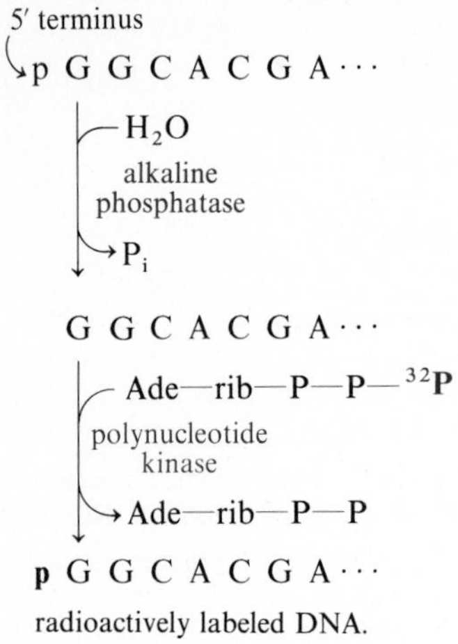

5′ terminus

p G G C A C G A ···

| — H$_2$O
| alkaline
| phosphatase
↓→ P$_i$

G G C A C G A ···

| — Ade—rib—P—P—^{32}P
| polynucleotide
| kinase
↓→ Ade—rib—P—P

p G G C A C G A ···

radioactively labeled DNA.

FIGURE 6–34 Preparation of 5′–^{32}P-labeled DNA strands for Gilbert-Maxam sequencing procedure.

a DNA with n bases the nested set has lengths of $n - 1, n - 2, n - 3, n - 4, \ldots,$ to the limit of $n - n$. Both methods use polyacrylamide gel electrophoresis to resolve the different-sized oligonucleotides and use radioactivity to detect their location on the gel. However, neither method produces a complete nested set in one operation but generates all possible lengths in four separate procedures.

Briefly, the Sanger method uses a DNA-replicating enzyme, *DNA polymerase,* to produce these different lengths by *copy synthesis* of the DNA target. By using one of four specific inhibitors of the DNA polymerase in each separate step, one can stop synthesis at all A locations, at all G locations, at all C locations, and at all T locations. Refer to the paper by Fiddes in the Literature of this chapter for more details of the Sanger method. We will focus on some of the details of the Gilbert-Maxam method, which generates oligonucleotides of different lengths by *chemical cleavage* of the DNA target.

The Gilbert-Maxam procedure begins by labeling one end of a single strand of DNA with radioactive phosphate (^{32}P) in order to later detect the fragments (see Figure 6–34). Labeling with ^{32}P at the 5′ end is done with two enzymes: (1) using *alkaline phosphatase* to remove the existing unlabeled phosphate from the 5′ end and then (2) treating the dephosphorylated DNA with *polynucleotide kinase* and ^{32}P(γ)-labeled ATP to transfer a ^{32}P phosphate to the 5′ terminus.

The labeled DNA is then divided into four batches. Each batch receives a different chemical treatment designed to modify the purine and pyrimidine bases in a different way. Then phosphodiester bonds involving only these modified bases are selectively cleaved to generate fragments. The treatments for G modification/cleavage are shown in Figure 6–35.

By controlling amounts of reacting materials and reaction conditions, one can approach a condition of having individual molecules of the target DNA

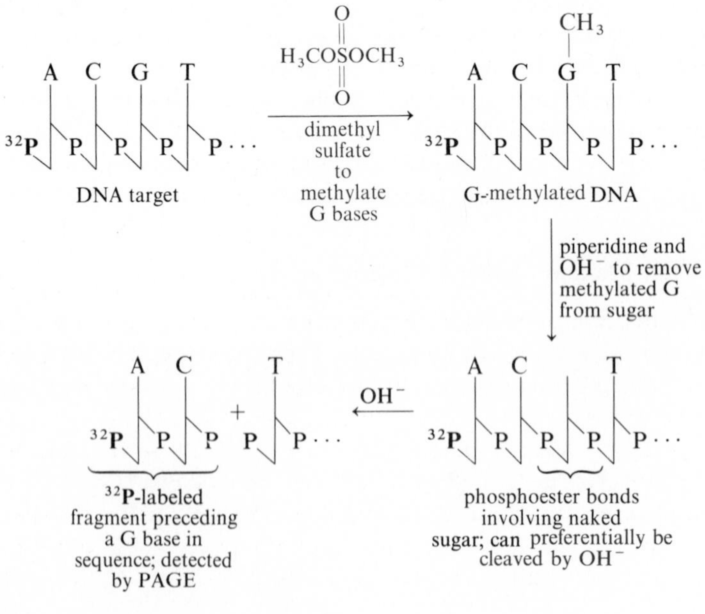

FIGURE 6–35 Treatments for G modification and cleavage.

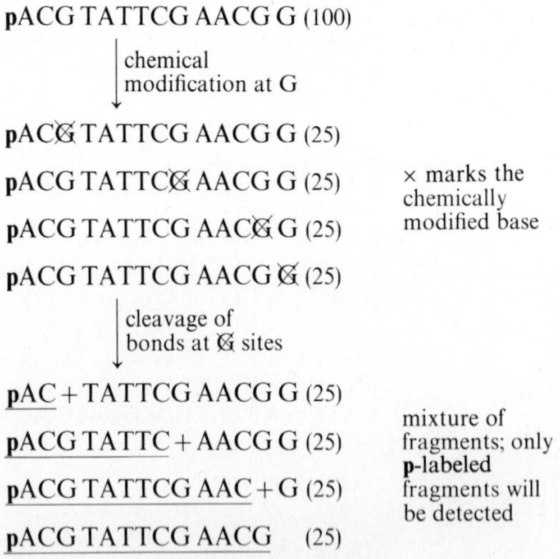

pACG TATTCG AACG G (100)

| chemical
| modification at G
↓

pACG̶ TATTCG AACG G (25)

pACG TATTCX̶ AACG G (25)

pACG TATTCG AACX̶ G (25)

pACG TATTCG AACG X̶ (25)

× marks the chemically modified base

| cleavage of
| bonds at X sites
↓

<u>**p**AC</u> + TATTCG AACG G (25)

<u>**p**ACG TATTC</u> + AACG G (25)

<u>**p**ACG TATTCG AAC</u> + G (25)

<u>**p**ACG TATTCG AACG</u>   (25)

mixture of fragments; only **p**-labeled fragments will be detected

FIGURE 6–36 G modification and cleavage to generate fragment lengths corresponding to all G positions in the sequence. Assume 100 molecules are present initially; $\mathbf{p} = {}^{32}P$.

being modified and cleaved at only one position (a single hit) but with nearly equal frequency at all such positions of the target base. This result is shown in Figure 6–36 for G modification/cleavage on a small oligonucleotide target with four G bases.

Note that each of the ^{32}P-labeled fragments terminates just before a G. The ideal scenario is to perform selective modification/cleavage at each of the four DNA bases. While this modification cannot yet be done, reagents and reaction conditions have been discovered that produce the following patterns of modification/cleavage: (1) only at G, as described, (2) at both A and G, (3) only at C, and (4) at both C and T. Thus fragmentation patterns at A bases and at T bases can be evaluated by difference (see Figure 6–37). (The details of A + G, C only, and C + T modification/cleavage can be found in the article by Maxam and Gilbert cited in the Literature of this chapter.)

Results are analyzed in the following way. The four fragment mixtures from each treatment are loaded side-by-side on a polyacrylamide gel slab. (The gel is cast in a buffer solution containing $6-8M$ urea, a denaturing agent, to prevent fragments from hydrogen-bonding with each other and with themselves.) During PAGE, fragments of different size will migrate differently because of sieving in the polyacrylamide matrix. After electrophoresis the pattern of resolution is recorded by exposing the gel to a piece of X-ray film. In the developed radioautograph the pattern of dark bands corresponds to the location on the gel of ^{32}P-labeled fragments. The unlabeled fragments from the other end of the DNA are also there, but they are not detected.

The complete sequence of the original target DNA is then read directly from the pattern of bands in the four lanes of the radioautograph. A G + A band appearing alongside a G band is read as G; a G + A band with no corresponding G band is read as A; a C + T band appearing alongside a C band is read as C; and a C + T band with no corresponding C band is read as T. Computer-interfaced electronic pens can be used to scan the film and compile the sequence.

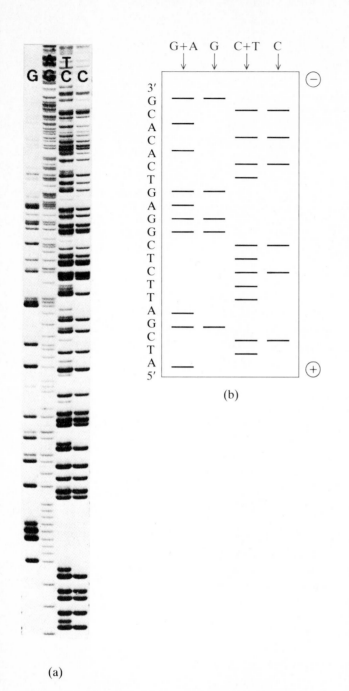

(a)

(b)

(5')p**ATCG**A̔TTCTCGGAGTCACACG(3')
p = ^{32}P in 5'-labeled DNA

p(N = 0 fragment; N = number of nucleotides)
p**ATC** (N = 3)
p**ATCG** (N = 4)
p**ATCGATTCTC** (N = 10)
p**ATCGATTCTCG** (N = 11)   from
p**ATCGATTCTCGG** (N = 12)   G + A
p**ATCGATTCTCGGA** (N = 13)
p**ATCGATTCTCGGAGTC** (N = 16)
p**ATCGATTCTCGGAGTCAC** (N = 18)
p**ATCGATTCTCGGAGTCACAC** (N = 20)

p**ATC** (N = 3)
p**ATCGATTCTC** (N = 10)
p**ATCGATTCTCG** (N = 11)   from
p**ATCGATTCTCGGA** (N = 13)   G only
p**ATCGATTCTCGGAGTCACAC** (N = 20)

p**A** (N = 1)
p**AT** (N = 2)
p**ATCGA** (N = 5)
p**ATCGAT** (N = 6)
p**ATCGATT** (N = 7)
p**ATCGATTC** (N = 8)   from
p**ATCGATTCT** (N = 9)   C + T
p**ATCGATTCTCGGAG** (N = 14)
p**ATCGATTCTCGGAGT** (N = 15)
p**ATCGATTCTCGGAGTCA** (N = 17)
p**ATCGATTCTCGGAGTCACA** (N = 19)

p**AT** (N = 2)
p**ATCGATT** (N = 7)
p**ATCGATTCT** (N = 9)   from
p**ATCGATTCTCGGAGT** (N = 15)   C only
p**ATCGATTCTCGGAGTCA** (N = 17)
p**ATCGATTCTCGGAGTCACA** (N = 19)

(c)

FIGURE 6–37 Gilbert-Maxam strategy of DNA sequencing. (a) Photo of
an actual radioautograph, different from the drawing. How much of this
sequence can you read? (b) Radioautograph pattern of the sequencing gel
corresponding to the fragments shown in part (c). Migration occurred from
− to +. (c) Original DNA and the resulting fragment mixture. Separate
portions of this original sample were subjected to four different modification/
cleavage treatments to yield the fragment mixtures shown.

An example of one application of DNA base sequencing is given in Figure 6–38. The sequence displayed shows 997 bases of a restriction fragment obtained from the chromosome of a virus responsible for forming sarcomas in mice. The tumor-causing action of the virus is due to a viral protein coded for by a gene on the viral chromosome called an *oncogene*. Previous studies established that the oncogene protein was a polypeptide of molecular weight 21,000—hence the designation p21.

When one scans a base sequence to find the gene, certain guidelines are used:

1. A triplet sequence of ATG is required to specify the start of the gene; ATG codes for methionine, which is the first amino acid put in place at the N terminus of a newly synthesized polypeptide chain (see Section 9–3).

2. Either TAA, TAG, or TGA is required as a stop signal downstream from the ATG start.

3. The ATG start signal should be preceded on the upstream side by various sequences recognized by the enzyme that synthesizes a messenger RNA molecule complementary to the gene.

Applying these criteria to the base sequence shown in Figure 6–38, researchers discovered that the base sequence contained three possible start signals coding for polypeptide chains having molecular weights of 30,000 (p30), 29,000 (p29), and 21,000 (p21). We will pass over the complex interpretation of what three different start signals may mean. The significant thing to emphasize is that one of the start signals can indeed account for the p21 oncogene protein.

The amino acid sequence displayed above the base sequence was evaluated from the known coding assignments of all triplets in the coding frame between the start and stop triplets. (The complete genetic code is displayed in Table 9–2.) The indirect determination of amino acid sequences from DNA base sequences of the coding gene agrees favorably with the traditional direct methods previously described in Chapter 4.

Sequencing RNA With Base-Specific Ribonucleases

One can now easily sequence the A, G, C, and U bases in RNA. After labeling the RNA with ^{32}P in much the same way as described earlier for DNA, base-specific ribonuclease enzymes (Table 6–5) can be used to generate a nested set of fragments in four separate treatments for analysis by PAGE. Ribonuclease T1 (RNase T1) catalyzes G-specific hydrolysis; RNase CL3 is C-specific; RNase U_2 is A-specific; and RNase PhyM is A + U–specific. Comparing fragments from RNase U_2 and RNase PhyM resolves A from U.

1

GCTCTAGTGGCAGTGTGTTGGTTGATAGCCAAAAGTTAATT<u>TTTAAAA</u>CATA<u>GTGTTT</u>

<u>TGGGGG</u>TTGGGGATTTAGCTCAGTGATAGAGCTCTTGCCTAGCACGCAAGCCCT<u>GGG</u>

<u>TTCGGTCCCC</u>CCAGCTCTGAAAAAAAGGAAAGAGAACAAAACAAAAACATATAGTGT

TTTATCTGTGCTT |ATG CCC GCA GCC CGA GCC GCA CCC GCC GCG
　　　　　　　　　met pro ala ala arg ala ala pro ala ala

GAC GAG CCC |ATG CGC GAC CCA GTC GCA CCC GTC CGC GCC CCC
asp glu pro met arg asp pro val ala pro val arg ala pro

GCC CTG CCC CGC CCC GCC CCG GGG GCA GTC GCG CCA GCA AGC
ala lev pro arg pro ala pro gly ala val ala pro ala ser

GGT GGG GCA AGA GCT CCT GGT TTG GAC GCC CCT GTA GAA GCG
gly gly ala arg ala pro gly lev ala ala pro val glu ala

|ATG ACA GAA TAC AAG CTT GTG GTG GTG GCC GCT AGA GGC GTG
met thr glu tyr lys leu val val val gly ala arg gly val

GGA AAG AGT GCC CTG ACC ATC CAG CTG ATC CAG AAC CAT TTT
gly lys ser ala leu thr ile gln leu ile gln asn his phe

GTG GAC GAG TAT GAT CCC ACT ATA GAG GAC TCC TAC CGG AAA
val asp glu tyr asp pro thr ile glu asp ser tyr arg lys

CAG GTA GTC ATT GAT GGG GAG ACG TGT TTA CTG GAC ATC TTA
gln val val ile asp gly glu thr cys leu leu asp ile leu

GAC ACA ACA GGT CCA GAA GAG TAT AGT GCC ATG CGG GAC CAG
asp thr thr gly gln glu glu tyr ser ala met arg asp gln

TAC ATG CGC ACA GGG GAG GGC TTC CTC TGT GTA TTT GCC ATC
tyr met arg thr gly glu gly phe leu cys val phe ala ile

AAC AAC ACC AAG TCC TTT GAA GAC ATC CAT CAG TAC AGG GAG
asn asn thr lys ser phe glu asp ile his gln tyr arg glu

CAG ATC AAG CGG GTG AAA GAT TCA GAT GAT GTG CCA ATG GTG
gln ile lys arg val lys asp ser asp asp val pro met val

CTG GTG GGC AAC AAG TGT GAC CTG GCT GGT CGC ACT GGT GAG
leu val gly asn lys cys asp leu ala gly arg thr val glu

TCT CGG CAG GCC CAG GAC CTT GCT CGC AGC TAT GGC ATC CCC
ser arg gln ala gln asp leu ala arg ser tyr gly ile pro

TAC ATT GAA ACA TCA GCC AAG ACC CGG CAG GGT GTA GAG GAT
tyr ile glu thr ser ala lys thr arg gln gly val glu asp

GCC TTC TAC ACA CTA GTA CGT GAG ATT CGG CAG CAT AAA CTG
ala phe tyr thr leu val arg glu ile arg gln his lys leu

CGG AAA CTG AAC CCG CCT GAT GAG AGT GGC CCT GGC TGC ATG
arg lys leu asn pro pro asp glu ser gly pro gly cys met

AGC TGC AAG TGT GTG CTG TCC TGA| CACCAGGTGAGGCAGGGACCAGCAAGACA
ser cys lys cys val leu ser STOP

TCTGGGGCAGTGGCCTCAGCTAGCCAGATGAACTTCATATCCACTTTGATGTCGCTCG₉₉₇

6–5 SECONDARY AND TERTIARY STRUCTURE OF DNA

Flexibility of Polynucleotide Chains

Previously, we explained (Section 4–3) that flexibility in a polypeptide chain is due to free rotation around the C^α—C and C^α—N axes of peptide bonds. In the repeat unit of polynucleotide chains (see Figure 6–39), there are 11 such torsional bonds. Thus there is considerable opportunity for geometric flexibility, particularly around the χ-glycoside bond axis connecting the base to the sugar.

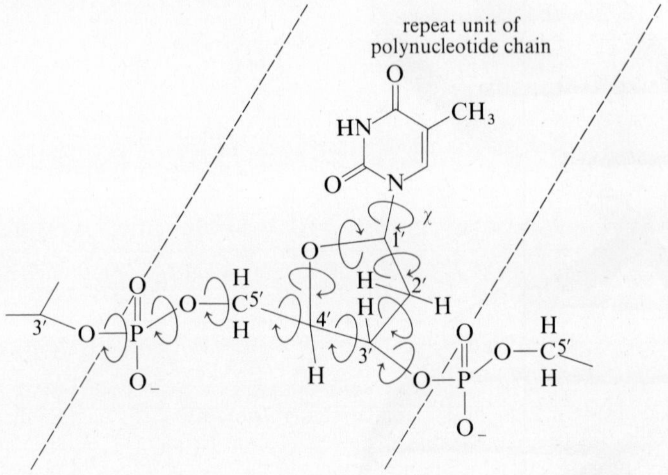

repeat unit of
polynucleotide chain

FIGURE 6–39 Flexibility in a polynucleotide. Each curved arrow identifies the possibility of rotation around a bond axis. Rotation at the sugar-base glycoside bond (χ, chi) is discussed later in the chapter.

However, because of constraints in the sugar ring structure and other steric constraints, totally free rotation is not possible. As we proceed, keep in mind that this element of structure is the reason for ordered conformations of nucleic acids and for the interconversion of conformations. We will refer back to this material at various times.

Double-Helix Structure of DNA

In the April 25, 1953, issue of *Nature,* James D. Watson and Francis H. C. Crick (working out of the Cavendish Laboratory in Cambridge, England) used only about a thousand words to describe their discovery of the three-

◄ FIGURE 6–38 Base sequence of a restriction fragment obtained from the Harvey murine sarcoma virus DNA chromosome. Three possible gene start signals (ATG) are denoted with ∟, and the TGA stop signal is denoted with ⌐. The third ATG (in color) specifies a polypeptide of molecular weight 21,000, in agreement with the p21 transforming protein encoded by the virus. Broken and solid underlining upstream from the ATG start signals are possible regulatory sequences governing the start of messenger RNA synthesis for this oncogene. The solid arrowheads (color) identify another possible regulatory signal, namely, the existence of two nearly perfect, extended repeat sequences (15 of 17 nucleotides are identical).

dimensional structure of DNA. They proposed the following (see Figure 6–40):

> The DNA molecule has a **double-helix structure** composed of two strands aligned with opposite polarity—one running in the $3' \rightarrow 5'$ direction and the other in $5' \rightarrow 3'$ direction—and intertwined with a right-handed twist. The purine and pyrimidine bases are inside the helical structure, where bases opposite each other on the two chains engage in hydrogen bonding along the entire length of the duplex.

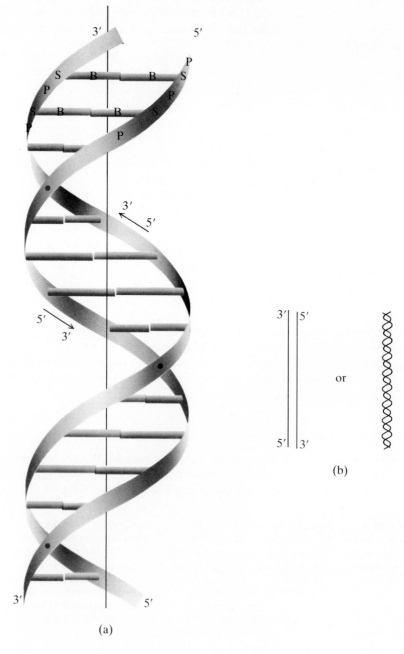

FIGURE 6–40 Double helix of DNA. (a) Diagrammatic representation of DNA according to Watson and Crick. The two ribbons symbolize the two phosphate—sugar chains, and the horizontal rods symbolize the pairs of hydrogen-bonded nitrogen bases holding the duplex together. The vertical line marks the central axis. (b) Schematic representations. On many occasions in this book we will represent the DNA structure in these simpler ways. [*Source:* Taken with permission from J. D. Watson and F. H. C. Crick, "Genetical Implications of the Structure of Deoxyribonucleic Acid," *Nature,* **171,** no. 4361, 964–967 (1953).]

Their proposal was proven correct and is generally recognized as the single achievement that revolutionized biological science in the twentieth century.

BASE PAIRING   The genius of Watson and Crick was in their interpretation of data available to them and in their construction of a model consistent with that interpretation. In addition, they happened to be at the right place at the right time. The critical empirical data came from Chargaff's earlier studies on the base composition of DNAs and from X-ray diffraction studies on DNA being performed at the time by Maurice Wilkins and Rosalind Franklin at King's College in London (see Note 6–4).

The pattern of interaction between nitrogen bases on opposing strands is in complete accord with Chargaff's observations. A purine is always hydrogen-bonded to a pyrimidine—hence the pur/pyr ratio of 1.0. More specifically, *adenine is always hydrogen-bonded to thymine* A═T (hence the A/T ratio of 1.0), and *guanine is always hydrogen-bonded to cytosine* G≡C (hence the G/C ratio of 1.0). Other base-pair (bp) combinations are incompatible with the geometry of the double helix. Accordingly, the A═T and G≡C combinations are called **complementary base pairs.** Moreover, the entire base sequence of one strand is totally complementary to the base sequence of the other strand.

The space-filling models in Figure 6–41 illustrate the *structural compatibility* of the planar bases for engaging in the dipole-dipole interactions of

NOTE 6–4
For their contributions Watson, Crick, and Wilkins were awarded the Nobel Prize in 1962. An extremely interesting and controversial narrative account of this period in scientific history and of the personalities involved is given by Watson himself in his best-selling book *The Double Helix.*

FIGURE 6–41 Purine-pyrimidine complementary base pairs in DNA. (a) Guanine-cytosine pairing. (b) Adenine-thymine pairing.

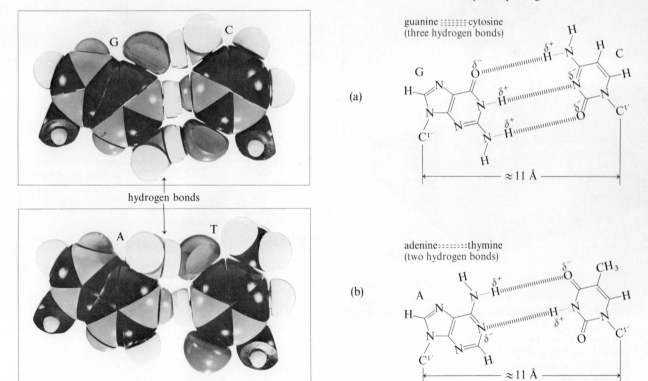

(a)

(b)

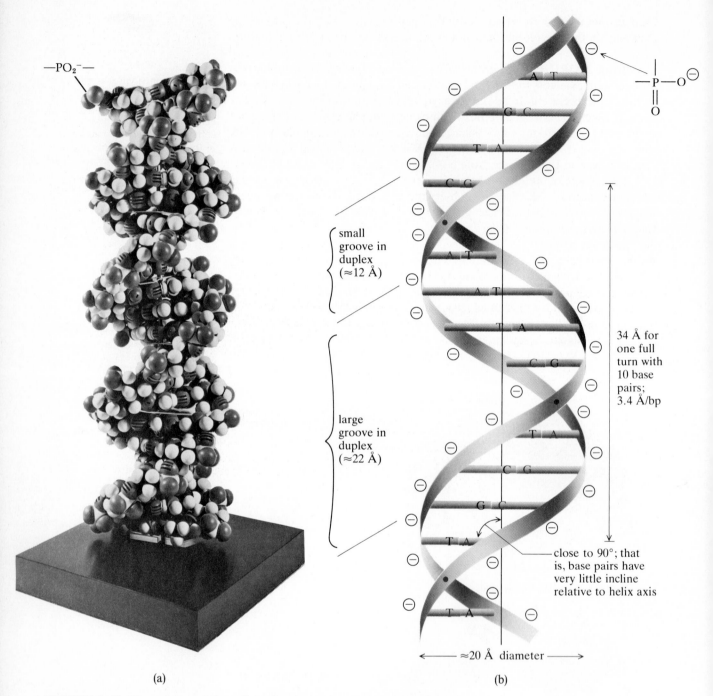

(a)   (b)

FIGURE 6–42 Dimensions and three-dimensional structure of the B form of DNA.
(a) Space-filling model, illustrating the hydrophobic core and the hydrophilic surface.
The model shows two complete turns of the duplex. (b) Grooves and molecular
dimensions. The width and the depth of the grooves are large enough to accommodate
the binding of proteins on the surface of DNA. (*Source:* Photograph provided by The
Ealing Corporation, Natick, Mass.)

hydrogen bonding. Without question, this bonding is one of the most striking examples of molecular logic that exists in nature. If you conclude that the bond strength of a G≡C pair with three hydrogen bonds should be greater than that of the A=T pair with two hydrogen bonds, you are correct. For example, this greater bond energy is reflected in the increase of the melting temperature T_m with increasing G + C content (refer to p. 244).

A more complete description of DNA structure appears in Figure 6–42, which shows molecular dimensions and the grooved surface of the duplex. With a diameter of about 20 Å, duplex DNA molecules are large enough to be clearly observed by electron microscopy. Although most DNAs have a *linear*, double-stranded (ds) structure, the micrograph in Figure 6–43 illustrates a *circular*, double-stranded structure of *plasmid* DNA, small DNA molecules found in pro-karyotes. The DNA chromosomes of several viruses and bacteria are also circular dsDNA molecules.

An exception to the duplex structure is found in some viruses that package the DNA chromosome as circular, single-stranded (ss) molecules. After infecting a cell, the complementary strand is synthesized to form the virulent duplex structure.

The sizes of some DNAs are given in Table 6–6. Figure 6–44 shows a specific DNA chromosome, illustrating its massive size.

STABILIZING FORCES   *Cooperative hydrogen bonding* obviously makes a significant contribution to stabilizing of duplex DNA. Additional stabilizing

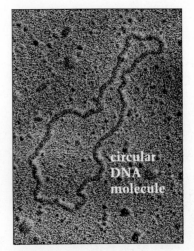

FIGURE 6–43 Electron micrograph of a plasmid DNA molecule obtained from *Escherichia coli.* Plasmids are small circular dsDNA molecules found in some bacteria in addition to the major chromosome. Plasmids carry genes conferring special attributes to the cell, such as genes that confer antibiotic resistance. Plasmids are also used as carrier molecules in genetic engineering experiments. (*Source:* Micrograph provided by Stanley N. Cohen, Department of Medicine, Stanford University.)

TABLE 6–6 Sizes of DNA molecules

ORGANISM	NUMBER OF BASE PAIRS (IN THOUSANDS; KILOBASES, kb)	TOTAL LENGTH[a] (μm)	STRUCTURE
Viruses			
Polyoma, SV40	5.2	1.7	circular ds
ϕX174	5.4	1.8 ⎫	circular ss to
M13 (fd, f1)	6.4	2.1 ⎭	ds on infection
P4	10.7	3.6 ⎫	
λ	48.6	16 ⎬	
T2, T4, T6	166	53	linear ds
Fowlpox	280	93 ⎭	
Bacteria			
Mycoplasma hominis	760	260 ⎫	
Escherichia coli	4,000	1,360 ⎭	circular ds
Eukaryotes		NO. OF CHROMOSOMES (HAPLOID)	
Yeast	13,500	4,600	17
Drosophila	165,000	56,000	4
Human	2,900,000	990,000	23
Lungfish	102,000,000	34,700,000	19

[a] Length = (kb) (0.34 μm).

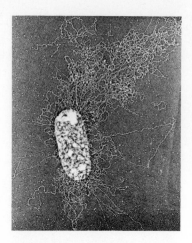

FIGURE 6–44 DNA chromosome seeping out of a disrupted *E. coli* cell. This photograph gives some perspective on the massive size of DNAs.

factors include the following. The tendency of the nonpolar aromatic purine and pyrimidine rings to self-associate in a stacked arrangement is the consequence of *hydrophobic influences*. Once the bases are stacked, an extensive network of van der Waals association among the bases would further stabilize the duplex. Finally, the existence (at pH 7) of the negatively charged phosphate groups along the outer surface of the hydrophilic sugar-phosphate backbone provides the opportunity for extensive *ion-ion associations* with other substances, such as inorganic metal ions (notably, Mg^{2+}) and positively charged organic substances. Examples of the latter include the *histones* and low-molecular-weight *polyamines*. Histones (see p. 267 in this chapter) are basic proteins containing (at pH 7) a large number of positively charged lysine and arginine side chains. Two examples of polyamines are

$$\overset{+}{H_3}N(CH_2)_4\overset{H}{N}(CH_2)_3\overset{+}{N}H_3 \qquad \overset{+}{H_3}N(CH_2)_4\overset{+}{N}H_3$$

$$\text{spermidine} \qquad\qquad\qquad \text{putrescine}$$

Histone/DNA interactions, occurring in the nucleus of eukaryotic cells, will be described shortly.

Polymorphism of DNA

Since the original Watson-Crick proposal, several different conformations for duplex DNA have been detected by X-ray diffraction. The conformation described by Watson and Crick is commonly called the B conformation, or simply B–DNA. It is obtained when fibrous DNA is prepared under conditions of high humidity. At low humidity a structure called A–DNA is obtained, which is shorter and broader than B–DNA. For A–DNA, 11 base pairs are packed in one complete turn that spans about 25 Å; the diameter is about 26 Å; and the base pairs are inclined about 20° to the helix axis. There are several other X-ray structures we will not describe.

Using model building, R. C. Hopkins established an alternative way of arranging the two strands in a double helix, namely, reversing the positions of the strands in B–DNA (see Figure 6–45). Aside from being able to form a complete right-handed helix, this alternative B′ configuration of strands can also support the formation of a left-handed, double-helix structure in which there is now extensive interest (see the next subsection).

These findings and many others give strong evidence that *in solution,* particularly in vivo, DNA does not exist in any one conformation, although various evidence suggests that the B form may be the major structure. This result raises the question about whether this *polymorphic character* of DNA has any biological significance. For example, do different segments of a DNA molecule exist in different conformations, and is this feature involved in the regulation of DNA functions? Preliminary evidence suggests that the answer is yes. Other questions relate to how local changes in the conformation of a DNA segment occur. Specific answers to such questions should be forthcoming.

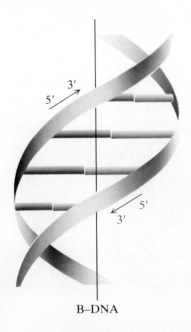

B–DNA

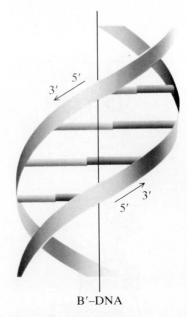

B'–DNA

FIGURE 6–45 Reversal of the pattern of interwinding strands. Reversal still allows for base pairing in a helical duplex.

Left-Handed Helical Geometry of DNA

In recent years crystals of synthetic oligonucleotides of precise size and base sequence have been studied by X-ray diffraction to obtain refined measurements on double-helix structures. True crystals give sharper diffraction patterns than fibers of DNA. When Alexander Rich and co-workers (in 1979) examined crystals of the self-complementary hexanucleotide

$$dCpdGpdCpdGpdCpdGp, \text{ or simply } d(CG)_3,$$

they observed that strands of $d(CG)_3$ did indeed form duplex structures that crystallized, but unexpectedly, the helix had a *left-handed twist*. The structure was dubbed the Z conformation because of the zigzag contour of the S—P—S—P—S backbone in the left-handed geometry (see Figure 6–46).

There is no definite proof yet that the Z form of DNA occurs in living cells. Rich and others have isolated antibodies that specifically recognize the left-handed twist of the Z structure and have shown that these Z antibodies did bind to certain regions of the salivary gland chromosome of the *Drosophila* fruitfly. However, this finding has been challenged by other researchers, who report that the Z antibody binding is an artifact resulting from the method used to prepare the chromosome for the assay. The isolation of other Z–DNA specific binding proteins from various cells also provides indirect evidence that Z–DNA occurs in cells. However, these Z-binding proteins are not yet known to have any function in living cells.

Even if it does occur naturally, still unclear at present is just what biological functions are associated with Z–DNA. Because the Z and B structures are

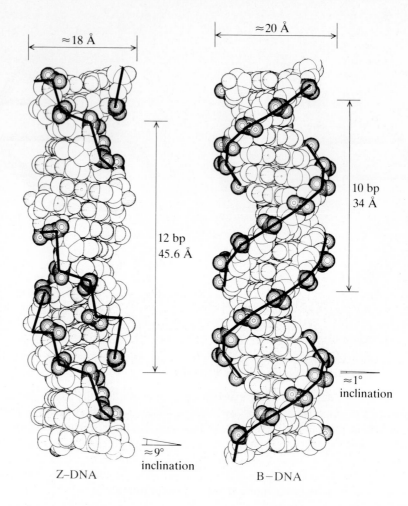

FIGURE 6–46 Comparison of Z–DNA (left-handed duplex) and B–DNA. In addition to having different handedness, Z–DNA is slightly slimmer and more elongated, and the topography of grooves is different. In Z the minor groove is deep, and the major groove has little or no depth, being almost flat to the surface. Both grooves in B–DNA show about the same intermediate depth. There is a very slight inclination of base pairs in B–DNA, while about a 9° inclination is seen in Z–DNA.

distinctly (not subtly) different, one obvious suggestion is that a local B ⇌ Z interconversion on a chromosome may serve to regulate the expression of genes. However, there is yet no proof of this hypothesis. Another possible function has been suggested by recent evidence that implicates regions of Z structure in the pairing of chromosomes in the process of genetic recombination.

Much has been learned about the chemistry of the Z duplex in the past five years. One noteworthy item is that the formation of the Z duplex is favored by—but not dependent on—an alternating GC pattern of base sequence. For this reason synthetic poly(CG)/poly(GC) substances are routinely used as model systems to study various aspects of the Z conformation: conditions that favor formation of Z, conditions that stabilize Z, and conditions that trigger the conversion of Z to B and vice versa.

Because of the difference in handedness, B ⇌ Z transitions in solution can be monitored by circular dichroism (see Figure 6–47). As indicated, B and Z structures give nearly mirror image CD spectra. Starting with B, any condition that causes a B → Z transition will be reflected in an altered CD spectrum,

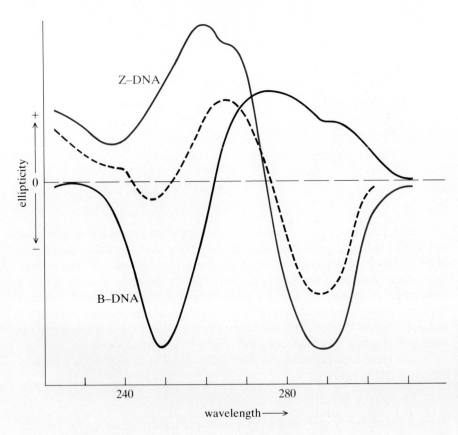

FIGURE 6–47 Circular dichroism (CD) spectra of Z and B forms of a double helix. The broken line represents a contribution of both B and Z during B ⇌ Z transitions. (If necessary, refer to Section 4–4 for a review of circular dichroism.)

becoming more like the Z spectrum as more Z structure is formed; the reverse is true for a Z → B transition.

One other interesting finding is that *methylation of C residues favors a B → Z transition.* Recall our previous discussion of the natural occurrence of m^5C in DNA and its possible role in regulating the expression of eukaryotic genes. Perhaps the existence of m^5C stabilizes a stretch of Z conformation, and the Z signal keeps genes in the region turned off.

While we cannot devote pages and pages to Z–DNA, before finishing let us consider how the B → Z transition might occur. It does not occur by a simple retwisting of the duplex in the opposite direction. Instead, it likely involves a complex series of internal rearrangements involving rotations around one or more bond rotation sites that exist in the nucleotide unit (refer to Figure 6–39).

The process probably begins by a local melting of a few base pairs; that is, bases pull away from each other, and the hydrogen bond associations are lost. If we consider the melt to occur in a region of $(GC/CG)_n$ structure, the C and G bases may change their orientation by 180°. For G bases this change can be achieved by the rotation of the G base around the χ bond axis connecting it to the sugar. In other words, G bases in Z have a conformation relative to the attached sugar that is different from that found in B (see Figure 6–48). The G conformation in Z is called *syn,* and in B it is called *anti.*

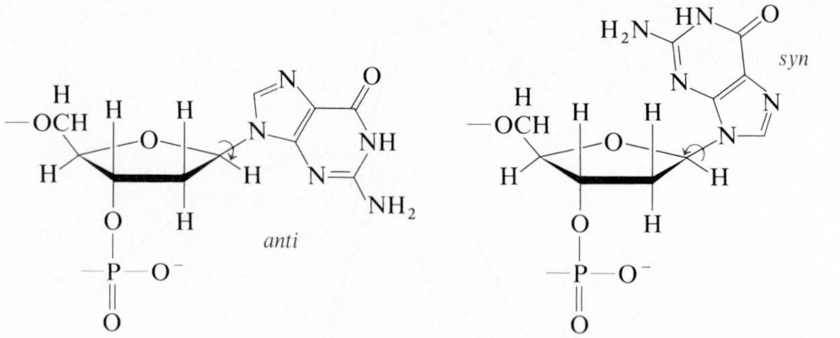

FIGURE 6–48 *Anti/syn*
conformations. In B–DNA all
bases have the *anti* orientation.
In Z–DNA the A, C, and T
bases are in the *anti* orientation,
but G is in the *syn* orientation.

However, for a C base, because the six-membered pyrimidine ring encounters steric hindrance to rotation around the χ bond axis, the 180° reorientation of C requires the complete rotation of the C sugar nucleoside unit. The orientation of a C base relative to the attached sugar is the same in both Z and B geometries—it is *anti*. All A and T bases in both B and Z structures also have the same *anti* orientation. After the G and C rotations are complete, the strands re-form a helical duplex, with the newly aligned G and C bases reengaging in hydrogen bonding but redirecting the twist with left-handedness.

The reason for pursuing this descriptive detail is to emphasize a broader aspect of DNA, namely:

The duplex molecular structure is not static.

Instead, the molecule can breathe, and as it opens, different things can happen. One is the transition of B → Z; another is the formation of loop-out structures to be described shortly; another is the interaction with DNA-binding proteins involved in various aspects of DNA biochemistry; another is the reaction of the bases with another substance (a mutagen, for example); another is the binding of substances to DNA by *intercalation*, where a substance slips into the core of the duplex, becoming wedged between the stacked base pairs (see Figure 6–49). Examples of intercalating substances are *ethidium bromide* (a DNA stain), *actinomycin* (an antibiotic), and some anticancer drugs.

Repetitive, Palindromic, and Inverted Repeat Sequences in DNA

FIGURE 6–49 Intercalation of a substance, represented by heavy color bars.

REPETITIVE SEQUENCES In descriptions of DNA emphasis is usually given to existence in DNA of single copies of genes, each having a unique base sequence coding for a messenger RNA, a transfer RNA, or a ribosomal RNA. A notable exception in many chromosomes is the existence of multiple copies (upward to several hundred in some cases) of genes for ribosomal RNA. However, in nuclear DNA of eukaryotes, depending on the species, unique gene sequences may comprise only about 10%–50% of the DNA information. Most of the rest of DNA contains *repetitive sequences,* the same sequence repeated many, many times. Some are short [100–500 base pairs (bp) in length], and others are known that have about 5000 bp. Moderately repetitive sequences may occur 1000–10,000 times; highly repetitive sequences occur 100,000–1,000,000 times. Neither the number of different repeat sequences nor their biological significance is yet known.

In human DNA about 3%–6% of the haploid genome (23 chromosomes) is contributed by just one family of repeat sequences, called the *Alu family* because the ≈300-bp sequence contains a single internal cleavage site for the restriction enzyme AluI. About 250,000–500,000 copies of this sequence are present.

PALINDROME SEQUENCES Sometimes, a segment of duplex DNA shows perfect twofold rotational symmetry owing to the occurrence of the same base sequence existing in opposite orientations on the two strands of the duplex. Such a base-pair sequence is called a *palindrome* (the term for a word that is spelled the same in either direction, such as madam). Perfect palindrome locations usually span 4–8 base pairs. Among other things, palindrome sites provide the *type of sequence recognized by most restriction enzymes*. For example (see Figure 6–50), for the three restriction enzymes listed in Table 6–5, the palindromic character of each sequence-specific site is clearly evident when the entire base-pair sequences of the duplex site are displayed.

Restriction enzymes may recognize the rotational symmetry directly, with individual enzymes having a preferred base sequence in the region for optimal binding. Alternatively, the linear duplex may breathe slightly, and some *intrastrand base pairing* may occur because half of a palindrome sequence in one strand is necessarily complementary to the other half. These tight *loop-outs* of each strand of the duplex are called *cruciform conformations* (see Figure 6–51), and these conformations may be what a restriction enzyme recognizes. Furthermore, DNA also contains interrupted palindrome sequences, which can provide the basis for larger and less strained cruciforms that may be conformational signals for the recognition of other DNA-binding proteins.

Remember (Table 6–5) that restriction enzymes not only recognize a specific palindromic site but also cleave specific phosphodiester bonds in *both strands*. Two types of cleavage patterns are known (see Figure 6–52). In *central cleavage* the same two bonds in the middle of the palindrome are cut, giving *blunt ends* for each of the two new termini. In *staggered cleavage* the same bonds

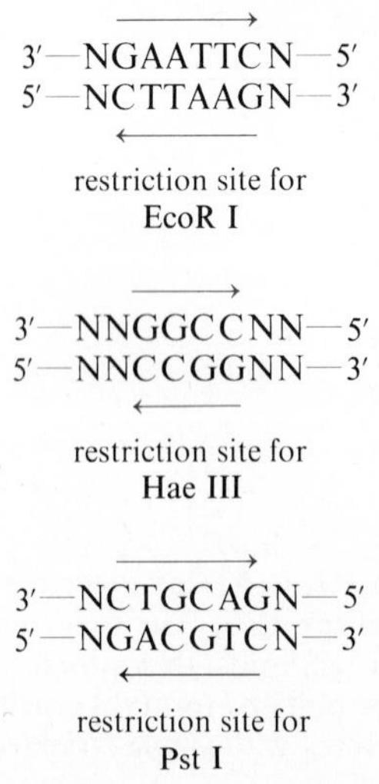

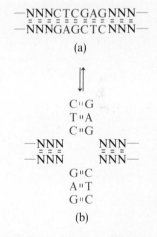

3′—NGAATTCN—5′
5′—NCTTAAGN—3′

restriction site for
EcoR I

3′—NNGGCCNN—5′
5′—NNCCGGNN—3′

restriction site for
Hae III

3′—NCTGCAGN—5′
5′—NGACGTCN—3′

restriction site for
Pst I

FIGURE 6–50 Palindrome sequences. The same sequence in opposite directions appears on both strands.

—NNNCTCGAGNNN—
—NNNGAGCTCNNN—

(a)

⇕

C G
T A
C G
—NNN   NNN—
—NNN   NNN—
G C
A T
G C

(b)

FIGURE 6–51 Reversible interconversion at palindrome locations. (a) Regular duplex. (b) Cruciform structures.

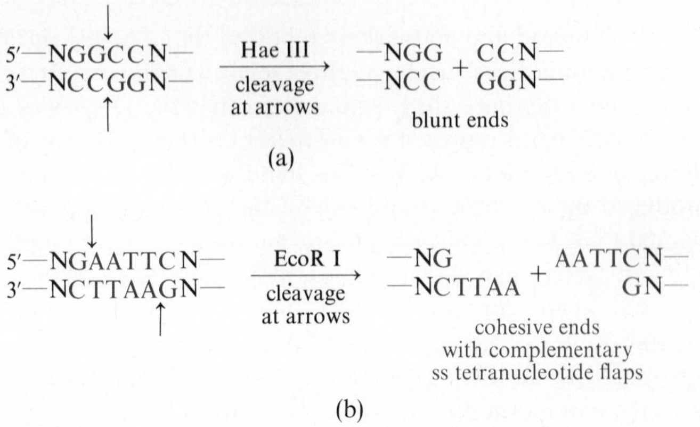

5′—NGGCCN—   Hae III   —NGG   CCN—
3′—NCCGGN—   ──────────→   —NCC + GGN—
                cleavage
                at arrows   blunt ends

(a)

5′—NGAATTCN—   EcoR I   —NG   AATTCN—
3′—NCTTAAGN—   ──────────→   —NCTTAA +   GN—
                 cleavage
                 at arrows   cohesive ends
                                   with complementary
                                   ss tetranucleotide flaps

(b)

◀**FIGURE 6–52** Cleavage patterns of restriction enzymes. (a) Central cleavage. (b) Staggered cleavage.

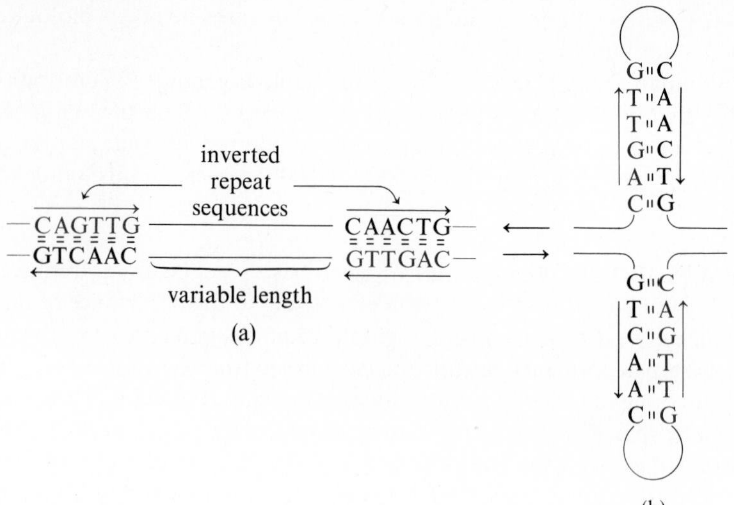

FIGURE 6–53 Inverted repeat sequences (IRS). The occurrence (a) of IRS in dsDNA provides an opportunity to form (b) cruciform structures with single-stranded loops.

on opposite sides of the center of symmetry are cut, giving ends with short single-stranded flaps that are complementary to each other and are called *cohesive ends*. Thus one EcoR I terminus of duplex DNA can recognize another EcoR I terminus and one Pst I terminus can recognize another Pst I terminus. The use of restriction enzymes that generate cohesive termini is beautifully exploited in the construction of recombinant DNAs (see Section 6–9).

INVERTED REPEATS   Examine Figure 6–53(a); you will see a base-pair sequence at the left that also appears at the right but is switched on the two strands and also oriented in the opposite direction. This sequence is called an *inverted repeat sequence* (IRS). The most notable significance of IRS regions is their occurrence at the end of *transposons*, segments of DNA that can move to a new position on the chromosome or to an entirely different chromosome. As shown in Figure 6–53(b), terminal IRS regions also allow the bracketed region of duplex DNA to assume a cruciform conformation with looped-out regions of single-stranded DNA.

Supercoiling of DNA

In 1963 J. Vinograd and co-workers reported that a highly purified DNA, extracted from polyoma virus, displayed three separate bands on ultracentrifugation: 20S, 16S, and 14S. Since they were certain that the DNA was pure, they suggested that each band represented a different structural form of the same DNA molecule (see Figure 6–54). The 14S band was shown to represent linear dsDNA produced by a double strand cut on the 16S material, itself shown to be a circular dsDNA. Being somewhat more compact, circular duplex molecules should sediment faster than linear duplex molecules. They further suggested that the fastest 20S band contains DNA in an even more compact structure—a twisted, circular duplex called *supercoiled DNA*. The coiled coil, or supercoiled, state was proposed to be the native structure, with the others being produced as a result of the extraction and handling procedures.

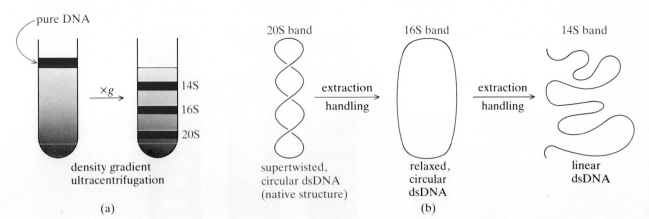

(a)

(b)

FIGURE 6–54 Original discovery of supercoiled DNA. (a) Three distinct zones observed from a highly purified DNA. (b) Explanation of (a). The three zones proved to be different states of the same molecule.

Proof of the supercoiled DNA structure has been provided from different areas, including visualization by electron microscopy (see Figure 6–55). Evidence for the existence of the supercoiled state in vivo has been provided by isolating enzymes from prokaryotes that are responsible for forming it—called *DNA gyrase*—and also for converting it back to relaxed structures—called *DNA swivelase*. Gyrase and swivelase are examples of enzymes that alter the topology of DNA. A general name for such an enzyme is *DNA topoisomerase*. A gyrase enzyme has not yet been detected in eukaryotes, but swivelase enzymes have been.

The DNA molecules with different degrees of supertwisting are called *topoisomers*. As shown in Figure 6–56, one can separate topoisomers via agarose gel electrophoresis (with ethidium bromide staining for detection). In lane 1 is an extract of native DNA, with the intense band at the bottom representing highly supercoiled dsDNA and the band at the top being some fully relaxed, circular dsDNA. Supercoiled molecules of DNA migrate further because they have a more compact structure and pass more easily through the pores of the agarose matrix. Lanes 2 and 3 illustrate the effects of treating the DNA with DNA swivelase for 5 min (lane 2) and 30 min (lane 3). Note the disappearance of the heavily supercoiled band at the bottom and the more intense banding at the top of the gel as molecules are relaxed. The intermediate bands represent DNA molecules caught with different degrees of supertwisting still present.

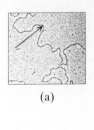

(a)

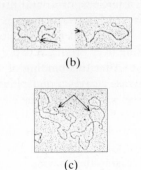

(b)

(c)

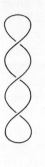

(d)

FIGURE 6–55 DNA duplex in different topological states. (a) Linear. (b) Highly supercoiled. (c) Two partially twisted structures. (d) Details of supercoiling. The line in the coil at the left represents the duplex DNA, which itself is coiled as shown in the drawing at the right. Thus the supercoiled structure is a *coiled coil*. (Source: Micrographs provided by Dr. Jack Griffith, Stanford University.)

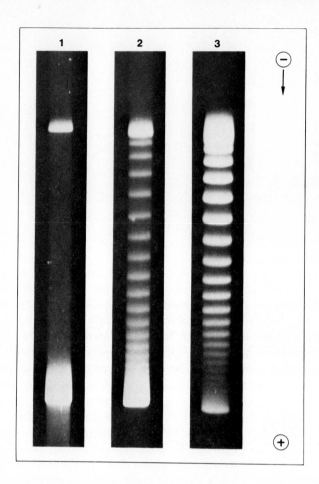

FIGURE 6–56 Separation of topo isomers via agarose gel electrophoresis. See text for description. (Source: Photograph provided by Walter Keller.)

In prokaryotes the action of DNA gyrase results in a supertwisted structure that is right-handed. This structure is called *negative supercoiled*. The extent of this twisting has been measured as occurring about once for every 200–250 base pairs, that is, one complete twist for every 20–25 turns of the double helix. Thus the circular DNA chromosome of *E. coli* with about 4 million base pairs could accommodate about 20,000 supertwists. Various studies have confirmed that the supercoiled structure of DNA is essential for numerous DNA processes. It also explains how the massive DNA molecule can get packaged inside the small volume of a bacterial cell, a nucleus, or a viral protein capsid.

How does DNA gyrase work? One suggestion is diagrammed in Figure 6–57. The DNA binds to gyrase, with duplex segments folding over each other. Topologically speaking, the folding over produces two different types of crossover nodes, positive and negative. After this binding of substrate to enzyme is complete, a nuclease action of gyrase cleaves a phosphodiester bond in each of the two strands of one duplex segment. The enzyme cleaves the bond only at the positive node. This double-strand cut is followed by the slippage of the uncut duplex strand through the cut strand. After this pass-through, the gyrase then acts as a ligase to rejoin the ends of the cut strand. The turnover of gyrase requires the hydrolysis of ATP, releasing the enzyme to start another cycle.

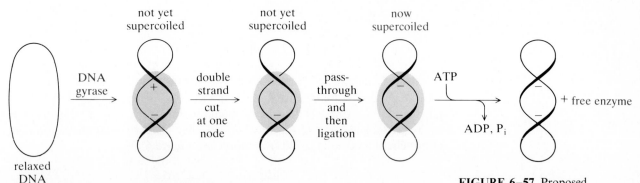

FIGURE 6–57 Proposed operation of gyrase.

Although they lack a gyrase enzyme, eukaryotes also contain supercoiled DNA occurring by virtue of the association of DNA with *histone proteins*. There are five principal types of histones (see Table 6–7): H1, H2A, H2B, H3, and H4; each one has a high content of the basic amino acids lysine and arginine. Because of their polycationic nature (many positive charges), histones are tailor-made to associate with the polyanionic backbone of DNA.

In addition to contributing to the stabilization of DNA structure, histones, according to some evidence, function in regulating the expression of genes. Histone-DNA interactions follow a regular pattern in which the DNA is wrapped around ordered clusters of histones (see Figure 6–58). Each histone cluster is an octamer composed of two molecules each of H2A, H2B, H3, and H4. Histone H1 binds to the outside of the coiled DNA. The length of DNA wrapped around one histone cluster spans about 150 base pairs, producing two turns around the cluster. Note that DNA coils around the histone clusters to yield a left-handed supertwist, called *positive supercoiling*. These aggregates are called *nucleosomes* or *v bodies* and serve as the elementary DNA-protein unit responsible for the organization of chromatin in the nuclei of eukaryotes. The article by Burlingame and co-workers listed in the Literature describes the recent success in solving the three-dimensional structure of the octameric histone core of the nucleosome.

TABLE 6–7 Lysine and arginine content of histone proteins

HISTONE[a]	% lys	% arg
H1	24.8	2.6
H2A	10.9	9.3
H2B	16.0	6.4
H3	9.6	13.3
H4	10.8	13.7

Note: A fifth to a third of amino acids are highly basic.
[a] Data are for histones isolated from calf thymus; most eukaryotic cells contain the same types.

FIGURE 6–58 Histone-DNA interactions in eukaryotes.

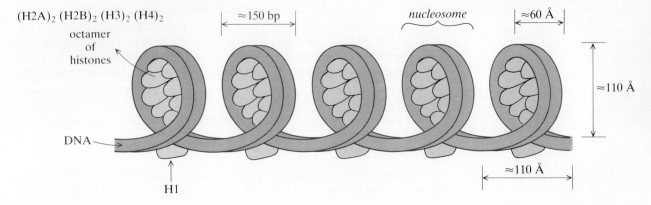

6–6   RNA STRUCTURE

The major RNAs found in all types of cells are *messenger RNA* (mRNA), *transfer RNA* (tRNA), and *ribosomal RNA* (rRNA). In *E. coli* the relative amounts of each are about 2% mRNA, 16% tRNA, and 82% rRNA. As a rough approximation, a similar distribution applies to all cells in general. All three types are *single-stranded structures* produced in the process of *transcription* from specific genes in DNA. Once formed, they all participate in the process of protein biosynthesis called *translation* (see Figure 6–59).

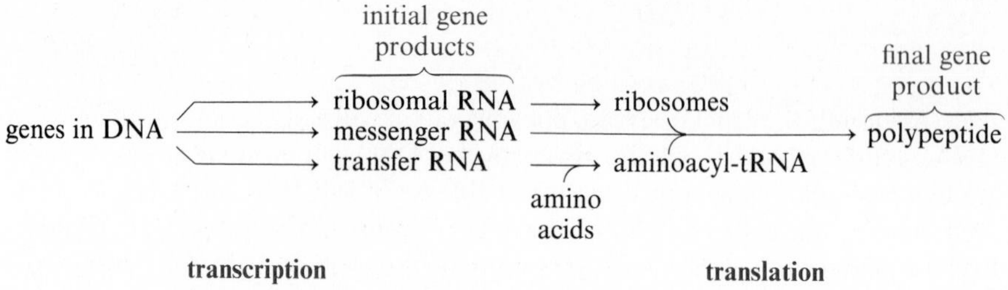

FIGURE 6–59 Flow of genetic information via transcription and translation. The flow DNA → RNA → protein is a common representation.

The rRNA gene transcripts are used in the assembly of ribosomes. The base sequence in the mRNA transcript provides the instructions for assembling polypeptide chains. In the translation process the instructions in mRNA are decoded by tRNA molecules, each carrying a specific amino acid to the ribosome surface where polypeptide assembly occurs. We will examine the complex details of transcription in the next chapter and of translation in Chapter 9. Our purpose now is to examine some highlights of RNA structure.

Messenger RNA

A living cell is capable of producing hundreds to thousands of different mRNA molecules of various sizes. Some degree of ordered structure probably exists in mRNA; but there is no conformation that applies to all mRNAs, since each has a unique sequence of A, G, C, and U bases.

The instructions in mRNA for directing polypeptide assembly are encoded in *triplet codons*—a succession of three bases—with each codon specifying one amino acid. For example, the codon UUU specifies phenylalanine, AUA specifies isoleucine, GAU specifies aspartic acid, and so forth. (The complete genetic code is shown in Table 9–2.) Thus a codon sequence in mRNA of . . . UUUGAUAUA . . . specifies the tripeptide segment of . . . phe-asp-ile . . . . In eukaryotes a single mRNA codes for a single polypeptide. In prokaryotes and in viral systems a single mRNA may code for one, two, or three polypeptides.

In the next chapter we will not only consider the details of transcription but also examine some special features of mRNA structure in eukaryotes.

Transfer RNA

A living cell may contain as many as 60 different tRNA molecules. They are the smallest nucleic acid (MW about 25,000), consisting of approximately

73–93 nucleotides. After attachment to an amino acid, with each tRNA attaching to a specific amino acid, the amino acid-tRNA adducts are positioned in the order specified by the codon sequence in mRNA. The reading of the mRNA codons is accomplished by the existence in each tRNA of a unique base sequence, called an *anticodon,* complementary in sequence to that of the codon.

Solutions of purified tRNA display a significant hyperchromic effect when heated, indicating a highly ordered native structure. The first insight of structure was provided in 1965 with the solution of the complete base sequence for a tRNA molecule. The sequence revealed that the tRNA chain could assume a variety of partially double-stranded arrangements by folding back on itself on the basis of *intrachain base pairing* involving GC and AU pairs. (*Note:* U and T are equivalent pyrimidines for base pairing with A. Both U and T differ only in a methyl group that has nothing to do with the base pairing to A.)

Of the various folding patterns suggested, the one that proved consistent with all tRNA sequences is the so-called *cloverleaf structure* (see Figure 6–60), consisting of four distinct double-stranded segments forming three major loops, one minor loop of variable size, and an open arm that includes the 3′ and 5′ ends of the chain. An X-ray diffraction analysis of tRNA crystals was completed in 1974. It revealed an L-shaped structure consisting of two distinct, partially double-helix stems aligned almost perpendicular to each other. Close inspection reveals that the L structure also contains three major loops and an open stem. Thus the cloverleaf pattern remains a valid two-dimensional representation of three-dimensional structure.

In addition to A, G, C, and U, the tRNA sequence displayed in Figure 6–60 indicates the presence of atypical (modified) nucleosides: D, mG, mC, m_2G, Ψ, T, mA. To date, more than fifty modified nucleosides have been detected. Many of these have methylated bases, but several others involve more complex modification. Some of the modifications—such as ribosylthymidine (T), pseudouridine (Ψ), and dihydrouracil (D)—are found in tRNAs of most organisms (see Figure 6–61, p. 271). Others are found only in prokaryotes, only in eukaryotes, or only in specific organisms. In any event, the presence of modified nucleosides in certain regions (and in some instances in a certain position) of the tRNA molecule clearly has important significance to both the structure and the function of rRNA.

Some of the established assignments of specific functions to specific regions of tRNA structure are as follows (see Figure 6–62):

1. The 3′ end of the open stem (with a —CCA sequence common to all tRNAs) is the site for the covalent attachment of the amino acid to tRNA.

2. The TΨC loop (loop I) is associated with the binding of the aminoacyl-tRNA species to ribosomes.

3. The D loop (loop III, containing dihydrouracil) is also implicated in the binding of aminoacyl-tRNA to ribosomes.

4. The anticodon is always contained in loop II. The presence of a modified nucleoside adjacent to the anticodon is also universal. The anticodon loop is also associated with ribosome binding.

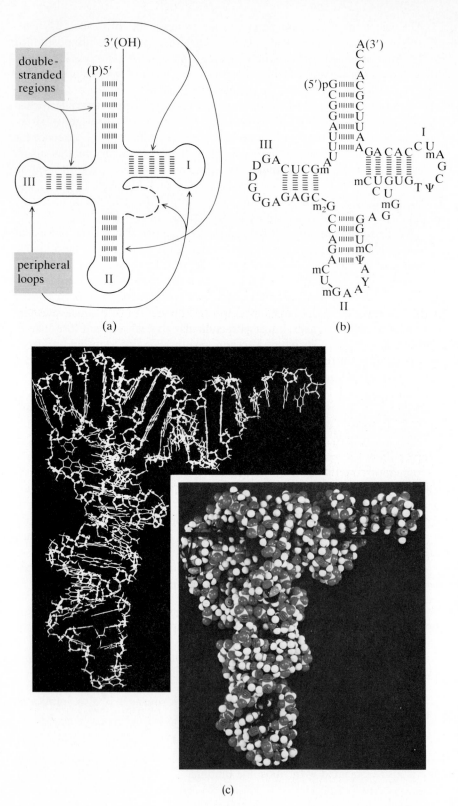

(a)

(b)

FIGURE 6–60 Transfer RNA.
(a) Cloverleaf representation. The
double-stranded regions are
caused by foldings and are
stabilized by hydrogen bonds
(ıııııııı) between complementary
base pairs. The peripheral loops
have no hydrogen bonding; there
are three major loops and one
minor loop of variable size. (b)
Phenylalanine tRNA from yeast,
with 76 nucleotides. (c) Molecular
wire model and space-filling
model of the three-dimensional
structure of tRNA. An upside-
down view from the rear would
represent the L shape. [*Source:*
Reproduced with permission from
S. H. Kim, "Three-Dimensional
Structure for Yeast Phenylalanine
Transfer-RNA," *Science,* **185,**
435–439 (1974). Copyright ©
1974 by the American Association
for the Advancement of Science.
See also F. L. Suddath, *Nature,*
248, 20–24 (1974).]

(c)

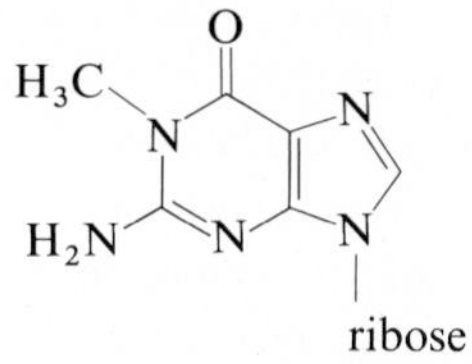

1-methylguanosine (mG) 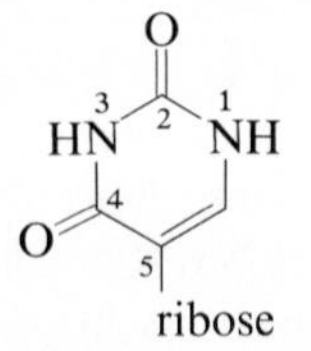 pseudouridine (Ψ) 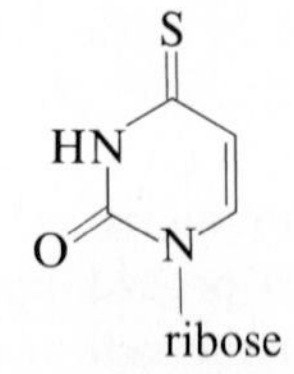 4-thiouridine

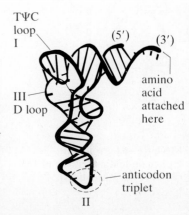

FIGURE 6–61 Examples of modified nucleosides in transfer RNA. In pseudouridine the modification involves the pyrimidine being linked to ribose at C5 rather than the usual N1.

Ribosomal RNA

Ribosomes are *multimolecular aggregates* of protein (about 35%) and RNA (about 65%); obviously, the RNA is called ribosomal RNA. An intact ribosome (70S in prokaryotes) is a complex of *two subunits* in noncovalent association: one heavy (50S) and one light (30S) (see Note 6–5). As shown in Figure 6–63, the intact complex can be dissociated to yield the subunits, which in turn can be dissociated further to yield rRNAs and various proteins. The heavy subunit contains two RNAs (a 23S rRNA and a 5S rRNA) and single copies of each of 30 different proteins and three copies of another protein. The light 30S subunit contains only one RNA (16S rRNA) and single copies of each of 21 different proteins. The total number of parts is 57, and 55 of them are different.

The 5S rRNA molecules (120 nucleotides) from many sources have been completely sequenced. More recently, the complete sequences of 16S (1542 nucleotides) and 23S (2904 nucleotides) species have also been solved. From a comparison of sequences and patterns of nuclease cleavage of 16S rRNAs from different sources, an elaborate pattern of ordered chain folding for the native three-dimensional structure has been proposed (see Figure 6–64). In addition to its obvious role as a molecular scaffold to support the association of 21 different proteins, certain domains of structure undoubtedly have specific functions, as in tRNA.

FIGURE 6–62 Functional domains of transfer RNA. See text for descriptions.

NOTE 6–5
Ribosomes from eukaryotic cells are larger (80S) and are composed of 60S and 40S subunits. The 40S subunit consists of an 18S RNA. The 60S subunit consists of a 28S RNA, a 5S RNA, and a 5.8S RNA.

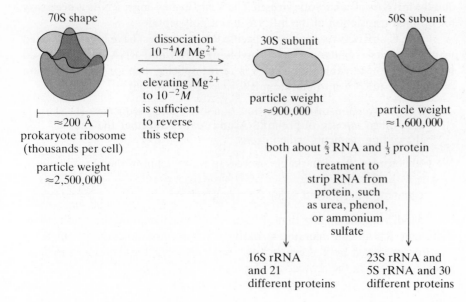

FIGURE 6–63 Molecular composition of ribosomes.

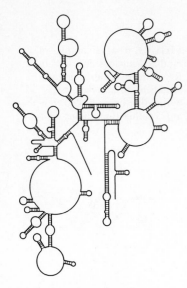

FIGURE–64 Proposed intrachain-folding pattern of 16S rRNA (1542 nucleotides).

TABLE 6–8 UMP-rich small nuclear RNAs

U-snRNA	CHAIN LENGTH[a]
U1	165
U2	188–189
U3	210–214
U4	142–146
U5	116–118
U6	107–108

[a] Range is for RNAs from different sources.

There is much evidence that the RNA and the protein molecules in each subunit interact with each other in a highly organized pattern. For example, one indication is provided by reconstitution experiments wherein active ribosome subunits are restored from their components previously isolated by dissociation procedures. Of course, the *self-assembly* of ribosomes is a natural event in the cell. Solving the intricate anatomy of the ribosome is difficult. Recent studies have revealed an image of overall shape and subunit association as was illustrated in Figure 6–63.

Small Nuclear RNAs

In addition to transfer RNAs and the 5S and 5.8S ribosomal RNAs, several new types of small RNAs (in the range of 4–8S) have been detected in eukaryotes, prokaryotes, and viruses. Some even occur in the free state (see Section 6–7). They were originally thought to be only fragments from the degradation of a larger mRNA or rRNA, but we now know that many are directly synthesized from genes in DNA as transcription products and then associate with several proteins to form *ribonucleoprotein particles* (RNPs). The type most studied is U-snRNA, a class of UMP-rich, small RNAs found in nuclei of eukaryotic cells.

At present six different species of U-snRNA have been detected (see Table 6–8). Species U1 occurs in highest concentration, with about a million copies present in the nucleoplasm of a single nucleus. The U1 sequence of 165 nucleotides has been determined for a variety of sources, from humans to insects. These sequences are very similar, suggesting that the structure and the function(s) of the molecule have been highly conserved over many years of evolution.

The first U1 sequence was solved in 1974. It revealed the presence of some Ψ (pseudouridine) residues, some methylated bases, and a novel element of structure at the 5′ end: a trimethylated G residue connected by a 5′—5′ triphosphate bond followed by one or two 2′-O-methylated sugar structures (see Figure 6–65). This *5′ cap modification* (also present in U2-, U3-, U4-, and U5-snRNAs) was subsequently found to be present also in eukaryotic mRNAs and the mRNAs of eukaryotic viruses. The 5′ cap in messenger RNAs is necessary for optimal translation of the mRNA into a polypeptide.

The U-snRNAs (probably as specific RNP particles) have been implicated in many diverse cellular processes involving DNA and RNA. One proposed function of considerable importance is in the **splicing step** for the conversion of precursor mRNA into mature mRNA (see Figure 6–66).

This conversion involves cutting out intervening sequences (IVS), also called *introns,* of pre-mRNA and sealing together the ends of the coding segments, called *exons.*

We will consider this remarkable process in more detail in the next chapter.

Catalytic Functions of Some RNAs

Modifying the dogma that all enzymes are proteins is a recent discovery that some RNAs can operate as *catalysts.* An enzyme function for an RNA was first proposed by T. Cech in 1982, who was studying the biosynthesis of ribosomal RNA in the protozoan *Tetrahymena thermophilia*. The objective was

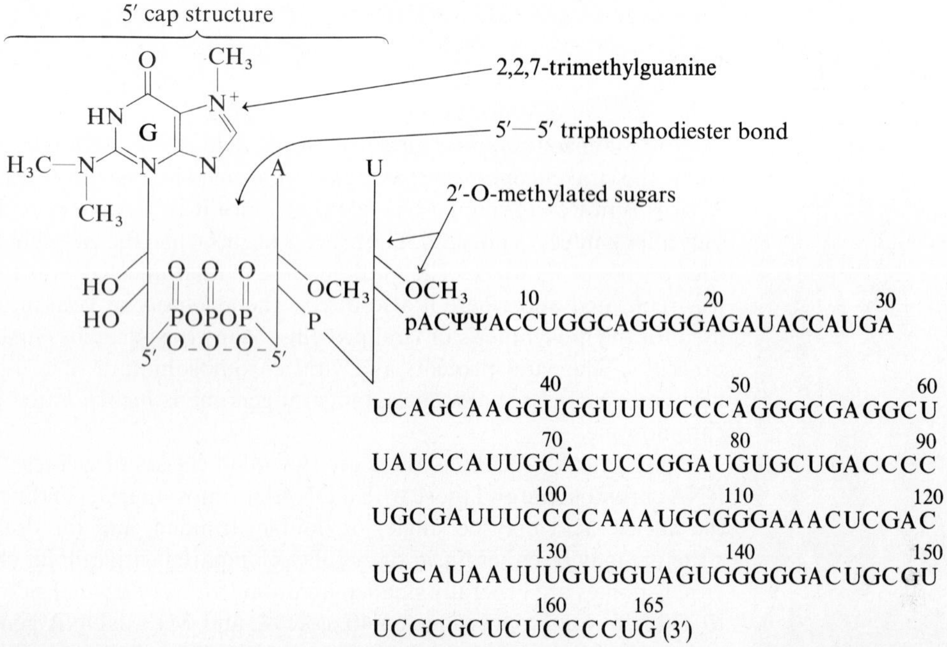

5′ cap structure

2,2,7-trimethylguanine

5′—5′ triphosphodiester bond

2′-O-methylated sugars

```
                                    10              20              30
pACΨΨACCUGGCAGGGGAGAUACCAUGA
          40              50              60
UCAGCAAGGUGGUUUUCCCAGGGCGAGGCU
       70 .          80              90
UAUCCAUUGCÁCUCCGGAUGUGCUGACCCC
          100            110            120
UGCGAUUUCCCCAAAUGCGGGAAACUCGAC
          130            140            150
UGCAUAAUUUGUGGUAGUGGGGGACUGCGU
          160      165
UCGCGCUCUCCCCUG (3′)
```

FIGURE 6–65 Primary structure of U1-snRNA. The · is a methylated base.

to determine how the precursor rRNA of the 26S rRNA was converted to the mature 26S rRNA, which is then assembled into the 60S ribosomal subunit. The conversion was known to require a splicing process involving the removal of a single intervening sequence 413 nucleotides in length. Because RNA splicing occurs at very specific nucleotide locations, researchers have always assumed that one or more specific enzymes are involved. However, Cech discovered that the splicing of pure pre-rRNA occurred in the total absence of any detectable protein, indicating that the catalytic action was in the RNA itself. In this type of self-splicing, however, the RNA molecule does not function as a true catalyst because the catalytic activity is lost in one step. That is, catalytic turnover—the same catalyst acting again and again on several substrate molecules—is not observed.

Then in 1983 S. Altman, N. Pace, and co-workers reported evidence for a true RNA catalyst associated with ribonuclease P (RNase P) activity. One of several nuclease enzymes responsible for cleaving RNAs in cells, RNase P is known to function in one of many steps occurring in the conversion of precursor tRNA molecules to the final mature tRNA molecules. Enzyme RNase P is specifically responsible for cleaving precursor molecules at the 5′ end to produce the correct 5′ ends in the mature molecules. This enzyme is found in all types of cells. Normally, RNase P is isolated as a small nucleoprotein aggregate composed of about 80% RNA and 20% protein by weight. They found that the catalytic action of intact RNase P could be performed by the RNA component alone, stripped of the protein, and further that the protein alone had no activity. The question now is, Are these examples of RNA catalysts only remnants of a more primitive age, or are they more widespread in nature, performing various roles of importance?

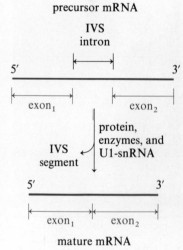

FIGURE 6–66 Conversion of precursor mRNA to mature mRNA in eukaryotes.

6–7 VIRUSES, VIROIDS, AND PRIONS

Viruses

Most *viruses are nucleoprotein particles* consisting of a nucleic acid molecule—the viral chromosome (genome)—packaged in a sheath of several proteins. A virus is not considered a true life form since it is capable of replicating itself only after it infects a host cell. The process begins when the virus binds to receptor sites on the cell surface. The viral genome is then injected into the cell, where it is replicated and where it also directs the assembly of viral messenger RNA used for the biosynthesis of viral proteins. Many host enzymes function in these processes. The viral proteins and viral chromosome are then assembled into new virus particles. Sometimes, the viral genome is incorporated into the host chromosome.

Genetically speaking, there are two main classes of viruses: those with an RNA chromosome and those with a DNA chromosome. Depending on the virus, the nucleic acid may be single- or double-stranded, and the double-stranded structure may be either linear or circular. Compared with cellular chromosomes, viral genomes are generally smaller, and some are very small. For example (refer to Table 6–6), the polyoma, SV40, ϕX174, and M13 dsDNA genomes have a molecular weight of about 3,500,000 and encode only about 10–12 genes for viral proteins.

Structurally speaking, the viruses comprise a heterogenous group, having various sizes (they are all small) and shapes. The unique shape of a virus results from the elegant geometric patterns of the associations among the protein and nucleic acid components. Some are rods, some spheres, some hexagons, some octagons, some icosahedrons, and some other shapes (see Figure 6–67). Biologically speaking, we have animal, plant, and bacterial viruses and several types of each.

On a practical level, the causal relationship of viruses to many diseases, including cancers, continues to guide research in virology. In addition, viruses have also proven to be immensely valuable as simple *model systems* for studying the complex biochemistry of molecular genetics.

Viroids and Prions

VIROIDS In 1967 T. O. Diener and co-workers reported evidence that the causative agent of spindle tuber disease in potatoes was not a typical virus but a *free,* "naked" RNA molecule. Subsequent studies established that this RNA did not require a helper virus but was capable of independent infectivity and replication. The term *viroid* was proposed for this subviral pathogenic agent. Viroids are much smaller than viral RNAs. For example, the potato spindle tuber viroid (PSTV) is an RNA molecule with 359 nucleotides (MW about 130,000). In the past decade about ten additional plant diseases have been established as also being caused by viroids, and five different viroids have been detected.

An interesting and unique characteristic of viroids is that the native structure appears to be that of a *circular single-stranded* RNA. Sequence studies of

(a)

(b)

FIGURE 6–67 Viruses.
(a) Tobacco mosaic virus.
(b) Bacterial virus (also called a bacteriophage).

viroids show extensive base-sequence homology and also support the possibility of extensive intramolecular base pairing to generate double-stranded character in the circular structure (see Figure 6–68). Four of the five viroids sequenced so far show identical sequence homology in one of these double-stranded segments.

Although the reasons for the disease-producing consequences of viroids are not yet understood, there is evidence rejecting one possibility. Apparently, viroid RNA does *not* function as an mRNA to direct the synthesis of a disease-causing viroid protein. Rather, it is a piece of foreign RNA that gets replicated in host cells by host enzymes and then somehow interferes with the transcription and/or translation of mRNAs in the host cell.

From where did viroids originate? There is as yet no answer to this question. Diener and others speculate that they are not evolutionary forerunners nor degenerate offshoots of classical viral RNAs but, rather, may have originated from intron segments of eukaryotic mRNA. An *intron* is an internal section of eukaryotic mRNA that is removed by nuclease action before the mRNA participates in the translation process (see Figure 6–66).

PRIONS   Although all viroids detected so far are pathogenic only for higher plants, there is no reason to believe that similar agents for other types of cells do not exist. In fact, the degenerative neurological disease of scrapie appears to be caused by a free, "naked" *protein* substance. Prusiner (see the Literature) proposes the name *prion* for such a "protein infectious particle." Recently, abnormal compounds found in the brains of deceased victims of Alzheimer's disease show extreme similarities to the scrapie prion.

6–8   NUCLEIC ACID HYBRIDIZATION

The base sequences of the two strands in duplex DNA are complementary, and the base sequence of an RNA transcript is complementary to one of the strands in the DNA gene coding for that RNA. The elegance of complementary base pairing has also been exploited in the laboratory for various purposes.

However, before we get to that technique, consider the following process: When a duplex DNA is heated, it *melts*. That is, stabilizing forces are weakened and disrupted, and the two strands begin to unravel; eventually, the strands completely separate. If the system is rapidly cooled, the single strands remain separated, and one could isolate one strand from the other. However, if the temperature is slowly lowered to about 20°–30° below the T_m and then held there for a time, most of the separated strands will reassociate to form the original duplex. The success of this *annealing process* (see Figure 6–69) is affected by temperature, by the size and the concentration of the nucleic acid, by pH, by solvent, and by the presence of countercations.

Annealing experiments have provided much information on the kinetics and the energetics of duplex structures. There are, however, variations on the process that extend its significance. For example, suppose that the original DNA solution contains DNAs from two different sources. Depending on the degree

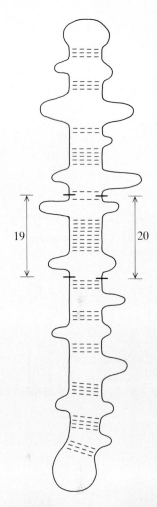

FIGURE 6–68 Proposed partial double-stranded structure of the circular, single-stranded RNA of PSTV. The two segments of sequence totaling 39 bases are identical in four different viroid RNAs. Dashed lines represent base-pair hydrogen bonding in complementary regions.

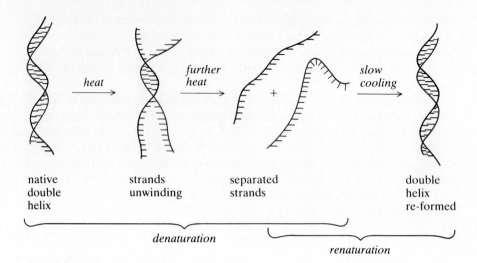

native
double
helix

native
double
helix

strands
unwinding

separated
strands

double
helix
re-formed

denaturation

renaturation

FIGURE 6–69 Annealing
process for DNA.

of base sequence homology in the two DNAs, annealing the mixture can result
in the formation of new *hybrid* duplex DNAs containing one strand of each of
the original DNAs (see Figure 6–70). For obvious reasons the process is called
nucleic acid hybridization. Extensive sequence homology will result in extensive
DNA/DNA hybrid associations. Thus a measure of the extent of hybridization
can be used to evaluate the genetic similarity of the two DNAs.

Other powerful applications of the hybridization scenario involve using
an RNA probe to isolate a specific piece of a DNA molecule or vice versa. Suppose, for example, you have isolated an mRNA molecule. With the strategy of
DNA/RNA hybridization (see Figure 6–71), you can use the mRNA as a probe
to isolate the segment of the DNA genome containing the gene coding for that
RNA.

In 1975 E. M. Southern combined gel electrophoresis, nucleic acid hybridization, and the different filtration characteristics of nucleic acids to design
a method for that purpose. The filtration properties that are exploited are as
follows: Under certain conditions dsDNA and ssRNA will pass through a paper

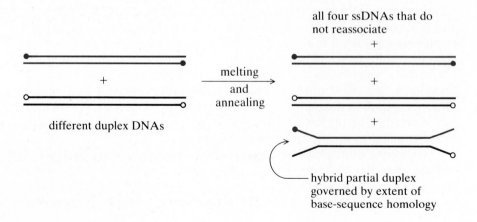

all four ssDNAs that do
not reassociate

+

+

+

different duplex DNAs

melting
and
annealing

hybrid partial duplex
governed by extent of
base-sequence homology

FIGURE 6–70 Nucleic acid
hybridization.

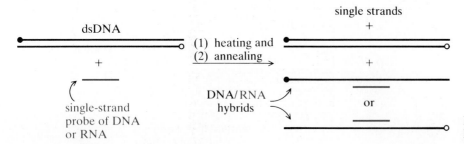

FIGURE 6–71 DNA/RNA hybridization.

filter, whereas ssDNA will be adsorbed and stick to the paper. This paper is not ordinary laboratory filter paper but a special grade composed of nitrocellulose and having very small pore sizes (0.025–0.045 μm). Modified nitrocellulose papers are now available that form very strong complexes with ssDNA.

Consider again the objective: to isolate a small piece of a very large DNA molecule. As shown in Figure 6–72(a), the first step is to cleave the large dsDNA molecule into several smaller pieces by using restriction enzymes. The next step is to isolate the right restriction fragment. Here is where the *Southern transfer technique* is utilized. After the mixture of double-stranded restriction fragments is subjected to gel electrophoresis, the gel is soaked in NaOH to denature the duplex fragments into single strands.

The gel with ssDNA and a piece of nitrocellulose paper are then sandwiched in an assembly [Figure 6–72(b)], which results in an upward flow of liquid through the gel into the nitrocellulose pad. Dissolved in the liquid, ssDNAs move from the gel to the nitrocellulose. Any dsDNA still remaining will be drawn through the nitrocellulose, but *ssDNA will bind* to the nitrocellulose. The nitrocellulose paper is then dried and remoistened with a solution containing the hybridization probe, such as a ^{32}P-labeled RNA. The RNA will permeate the nitrocellulose matrix but not bind to it unless it finds a bound ssDNA with a complementary base sequence, in which case it will form a DNA/RNA hybrid.

Afterward, the nitrocellulose is gently washed to remove any unhybridized RNA, dried, and exposed to X-ray film. The radioautograph will identify the location on the gel of the ^{32}P-labeled DNA/RNA hybrid, which can then be cut out. The ssDNA of the hybrid can be recovered after strand dissociation, and it in turn can then be used as a probe in the same process with the same restriction fragments to recover the intact dsDNA segment.

6–9 CONSTRUCTING RECOMBINANT DNAs

Researchers can now insert a natural or synthetic gene into a carrier DNA that is then used to infect a host cell where the inserted gene is normally expressed. The assembly of a DNA with new and novel genetic instructions—a *recombinant DNA*—is genetic engineering performed at the molecular level. It is the newest and most exciting technology of the twentieth century. In less

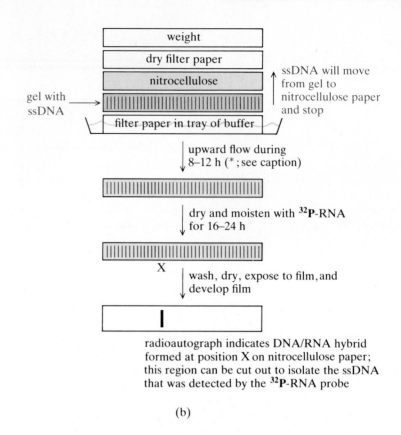

(b)

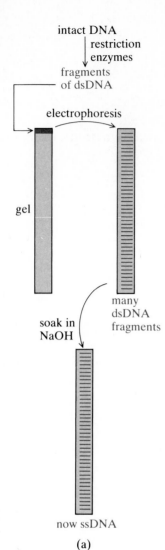

FIGURE 6–72 Southern transfer technique to recover DNA from electrophoresis gel. (a) Initial steps. (b) Final steps and results. The time specified in * can be shortened by carrying out the transfer electrophoretically in an appropriate apparatus. Variations of this procedure are also available to transfer RNAs (the *Northern* transfer technique) and proteins (the *Western* transfer technique) from electrophoresis gels to modified nitrocellulose papers.

than a decade developments have been rapid and remarkable, impacting not only on basic science but also on medicine, animal breeding, agriculture, law, government, finance and business, and ethics. Indeed, most elements of society will eventually be affected.

Although very complex and tedious in actual operation, in general design the preparation of new life forms is simple. Here are the general steps.

1. With the use of the techniques of restriction cutting and of DNA/RNA hybridization, a gene-carrying segment of a natural chromosome is isolated. Alternatively, a gene of defined base sequence is chemically synthesized in the laboratory. Procedures for these difficult processes are constantly being improved.

2. Infective but nonpathogenic DNAs, such as certain viral DNAs or plasmid DNAs, are used as carriers (*vectors*) of the foreign gene. The vector DNA is cut at specific locations with restriction enzymes, and the foreign gene is inserted precisely at the location of the cut, sealed in place by using DNA ligase. This step is the most critical part of the process because one must insert the gene in an optimum position for its subsequent expression. Various plasmid DNAs (see Figure 6–43) and bacterial virus DNAs are used as vectors for transforming bacterial cells; a special plasmid called Ti can be used to transform plant cells; and nonpathogenic animal viruses such as SV40 are used to transform animal cells.

3. Host cells are then exposed to the recombinant DNA vector, which enters the cell, where it is independently replicated and expressed as an extra-chromosomal element or in some cases is actually incorporated into the host chromosome.

Rather than proceed with further generalities, let us consider a specific example to partially illustrate the state of the art (see Figure 6–73). The example is the construction of a recombinant *E. coli* strain capable of producing human interleukin-2 (IL–2), a key protein (MW about 15,000) produced in small amounts by human lymphocytes during an immune response. Protein IL–2 also stimulates the activity in the body of natural killer cells that are capable of destroying tumor cells. Exploring the use of IL–2 in the therapy of immune disorders and as a potentially effective and safe anticancer treatment is obvious, but IL–2 is not readily available. Recently, however, the human IL–2 gene was cloned and expressed in bacteria.

We will not describe how it was done, but a 706-bp fragment of dsDNA containing the human IL–2 gene was isolated. The IL–2 fragment had one blunt end and one Pst I cohesive end. Another critical duplex fragment (120 bp in length) was cut out of a previously engineered plasmid containing the fragment. This short fragment contains base sequences that function to mark the start of a gene and promote the optimal transcription of the gene. (Such pre-gene regulatory regions are described in the next chapter.) Because of the manner in which the plasmid was engineered, this control fragment could be excised with restriction enzymes, yielding one EcoR I cohesive end and one blunt end with ATG—the universal codon specifying the start of a gene. An additional restriction site for Hind III also existed just before the blunt end.

The plasmid vector used was pBR322, a circular DNA (5995 bp) containing single restriction sites for EcoR I and Pst I and two genes that render a host bacterial cell resistant (^{r}) to each of two antibiotics, ampicillin (amp) and tetracycline (tet). Both EcoR I and Pst I were used to cut the plasmid into two linear pieces, and the large fragment was then isolated by gel electrophoresis. The IL–2 fragment, the control fragment, and the large pBR322 fragment were then incubated in the presence of DNA ligase to seal the three together.

Since several ligation patterns are possible, the right one must be isolated. This isolation was done by exposing *E. coli* host cells to the ligation mixture and growing the treated cells in the presence of tetracycline. Any cells that grow must carry the tet-resistant gene. From tet-resistant colonies the experimenter must find one that contains a reconstituted plasmid also carrying the IL–2 and control fragments ligated in the right positions. This colony was found by doing restriction mapping on plasmids from tet-resistant cells. The properly engineered plasmid was cut with EcoR I and Pst I to yield two fragments of specific sizes, and the small control fragment was further cut with Hind III to give two subfragments also of specific sizes. The base sequences of these fragments were also checked with the base sequences of the original spliced fragments.

Properly transformed cells were grown and harvested, and a cell-free extract was prepared. An SDS–PAGE analysis of the extract revealed the presence of a new bacterial protein (MW about 15,000), which was subsequently

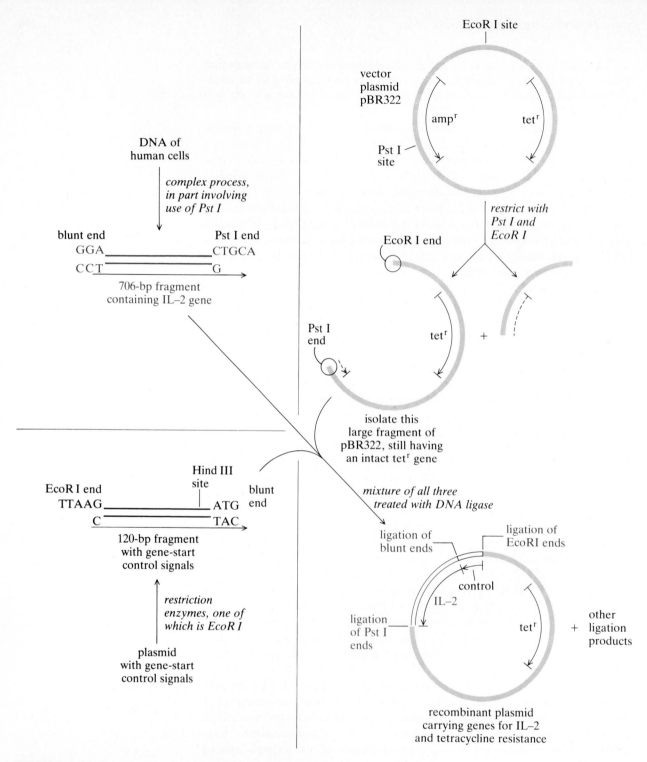

FIGURE 6–73 Construction of a recombinant DNA plasmid carrying a human gene for interleukin-2 (IL–2). The final recombinant plasmid is used to infect normal *E. coli* cells and screen for tet-resistant transformed cells that produce IL–2. [For complete details of the construction and the biological assays of the bacteria-produced interleukin-2, see S. A. Rosenberg et al., *Science*, **223**, 1412–1414 (1984).]

shown to be virtually identical in terms of biological activity to human inter-leukin-2. Remarkably, the amount of IL–2 protein produced was very high, representing about 5% of the total cellular protein.

6–10 CHEMICAL SYNTHESIS OF DNA

Owing to the pioneering efforts of G. Khorana (Nobel Prize, 1968), effi-cient methods of organic synthesis have been developed to assemble DNA in the laboratory. In the past five years techniques have been perfected to provide better yields in faster times, and an *automated, solid-phase procedure* has been developed. With microprocessor-controlled DNA synthesizers one can assemble a piece of duplex DNA with 500–600 bp in about two months. Done manually, the task could take years. Unlike the procedure for successive condensation of amino acids into a long polypeptide chain, in duplex DNA synthesis sets of com-plementary oligonucleotide segments (10–20 bases long) are separately assem-bled. Hybridization and DNA ligase are then used to assemble the DNA duplex structures.

Phosphite triester synthesis (see Figure 6–74) is currently the best method for oligonucleotide synthesis. The method involves the condensation of highly reactive deoxynucleoside-3′-phosphoramidites, wherein the phosphorus atom is present in a trivalent oxidation state. After the connecting phosphoester bond is formed, the trivalent phosphorus atom is easily oxidized to the pentavalent state. The most stable deoxynucleoside-3′-phosphoramidites are prepared as indicated in Figure 6–75. The ⊓ symbols indicate previous chemical modifi-cations that must be made on a nucleoside to block the 5′—OH and the —NH$_2$ group of G, C, and A bases, preventing these functions from participating in competing side reactions.

The first cycle in the sequence of oligonucleotide synthesis (refer to Figure 6–74) begins [step (a)] with condensing a nucleoside-3′-phosphite with the 5′—OH of the first nucleoside that has been previously attached by covalent linkage to the solid-phase polymeric support **S.** Requiring the presence of an activating agent, this step produces a stable phosphotriester species in nearly quantitative yield. Then [step (b)] a brief reaction with I$_2$ oxidizes the trivalent phosphorus to the more stable pentavalent state. Next [step (c)], the 5′-blocking group is removed to expose the next reactive 5′—OH. A second cycle of steps (a), (b), and (c) will give a trinucleotide, and so on. About fifteen cycles can be performed in 5–6 hours.

After the desired length is achieved, the **S**—oligonucleotide anchor bond is cleaved; all other blocking groups are removed; and finally, the methyl phos-phates are hydrolyzed to yield the finished product (plus any incompletely extended oligonucleotides from any cycle). The crude product mixture is sub-jected to column chromatography or gel electrophoresis to purify the product of desired length. Overall yields in excess of 90% have been reported for 10 cycles, and 70% for 14 cycles.

The specific oligonucleotide segments synthesized are dictated by a grand design previously mapped out for the assembly of a duplex DNA. For a DNA

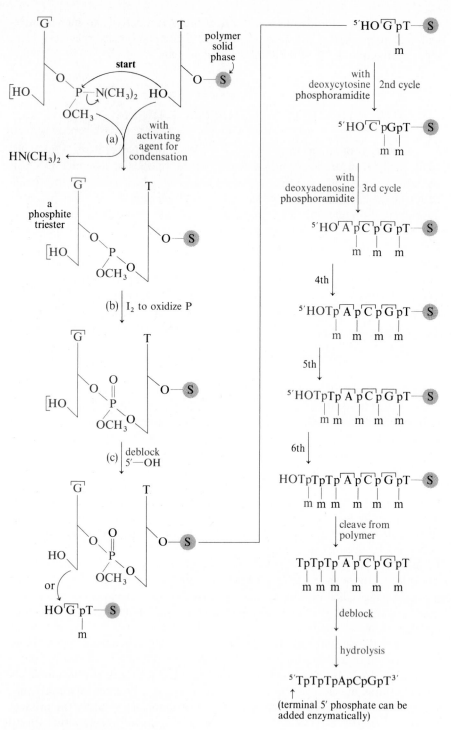

FIGURE 6–74 Solid-phase chemical synthesis of oligonucleotides by a phosphotriester method. The ⌐¬ symbol represents the presence of blocking groups to prevent unwanted reactions.

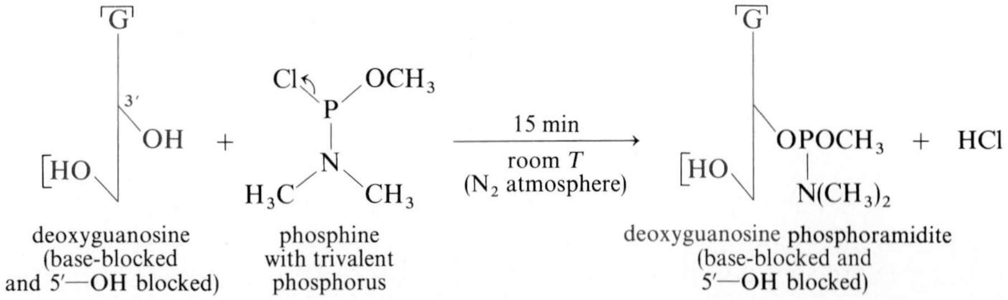

FIGURE 6–75 Preparation of a deoxynucleoside-3'-phosphoramidite. The phosphine by name is chloro-N,N-dimethylamino-methoxyphosphine.

encoding a gene for a polypeptide, the known amino acid sequence of the polypeptide determines what the sequence of codons must be. Since several amino acids are coded for by more than one codon (see Table 9–2), numerous base sequences are possible. Selection of the one sequence to be synthesized is based on avoiding competitive hybridizations when the oligonucleotides are annealed for ligation.

Consider the example illustrated in Figure 6–76 for the planned synthesis of a gene coding for a human interferon. For the known sequence of 166 amino acids, 166 successive codons ($3 \times 166 = 498$ bases in each strand) are required. The N terminus codon was preceded by an ATG codon (for methionine) to specify the start of the gene, and the C terminus codon was followed by a TAA codon to specify the signal for ending the gene (for 6 more bases).

Since the plan was to later splice the duplex into a plasmid for cloning, each end was designed to have single-strand flaps complementary to flaps generated by restriction enzyme treatment of the plasmid—a BamH I flap at one end and a Sal I flap at the other end (for 10 more bases, or a grand total of 514 bases in each strand). As indicated, their sequence was carved up into 67 different oligonucleotides, each of about 15 residues.

After their synthesis, groups of 5, 6, 7, or 8 were annealed and ligated to give short duplex fragments with cohesive ends. Groups of these fragments were then annealed and ligated to assemble four larger segments, which in turn were annealed and ligated to give the final product. The synthetic gene was successfully spliced into a plasmid, the plasmid was cloned in *E. coli* cells, and the bacteria produced interferon.

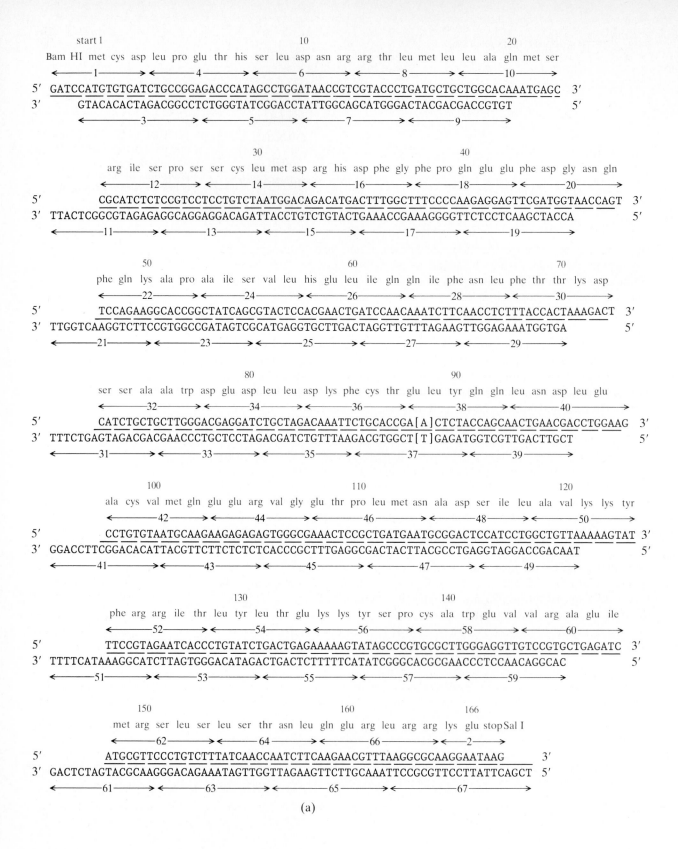

(a)

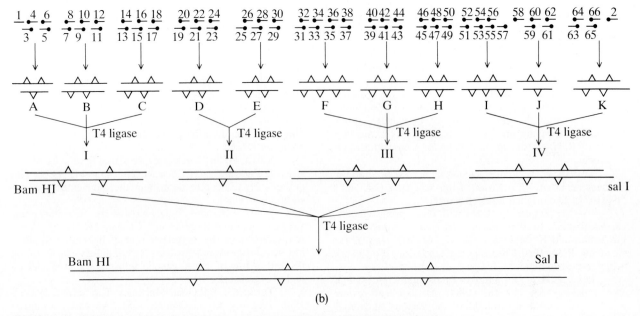

(b)

◀ FIGURE 6–76 Design of the synthesis for an interferon gene. (a) Amino acid sequence of interferon. This sequence was used to define a codon (base) sequence for the top strand of DNA. This definition fixes the base sequence of the other strand, which must be complementary. Cohesive termini and start/stop signals were also planned. (b) Grand design of the stepwise annealing and ligation assembly of the duplex from 67 oligonucleotide fragments. Solid dots identify phosphorylated 5' ends of oligonucleotides. Carets (∧) identify locations of ligation. For details, see M. D. Edge, et al., *Nature*, **292**, 756–761 (1981).

LITERATURE

ADAMS, R. L. P., R. H. BURDON, A. M. CAMPBELL, and R. M. S. SMELLIE. *The Biochemistry of Nucleic Acids.* 8th ed. New York: Academic Press, 1976. An excellent introductory source.

BAGLIONI, C., and T. W. NILSEN. "The Action of Interferon at the Molecular Level." *Am. Sci.,* **69**, 392–399 (1981). A review article.

BURLINGAME, R. W., W. E. LOVE, B. WANG, R. HAMILIN, N. XUONG, and E. N. MOUDRIANAKIS. "Crystallographic Structure of the Octameric Histone Core of the Nucleosome at a Resolution of 3.3 Å." *Science,* **228**, 546–553 (1985). Detailed three-dimensional models of the octameric histone aggregate and associated DNA are presented.

CARUTHERS, M. H. "Gene Synthesis Machines: DNA Chemistry and Its Uses." *Science,* **230**, 281–285 (1985). Good review article of the automated, solid phase phosphite triester method to synthesize DNA.

COLD SPRING HARBOR LABORATORIES. *Structures of DNA.* Cold Spring Harbor Symposia on Quantitative Biology, vol. 47 (parts I and II). New York: Cold Spring Harbor Laboratories, 1983. A collection of 135 research papers on numerous aspects of DNA structure.

COZZARELLI, N. R. "DNA Gyrase and the Supercoiling of DNA." *Science,* **207**, 953–960 (1980). A review article.

DIENER, T. O. "The Viroid—A Subviral Pathogen." *Am. Sci.,* **71**, 481–489 (1983). An excellent review article for the beginner.

GILBERT, W. "DNA Sequencing and Gene Structure." *Science,* **214**, 1305–1312 (1981). The author's Nobel lecture describing the chemical method of sequencing DNA.

GILBERT, W., and L. VILLA-KOMAROFF. "Useful Proteins from Recombinant Bacteria," *Sci. Am.,* **242**, (4) 74–97 (1980). A description of how prokaryotes can be engineered to produce the eukaryote proteins insulin and interferon.

GOLDBERG, M. A. "Cyclic Nucleotides and Cell Function." *Hosp. Pract.,* **9**, 127–142 (1974). An excellent introductory-level review article summarizing the regulatory functions of cyclic AMP and cyclic GMP. The primary focus is on mammalian cells.

ITAKURA, K. J., T. ROSSI, and R. B. WALLACE. "Synthesis and Use of Synthetic Oligonucleotides." *Annu. Rev. Biochem.,* **53**, 323–356 (1984). A thorough review article.

KORNBERG, R. P. "The Structure of Chromatin." *Annu. Rev. Biochem.,* **46,** 931–954 (1977). A review article of nucleosome structure.

LAKE, J. A. "The Ribosome." *Sci. Am.,* **245,** (2) 84–97 (1981). Current knowledge of ribosome structure and function is described.

MARX, J. L. "Z–DNA: Still Searching for a Function." Research News article in *Science,* **230,** 794–796 (1985). Brief account of the current state of knowledge about the biological significance of left-handed DNA. See the article by Rich, Nordheim, and Wang for a more thorough treatment.

MAXAM, A. M., and W. GILBERT. "A New Method for Sequencing DNA." *Proc. Natl. Acad. Sci., USA,* **74,** 560–564 (1977). The original research paper.

———. "Sequencing DNA with Base-specific Chemical Cleavages." In *Methods in Enzymology,* edited by L. Grossman and K. Moldave, vol. 65, 499–560. New York: Academic Press, 1980. Chemistry and laboratory procedures are described in detail.

NOLLER, H. F. "Structure of Ribosomal RNA." *Annu. Rev. Biochem.,* **53,** 119–162 (1984). An excellent review article with several illustrations of the various ribosomal RNA species with emphasis on the large 16S and 23S structures.

NOMURA, M. "The Assembly of Bacterial Ribosomes." *Science,* **179,** 864–873 (1973). A review article describing the molecular anatomy of ribosomes, the pattern in which they may be assembled, and their cellular function.

NOVICK, R. P. "Plasmids." *Sci. Am.,* **243,** (12) 102–127 (1980). A review of plasmid biology, structure, and replication.

PRUSINER, S. B. "Novel Proteinaceous Infectious Particles Cause Scrapie." *Science,* **216,** 136–144 (1982). A description of the evidence for and characteristics of this subviral prion pathogen.

RICH, A., A. NORDHEIM, and A. H. J. WANG. "The Chemistry and Biology of Left-handed Z–DNA." *Annu. Rev. Biochem.,* **53,** 791–846 (1984). A thorough review article.

ROBERTS, R. J. "Directory of Restriction Endonucleases." In *Methods in Enzymology,* edited by L. Grossman and K. Moldave, vol. 65, 1–15. New York: Academic Press, 1980. A description of about 200 enzymes.

SANGER, F. "Determination of Nucleotide Sequences in DNA." *Science,* **214,** 1205–1214 (1981). The author's Nobel lecture describing the enzymatic methods of sequencing DNA.

SMITH, H. O. "Nucleotide Sequence Specificity of Restriction Endonucleases." *Science,* **205,** 455–462 (1979). The author's Nobel lecture describing the biochemistry of restriction enzymes.

STEIN, J. S., and L. J. KLEINSMITH. "Chromosomal Proteins and Gene Regulation." *Sci. Am.,* **232,** 46–57 (1975). A discussion of the suggested role of histones and nonhistone chromosomal proteins as regulatory elements for the expression of DNA genetic information in nuclei of higher organisms.

TUCKER, J. B. "Gene Machines: The Second Wave." *High Technol.,* 50–59 (March 1984). An excellent introductory review article on the automated synthesis of DNA.

WANG, J. C. "DNA Topoisomerases." *Sci. Am.,* **247,** 94–109 (1982). Introductory-level description of enzymes that supercoil DNA.

———. "DNA Topoisomerases." *Annu. Rev. Biochem.,* **54,** 665–696 (1985). A detailed review of these enzymes covering mechanism of action and biological processes that involve supercoiling of DNA.

WATSON, J. D. *The Double Helix.* New York: Atheneum, 1968. An interesting and revealing narrative account of the events and persons associated with the discovery of DNA structure. It became a best-seller. An updated edition (1980) has been prepared by Norton Press.

EXERCISES

6–1. Draw the structural formula for each of the following substances: (a) guanosine-3′-monophosphate; (b) deoxyadenosine-5′-diphosphate; (c) 5′-dADP; (d) 5′-dTMP; (e) thymidine; (f) cytosine-5′-triphosphate; (g) $\cdots$ pUpGp $\cdots$; (h) guanylic acid.

6–2. Draw the structures illustrating the keto-enol tautomerism for (a) guanine and (b) cytosine. Why does adenine not exhibit tautomerism?

6–3. A purified DNA preparation was found to contain 30.4% adenine and 19.6% cytosine. The adenine-thymine ratio was 0.98, and the guanine-cytosine ratio was 0.97. Calculate the amount of guanine and thymine in this DNA and also the ratio of purine bases to pyrimidine bases.

6–4. Draw a representation of the complementary base pair found in RNA that corresponds to the adenine-thymine pair in DNA.

6–5. A pure DNA preparation was shown to have a T_m of 85°C. What is the percentage of A + T in this DNA? (Assume the measurement was made on a solution of the DNA containing $0.2M$ Na$^+$.)

6–6. Estimate the buoyant density of the DNA referred to in Exercise 6–5.

6–7. Shown below is an oligonucleotide segment of a ribonucleic acid (RNA) molecule. What type of cleavage pattern (if any) will result in this segment by treating the RNA with

(a) sodium hydroxide; (b) pancreatic ribonuclease A; (c) ribonuclease T1; (d) nuclease S1; (e) exonuclease III?

pGpGpCpUpApCpGpUpApGpApUpCpA

6–8. Sketch the four patterns of radioactive zones expected if the top strand in the duplex shown below was subjected to the Gilbert-Maxam sequencing technique. What will the patterns look like if the other strand is sequenced?

pCACTTACTTGTTAACCGCC
GTGAATGAACAATTGGCGGp

6–9. A section of an RNA-sequencing radioautograph pattern is shown in Figure 6–77. What nucleotide sequence in RNA is represented by this portion of the radioautograph?

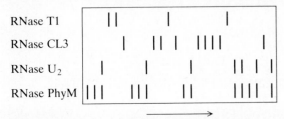

direction of migration

FIGURE 6–77

6–10. The numbers to the right of the photograph in Figure 6–31 identify the size of each restriction fragment of dsDNA, in numbers of base pairs. Construct a graph by plotting the logarithm (base 10) of base-pair length versus distance of migration for each fragment. Using a ruler scaled in millimeters, measure the distances migrated by each fragment,

using the top of the photograph as the starting point. What do you suppose this type of experiment can be used for?

6–11. The first three nucleotides of a hexanucleotide palindrome sequence in one strand of dsDNA are CTG. What is the complete palindrome sequence of base pairs?

6–12. When a restriction enzyme yields staggered cleavage at a hexanucleotide palindrome sequence, does the single-stranded tetranucleotide segment terminate with a free $3'$—OH, a free $5'$—OH, a $3'$—OPO_3^{2-}, or a $5'$—OPO_3^{2-}?

6–13. The plasmid pBR322 (see Figure 6–73) is cleaved by several restriction enzymes at only single locations. The EcoR I site identified in Figure 6–73 is one. Another is a Hind III site that is located about 30 bp away from (clockwise) the single EcoR I site. Close study of Figure 6–73 will provide some additional information about a Hind III restriction site in the gene-start control segment used in the preparation of the pBR322/IL–2 recombinant plasmid. There is no Hind III site in the DNA fragment containing the IL–2 gene. Using this information about Hind III sites, suggest a simple experiment that can be used to confirm that the tetracycline-resistant, IL–2–producing cells contained the complete recombinant plasmid that was designed as diagramed in Figure 6–73.

6–14. The total amount of dsDNA in the adult human body is approximately 0.5 g. Using information in Table 6–7 and the relationship that a single base pair of dsDNA has a mass of about 10^{-21} g, calculate the distance that would result if all of the DNA molecules in the body were placed end to end. To get some perspective, you might wish to refer to a source book (a dictionary) and check on the distance from the earth to the sun.

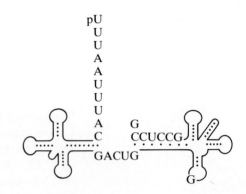

CHAPTER SEVEN

TRANSCRIPTION: RNA BIOSYNTHESIS

All of the instructions for the chemical processes of an organism are provided by DNA, the molecular reservoir of genetic information. A popular way of summarizing the flow of information is

$$DNA \rightarrow RNA \rightarrow protein$$

representing the conversion of the genetic language of a nucleotide sequence into an amino acid sequence. A preliminary introduction was given in the preceding chapter.

The DNA → RNA symbolism represents **transcription,** the biosynthesis of RNA molecules having nucleotide sequences specified by genes in DNA (see Figure 7–1). The transcript RNA may be transfer RNA, ribosomal RNA, or messenger RNA, and in eukaryotes it also may be small nuclear RNA. The

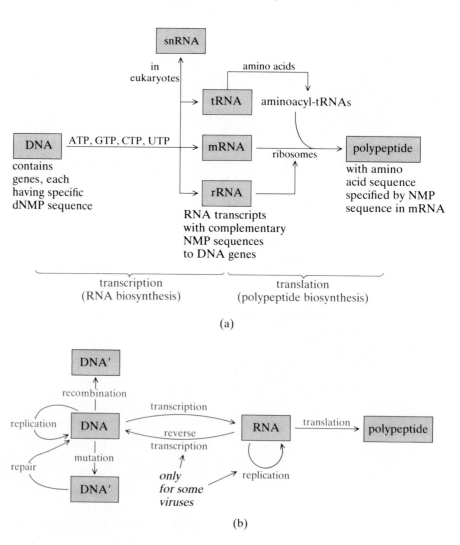

(a)

FIGURE 7–1 Molecular genetics. (a) Summary of the DNA → RNA → protein processes responsible for the nucleotide (NMP) sequence in genes of DNA providing instructions via RNA that specify the sequence of amino acids in a polypeptide. (b) Other events occurring that involve the processing of genetic information. Some of these events apply only to certain viruses. The symbolism DNA′ represents DNA that has been modified to have different genetic instructions.

(b)

RNA → protein symbolism represents **translation,** the biosynthesis of polypeptide chains having an amino acid sequence dictated by the nucleotide sequence of mRNA with auxiliary participation of tRNAs and ribosomes.

These events, however, are only two of many comprising the whole of molecular genetics. Others include the following:

1. *DNA replication:* The biosynthesis of a duplicate copy of DNA prior to cell division (DNA → DNA).

2. *DNA repair:* The removal and resynthesis of short segments of DNA damaged by chemical or physical agents or of DNA synthesized with errors during replication.

3. *DNA recombination:* The exchange of gene segments between different DNA molecules.

4. *DNA transposition:* A nonclassical type of genetic recombination involving the movement of a gene from one location to another on the same chromosome or to a different chromosome.

In cells infected with viruses having an RNA chromosome, the events include the following:

5. *Reverse transcription:* The biosynthesis of a DNA molecule complementary in sequence to the RNA chromosome (RNA → DNA).

6. *RNA replication:* The biosynthesis of a duplicate copy of the viral RNA chromosome prior to the formation of new particles (RNA → RNA).

In this chapter we will focus on the processes of transcription and reverse transcription. Two related topics are also examined, *gene regulation* and *oncogenes.* Replication, repair, and recombination processes are considered in Chapter 8, and translation is considered in Chapter 9. In addition to describing the key enzymes involved in these processes—*polymerases, nucleases, ligases,* and others—this chapter will also focus on various *noncatalytic binding proteins* that participate in many crucial steps.

7-1 TRANSCRIPTION OF DNA via RNA POLYMERASE

General Features and Terminology

All organisms use the same type of enzyme—a *DNA-directed RNA polymerase*—to assemble RNA according to DNA instructions (see Figure 7–2).

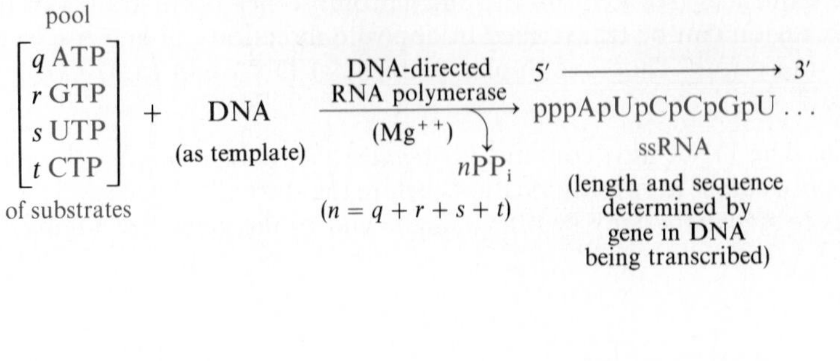

$$
\begin{bmatrix} q\,\text{ATP} \\ r\,\text{GTP} \\ s\,\text{UTP} \\ t\,\text{CTP} \end{bmatrix} + \underset{\text{(as template)}}{\text{DNA}} \xrightarrow[\substack{\text{(Mg}^{++}) \\ \downarrow \\ n\text{PP}_i}]{\substack{\text{DNA-directed} \\ \text{RNA polymerase}}} \text{pppApUpCpCpGpU}\ldots
$$

pool
of substrates

$(n = q + r + s + t)$

5′ —————→ 3′

ssRNA
(length and sequence
determined by
gene in DNA
being transcribed)

FIGURE 7–2 Overall reaction catalyzed by the DNA-copying enzyme of transcription.

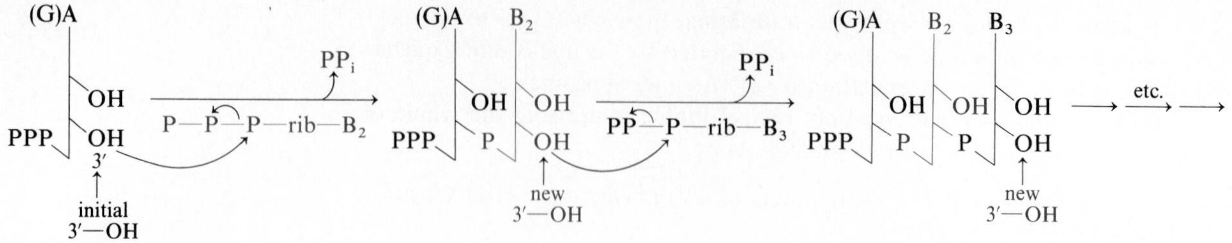

FIGURE 7–3 Illustration of $5' \to 3'$ growth for a polynucleotide chain (B is any base).

It is called DNA-directed because the DNA is necessary to act as a *template* for linking nucleotides in a sequence complementary to one of the strands in the duplex. In other words, the RNA polymerase for transcription is a true *copy enzyme*. Much evidence of this copying capability has been provided, with hybridization of the product RNA and the template DNA providing convincing proof. (Except for times when we must distinguish among different types of RNA polymerases, we will routinely refer to DNA-directed RNA polymerase simply as RNA polymerase.)

Whatever its source (animal, plant, bacteria), RNA polymerase has the following characteristics of operation:

1. Triphosphonucleotides (NTPs) are required as the monomeric substrates; monophospho- and diphosphonucleotides are inactive as substrates.

2. Mg^{2+} is a required cofactor for optimal activity.

3. Successive NTPs are condensed (with elimination of pyrophosphate, PP_i) so that the growth of the RNA chain is from the $5'$ end to the $3'$ end (see Figure 7–3). This activity is called $5' \to 3'$ *polymerization* or $5' \to 3'$ *chain growth*.

4. No *primer piece* of RNA is required to start the copy synthesis. The enzyme positions the first triphosphonucleotide (usually ATP or GTP) and uses its $3'$ —OH to condense with the next nucleotide brought into position. After the phosphodiester bond is formed, the $3'$ —OH of the dinucleotide then condenses with the next nucleotide brought into position; and so on. This activity is called *self-priming polymerization*.

5. During operation the bound enzyme copies only from one strand of the duplex DNA, with the enzyme being displaced in the $3' \to 5'$ direction of the template. For a given gene duplex location either strand can serve as the template (see Note 7–1), with strand selection being governed by specific pregene signal sequences (see item 6). In some chromosomes both strands of the same duplex region can be transcribed in opposite directions, giving rise to different RNA transcripts. Thus one duplex segment of DNA can carry two genes (see Figure 7–4).

6. The DNA also contains (a) *sequence signals to initiate* the binding of RNA polymerase at specific locations before the start of the gene and (b) *sequence signals to terminate* copy synthesis at the end of the gene. See Figure 7–5.

NOTE 7–1
The strand of dsDNA that has the same nucleotide sequence as the RNA transcript will be called the *sense strand* and will be designated (+). The other strand, serving as template and having a nucleotide sequence complementary to the (+) RNA transcript, will be called the *sense-template strand* and will be designated (−).

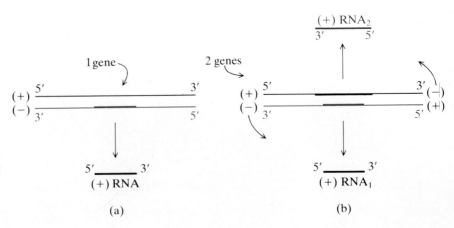

FIGURE 7–4 Origins of RNA transcripts. (a) Usual event: Only one strand of dsDNA providing the template instructions. (b) Occasional event: Both strands acting as the template for different transcripts. Complementary sequences are coded by color with black.

Note: The in vivo events described here also require the supercoiled structure of DNA. Supercoiling may assist in the recognition of start and stop signals and/or in the initial local deformation of duplex DNA to accommodate the binding of the polymerase to the template strand. See Figure 7–6.

RNA Polymerases and Their Binding to Promoters

Prokaryotes contain only a single RNA polymerase for transcription. Eukaryotes contain four different enzymes: three located in the nucleus and one in the mitochondrion (see Table 7–1). All are large, heterogeneous oligomeric proteins. The most extensively studied and best understood is the pentameric $\alpha_2\beta\beta'\sigma$ RNA polymerase in prokaryotes. Proper binding of polymerase requires the intact pentamer, but the activity of processive copy synthesis can be sustained by the tetrameric complex of $\alpha_2\beta\beta'$ called the *core enzyme*.

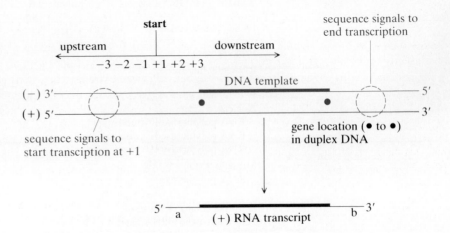

FIGURE 7–5 Gene signals. The base positions before the start of a gene are referred to as being *upstream,* with individual positions designated as . . . , $-5, -4, -3, -2, -1$. Base positions in the transcribed region are designated 1, 2, 3, 4, 5, . . . , proceeding in the *downstream* direction to the termination site. The final RNA transcript contains the entire information from ● to ● plus short leader (a) and trailer (b) segments.

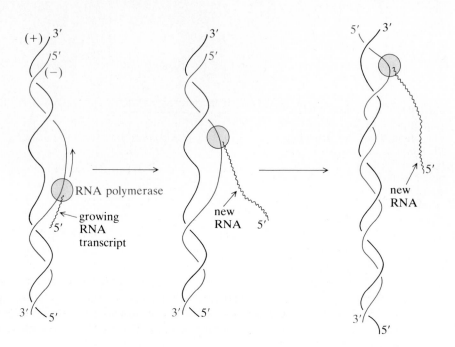

FIGURE 7–6 RNA polymerase advancing on the template strand after a local unraveling of the DNA duplex. The growing RNA transcript remains attached to the polymerase until the termination site is reached.

START RECOGNITION   During RNA assembly the α and β' subunits apparently provide binding sites for holding the enzyme to the template strand, and the β subunit provides the active site for forming new phosphodiester bonds in RNA. The σ subunit also performs a DNA-binding function and a very critical one: σ scans the DNA duplex searching for the *promoter site,* the *sequence signal specifying the start of the gene.* Once copy synthesis is underway, the subunit dissociates, leaving the core enzyme to continue. The affinity of σ for the promoter is extremely high ($K_{association} = 2 \times 10^{11}$), reflecting the specificity of recognition between σ and the promoter sequence.

TABLE 7–1 DNA-directed RNA polymerases in living cells

ENZYME	COMPOSITION (MW)
Prokaryotes	
RNA polymerase	$\alpha_2\beta\beta'\sigma$ (480,000) with bound Zn^{2+}
Eukaryotes	
Nuclear	
RNA polymerase I[a] (nucleolus)	Octamer of six subunits (500,000)
RNA polymerase II[b] (nucleoplasm)	Five subunits
RNA polymerase III[c] (nucleoplasm)	Nine subunits
Mitochondrion	
RNA polymerase	Variable in different species

[a] For transcribing rRNA genes.
[b] For transcribing mRNA genes.
[c] For transcribing tRNA genes.

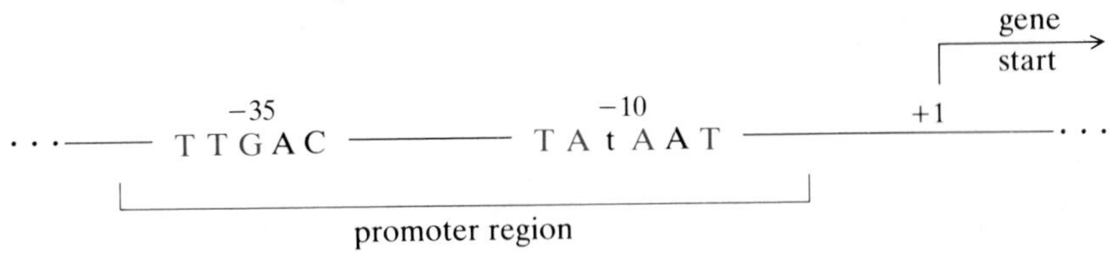

gene
start

-35 -10

$\cdots$ —— T T G A C ———————— T A t A A T —— $+1$ —— $\cdots$

promoter region

FIGURE 7–7 Summary of consensus sequence features of prokaryotic promoters recognized by *E. coli* RNA polymerase. Only one strand of DNA is shown. Color capital symbols identify strongly conserved residues; black capitals for moderately conserved residues; lowercase symbols for weakly conserved residues.

With the use of the techniques of restriction, hybridization, cloning, and base sequencing, about one hundred promoter-containing regions of DNA have been characterized from prokaryotes. These studies have established that although there is no universal nucleotide sequence in every promoter, two common features of sequence homology exist: (1) a **TTGAC** sequence located at about -35 and (2) a **TAtAAT** sequence located at about -10 (see Figure 7–7 and Note 7–2). A highly conserved sequence discovered by comparing the sequences of the same region isolated from different sources is called a *consensus sequence*. In a consensus sequence, as opposed to a universal sequence, the frequency of homology can vary for each residue. For example, in the promoter regions described here, boldface capital symbols represent strongly conserved residues ($>75\%$ homology in all promoters), regular capital symbols represent moderately conserved residues (50%–75%), and lowercase symbols represent weakly conserved residues (40%–50%).

When mutations occur in these regions of the promoter, the rate of transcription is greatly diminished, providing good evidence that they have an indispensable role in binding RNA polymerase. Promoters in eukaryotes are described in the next subsection.

NOTE 7–2
The -10 region, rich in T and A, is sometimes called the TATA region. It is also called the Pribnow box (in prokaryotes) and the Goldberg-Hogness box (in eukaryotes) after the individuals who first discovered it.

TERMINATION SIGNALS The action of RNA polymerase ceases when the enzyme dissociates from the template DNA. The point at which dissociation occurs appears to be specified in different ways. In prokaryotes an auxiliary protein factor called *rho* (ρ) is implicated in recognizing termination sequences in DNA. The ρ–DNA complex obstructs further movement of the advancing polymerase on the template. The formation of secondary-structure characteristics in the RNA being assembled may also promote the dissociation. The sequence details of a possible consensus termination site are not yet discovered for either eukaryotes or prokaryotes.

Additional Signals of the Promoter-Polymerase Interaction in Eukaryotes

In prokaryotes the promoter site is a compact segment, with the enclosed TTG and TATA regions providing signals for recognition by RNA polymerase. By comparison (see Figure 7–8), in eukaryotes promoter regions are apparently more extended and have several signal elements differing slightly for different polymerases. The extended promoter region for RNA polymerase II, which transcribes mRNA genes, appears to have the following elements: (1) a PyPyAPyPy cap site specifying the location of starting the 5′ terminus of mRNA, (2) a TATA region at about the -30 position, necessary to accurately position the polymerase at the cap site, (3) one or more additional sequences

FIGURE 7–8 Comparison of promoter signals in prokaryotes and eukaryotes.

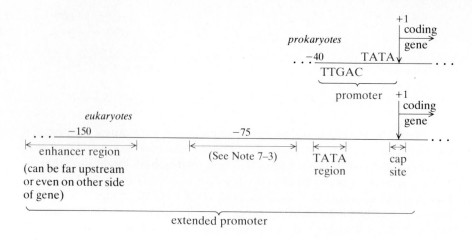

NOTE 7–3
At present five different sequences in the −30 to −100 upstream region have been implicated as contributing to the extended promoter region of eukaryotic genes. They are:

GGGGCGG
CCATT
GCCACACCC
GGCCACGTGACC
ATGCAAAT

Different promoters have different combinations of one .or more of these elements.

located between −30 and −100 (see Note 7–3), and (4) other signals even further upstream, further than −100.

In 1981 a sequence signal of the fourth category—that is, a signal far removed from the start of transcription—was discovered in the DNA chromosome of the eukaryotic virus SV40. It has since been found in the DNA of other eukaryotic viruses and in the DNA of eukaryotic cells, including human cells. The sequence signal is called a *transcription enhancer*. In the SV40 genome the enhancer element is 72 bp long, and two copies are present in tandem (see Figure 7–9). Removing one of the repeats with restriction enzymes does not affect transcription of the SV40 genome. Removing both, however, significantly reduces the rate of viral transcription.

FIGURE 7–9 Segment of the SV40 chromosome. The intact dsDNA has 5243 bp. The segment shown here highlights the transcription enhancer region between two transcription-start regions, proceeding in opposite directions and at different times in the cycle of viral expression. The region also contains a TATA signal and some other interesting sequence characteristics of possible significance. Boxes identify specific regions on the dsDNA represented by the thick line.

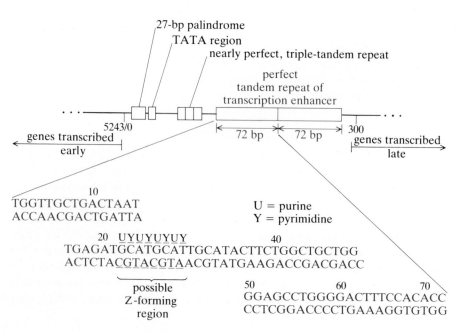

Other studies have established another interesting feature of the enhancer element: Moving the enhancer even further upstream (> -2000) or even on the other side of the gene does not affect its function. Obvious questions apply to what distinguishes the enhancer sequence and how the enhancement of transcription is achieved. The answers are not yet known.

The SV40, 72-bp enhancer contains no palindrome sequences. However, it does contain an uninterrupted 8-bp stretch composed of a tetranucleotide repeat: GCATGCAT. This octanucleotide sequence has the pattern PuPyPuPyPuPyPuPy, that is, an alternating purine-pyrimidine sequence, and half of it is GC—GC—. The significance is that poly(dGdC) oligonucleotides readily form the left-handed Z conformation of DNA. Confirming the obvious implication, recent studies have shown that local regions of Z–DNA do in fact exist in the SV40 enhancer.

What might this result mean to the influence of an enhancer on transcription? One hypothesis is that a transcription enhancer region having some local Z structure is tightly bound to and perhaps stabilized by a specific Z-binding protein. The Z–DNA/protein complex may in turn be the signal recognized by RNA polymerase or some other transcriptional-protein factor. The polymerase binding may in turn dislodge the Z-binding protein, prompting a Z → B transition. When this change occurs, a local increase in duplex supercoiling may arise, generating a local stress that enhances the entry and/or stronger binding of polymerase. (It wasn't mentioned in the previous chapter, but the degree of supercoiling in Z–DNA is less than in B–DNA.) Much additional study is necessary to support and/or modify these speculations.

In recent years transcription enhancer regions have been detected in other viruses and also in cellular DNA of eukaryotes. They are not just a viral peculiarity. Although tandem repeats of 50–130 bp are common to enhancer elements, they are not observed in all. Indeed, many of the cellular enhancers detected to date do not show obvious repeats. Nevertheless, even though sequence repetition is not an essential feature of enhancers, it does potentiate the effect. Those interested in pursuing this topic further can refer to the Gluzman references given in the Literature.

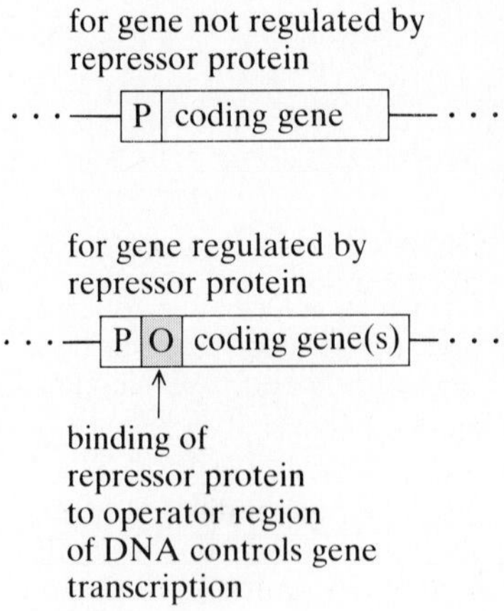

Control of Some Promoters by Repressor Proteins

GENERAL PRINCIPLES The regulation of transcription in eukaryotes by enhancers was not the first discovery of an in vivo, gene-regulating process. Twenty years earlier (1961) F. Jacob and F. Monod proposed that some genes are regulated by *repressor proteins*. By 1970 various genetic and biochemical studies verified the hypothesis, and in recent years the structures of several repressors have been solved and proposals for their mode of action have been formulated.

A repressor protein is itself coded for by a gene, and of course, each different repressor is coded for by a different gene. When present in vivo, the function of a repressor is to bind with DNA at a specific pregene location called an *operator* (O) *site*. The operator is commonly located between the promoter (P) site and the gene start (see Figure 7–10); often the promoter and operator sites overlap slightly. The entire contingent of promoter, operator, and the immediate tandem of coding genes is called an *operon*.

for gene not regulated by repressor protein

$\cdots$ —[P | coding gene]— $\cdots$

for gene regulated by repressor protein

$\cdots$ —[P | O | coding gene(s)]— $\cdots$

binding of repressor protein to operator region of DNA controls gene transcription

FIGURE 7–10 Some genes being controlled by repressor proteins that bind to an operator (O) site. Additional descriptions are given in Figure 7–11.

When a repressor is bound to its operator, the genes in the operon are *off*—that is, they are not being transcribed, for the following reason:

The binding of RNA polymerase at the promoter is obstructed by the bound repressor.

This state of negative control can be reversed by the process of *induction* (also called *derepression*), wherein the repressor is deactivated for binding to the operator. A vacant operator poses no obstruction for polymerase binding at the promoter, and the gene is turned *on*—that is, the gene is now transcribed. The operon genes are called inducible genes, and their products are called inducible proteins.

Induction usually occurs as the result of the repressor protein binding to another substance called an *inducer*. As a classic example of allosterism, the inducer-repressor binding changes the conformation of the repressor protein with a high affinity for DNA to a low-affinity conformation. In some cases induction also occurs as a result of the repressor protein being degraded in the cell.

For some repressors the scenario just described operates in a reciprocal fashion: In the absence of inducer the free repressor has an inactive conformation, the operator is vacant, and the gene is on. When the inducer—now more properly called a *co-repressor*—binds to the repressor, the repressor is activated for operator binding, and the gene is off.

Two of the most studied systems of repressor gene regulation are the biosynthesis of β-galactosidase in *E. coli* and the activation of the lytic cycle of the lambda (λ) bacterial virus. Both systems also involve the participation of auxiliary regulatory proteins and represent interesting examples of the complexity of gene regulation.

LAC REPRESSOR AND CAP   Grown in the presence of the sugar glucose, *E. coli* does not produce three proteins involved in the metabolism of the sugar lactose: β-galactosidase for catalyzing the hydrolysis of lactose to glucose and galactose, a permease for controlling the entry of lactose (lac) into the cell, and a transacetylase enzyme of yet unknown function. In the *lac* operon all three genes are adjacent to each other and under the control of a single promoter/operator tandem (see Figure 7–11). The active lac repressor—an oligomeric protein composed of four identical subunits [MW = $(37,500)_4$]—keeps the entire *lac* operon off.

In addition to providing the recognition site for RNA polymerase, the *lac* promoter provides a binding site for another regulatory protein called *catabolite activator protein* (CAP). A CAP binding site occurs in other promoters but is not a common element of all prokaryote promoters. CAP is a dimer composed of two identical subunits [MW = $(22,500)_2$] and is only active when cyclic AMP is bound to it. When cells are growing on glucose, the cellular level of cyclic AMP is low, and CAP is inactive.

If lactose replaces glucose as the nutrient, two things occur: (1) Lactose acts as an inducer, deactivating the lac repressor, and (2) the absence of glucose causes a shift in cellular metabolism, increasing the cellular level of cyclic AMP, which activates CAP. An interesting dichotomy of effects now results. The dis-

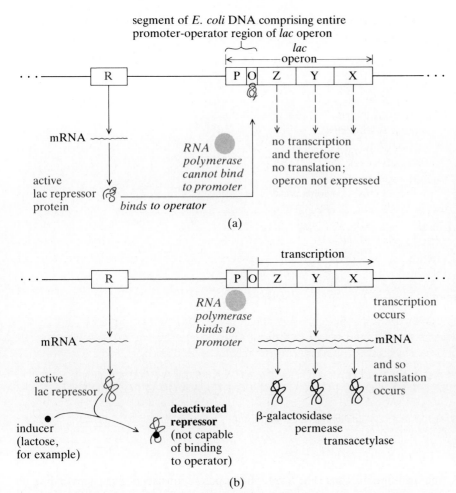

FIGURE 7–11 Diagrammatic summary of gene control by induction, using the *lac* operon of *E. coli* as an example. Gene segments are not drawn to scale. The Z, Y, and X tandem of genes code for separate enzymes; R is the gene coding for the lac repressor; P is the promoter; O is the operator. (a) No inducer. In the absence of an inducer, gene transcription is blocked by the repressor/operator binding. The segment of *E. coli* that comprises the entire promoter-operator region of the lactose operon has been isolated, and its sequence has been determined. The total length is 122 bp, with the operator locus comprising about 40 (one-third) and the promoter comprising about 80 (two-thirds). The active lac repressor protein binds to the operator, preventing the binding of the RNA polymerase at the promoter locus and thus preventing transcription. (b) Inducer present. In the presence of inducer the repressor is deactivated, and genes are expressed.

sociation of one DNA-binding protein (the lac repressor) on one side of the promoter and the association of a second DNA-binding protein (the cylic AMP–CAP complex) to the other side of the promoter stimulate the binding of RNA polymerase. So the *lac* operon is turned on, exhibiting about a thousand-fold increase in the rate of transcription.

Sequence studies of the operator and the CAP site in the promoter have shown (see Figure 7–12) that each contains a nearly perfect palindrome region involving 29 of 35 bp in the operator and 14 of 16 bp in the CAP site. Recent studies strongly suggest that these locations of twofold rotational symmetry provide the structural information recognized by these two proteins. The structures of the proteins are consistent with such a sequence signal. Being a homogeneous dimer and homogeneous tetramer, both proteins can themselves assume a symmetrical structure with a twofold axis of symmetry. We will develop this feature further as we proceed.

HISTIDINE AND TRYPTOPHAN OPERONS Two examples of transcription control in prokaryotes involving a co-repressor are the tryptophan

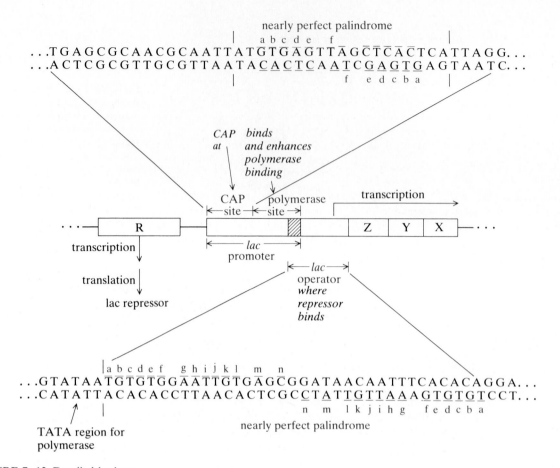

FIGURE 7–12 Detailed look at the regulatory region of the *lac* operon. Gene segments are not drawn to scale. Hatching signifies the slight overlapping of the operator and promoter regions. Base-pair sequences of the operator and CAP site are identified. The Z, Y, and X genes are identified in Figure 7–11.

and histidine operons. The tryptophan operon consists of five genes coding for separate enzymes involved in the biosynthesis of tryptophan. Twice as large, the histidine operon contains ten genes coding for separate enzymes involved in the biosynthesis of histidine. When the growth medium provides sufficient histidine and tryptophan to sustain the needs of the cell, each amino acid acts as a co-repressor, turning off the respective operons. Histidine activates the his repressor; tryptophan activates the trp repressor. Consult other sources in molecular genetics for further aspects of these systems, particularly the trp operon, which has an additional element of control called *attenuation,* involving an interesting regulatory effect mediated by a transfer RNA molecule.

LAMBDA AND CRO REPRESSORS The genome of the lambda (λ) bacteriophage is a linear dsDNA with 48,502 bp coding for about forty genes. When it enters a host cell, the linear molecules are circularized, and then either of two events occurs (see Figure 7–13): (1) A *lytic condition* results where viral DNA is transcribed, translated, and replicated to produce new phage particles, and eventually the cell bursts; or (2) a *lysogenic condition* results where the λ–DNA is incorporated into the host chromosome, where it has a dormant existence. At any time the lysogenic state can spontaneously revert to the lytic state. The frequency of this spontaneous process is very low. The reversion can also be induced by exposing lysogenic cells to UV radiation.

The lysogenic versus the lytic state is controlled by two viral genes, each coding for a different repressor protein. The two genes have the designations *cI* and *cro*. The *cI* gene codes for the *lambda repressor;* the *cro* gene codes for the *cro repressor*. The interaction of these two repressors is one of the most remarkable illustrations of the complexity and the beauty of gene control. A simple explanation is as follows: In the lysogenic state the *cI* gene is on and the *cro* gene is off; in the lytic state just the reverse applies—the *cI* gene is off and the *cro* gene is on.

The *cI* and *cro* genes are located between two oppositely oriented transcription-start regions [see Figure 7–14(a)]. Both early and late transcription starts are preceded by promoter regions, and these regions are preceded by two operator sites for the lambda repressor—the left operator O_L and the right operator O_R. The *cro* gene is located just to the right of O_R. Each O site is about 100 bp long (10 helical turns) and is able to accommodate three repressor molecules. Sequence studies have shown that each O site contains three similar stretches about 20 bp in length (for example, O_R1, O_R2, and O_R3), each showing some palindrome character. Though similar, the three operator segments do have some important subtle differences.

Where is the promoter for the *cro* gene? It is in the right half of the O_R site. And where is the promoter for the *cI* gene? It is in the left half of the O_R site. [See Figure 7–14(b).] Yes, you are reading correctly: The O_R site, which was initially described as the lambda repressor binding site, also serves two other functions. And there is more: The O_R3 segment of O_R is also a high-affinity binding site for the cro repressor. Thus a single 100-bp segment of DNA performs four functions. The key feature is that the cro and lambda repressors have different binding affinities for the operator sites. Of particular significance is that the O_R3 site has a low affinity for the lambda repressor but a high affinity for the cro repressor. Let us now sort all of this out in terms of lysogenic versus lytic states.

The lysogenic state [see Figure 7–14(c)] requires that no cro repressor be produced (the *cro* gene is off) and the lambda repressor be produced (the cI gene is on). Turning off the *cro* gene is explained by the binding of the lambda repressor to the O_R1 and O_R2 segments of O_R. Since these two segments also serve as the *cro* promoter, the binding of the lambda repressor blocks the *cro* promoter, and late transcription is shut down. The viral genes that are transcribed early are also shut down by the lambda repressor binding to O_L. With massive amounts of repressor the O_R3 site can also be flooded by lambda repressor. However, O_R3 is not a high-affinity site for the lambda repressor, and in the presence of repressor levels characteristic of the lysogenic state, the lambda repressor is not bound to O_R3. Rather, when O_R1 and O_R2 are occupied with the lambda repressor, the event proposed for O_R3 is *binding of RNA polymerase* [see Figure 7–14(c)]. Since O_R3 doubles as the promoter for *cI*, the *cI* gene is kept on. In other words, the *lambda repressor suppresses the expression of all viral genes except its own, for which it has an activating effect.*

The switch from the lysogenic state to the lytic state [Figures 7–14(c) and 7–14(d)] is triggered by the *degradation of the lambda repressor* by an enzyme

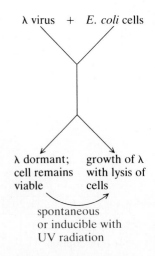

FIGURE 7–13 Lytic versus lysogenic conditions.

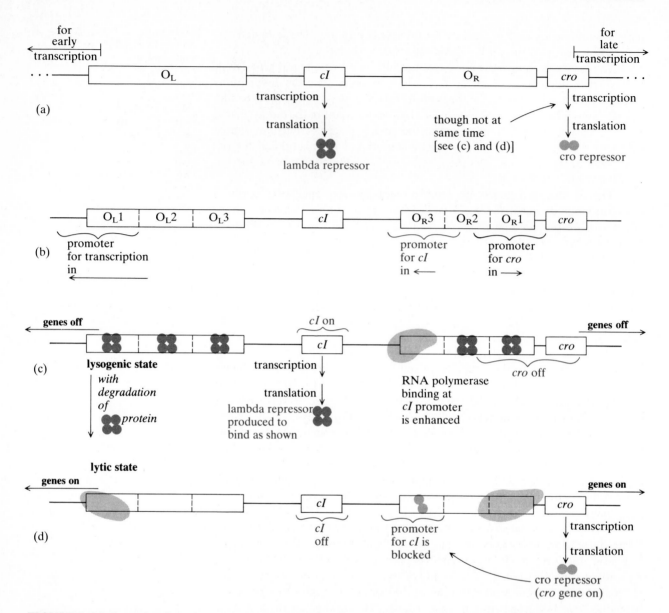

FIGURE 7–14 Control region of lambda (λ) DNA, displayed in steps of increasing schematic complexity. Gene segments are not drawn to scale. (a) and (b) Identification of the regions. (c) Lysogenic state with *cro* off and *cI* on. (d) Lytic state with *cro* on and *cI* off.

in the host cell—an enzyme that is activated by UV light. When the lambda repressor is removed, the promoter sites for the early viral genes and for the *cro* gene become vacant, giving rise to the cro repressor. When cro repressor binds to its high-affinity O_R3 site, the *cI* promoter is obstructed. So *cI* is turned off, and all viral genes can be transcribed. Thus in the lytic state the cro repressor represses the synthesis of the lambda repressor. In the lysogenic state the situation is just the reverse.

In recent years the three-dimensional structures of the cro repressor, the lac repressor, the CAP protein, and the equivalent of half the lambda repressor have been solved by X-ray diffraction. All of these DNA-binding,

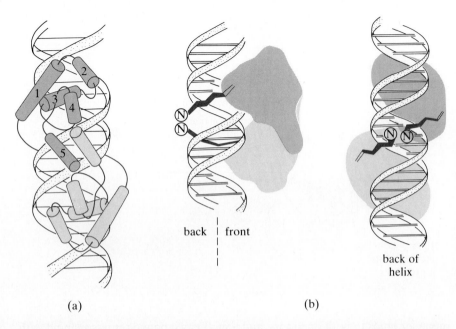

back | front

back of
helix

(a)   (b)

FIGURE 7–15 Lambda repressor/DNA binding. The models are based on crystallographic studies of a dimeric form of lambda repressor and the O_R1 segment of lambda DNA. (a) Symmetric binding of the repressor, which is shown in color. Cylinders represent α-helix segments of secondary structure. Binding contacts involving the helix 2–helix 3 region are particularly critical (see Figure 7–16). (b) Side and rear views of (a), showing how the N terminal arms from each helix 1 region are proposed to reach around the DNA to contact sites on the back side of the binding site. Shaded areas of color represent the bulk of the repressor protein. (*Source:* Adapted from photographs provided by Carl Pabo.)

gene-regulating proteins are different, but they are all symmetrical oligomers. Model-building studies have shown that each can recognize palindromic elements of twofold rotational symmetry in extended DNA. Cruciform structures (hairpins and loops) are not involved. A model proposed for the binding of the lambda repressor to the lambda operator is displayed in Figure 7–15.

Although each protein has unique features of secondary and tertiary structure, there are some strong similarities. In fact, in one region of structure, where the proteins are proposed to make important binding contacts with extended duplex DNA, the proteins are nearly identical. This consensus element of structure consists of two short α-helix segments connected by a short, tight turn (see Figure 7–16). In Figure 7–15 these segments are helix regions 2 and 3. Although the entire protein structure is bound primarily on one side of the DNA, the lambda repressor makes an interesting rear-side contact (see Figure 7–15). This contact results from flexible nonhelical arms at the amino end of each polypeptide subunit that wrap around the DNA. As the structures of other DNA-binding proteins are solved, it will be interesting to see how widespread this pattern of interaction is in nature.

For many more details than space allows here, see the articles by Ptashne, Johnson, and Pabo and by Pabo and Sauer cited in the Literature.

NEW TRANSCRIPTION CONTROL   Indicating that other regulatory devices for transcription undoubtedly exist, a new and novel element of control was recently detected in *E. coli* for the *gal operon*—a group of genes coding for proteins that participate in the metabolism of the sugar galactose. Contrary to the usual promoter/operator tandem in a classical operon, the *gal* operon has *two* operators, and both are located in unexpected positions (see Figure 7–17). One is positioned upstream from the promoter, and the second is located inside the first coding gene of the operon. Researchers propose that the gal

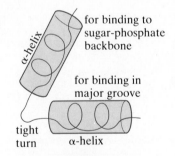

for binding to
sugar-phosphate
backbone

for binding in
major groove

tight
turn   α-helix

FIGURE 7–16 Consensus feature of structure detected in several DNA-binding proteins. In Figure 7–15 this feature corresponds to the helix 2–helix 3 region in the lambda repressor.

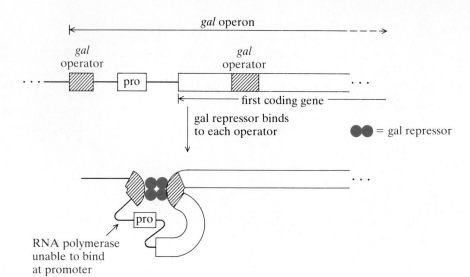

FIGURE 7–17 Proposed action of gal repressor at two *gal* operator sites (pro symbolizes promoter).

repressor binds to both operators, and then the two bound repressors bind with each other, looping the DNA in the promoter region. Perhaps in this loop the promoter is less accessible for binding RNA polymerase. The possibility that other operons are controlled in this manner will require further study.

This type of interaction suggests that DNA-binding proteins can act at widely spaced locations on DNA, with the event at one location communicating with the event at the second location. In other words, we have *action over distances* on DNA.

Inhibition of Transcription by Antibiotics

The mode of action of about thirty antibiotics involves an inhibition of transcription in one of three ways. Depending on the antibiotic, there may also be some prokaryotic versus eukaryotic differences.

Type I antibiotics bind to RNA polymerase, causing a direct inactivation of the enzyme. *Rifampicin* binds to the β subunit of prokaryotic RNA polymerase, inhibiting the initiation of chain synthesis involving pppApN formation. *Streptolydigin* also binds to the β subunit but inhibits the progressive events of elongation after initiation of chain growth. For polymerases in eukaryotes the mitochondrial enzyme is also rifampicin-sensitive, but the nuclear polymerases are unaffected by rifampicin. α-*Amanitin*, a toxic poison in certain mushrooms, is also a potent RNA polymerase inhibitor but specifically for eukaryotic RNA polymerase II.

Type II antibiotics cause an indirect inactivation of RNA polymerases by binding to the DNA template, thus blocking the binding of polymerase to the template or blocking the movement of polymerase along the template. All polymerases are affected, regardless of source. The classic example is *actinomycin D*, which noncovalently binds to duplex DNA by *intercalation*—wedging itself into the core of the duplex between the closely stacked base pairs.

Type III antibiotics are represented by *coumermycin, novobiocin,* and *nalidixic acid.* These antibiotics are also indirect inhibitors that act in prokaryotes by inhibiting the activity of *DNA gyrase,* the enzyme responsible for forming (see Figure 6–57) the supercoiled structure of DNA. This type of inhibitor provides evidence that DNA supercoiling is required for transcription in vivo, at least in prokaryotes.

7–2 PROCESSING OF INITIAL RNA TRANSCRIPTS INTO MATURE RNAs

Most of the initial RNA transcription products formed in vivo do not represent a biologically active tRNA, mRNA, or rRNA. Rather:

They are inactive precursors that must be converted to the mature active forms via various RNA modification steps.

Called *RNA processing,* the alterations include the following types:

1. *Cleaving* the large precursors into fragments.

2. *Trimming* the 5′ and/or 3′ ends of the precursor or of a fragment.

3. *Extending* the 5′ and/or 3′ ends with additional nucleotides.

4. *Splicing* a precursor by removing one or more internal segments and sealing the other pieces together.

5. *Modifying* bases in specific positions.

All types of cellular RNAs are processed in one or more ways, each in a characteristic manner. There are both similarities and differences in eukaryotic and prokaryotic cells.

Ribosomal RNA

The *E. coli* chromosome contains at least seven ribosomal RNA genes that have been mapped in various locations on the chromosome. Each gene encodes all of the information for the three prokaryotic rRNAs: 23S rRNA, 16S rRNA, and 5S rRNA. These genes also encode information for one, two, three, or four different transfer RNAs, depending on the gene. Initially, the ribosomal RNA gene is transcribed into a single RNA transcript about 7000 nucleotides in length. These initial transcripts contain the individual rRNA sequences in the order 16S, 23S, 5S, from the 5′ to 3′ direction. The rest of the transcript consists of leader, spacer, trailer, and tRNA sequences. The map of one such rRNA transcript is shown in Figure 7–18.

As indicated in the figure, the initial cuts are proposed to be made at specific positions by a cellular endoribonuclease called *RNase III.* This recognition and specificity are undoubtedly determined by unique elements of loops and hairpins in the secondary and tertiary structures of the initial transcript. Refer to Figure 6–64 to refresh your understanding about folded structures of ssRNA. The

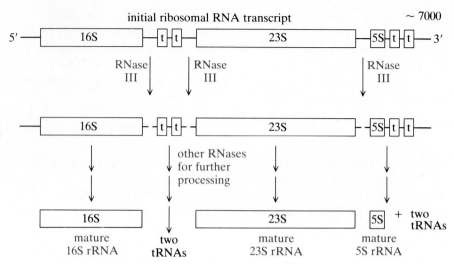

FIGURE 7–18 Processing of the primary ribosomal RNA transcript in *E. coli* into mature rRNAs. The thick line represents single-stranded RNA, with boxes identifying specific RNA sequences designated with S values. The symbol t identifies four different transfer RNAs also contained in the primary transcript. The actual in vivo structure of this precursor RNA will have a unique folded conformation, with many loops, bulges, and double-stranded stems.

cuts made by RNase III release smaller precursor fragments: pre-16S rRNA, pre-23S rRNA, and pre-5S rRNA. Other specific endonucleases and exonucleases then further cleave and trim these fragments until the mature rRNAs and mature tRNAs are formed. About twelve different RNases have been detected in *E. coli* to date, and others probably exist.

Although Figure 7–18 diagrams the processing steps as occurring on an intact pre-rRNA transcript, these events in vivo probably occur while the initial transcript is being assembled and still attached to RNA polymerase. And finally another item: Whether any of the bits and pieces of scrap RNA produced have any biological function is not known.

Similar events occur in the nuclei of eukaryotes, where a single pre-rRNA transcript is cut and trimmed to yield 28S rRNA, 18S rRNA, and 5.8S rRNA. The 5S rRNA in eukaryotes originates from a separate gene and RNA transcript. In some eukaryotic cells the processing of the primary rRNA transcript also involves a splicing operation.

Transfer RNA

The rRNA processing described above included reference to tRNAs encoded in the rRNA genes. However, most transfer RNAs are coded for by tRNA genes not rRNA genes. A given tRNA gene may encode one tRNA or two or more different or identical tRNAs. In either case, the processing of the initial pre-tRNA transcript also involves specific events of cutting and trimming, as previously described. In prokaryotes the enzyme *RNase P* participates in many of the events occurring at the 5′ end. (Recall that the enzymatic activity of RNase P is apparently due an RNA component. See Section 6–6, p. 272.)

Other important processing steps for tRNA include adding nucleotides as necessary at the 3′ end to form the universal —CCA sequence at the 3′ terminus and the modification of about 6–12 bases in specific positions of each tRNA. The base modifications, for which specific enzymes are involved, include methylation, hydrogenations, deaminations, the conversion of the normal $C^{1'}$—N^1 glycoside bond between ribose and uracil to a $C^{1'}$—C^5 linkage, yielding pseudouridine Ψ, and the incorporation of sulfur atoms. An interesting recent discovery is the incorporation of selenium atoms in some prokaryotic tRNAs. Most (not all) of these base modifications occur prior to the events of cutting and trimming.

The display in Figure 7–19 summarizes a sequence of cutting, trimming, and extending events for a proline-tRNA/serine-tRNA precursor. Except for the one that is shown in the diagram, the events of base modification are not illustrated because the order in which they occur is not known for any tRNA.

Messenger RNA in Prokaryotes

The only exception to posttranscriptional processing is messenger RNA in prokaryotes. All mRNA transcripts in prokaryotes are used in translation as is. (There is one exception known so far: A viral mRNA transcribed in a host bacterial cell from the T7 bacteriophage DNA genome is cut by RNase III into five smaller viral mRNAs.) Nuclease cleavage of prokaryotic mRNA occurs only as part of the natural degradation of mRNA. For most prokaryotic mRNAs this degradation occurs within about 1–5 min; in eukaryotes mRNA is also unstable but has a longer lifetime—up to several hours. This biological instability of mRNA is in contrast to the much more stable condition of rRNAs and tRNAs.

Messenger RNA in Eukaryotes

There are three processing stages for most of the primary mRNA transcripts produced in the nucleus of eukaryotes: (1) *capping* the 5′ terminus, (2) *extending* the 3′ terminus with a *poly(A)* segment, and (3) *splicing.* The 5′ cap and the 3′ poly(A) modifications occur very early (within 1–2 min) in the biogenesis of mRNA. Splicing is a much slower process, spanning 10–20 min. All take place in the nucleus before the mature mRNA exits into the cytoplasm.

CAPPING THE 5′ TERMINUS   Operating in the sequence *oligonucleotide triphosphatase, guanyl transferase,* and a *methyl transferase,* these nuclear enzymes convert the 5′ pppApN · · · terminus of the initial mRNA transcript to ^{7m}GpppApN · · · (see Figure 7–20). The most novel feature of this structure is the 5′ → 5′ triphosphate linkage. Later, when the processed mRNA enters the cytoplasm, the 2′—OH of the A nucleotide (and sometimes the 2′—OH of the next N nucleotide) is further methylated. Recall that 5′ capping was described previously in Figure 6–65 as a structural feature of small nuclear RNAs, where it was first discovered. Whereas the G base is trimethylated in snRNAs, it is only monomethylated in mRNA. The significance (if any) of this difference is unknown. There is strong evidence that the 5′ capping of mRNA is indispensable for the optimum binding of the mRNA to the light 40-S subunit of eukaryotic ribosomes.

METHYLATION OF INTERNAL A RESIDUES   Methylation is not confined solely to the 5′ cap structure. Small amounts of N^6-methylated adenine (m^6A) have also been detected from internal positions in mRNA. Although

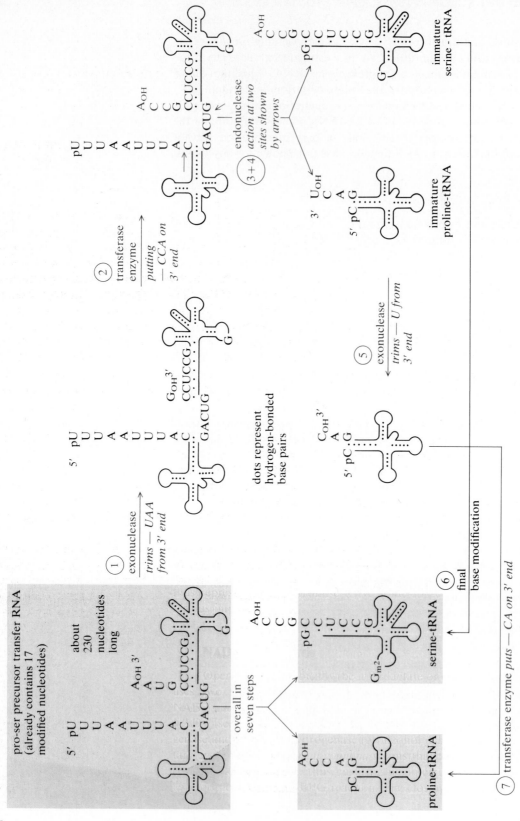

FIGURE 7–19 Partial summary of the production of two different tRNAs from a mixed dimeric precursor tRNA. Although not shown here, the precursor is itself a fragment from a larger pre-tRNA primary transcript. Dots identify hydrogen bonding in duplex stems.

internal methylation of adenine occurs early in the development of mRNA, and the m^6A residues are retained during all processing steps, there is as yet no understanding of what role (if any) they may have.

POLYADENYLATION OF THE 3′ TERMINUS   Also in the nucleus, the mRNA is modified at the 3′ end owing to the presence of a nuclear enzyme called *poly(A)-polymerase*. This polymerase uses the 3′—OH of the initial mRNA transcript as a primer for adding several more (≈ 150–200) AMP units:

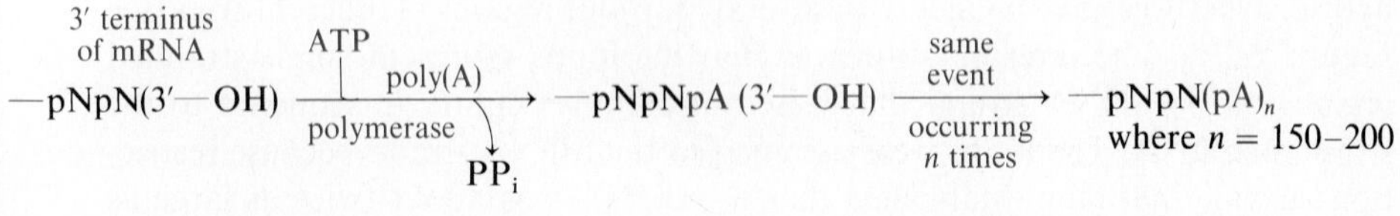

$$\overset{\substack{3'\ terminus \\ of\ mRNA}}{-pNpN(3'-OH)} \xrightarrow[\substack{poly(A) \\ polymerase}]{\substack{ATP \\ \searrow PP_i}} -pNpNpA\ (3'-OH) \xrightarrow[\substack{occurring \\ n\ times}]{\substack{same \\ event}} \underset{where\ n = 150-200}{-pNpN(pA)_n}$$

The poly(A)-polymerase uses no template for this operation. The major role of the 3′—poly(A) tail appears to be in conferring some stability to mRNA by protecting it from the degradative action of nucleases in the nucleus and in the cytoplasm. Poly(A) may also assist in the transport of mRNA out of the nucleus.

Poly(T) affinity chromatography is a useful laboratory procedure exploiting the 3′—poly(A) tail. By packing a column with a solid matrix having covalently attached poly-thymidylate chains, one can separate mRNAs from a heterogeneous mixture of other RNAs and proteins. After the mRNAs are washed from the column, individual mRNAs can be isolated by electrophoresis and hybridization techniques.

SPLICING   The release of bioactive fragments from inactive precursor proteins and precursor RNAs by precise cutting operations performed by peptidase and nuclease enzymes is exemplary of the specificity and the complexity of biochemical processes. In 1977 Chambon and co-workers reported an even more remarkable event for the processing of nuclear eukaryotic mRNA—*splicing*. It is also now known to occur with certain rRNAs and tRNAs in some eukaryotic cells.

First, let us consider some background material. As part of an overall investigation into the nature of gene regulation, the Chambon group decided to work on the ovalbumin gene in chickens. Ovalbumin (386 amino acids in a single chain) is the major protein in egg white. It is produced in large amounts only by specialized tubular gland cells of the oviduct and only when a hen is laying. Although the same gene is present in every chicken cell, ovalbumin is not produced in any other cell type.

The researchers decided to compare the ovalbumin gene in the oviduct with the ovalbumin gene from other cells. Perhaps the oviduct gene had key pre-gene regulatory signals not present on other cells. First, the mRNA for ovalbumin was isolated and purified from the cytoplasm of oviduct cells of laying hens. Next, with the enzyme RNA-directed DNA polymerase (see Section 7–3; reverse transcription), a complementary DNA copy (cDNA) of the ovalbumin gene was made in vitro. The cDNA was then inserted into a plasmid for cloning in order to increase the supply. The cDNA was next used as a hybridization

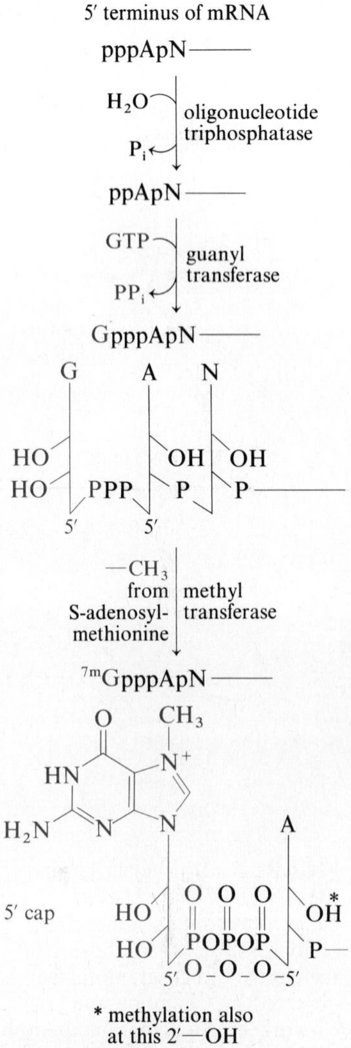

FIGURE 7–20 Process of 5′ capping for eukaryotic mRNA. All events shown occur in the nucleus. The subsequent methylation of the 2′—OH occurs in the cytoplasm (^{7m}G is N^7-methylguanosine).

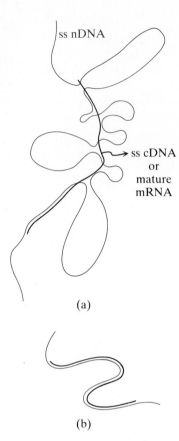

ss nDNA

ss cDNA
or
mature
mRNA

(a)

(b)

FIGURE 7–21 Evidence from hybridization that genes in eukaryotes contain noncoding regions. (a) Idealized drawing of the electron micrograph image observed by Chambon and co-workers. An actual micrograph of this type is shown in Figure 7–23. The black line represents ssDNA carrying a nucleotide sequence from the original nuclear gene. The color line represents ssDNA carrying a nucleotide sequence corresponding to the processed mRNA initially transcribed from the original nuclear gene. Identical results would be obtained by using the mature mRNA directly. Complementary hybridized regions occur where the two lines are together. (b) Type of image one would expect if the two strands of ssDNA were fully complementary.

probe to isolate the ovalbumin gene from fragments of the native nuclear DNA chromosome (nDNA) of various chicken cells. What they found didn't answer their original question of what regulates the ovalbumin gene (at least not yet), but it did have a revolutionary effect on our knowledge of eukaryotic mRNA.

By hybridization and restriction enzyme mapping, the comparison of c(ovalbumin)DNA to an n(ovalbumin)DNA fragment gave unexpected results. Electron microscopy of cDNA/nDNA hybrid duplexes did not reveal an image of an extended duplex structure. This structure is what you would expect if the two strands were perfect or near-perfect complements of each other. Rather, the hybrid structures gave images with several loop-out regions of different sizes (see Figure 7–21). The interpretation was that the loops represent single-stranded regions of the nDNA strand that were not complementary in sequence to the strand of cDNA. The loops were assigned to the nDNA strand because restriction enzyme mapping established that the nDNA was about twice as large as the cDNA strand.

The meaning of this observation was not immediately apparent. Additional studies confirmed one speculation that is now accepted as applying to nearly all mRNA coding genes in eukaryotic DNA: They are **split genes.** In what sense are they split?

> The base-pair sequence in the gene consists of separate segments coding for amino acids interrupted by separate segments that have a noncoding function.

The coding segments are called *exons;* the noncoding segments are called *introns* or *intervening sequences* (IVS).

The transcription of a split gene by RNA polymerase II gives an initial mRNA transcript containing both exons and introns (see Figure 7–22). After 5′ capping and 3′ poly(A) extension are complete, the process of *splicing* occurs:

> Precise cuts are made at exon-intron and intron-exon junctions to excise the intron. Then the ends of the two flanking exons are joined.

The process occurs in the nucleus. When several introns occur, they are not all excised simultaneously but in stages. Also, the exon-exon fusion occurs without scrambling the order of exons in the final spliced product. With the introns removed, the start-to-end coding frame is now continuous; the mRNA is fully matured and is ready for transport to the cytoplasm for translation into a polypeptide.

So far, the only known exceptions to split mRNA genes in eukaryotes are the genes for interferon (animals) and histone polypeptides. All other genes examined have one, two, several, or many introns: 1 in the actin gene, 2 in both the α and β genes for hemoglobin, 2 for the preproinsulin gene, 3 in the lysozyme gene, 7 in the ovalbumin gene, 16 in the conalbumin gene (this gene is shown in Figure 7–23), and 51 in the α-2 collagen gene. The size of intron sequences is also variable. For example, the two introns of the β globin gene contain 120 bp and 550 bp, representing approximately 40% of the entire gene. An extreme example is provided by the folate reductase gene, of which 95% is composed of introns.

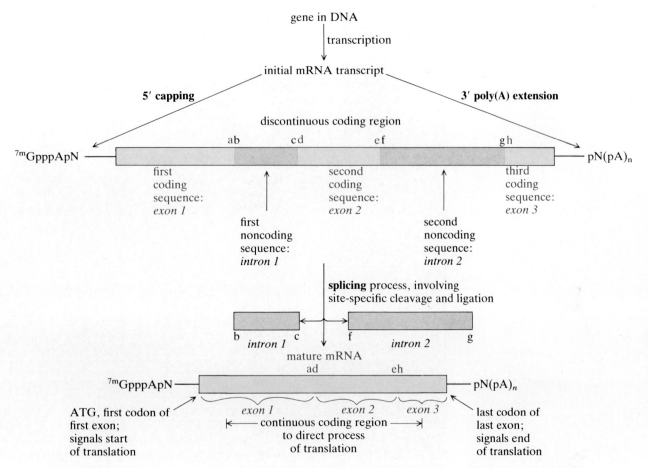

FIGURE 7–22 Diagrammatic summary of processing for eukaryotic mRNA, highlighting the splicing of mRNA. Exon segments are shaded in color; introns are in grey. The lowercase letters a through h are used to mark the last and first nucleotides at each exon-intron and intron-exon junction, emphasizing the precision of the splicing process.

There are three obvious questions regarding split genes: (1) How did they evolve? (2) What functions, if any, do introns perform? (3) How does the splicing occur? There is very little insight into either of the first two questions. Those interested can refer to the Chambon article cited in the Literature and other sources for some speculative views.

Steady progress is being made in defining the molecular mechanism of splicing, although it is not yet definitely understood. Because of the precision in which the cuts and the seals are made, nuclease and ligase enzyme functions are clearly involved. Some people estimate that there may be 10–100 different enzymes that function in splicing events, rather than just a single enzyme. However, none of these enzymes have yet been isolated.

A more definitive finding comes from a determination of base sequences of about 200 different introns from various mRNA genes. Although each intron has a unique sequence, there are two elements of sequence common to nearly all of them: The sequence of an intron (1) begins on the 5′ side with GU and (2) ends on the 3′ side with AG (see Figure 7–24). Though not as universal as the (5′)GU . . . AG(3′) pattern, other common features are (3) the 3′ AG end of the intron is preceded by a pyrimidine-rich region, and (4) the last nucleotide at the 3′ end of an exon is G.

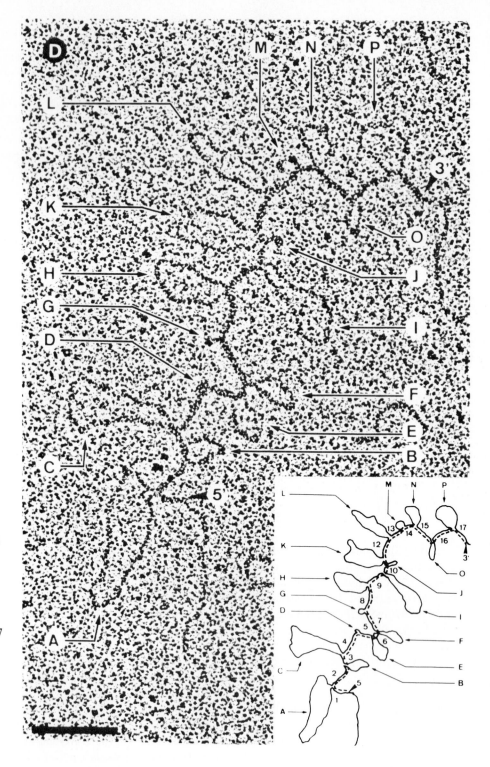

FIGURE 7–23 Electron micrograph of a DNA/RNA hybrid containing a DNA strand of the gene coding for the conalbumin protein and the mature mRNA for conalbumin. The insert is an interpretative drawing, with the solid line representing ssDNA and the dashed line being mRNA. The 17 exon sequences are numbered, and the looped-out introns are lettered. The bar represents 0.1 μm. The DNA/RNA image may be more discernible by viewing the micrograph at arm's length. (*Source:* Photograph provided by P. Chambon.)

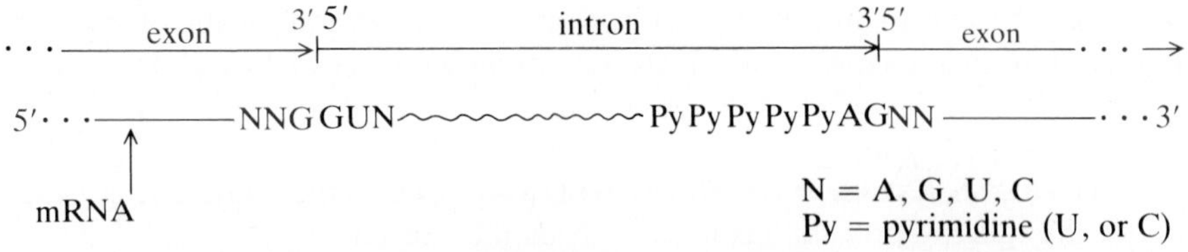

FIGURE 7–24 Current proposal of sequence signals that identify exon/intron junctions.

N = A, G, U, C
Py = pyrimidine (U, or C)

If simple signals like these are involved, how is it that they are recognized? There is some evidence for the participation (see Figure 7–25) of *small nuclear RNAs* (refer to Section 6-6, p. 272) to provide specially folded, partially duplex mRNA structures for recognition by enzymes that certainly participate in splicing. From sequence comparisons of snRNAs and the consensus splice junction, researchers have hypothesized that parts of the snRNA sequence may be sufficiently complementary to a combined UGGA sequence from the junctions on both ends of the introns. An snRNA–mRNA base pairing could bring the two junctions into close proximity as part of a region with characteristic secondary and tertiary structure. Nuclease and ligase enzymes may specifically

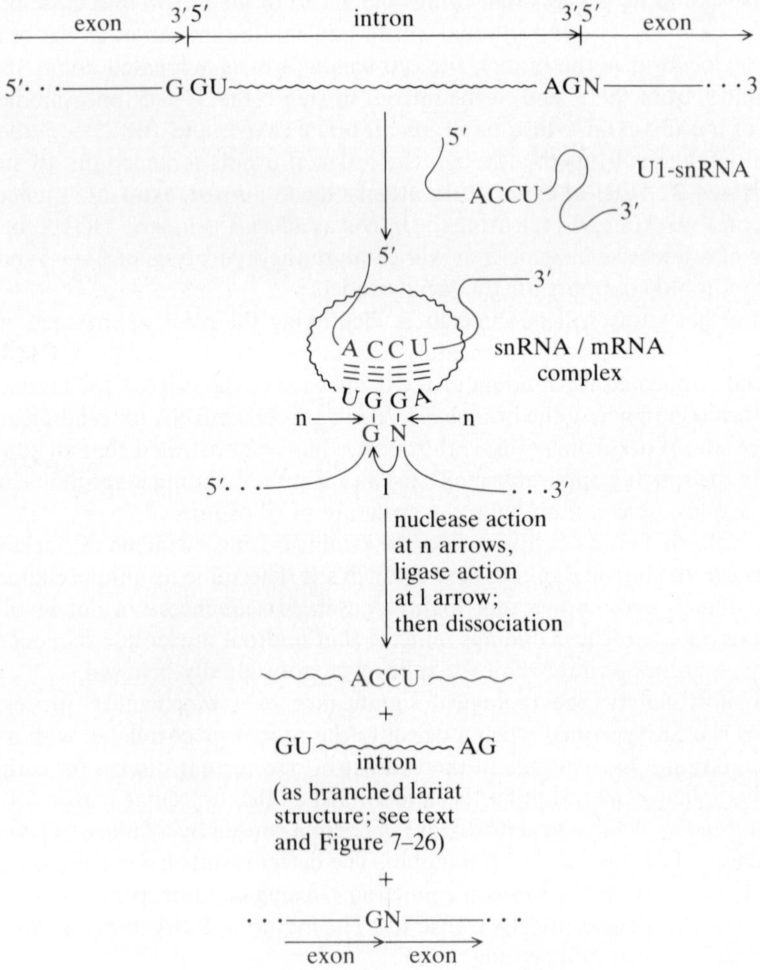

FIGURE 7–25 Hypothesis for participation of small nuclear RNA in the splicing of mRNA.

recognize this region, bind to it, and perform the cutting and sealing of splicing. Refer to Figure 6–65 and note that the minimum sequence of ACCU does occur near the 5' terminus in U1-snRNA.

INTRONS EXCISED AS BRANCHED LARIATS   One interesting fact about the splicing process has recently been discovered:

Branched RNA structures are formed as intermediates.

Nucleic acids with branching had never been detected until 1983, when Edmonds and Wallace discovered that the 2'—OH of ribose from an internal nucleotide can form a 2' → 5' phosphodiester bond. Shortly thereafter, two research groups independently established that the formation of branched RNA naturally occurs during the splicing of messenger RNA, with the branch site occurring in excised intron segments.

A suggestion for the excision of a branched intron is shown in Figure 7–26. No details of enzyme action and auxiliary participants, such as small nuclear RNA, are shown because such details are still unknown. Step a of Figure 7–26 identifies a folding process that brings the 5' end of the intron into close proximity to the 2'—OH of an internal intron nucleotide that serves as the branch site. The location of the branch site is a *unique A residue* located about 18–40 nucleotides from the 3' end of the intron. In step b the 3' → 5' phosphodiester bond of the (3')-exon 1/intron-(5') junction is cleaved, and the 2' → 5' diester branch linkage is formed, although the order of events is uncertain. In step c the released 3'—OH of exon 1 then attacks the (3')-intron/exon 2-(5') junction, fusing the two exons and releasing the intron as a *lariat structure*. The recent discovery of a nuclease enzyme that will catalyze the hydrolysis of 2' → 5' bonds offers additional support for the lariat model.

Further study will be directed at identifying the roles of enzymes, small nuclear RNAs, and possibly noncatalytic binding proteins. To date, U1-snRNA has been implicated for binding at the 5' splice site, U2-snRNA for binding in the intron region where the branch is formed, and U5-snRNA for binding at the 3' splice site. Work from several laboratories has demonstrated that all components of the splicing apparatus may exist as a complex ribonucleoprotein cluster called a *spliceosome*, resembling the structure of ribosomes.

Additional study is also needed to establish what elements of nucleotide sequence in the intron flanking the A branch site determine its unique character. Initial studies have shown that highly conserved sequences are not involved. Whatever the case, these findings indicate that internal nucleotide sequences of the intron are more important for splicing than originally believed.

Unfortunately, the biological significance of a biochemical process in humans is clearly verified when a defect in the process is correlated with a disease condition. One example in the context of our current discussion is that a defective splicing operation has been identified as the molecular reason for one type of *β-thalassemia*, a genetic disease of anemia caused by a failure to produce the β chain of normal adult hemoglobin. The defect results from a single nucleotide change at the exon 2/intron 2 junction, causing an improper splicing event to occur in the β chain mRNA transcript. The mature mRNA does not code for the normal β chain polypeptide.

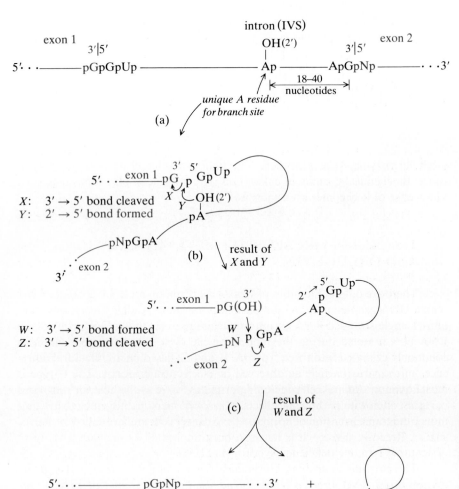

FIGURE 7–26 Introns excised as branched lariat structures. See text for descriptions of steps a through c. Enzymes are certainly involved. Noncatalytic binding proteins and small nuclear RNAs may also function in forming and stabilizing one or more of the alignments.

PROCESSED GENES The discovery of split genes was a surprise. There is now evidence of still another surprise: An intron-containing gene and an intron-lacking gene, *both coding for the same protein,* are present in the same DNA genome. The intron-lacking genes have been called *pseudogenes* or *processed genes.* So far, they have been detected for only a small number of protein-coding genes, but others probably exist. The sketch in Figure 7–27 shows both genes on the same chromosome, but they have also been located on different chromosomes.

Processed genes may arise in vivo by the action of an enzyme having reverse-transcription activity to use the mature mRNA as a template to yield a complementary piece of DNA, which is then reintegrated into the host chromosome. The reverse-transcription enzyme may be of viral origin, or it may be

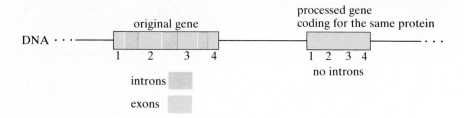

FIGURE 7–27 Split and processed genes.

a cellular enzyme. At present there is no understanding of the biological significance. Biochemical genetics in eukaryotes is already a complex subject, and our knowledge of it becomes even more so.

7–3 RETROVIRUSES, REVERSE TRANSCRIPTION, AND ONCOGENES

There are different families of viruses that contain an RNA genome rather than a DNA genome in the virion particle. One family, called *retroviruses,* contains a single-stranded RNA chromosome and requires that a DNA copy of the RNA be assembled during the replication life cycle. The retrovirus family is commonly classified into three types (A, B, and C) based on architectural differences in the virus particle as observed by electron microscopy. The C-type is most common. Animal cells, including primates, serve as the host for retrovirus particles, often with pathogenic effects. Moreover, many pathogenic retroviruses induce the transformation of normal cells to cancer cells and are called *oncogenic viruses.* Recently, a retrovirus has also been implicated as the causative agent of acquired immune deficiency syndrome (AIDS).

The most distinguishing biochemical feature of a retrovirus is the participation of an *RNA-directed DNA polymerase* (first discovered in 1970) to assemble a double-stranded DNA according to the template instructions of the single-stranded RNA. On the basis of a comparison with the usual process of DNA → RNA, this process of RNA → DNA is called *reverse transcription,* and the polymerase is sometimes called *reverse transcriptase.* This enzyme is a true viral protein, coded for by a gene in the viral chromosome. A cellular counterpart has not yet been detected.

A mature retrovirus particle contains two copies of the ssRNA chromosome, (+)RNA, and several copies of the reverse transcriptase enclosed in a protein core, which in turn is surrounded by an outer envelope containing other viral proteins. After the virus binds to a receptor site in the membrane of a target cell, the (+)RNA chromosome and the reverse transcriptase enter the cell (see Figure 7–28). In the cytoplasm of the host cell the reverse transcriptase uses the (+)RNA to assemble a ss(−)DNA and then uses the ss(−)DNA to assemble the complementary ss(+)DNA strand. The two ssDNA strands then associate to yield a viral ds(±)DNA. (See the Kornberg source cited in the Literature of Chapter 6 for some of the details proposed for the mechanism of this double-copying operation.) The viral dsDNA then enters the nucleus of the host cell, where it becomes integrated into a host chromosome. The integration process

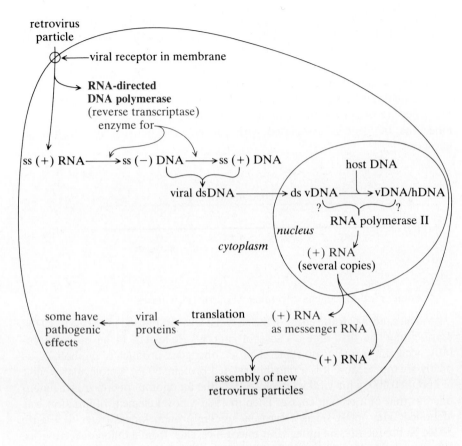

FIGURE 7–28 Replication cycle of a retrovirus characterized by the process of reverse transcription (RNA → DNA). Question marks indicate uncertainty about whether or not integration of viral DNA is required for its transcription.

requires a circularized form of the viral dsDNA. Whether this integration into host DNA is necessary for viral replication is still another unresolved question.

In the nucleus the remainder of the retrovirus replication cycle continues, with assistance from one of the nuclear transcription enzymes (probably DNA-directed RNA polymerase II), which transcribes the integrated viral dsDNA to yield several copies of single-stranded (+)RNA. The (+)RNA exits the nucleus and is used as a viral messenger RNA to translate the viral genes into viral proteins. These proteins include structural proteins for assembly [with (+)RNA] of new viruses, the RNA-directed DNA polymerase, and in many cases at least one protein that is pathogenic to the host cell.

Retrovirus RNA Chromosome

Apparently, all retroviruses have similar ssRNA chromosomes—similar in structural features and in the composition and arrangement of genes. The RNA is 9000–10,000 nucleotides in length (2–3 million in molecular weight) and resembles eukaryotic mRNA in that the 5′ terminus is capped and the 3′ terminus is polyadenylated. The genetic organization of avian sarcoma virus (ASV) (see Note 7–4) is known in some detail and serves as a prototype for other retroviruses.

NOTE 7–4
The avian sarcoma virus was first described in 1911 by Peyton Rous and is also called the Rous sarcoma virus (RSV).

The ASV chromosome contains four genes, identified with three-letter names symbolic of the proteins they code for: (1) the *gag* gene coding for a polypeptide precursor that is carved into various proteins present in the viral core, (2) the *pol* gene coding for the RNA-directed DNA polymerase, (3) the *env* gene coding for proteins that appear in the viral envelope, and (4) the *src* gene coding for a protein that *causes the formation of sarcomas.* As indicated in Figure 7–29, the genes are organized in a *gag/pol/env/src* sequence in the 5′ to 3′ direction. Since the *src* gene is associated with cancer-using activity, it is called an **oncogene.**

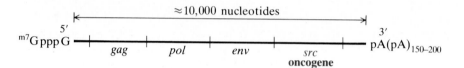

FIGURE 7–29 Organization of the ssRNA chromosome of ASV.

Some Viral Oncogenic Proteins Are Protein Kinases

The cancer-causing effect of the *src* gene was established in the early 1970s, representing the first time a cause of cancer was related to a specific naturally occurring protein molecule. The *src* oncogene product is symbolized as pp60v-src (pp because it is a phosphorylated protein; 60 for a molecular weight of about 60,000; v for viral source). In 1978 the function of pp60v-src was identified as that of a *protein kinase.* Then in 1980 a novel element of function was identified: The pp60v-src is a *tyrosine-specific* protein kinase (refer to Figure 5–36). Subsequently, six other viral oncogenes, each from a different retrovirus, have also been shown to code for tyrosine-specific protein kinases.

These findings prompted an obvious working hypothesis to explain the viral transformation of normal cells to cancer cells: (1) The viral protein kinase catalyzes the conversion of one or more cellular proteins from a nonphosphorylated form to a phosphorylated form, and (2) in this altered state the phosphorylated cellular protein(s) acquires different functions, triggering a single event or cascade of events that eventually destroy the growth control of the cell.

The first part of the hypothesis has been verified by findings that ASV-infected cells do contain some phosphorylated proteins not detected in uninfected cells. One of these proteins is *vinculin,* a component of the cytoskeletal structure in eukaryotes. In ASV-infected cells the level of tyrosine-phosphorylated vinculin is increased about twentyfold. Researchers have suggested that the phosphorylation of vinculin may alter its interaction with other cytoskeletal proteins, causing a change in the entire cytoskeletal network. This change may in turn alter the shape of the cell and its physical contact with other cells. To the extent that cell shape and cell/cell contacts are responsible for the regulation of the normal growth and development of cells, such effects would contribute to tumorigenesis.

As explained in the next subsection, altering normal growth control and development by protein kinases is only one aspect of oncogenic transformation.

Cellular Oncogenes

A gene theory of cancer was first suggested in the 1960s, and it is now proven. The general principle is as follows: Cellular chromosomes naturally contain cancer-causing genes that are normally dormant, that is, they are not expressed because of some type of negative control that represses transcription. But when activated, they are actively transcribed into protein products, which cause the transformation of normal cells to cancer cells.

When viral oncogenes were discovered, the question immediately was raised about their origin. Do viral oncogenes (v *gene*) arise from cellular oncogenes (c *gene*)? The answer is definitely yes. The evidence comes from hybridization studies, and the first success (1975) was provided with the finding that the v-*src* gene has a nearly identical counterpart, c-*src,* in the DNA chromosomes of chickens. The c-*src* gene is a split gene with six introns; the v-*src* gene is a fully processed gene lacking the introns. The cellular oncogenes are also called *proto-oncogenes,* reflecting the suggestion that retroviruses acquired the oncogene from the host during a previous infection. How this acquisition may have occurred is not yet understood.

The exact number of cellular oncogenes existing in the genome of a eukaryotic cell is unknown. Of the estimated 30,000 human genes, about 50–100 may be oncogenes. At the time of this writing, 24 oncogenes had been detected in vertebrate sources, and 22 of these are known to be carried as viral oncogenes in retroviruses. The other two have been detected in cancer cells not infected by any virus. As described in the next subsection, the functions of oncogenes encompass various protein activities.

Other Types of Oncogenic Proteins

The type of protein function coded for by about half of the known oncogenes has been identified. Seven of the oncogenes (*src, ros, yes, fgr, abl, fps,* and *ras*) code for proteins that have *protein kinase* activity, and all but *ras* are tyrosine-specific. This function appears to be the major class of oncogene proteins. Recently, other interesting functions of oncogene-coded proteins have been detected. In 1983 researchers found that the p28v-sis protein coded for by the *sis* gene in the simian sarcoma virus was very similar to the *plasma-derived growth factor* (PDGF) (refer to Section 3–7). Linking an oncogene to a growth factor protein has obvious implications in explaining how a healthy cell with normal growth control is converted into a cell lacking growth control. At a time when it should not be, a cell is bombarded by excessive amounts of a substance that stimulates growth and/or cell division.

The protein of the *erb-B* oncogene has also been linked to the biochemistry of growth factors. However, the erb-B protein is not itself a growth factor but, rather, is a *receptor protein* for a growth factor, specifically, the *epidermal growth factor* (EGF) receptor (refer to Section 3–7).

Three other oncogene proteins express proteins that *bind strongly to GTP.* Proteins that bind GTP are sometimes called *G proteins.* How these G proteins cause transformation is not yet clear. Perhaps the binding of GTP (and subsequent hydrolysis of GTP to GDP) is necessary to activate the oncogene protein for some other function it may have. Another possibility is that the oncogene-coded G protein simulates the action of one or more normal G proteins, some of which are known to regulate key cellular processes. This latter hypothesis has support from recent research in yeast cells that established that the *ras*-coded G protein appears to simulate the normal G protein responsible for regulating the activity of adenylate cyclase, the enzyme responsible for producing cyclic AMP, a major regulator of many cell processes. In yeast cells the activation of the *ras* oncogene results in a strong activation of adenylate cyclase. A similar *ras* effect in mammalian cells is likely but not yet established.

Representative of other oncogene proteins that act in the nucleus of cells, one protein has been shown to have strong *DNA-binding activity*. Although there is no evidence yet, such an event could result in the activation of a cellular transforming gene that is normally repressed.

This diversity of oncogene function clearly indicates that oncogenic transformations can occur by various mechanisms. Moreover, recent studies indicate that such transformations require the *cooperation of at least two oncogenic proteins*. For example, normal fibroblast cells produce very little of the protein coded for by the cellular oncogene *myc* (so designated because it was first detected in the retrovirus MC29). However, when exposed to PDGF, the fibroblasts produce 40 times the amount of the myc protein. Another study has shown that fibroblast cells are completely transformed to malignant cells only if infected with both a *ras* gene and a *myc* gene. Separate transfections with either gene give only incomplete transformation. These results strongly suggest that cellular transformation is a multistep process with greater complexity than once envisioned.

Expression of Cellular Oncogenes

Given the natural existence of oncogenes, an obvious question is, What sort of activation mechanisms operate to turn these genes on and start the transformation process? There are several possibilities.

First, in the case of retrovirus activation the expression of the *oncogene carried by the virus* will obviously *increase the intracellular concentration* of the oncogenic protein. In this case the cellular oncogene is not directly activated but is expressed as the viral oncogene counterpart. Second, nonviral activation of a dormant cellular oncogene may occur because of a *mutation* occurring in the transcription control region of a cellular oncogene that turns the gene on or a mutation in the oncogene itself giving rise to an altered oncogene product but with greatly increased oncogenic activity. Third, nonviral activation may occur because of a *movement* (transposition) of an oncogene to a different position on a chromosome, where it comes under the influence of a strong positive control signal such as a transcription enhancer element. Fourth, nonviral activation may be caused by *gene amplification*—an increase in the number of oncogene copies due to a localized replication of DNA.

The formation of a superactive oncogene protein by mutation (the second possibility) was recently proven by comparing a nonviral oncogene obtained from human bladder carcinoma cells with its protooncogene in normal cells. Both genes are expressed at nearly the same levels in the two cell populations.

If overexpression is not involved, then the only apparent conclusion is that the oncogene protein in the cancer cell is an altered form of the normal oncogene protein. In a brilliant piece of modern research in molecular genetics, this reasoning was proven by isolating the segment of the oncogene containing the mutation. Base-sequencing procedures then revealed that the sequence of amino acid codons differed in only one codon involving a single nucleotide change—a *point mutation*. The coding information changed from a codon of GGC for glycine in the normal protooncogene to the codon of GTC for valine in the carcinoma oncogene. This was the first time a genetic lesion had been established as causing the change of a benign gene into an active gene for the growth of a human tumor. For details of this remarkable finding, see the paper by Weinberg listed in the Literature.

LITERATURE

BISHOP, J. M. "Oncogenes." *Sci. Am.,* **246,** (3) 80–92 (1982). The discovery and the nature of oncogenes are described.

BUJARD, H. "The Interaction of *E. coli* RNA Polymerase with Promoters." *Trends Biochem. Sci.,* **5,** 274–278 (1980). A brief review of the recognition of promoter sequence by the sigma subunit of RNA polymerase.

CHAMBON, P. "Split Genes." *Sci. Am.,* **244,** (5) 60–71 (1981). The discovery and the nature of split genes are described.

DICKSON, R. C., J. ABELSON, W. M. BARNES, and W. S. REZNIKOFF. "Genetic Regulation: The *Lac* Control Region." *Science,* **187,** 27–35 (1975). A description of the nucleotide sequence of the promoter-operator region of *E. coli* DNA and a discussion of how the *lac* operon is regulated.

GLUZMAN, Y., ed. *Eukaryotic Transcription.* Cold Spring Harbor: Cold Spring Harbor Laboratories, 1985. Collection of brief papers discussing advances in this field since 1983 (see Gluzman and Shenk).

GLUZMAN, Y., and T. Shenk, eds. *Enhancers and Eukaryotic Gene Expression.* Cold Spring Harbor: Cold Spring Harbor Laboratories, 1983. Collection of brief papers discussing the advances in this field.

HUNTER, T. "The Proteins of Oncogenes." *Sci. Am.,* **251,** (2) 70–79 (1984). The various functions of oncogene proteins are described, with speculations on how they transform normal cells to tumor cells.

JACOB, F., and J. MONOD. "Genetic Regulatory Mechanisms in the Synthesis of Proteins." *J. Mol. Biol.,* **3,** 318–356 (1961). The original paper proposing the messenger RNA hypothesis and the nature of gene control by induction and repression of operons by regulatory genes.

PABO, C. O., and R. T. SAUER. "Protein DNA Recognition." *Annu. Rev. Biochem.,* **53,** 293–321 (1984). A detailed review article of DNA binding with *Cro* and lambda repressors and CAP, with general implications for other DNA-binding proteins.

PADGETT, R. A., M. M. KONARSKA, P. J. GRABOWKI, S. F. HARDY, and P. H. SHARP. "Lariat RNA's as Intermediates and Products in the Splicing of Messenger RNA Molecules." *Science,* **225,** 898–903 (1984). Discovery (also see Ruskin, Krainer, Maniatis, and Green) of a branched intermediate during RNA splicing.

PTASHNE, M., A. D. JOHNSON, and C. O. PABO. "A Genetic Switch in a Bacterial Virus." *Sci. Am.,* **247,** (5) 128–140 (1982). The intricate interplay of the cro and lambda repressors at the lambda operator is described. The structure of DNA-binding proteins is also emphasized.

REED, R., and T. MANIATIS. "Intron Sequences Involved in Lariat Formation During pre-mRNA Splicing." *Cell,* **41,** 95–105 (1985). Research into the nature of sequence signals that affect the splicing process.

RUSKIN, B., A. R. KRAINER, T. MANIATIS, and M. R. GREEN. "Excision of an Intact Intron as a Novel Lariat Structure During pre-mRNA Splicing in Vitro." *Cell,* **38,** 317–3313 (1984). Discovery (also see Padgett, Konarska, Grabowski, Hardy, and Sharp) of a branched intermediate during RNA splicing.

TEMIN, H. M. "The DNA Provirus Hypothesis." *Science,* **192,** 1075–1080 (1976). The Nobel Prize lecture describing current thoughts on the significance of reverse transcription and the infectivity of RNA viruses.

WALLACE, J. C., and M. EDMONDS. "Polyadenylated Nuclear RNA Contains Branches." *Proc. Natl. Acad. Sci., USA,* **80,** 950–954 (1983). The initial discovery of branched RNA.

WEINBERG, R. A. "The Action of Oncogenes in the Cytoplasm and Nucleus." *Science,* **230,** 770–776 (1985). A current review article.

WEINBERG, R. P. "A Molecular Basis of Cancer." *Sci. Am.,* **249,** (5) 126–142 (1984). How a dormant oncogene can be activated by a mutation is described.

EXERCISES

7–1. What relationship does the nucleotide sequence in an RNA transcript have to the nucleotide sequence of the sense strand in dsDNA?

7–2. Consider Figure 7–30, showing a gene isolated from DNA having one intron (wavy line) and the arrows identifying two sites for the hypothetical restriction enzyme Rcb. Obviously, one could use such a restriction pattern to isolate a fragment containing the intact intron. By partial DNA-sequencing procedures, how would you determine which of the restriction fragments contained the intron?

Rcb   Rcb

FIGURE 7–30

7–3. Assume you have isolated the mature mRNA corresponding to the gene in Figure 7–31, where the wavy lines represent introns. You then apply the hybridization technique and view the results via electron microscopy. Draw the type of image you would expect to observe.

FIGURE 7–31

7–4. A 30-bp segment from ds λDNA containing the 17-bp sequence of O_R1 is shown in Figure 7–32. How much two-fold rotational symmetry exists in the O_R1 site: 100%, about 60%, about 30%, or none at all? Use the base pairs in the center of the O_R1 region to make the evaluation.

7–5. Assume that each strand of the O_R1-containing duplex fragment in Figure 7–32 is end-labeled with ^{32}P. The labeled DNA is then divided into two portions: First, you react one portion with dimethyl sulfate under conditions that give methylation only at G bases. Assume that the reaction conditions will result in all G bases in the duplex being methylated. Then you separate the methylated, ^{32}P-labeled

```
         5        10       15       20       25       30
        ACTATTTTACCTCTGGCGGTGATAATGGTT
        TGATAAAATGGAGACCGCCACTATTACCAA
             |←                          →|
                     O_R1 of λDNA
```

FIGURE 7–32 Segment of DNA used for Exercises 7–4 and 7–5.

strands. The upper strand is recovered and treated under reaction conditions that will cleave phosphodiester bonds at the sites of G-methylation, yielding a mixture of ^{32}P-labeled fragments of different lengths, depending on which methylated G site is cleaved. This fragment mixture is then examined by PAGE and a radioautograph is prepared. Second, to the other portion, you add a pure sample of the lambda repressor and incubate the mixture, allowing sufficient time for saturation binding of the repressor to the DNA. Then you carry out the same methylation and cleavage reactions and perform PAGE under the same conditions as given for the first portion. The results for portions 1 and 2 are summarized in Figure 7–33, showing a drawing of the comparative radioautographs. What can be concluded from these results about the contact sites between bound lambda repressor and G residues in the upper strand of O_R1?

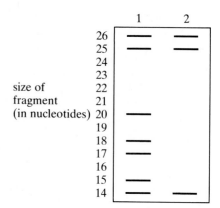

FIGURE 7–33

7–6. Most of the descriptive material covered in this chapter is embodied in the following list of terms and symbols. Define or describe each term: upstream/downstream, $5' \rightarrow 3'$ synthesis, promoter, sigma factor, operator, template, TATA, lytic versus lysogenic, enhancer, *cI* and *cro* genes, repressor, operon, induction, hybridization, $5'$ cap, retrovirus, poly (A), oncogenes, intron, oncogene proteins, exon, conserved sequences, pre-RNAs, spliceosome, lariat.

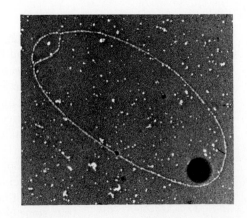

CHAPTER EIGHT

DNA BIOSYNTHESIS, REPAIR, AND RECOMBINATION

In announcing their discovery of the double-helix structure of DNA, Watson and Crick suggested that it was obvious—at least in general terms—how duplicate copies of DNA are assembled. Each strand of the duplex can guide the assembly of a new complementary strand according to base-pairing principles. As it happens, nature has evolved very elaborate mechanisms in order to direct this base-pairing recognition, and even now, some thirty years later, research still continues to unravel the intricate details of this very complex process.

The objectives of this chapter are to examine the main features (general and specific) of the DNA → DNA replication and to also consider the related processes of mutation, DNA repair, and DNA recombination. These topics are truly some of the most fascinating biochemical events that occur in living cells.

8–1 DNA REPLICATION

General Principle: Semiconservative Replication

In the late 1950s M. Meselson and F. W. Stahl experimentally established (see Figure 8–1) the in vivo pattern of how the duplex structure of DNA is duplicated. The *E. coli* bacterial cells were first grown in a medium containing $^{15}NH_4Cl$ as the sole nitrogen source. Nitrogen ^{15}N is a stable, heavy isotope of ^{14}N, the latter being the most abundant form of nitrogen. Growth in such a medium will label all nitrogen-containing compounds with ^{15}N. The DNA of the cells will be extensively labeled, because each DNA molecule will contain millions of ^{15}N atoms in the many purine and pyrimidine bases. Accordingly, ^{15}N–DNA will have a greater density (it will be heavier) than ^{14}N–DNA or any hybrid of DNA containing a portion of ^{14}N and a portion of ^{15}N, and it should be separable by density gradient ultracentrifugation.

After ^{15}N labeling was complete, the cells were washed and transferred to a similar medium containing only $^{14}NH_4Cl$, and growth was continued. Every time the population doubled—that is, with the production of every new generation of cells—a sample was removed and the DNA assayed for its content of ^{14}N and ^{15}N. Ultracentrifugal analysis of the DNA obtained after one complete generation revealed only one zone in the tube: 50% lighter than completely ^{15}N–DNA and 50% heavier than the completely ^{14}N–DNA. No other bands of DNA were present. This result clearly indicated that after one round of cell division the replicated DNA molecules were 50–50 *hybrids* of ^{14}N and ^{15}N.

The DNA extracted after two cell doublings (second-generation cells) gave only two zones containing equal amounts of DNA. One zone had the same sedimentation as the 50–50 ^{14}N–^{15}N hybrid from the first-generation cells, and the second zone was equivalent to 100% ^{14}N–DNA. Similar experiments since performed with many other prokaryotic and eukaryotic cells have given the same results.

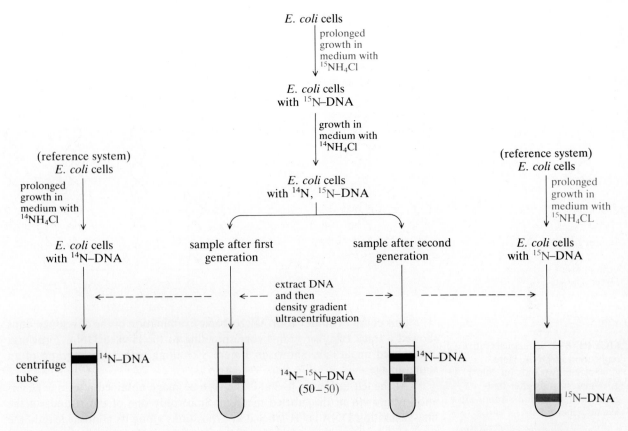

FIGURE 8–1 Diagrammatic summary of the Meselson-Stahl experiment proving the semiconservative mechanism of DNA replication. Also see Figure 8–2.

The data clearly reveal that DNA replication must proceed by a process (see Figure 8–2) wherein each strand of the parent DNA duplex acts as a template for the synthesis of a new complementary strand. Furthermore:

Each template and its new complement remain associated to give the next generation of DNA. This pattern is called semiconservative replication.

General Principle: Simultaneous Synthesis of Both Strands

The next important insight into the molecular nature of DNA replication was provided by J. Cairns and co-workers in 1963, who succeeded in *observing* the DNA molecule of *E. coli* during stages of its replication. This feat was accomplished by growing *E. coli* cells in a growth medium to which was added some radioactive thymidine (tritium-labeled; ^{3}H). After entering the cell, thymidine was converted to TTP and was utilized in the synthesis of DNA as it was replicated. Obviously, the DNA was then radioactively labeled owing to the thymine positions being ^{3}H-labeled. After the addition of (^{3}H)-thymidine, cells were removed at various times; the cells were lysed; and the DNA was recovered by ultrafiltration. The filter was mounted on a glass surface and exposed (about two months) to X-ray film to detect the weak β radiation emitted by the

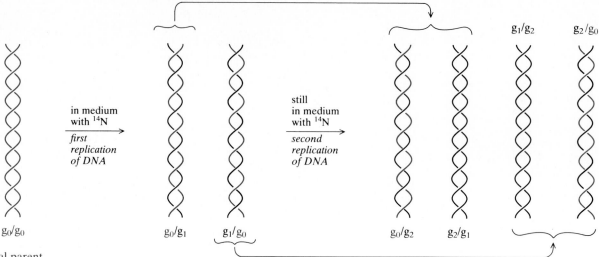

initial parent
DNA labeled
in both
strands
with ^{15}N

FIGURE 8–2 Semiconservative
replication of DNA. Here
g_0 = original strands (in color);
g_1 = new strands after first
generation; g_2 = new strands after
second generation.

^3H atoms of the thymine bases. Microscopic examination of the developed films showed a track of silver grains corresponding to the labeled DNA. Drawings of idealized images are shown in Figure 8–3 along with a photograph of an actual image.

At the left in Figure 8–3(a) is the type of image obtained after one generation of growth in the labeled medium. Since only one of the strands of the first-generation DNA is ^3H-labeled at T positions along its contour length, the track of silver grains is of uniform density. When DNA was extracted from cells *during* the second-generation growth period in the same labeled medium, an image of the type shown at the right was observed. It showed two circular regions, with part of the silver grains being about twice as intense as the other. The more intense part of the image was interpreted as representing the copying of the already labeled strand of the first-generation duplex, which would now contain twice as much ^3H-thymine, since both strands now contain the radioactive label. The less intense parts were interpreted as the copying of the unlabeled strand of the first-generation duplex and the segment of the first-generation duplex that had not yet been replicated.

Cairns concluded that the image of this DNA, caught in the course of replication, must mean not only that replication was semiconservative but also that the *synthesis* of both strands was *simultaneous*. Cairns went on to propose a model of DNA replication, a key feature of which was the suggestion that DNA contains a specific location, called the *replication origin,* where copy synthesis begins. Beginning at the origin, some type of *swivel mechanism* is necessary to locally unwind the helical structure of duplex DNA to expose both strands for replication.

Progressing away from the origin, copy synthesis can be *unidirectional* or *bidirectional* (see Figure 8–4). Various studies have established that bidirectional synthesis is the major strategy used in all types of cells. Unidirectional synthesis does occur with the replication of many viral DNAs. In either case, the *repli-*

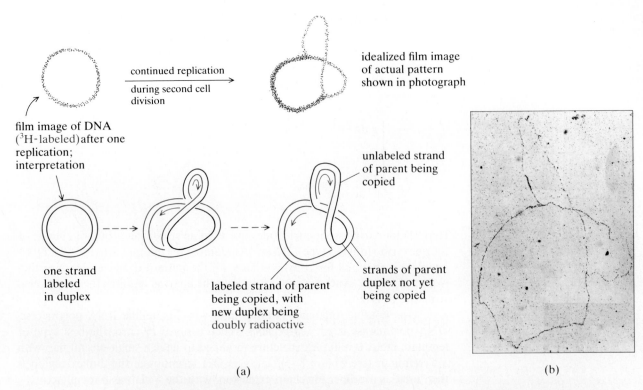

idealized film image of actual pattern shown in photograph

film image of DNA (³H-labeled) after one replication; interpretation

one strand labeled in duplex

labeled strand of parent being copied, with new duplex being doubly radioactive

unlabeled strand of parent being copied

strands of parent duplex not yet being copied

(a)

(b)

FIGURE 8–3 Simultaneous synthesis of dsDNA. (a) Idealized images and interpretations. In the lower line drawings color represents the ³H-labeled strand and black represents the unlabeled strand. (b) Photograph of an actual image.

cation fork (Cairns's term) will continue to advance via a swivel mechanism along the template duplex until both strands are replicated.

Enzymology of DNA Replication: Basic Features

Although many proteins participate, the central enzyme involved in DNA replication is *DNA-directed DNA polymerase*. (Hereafter, we will simply refer to DNA polymerase; but remember that this term should not be confused with the RNA-directed DNA polymerase operating in reverse transcription.) A. Kornberg (Nobel Prize, 1959) and co-workers pioneered in the initial isolation and characterization of DNA polymerase from *E. coli* in 1958. Continued studies by Kornberg and numerous other investigators over the past 25 years have unraveled many of the details of DNA replication. Complete understanding is elusive because of the enormous complexity of the process. The coverage here will highlight only a small fraction of current knowledge. Those interested in

FIGURE 8–4 Two patterns of duplex copy synthesis. (a) Unidirectional synthesis from the origin. (b) Bidirectional synthesis from the origin. (c) Simple diagrammatic representation of a replication fork: one fork in unidirectional, two forks in bidirectional.

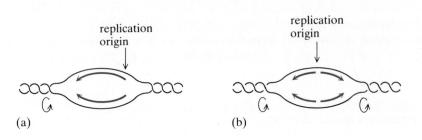

replication origin

replication origin

(a)

(b)

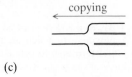

copying

(c)

further study should consult the book *DNA Replication* by A. Kornberg, considered as the most thorough and authoritative treatise on the subject.

In many respects the operation of DNA polymerase in the replication of DNA is similar to that described in Chapter 7 for the operation of RNA polymerase in the transcription of DNA:

$$
\begin{bmatrix} q\ \text{dATP} \\ r\ \text{dGTP} \\ s\ \text{dTTP} \\ t\ \text{dCTP} \end{bmatrix}_{\substack{\text{pool} \\ \text{of subtrates}}} + \underset{\text{(as template)}}{\text{DNA}} \xrightarrow[\underset{n\text{PP}_i}{(\text{Mg}^{++})}]{\substack{\text{DNA-directed} \\ \text{DNA polymerase}}} \underset{\substack{\text{(copy of} \\ \text{template)}}}{\text{DNA}}
$$

Here (1) four triphosphonucleotides are used as substrates—but, of course, of the deoxyribo (dNTP) type, (2) Mg^{2+} is required for optimal activity, (3) a DNA template is required for copy synthesis, (4) the pattern of nucleotide insertion is that of $5' \rightarrow 3'$ synthesis, and (5) in vivo the enzyme requires the supercoiled structure of DNA.

One notable difference (among many) is that unlike RNA polymerase, DNA polymerase is *not self-priming*. When assayed in vitro, the best type of template DNA is a duplex structure modified to have a single-strand flap with a 5' terminus (see Figure 8–5). The 3'—OH terminus of the duplex region is then used as a primer for chain extension, with the 5' flap as a template.

Another significant distinction of DNA polymerase (in prokaryotes) is that it also exhibits *exonuclease activity*. The most common type is $3' \rightarrow 5'$ exonuclease activity, removing nucleotides from a 3' terminus. Some also have $5' \rightarrow 3'$ exonuclease activity, removing nucleotides from the 5' terminus. That a DNA-polymerizing enzyme should also be a DNA-degrading enzyme may not seem to make any sense, and when first discovered, it didn't. However, we now understand that the exonuclease operations of DNA polymerase are crucial to various aspects of normal DNA replication in vivo.

A particularly important role of the $3' \rightarrow 5'$ exonuclease activity allows DNA polymerase to correct a base-pairing mistake during copy synthesis. The process is called *proofreading*. For example, when the template instruction is G, the incoming nucleotide should be dCTP to yield a GC pair. If dATP, dGTP, or dTTP is used, the template is copied incorrectly to yield either GA, GG, or GT. In quick order DNA polymerase can apparently recognize such an error and correct it. The recognition probably depends on improper base-pair interactions, resulting in a slightly frayed 3' end on the primer strand. The correction

FIGURE 8–5 DNA-directed DNA polymerase, which requires a 3'—OH terminus to prime copy synthesis of template strand.

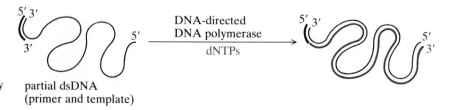

partial dsDNA
(primer and template)

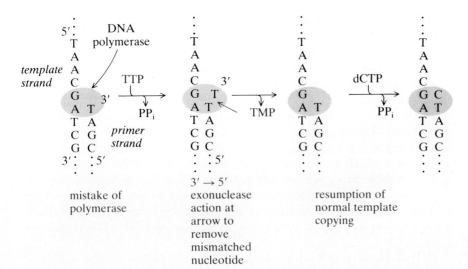

FIGURE 8–6 Proofreading by DNA polymerase.

is made via the $3' \rightarrow 5'$ exonuclease action of the polymerase to remove the mismatched nucleotide from the 3' terminus (see Figure 8–6). Then copy synthesis resumes. Some other functions of the exonuclease activity of DNA polymerase will be explained later.

Supplemented by the proofreading capability and other repair devices, which we will discuss later, the template-copying action of DNA polymerase has the ability to replicate DNA with a very high degree of fidelity. Excluding the presence of mutagenic conditions, the normal error frequency is about 1 in 200 million—truly remarkable.

Multiple Polymerases in Cells

The original DNA polymerase isolated from *E. coli* and characterized by Kornberg and others is commonly referred to as DNA polymerase I (or simply pol I). Subsequent studies detected two other DNA polymerases in *E. coli*, DNA polymerases II and III (pol II and pol III), so designated for the order of their discovery. The existence of three enzymes is now known to be true of most prokaryotes. In addition, most eukaryotic cells also contain at least three different DNA polymerases, designated α, β, and γ. Numerous other DNA polymerases have been isolated from cells infected with certain viruses owing to the expression of a viral gene coding for a viral DNA polymerase.

Although all DNA polymerases have the same general features of operation previously described, there are some specific differences in operation. For example, the eukaryotic DNA polymerases lack nuclease activity. In addition different polymerases participate in different aspects of DNA synthesis in vivo. In eukaryotes pol α is the major enzyme associated with chromosome replication, pol β appears to be involved in repair and recombination processes, and pol γ is localized primarily in mitochondria, where it probably functions in the replication of mitochondrial DNA. Polymerases pol α and pol β are localized in the nucleus. The complexity of DNA-polymerizing enzymes is compounded by observations that the relative amounts of the different enzymes can change with

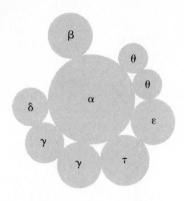

FIGURE 8–7 DNA polymerase III aggregate. See also Table 8–1. The minimum pol III holoenzyme core consists of α, ε, and θ; α is the polymerase component.

TABLE 8–1 Composition of DNA polymerase III

SUBUNIT	MOLECULAR WEIGHT
α	140,000
τ	83,000
γ	52,000
β	37,000
δ	32,000
ε	25,000
θ	10,000

the nutritional state and age of a cell and the possibility that under certain conditions they can substitute for each other.

We will focus on DNA replication in prokaryotes, since the process is better understood in prokaryotes than in eukaryotes. In prokaryotes pol III (a single polypeptide chain designated α, with a molecular weight of about 140,000) is the major polymerizing enzyme. In vivo the active holoenzyme form of pol III is a multisubunit complex consisting of α in association with other polypeptides. The largest aggregate (see Figure 8–7 and Table 8–1) consists of one copy of a pol III and six other subunits. The minimum holoenzyme consists of a pol III and just two other subunits. At different stages of DNA replication different aggregates may participate. Specific functions for each of the subunits have not yet been clearly defined. There is some evidence that the β subunit may perform a function for pol III similar to the sigma (σ) subunit in RNA polymerase, that is, recognizing the location on DNA where replication starts.

DNA polymerase I, also a large polypeptide with molecular weight of 109,000, functions primarily as an auxiliary enzyme to pol III in one of the crucial steps of DNA replication and also in the repair of damaged DNA. We will discuss both processes later. Unlike pol III, the in vivo active form of pol I does not involve any association with other protein subunits. Polymerase pol II (an oligomeric protein) has not yet been implicated in any crucial aspect of DNA replication. It may be merely a reserve enzyme capable of replacing pol I or pol III under certain conditions.

DNA Replication in Vivo Requires Many Auxiliary Proteins

Several proteins, many of which are enzymes, supplement the action of DNA polymerase in DNA replication. In prokaryotes the participation of at least a dozen different auxiliary proteins has been established by the ultimate proof, namely, isolating mutants that exhibit abnormal DNA replication owing to a gene defect for the production of a specific protein. In view of the many proteins that participate, a more realistic procedure is to refer to a DNA replication system rather than a single enzyme.

These auxiliary proteins assist in eight activities: (1) the recognition of origin sites, (2) the local unwinding of the DNA duplex to expose single strands for template copying, (3) the stabilization of the unwound structure, (4) the provision of primer strands to initiate the action of DNA polymerase, (5) the swivel process of advancing the replication fork, (6) the final steps of assembling two complete strands, (7) the recognition of termination sites, and (8) the supercoiling of the two new DNA molecules. A few of these auxiliary proteins are identified in the following sections.

Origin of DNA Replication

Replication of the DNA chromosome of most prokaryotes, plasmid DNAs, and viral DNA chromosomes begins at a single specific site on DNA called the *replication origin,* or *ori* site. The DNAs of eukaryotes appear to have multiple origin sites providing multiple start locations. Isolation and base se-

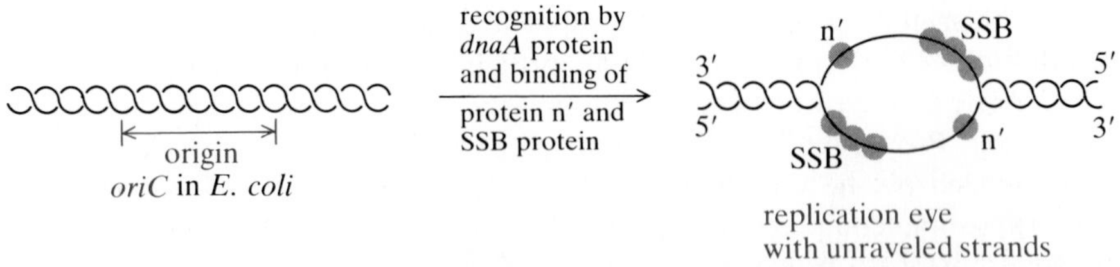

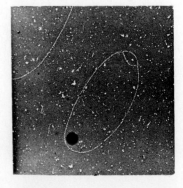

FIGURE 8–8 Unraveling of strands at the origin of DNA replication. A micrograph is shown in Figure 8–9. The next step in replication is shown in Figure 8–10.

FIGURE 8–9 Electron micrograph of a DNA replication eye in a replicating DNA plasmid, shown here in a relaxed, circular state. (*Source:* Photograph provided by Arthur Kornberg.)

quencing of *ori* sites from various sources have failed to reveal any universal *ori* sequence. Different DNAs apparently have different signals. We will briefly consider just one example: the origin site in *E. coli* (called *oriC*) and several other related gram-negative bacteria. The sequence of the *oriC* location (about 250 bp long) has some interesting features, such as a palindrome sequence of GATC that is repeated 14 times and two sets of inverted repeat sequences each 9 bp in length. Is there a special element of looped-out secondary structure? Do the GATC repeats mean anything? As yet, there is no understanding of how these and/or other features are used as signals to identify the *oriC* location.

The *oriC* recognition is known to involve various auxiliary proteins designated as the *dnaA* protein, protein n′, and the SSB protein. In ways not yet known, these proteins bind to the *ori* site and cause a local unraveling of the duplex strands to form a replication *eye* (also called a replication *bubble*); see Figures 8–8 and 8–9. The function of the SSB protein is to stabilize this localized site of unwound strands by acting as a <u>s</u>ingle-<u>s</u>trand <u>b</u>inding protein. In the replication eye each of the unraveled strands is now accessible to provide template instructions for new DNA synthesis. The formation of the replication eye is quickly followed by the process of priming.

RNA Provision of a 3′—OH Primer Terminus

Unable to start its own chain, DNA polymerase depends on another enzyme to provide a primer oligonucleotide chain. The enzyme is called *DNA primase* and operates as a DNA-directed RNA polymerase, but it is not the same protein that functions in the transcription of DNA. The DNA primase copies part of a template DNA strand into a short piece of ssRNA (2–10 nucleotides long), which remains annealed to the template strand (see Figure 8–10). Originally discovered for *E. coli* replication, the strategy of RNA priming is now known to be common for both prokaryotes and eukaryotes.

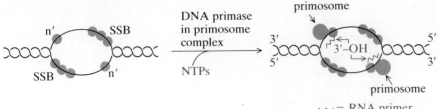

⌇⌇ = RNA primer, 1–10 nucleotides long

FIGURE 8–10 RNA priming at the replication eye. Subsequent events are shown in Figure 8–11.

The priming process is quite complicated, using at least five other proteins in addition to DNA primase: protein i, protein n, protein n'', *dnaC* protein, and *dnaB* protein. All components function as a multisubunit complex termed a *primosome*. Proper binding of the primosome at the replication eye is proposed to be guided by the previously bound protein n'. For bidirectional synthesis the primosome complex binds at opposite ends of the replication eye, laying down a short RNA primer on each template strand. The DNA is now prepared for DNA polymerase.

Processive Copying via DNA Polymerase Requires an Advancing Replication Fork

Binding of a pol III holoenzyme complex begins the copy synthesis of the template single strands, using the RNA primer for chain extension (see Figure 8–11). Note that the two template strands are copied in *opposite directions*. This feature is consistent with the fact that DNA polymerase is only capable of $5' \rightarrow 3'$ synthesis, and so it cannot copy the two strands in the same direction. For bidirectional synthesis two separate pol III complexes are involved at opposite ends of the replication eye.

As pol III advances on the templates, the replication forks must continue to advance in opposite directions in order to continually expose more of the templates. The unwinding of duplex DNA at the fork is known to require the participation of auxiliary proteins called *helicases*. The process is ATP-dependent, that is, energy-dependent. After helicase proteins bind to the fork structure, ATP binds to the helicase, which catalyzes the hydrolysis of ATP to ADP and P_i, a process called *ATPase activity*. In a way not yet understood, the ATP/helicase binding and the subsequent hydrolysis of ATP cause the melting of one or two base pairs at the fork to permit some strand separation. As single-strand regions develop, the binding of SSB proteins stabilizes the expanded structure of the replication eye.

Semidiscontinuous Synthesis

The advancing replication fork allows pol III to continue copy synthesis on the 3'—OH of the new DNA just assembled. The other segments of newly exposed template require additional priming by the primosome complex. The pol III uses these primers for additional copy synthesis, as permitted by the length of the template not yet copied, stopping as the 5' end of a previously used primer is approached. Continual fork advancement and continual pol III and primosome operation yield a replicating intermediate, as shown in Figure 8–11. Note that in each direction from the initial origin, one of the template strands is being copied continuously (called the *leading strand*), whereas the other template strand is copied discontinuously *in segments* (called the *lagging strand*). This pattern of DNA replication is termed *semidiscontinuous synthesis*.

The discontinuous assembly of the lagging strand was first indicated by the research of R. Okazaki. Using special growth conditions to slow down the rate of the DNA replication process in *E. coli,* Okazaki found that most of the newly formed DNA sedimented at 10S–12S, corresponding to polynucleotide chain lengths in the range of 1000–2000 nucleotides. The significance of these *Okazaki fragments* is evident from the diagram in Figure 8–11.

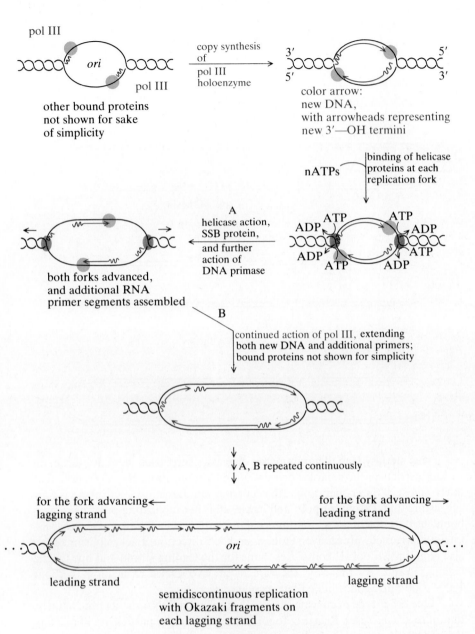

FIGURE 8–11
Semidiscontinuous synthesis of DNA in a bidirectional pattern. Although not shown throughout the diagram, most of the proteins remain bound, and all operations occur simultaneously. This action is called a *processive operation*. The completion of synthesis involving the connection of the lagging-strand fragments is shown in Figure 8–12.

During copy synthesis the Okazaki fragments are assembled into a complete copy of the lagging strand by the action of other enzymes, primarily *DNA polymerase I* (pol I) and *DNA ligase* (see Figure 8–12). Polymerase I performs two functions: (1) The $5' \rightarrow 3'$ exonuclease activity removes the RNA primer, and (2) the copy synthesis activity fills the gaps, stopping short of the $5'$ end of the next fragment. The DNA ligase activity connects the $3'$ and $5'$ ends of adjacent fragments. Experiments showing that Okazaki fragments accumulate in mutant cells deficient in DNA ligase or pol I activity provide conclusive proof of this aspect of DNA replication.

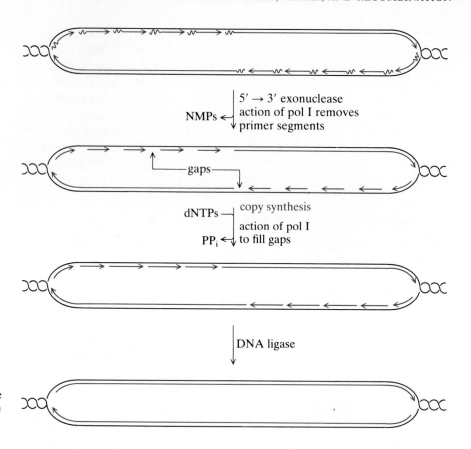

FIGURE 8–12 DNA polymerase I and DNA ligase completing the assembly of the lagging strands.

The step-by-step diagrammatic displays were used here to focus on various aspects of the complex process of DNA replication. However, note that in vivo most of these operations are occurring simultaneously. Kornberg is now suggesting that, except for pol I and ligase, all other replicating proteins remain bound as part of a massive multimolecular complex, which he terms a *replisome*, for uninterrupted, processive synthesis in each direction. Each replisome may also contain two pol III complexes—one for the leading strand and one for the lagging strand. By the way, the rate of in vivo DNA replication is about 1000–1500 nucleotides linked per second.

One other type of enzyme activity is involved in DNA replication, namely, that of *topoisomerases*. Relaxing topoisomerases initially participate in relaxing the negatively supercoiled state of DNA to provide access to the *ori* site. A supertwisting topoisomerase (DNA gyrase) can function to relieve the strain of positive supercoiling that will result ahead of the advancing replication forks and also convert the final two new daughter duplexes into the negatively supercoiled state.

Further Normal Modification of DNA

Newly replicated DNA is further processed by *methyltransferase* enzymes, using S-adenosylmethionine (SAM) as a source of —CH$_3$ groups for the methylation of some adenine and cytosine residues in prokaryotes and primarily cytosine residues in eukaryotes. The only specific function associated with base methylation in prokaryotes is protection of the host chromosome from the degradative action of its own restriction enzymes. In eukaryotes, as mentioned in a previous discussion, the pattern of methylation may determine which genes are transcribed and which are not.

In the nuclei of eukaryotes newly replicated DNA becomes associated with the various histone proteins (refer to Figure 6–58) and some nonhistone proteins to form the mature chromosomes. Some proteins and small basic peptides and amines bind also to the DNA in prokaryotes.

Inhibition of DNA Replication

The list of DNA replication inhibitors is long and varied in regard to both type of function and mode of action. Some bind to and deactivate DNA polymerase; some bind to and deactivate a specific auxiliary protein; some intercalate with the template DNA, interfering with its ability to be copied; and some are substitute analogs of the normal dNTP substrates, causing (in many instances) an inhibition of further chain extension or acting as competitive inhibitors of the normal dNTPs. Some are antibiotics; some are mutagens; some are chemical toxins; some are effective as antiviral drugs; some are used as drugs in the treatment of cancer.

Some specific examples are *arabinosyl adenine* (ara-A) and *arabinosyl cytosine* (ara-C), two promising antiviral drugs (see Figures 6–14 and 6–15), which inhibit DNA replication as nucleotide analogs; the antibiotic *actinomycin,* which intercalates with DNA; and the antibiotics *nalidixic acid* and *coumermycin,* which inhibit the action of DNA gyrase. Lacking both 2′—OH and 3′—OH functions, synthetic *2′,3′-dideoxy-NTP analogs* are a special class of inhibitors used in the Sanger method of base sequencing. When a 2′,3′-dideoxy-dNMP unit is positioned in the growing DNA chain, further extension is impossible because the 3′ terminus now lacks a 3′—OH function.

Extensive discussion and structure displays are beyond the scope of this book. The Waring article cited in the Literature is a good source for information on anticancer drugs.

8–2 MUTATION AND DNA REPAIR

Mutation Classifications

Mutation is a heritable change in the genetic material. The simplest alteration, a *point mutation,* involves a change in a single base pair of DNA. Depending on the type of base-pair alteration, the change is considered a *substitution mutation* or a *frame shift mutation,* each occurring in either of two ways (see Figure 8–13). A substitution mutation may be a *transition* or a *transversion.* (In describing each, we will use + and − to designate the complementary strands of DNA.) A transition defines a change of one (+)purine-pyrimidine(−) pair to another, such as (+)GC(−) → (+)AT(−) or

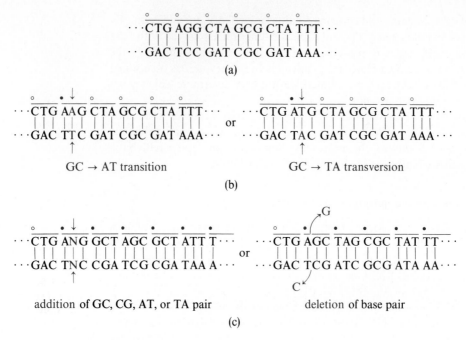

FIGURE 8–13 Point mutations. (a) Normal (wild-type) sequence of triplets (o). (b) Substitution mutations. (c) Frame shift mutations. The • identifies triplets affected by base-pair mutation.

$(+)AT(-) \rightarrow (+)GC(-)$. A transversion defines a change of a $(+)$purine-pyrimidine$(-)$ pair to a $(+)$pyrimidine-purine$(-)$ pair, such as $(+)GC(-) \rightarrow (+)TA(-)$ or $(+)GC(-) \rightarrow (+)CG(-)$. The two types of frame shift mutations are *additions* (an extra GC or AT pair) and *deletions* (loss of a GC or AT pair).

Mutations in protein-coding genes are also classified as *lethal* or *silent*. Lethal mutations are characterized by a defective cellular function due to the failure to produce a particular protein or the production of a defective protein. Silent mutations do not give rise to defective cellular functions. They are characterized by the production of a slightly different protein that still exhibits virtually normal activity.

Lethal mutations are almost always the consequence of frame shift alterations that change the entire sequence of coding triplets, starting at the location of the base-pair addition or deletion. Silent mutations are more likely with base-pair substitutions that change only one coding triplet in the gene, thus potentially changing only one amino acid in the protein. Because of codon degeneracy (see Section 9–7), the new codon may even code for the same or a similar amino acid, and thus the protein will be unaffected. When the new codon does code for a different amino acid, the activity of the modified protein may still be unaffected if the amino acid involved is not a critical structural or active-site residue. For example, at the present time about two hundred seventy-five different hemoglobin variants have been discovered, and the great majority of them involve single amino acid replacements in either the α or β chain. Only four of these variants are seriously defective and associated with disease conditions; one of these is HbS, the sickle-cell hemoglobin.

Various Causes of Mutation

Mutations arise for three reasons: (1) Normal mistakes occur during the replication of DNA; (2) spontaneous alterations under normal in vivo conditions occur in DNA after its replication; or (3) alterations in DNA occur owing to exposure to mutagenic chemicals or mutagenic radiation. Because of the ability of DNA polymerase to accurately copy a template and to instantly proofread and correct mismatches, the normal error rate in DNA replication is very low—about one base-pair substitution occurring for every 10^7–10^{11} base pairs replicated. This low rate of misincorporation is also attributed to the operation of enzyme systems that can recognize and correctly edit base mismatches after DNA synthesis and other enzymes that can repair alterations in DNA. We will describe some of these repair systems shortly.

One type of spontaneous alteration is the nonenzymatic cleavage of sensitive bonds in DNA by hydrolysis. Particularly susceptible are the glycoside bonds connecting the purine and pyrimidine bases to the $C^{1'}$ of deoxyribose. The loss of a base is called *depurination* or *depyrimidination,* and the location is designated an *AP site* (for apurinic or apyrimidinic); see Figure 8–14. When DNA with AP sites replicate, base-pair deletions can arise. Again, the operation of DNA repair enzymes minimizes the chance of mutation by correcting the AP lesions.

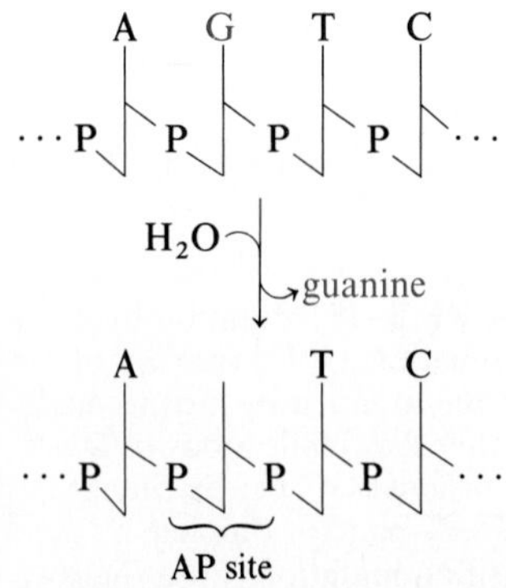

FIGURE 8–14 Formation of an AP site. The example shown is an apurinic site.

Mutagenic Chemicals and Radiation

CHEMICALS A list of mutagenic chemicals would be massive. Many are part of the normal biosphere, others are pollutants of modern society, and numerous others are merely nonbiological, synthetic materials that have become part of the chemical library in our society. Whatever the type, mutagenicity is caused by a reaction with DNA, altering the structure of DNA in such a way that a mutation occurs when the altered DNA is replicated.

Some mutagens, like nitrous acid (HNO_2), formaldehyde (H_2CO), and bisulfite (HSO_3^-), cause *deaminations* to occur in G, C, and A bases. In deamination a $C—NH_2$ function is changed to a $C{=}O$ function in the base. Mutations occur because the deaminated bases have different hydrogen bonding patterns, directing the insertion of different bases during subsequent replications. For example, HNO_2 deaminates a cytosine ring to yield uracil (see Figure 8–15). Thus a GC pair will be initially converted to a GU pair. Then in the first replication of this altered DNA, the template strand with U will specify for A to yield an AU pair. In the second replication the template strand with A will specify for T to yield an AT pair. The overall effect is a $(+)GC(-) \rightarrow (+)AT(-)$ transition.

Dimethylnitrosamine (DMNA) is typical of a large group of mutagens that act as *methylating agents* (or generally, *alkylating agents*). The sites of methylation are the nucleophilic O and N atoms in the bases, giving rise to various $O—CH_3$ and $N—CH_3$ derivatives. For example, cytosine can be converted to an N^3-methyl, O^2-methyl, or N^4-methyl derivative (see Figure 8–16). Mutations arise because the methylated bases also have different base-pairing properties, causing base-pair substitutions after two rounds of DNA replication.

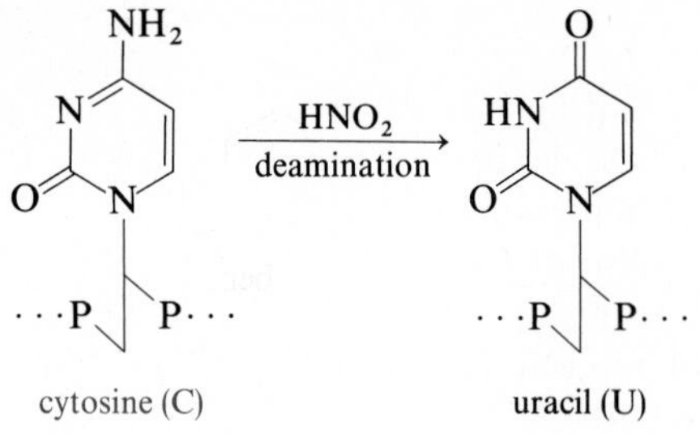

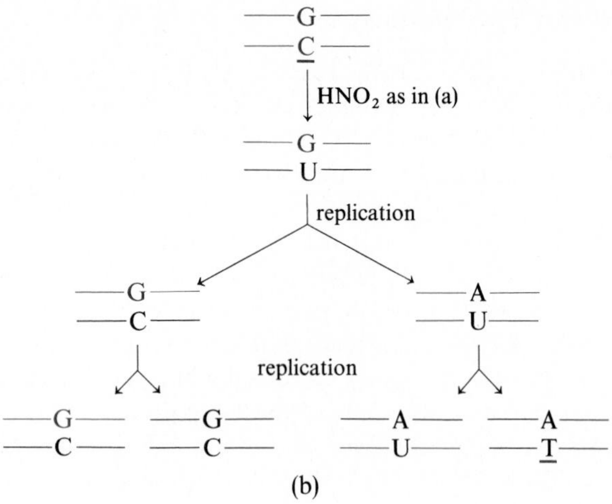

FIGURE 8–15 Mutation by deamination. (a) Conversion of cytosine to uracil by nitrous acid. (b) After two replications of DNA. The original GC pair is changed to an AT pair to complete a transition mutation. Color lines represent the original strands of DNA.

FIGURE 8–16 Mutation by an alkylating agent. The example shows possibilities for cytosine methylation by dimethylnitrosamine (DMNA). All of the methylated forms of cytosine will have altered base-pairing properties and increased susceptibility for depyrimidination to yield an AP site.

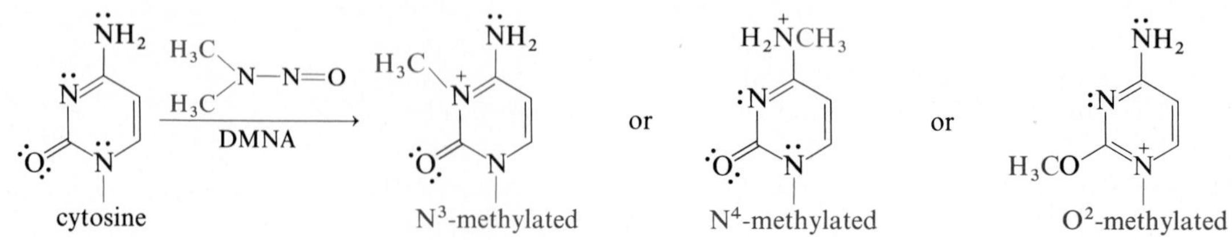

Many mutagenic substances become mutagenic only *after partial metabolism* in the host cell. A notable example of this metabolic activation of a mutagen is *benzo[a]pyrene* (BAP), the most common polycyclic aromatic hydrocarbon in the atmosphere, produced in the incomplete combustion of fossil fuels. In various cells BAP is metabolized by two normal enzymes [an oxidase called cytochrome P–450 (see Section 15–6) and an epoxide hydratase] to yield different diol-epoxide derivatives such as the one shown in Figure 8–17. The tendency of the diol-epoxide to yield a highly reactive carbonium ion species that can react with :N atoms of purines and pyrimidines (particularly the N^7 of G) accounts for the mutagenicity (and carcinogenicity) of the hydrocarbon. A common final consequence of this bulky alkylation is the formation of an AP lesion. Recent studies have linked benzo[a]pyrene as a particularly effective agent for causing human breast cancers.

Other mutagenic chemicals react with DNA to *form cross-links* between bases on the same strand or on opposite strands and others *intercalate* with DNA. Both processes interfere with DNA polymerase action, causing a "skip" or a "stutter" of the enzyme on its template and yielding frame shift mutations. Other mutagens react in such a way as to cause a base in the normal anti conformation to change rotation to the syn conformation—a change that may affect the normal base-pairing recognition of the rotated base. Still others may cause the stable keto tautomer of a base to change to an enol tautomer with different base-pairing properties. In other words, mutations can be caused by a myriad of chemical effects.

RADIATION Exposure to *ultraviolet* (UV) *radiation* causes cyclobutyl cross-links to form between adjacent thymine bases. The adduct is called a *thymine dimer* (see Figure 8–18). Earth organisms are protected from the most harmful UV emission of the sun by the ozone layer in the outer atmosphere, which absorbs a large percentage of solar UV light.

Exposure to *ionizing gamma radiation* causes various types of DNA alterations, such as base modifications including thymine dimers, cleavage of sugar-base bonds, and cleavage of phosphodiester bonds.

DNA Repair Processes

All types of organisms contain enzymes that function in repairing damaging alterations in DNA. Most are inducible enzymes, produced in response to some signal identifying the presence of a mutagen and lesions in DNA. The most abundant repair enzymes are *DNA glycosylases* (about fifteen are known so far), which remove altered bases by catalyzing the cleavage of base-sugar bonds to generate AP sites. The AP sites in turn are recognized by specific *AP endonucleases* that catalyze cleavage of phosphodiester bonds on both sides of the AP site. The nicked AP site is probably enlarged by the $5' \rightarrow 3'$ exonuclease action of DNA polymerase, then the gap is filled by the template-copying activity of DNA polymerase, and finally DNA ligase seals the repaired strand.

The example in Figure 8–19 illustrates the action of uracil-DNA glycosylase, removing a uracil base that arises from the deamination of cytosine we highlighted earlier. Other examples include hypoxanthine-DNA glycosylase for repairing an adenine → hypoxanthine deamination, xanthine-DNA glycosylase for repairing a guanine → xanthine deamination, 3-methyladenine-DNA glycosylase for removing methylated A bases, and 7-methylguanine-DNA glycosylase for removing methylated G bases.

The AP endonucleases include two types, one acting on the 3′ side of the AP site and the other acting on the 5′ side. Perhaps both enzymes act in concert to completely remove the sugar phosphate moiety, leaving the site with 3′—OH and 5′ phosphorylated termini suitable for subsequent action of DNA polymerase and DNA ligase.

An obvious way to repair an AP site is direct reattachment of the proper free base. Enzyme extracts exhibiting this type of *insertase activity* have been recently reported, but complete purification and characterization have not yet been achieved.

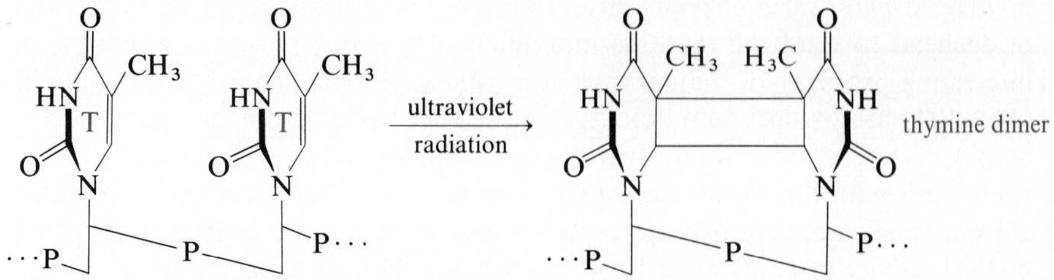

benzo[a]pyrene

diol-epoxide

reactive carbonium ion species toward DNA

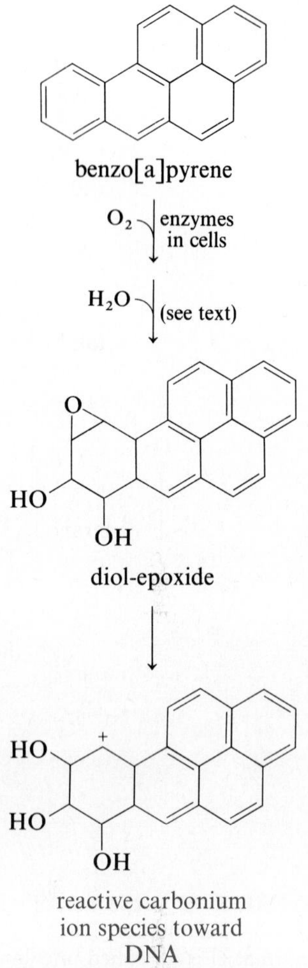

FIGURE 8–17 Mutagenicity of polycyclic hydrocarbons. It requires in vivo action of cellular enzymes.

thymine dimer

FIGURE 8–18 Thymine dimer formation. Two adjacent thymine rings on the same strand are involved.

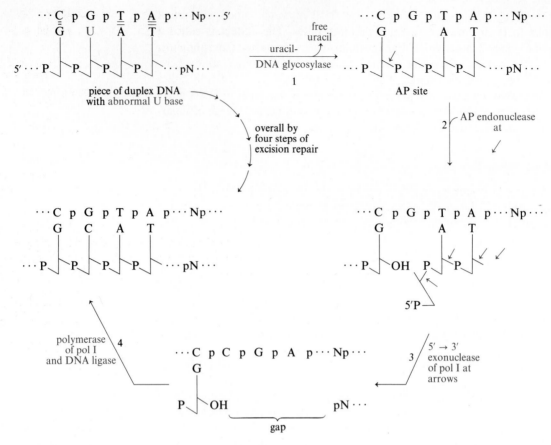

FIGURE 8–19 Correction of a C → U deamination. The abnormal U is identified in color. The complementary strands are represented differently to illustrate the action of repair enzymes in steps 1 through 4.

One of the most novel repair enzymes is O^6-*methylguanine methyltransferase.* The novel aspect of the enzyme is that it has a turnover number of 1, meaning that it does not really turn over at all, but a single molecule functions as a catalyst only once. The chemistry of the reaction involves a direct transfer of the methyl group from an O^6-methylguanine base to an —SH group of the methyltransferase. Apparently, the —SCH$_3$ modification destroys the activity of the enzyme. This type of enzyme—rare in nature—is called a *suicide enzyme.* Whether the active form of the enzyme can be regenerated by some auxiliary process is not yet known.

Two different repair processes for thymine dimer lesions are known. One involves a *photoylase* enzyme catalyzing the direct cleavage of the cyclobutyl bridgehead to yield the thymine monomers. The name photoylase reflects an interesting property of the enzyme: It is *photoactivated,* requiring visible light (340–400 nm) for optimum activity. The second thymine dimer repair process involves excising the lesion with specific endonucleases, followed by gap resynthesis with DNA polymerase (see Figure 8–20). The *E. coli* enzyme, called *uvr* endonuclease, can also act at DNA lesions where the duplex is distorted by the presence of a bulky substituent such as a polycyclic aromatic hydrocarbon

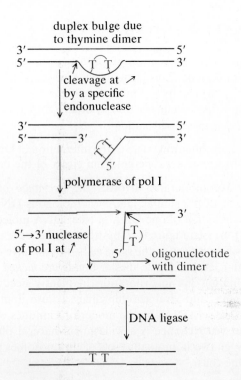

FIGURE 8–20 Repair of thymine dimers.

or a cross-linking reagent. Strand cleavage may occur on either side of the dimer site.

DNA Repair Defects in Disease and Aging

Defects in DNA repair mechanisms are known to be associated with certain disease conditions in humans. A specific case is the rare skin disorder called *xeroderma pigmentosum*. Patients with this disease are extremely sensitive to sunlight and show a high susceptibility to skin cancer. The biochemical defect is the absence of an excision-repair enzyme similar to the *uvr* endonuclease of *E. coli*.

Although there are numerous suggestions explaining the progressive loss of cellular functions with age, there is no consensus explanation. Generally speaking, the aging process and some associated diseases certainly involve increased damage to various sensitive structures with an associated loss in function. Proteins, nucleic acids, and membranes are obvious candidates. One popular theory proposes that much of this damage is caused by increased cellular levels of highly reactive free-radical species that are normally detoxified (see Section 15–5). In addition to the direct damage to DNA, one or more DNA repair enzymes can be damaged, making a bad situation even worse. As yet there is no firm evidence that such an explanation applies to the dramatic aging disease of *progeria*, in which children ten years of age appear to be 60–80 years of age.

8–3   DNA RECOMBINATION

The high-fidelity replication of DNA determines genetic stability by reproducing the same genetic instructions when cells divide. Although new phenotypes are rarely formed by spontaneous mutations, a tremendous degree of genetic variation does occur because of other DNA processes, grouped under the heading of *genetic recombination:*

> Genetic recombination refers to the interaction of DNA molecules to yield DNA molecules different from either of the originals.

Classical recombination involves the *exchange of genetic information* between two genetically similar DNAs (homologous DNAs). Another type involves the *incorporation* of one DNA (a total DNA molecule or just part of it) into another DNA (see Figure 8–21). Although there are various ways in which these processes occur (even in the same cell), they all have one thing in common: At the molecular level of DNA, diester bonds are *nicked and sealed.*

Recombination via incorporation frequently occurs in nature between host chromosomes and viral chromosomes. It also applies to the laboratory construction of novel DNAs by the modern techniques of genetic engineering. Recombination via exchange is a widespread natural phenomenon occurring primarily between two homologous sets of chromosomes provided by male and female germ line cells.

Recombination via Strand Exchange

A detailed explanation for classical exchange recombination between two homologous DNAs was originally suggested by R. Holliday in 1964 from extensive genetic studies with fungi. In 1977 H. Potter and D. Dressler obtained supporting electron microscopic evidence. We will highlight the mechanism of the Holliday model, which explains recombination without any need for DNA replication. Note, however, that other models have also been proposed, one of which does interface recombination with DNA replication.

The Holliday model is described in two stages: (1) the formation of a crossover intermediate between two dsDNAs and (2) the processing of the crossover intermediate to complete the exchange of single-strand segments. Although

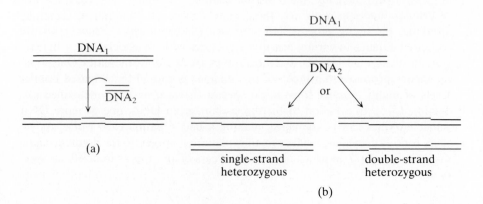

FIGURE 8–21 Recombination of DNA. (a) Foreign piece of DNA incorporated into a host DNA. (b) Exchange of DNA.

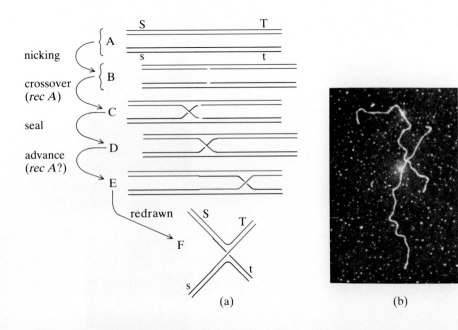

(a) (b)

FIGURE 8–22 Holliday model for classical genetic recombination via strand exchange between homologous chromosomes. (a) Phase 1. The crossover formation and advance to form a chi intermediate is shown here. See Figure 8–23 for phase 2. The ST and st notations identify gene locations on two homologous chromosomes, one in color and one in black. The E and F structures represent the Holliday intermediates for recombination via exchange. In the F structure the two long arms are cis, and the two short arms are cis. *Reminder:* The parallel lines represent double-helix structures. Cis means the same side of the crossover junction. (b) Structure F as seen via electron microscopy.

we do not yet know exactly how these events occur, we can describe some catalytic and noncatalytic proteins that have been implicated from studies with prokaryotic bacteria. Similar proteins undoubtedly operate in eukaryotes. In *E. coli* at least four separate proteins are considered to participate in exchange recombination: the *recA* protein, a protein composed of *recB* and *recC* subunits, a single-strand binding (SSB) protein, and a topoisomerase. There is much evidence implicating *recA* as the major recombinant protein.

The occurrence of crossing-over is envisioned as shown in Figure 8–22. The process may begin with the *recBC* protein sliding along a double helix, causing local unraveling as it moves, with unraveled strands being momentarily stabilized by the binding of SSB. The *recA* protein, having a strong affinity to bind with ssDNA, may also bind at this time and promote additional unwinding. A specific endonuclease or topoisomerase may then cleave a diester bond in one strand of each duplex (A → B). Another function of *recA* may be to direct the location of the single-strand nicks.

The critical crossover step (B → C) involves the reciprocal invasion of a single-stranded segment from one DNA into the duplex structure of the other DNA—a process proposed to be mediated also by *recA* and maybe requiring ATP. One speculation is that the ssDNA/*recA* complex may associate with the other dsDNA to yield a triple-stranded ssDNA/*recA*/dsDNA complex. In this complex the *recA* may promote the unraveling of dsDNA until a region of homology is encountered with the invading ssDNA segment. When homology is found, the invading ssDNA will displace the other ssDNA, which in turn will then invade the other DNA by associating with its complementary region previously vacated by the first invading ssDNA. (Nicking may occur at this time rather than before the participation of *recA*.) This reciprocal invasion completes the crossover, which will be stabilized by the sealing action (C → D) of a topoisomerase or DNA ligase. (Remember that topoisomerases have both nicking and sealing activities, that is, both endonuclease and ligase functions.)

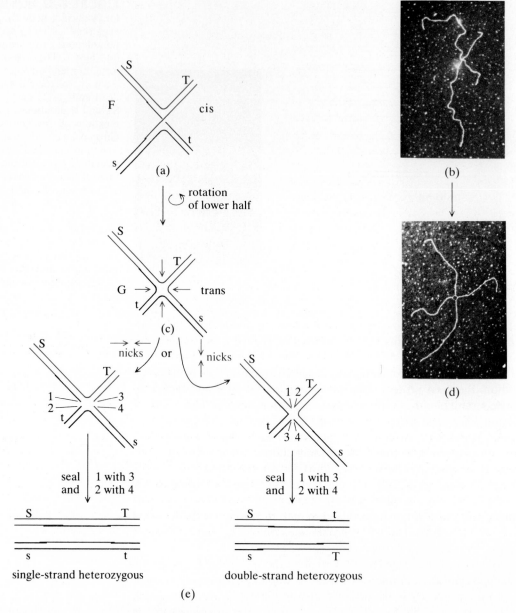

FIGURE 8–23 Phase 2 of the Holliday mechanism.
(a) Structure F from Figure 8–22. The process begins with
a rotation of the cis form of the crossover intermediate
to yield a trans form of the same intermediate, where
trans means opposite sides of the crossover single-stranded
ring. (b) Structure F as seen via electron microscopy.
(c) Structure G, the trans form. Now the two long arms
are opposite each other, as are the two short arms.
(d) Structure G as seen via electron microscopy. (e) Two
possible nicking/sealing patterns yielding recombinant
DNAs.

The strand exchange is continued—that is, the crossover is advanced (D → E)—as a result of the four duplex regions rotating around their axes. This advance may also be mediated by *recA* and ATP and perhaps also by the action of a topoisomerase. The last structure (E) is called a *Holliday intermediate* of recombination, redrawn (E → F) to more clearly illustrate the final steps of strand exchange and to correlate with images in electron micrographs.

The completion of strand exchange is shown in Figure 8–23, beginning with a rotation of the initial crossover structure to produce a structure G consisting of four double-helix arms protruding from a ring of primarily single-stranded DNA. Because of the resemblance to the Greek letter χ, F and G are called chi intermediates. Cleaving bonds opposite each other in the ring and sealing the ends yield the final recombinant DNAs. Researchers are not yet certain whether these operations involve a single nicking-and-sealing topoisomerase or a chi-specific endonuclease followed by DNA ligase. In any case, note that two different nicking/sealing patterns are possible, to yield recombinant DNAs heterozygous in only one strand or heterozygous in both strands.

DNA Transposition

Twenty-five years ago Barbara McClintock collected genetic evidence that she argued could only be explained by the *movement of genes* along a chromosome or from one chromosome to another. Despite the skepticism and rejection of others, she never abandoned her hypothesis; recently, she was proven to be absolutely correct (Nobel Prize, 1983). Genes *do* change position. Indeed, whole clusters of genes can change position (see Figure 8–24). This phenomenon of nonclassical recombination is called *transposition* and is known to occur in both prokaryotes and eukaryotes. Its total involvement in the flow of genetic information is not yet known, but there is evidence that transposition is an important event in determining which genes of a chromosome are expressed and which are not. Thus it may have an important role in cellular differentiation. In bacteria, transposition does explain the rapid evolution of bacterial strains having multiple resistance to antibiotics, a development of great concern to modern medicine.

Movable segments of DNA are called *transposons,* and some transposons have been isolated from plasmids of bacteria. Base-sequence studies have re-

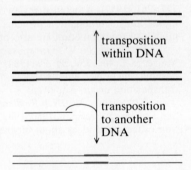

FIGURE 8–24 DNA transposition.

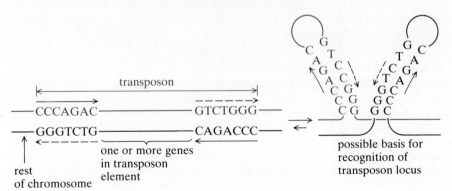

FIGURE 8–25 Transposons, marked by inverted sequences at each end.

vealed an aspect of structure undoubtedly related to the recognition of transposons by proteins involved in the process: The ends of a transposon element have *inverted sequences* (see Figure 8–25, p. 345). This spaced palindromic feature may give rise to a looped-out structure of the transposon locus, providing recognition signals for the nicking-and-sealing enzymes that obviously participate to excise the transposon and reinsert it elsewhere.

L I T E R A T U R E

CAIRNS, J. "The Bacterial Chromosome." *Sci. Am.,* **214**, 36–44 (1966). A description of the original in vivo experiments establishing the simultaneous replication of duplex DNA.

COHEN, S. N., and J. A. SHAPIRO. "Transposable Genetic Elements." *Sci. Am.,* **242**, 40–49 (1980). A description of the discovery and nature of transposons. A general discussion of the significance of movable genetic elements is also included.

DRESSLER, D., and H. POTTER. "Molecular Mechanisms in Genetic Recombination." *Annu. Rev. Biochem.,* **51**, 727–761 (1982). A detailed review article of the Holliday model and recombination proteins in prokaryotes. Other models of recombination are also briefly described.

FEDOROFF, N. V. "Transposable Genetic Elements in Maize." *Sci. Am.,* **250**, 84–99 (1984). A description of transposable genetic elements and some molecular details on how transposition occurs are given.

KORNBERG, A. *DNA Replication.* (and *Supplement*). San Francisco: Freeman, 1980, 1982. An excellent presentation of facts and ideas about biochemical aspects of DNA

biosynthesis. Polymerases, repair, recombination, restriction, and transcription are covered. Suitable for students.

LINDAHL, T. "DNA Repair Enzymes." *Annu. Rev. Biochem.,* **51**, 61–87 (1982). A detailed review article of this interesting class of enzymes.

MESELSON, M., and F. W. STAHL. "The Replication of DNA in *Escherichia coli.*" *Proc. Natl. Acad. Sci., U.S.A.,* **44**, 671–682 (1958). The original paper describing the experimental approach used to establish the semiconservative scheme of DNA replication.

SCOVASSI, A. I., P. PLEVANI, and U. and U. BERTAZZONI. "The Eukaryotic DNA Polymerases." *Trends Biochem. Sci.,* **5**, 335–337 (1980). A brief review of the multiple DNA polymerases in eukaryotes.

SOBELL, H. M. "How Antimycin Binds to DNA." *Sci. Am.,* **231**, 82–92 (1974). The phenomenon of intercalation is described and clearly illustrated.

WARING, M. J. "DNA Modification and Cancer." *Annu. Rev. Biochem.,* **50**, 159–192 (1982). Structures and mode of action of anticancer drugs are reviewed.

E X E R C I S E S

8–1. Assume that the chromosome of *E. coli* replicates according to a *conservative scheme.* Which of the drawings in Figure 8–26, numbered 1 through 6, represent the density gradient ultracentrifugation pattern of the cellular DNA obtained in the Meselson-Stahl experiment from (a) the first generation of cells, that is, after one replication, and (b) the second generation of cells, that is, after two replications? Explain.

8–2. In the Meselson-Stahl experiment proving the *semiconservative scheme* for the replication of DNA, which of the patterns in Figure 8–26 corresponds to the density gradient pattern obtained from the third generation of cells, that is, after three replications? Describe the relative proportions of each type of DNA that are present in the extract.

8–3. In the Cairns experiment (Figure 8–3) the micrograph of replicating DNA was interpreted in terms of unidirectional copy synthesis. For assistance with this question the replicating DNA is shown again in Figure 8–27. (a) Identify where the *ori* location would be located according to the way the image is drawn. Now in responding to (b) and (c), all you need to do is identify how the small curved arrows should be repositioned. (b) How would you modify the diagram to indicate unidirectional synthesis proceeding from the *ori* site being at a different location? (c) How would you modify the diagram to indicate that bidirectional copy synthesis was operating?

8–4. The result of deamination in the presence of HNO_2 was shown in Figure 8–15(a) for the conversion of cytosine

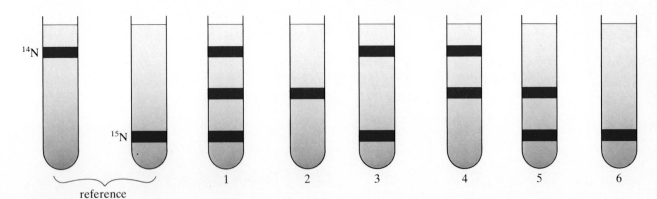

FIGURE 8–26 Setups for Exercise 8–1.

to uracil. Using this diagram as a guide, draw the necessary structures to illustrate the adenine → hypoxanthine and guanine → xanthine modifications. You know the structures of adenine and guanine; the structures of hypoxanthine and xanthine will be evident after you illustrate the deamination.

8–5. Suppose the base hypoxanthine (I) uses cytosine as its complementary base. Identify the substitution mutation that would result for a (+)AT(−) base pair if the A in the (+) strand were deaminated to hypoxanthine (I). Is this mutation a transversion or a transition mutation?

8–6. Suppose the base xanthine (X) uses thymine as its complementary base. Identify the substitution mutation that would result for a (−)CG(+) base pair if the G in the (+) strand were deaminated to xanthine (X). Is this mutation a transversion or a transition mutation?

8–7. Much of the descriptive information in this chapter is embodied in the following terms and symbols. Define or describe each term: semiconservative replication, semidiscontinuous copy synthesis, replication fork, primosome, replisome, SSB proteins, lagging/leading strands, pol III versus pol I roles, frame shift/substitution, transition/transversion, proofreading, AP site, AP endonuclease, intercalation, DNA recombination, DNA transposition, crossover, chi intermediate, thymine dimer, single-strand heterozygous DNA, double-strand heterozygous DNA, inverted sequence elements.

FIGURE 8–27 Replicating DNA for Exercise 8–3.

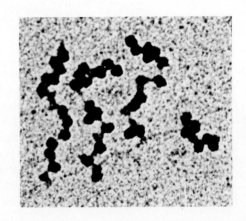

CHAPTER NINE

TRANSLATION: PROTEIN BIOSYNTHESIS

The vast majority of genes in a chromosome code for the biosynthesis of a specific polypeptide (protein). After transcription (and mRNA processing in eukaryotes), the base sequence of mRNA is used to provide instructions for the sequential assembly of amino acids into a polypeptide. The process is called translation because of the conversion in the "language" of primary structure from base sequence into amino acid sequence.

As stated in previous chapters, the instructions in mRNA are provided by sets of three successive bases called triplet *codons*. The coding assignments of all 64 codons (see Note 9–1) have been known for about twenty years. During this time research has continued to unravel the detailed mechanism of the translation process. Although much has been learned, some aspects are at best only slightly understood.

Because the translation process is so complex, the process is customarily considered as occurring in four steps:

1. *Activation* and *selection* of amino acids.
2. *Initiation* of polypeptide chain formation.
3. *Elongation* of the polypeptide chain.
4. *Termination* of polypeptide chain formation.

The assembly steps (2, 3, and 4) actually occur on the surface of *ribosomes,* with some individual ribosomal proteins performing specific functions. In addition to messenger RNA and ribosomes, other participants include *transfer RNA,* various *noncatalytic proteins,* and, as sources of energy, *ATP* and *GTP*.

One or more of the following steps also apply to a newly assembled polypeptide (protein):

5. *Chemical modifications* of specific amino acid R groups.
6. *Chain association* to yield oligomeric structures.
7. *Prosthetic groups* that become bound.
8. *Concentrated packaging* of proteins into lysosomes or into granules for extracellular secretion.
9. *Assimilation* into the matrix of a biomembrane.
10. *Transport across membranes,* accounting for the entry of a protein into another subcellular compartment (for example, into the mitochondrion).

Before examining some of these events, we will focus on the structure of ribosomes, extending the preliminary description given in Chapter 6.

9–1 RIBOSOMES

General Architecture

Only in the past ten to fifteen years have intensive studies been directed at resolving the intricate structure and architecture of the ribosome—a complex ribonucleoprotein aggregate (see Table 9–1). The smaller prokaryote ribosomes

TABLE 9–1 Summary of ribosome components

SUBUNIT	SIZE (daltons)[a]	SINGLE COPIES OF RNAs PRESENT		DIFFERENT PROTEINS PRESENT		
		SPECIES	NUCLEOTIDE LENGTH	SINGLE COPIES	MULTIPLE COPIES	TOTAL
Prokaryotes	(70S; *E. coli*)					
30S	900,000[b]	16S	1542	21	None	21
50S	1,600,000[b]	5S	120	30	1 present in 4 copies	34
		23S	2904		(Total of 52 different proteins)	
Eukaryotes	(80S; rat liver)					
40S	1,400,000[c]	18S	1874	About 30–35 proteins[e]		
60S	2,900,000[d]	5S	≈120	About 45–50 proteins[e]		
		5.8S	≈160			
		28S	4718			

[a] The dalton is a unit of mass nearly equivalent to the mass of a single hydrogen atom; the terms *dalton* and *molecular weight* are used interchangeably.

[b] About 66% of the mass of each subunit is contributed by RNA.

[c] About 40% is contributed by RNA.

[d] About 60% is contributed by RNA.

[e] The precise population is not yet known.

have been studied longer and are better understood. In fact, all 55 components of the prokaryote ribosome have been isolated. All of the rRNAs have been sequenced, and hypothetical secondary structures have been suggested (see Figure 6–64). Most of the ribosomal proteins have also been sequenced. Scenarios for the in vivo assembly of these components into a mature ribosome have also been proposed.

More recently, various techniques—such as electron microscopy, probing with antibodies directed against individual ribosomal proteins, X-ray and neutron scattering, and treatment with chemical reagents that form cross-links between neighbor ribosomal proteins and between ribosomal proteins and RNA—have been used to evaluate the overall ribosome shape and specific topographical features of the ribosome surface. The many details of the progress made in advancing the knowledge of ribosomes are available in various review articles, such as those by Wittmann and by Brimacombe, Stoffler, and Wittmann cited in the Literature. The article by Lake is a good introductory source for some of these details.

Current evidence suggests that there are four subtly different ribosome structures occurring in nature: the prokaryote ribosome, the eukaryote ribosome, and two structural types in the archaebacteria family. Sketches of one proposal for the general shapes of the subunits of the prokaryote ribosome are shown in Figure 9–1. The 30S subunit apppears to have two domains (called a "head" and a "body") with an intermediate "collar" or "neck." The lower body has a distinct protrusion at the collar region. The 50S subunit appears to have

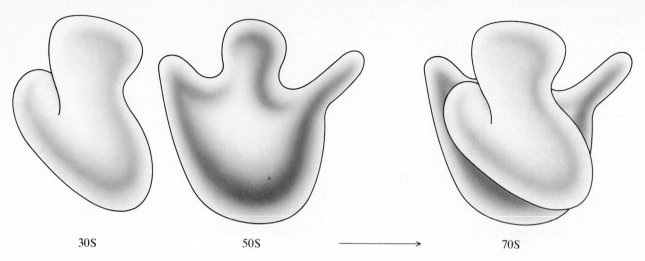

30S   50S   ⟶   70S

FIGURE 9–1 General shapes of prokaryote ribosome subunits and intact ribosome.

a hemispherical-shaped body with three distinct protrusions that differ in size and shape. In the intact 70S complex the 30S subunit is partially cradled in the shallow pocket of the 50S subunit.

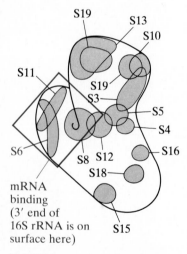

FIGURE 9–2 Topographical locations of some of the proteins in the 30S subunit when viewed as drawn. The domain boxed in a bold line is implicated as the mRNA-binding site. In the notation S19, the S is for small subunit; the number designates the position of a protein after gel electrophoresis.

Locations of Proteins and RNAs

All available evidence suggests that each subunit is a tightly packed complex, with the ribosomal RNAs being located primarily (though not exclusively) in the core of each subunit. The surface is primarily protein in character, with individual ribosomal proteins having various degrees of surface exposure. The RNA-protein architecture is anything but a mishmash of jumbled components. Rather, all components occupy preferred "native" positions, with certain protein-protein and protein-RNA associations responsible for forming specialized functional domains on the surface of each subunit. These domains serve to bind mRNA, to bind tRNAs, to bind auxiliary proteins that operate in translation, to catalyze the formation of peptide bonds, to effect the translocation of bound tRNAs from one binding site to another, and perhaps to recognize receptor sites in the membrane of the endoplasmic reticulum.

Progress continues to be made in locating positions occupied by individual proteins in each subunit. A partial topographical map for the 30S subunit is illustrated in Figure 9–2. The drawing also attempts to convey the message that while some of the ribosomal proteins are compact globular proteins, some are very elongated, and some are intermediate in shape. The drawing also identifies the proposed location of an important functional domain of 30S—a binding site for mRNA. Note that a segment of the 3′ terminus of 16S rRNA is proposed to be exposed at the surface in this region.

Phosphorylation of Ribosomal Proteins in Eukaryotes: Proposed Regulation of Ribosome Function

There is mounting evidence that the activity of cellular ribosomes (particularly in eukaryotic cells) can change in response to various signals such as viral infections, a few hormones, and various modifications in growth conditions of cells in culture. A major strategy of directly regulating eukaryote ribosome activity appears to be *phosphorylation* of one or more ribosomal proteins, with increased activity closely associated with the occurrence of phosphorylation. The major target of phosphorylation appears to be the S6 protein in the 40S subunit, suggesting a crucial role of S6 in stimulating ribosomes to operate in protein synthesis. Although not yet conclusive, some evidence implicates S6 phosphorylation with enhancing the binding of 40S to mRNA.

The state of S6 phosphorylation in ribosome-stimulated cells has been shown to involve not just one but several serine and threonine residues in the S6 protein. At least ten serine —CH_2OH and two threonine —$CH(OH)$— groups get phosphorylated and apparently in a specific order. These phosphorylations occur on S6 in an intact ribosome and involve the catalytic participation of *protein kinases* (see Figure 9–3). Protein phosphatases would operate in the dephosphorylation of S6.

The reason for presenting this information at this time is to emphasize that ribosomes are not static, stonelike structures that provide only a fixed binding surface for the events of polypeptide biosynthesis. Rather, most evidence points to a more dynamic structure capable of oscillating between different activity states owing to subtle conformational rearrangements and realignments of the component parts. Local conformational changes of one or more key ribosomal proteins may set in motion new protein-RNA and protein-protein interactions, restructuring one or more functional domains in the respective subunits. Other than the suggested involvement of S6 phosphorylation as a possible trigger of such changes, as yet very little is known about this aspect of ribosomes.

Later in this chapter (Section 9–6), another aspect of regulating translation in eukaryotes is described, involving an auxiliary nonribosomal protein, which may be related in part to the antiviral activity of interferon. We will also describe a newly discovered mode of regulating translation in prokaryotes.

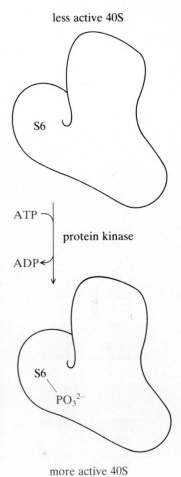

less active 40S

S6

ATP

protein kinase

ADP

S6

PO_3^{2-}

more active 40S

FIGURE 9–3 Phosphorylation of the 40S S6 protein. It is associated with regulating ribosome function in eukaryotes.

9–2 ACTIVATION AND SELECTION

Aminoacyl Transfer RNA Adducts

In all types of cells the first event of translation is the ATP-dependent conversion of each amino acid to an *aminoacyl-tRNA species*. This conversion accomplishes two things: (1) The reactivity of the amino acid is enchanced for peptide bond formation (*activation*), and (2) the amino acid is matched with a specific transfer RNA (*selection*) (see Note 9–2). The enzyme, an *aminoacyl-tRNA synthetase,* operates in two steps. In forming an aminoacyl-AMP species (step 1), ATP energy is expended to form a mixed anhydride intermediate wherein the C of the C=O now has a greatly increased reactivity. In step 2 a terminal —OH of the 3′ terminus of transfer RNA attacks this carbonyl (displacing AMP) to yield the aminoacyl-tRNA adduct. Since this adduct is an ester, the increased reactivity of the aminoacyl C=O group is conserved.

NOTE 9–2
A common way to identify a specific tRNA carrying the genetic correspondence for a specific amino acid is to identify the correspondence with a superscript of the amino acid abbreviation. For example, the tRNA for alanine is designated as tRNA[ala].

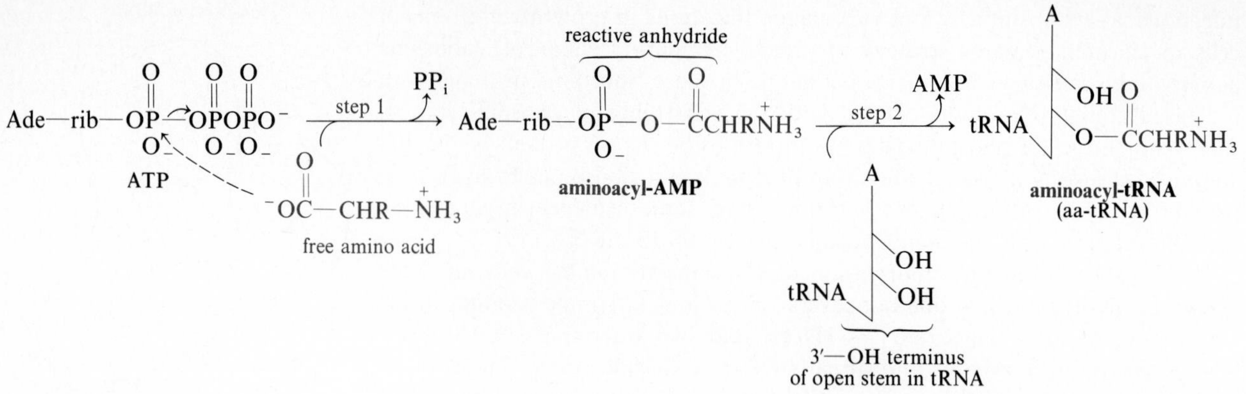

FIGURE 9–4 Activation and selection of amino acids.

An interesting variation in the family of aminoacyl-tRNA synthetases is that some enzymes use the 3′—OH of tRNA (shown in Figure 9–4), others use the 2′—OH, and others use either the 2′—OH or 3′—OH for attaching the aminoacyl moiety.

The synthetase enzyme is highly specific in its recognition of both amino acid and transfer RNA substrates.

Indeed, living cells contain at least 20 different synthetases, one for each of the 20 amino acids. The number of different tRNA molecules per cell is even greater, since many of the amino acids can be matched with at least two different tRNA molecules and some with three. A minimum of 32 different tRNAs can be predicted (see the wobble hypothesis discussed in Section 9–7).

The basis for this enzyme specificity is, of course, the highly ordered recognition of the active site for a particular combination of amino acid and transfer RNA. A logical question in this regard is, What are the features of substrate structure recognized for binding by the synthetase? With the amino acid substrates each has its own unique structure conferred by the side chain. With transfer RNA the issue is much more complex. (See Figure 9–5 for a review of transfer RNA structure.) Indeed, despite years of investigation with many different approaches, a clear picture of synthetase/tRNA recognition is not yet available. However, these studies have established that the one unique element of structure in each tRNA, namely, the anticodon loop, is not a critical recognition site for all synthetases—only for some.

The current proposal is that an aminoacyl-tRNA synthetase makes several contacts with a tRNA, primarily along the inner bend region of the L-shaped structure (see Figure 9–6). Contact of the active site with the 3′ terminus of the open stem is, of course, an essential contact region. Moreover, two or more residues in the open-stem region appear to confer some of the tRNA specificity recognized by each synthetase. Depending on the individual enzyme, additional contacts extend either partially or totally to the anticodon loop. Individual contacts in these extended regions that might contribute to the specificity of tRNA/synthetase recognition are not yet understood.

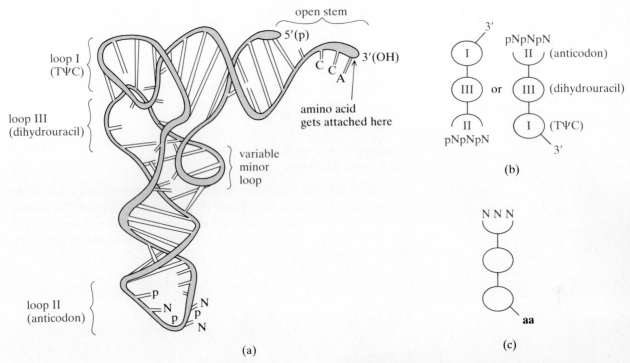

(a)

(b)

(c)

FIGURE 9–5 Review of transfer RNA structure.
(a) Major domains of structure/function. Each tRNA has
a unique pNpNpN anticodon triplet. (b) Shorthand
representations. In later diagrams depicting events of
translation, tRNA structure will be symbolized in shorthand
form as explained. (c) Symbolism for an aminoacyl-tRNA
adduct; **aa** identifies an attached aminoacyl group.

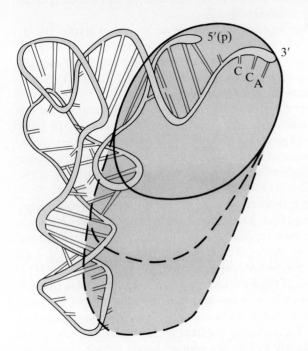

FIGURE 9–6 Specific binding of tRNA/aminoacyl
synthetase. Different synthetases (color shading) may make
different degrees of contact with tRNA; all contact the
open stem.

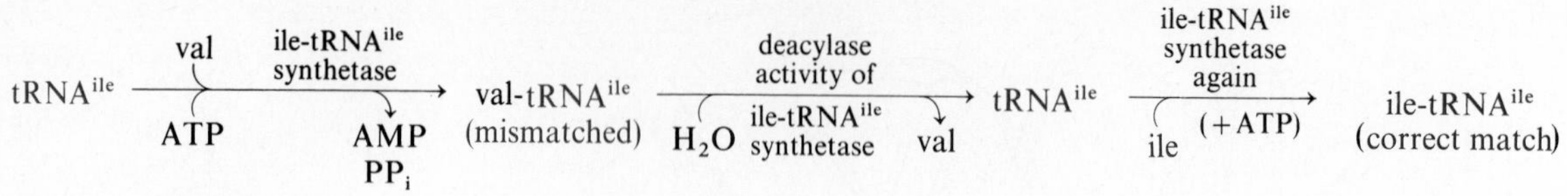

$$\text{tRNA}^{\text{ile}} \xrightarrow[\text{ATP}]{\overset{\text{val}}{} \quad \overset{\text{ile-tRNA}^{\text{ile}}}{\text{synthetase}}} \text{AMP} \atop \text{PP}_i} \rightarrow \underset{\text{(mismatched)}}{\text{val-tRNA}^{\text{ile}}} \xrightarrow[\text{H}_2\text{O} \quad \overset{\text{ile-tRNA}^{\text{ile}}}{\text{synthetase}} \quad \text{val}]{\overset{\text{deacylase}}{\text{activity of}}} \text{tRNA}^{\text{ile}} \xrightarrow[\text{ile} \quad (+\text{ATP})]{\overset{\text{ile-tRNA}^{\text{ile}}}{\text{synthetase}} \atop \text{again}} \underset{\text{(correct match)}}{\text{ile-tRNA}^{\text{ile}}}$$

Figure 9–7 Self-correcting action of aminoacyl-tRNA synthetases.

Aminoacyl-tRNA Synthetases Can Correct Mistakes

The critical matching of an amino acid with the right tRNA is not left to a single event. If a mismatch does occur (for example, val-tRNA^{ile} is formed rather than ile-tRNA^{ile}), the synthetase can also function as an aminoacyl-tRNA *deacylase,* removing the mismatched amino acid and then operating again as a synthetase. See Figure 9–7.

9–3   INITIATION OF TRANSLATION

Auxiliary Proteins As Initiation Factors

In both prokaryotes and eukaryotes all phases of the translation process depend on the participation of nonribosomal proteins. Proteins operating only in specific steps of the initiation phase are called *initiation factors,* designated IF in prokaryotes and eIF in eukaryotes.

The three initiation factors essential for translation in prokaryotes (IF–1, IF–2, and IF–3) have all been purified and extensively characterized in terms of protein structure. Much has also been learned about their individual and collective functions. The situation is more complex and less resolved in eukaryotes, which contain about a dozen initiation factors.

Initiation factors exert their effect by binding to specific regions on the ribosome surface or to specific domains on aminoacyl-tRNA adducts. In addition, a bound factor may itself provide a new binding location for another factor. A more thorough examination of these effects will be given shortly (see Figure 9–11). All of these highly specialized protein/protein and protein/RNA interactions guide a complex sequence of events resulting in the formation of an *initiation complex*—an aggregate composed of mRNA, an intact ribosome, and the initial aminoacyl-tRNA adduct. A simplified sketch of this complex is shown in Figure 9–8, with each subunit represented only as a rectangle. The existence of an mRNA-binding site in 30S (or 40S) is implied. The representation does draw specific attention to three other functional domains in 50S (or 60S): (1) a *P site* involved in the binding of the TΨC loop region of the initiator fmet-tRNA adduct (see the next section), (2) an *A site* involved in binding the TΨC loop of later incoming aminoacyl-tRNA adducts, and (3) a *peptidyl transferase* domain, composed of a surface-oriented segment of 23S RNA and ribosomal protein(s), that acts catalytically in forming peptide bonds during elongation.

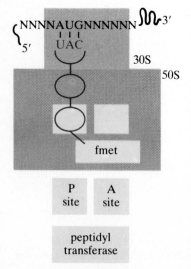

FIGURE 9–8 Initiation complex of 70S|mRNA|fmet-tRNA$^\text{f}$.

AUG Codon in mRNA for Specifying the Start of Translation

With the exception of some novel start signals recently discovered in some genes of mitochondrial DNA and chloroplast DNA, the DNA sense sequence of ATG signals a gene start, coding for the first amino acid at the N terminus. When the sense template of ATG is transcribed into mRNA, the *AUG codon* is formed—a codon specifying *methionine*. This initiation codon is located near the 5' end of the mRNA, preceded by a nontranslated leader segment (see Figure 9–9). Thus a methionine-tRNA adduct with a *UAC anticodon* in the tRNA is required for complementary recognition.

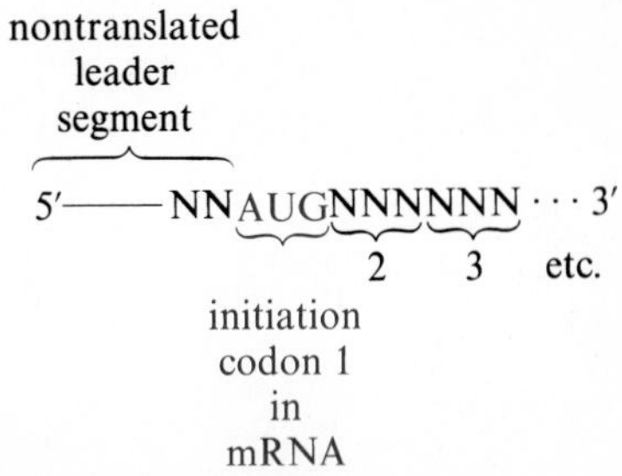

This AUG initiation signal was first discovered in the prokaryote *E. coli,* also revealing another interesting aspect of initiation: The N terminal amino acid of a newly synthesized polypeptide chain is always *N-formylmethionine* (fmet). Subsequent discoveries that *E. coli* contains two distinct tRNAs for methionine and also a specific *formyltransferase* enzyme provided the basis for understanding the origin and the role of N-formylmethionine.

Although the two tRNAs for methionine (designated as tRNA^{fmet} and tRNA^{met}) both form met-tRNA adducts in the presence of the same synthetase, only the met-tRNA^{fmet} adduct can serve as substrate for the reaction catalyzed by the formyltransferase (see Figure 9–10). This reaction involves the transfer of a one-carbon formyl group

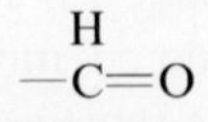

$$
\begin{array}{c}
\text{H} \\
| \\
-\text{C}=\text{O}
\end{array}
$$

from formyl-tetrahydrofolic acid (formyl-FH$_4$) to the alpha amino group of methionine in met-tRNA^{fmet}. (The formation of various one-carbon adducts of tetrahydrofolic acid and their crucial role in several areas of cellular metabolism will be examined in Chapter 18.)

Thereafter the fmet-tRNA^{fmet} adduct, having a UAC anticodon, is used only in the initiation phase to recognize the AUG initiation codon near the 5' end of the mRNA. The nonformylated met-tRNA^{met}, also having the same UAC anticodon, is used only in the elongation phase to recognize internal AUG codons of the mRNA-coding frame.

A similar situation operates in eukaryotes. Two different tRNAs for methionine exist, and only one met-tRNA adduct recognizes the start AUG codon; the other recognizes only internal AUG codons. However, there is a distinct

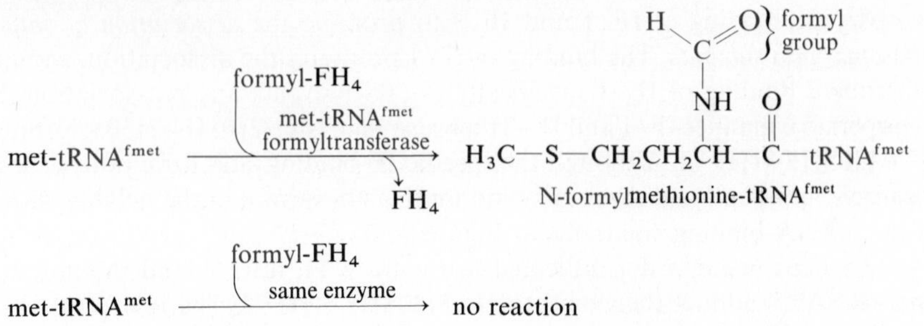

nontranslated
leader
segment

5'———NNAUGNNNNNN···3'
                2   3   etc.

initiation
codon 1
in
mRNA

FIGURE 9–9 AUG codon, the start of the transcript message. In the sense strand of the DNA gene duplex, this start signal is coded as ATG.

met-tRNA^{fmet} $\xrightarrow[\text{FH}_4]{\substack{\text{formyl-FH}_4 \\ \text{met-tRNA}^{fmet} \\ \text{formyltransferase}}}$ $\text{H}_3\text{C}-\text{S}-\text{CH}_2\text{CH}_2\text{CH}-\overset{\text{O}}{\overset{||}{\text{C}}}-\text{tRNA}^{fmet}$

N-formylmethionine-tRNA^{fmet}

met-tRNA^{met} $\xrightarrow[]{\substack{\text{formyl-FH}_4 \\ \text{same enzyme}}}$ no reaction

FIGURE 9–10 Selective formylation.

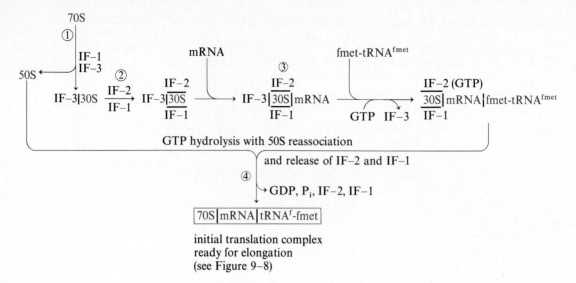

FIGURE 9–11 Summary of the events proposed for forming the initiation complex of translation in prokaryotes. The | and — symbols identify numerous protein/protein- and protein/RNA-binding interactions. See text for descriptions of steps 1 through 4.

difference: Eukaryotes do not use formylation to differentiate the two adducts. The individuality of each met-tRNA (the initiator met-tRNA^{imet} and the internal met-tRNA^{met}) is conferred by structural differences in the two tRNAs. Perhaps the most significant distinction is that the tRNA^{imet} species lacks an element of structure that is considered universal of all tRNAs. Specifically, the universal TΨC loop sequence is replaced by AUC in tRNA^{imet}.

Formation of the Initiation Complex in Prokaryotes

The statement

$$70S + \text{met-tRNA}^{fmet} + \text{mRNA} \xrightarrow[\substack{\text{GTP} \quad \text{GDP, P}_i}]{\text{IF–1, IF–2, IF–3}} [\text{fmet-tRNA}^{fmet}_{UAC}|_{AUG}\text{mRNA}]70S \quad \text{ternary initiation complex}$$

summarizes the initial step in translation: an energy-(GTP) dependent formation of a ternary complex wherein the AUG codon of mRNA is recognized by the UAC anticodon of fmet-tRNA^{fmet} on the ribosome surface. The steps in this assembly and the participation of IF proteins are currently understood as depicted in Figure 9–11.

There is conclusive evidence that the first step, occurring soon after a 70S ribosome is released from a previous participation in the translation process, involves the binding of IF–1 and IF–3 to promote the *dissociation of intact ribosomes into subunits*. The binding of IF–1 promotes the dissociation, and the subsequent binding of IF–3 specifically to 30S prevents any reassociation. In a cooperative manner IF–1 and IF–2 then also bind (step 2) to IF–3|30S, forming IF–3|30S|IF–1|IF–2. Although the specific IF-binding sites have not yet been localized, apparently all three IFs bind on the 30S surface in the neighborhood of the mRNA-binding site (refer to Figure 9–2).

Perhaps in the order indicated in Figure 9–11, mRNA and the initiator fmet-tRNA^{fmet} adduct then bind to IF–3|30S|IF–1|IF–2. The release of IF–3 and the binding of GTP are also proposed to occur at this time. In vitro studies with highly purified IF–2 have established that the presence of IF–2 is essential for the binding of fmet-tRNA^{fmet}. Finally, in step 4, a step dependent on the hydrolysis of GTP, the 50S reassociates, and IF–1 and IF–2 are released to yield the completed initiation complex. The source of the GTPase activity for GTP hydrolysis appears to be IF–2 (see Note 9–3). The exact sequence of events and the changes associated with GTP hydrolysis are not yet certain.

In this initiation process a crucial step is that the 5′ end of mRNA bind to 30S for a precise delivery of the AUG start codon to the initiator fmet-tRNA^{fmet}. In prokaryotes the signal directing this pattern of mRNA binding is provided by a short AG-rich (purine-rich) sequence in the nontranslated leader segment upstream from the AUG start codon. A proposed consensus sequence is (5′) ⋯ <u>AGGAGGU</u> ⋯ (3′); see Figure 9–12. It is sometimes called the *Shine-Dalgarno (SD) sequence,* after those who first discovered it.

What is so special about this sequence? It so happens that the 3′ end of the 16S rRNA, located at the 30S surface in the region of the mRNA-binding site, has a pyrimidine-rich sequence complementary to the SD sequence: (3′)AU<u>UCCUCCACUAG</u> ⋯ (5′). Thus the recognition of the 5′ end of prokaryote mRNA is proposed to involve base pairing between the mRNA leader and the 3′ terminus of 16S rRNA.

Formation of the Initiation Complex in Eukaryotes; Special Importance of eIF–2

Some of the known details of the initiation process in eukaryotes are shown in Figure 9–13. A comparison with Figure 9–11 will reveal an overall similarity to the prokaryote system, with eIF–2, eIF–3, and eIF–5 similar in action to the prokaryote IF proteins.

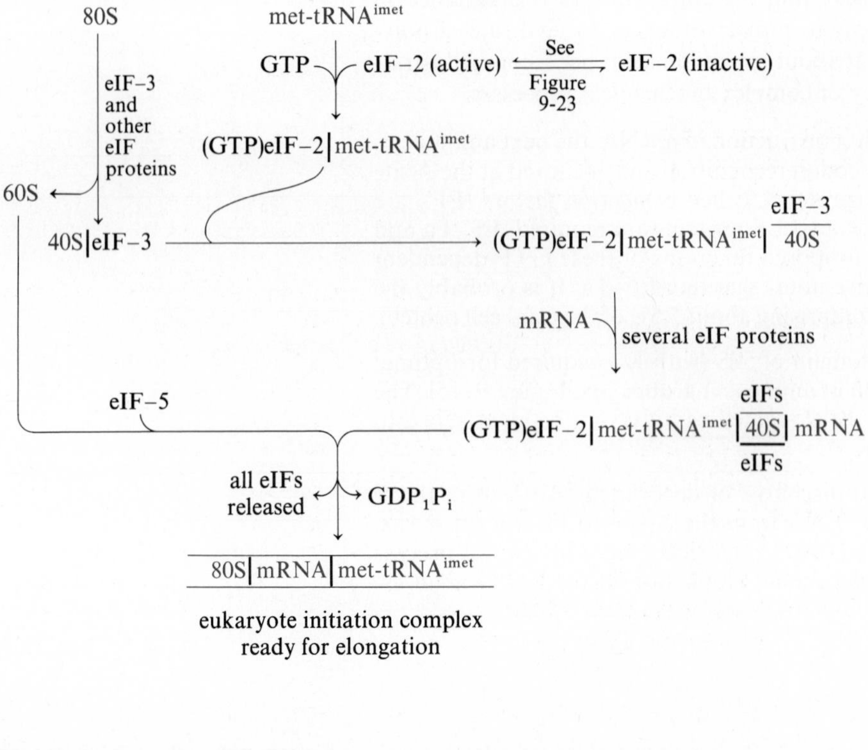

NOTE 9–3
Protein IF–2 is an example of a G protein, previously described in Section 7–3. The binding of GTP and its subsequent hydrolysis to GDP + P$_i$ are associated with conformational changes in the G protein.

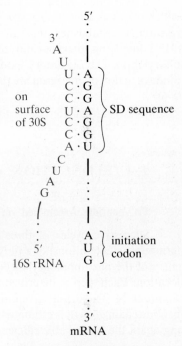

FIGURE 9–12 Initial binding of mRNA, guided by the 5′ leader region of mRNA and the 3′ terminus of 16S rRNA.

FIGURE 9–13 Partial summary of the proposed events for forming the initiation complex of translation in eukaryotes. The regulatory control involving eIF–2 is described in Section 9–6.

80S   met-tRNA^{imet}

GTP — eIF–2 (active) ⇌ (See Figure 9–23) eIF–2 (inactive)

eIF–3 and other eIF proteins

(GTP)eIF–2|met-tRNA^{imet}

60S ←

40S|eIF–3 ⟶ (GTP)eIF–2|met-tRNA^{imet}| eIF–3 / 40S

mRNA — several eIF proteins

eIF–5

(GTP)eIF–2|met-tRNA^{imet}| eIFs / 40S / eIFs | mRNA

all eIFs released ← GDP$_1$P$_i$

80S|mRNA|met-tRNA^{imet}

eukaryote initiation complex
ready for elongation

Significant differences between the eukaryote and prokaryote systems already mentioned include the large number of eIF proteins that participate and the nonformylation of the initiator met-tRNA^{imet} adduct. Another difference involves the basis for finding the AUG start codon at the 5′ end of mRNA. Sequence studies of leader segments in eukaryote mRNA have not revealed a consensus-type, purine-rich, Shine-Dalgarno signal as in prokaryotes. One proposal is that perhaps the $^{7m}GpppX^m$ cap at the 5′ end of eukaryote mRNA may provide the necessary signal. This proposed role of the 5′ cap is subtly different from another well-established role of capping, namely, enhancing the efficiency of translation by stabilizing the mRNA/40S association.

Still another difference centers on the biochemistry of the eIF–2 protein, which serves as a *regulatory protein* for translation in eukaryotes. In 1976–1977 researchers established that an active $\rightleftharpoons$ inactive interconversion applies to eIF–2 involving enzyme-catalyzed modifications of phosphorylation/dephosphorylation. The regulatory biochemistry of eIF–2 and a proposed link to the mode of action of interferon are described more fully later in the chapter (Section 9–6).

9–4   ELONGATION AND TERMINATION

Polypeptides Assembled via a Repetitive Process

Successive cycles of three events result in the progressive growth (N terminus → C terminus) of a polypeptide chain from the initiation complex: (1) entry of the next aminoacyl-tRNA, (2) peptide bond formation, and (3) translocation. Each step is described next, and the entire process is diagramed in Figure 9–14. The details given apply to prokaryotes. The in vivo rate of polypeptide chain growth is estimated at about ten amino acids per second, illustrating again the remarkable efficiency of complex biochemical processes.

1. According to the next codon instruction of mRNA, the next aminoacyl-tRNA adduct is positioned via anticodon recognition and anchored at the A site on 50S. Both GTP and auxiliary proteins, called *elongation factors* (EF), are required in this step. In prokaryotes two EF proteins are required, EF–Tu and EF–Ts. (We will not examine the proposed functions of these GTP-dependent proteins.) One interesting descriptive note regarding EF–Tu: It is probably the most abundant protein in *E. coli*, comprising about 5.5% of the total cell protein.

2. The *peptidyl transferase* domain of 50S (with K^+ required for optimal activity) catalyzes the condensation of aminoacyl groups (see Figure 9–15). The pattern of condensation results in the elongated peptidyl unit being attached to the tRNA bound at the A site.

3. The final step involves (a) discharge of deacylated tRNA from the P site, (b) movement of the peptidyl-tRNA from the A site to the vacant P site, and (c) movement of the mRNA relative to the ribosome. The overall process is called *translocation*. The GTP-dependent elongation factor EF–G is an essential participant.

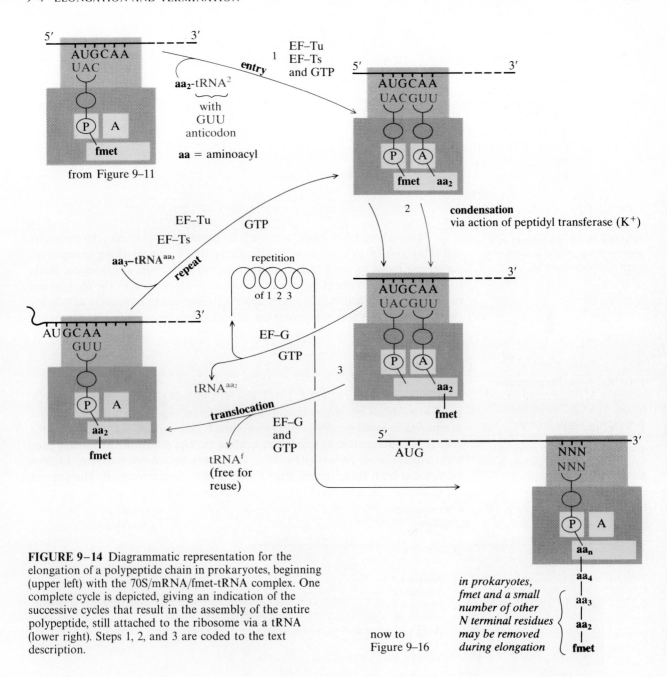

FIGURE 9–14 Diagrammatic representation for the elongation of a polypeptide chain in prokaryotes, beginning (upper left) with the 70S/mRNA/fmet-tRNA complex. One complete cycle is depicted, giving an indication of the successive cycles that result in the assembly of the entire polypeptide, still attached to the ribosome via a tRNA (lower right). Steps 1, 2, and 3 are coded to the text description.

now to Figure 9–16

The process will recycle (steps 1, 2, and 3 again) according to instructions provided by the next codon. And the recycling will continue until instructions to terminate are encountered.

Although there are several differences involving eukaryote-specific EF proteins, the same general strategy applies to the elongation phase in eukaryotes.

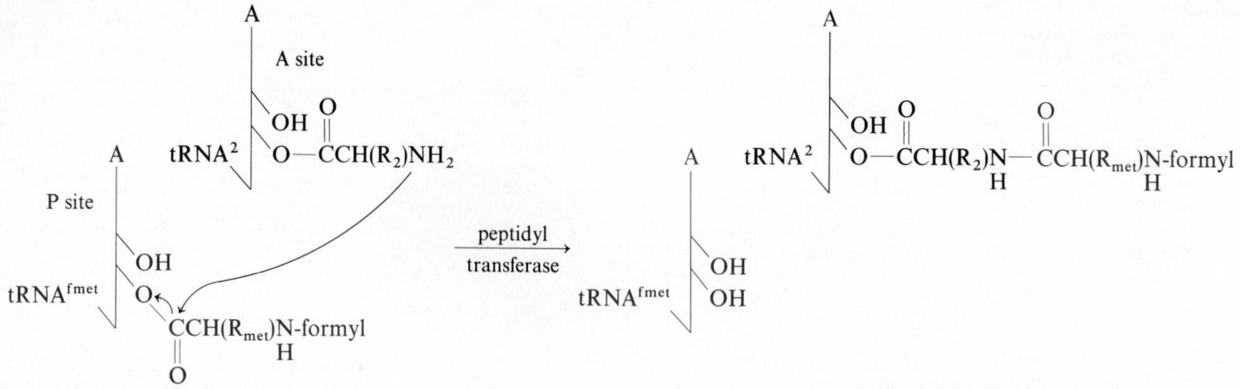

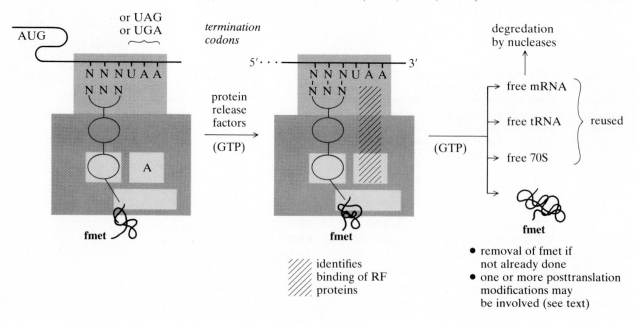

FIGURE 9–15 Formation of a peptide bond on the surface of a large ribosome subunit.

Various *antibiotics inhibit* protein biosynthesis. For example, *chloramphenicol* (also known as chloromycetin) binds at the A site, blocking elongation; *puromycin* binds to the P site, thus blocking translocation; *streptomycin* binds to a ribosomal protein of the small subunit and interferes with the recognition of codon and anticodon. The many specific differences in molecular details between prokaryotes and eukaryotes are indicated by the fact that eukaryotes are insensitive to chloramphenicol inhibition.

One of Three Codons for Stopping the Process of Elongation

FIGURE 9–16 Diagrammatic summary of the termination and release of a fully assembled polypeptide chain from the 70S/mRNA/polypeptidyl-tRNA complex. The precise mode of action of the release factors is not known.

The appearance of UAA, UAG, or UGA in the codon sequence of mRNA gives instructions for termination of chain extension. (Again, there are some novel exceptions in mitochondria and chloroplasts.) Recognition of these *terminator codons* is mediated not by anticodon-bearing tRNAs but rather by specific proteins called *release factors* (RF). Also, GTP may be required in vivo. In prokaryotes at least three factors (RF–1, RF–2, RF–3) are involved. The current

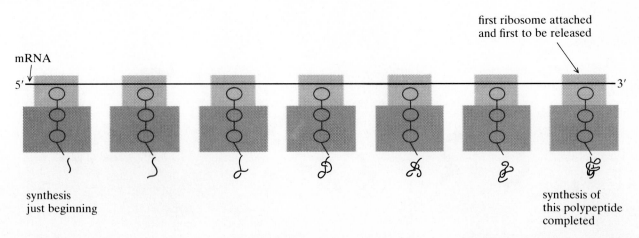

mRNA

first ribosome attached
and first to be released

5′ ————————————————————————————————— 3′

synthesis
just beginning

synthesis of
this polypeptide
completed

FIGURE 9–17 Messenger RNA/
polysome complex depicting seven
copies of the same polypeptide in
different stages of development on
each of seven ribosomes.

view (Figure 9–16) is that with assistance from RF–3, either RF–1 or RF–2 specifically recognizes the termination codon, and the interaction binds RFs to the ribosome near the vacant A site. The binding of RF proteins is also involved in cleaving the completed polypeptide chain from the tRNA at the P site and in promoting the release of free tRNA and mRNA from the ribosome.

Simultaneous Synthesis of Polypeptides on Polysomes

So that we could focus attention on details, the preceding descriptions were limited to the growth of a single polypeptide on a single ribosome attached to a single mRNA. However, in vivo the process is much more efficient, with a single mRNA being translated simultaneously on a cluster of ribosomes, called *polysomes.* Initiation, elongation, and termination apply to a growing polypeptide chain on each ribosome in the cluster (see Figure 9–17). Obviously, the advantage of this arrangement is that several copies of the polypeptide can be made before the mRNA undergoes degradation by the action of cellular nucleases.

The size of polysomal complexes varies widely and is generally a function of the length of the mRNA molecule. Extremely large mRNAs may be complexed to as many as 50–100 ribosomes. Generally, however, polysomal clusters contain 3–20 ribosomes.

Electron micrographs such as those in Figure 9–18 provide visual evidence. A polysomal cluster of about 12 ribosomes (heavy black bodies) in association with mRNA (the thin thread that appears to contact the ribosomes) is clearly visible at the extreme right. Progressively smaller polysome clusters, also in association with mRNA, are evident toward the left. The long thread running across the field is DNA.

Where a ribosome is in close proximity to the DNA, note the presence of a smaller granule that is less electron dense than the ribosome. Presumably, this granule represents a molecule of DNA-directed RNA polymerase attached to its DNA template. Thus, the entire image is a view of both *transcription and translation,* indicating that the two processes are *tightly coupled in both space and time* (at least in prokaryotes). In other words, soon after the tran-

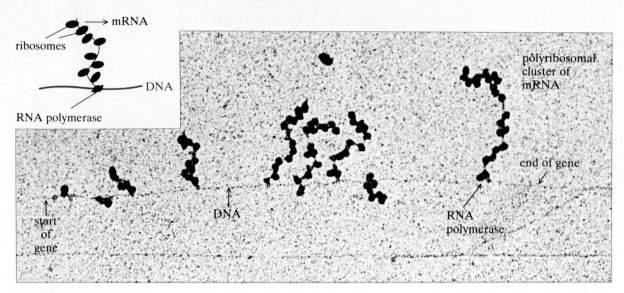

FIGURE 9–18 Electron microscopic evidence of genes in action. This micrograph is interpreted as depicting DNA, RNA polymerase, messenger RNA, and ribosomes caught in the acts of transcription and translation. Polypeptide chains attached to ribosomes are not visible in electron microscopy. The specimen was obtained from a cell-free extract (produced by osmotic shock) of rapidly dividing *E. coli.* (*Source:* Reproduced, with permission, from "Electron Microscopy of Genetic Activity" by B. A. Hamkalo and O. L. Miller, *Annual Review of Biochemistry,* Volume 42. Copyright © 1973 by Annual Reviews, Inc. All rights reserved. Photograph provided by Dr. O. L. Miller.)

scription of messenger RNA is begun, its translation begins. Then as the size of the mRNA increases with continued transcription, more ribosomes become associated with it. Note how the polysome cluster gets larger as transcription proceeds along the gene from left to right.

9–5   ADDITIONAL PROCESSING OF POLYPEPTIDES

A Survey

Nearly all polypeptide chains undergo one type of processing, namely, the removal of the fmet or met residue from the N terminus. This exopeptidase-catalyzed process may occur during elongation or after complete assembly. As described later, sometimes additional N terminal residues are removed, particularly for polypeptides that move across a membrane or are assimilated into a membrane.

For many proteins this N terminal trimming completes the formation of the mature bioactive form: It is a single polypeptide chain, requiring no further changes. Moreover, it may be released from a ribosome directly to the cytoplasm, where it starts to function as a soluble protein. However, the assembly of the mature and properly compartmentalized form of many other proteins require one or more additional processing events. Depending on the protein, these events include the following:

1. The oxidation of cysteine —SH groups to form intrachain and/or interchain disulfide —S—S— bonds (probably the most common type of R group chemical modification characteristic of proteins).

2. Various other R group modifications, such as hydroxylation, amidation, methylation, iodination, carboxylation, and phosphorylation.

3. The covalent attachment or strong noncovalent association of metal ion cofactors (Zn^{2+} or Fe^{2+}, for example), of coenzymes (such as biotin), and of prosthetic groups (such as heme and carbohydrate groupings).

4. The cleavage of one or more specific internal peptide bonds in a precursor protein to release bioactive fragments.

5. The noncovalent association of identical or different polypeptide chains to produce a bioactive oligomeric structure.

6. The assimilation of a protein into a membrane.

7. The *export* of a protein from a cell; that is, its secretion as an extracellular protein.

8. The *import* of a protein, assembled on ribosomes in the cytoplasm, into a different cellular compartment.

Examples of events 1 through 5 have been encountered in previous chapters and much more will come in future chapters. At this time attention will be directed to the last three membrane-involved processes.

Signal Peptide Sequences

About twenty years ago researchers proposed that proteins, destined for secretion from the cells in which they are produced, are assembled on ribosomes bound to the endoplasmic reticulum. These proteins are then transferred into the inner compartment of the endoplasmic reticulum (called the lumen or cisterna) for passage to the Golgi body, where concentrated protein vesicles are assembled for release from the cell. Researchers then established that the transfer across the reticulum membrane begins before the synthesis of the polypeptide is completed. Because it occurs during translation, this transfer is termed *cotranslational transfer*.

Later, evidence was obtained demonstrating that some proteins can also move across a membrane after their synthesis is complete. Such *posttranslational transfer* applies to the *import* of a protein to the mitochondrion from the cytoplasm. This type of transfer is distinct from the entry of cytoplasmic proteins into the nucleus through actual pores (openings) in the nuclear membrane. Still other proteins only penetrate a membrane, partially or totally, remaining lodged in the membrane as a functional component of that membrane. This transfer is termed *assimilation transfer* and can occur cotranslationally or posttranslationally.

In 1975 the biochemical basis of the way proteins are selected for cotranslational transfer was discovered, and similar mechanisms have been found to apply for posttranslational and assimilation transfer processes. In cotranslational transfer the synthesis of a polypeptide begins on a free ribosome (not membrane bound). While engaged in elongation, the ribosome then becomes attached to the endoplasmic reticulum, where assembly of the polypeptide is completed (see Figure 9–19).

The attachment is guided by the recognition of a *signal peptide*— a short sequence at the N terminus region of the growing polypeptide.

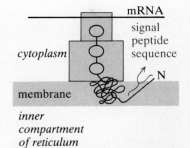

FIGURE 9–19 Signal peptide theory. This diagram shows membrane recognition occurring during assembly of the protein. See Figures 9–20 and 9–21 for additional features. A similar recognition can also occur after the polypeptide is released from the ribosome.

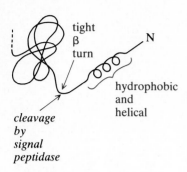

FIGURE 9–20 Proposed elements of the signal peptide structure for membrane recognition and action by signal peptidase.

As it is being assembled, the polypeptide permeates the membrane, "tunneling" its way to the other face of the membrane, where it is released into the lumen of the reticulum. However, research has clearly established that during transfer and before release the following event occurs:

A segment of the N terminus containing the signal peptide is removed by a proteolytic enzyme.

Designated *signal peptidase,* the enzyme resides in the membrane.

Although this *signal peptide theory* is now widely accepted, the signal has not been defined in terms of a specific, highly conserved, N terminal consensus sequence—though not because of a lack of effort. Numerous signal peptide regions have been sequenced, but they show considerable variation in both length (in the range of 15–30 residues) and amino acid sequence. Rather, the important elements of structure appear to be contributed by a combination of other factors: possibly (1) a stretch of 9–12 hydrophobic (apolar) residues in the center of the signal region, (2) a tendency for this region to be set off by a tight β turn near the site of cleavage, and (3) a tendency to form α-helix geometry in the hydrophobic region. See Figure 9–20.

Signal Recognition Particle

How is the signal peptide region recognized? Is there a receptor protein or a receptor/lipid cluster in the membrane? Recent studies implicate a mechanism involving a newly discovered 11S ribonucleoprotein particle, called the *signal recognition particle* (SRP). The SRP is composed of single copies of seven different proteins and a small 7S RNA (about 250 nucleotides in length). The 7S RNA of SRP is of further interest because it shows considerable homology in base sequence to the Alu repeat sequence (refer to p. 263 in Chapter 6) and may represent the first function discovered for DNA repeat sequences.

A complete model for the mechanism of cotranslational transfer is summarized in Figure 9–21. In addition to showing aspects already described, the model represents the participation of separate membrane receptor proteins for binding the 60S subunit of the eukaryote ribosome and for binding the SRP.

In closing this section, we note that the discovery of the 11S SRP modifies the classical description of a functional ribosome as being composed of two ribonucleoprotein subunits. Ribosomes engaged in the translation of mRNAs to produce proteins for transport into the reticulum appear to consist of three operational subunits, 40S, 60S, and 11S, with the 11S subunit having a transient association.

9–6   REGULATION OF PROTEIN BIOSYNTHESIS

Until about 1973 the only established process for regulating protein biosynthesis was the classical *induction-repression mechanism,* which we examined in Section 7–1. Recall that induction/repression directly controls the biosynthesis of RNA (transcription) and thus *indirectly* controls the biosynthesis of

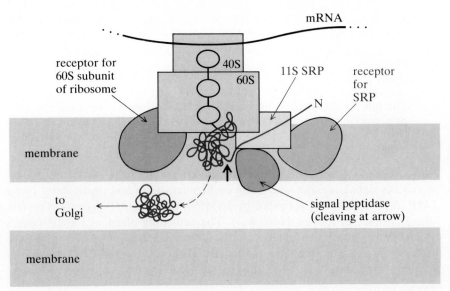

FIGURE 9–21 Signal peptide model for the cotranslational transfer of proteins into the inner compartment of endoplasmic reticulum. At some point after translation begins, the ribosome becomes membrane-bound. The SRP is the signal recognition ribonucleoprotein particle proposed to bind with the N terminal signal peptide region of a growing polypeptide.

proteins (translation). In the past decade, however, some new regulatory aspects of translation have been discovered, aspects involving mechanisms that *directly* regulate the translation process by controlling the activity and the synthesis of ribosomes. One was mentioned earlier in this chapter, namely, in eukaryotes the proposed stimulation of ribosome activity linked to the phosphorylation of the S6 ribosomal protein. Further research is still needed to obtain definite proof and to evaluate whether other ribosomal proteins fall into the same category. At this time we will examine two other direct-control mechanisms—one operating in eukaryotes and the other definitely operating in prokaryotes and maybe in eukaryotes.

Phosphorylation of eIF–2 for Regulation of Translation in Eukaryotes: Introduction to Protein Kinase Cascades

Reticulocytes are bone marrow–produced cells that are released into the blood for eventual maturation into erythrocytes (red blood cells). Reticulocytes are specialized cells engaging in extensive levels of protein biosynthesis—particularly in the production of the α and β polypeptide chains of hemoglobin. Because they are geared for a high rate of translation, reticulocytes have been very useful in studying various aspects of the translation process.

A significant finding in the early 1970s was that the biosynthesis of the globin chains (and other proteins) is greatly stimulated in reticulocytes by the presence of the iron-containing heme prosthetic group. Subsequent studies established that the enchancement effect of heme was due to its ability to short-circuit a cellular process responsible for inhibiting translation. In other words, *heme activates by inhibiting the operation of an inhibitor.*

Eventually, the heme effect was explained in precise biochemical events involving *eIF–2* and *protein kinases.* Applicable in part to all eukaryote cells, the mechanism ultimately centers on the phosphorylation of the α subunit of the eIF–2, converting this initiation factor into an inactive form that is unable to promote the binding of met-tRNA^{imet} to the 40S subunit. This phosphorylation involves a highly specific protein kinase, called *eIF–2 kinase I*. In fact, the eIF–2 kinase I is the *translation inhibitor* referred to above and is the ultimate target of the heme effect.

> The link to the heme enhancement of translation is accounted for by a *protein kinase cascade*—a fascinating process involving the activity of one protein kinase being regulated by another protein kinase.

Such a regulatory cascade begins with a protein kinase that is under the control of another signal, a major one being cyclic nucleotides and particularly cyclic AMP. A protein kinase activated by cAMP is called *cAMP-dependent protein kinase* (cAMP–dPK) (see Note 9–4). The activating effect of cAMP occurs by cAMP binding to an inactive R_2C_2 tetrameric form of cAMP–dPK (see Figure 9–22), causing the tetramer to dissociate into R_2 and C subunits. The active protein kinase is the *free C subunit,* having a different conformation than in R_2C_2.

One of the protein substrates for the protein kinase C subunit is eIF–2 kinase I. As shown in Figure 9–23, this phosphorylation of eIF–2 kinase I is responsible for activating this second protein kinase, which then uses eIF–2 as a protein substrate to form the inactive phosphorylated form of eIF–2. The presence of heme in reticulocytes prevents the activation of cAMP–dPK—thus preventing the activation of eIF–2 kinase I, preventing the inactivation of eIF–2, allowing the active form of eIF–2 to accumulate (owing to the participation of a protein phosphatase), and enhancing the initiation phase of translation. We will not discuss the process here, but we note that in phosphorylation/dephosphorylation processes the protein phosphatase enzymes may also be regulated in a reciprocal manner to the protein kinases. In other words, a kinase-activating signal can also serve as a phosphatase-inhibiting signal and vice versa.

The protein kinase cascade regulating eIF–2 activity is only one example of this type of regulatory mechanism. Another important example, responsible for regulating carbohydrate metabolism, will be described in Chapter 13.

Antiviral Action of Interferon Associated in Part to Control of eIF–2 Activity

There was considerable skepticism for the proposal of Isaacs and Lindenmann in 1957 that animal cells can protect themselves from viral infection by producing a special type of protein (they called it *interferon*) that prevents viral growth. Subsequent studies not only proved them right but also extended the importance and, unfortunately, the complexity of the original hypothesis (see Figure 9–24). For example, we now know that an animal can produce several different interferons from different cells. Moreover, the antiviral function of interferon is but one of several cellular effects now associated with the family of interferon proteins. Other important biological effects of interferons include the inhibition of cell growth (normal and tumor) and a regulatory role in some aspects of the immune response.

NOTE 9–4
Those protein kinases not directly responsive to cyclic AMP are called cAMP-independent protein kinases (cAMP–iPK). Two examples of this class are eIF kinase I and eIF kinase II.

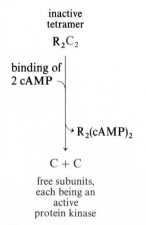

FIGURE 9–22 Activation of cyclic AMP–dependent protein kinase (cAMP–dPK). The R represents regulatory subunit; C is for catalytic subunit.

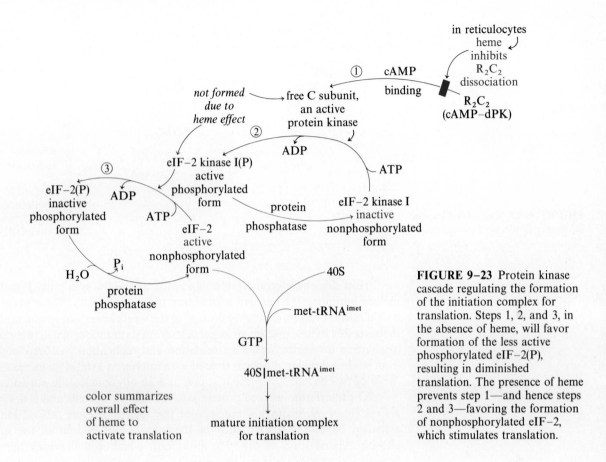

FIGURE 9–23 Protein kinase cascade regulating the formation of the initiation complex for translation. Steps 1, 2, and 3, in the absence of heme, will favor formation of the less active phosphorylated eIF–2(P), resulting in diminished translation. The presence of heme prevents step 1—and hence steps 2 and 3—favoring the formation of nonphosphorylated eIF–2, which stimulates translation.

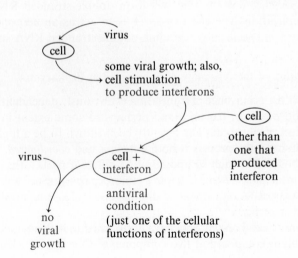

FIGURE 9–24 Simplified summary of interferon processes.

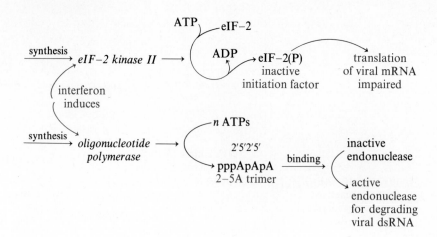

FIGURE 9–25 Two of the known biochemical effects of interferon.

How does interferon establish an antiviral condition in cells? Possibilities include inhibiting the entry of a virus into the cell, inhibiting the translation of a viral mRNA, inhibiting the replication of the viral chromosome, and inhibiting the assembly of new mature virus particles from its components. A two-pronged mechanism for inhibiting both translation and replication has been suggested from studies showing that cells treated with interferon exhibit an increased synthesis of two enzymes, a protein kinase and an oligonucleotide polymerase.

The interferon-induced protein kinase is eIF–2–specific and is designated *eIF–2 kinase II,* distinguishing it from the related enzyme (eIF–2 kinase I) activated by cAMP–dPK. Thus in the presence of interferon, the inactive phosphorylated form of eIF–2 is produced, which will diminish the translation of viral mRNA (see Figure 9–25). The oligonucleotide polymerase is, specifically, *(2′ → 5′) oligoadenylate polymerase,* responsible for forming the pppA(2′)p(5′)A(2′)p(5′)A trimer (refer to Figure 6–27). There is evidence that the presence of the 2–5A trimer enables a cell to combat a viral infection by preventing the replication of the virus. Specific targets are viruses having a single-stranded RNA chromosome that will form double-stranded RNA structures during its replication. The effect of 2–5A is to act as an *activator of a latent RNase* capable of specifically degrading double-stranded RNA structures.

Regulation of the Assembly of Ribosomes in Prokaryotes

In a cell the occurrence of translation is obviously dependent on the availability of ribosomes. In bacteria (and perhaps to some extent in eukaryotes), the production of ribosomes has recently been shown to be a highly regulated process. Moreover, the process is both *balanced* and *coordinated.* Thus in producing N ribosomes—each composed of 55 different parts—the cell produces only N(55) total parts. Also, if fewer or more parts are needed, the cell can make the appropriate adjustment to diminish or increase its production of the precise number of parts.

M. Nomura and co-workers have established that this regulation of ribosome assembly is composed of two components. One control element involves regulating the biosynthesis of ribosomal RNAs, and the second involves regulating the biosynthesis of ribosomal proteins. Both controls operate via a *feedback inhibition strategy* with two steps: (1) The transcription of rRNA genes is inhibited by the free ribosomes present in excess, and (2) the translation of the mRNA for ribosomal protein is inhibited by the free ribosomal proteins present in excess.

Under most growth conditions bacteria produce a population of ribosomes proportional to the growth rate, that is, the total rate of protein biosynthesis. Moreover, in a phase of active growth, cells are using nearly all of the ribosomes. If conditions change to reduce the growth rate, the prokaryote cell responds not by decreasing the activity of ribosomes but, rather, by *producing fewer ribosomes.*

The first signal of a reduced growth rate is an increase in the number of nonfunctioning ribosomes, that is, ribosomes no longer involved in translation. The Nomura group and others have demonstrated the following result:

> This increase in free ribosomes inhibits further transcription of the ribosomal RNA genes.

This inhibition is proposed to occur by free ribosomes acting as "repressor particles" that bind to the promoter regions of rRNA genes, blocking RNA polymerase action. Although the proposed model is more complex than that shown in Figure 9–26, we will not touch upon further details.

Regulating the translation of the mRNAs coding for ribosomal proteins involves a direct-feedback inhibition by excess ribosomal proteins.

> Here we have the first example of a protein inhibiting its own biosynthesis at the level of translation.

This feedback inhibition is explained by a free ribosomal protein binding to its mRNA near the AUG start codon, preventing the mRNA from forming the initiation complex with 70S ribosomes and fmet-tRNA^{fmet}. Excess ribosomal proteins arise when enough ribosomes are assembled to satisfy the growth rate and as free ribosomes inhibit the production of any further rRNA, as just described. A decrease in the production of rRNA will cause an increase in the cellular level of free ribosomal proteins.

However, in the case of ribosomal proteins each of the 52 different proteins does not function as a repressor only of its own synthesis. Rather, only a few of the proteins act as repressors, inhibiting not only their own synthesis but also the synthesis of other proteins. They act in this way because many of the ribosomal protein genes in prokaryote chromosomes are arranged in tandem, and the entire tandem (an operon) is under the influence of a single promoter (see Figure 9–27). Thus an entire group of genes is transcribed as a unit into a single mRNA. For example, one tandem consists of genes for the following ribosomal proteins, arranged in the order listed: S10, L3, L4, L23, L2, (L22, S19), S3, L16, L29, and S17. In this group of proteins only one—namely, L4—serves as a repressor, but it inhibits the translation of the *entire set* of 11 proteins. Other regulatory ribosomal proteins identified so far include L1, L10, S7, S8, S4, and S20, each regulating a different mRNA. The collective effect is the regulation of translation for a total of 32 ribosomal proteins.

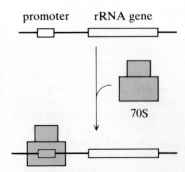

RNA polymerase cannot bind to promoter; thus no transcription of rRNA gene

FIGURE 9–26 Excess free ribosomes, which can repress rRNA genes.

FIGURE 9–27 Organization of one tandem of ribosome protein genes in the *E. coli* chromosome and the proposed mode of action of the L4 protein as a repressor of translation. (The exact sequence of the L22 and S19 genes is not yet certain.) (a) Operon of 11 ribosomal protein genes transcribed into a single mRNA unit; P is promoter. (b) Result when ribosome parts are being assembled. (c) Result when sufficient ribosomes are available to support the growth rate of the cell. The L4 protein binds near the 5′ terminus of mRNA, and for reasons not yet completely understood, none of the coding regions in the mRNA are translated. Possibly, one or two other proteins supplement the L4 effect by binding at other locations on the mRNA.

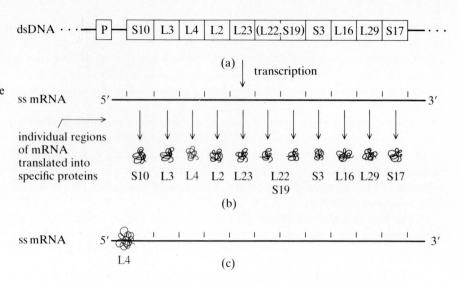

This description only scratches the surface of what is emerging as a complex group of mechanisms that appear to contribute to the regulation of ribosome assembly. Moreover, one or more of these systems may apply to the regulation of protein biosynthesis in general and not just to the biosynthesis of ribosomal proteins. For example, consider the phenomenon of "stringent control" occurring under growth conditions where the supply of aminoacyl-tRNA adducts is severely limited owing to a nutritional deprivation where cells are starved for amino acids. Cells adapt not only with a rapid decrease in ribosome production but also with a large increase in the production of novel guanosine polyphosphates such as *guanosine tetraphosphate,* ppGpp, and *guanosine pentaphosphate,* pppGpp. Do these compounds serve in some fashion as inhibitors for the synthesis of ribosomal RNAs and/or ribosomal proteins? As yet, this question is not definitely answered.

The regulatory aspects of ribosome assembly have been characterized in prokaryote cells but as yet have not been established as operating in eukaryotes. However, the prokaryote system may prove to be a useful model system, guiding future investigations into the nature of assembling ribosomes in eukaryotes, where the transcription and translation processes are much more complex.

9–7 GENETIC CODE

Codon Assignments

That genetic information flows from DNA through a messenger RNA molecule was first proposed in 1961 by the French biochemists F. Jacob and J. Monod (Nobel Prizes in 1965). The later work of M. Nirenberg, H. C. Khorana, and R. Holley (Nobel Prizes in 1968) served four purposes: (1) It proved the messenger RNA hypothesis; (2) it proved that the genetic language was written in sequences of three bases, called *triplet codons;* (3) it established

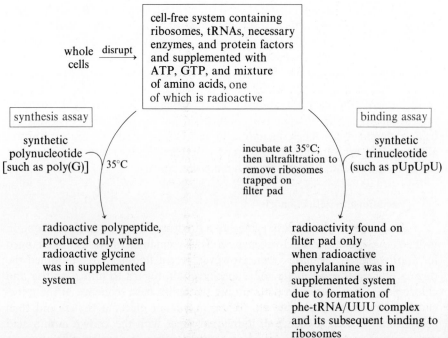

FIGURE 9–28 In vitro assays used to identify coding assignments for mRNA codons.

that the mRNA code is read by the complementary recognition of *anticodon triplets* in tRNA; (4) it established the amino acid–coding assignments for most of the 64 possible codons.

Nirenberg pioneered in using cell-free systems to study the synthesis of polypeptides according to instructions provided by synthetic RNAs and to study the binding of aminoacyl-tRNAs to ribosomes in response to the presence of synthetic trinucleotides of known sequence (see Figure 9–28). Khorana pioneered in the development of procedures for the chemical synthesis of poly- and oligonucleotides; Holley was responsible for the solution of cloverleaf tRNA structure with a distinct anticodon region.

In Nirenberg's *synthesis assays* homogeneous polyribonucleotides resulted in the production of homogeneous polypeptides: poly(U) as mRNA gave poly(phe) as product; poly(A) gave poly(lys); poly(C) gave poly(pro); and poly(G) gave poly(gly). The triplet nature of the code was categorically established by the use of heterogeneous polyribonucleotides composed of two nucleotides in alternating sequence. For example, a poly(UG) message (... UGUGUGUGUGUGUGUGUGUG...) gave a polypeptide composed of only cysteine and valine in alternating sequence (... cys-val-cys-val...). Thus the reading frame of the synthetic mRNA had to be ... UGUGUGUGUGUG.... A doublet code would have given two homogeneous polypeptides—one for a ... GUGUGU ... reading frame and another for a ... UGUGUG ... reading frame. Codon assignments based on the homogeneous mRNAs were obvious: UUU = phe, AAA = lys, CCC = pro, and GGG = gly. The poly(UG) result could not be used to make any definite coding assignments. (Why?)

Although the results using several other heterogeneous mRNAs did provide a basis for many either/or assignments and a few definite assignments, most were identified by the *binding assay,* using synthetic trinucleotides as "messages" to direct the attachment of specific aminoacyl-tRNA species to ribosomes. In this way nearly 50 codons were unequivocally assigned to certain amino acids. Experimental ambiguities prevented complete success, although probable assignments were made in several cases. Eventually (by 1966), with improvements in the binding assays and with the work of other investigators using different approaches, including in vivo studies with mutant organisms, all 64 coding assignments were defined (see Table 9–2): There are 61 *codons for amino acids* and 3 *that serve as termination signals.*

Variations in Mitochondria

All prokaryotes and eukaryotes and all viruses use the same set of codon instructions, indicating that the genetic code has remained essentially unchanged for approximately 3 billion years. However, recent findings have modified the dogma of a universal genetic code, specifically in regard to mitochondria and chloroplasts. The success in determining the entire base sequence of the mitochondrial DNA chromosome (16,569 bp in human placenta tissue) and then comparing the base sequence of individual genes with the known amino acid sequences of some the corresponding polypeptides revealed the differences in coding instructions listed in Table 9–3.

TABLE 9–2 Summary of the coding assignments of the 64 triplet codons

	$(5') \cdots pNpNpN \cdots (3')$ in mRNA				
BASE AT 5′ END OF CODON↓	MIDDLE BASE OF CODON →				**BASE AT 3′ END OF CODON↓**
	U	C	A	G	
U	phe (UUU)	ser	tyr	cys	U
	phe	ser	tyr	cys	C
	leu	ser	termination	termination·	A
	leu	ser	termination	trp	G
C	leu	pro	his	arg	U
	leu	pro	his	arg	C
	leu	pro	gln	arg	A
	leu	pro	gln	arg	G
A	·ile	thr	asn	ser	U
	ile	thr	asn	ser	C
	·ile	thr	lys	arg·	A
	met (and initiation)	thr	lys	arg·	G
G	val	ala	asp	gly	U
	val	ala	asp	gly	C
	val	ala	glu	gly	A
	val	ala	glu	gly	G

Note: The · symbolizes those codons having different instructions in the mRNA of mitochondria (see Table 9–3).

TABLE 9–3 Differences in the genetic code in mitochondria

CODONS	UGA	AUA	AGU	AGG	AUU
Universal code	termination	ile	arg	arg	ile
Mitochondrial code	trp	met and initiation	termination	termination	ile and possibly initiation

Degeneracy of the Genetic Code

Eighteen of 20 amino acids are coded for by more than one codon, a feature called *degeneracy*. Nine amino acids exhibit twofold degeneracy; one has threefold degeneracy; five have fourfold degeneracy; three have sixfold degeneracy. The phenomenon is of immense significance to living organisms. Because of codon degeneracy, certain errors in DNA replication or DNA transcription can occur without a corresponding change in the genetic information or in its expression.

For example, consider the alanine codon GCU, which arises from the transcription of CGA in the antisense-template strand of DNA. If GCU were the only code for alanine, then any alteration during replication or transcription of the CGA sequence in DNA would change the information. However, because of the fourfold degeneracy existing in the third position of the codon, only an error involving either of the first two bases will change the information (see Figure 9–29).

In all cases of twofold, threefold, and fourfold degeneracy, the variation involves the *third base* of the triplets. On still closer inspection, note that where only twofold degeneracy exists, the pattern of the third-base change always involves a purine for a purine or a pyrimidine for a pyrimidine: $NNpur_a$ and $NNpur_b$, or $NNpyr_a$ and $NNpyr_b$. Both characteristics suggest that the genetic code may originally have been based in a language using doublet codons and eventually evolved into using triplet codons.

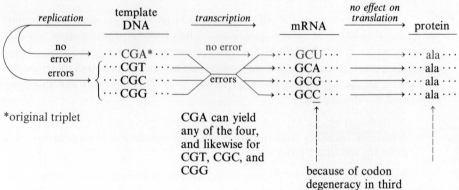

*original triplet

CGA can yield any of the four, and likewise for CGT, CGC, and CGG

because of codon degeneracy in third base (all code for alanine), none of these errors will change genetic instruction

FIGURE 9–29 Codon degeneracy which can compensate for substitution mutations. Similar effects apply to all sets of degenerate codons, sometimes involving the middle base.

TABLE 9–4 Degenerate codons of aliphatic, nonpolar amino acids

ALANINE	VALINE	LEUCINE	ISOLEUCINE
GCU	GUU	CUU	AUU
GCA	GUA	CUA	AUA
GCC	GUC	CUC	AUC
GCG	GUG	CUG	
		UUA	
		UUG	

Further inspection of the genetic code reveals that it is characterized by another type of degeneracy. To illustrate, the sets of degenerate codons for alanine, valine, leucine, and isoleucine are listed in Table 9–4. We have already noted that a code for alanine will survive errors in DNA replication and DNA transcription involving the third-base position because of the four-codon degeneracy for alanine. If errors involve the first and second bases, the code for alanine is lost. However, certain changes in the first or second base may occur without losing the code for the *type* of amino acid that alanine represents.

Look at the columns of codons in Table 9–4. Codons GCU and GUU (differing in the second-base position) code for alanine and valine, respectively, but each is an *aliphatic, nonpolar amino acid*. Likewise, CUU and AUU (differing in the first-base position) code for leucine and isoleucine, respectively, but again, both are aliphatic, nonpolar amino acids. Indeed:

> Any one of the 17 codons listed will code for an aliphatic, nonpolar amino acid.

This sort of pattern also applies to other groups of amino acids, although not to the same extent. But don't let this result mislead you; most changes in the first- or second-base position, or both, will result in significant alterations in the coding instructions.

Overlapping Codons and the Reading of Codons in Different Frames

The chromosome of the ϕX174 bacterial virus is a single-stranded circular molecule, consisting of 5386 nucleotides, coding for the production of ten different proteins: A, A*, B, C, D, E, F, G, H, J. However, various genetic and physical studies established that 5386 nucleotides do not contain enough information to code for all these proteins. The sequencing (in 1977) of the entire chromosome explained the discrepancy:

> At different places the same stretch of DNA can code for different functions, meaning that transcription and translation can occur in different reading frames.

As shown in Figure 9–30, proteins A, A*, and B are all encoded in the same region. The same applies to proteins D and E. The A and A* proteins are different only because different start locations are used. The codons are read in the same reading frame. However, not only does the B protein in the A region have

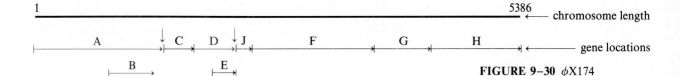

FIGURE 9–30 φX174 chromosome (shown here in a linear form for easier sight evaluation). In addition to the AB and DE gene overlap, initiation and termination codons overlap at locations indicated by ↓.

a different start codon, but the codons are read in a different reading frame. The same applies to the E gene within the D gene. Two other related examples occur at adjacent genes where initiation codons overlap with termination codons of the preceding gene. The reading frame changes from one gene to the next. Overlaps of this type occur at the A/C junction and the D/J junction.

The multifunctional character of the φX174 chromosome is not a unique property displayed by only a few very small DNA viruses. Overlapping initiation and termination codons have also been detected in much larger chromosomes of other bacterial viruses, in chromosomes of animal viruses, in the chromosome of *E. coli,* and in mitochondrial DNA. Thus the possibility that some of these tactics of compressing information into DNA may be of more general significance.

Novel Base Pairing in the Codon-Anticodon Recognition

The key to the fidelity of the stepwise assembly of a polypeptide chain is the complementary base-pair recognition of a $5' \rightarrow 3'$ codon sequence in mRNA by a $3' \rightarrow 5'$ anticodon sequence in tRNA. See Figure 9–31. Although some tRNAs recognize and bind with only one specific codon, others have the ability to do so with two or three different codons. The biological significance of this characteristic is related to the degeneracy of the genetic code and was first expressed by Francis Crick in what he termed the *wobble hypothesis* of codon-anticodon recognition.

Wobble refers to the ability of a base in the anticodon to assume different spatial orientations; the other two bases exist in a more fixed orientation. For reasons we will not pursue, Crick argued that the *wobble position* is located at the 5'end of the anticodon.

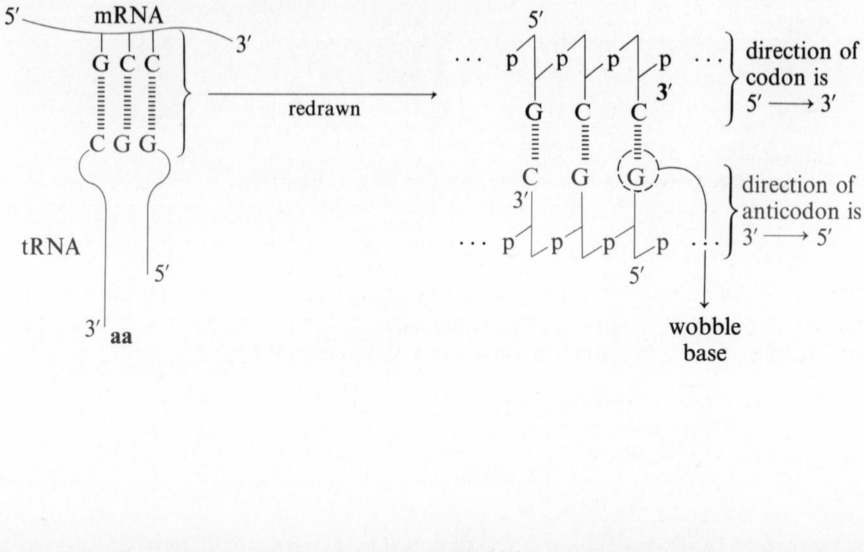

FIGURE 9–31 Codon-anticodon recognition.

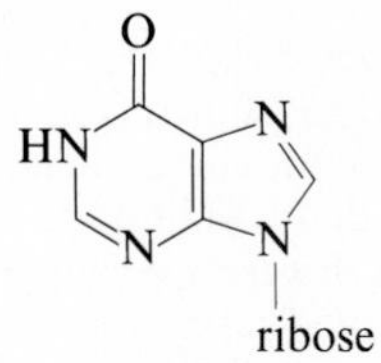

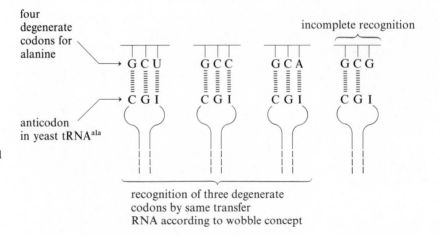

FIGURE 9–32 Inosine with the purine hypoxanthine. The original evaluation of novel base pairing at the 5′ end of the anticodon included hypoxanthine because it had been detected in the anticodon of some tRNAs that had been sequenced at the time.

TABLE 9–5 Base-pair combinations predicted by the wobble hypothesis

BASE AT 5′ END OF ANTICODON	BASE AT 3′ END OF CODON
I	A, C, or U
G	C or U
U	A or G
A[a]	U
C[a]	G

[a] Wobble does not permit any novel combinations with these bases when they are in the anticodon.

He further argued:

> Wobble will permit the same anticodon to recognize more than one codon because novel base pairs may be produced between the 5′ end of the anticodon and the 3′ end of the codon.

These base pairs are novel in the sense that they do not follow the classical pairings of A–T, G–C, and A–U known to exist in DNA and RNA.

On the basis of theory and model building, Crick further proposed that wobble will only be possible with certain bases and, furthermore, that novel base pairings will be restricted to certain combinations. The wobble bases are U, G, and I where I represents inosine (see Figure 9–32), a ribonucleoside containing the purine hypoxanthine. Wobble is not permitted with A and C. Table 9–5 summarizes the base-pairing possibilities predicted by Crick: three if the base at the wobble position is I and two if the base at the wobble position is G or U.

Let us now briefly examine how the wobble concept is related to the degeneracy of the genetic code. As a specific example, consider the sequence of yeast tRNA^{ala}, which has the anticodon sequence (3′)CGI(5′). Since it is a tRNA for alanine, it obviously recognizes an alanine codon. However, the genetic code exhibits a fourfold degeneracy for alanine, the four codons being GCU, GCC, GCA, and GCG (all 5′ → 3′). See Figure 9–33. This degeneracy can be accounted for if four different tRNAs for alanine exist in the cell. But if the wobble concept applies, then possibly all four codons can be recognized by a minimum of two different tRNAs. Furthermore, the tRNA described here can recognize three of the four degenerate codons. That is, the 5′ position of the anticodon is I, which means that the CGI anticodon can recognize GCU, GCC, and GCA but not GCG. As it happens, this pattern is precisely the pattern observed in binding studies made with purified tRNA^{ala} and synthetic preparations of these four codons.

Although no one has ever determined exactly how many different tRNAs are present in a cell, wobble predicts that a minimum of 32 different tRNAs are required. The precise number (be it 32, 33, 34, or whatever) may vary from or-

FIGURE 9–33 Wobble concept and degeneracy. The GCG codon would have to be recognized by a tRNA with a regular anticodon (CGC) or with a wobble anticodon (CGU).

ganism to organism. In any event, the wobble phenomenon provides another safety valve for minimizing errors in the translation readout of genetic information.

The DNA chromosome of mitochondria contains only 22 different tRNA genes. If some additional nucleus-produced tRNAs are not imported into mitochondria, then the anticodon-codon recognition in mitochondria involves either (1) more wobble than is defined in Table 9–5 or (2) the interaction of some triplets via only two bases rather than three. Whatever the case, mitochondria are clearly more versatile than was originally thought.

LITERATURE

BRIMACOMBE, R., G. STOFFLER, and H. G. WITTMANN. "Ribosome Structure." *Annu. Rev. Biochem.*, **47**, 217–250 (1978). A thorough review article focusing on the structure and the location of ribosomal RNAs and proteins and the three-dimensional shape of ribosomes.

CLARK, B. "The Elongation Step of Protein Biosynthesis." *Trends Biochem. Sci.*, **5**, 207–209 (1980). A brief review article.

EIGEN, M., W. GARDINER, P. SCHUSTER, and R. WINKLER-OSWATITSCH. "The Origin of Genetic Information." *Sci. Am.*, **244**, 88–118 (1981). A description of the hypothesis that RNA was the prebiotic (before-life) carrier of genetic information.

FIDDES, J. C. "The Nucleotide Sequence of a Viral DNA." *Sci. Am.*, **237**, 54–67 (1977). The overlapping genes in ϕX174 DNA are described.

HUNT, T. "The Initiation of Protein Biosynthesis." *Trends Biochem. Sci.*, **5**, 178–181 (1980). A brief review of the initiation stage of translation in eukaryotes.

KREIL, G. "Transfer of Proteins Across Membranes." *Annu. Rev. Biochem.*, **50**, 317–348 (1981). A thorough review article highlighting the signal peptide hypothesis.

LAKE, J. A. "The Ribosome." *Sci. Am.*, **245**, 84–97 (1981). A detailed description of the translation process, with emphasis on the three-dimensional shape of a ribosome and the participation of ribosomal proteins.

LEDER, P. "The Genetics of Antibody Diversity." *Sci. Am.*, **246**, 102–115 (1982). New discoveries are described, explaining how a few hundred genes can give rise to possibly billions of different proteins by elaborate processes involving gene recombination and mRNA splicing.

LENGYEL, P. "Biochemistry of Interferons and Their Actions." *Annu. Rev. Biochem.*, **51**, 251–282 (1982). A review article with a self-explanatory title.

MAITRA, U., E. A. STRINGER, and A. CHAUDHARI. "Initiation Factors in Protein Biosynthesis." *Annu. Rev. Biochem.*, **51**, 869–900 (1982). A thorough review article with a self-explanatory title.

MARSHALL, R. E., C. T. CASKEY, and M. NIRENBERG. "Fine Structure of RNA Codewords Recognized by Bacterial, Amphibian, and Mammalian Transfer RNA." *Science,* **155**, 820–825 (1967). Conclusive evidence for the universality of the basic language of the genetic code on the basis of in vitro binding studies with 50 synthetic codons, with a discussion of specific variations in terms of their phylogenetic and evolutionary significance.

NIRENBERG, M. W. "The Genetic Code: I." *Sci. Am.*, **208**, 80–94 (1963). A description of (a) classical experiments with synthetic oligo- and polyribonucleotides proving that the sequence of bases in an RNA molecule specifies the order of insertion of amino acids during the assembly of a polypeptide chain; (b) the nature of the genetic code; and (c) how many of the triplet codes were determined by laboratory studies.

NOMURA, M. "The Control of Ribosome Synthesis." *Sci. Am.*, **250**, 102–115 (1984). The control of ribosomal RNA and ribosomal protein biosynthesis via feedback inhibition of excess ribosomes is described and illustrated.

OCHOA, S., and C. DE HARO. "Regulation of Protein Synthesis in Eukaryotes." *Annu. Rev. Biochem.*, **48**, 549–580 (1979). The phosphorylation of eIF–2 is highlighted.

UY, R., and F. WOLD. "Posttranslational Covalent Modification of Proteins." *Science,* **198**, 890—896 (1977). A good review article.

WEISSBACH, H., and S. OCHOA. "Soluble Factors Required for Eukaryotic Protein Synthesis." *Annu. Rev. Biochem.*, **45**, 191–216 (1976). A thorough review article.

WICKNER, W. T., and H. F. LODISH. "Multiple Mechanisms of Protein Insertion into and Across Membranes." *Science,* **230**, 400–407 (1985). A current review article of the signal peptide theory comparing and contrasting protein/membrane interactions for the endoplasmic reticulum, mitochondria, and the bacterial plasma membrane.

WITTMANN, H. G. "Components of Bacterial Ribosomes." *Annu. Rev. Biochem.*, **51**, 155–184 (1982). Structure and function of the individual RNA and protein components of the ribosome are reviewed.

———. "Architecture of Prokaryotic Ribosomes." *Annu. Rev. Biochem.*, **52**, 35–66 (1983). An updated revision of the review cited above by Brimacombe, Stoffler, and Wittmann.

EXERCISES

9–1. The polypeptide chain of myoglobin contains 153 amino acid residues. Theoretically, then, how many base pairs will be present in the gene that specifies the configuration of this polypeptide?

9–2. Using the genetic code and the description of the difference in the amino acid sequence in the β chain of sickle-cell hemoglobin (see p. 133), prove that a change in a single base pair in DNA can result in a defective protein.

9–3. Evaluate whether the following nucleotide sequence of the sense-template strand in a nuclear DNA corresponds to (a) the *start* of an mRNA gene, (b) the *end* of an mRNA gene, or (c) a *midregion* of an mRNA gene. Explain your reasoning. Assume no transcriptional errors occur.

$$3' \cdots \text{ATGTCGCATGAGGCTAGCTCAAAC} \cdots 5'$$

9–4. How many different tRNA molecules are required to read the information in the nucleotide sequence shown in Exercise 9–3? Assume the reading frame begins with the A nucleotide at the 3' end.

9–5. Verify that a living cell must contain a minimum of 32 different tRNA molecules.

9–6. (a) Write a sequence of nucleotides within a gene on the sense strand (+) of DNA that will specify the following amino acid sequence in the polypeptide coded for by the gene:

$$\cdots \text{ala-asp-trp-gly-pro} \cdots$$

(b) Focusing only on A positions in each DNA triplet of the sense strand, how many AT → GC substitutions can occur in your sequence without drastically altering the functional sense of this segment of genetic information?

9–7. A leucine-tRNA molecule (yeast) having the anticodon (3')GAA is capable of binding with only one of the six degenerate codons assigned to leucine. Given that a second leucine-tRNA (yeast) having the anticodon (3')AAC has been isolated, predict (a) what codon or codons it will bind with, and (b) the *minimum* and the *maximum* number of additional leucine-tRNAs that will be required to recognize the codons that still remain unaccounted for.

9–8. In Chapter 8 we stated that substitution mutations may be silent or lethal (see Section 8–2). Explain why one would expect a high frequency for lethal mutations to arise when a substitution mutation affects the base at the 3' end of either of the degenerate codons for tyrosine.

9–9. Using the style shown in Figure 9–29, illustrate how codon degeneracy can overcome certain substitution mutations that might occur in DNA template triplets (a) for phenylalanine and (b) for arginine.

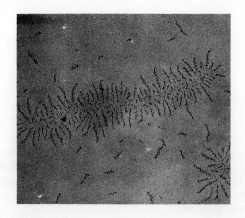

CHAPTER TEN

CARBOHYDRATES

Carbohydrate is a very general term that applies to a very large number of materials covering a wide spectrum of chemical structure and biological function. It was suggested as a name almost a hundred years ago to refer to those naturally occurring substances having a composition according to the formula—$(C \cdot H_2O)_n$—, that is, carbon·hydrate (see Note 10–1).

In most organisms carbohydrate material—largely in the form of the simple sugar glucose—is the primary foodstuff, providing most of the *energy* and *carbon* required in the biosynthesis of proteins, nucleic acids, lipids, and other carbohydrates. Many of the polymeric carbohydrates belong to one of two categories: (1) those that have a *structural role,* such as cellulose in plants, and (2) those that have a *carbon and energy storage role,* such as starch in plants and glycogen in animals and bacteria.

Among various other roles of naturally occurring carbohydrate materials, we can list gellike substances that lubricate bone joints, determinants of the different blood groupings, components of some antibiotics, gum secretions that assist in the healing of plant wounds, protective coatings of bacteria, and components of the bacterial cell wall.

Often the substance in question is not a pure carbohydrate but rather one that is partially composed of carbohydrate. The two main examples are *glycoproteins* and *glycolipids.* (The prefix *glyco-* is commonly used to designate the presence of carbohydrate.)

There is no way that the structure and the function of carbohydrates can be thoroughly covered in a single chapter. What we will try to do is introduce important principles that apply to carbohydrates in general, highlighting only a few of the specifics mentioned above. This chapter covers the following classes of carbohydrates: (1) the *monosaccharides* (monomeric units), sometimes referred to as simple sugars, (2) the *oligosaccharides* (two to several monomers linked together), and (3) the *polysaccharides,* which may have as many as a few thousand monomeric units. (The term *saccharide* is derived from the Greek *sakchar,* meaning "sugar or sweetness," and is related to the characteristic taste of many of the simple carbohydrates.) The chapter concludes with a discussion of carbohydrate structures that are conjugated with (poly)peptide components.

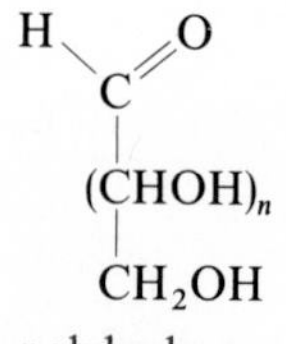

H—C=O
|
$(CHOH)_n$
|
CH_2OH

polyhydroxy
aldehyde

(a)

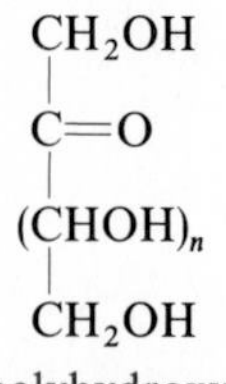

CH_2OH
|
C=O
|
$(CHOH)_n$
|
CH_2OH

polyhydroxy
ketone
(2-keto)

(b)

FIGURE 10–1 General classes of monosaccharides. (a) Aldose. (b) Ketose.

10–1 MONOSACCHARIDES

Basic Structure

The *monosaccharides* can be described as polyhydroxy aldehydes, polyhydroxy ketones, and derivatives thereof. The derivatives will be discussed later. All simple monosaccharides have the general empirical formula $(CH_2O)_n$, where *n* is a whole number ranging from 3 to 9. Regardless of carbon number, all monosaccharides can be grouped into one of two general classes: *aldoses* or *ketoses* (see Figure 10–1). The *-ose* ending is characteristic in carbohydrate nomenclature. The ending *-ulose* is irregularly used to designate a simple ketose.

(a)

```
                                                                      H
                                                                      |
                                                   H                 ¹C=O
                                                   |                  |
                                H                 ¹C=O           H—²C—OH
                                |                  |                  |
              H              ¹C=O           H—²C—OH           H—³C—OH
              |               |                  |                  |
            ¹C=O         H—²C—OH          H—³C—OH           H—⁴C—OH
             |                |                  |                  |
        H—²C—OH          H—³C—OH          H—⁴C—OH           H—⁵C—OH
             |                |                  |                  |
          ³CH₂OH           ⁴CH₂OH           ⁵CH₂OH            ⁶CH₂OH

          C₃H₆O₃            C₄H₈O₄           C₅H₁₀O₅           C₆H₁₂O₆
         aldotriose        aldotetrose      aldopentose       aldohexose
       (glyceraldehyde)
```

(b)

```
                                                                  ¹CH₂OH
                                                                   |
                                                 ¹CH₂OH           ²C=O
                                                  |                |
                                ¹CH₂OH           ²C=O         H—³C—OH
                                 |                |                |
              ¹CH₂OH            ²C=O         H—³C—OH         H—⁴C—OH
               |                 |                |                |
             ²C=O           H—³C—OH         H—⁴C—OH         H—⁵C—OH
               |                 |                |                |
             ³CH₂OH            ⁴CH₂OH           ⁵CH₂OH            ⁶CH₂OH

           ketotriose        ketotetrose      ketopentose       ketohexose
      (dihydroxyacetone)
```

FIGURE 10–2 Aldose and ketose families. (a) Aldose family. (b) 2-Ketose family. These linear structural formulas are referred to as Fischer projection formulas. The accepted convention of numbering carbons is as shown here; asymmetric carbons are in color.

Aldoses contain a functional aldehyde grouping —(H)C=O whereas ketoses contain a functional ketone grouping >C=O (see Figure 10–2). Subclasses are then distinguished on the basis of carbon content according to the following terms: aldotriose, ketotriose, aldotetrose, ketotetrose, and so on. *Glyceraldehyde* is the simplest aldose, and *dihydroxyacetone* is the simplest ketose. Each is considered the parent compound of higher >C(H)OH homologues in each class.

Stereoisomerism: Enantiomers, Diastereomers, and Epimers

With the exception of dihydroxyacetone, all other simple sugars are chiral because they contain asymmetric carbon atoms, the total number being equal to the number of internal >C(H)OH groups. The number of stereoisomers corresponds to 2^n, where n equals the number of asymmetric carbons. For example, an aldohexose with a general formula of $C_6H_{12}O_6$ and four asymmetric carbons, that is, four >C(H)OH groups, can exist in any one of 16 possible isomeric forms, with 8 D forms and 8 L forms.

With D-glyceraldehyde as the parent compound, one can construct a chart (see Figure 10–3) illustrating the structures of all D-aldoses through the aldohexose group. The figure illustrates sugars of one homologous family as originating from the next lower homologous group, with the chain extended

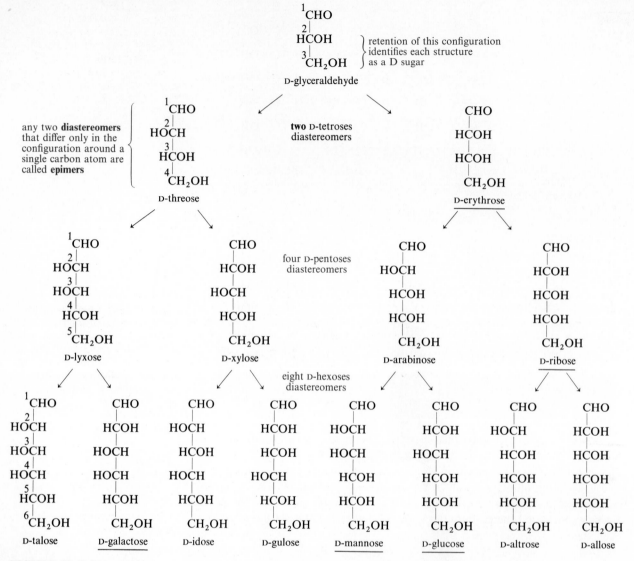

any two **diastereomers** that differ only in the configuration around a single carbon atom are called **epimers**

{ retention of this configuration identifies each structure as a D sugar

two D-tetroses diastereomers

four D-pentoses diastereomers

eight D-hexoses diastereomers

FIGURE 10–3 Structural relationships among D-aldoses. Those that are underlined occur more commonly.

by the generation of a new $>$C(H)OH at position 2. Note that every time this extension occurs, the hydroxyl group at C2 can assume two possible orientations, while all other $>$C(H)OH groupings remain unchanged.

> The overall configuration of each sugar (D in this case) is fixed by the orientation of the $>$C(H)OH position most distant from the functional carbonyl group.

This position is C5 in hexoses, C4 in pentoses, and C3 in tetroses. A similar diagram can be constructed for the L series, beginning with L-glyceraldehyde. The L sugars are the *enantiomers* of the D sugars—nonsuperimposable mirror images. For example, in L-glucose the configuration of every asymmetric

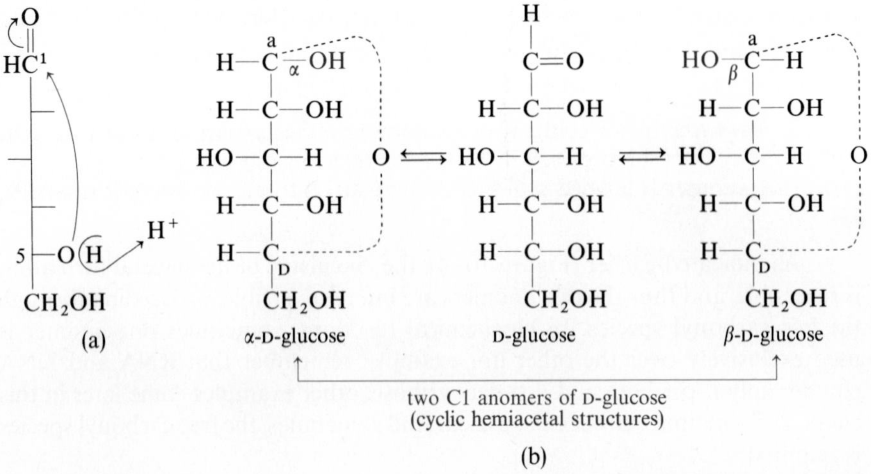

(a)

α-D-glucose   D-glucose   β-D-glucose

two C1 anomers of D-glucose
(cyclic hemiacetal structures)

(b)

FIGURE 10–4 Anomers, which arise by intramolecular condensation. Some D sugars are illustrated here. (a) Cyclic hemiacetal formation via C1—O5 interaction in aldohexose. (b) Some D sugars. The dashed line represents a distorted bond projecting toward the rear. The C^a is the anomeric carbon.

$>$C(H)OH is opposite that in D-glucose. Sugars with the D configuration predominate in nature.

Any two isomers that are nonsuperimposable, nonmirror images are called *diastereomers*. When two diastereomers differ only in the configuration on one asymmetric carbon, they are called *epimers* (see Figure 10–3). For example, D-glucose and D-galactose are C4 epimers. Table 10–1 lists the simple sugars that most frequently occur in nature. The structure of a common ketohexose, D-fructose, is also shown. The other ketose structures will be encountered in Chapters 13 and 16.

Stereoisomerism: Anomers

Each D or L sugar having five or more carbons can exist as either of two different diastereomers called α and β *anomers*. They arise from an additional asymmetric $>$C(H)OH carbon formed by the intramolecular condensation of the carbonyl group and a distant —OH (see Figure 10–4). The resultant structure is called a *cyclic hemiacetal* (for aldoses) or a *cyclic hemiketal* (for ketoses). Generally, aldohexoses form six-membered rings via a C1—O5 interaction; aldopentoses form five-membered rings via a C1—O4 interaction; ketohexoses form five-membered rings via a C2—O5 interaction; ketopentoses also form five-membered rings via a C2—O5 interaction.

The new asymmetric center, involving what was the original carbonyl carbon, is called the *anomeric carbon,* C^a, and the two possible diastereomeric configurations are called *anomers*. One is designated as the α anomer, and the other is the β anomer. (Anomers are a special class of epimers.) Two conventions used for assigning the α or β label in the linear Fischer representation are explained in the following list. They are illustrated for the D-glucose anomers shown in Figure 10–4 and also for the L configuration of a general aldohexose in Figure 10–5. The same rules apply to pentoses. Of course both rules result in the same assignment of α or β.

1. For a D sugar the α anomer has the anomeric OH group directed to the right (C^a—OH), and the β anomer has the OH directed to the left (OH—C^a).

TABLE 10–1 Commonly occurring monosaccharides

ALDOSES
D-Glyceraldehyde
D-Erythrose
D-Ribose
D-Galactose
D-Mannose
D-Glucose

KETOSES
Dihydroxyacetone
D-Xylulose $(C_5)^a$
D-Ribulose $(C_5)^a$
D-Sedoheptulose $(C_7)^a$
D-Fructose (C_6) (structure below)

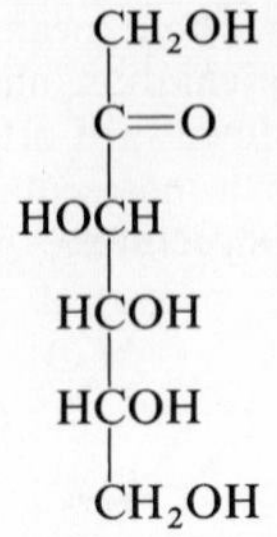

^{a} See Section 13–9 for structures.

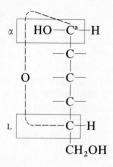

FIGURE 10–5 Anomer labeling for an L configuration. Here the anomeric OH is to the left; therefore it is α. The boxed orientations are cis (both on the left side of the vertical carbon chain). Again, therefore, it is α.

For an L sugar the convention is just the reverse; that is, C^a—OH is β and OH—C^a is α. (Reversing the assignments should make some sense in view of the mirror image relationship between D and L sugars.)

2. Be it a D or an L sugar, if the anomeric OH group and the would-be OH of the D/L-conferring carbon have a cis relationship relative to the carbon chain, the anomer is labeled α. If the relationship is trans, the anomer is labeled β.

As indicated earlier (Figure 10–4), the chemistry of hemiacetal formation is reversible, and thus the two anomers are interconvertible, proceeding through the free carbonyl species. In biochemical reactions, sometimes one anomer is used exclusively over the other (for example, remember that RNA and DNA contain only β-D-ribose and β-D-deoxyribose; other examples come later in this chapter). Sometimes, either can be used. And sometimes, the free carbonyl species is required.

Haworth Projection Formulas

Obviously, a Fischer projection formula of the hemiacetal structure is inaccurate in what it represents. Chemical bonds do not have right-angle bends. In 1929 Haworth suggested a more realistic representation in which both five-membered and six-membered cyclic structures are depicted as planar ring systems, with the hydroxyl groups on each C oriented either up or down from the plane of the ring. Although it, too, does not really represent the actual three-dimensional structure of a sugar, the Haworth representation has been used as an easy-to-draw formula that also permits a quick evaluation of the relative orientations of the —OH groups in the structure. Because of the structural similarity to the organic compounds called *furan* and *pyran*, a five-membered cyclic hemiacetal is labeled a *furanose*, and a six-membered hemiacetal ring is called a *pyranose* (see Figure 10–6). The simplified Haworth form with carbons omitted from the ring is routinely used.

For any D sugar the conversion of a linear Fischer formula into a Haworth formula proceeds as follows:

1. An H or OH group directed to the right of the carbon chain in the Fischer structure is given a downward orientation in the Haworth formula. If directed to the left in Fischer, it has an upward orientation in Haworth.

2. The terminal —CH₂OH grouping is given an upward orientation in the Haworth formula.

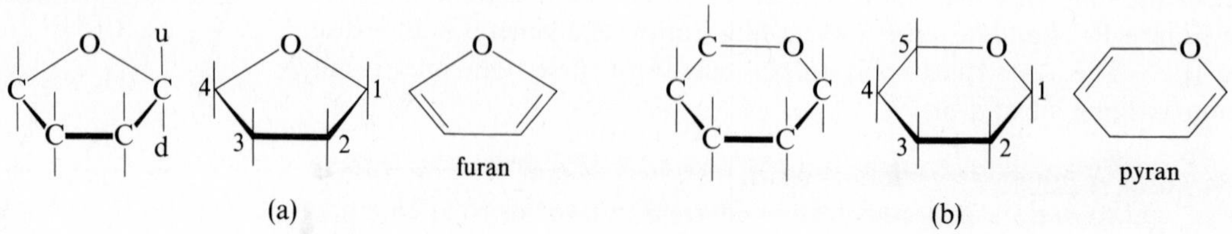

FIGURE 10–6 Haworth projection formulas. (a) Haworth representations of furanose structures. The u means an upward orientation, and the d means a downward orientation. (b) Haworth representations of pyranose structures.

(a)   (b)

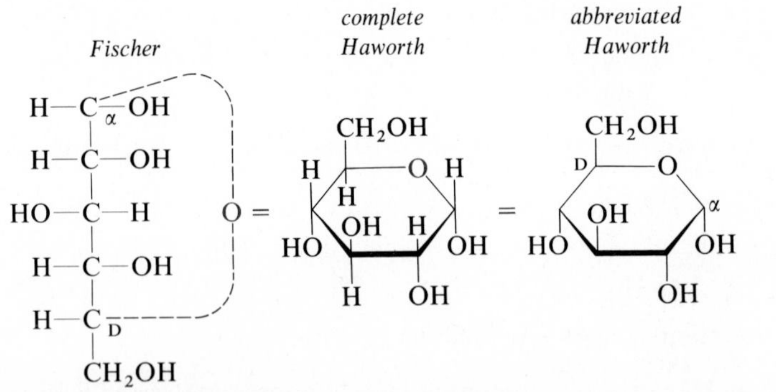

(a) α-D-glucopyranose

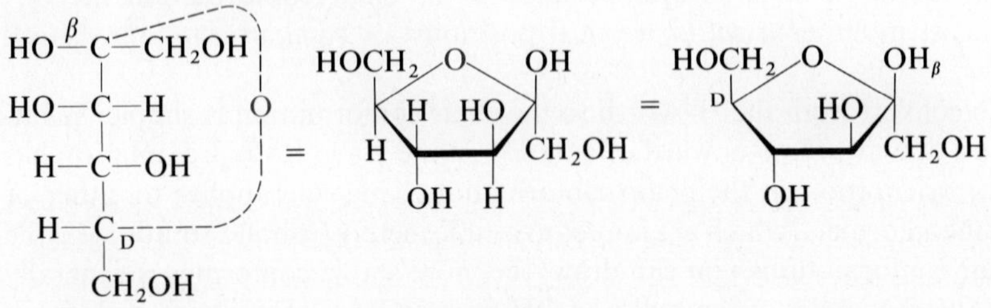

(b) β-D-fructofuranose

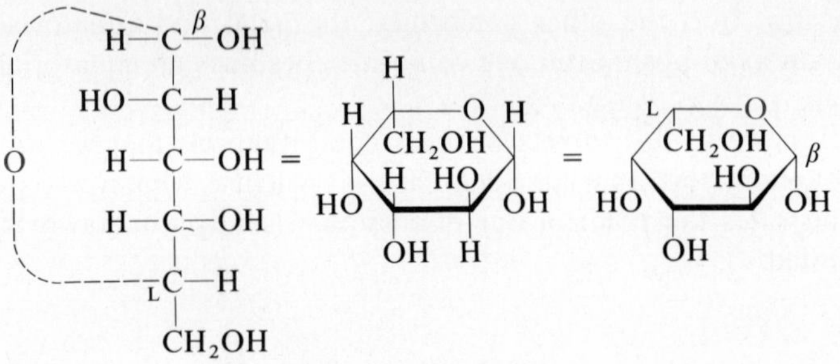

(c) β-L-galactopyranose

FIGURE 10–7 Three examples of converting Fischer representations to Haworth representations.

For an L sugar the rule in step 1 is the same, but the terminal —CH$_2$OH grouping is projected downward in the Haworth formula.

The structures of α-D-glucopyranose, β-L-galactopyranose, and β-D-fructofuranose illustrate the conversion (see Figure 10–7). Note the abbreviated, shorthand forms, in which only dashes are used to represent the position of the —OH group and all H's are omitted. Also note that the reversal of the α/β designations defined in the Fischer projections carries over to the Haworth representations. In the Haworth formula for D sugars the orientation of the α configuration at the anomeric carbon is down and that of the β configuration is up. For L sugars the α and β orientations in Haworth are just the reverse.

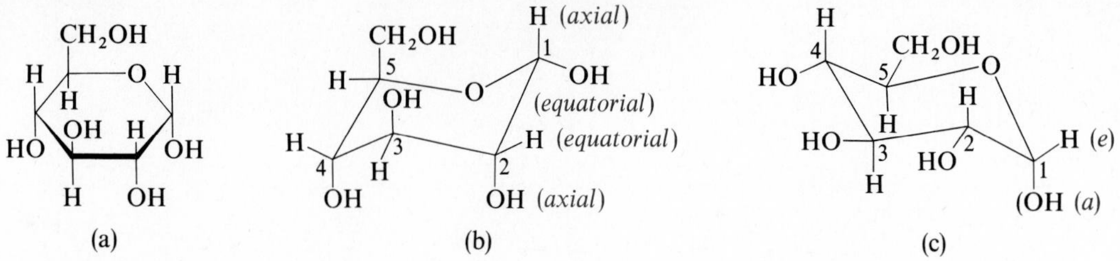

(a)   (b)   (c)

FIGURE 10–8 Structures for α-D-glucopyranose. (a) Haworth representation. (b) Chair conformation 1: *eaaa* for 12345. (c) Chair conformation 2: *aeeee* for 12345. This form is the more stable form.

Conformation of Sugars in Solution

A planar Haworth representation closely approximates the actual shape of the furanose ring structure, which is nearly flat. The preferred conformation of pyranose rings is nowhere near being flat but is the *chair structure* that is common to such things as cyclohexane. In the chair conformation the OH groups exist in either *axial* (*a*, vertical positions) or *equatorial* (*e*, nonvertical positions).

To convert from the Haworth to the chair conformation is simple: Whatever has a downward or upward orientation in the Haworth representation has a similar orientation in the chair conformation. This rule applies to either of two chair conformers, which are interconvertible merely by bond rotation. Of the two chair conformations you can draw, the more stable conformer is generally the one that has most of the bulky —OH and —CH$_2$OH in equatorial-group positions. In the case of α-D-glucopyranose (see Figure 10–8), the *aeeee* isomer is more stable. (From your study of organic chemistry, remember that when a chair conformer flips into the other conformer, the axial and equatorial orientations invert; an axial position in one conformer becomes an equatorial position in the other, and vice versa.)

There is much more to this subject of sugar conformations that we will not consider. Throughout this and later chapters we will use whatever representation best illustrates the point of our discussion. Usually the Haworth forms will be presented.

10–2   DERIVATIVES OF SUGARS

Owing to the presence of the functional OH groups and the possible existence of a

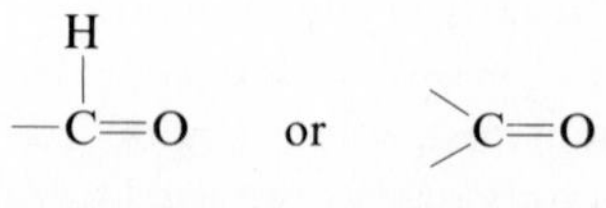

$$\underset{\text{H}}{\overset{\displaystyle |}{-\text{C}}}=\text{O} \quad \text{or} \quad {>}\text{C}=\text{O}$$

groupings, sugars can undergo the spectrum of reactions that are common to alcohols, aldehydes, and ketones. Then there are some class reactions that are a consequence of the cyclic polyhydroxy ring structure. Thus sugars undergo a large number of reactions. However, the scope of our consideration will be limited to only a few, with a particular focus on reactions of biological significance.

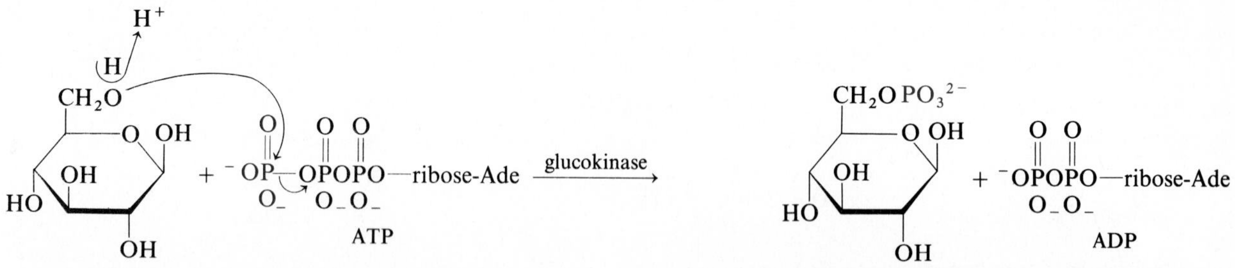

β-D-glucose

β-D-glucose-6- phosphate
(*enzyme specifies* interaction with —OH of C6)

FIGURE 10–9 Kinase-catalyzed formation of sugar phosphates.

Esterification: Phosphate Esters and Nucleoside Diphospho Sugars

Alcohols readily form esters when reacted with acids, anhydrides, or acyl halides. The most important types of sugar esters that occur in living cells are the *phosphate esters* (phosphoesters) and the *nucleoside diphosphate esters.*

SUGAR PHOSPHATE ESTERS The biosynthesis of sugar phosphate esters usually occurs via a kinase-catalyzed phosphoryltransfer process, with ATP as the phosphate donor (see Figure 10–9). A variety of kinase enzymes exist, producing monophosphoesters such as glucose-6-phosphate, ribose-5-phosphate, fructose-6-phosphate, and galactose-1-phosphate. Diphosphoesters also exist, such as fructose-1,6-bisphosphate. Some sugar phosphates are produced differently. For example, glucose-1-phosphate is formed by the stepwise cleavage of glycogen or starch polysaccharides with inorganic pyrophosphate, or by the isomerization of glucose-6-phosphate (see Figure 13–3).

The biological significance of the sugar phosphates is that they represent the *metabolically active form* of sugars. In other words, in most instances when a sugar participates as a substrate in an enzyme-catalyzed reaction, it does so as a phosphate ester, with the charged phosphate group engaging in binding and/or catalytic events. In addition to enhancing the reactivity of sugars, phosphorylation confers another benefit to a cell because the cell membrane is not very permeable to the passage of sugar phosphates. In effect, then, once they are formed within the cell, the sugar phosphates are more or less trapped inside.

NUCLEOSIDE DIPHOSPHATE ESTERS One of the most important reactions that the sugar-1-phosphates undergo is with another unit of a nucleoside triphosphate to yield *nucleoside diphospho derivatives.* The general reaction, catalyzed by an enzyme referred to as either an NDP-sugar pyrophosphorylase or a nucleotidyl transferase, is shown in Figure 10–10. Although all types of nucleoside diphospho sugars are found in nature, the *UDP sugars* are the most abundant.

The NDP sugars participate in many reactions. Basically, however, all of these reactions can be grouped under two types, involving (1) the biosynthesis of oligomeric and polymeric carbohydrates and (2) various chemical transformations of simple sugars, such as the isomeric interconversion of galactose and

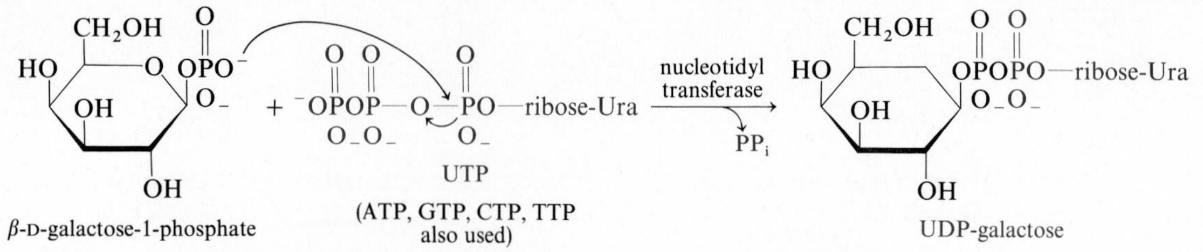

β-D-galactose-1-phosphate

(ATP, GTP, CTP, TTP also used)

UTP

UDP-galactose

FIGURE 10–10 Formation of NDP sugars.

glucose. Both types are discussed in Chapter 13. A few examples of each type are given in the next few pages.

1. $(glucose)_n$ + UDP-glucose → $(glucose)_{n+1}$ + UDP

2. UDP-glucose ⇌ UDP-galactose

Oxidation of Sugars to Sugar Acids and CO_2

Three different acid derivatives of aldoses can be produced (in the laboratory and in living cells) by oxidizing terminal groupings to COOH groups. Acids arising from the oxidation of the terminal —CHO group are called *glyconic acids*. If the terminal —CH_2OH group is oxidized, a *glycuronic acid* is produced; and if both terminal groups are oxidized, a *glycaric acid* is produced. The three acids of D-glucose are shown in Figure 10–11; also indicated is their tendency to undergo an intramolecular elimination of H_2O to yield six-membered *cyclic lactones*.

In plants and most vertebrates—except primates (including humans) and guinea pigs—*L-ascorbic acid* (vitamin C) can be produced from D-glucose. Involving the intermediate formation of sugar acids, this interesting sequence of enzyme-catalyzed reactions is outlined in Figure 10–12. After UDP-D-glucose is formed, the —C(6)H_2OH of glucose is oxidized to yield D-glucuronic acid. This acid, in turn, is reduced at the terminal —CHO to yield a sugar acid having terminal —CH_2OH and —COOH groups, that is, a glyconic acid. The specific glyconic acid, namely, L-gulonic acid, is evident by rotating the structure 180° in the plane of the paper. After a lactone is formed, a final dehydrogenation (oxidation) step yields L-ascorbic acid. Lacking the normal enzyme that catalyzes the last dehydrogenation step, humans are unable to synthesize ascorbic acid. Without a dietary supply humans are prone to the disease of scurvy.

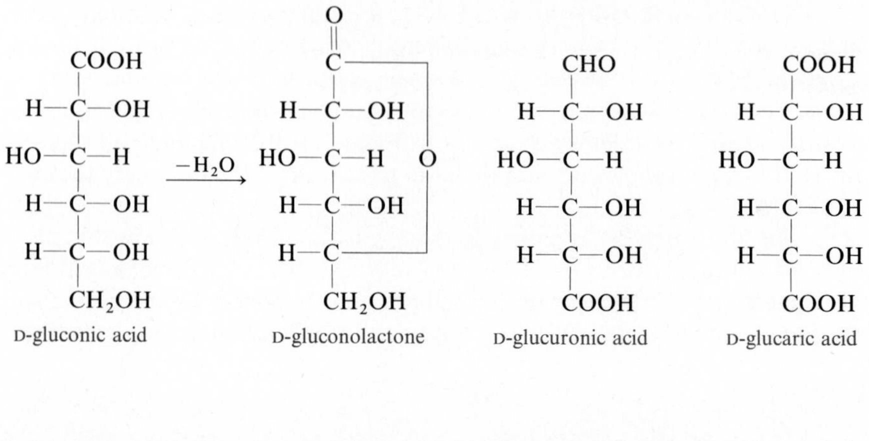

FIGURE 10–11 Acids derived from D-glucose

D-gluconic acid   D-gluconolactone   D-glucuronic acid   D-glucaric acid

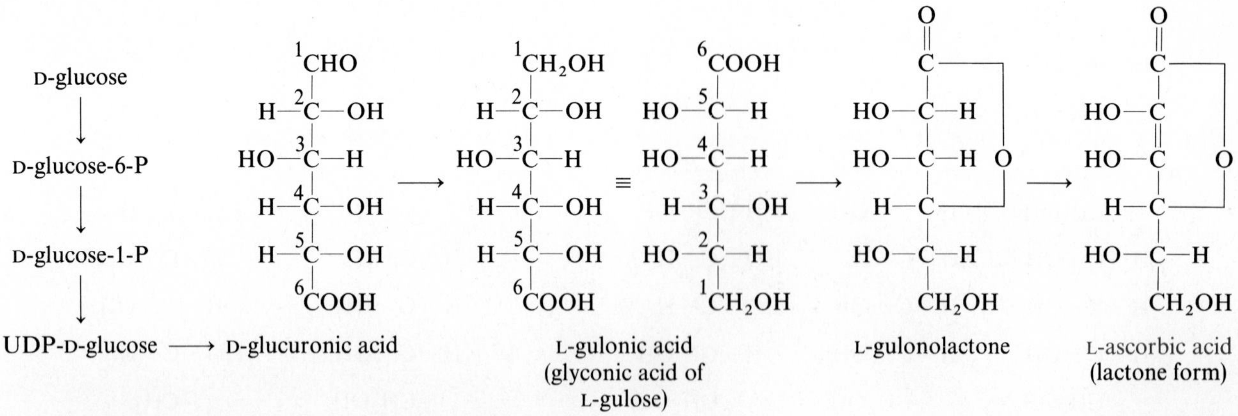

D-glucose
↓
D-glucose-6-P
↓
D-glucose-1-P
↓
UDP-D-glucose ⟶ D-glucuronic acid

$$
\begin{array}{c}
^1\text{CHO} \\
\text{H}-^2\text{C}-\text{OH} \\
\text{HO}-^3\text{C}-\text{H} \\
\text{H}-^4\text{C}-\text{OH} \\
\text{H}-^5\text{C}-\text{OH} \\
^6\text{COOH}
\end{array}
$$

$$
\begin{array}{c}
^1\text{CH}_2\text{OH} \\
\text{H}-^2\text{C}-\text{OH} \\
\text{HO}-^3\text{C}-\text{H} \\
\text{H}-^4\text{C}-\text{OH} \\
\text{H}-^5\text{C}-\text{OH} \\
^6\text{COOH}
\end{array}
\equiv
\begin{array}{c}
^6\text{COOH} \\
\text{HO}-^5\text{C}-\text{H} \\
\text{HO}-^4\text{C}-\text{H} \\
\text{H}-^3\text{C}-\text{OH} \\
\text{HO}-^2\text{C}-\text{H} \\
^1\text{CH}_2\text{OH}
\end{array}
$$

L-gulonic acid
(glyconic acid of L-gulose)

L-gulonolactone

L-ascorbic acid (lactone form)

(*this enzyme absent in humans)

FIGURE 10–12 Formation of L-ascorbic acid.

Ascorbic acid is one of a small group of substances, referred to as *antioxidants*, that protect living cells from the destructive action of powerful oxidizing agents that can be formed from O_2 (see Section 15–6). Remember that glutathione also performs a similar antioxidant function. Further research is still necessary to substantiate the claims that large daily doses of vitamin C can provide protection against the common cold and perhaps some types of cancer.

Reduction of Sugars to Sugar Alcohols and Deoxy Sugars

Two important reduced forms of the sugars are the *polyhydroxy alcohols* and the *deoxy structures*. In cells the sugar alcohols are usually formed in reactions catalyzed by specific dehydrogenases, using the coenzyme NADH or NADPH as a hydrogen (electron) donor. Some specific examples of reduced sugars are shown in Figure 10–13, and a brief statement of biological role is also given.

The formation of deoxy sugars involves the substitution of an OH group by an H atom. The most important biological example is D-2-deoxyribose, the sugar of DNA; the complex details of the D-ribose → D-2-deoxyribose conversion are discussed on p. 705. Another example is L-6-*deoxygalactose* (called L-*fucose*), which is found in the carbohydrate grouping of several glycoproteins in the surface coat of animal cells.

In all types of organisms glucose-6-phosphate is, among other things, a precursor in the biosynthesis of *inositol,* a six-membered cyclic polyhydroxy alcohol. The most common isomer (there are nine different stereoisomers) is *myoinositol:*

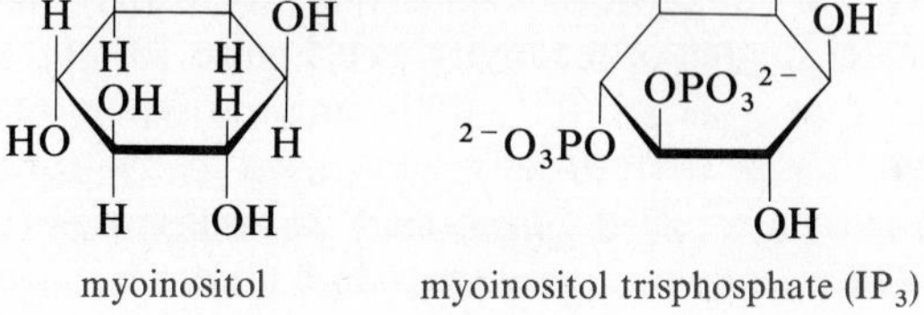

myoinositol

myoinositol trisphosphate (IP₃)

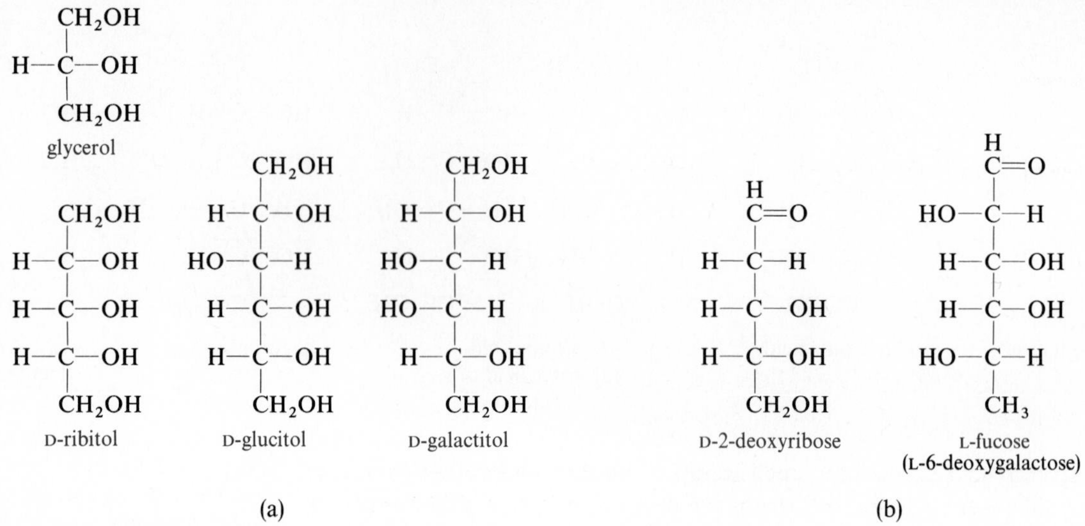

$$
\begin{array}{c}
\text{CH}_2\text{OH} \\
| \\
\text{H}-\text{C}-\text{OH} \\
| \\
\text{CH}_2\text{OH}
\end{array}
$$
glycerol

D-ribitol   D-glucitol   D-galactitol

D-2-deoxyribose   L-fucose
(L-6-deoxygalactose)

(a)   (b)

FIGURE 10–13 Reduced sugars.
(a) Polyhydroxy alcohols.
Glycerol is a component of lipids.
D-Ribitol is a component of
riboflavin and the coenzymes
FAD and FMN. D-Glucitol, also
called sorbitol, has various commercial uses, such as a sweetening
agent; high levels of it are found
in semen. D-Galactitol is also called
dulcitol. Cataract formation in the
lens of the eye may be related
to an accumulation of this sugar
alcohol. (b) Deoxy sugars.
D-2-Deoxyribose is found in
nucleotides of DNA. L-Fucose is
a key sugar in several animal
glycoproteins.

reduced
carbon

$(\text{CH}_2\text{O})_n + n\text{O}_2$

(photosynthesis)   (respiration)
energy   energy

$n\text{CO}_2 + n\text{H}_2\text{O}$

oxidized
carbon

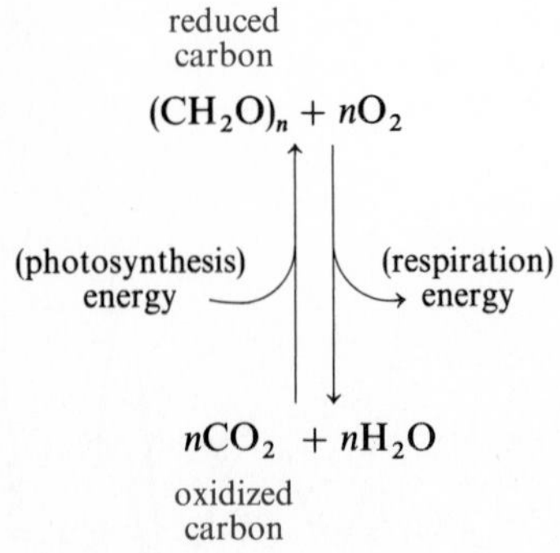

FIGURE 10–14 Photosynthesis
and respiration.

In plants a large portion of inositol occurs as a fully phosphorylated form (that is, a hexophosphoester), which is believed to be a storage form of phosphate in plants. A similar phosphate storage role is not found in animals. In animals most of the inositol is found in cell membranes as a component of the lipid *phosphatidyl inositol* (see Section 11–3). Also in Chapter 11, we will highlight a recent discovery that *inositol trisphosphate* may be an important bioregulator linked to the operation of some hormones and perhaps even to the operation of some oncogene proteins.

$CO_2 \rightleftharpoons (CH_2O)$ Redox Chemistry as the Foundation of Our Biosphere

The complete oxidation of a sugar results in its total degradation to CO_2 and H_2O and is accompanied by the release of a large amount of useful energy (see Figure 10–14). This overall process, of course, represents the oxidative and energy-yielding process of *respiration*. Respiration is complemented by the reductive, energy-requiring process of *photosynthesis,* which results in the fixation of atmospheric CO_2 for conversion to plant carbohydrate material. The many details of these metabolic processes are presented in Chapters 13 through 16.

Glycosides

When an alcohol (ROH) reacts with another alcohol (R′OH), the product is an ether (R—O—R′). Sugars can react in this way, with the *anomeric* C^a—OH site having a greater degree of reactivity (see Figure 10–15). When reaction is limited to the anomeric carbon (C^a), the structure that results is a full *acetal* called a *glycoside*. Those derived from pyranoses are called *pyranosides,* and those from furanoses are called *furanosides*. In either case the newly formed linkage of C^a—OR is called a *glycosidic bond*. Under stronger conditions the

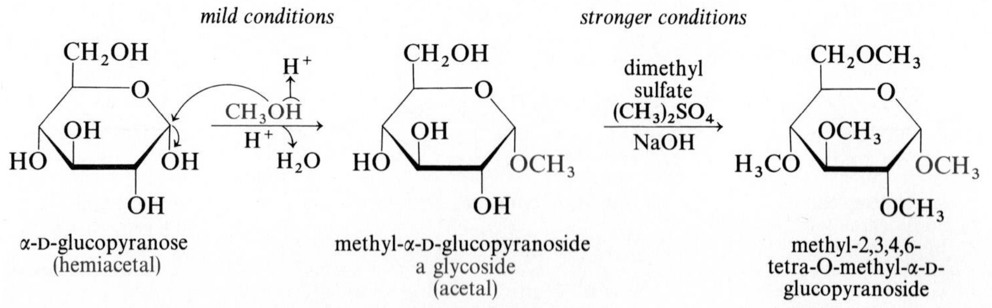

FIGURE 10–15 Glycoside formation.

other C—OH positions will react to give a completely substituted sugar.
The glycosidic linkage is of extreme biological significance:

> It represents (neglecting some obscure exceptions) the co-valent linkage between successive monosaccharides in oligo- and polysaccharides.

Different glycoside bonds can result from different combinations of the α or β carbon of one sugar and the various OH groups in the other sugar. The structures in Figure 10–16 illustrate two of the five different ways α-D-glucose molecules can be linked with each other. Note the symbolism used in identifying the specific pattern of each bond.

All types of combinations are found in nature, such as $\alpha(1 \to 3)$, $\alpha(1 \to 2)$, $\beta(1 \to 6)$, $\beta,\beta(1 \to 1)$, and so on. However:

> Each particular oligosaccharide and polysaccharide contains a specific pattern of glycoside bond(s) between the monomeric residues.

Indeed, in some instances this specific pattern of bonding is the primary structural difference between otherwise identical oligomers or polymers. With the exception of disaccharides, all linear oligomers or polymers will contain monomeric residues involved in two glycosidic linkages, except for the two residues at each end of the chain, which are involved in only one (see Figure 10–17). Some residues may be involved in three glycosidic bonds, a situation common to *branched* oligomers and polymers.

A variety of glycoside materials are found in nature, particularly in the plant kingdom. Several have therapeutic uses. *Digitoxin,* a potent stimulant of heart muscle, is but one of several examples. Several antibiotics are glycosides; some examples are *erythromycin, streptomycin,* and *puromycin.* You should also recall that a nucleoside contains a C—N glycoside bond.

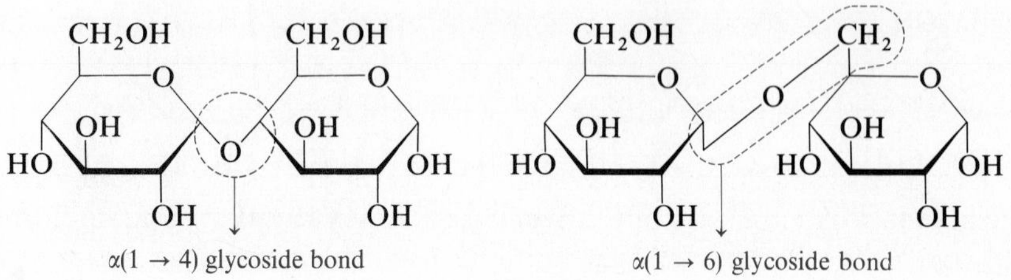

α(1 → 4) glycoside bond   α(1 → 6) glycoside bond

FIGURE 10–16 Two glycoside linkages between α-D-glucose. These two disaccharides of α-D-glucose are different chemical compounds with many different properties. What three other different linkages are possible?

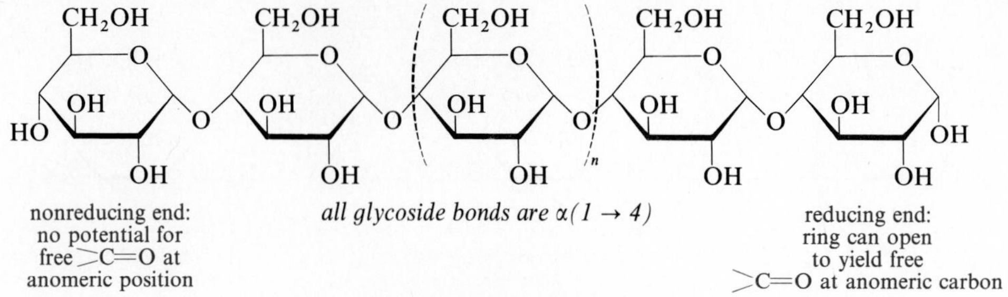

nonreducing end:
no potential for
free $>C=O$ at
anomeric position

all glycoside bonds are α(1 → 4)

reducing end:
ring can open
to yield free
$>C=O$ at anomeric carbon

FIGURE 10–17 Linear polyglucose chain.

Other Sugar Derivatives

Without examining the chemistry of their formation in living cells, let us conclude this section by merely identifying the structures of other important sugar derivatives found in many oligo- and polysaccharides. *Amino sugars* contain an amino ($-NH_2$) group in place of an $-OH$ group, and *N-acetyl amino sugars* contain the amino group with an acetyl

$$-\overset{\overset{\textstyle O}{\|}}{C}-CH_3$$

substituent. Hexose derivatives are most common, with the replacement position usually being C2. 2-Amino-D-glucose (also called D-*glucosamine*), 2-amino-D-galactose (also called D-*galactosamine*), and the corresponding acetylated forms, N-acetyl-D-glucosamine and N-acetyl-D-galactosamine, are the most abundant examples (see Figure 10–18).

N-acetyl-D-glucosamine is also a precursor of two other specific sugar derivatives: *muramic acid* (in bacteria), a component of bacterial cell walls, and *sialic acid* (in many types of cells), a component of many glycoproteins and glycolipids.

Certain types of polysaccharides found in animals contain sugar residues having a *sulfate group* ($-OSO_3^-$) in the structure. Specific examples are illustrated later in this chapter.

10–3 DISACCHARIDES

Among the many *disaccharides* of natural origin, *sucrose* and *lactose* are the most abundant and most important. Sucrose is composed of *α-D-glucose* and *β-D-fructose* linked via the α1 of glucose and the β2 of fructose—an

FIGURE 10–18 Amino sugar derivatives.

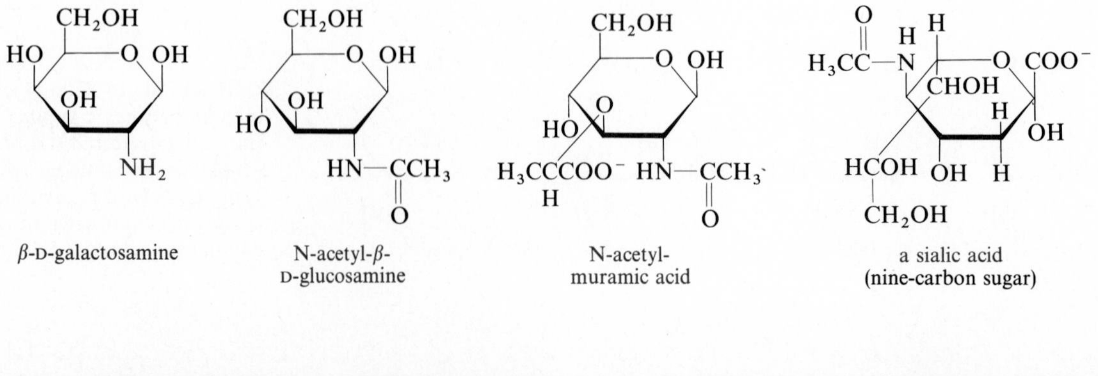

β-D-galactosamine

N-acetyl-β-
D-glucosamine

N-acetyl-
muramic acid

a sialic acid
(nine-carbon sugar)

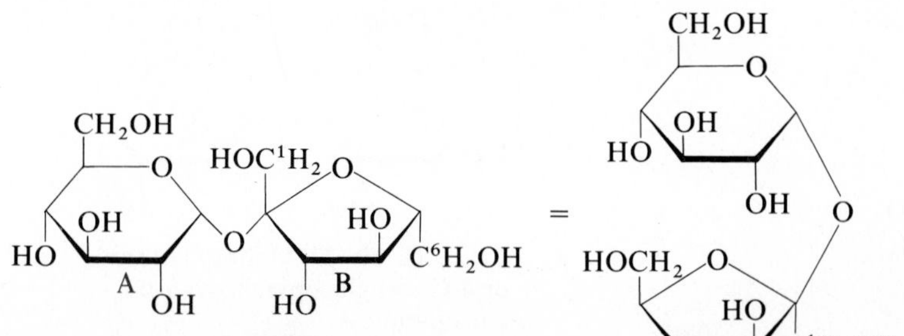

sucrose

2-O-α-D-glucopyranosyl-β-D-fructofuranoside

A

B

FIGURE 10–19 Haworth formula for sucrose. Since both anomeric carbons are involved in the glycoside bond, a free carbonyl group ($\diagup C{=}O$) is not possible. So sucrose is a nonreducing sugar.

$\alpha,\beta(1 \rightarrow 2)$ glycosidic linkage. Found throughout the plant kingdom, sucrose is most abundant in sugarcane, sugar beets, and maple syrup. It is the primary granulated product obtained from the processing of these materials and is commonly known as table sugar.

After being synthesized in the green leaves, sucrose is transported to various other parts of the plant, primarily for storage. When a carbon and an energy source are needed, the sucrose is then hydrolyzed to glucose and fructose, which enter the mainstream of metabolism. The same hydrolytic degradation occurs during digestion in animals that consume plants. Sucrose provides one of the major dietary supplies of hexoses for the animal kingdom. Everyone is familiar with its sweetening and flavor-enhancing properties, as well as the fact that an excess intake can be harmful. The Haworth formula for sucrose is shown in Figure 10–19. Note the explanation in the caption of Figure 10–19 for the classification of sucrose as a nonreducing sugar.

In plants the biosynthesis of sucrose proceeds by either of two routes, both of which involve the participation of UDP-glucose:

$$\text{UDP-D-glucose} + \text{D-fructose} \rightarrow \text{UDP} + \text{sucrose} \qquad \text{(major route)}$$

$$\text{UDP-D-glucose} + \text{D-fructose-6-phosphate} \rightarrow \text{UDP} + \text{sucrose phosphate} \rightarrow \text{sucrose} + P_1$$

Lactose (see Figure 10–20) is composed of *β-D-galactose* and *D-glucose,*

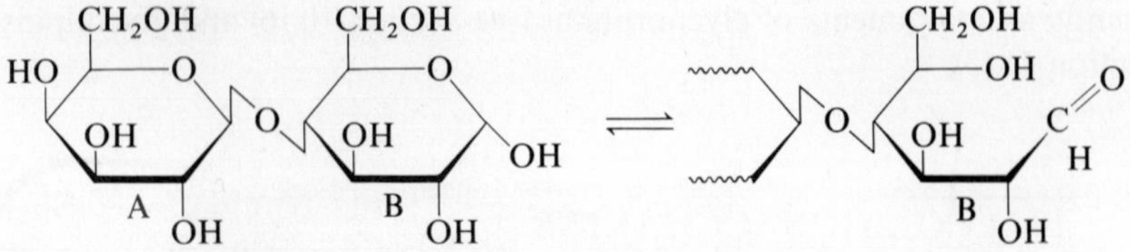

lactose (α form)

4-O-β-galactopyranosyl-α-D-glucopyranose

A

B

FIGURE 10–20 Lactose. Since the anomeric carbon of the glucose residue is not involved in the glycoside bond, a free carbonyl group is possible via ring opening. So lactose is a reducing sugar.

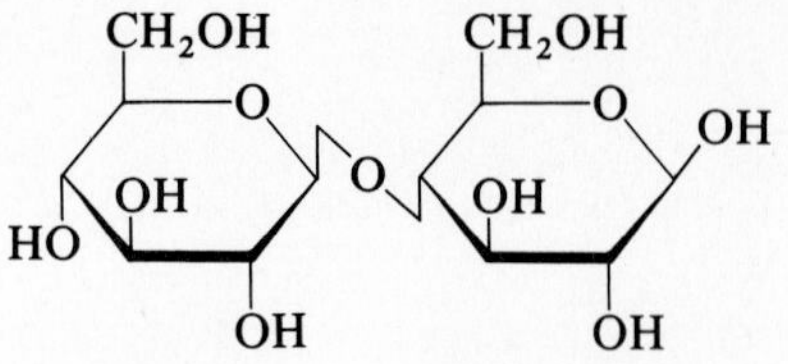

FIGURE 10–21 Cellobiose (β form), or 4-O-β-D-glucopyranosyl-β-D-glucopyranose.

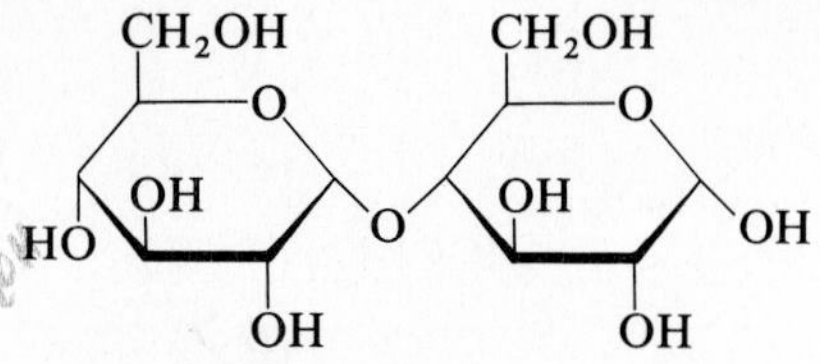

FIGURE 10–22 Maltose (α form), or 4-O-α-D-glucopyranosyl-α-D-glucopyranose.

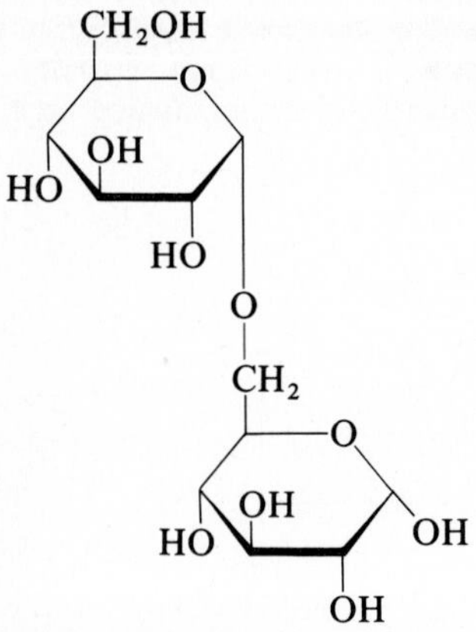

FIGURE 10–23 Isomaltose (α form), or 6-O-α-D-glucopyranosyl-α-D-glucopyranose.

linked via the β1 of galactose and position 4 of glucose, a β(1 → 4) glycosidic linkage. It is the most abundant carbohydrate material in the milk of mammals (milk is about 5% lactose).

After its formation from lactose during digestion, the metabolic utilization of D-galactose is preceded by conversion to D-glucose in a multistep, UTP-dependent process. Normal operation of this conversion is critical, since high levels of blood galactose contribute to severe physiological disturbances that can be fatal, especially to infants. The abnormality is a genetic, metabolic disorder involving a malfunctional enzyme in one of the steps in the D-galactose → D-glucose conversion. The condition is termed *galactosemia* (see Section 13–5). The biosynthesis of lactose in the mammary gland proceeds as follows:

UDP-D-galactose + glucose → UDP + lactose

Cellobiose (Figure 10–21), *maltose* (Figure 10–22), and *isomaltose* (Figure 10–23) are examples of homogeneous disaccharides: All are composed only of D-glucose. However, they differ in the nature of the glycosidic linkage. Cellobiose with β(1 → 4) is the sole repeating unit in cellulose, maltose with α(1 → 4) is the sole repeating unit of the amylose fraction of starch; isomaltose with α(1 → 6) is found in the amylopectin fraction of starch and glycogen. The notion of a repeating unit is developed further in the section on polysaccharides. As the structures here indicate, all three are reducing sugars (refer again to the captions of Figures 10–19 and 10–20).

Disaccharides are the most common oligomeric carbohydrates, occurring as free structures in living cells. Larger oligomeric groupings occur more frequently as components of glycoproteins (see Section 10–6) and glycolipids (see Section 11–3).

10–4 OLIGOSACCHARINS

A new and important function of carbohydrates has been recently discovered in plants:

Oligosaccharide materials, called *oligosaccharins,* function as plant hormones.

The regulatory functions include control of all the major plant processes: growth, development, reproduction, and defense against disease.

The first oligosaccharin discovered was a heptaglucoside composed of a pentaglucoside chain having all $\beta(1 \rightarrow 6)$ linkages, with the second and fourth glucose residue in the chain each having a single glucose branch residue linked by a $\beta(1 \rightarrow 3)$ bond:

$$\text{Glc} \xrightarrow{\beta(1 \rightarrow 6)} \text{Glc} \xrightarrow{\beta(1 \rightarrow 6)} \text{Glc} \xrightarrow{\beta(1 \rightarrow 6)} \text{Glc} \xrightarrow{\beta(1 \rightarrow 6)} \text{Glc}$$

$$\uparrow \beta(1 \rightarrow 3) \qquad\qquad \uparrow \beta(1 \rightarrow 3)$$

$$\text{Glc} \qquad\qquad\qquad \text{Glc}$$

(Glc)$_7$ oligosaccharin

The sites of branching are critical. Indeed, any other combination of branching on the pentaglucoside chain gives an oligosaccharide with no regulatory activity.

Oligosaccharins are produced as fragments from the various complex polysaccharides that comprise the cell wall structures of plants. The formation of an oligosaccharin may occur in response to the well-known plant hormones (auxin, abscisic acid, cytokinin, ethylene, and gibberellin), which may stimulate specific enzymes that degrade specific cell wall polysaccharides to give specific fragments. For an introductory review of this important finding, see the article by Albersheim and Darvill cited in the Literature.

10–5 POLYSACCHARIDES

The three elements of structure that define a *polysaccharide* are (1) the identity of the constituent glycosyl monomers, (2) the nature of the glycosidic linkages between them, and (3) when appropriate, the sequence of glycosyl residues. Relative to the first point, polysaccharides can be classified as *homopolysaccharides* (all glycosyl residues identical) or *heteropolysaccharides* (different glycosyl residues). Since heteropolysaccharides are usually composed of only two different glycosyl monomers in a *repetitive sequence*, they are non-informational molecules. Polysaccharides of natural origin vary greatly in size, ranging up to several thousand glycosyl residues. In addition, the residue count frequently varies among samples of the same material. Thus the molecular weight of polysaccharides has limited physicochemical value.

The following abbreviations will be used to identify glycosyl residues in oligo- and polysaccharides:

Glc for D-glucose	GlcNAc for N-acetyl-D-glucosamine
Gal for D-galactose	GalNAc for N-acetyl-D-galactosamine
Man for D-mannose	GlcUA for D-glucuronic acid
Fuc for L-fucose	IdoUA for L-iduronic acid
Xyl for D-xylose	NAN for N-acetyl-neuraminic acid
GalNH$_2$ for D-galactosamine	(sialic acid)
GlcNH$_2$ for D-glucosamine	

Homopolysaccharides: Cellulose

NOTE 10–2
Cellulose is sometimes called a
β-glucan, and amylose is called
an α-glucan.

Cellulose (see Note 10–2) is the most abundant organic compound of natural origin on the face of the earth. It occurs throughout the plant kingdom as a structural component of the cell wall, and frequently, it is the major component. In parts of some plants it is present in almost pure form. For example, the seed hairs of the cotton plant are particularly rich in cellulose, containing 98%–99%. Although plants are its major source, cellulose is also produced (in small quantity) by some bacteria and animal organisms.

Microscopic cellulose fibers are aggregates of a variable number of unbranched polyglucose chains in parallel alignment with each other. If your eyes had the resolving power of a microscope, the pages these words are printed on would serve as a good visual aid for studying such fibrils. In each chain *all residues are β-D-glucose,* and all are linked via $\beta(1 \rightarrow 4)$ glycosidic bonds, as illustrated in Figure 10–24. The notion of a *repeating disaccharide unit* in a polysaccharide chain should be obvious at this point, with β-cellobiose repeating in cellulose.

The structure of cellulose is well suited for its biological role. The hydrogen bonding capacity between individual chains is quite high, with each residue contributing three OH groups that may participate. This feature confers a high degree of strength to the intact fiber and is also the basis of its water insolubility. In the cell walls of plants these cellulose fibers are densely packed in layers that are further strengthened by the presence of other substances such as *hemicellulose, pectin,* and *lignin,* which function as cementing materials. The strength of this aggregate should be obvious when you gaze at a 100-ft oak tree standing erect even in a heavy wind. Hemicellulose is a polysaccharide composed primarily of D-xylose; pectin is a polysaccharide composed primarily of D-galacturonate. Lignin is a complex polyphenolic substance.

The nutritional value of cellulose is virtually nil for the higher animals, with the exception of ruminants. The reason is that the combined digestive secretions of the mouth, stomach, and intestine do not contain a *cellulase* that would cleave the $\beta(1 \rightarrow 4)$ glycosidic bonds and yield free glucose units. However, cellulases are found in nature, most commonly in various insects, snails, fungi, algae, and bacteria, including bacteria found in the rumen stomach of ruminant animals. For obvious reasons such organisms are referred to as being *cellulolytic.*

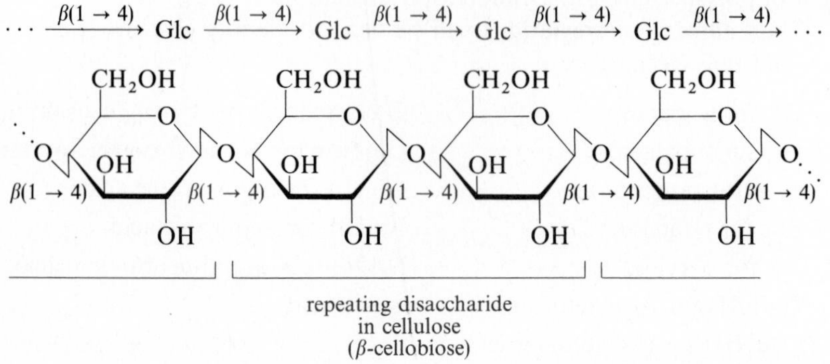

FIGURE 10 – 24 Cellulose, poly[$\beta(1 \rightarrow 4)$]D-glucose.

repeating disaccharide
in cellulose
(β-cellobiose)

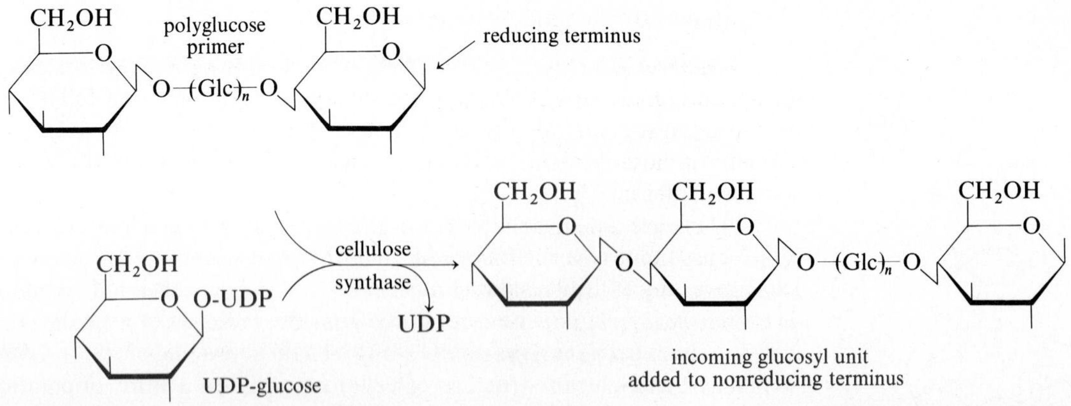

FIGURE 10–25 Primer extension biosynthesis of cellulose.

The biosynthesis of cellulose involves the successive condensation of β-D-glucosyl residues to the nonreducing terminus of a *primer polyglucose fragment*. The single enzyme involved is called *cellulose synthase*, using a nucleoside diphospho glucose derivative as substrate for the donation of the incoming glucosyl unit (see Figure 10–25). Enzymes of this sort are classified as transferases, specifically, *glycosyltransferases*.

Chitin

A subtle variation on the structure of cellulose is found in *chitin*, a linear homopolysaccharide consisting only of N-acetyl-D-glucosamine residues linked by $\beta(1 \rightarrow 4)$ bonds (see Figure 10–26). Because it is a close structural relative of cellulose, we might argue that it should have a biological function similar to that of cellulose. And indeed, it does, with chitin being the major organic structural component of the exoshells of some invertebrates (lobster, for example). It is also found in most fungi, many algae, and some yeasts as a cell wall component. As in cellulose, individual chains are bundled together via hydrogen bonding. Chitin biosynthesis is similar to that of cellulose.

repeating disaccharide in chitin

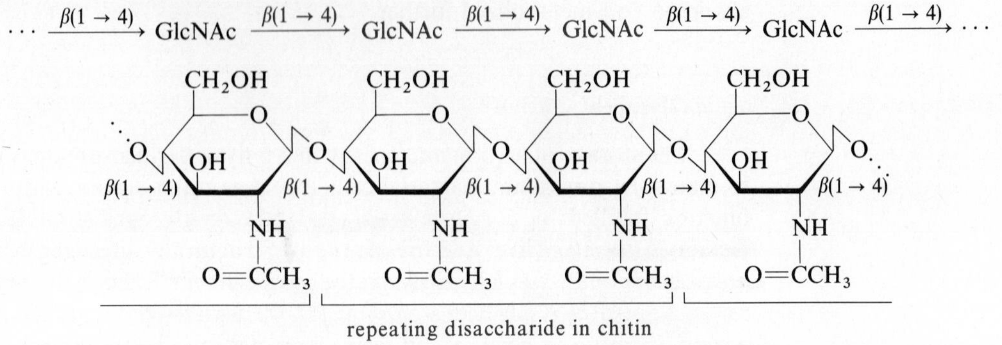

FIGURE 10–26 Chitin, poly[$\beta(1 \rightarrow 4)$]N-acetyl-D-glucosamine.

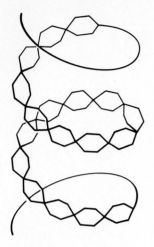

FIGURE 10–27 Helical conformation of amylose, poly[α(1 → 4)]D-glucose.

Starch Amylose and Starch Amylopectin

Starch is used to refer to a group of materials of varying size and shape. Occurring exclusively in plants, starch is found inside the plant cell as granules in the cytoplasm and also in plastids, including the chloroplast. Like cellulose, starch is a homogeneous polymer of D-glucose, but in starch the α-D-glucose isomer is present.

There are two distinct poly(α-D-glucose) structures in what is called starch. One component is termed *amylose,* the other *amylopectin.* Amylose is a *linear-chain structure* with all residues linked via α(1 → 4) glycoside bonds. Amylopectin is a *branched-chain structure,* resulting from the presence of a small number of α(1 → 6) linkages at various points along a chain consisting of α(1 → 4) linkages. Whereas the β-glucan structure of cellulose assumes a fibril orientation, the α-glucan structure of amylose prefers a *helical, coiled conformation* (see Figure 10–27). No preferred conformation of amylopectin has been suggested. A diagrammatic representation of each material is shown in Figure 10–28.

The biological role of starch is *food storage* in plants. When a source of carbon and energy is needed, starch is released from granules and then degraded by enzymes. Most plants contain two distinct hydrolyzing enzymes, traditionally named *α-amylase* and *β-amylase.* Both attack the amylose and amylopectin fraction at α(1 → 4) sites but in a different pattern. Cleavage with α-amylase is random, occurring at different loci to yield a mixture of glucose and maltose. The action of β-amylase is more ordered; it is characterized by successive removal of only maltose units, beginning at a nonreducing terminus (see Figure 10–28). (Note that amylose has a single nonreducing terminus and a single reducing terminus, but amylopectin has several nonreducing ends and a single reducing terminus.) Neither enzyme is capable of hydrolyzing the α(1 → 6) linkages. Thus, whereas the combined action of the two enzymes will completely degrade amylose to glucose and maltose, amylopectin is only partially degraded. However, other catalysts, called *debranching enzymes,* specific for hydrolyzing the α(1 → 6) linkage, do exist in cells.

Unlike cellulose, starch is digestible by humans (and most other organisms) owing to the presence of *salivary amylase* and *pancreatic amylase* in the digestive secretions. Both enzymes are similar in action to the α-amylase of plants. In combined action with other digestive enzymes (notably *maltase* and debranching enzymes), starch is completely degraded to α-D-glucose, which is absorbed and metabolized further.

Glycogen

The storage of carbon and energy in a poly-α-D-glucose state is not unique to plants. In animal and bacterial cells the same thing occurs, with the storage function fulfilled by *glycogen.* In higher animals glycogen granules are most abundant in cells of liver and muscle tissue. Structurally, glycogen is a branched polyglucose molecule identical to the amylopectin fraction of starch in all respects except that glycogen is more highly branched (see Figure 10–28). Branch points on glycogen occur about every 8–10 residues along the α(1 → 4) chain. In starch amylopectin, branch points occur about every 25–30 residues. Within

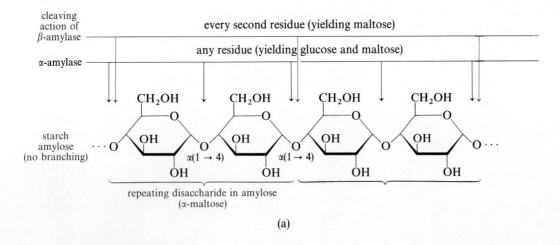

(a)

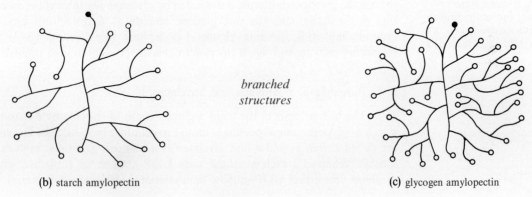

(b) starch amylopectin   *branched structures*   (c) glycogen amylopectin

(d)

FIGURE 10–28 Diagrammatic representations of starch. (a) Starch amylose. The action of α- and β-amylases is also indicated. (b) Starch amylopectin. (c) Glycogen amylopectin. In (b) and (c) an α(1 → 6) linkage occurs wherever there is a junction of two lines. Along each line all glucose residues are linked via α(1 → 4) bonds. These representations do not depict any three-dimensional structure; they depict only the branching and the greater degree of it in glycogen. Only one of the ends corresponds to a reducing terminus (●); all others are nonreducing (○). (d) Glycoside bonding at a branch site.

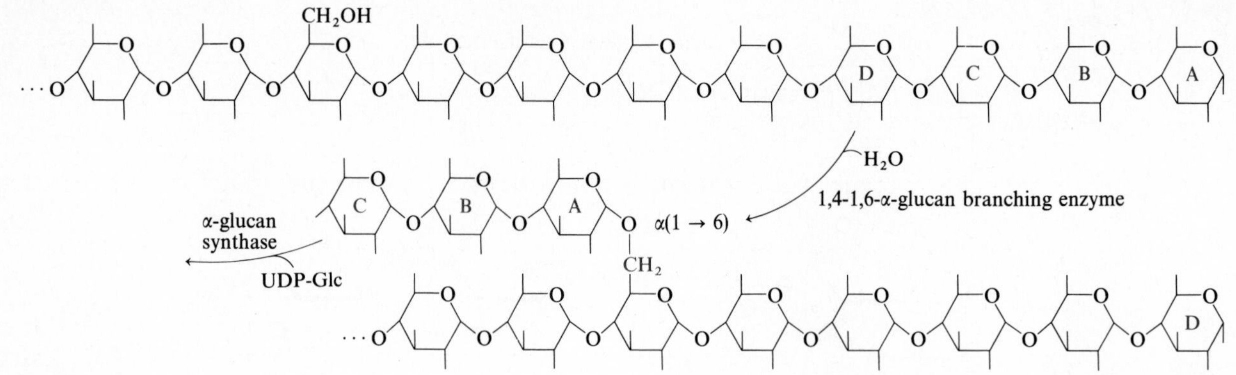

FIGURE 10–29 Formation of branch sites in glycogen and starch amylopectin. After a branch site is formed, the nonreducing end of the branch will be further extended by α-glucan synthase, as indicated.

the cell the glycogen molecule is degraded by *glycogen phosphorylase,* an enzyme that sequentially removes one glucose residue at a time from any of the nonreducing ends, yielding glucose-1-phosphate (see Section 13–3). A debranching enzyme will again operate to cleave $\alpha(1 \rightarrow 6)$ branch points.

Biosynthesis of Amylose and Amylopectin

The biosynthesis of the main α-glucan chain of starch amylose and either starch or glycogen amylopectins is similar to cellulose biosynthesis. The enzymes are called *starch synthase* and *glycogen synthase* (or, generally, α-glucan synthases). Whereas starch synthase uses UDP-glucose as substrate, glycogen synthase uses either UDP-glucose (animals) or ADP-glucose (bacteria):

$$\text{UDP-Glc} + \text{Glc} (\text{Glc})_n \text{Glc} \xrightarrow[\text{synthase}]{\alpha\text{-glucan}} \text{Glc}\!-\!\text{Glc}\!-\!(\text{Glc})_n\!-\!\text{Glc} + \text{UDP}$$

primer

The formation of $\alpha(1 \rightarrow 6)$ branch sites in amylopectins occurs in an interesting manner. The C6 branch site is *not* formed by a direct transfer from a single NDP-glucosyl donor. Rather, the enzyme (*1,4-1,6-α-glucan branching enzyme*) cleaves a short oligonucleotide fragment from a nonreducing end of an α-glucan chain and transfers it to a C6 hydroxyl (see Figure 10–29). The nonreducing end of this branch fragment can then be further extended by the action of an α-glucan synthase.

Glycosaminoglycans

Heteropolysaccharide substances are widely distributed in the animal kingdom, with the *glycosaminoglycans* comprising a particularly important class in vertebrates. The glycosaminoglycans are so-named because of the presence of amino sugars, usually the N-acetyl forms of D-glucosamine and D-galactosamine. Another common characteristic is the presence of many negatively charged carboxylate ($-\text{COO}^-$) and sulfate ($-\text{OSO}_3^-$) groups.

Glycosaminoglycans do not occur in the free state. Rather, they are found in association with other substances, notably proteins and protein/lipid aggre-

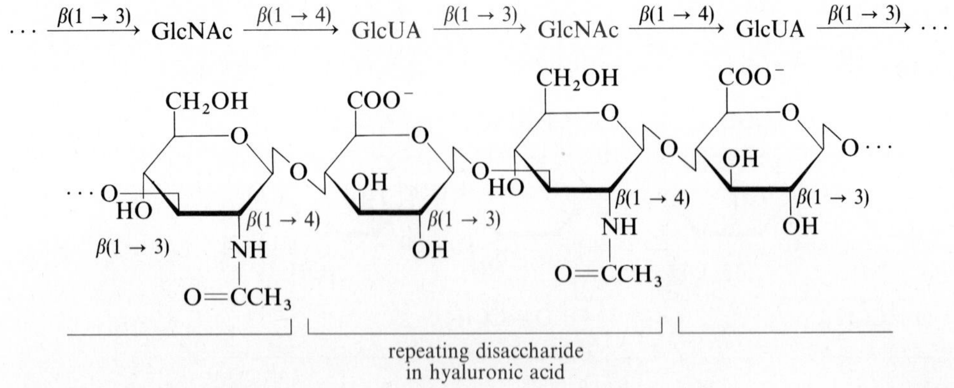

$$\cdots \xrightarrow{\beta(1 \rightarrow 3)} \text{GlcNAc} \xrightarrow{\beta(1 \rightarrow 4)} \text{GlcUA} \xrightarrow{\beta(1 \rightarrow 3)} \text{GlcNAc} \xrightarrow{\beta(1 \rightarrow 4)} \text{GlcUA} \xrightarrow{\beta(1 \rightarrow 3)} \cdots$$

repeating disaccharide
in hyaluronic acid

FIGURE 10–30 Hyaluronic acid. The structure is an alternating copolymer of D-glucuronic acid and N-acetyl-D-glucosamine. As indicated, the composition of an alternating copolymer can be described in terms of a repeating disaccharide with specific intraglycoside and interglycoside linkages.

gates. The term *proteoglycan* is used to refer to a protein/glycosaminoglycan association involving a covalent attachment.

In this section we will describe the structures of some important glysoaminoglycans in vertebrates, namely, hyaluronic acid, chondroitin sulfate, keratan sulfate, and heparin. The first three are part of the proteoglycan composition of connective tissue, which is described later in the chapter (see Section 10–7).

HYALURONIC ACID   In connective tissue *hyaluronic acid* is the primary component of *ground substance,* a gelatinous matrix with embedded collagen fibrils, filling the extracellular spaces of tissue. Another abundant source is synovial fluid, a viscous packing around bone joints serving as a lubricant and a shock absorber. The vitreous humor and umbilical cord are also rich in hyaluronic acid.

The structure is a linear polymer with a disaccharide repeating unit of *D-glucuronic acid* and *N-acetyl-D-glucosamine*. These residues are linked by a $\beta(1 \rightarrow 3)$ bond, and successive disaccharide repeating units are linked by $\beta(1 \rightarrow 4)$ bonds (see Figure 10–30). The high viscosity of hyaluronic acid is in part related to its polyanionic character at physiological pH (because of the many negatively charged —COO$^-$ groups), which favor excessive hydration and interchain hydrogen bonding.

CHONDROITIN SULFATE AND KERATAN SULFATE   Two other glycosaminoglycans associated with connective tissue are *chondroitin sulfate* and *keratan sulfate*. Each has the following similarities to hyaluronic acid: an alternating $\beta(1 \rightarrow 3)$ and $\beta(1 \rightarrow 4)$ pattern of glycoside bonds and the presence of an N-acetyl amino sugar. In addition to the identity of the specific glycosyl residues, another difference is the presence of negatively charged sulfate (—OSO$_3^-$) groups.

The repeating disaccharide of chondroitin sulfate is composed of *D-glucuronic acid and N-acetyl-D-galactosamine*, linked as $\beta(1 \rightarrow 3)$. Sulfate groups are inserted at either C4 or C6 positions of GalNAc residues (see Figure 10–31). The presence of 4-O-SO$_3^-$, 6-O-SO$_3^-$, and —COO$^-$ groups contributes to a very polyanionic structure. The length of a chondroitin sulfate chain can

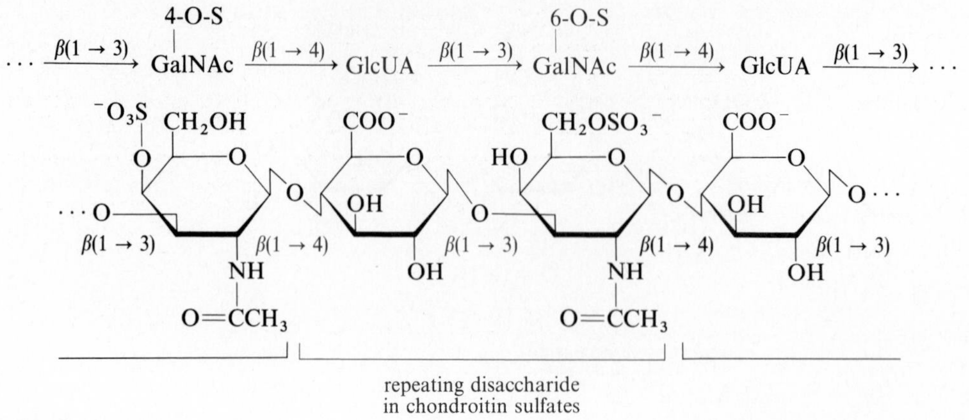

FIGURE 10–31 Chondroitin sulfate.

vary considerably, ranging from 15 to 150 disaccharide units in different tissues. Chain length and the extent of sulfation may be affected by the age of the tissue and may also be related to some disease conditions.

The repeating disaccharide of keratan sulfate is composed of *D-galactose* and *N-acetyl-D-glucosamine,* linked as $\beta(1 \to 3)$. Sulfate groups are added principally at C6 positions of GlcNAc residues (see Figure 10–32). Chain lengths, ranging from 10 to 50 disaccharide units, are typically less than for chondroitin sulfates.

The biosynthesis of chondroitin sulfate, keratan sulfate, and other glycosaminoglycans occurs in the Golgi body. Throughout the entire assembly process the growing polysaccharide chains are attached to serine residues in the protein component of the final proteoglycan structure. The protein is formed on ribosomes associated with the endoplasmic reticulum, moved into the lumen of the reticulum, and transported to the Golgi body for modification into the proteoglycan.

Highly specific *glycosyltransferases* and *sulfotransferases* in the Golgi body function in glycosaminoglycan formation. The process begins with the assembly of short oligosaccharide segments that serve as specific primers for individual glycosaminoglycans. In the case of chondroitin sulfate biosynthesis (see Figure

FIGURE 10–32 Keratan sulfate.

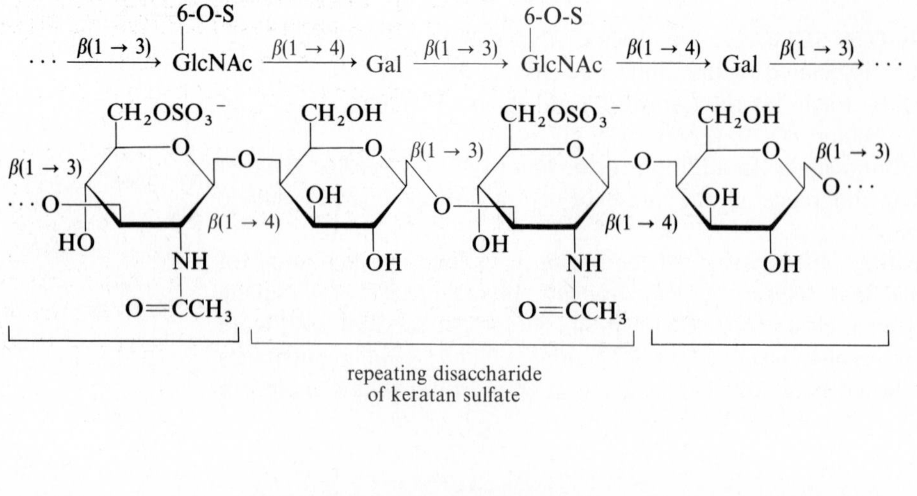

repeating disaccharide
of keratan sulfate

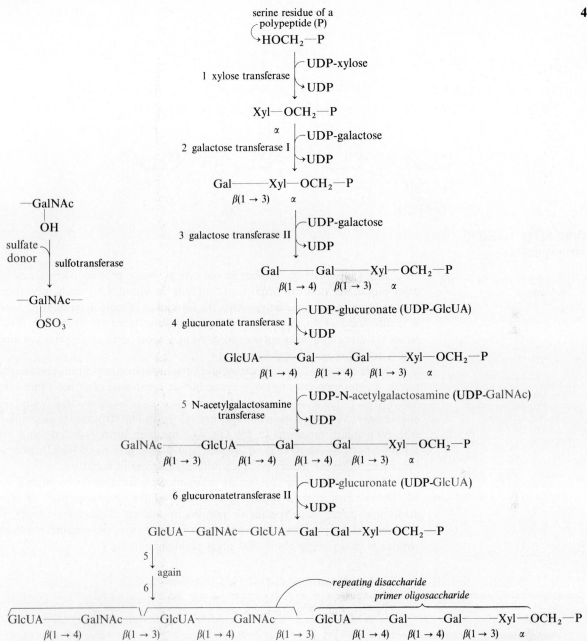

FIGURE 10–33 Biosynthesis of a chondroitin sulfate. The source of the sulfate is described in Chapter 18. After the final step shown here, there are 15–150 more cycles of steps 5 and 6 to extend the chain further, and there is random action of sulfotransferase enzymes to attach —SO_3^- groups to C4 and C6 of GalNAc residues.

10–33), a tetrameric —Xyl—Gal—Gal—GlcUA linker is assembled on serine side chains. The serine—xylose linkage is a glycoside bond involving the anomeric carbon of xylose. The successive alternating action of two glycosyltransferases, one specific for UDP-GalNAc and the other for UDP-GlcUA, extends the primer with the GalNAc—GlcUA repeating disaccharide. Sulfate groups are added as indicated in the caption of Figure 10–33.

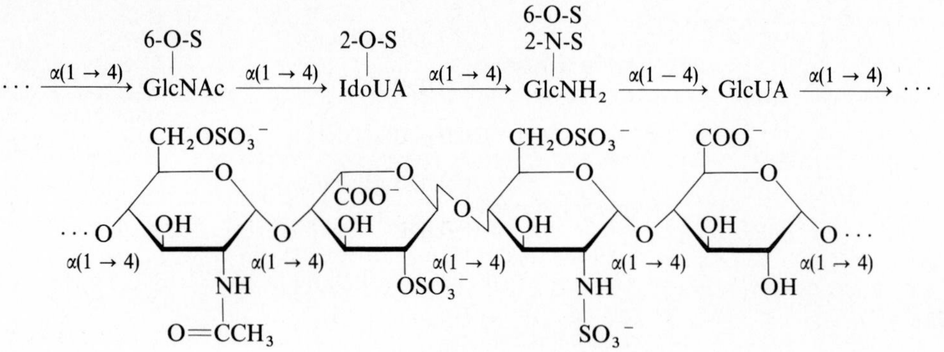

$$\cdots \xrightarrow{\alpha(1 \to 4)} \overset{\overset{\displaystyle 6\text{-}O\text{-}S}{|}}{\text{GlcNAc}} \xrightarrow{\alpha(1 \to 4)} \overset{\overset{\displaystyle 2\text{-}O\text{-}S}{|}}{\text{IdoUA}} \xrightarrow{\alpha(1 \to 4)} \overset{\overset{\displaystyle \overset{6\text{-}O\text{-}S}{2\text{-}N\text{-}S}}{|}}{\text{GlcNH}_2} \xrightarrow{\alpha(1 - 4)} \text{GlcUA} \xrightarrow{\alpha(1 \to 4)} \cdots$$

FIGURE 10–34 Typical sequence in heparin.

HEPARIN   Several tissues in the animal body produce substances that enter blood and perform important functions in blood. One such substance is *heparin,* a proteoglycan that prevents the formation of clots in circulating blood. It is also used as an anticoagulant agent, administered intravenously (IV) to patients during and after surgery to prevent a short-term impairment of circulation during these critical times.

More complex than other structures, the heparin glycosaminoglycan structure is a heterogeneous, linear polysaccharide composed of at least four sugar derivatives: *D-glucuronic acid, L-iduronic acid, D-glucosamine,* and *N-acetyl-D-glucosamine.* Most glycoside linkages are $\alpha(1 \to 4)$. The structure is also sulfated with 6-O-sulfate functions in the GlcNAc residues, both N-sulfate and 6-O-sulfate groups in GlcNH$_2$ residues, and 2-O-sulfate groups in IdoUA residues. Moreover, different tissues produce different degrees of sulfation.

A sample structure illustrating many of these features is shown in Figure 10–34. The disaccharide repeating unit involves the general pattern of uronic acid/amino sugar, with specific derivatives in different disaccharide units. The heparin proteoglycan aggregate has about 5–15 of these polysaccharide chains, with each chain being about 500 sugar residues in length.

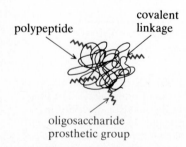

FIGURE 10–35 General representation of glycoprotein structure. Carbohydrate groups are exposed on the surface.

10–6 GLYCOPROTEINS

There are different types of carbohydrate and (poly)peptide conjugates that occur naturally. The term *glycoprotein* is generally used in reference to a "true" molecule of specific size composed of one or more oligosaccharide units covalently attached to specific amino acid side chains (see Figure 10–35). A glycoprotein usually has a higher percentage of protein than carbohydrate. The terms *proteoglycan* and *peptidoglycan* designate massive aggregates, composed of carbohydrate and proteins or small peptides, for which the word *molecule* has no precise meaning. Proteoglycan particles have a higher percentage of carbohydrate than protein. Glycoproteins and some related topics are examined in this section; proteoglycans and peptidoglycans are described in the next section.

General Features of Glycoproteins

Glycoproteins occur in all types of organisms, but they are especially prevalent in the fluids and cells of animals, where they are associated with many functions (see Table 10–2). They are found extensively in the membranes of cells or in association with the membrane as a component of the surface coat. The list in Table 10–2 reveals that many of these functions relate to communication—cell/cell recognition, cell/molecule recognition, organelle/molecule recognition, and molecule/molecule recognition. Clearly, all of these processes are important normal processes in all living organisms. Abnormalities of these processes are most probably associated with many diseases, including cancer, and with other conditions such as rejection of tissue transplants.

An obvious question concerns the reason for the presence of carbohydrate. One proposal is that the attachment of sugars to a protein is the identifying chemical label used to tag proteins that are destined to be utilized outside of the cell or in the membranous network of the cell. Thus those proteins to be retained and used in the cytoplasm of the cell are nonglycosylated. Although this hypothesis is still being debated, there is clear evidence that its basic premise is valid, namely, that the carbohydrate moiety does confer an *additional recognition factor* to the protein. In the simplest sense this premise means that the specific binding of a glycoprotein to another molecule or molecular aggregate may be based on the specific binding of the carbohydrate moiety.

Glycoproteins as a group exhibit great differences in their carbohydrate content, which ranges from less than 1% to as high as 80% of total weight. Glycoproteins with 4% or more carbohydrate are sometimes termed *mucoproteins* because they exhibit a very high viscosity.

> The covalent linkage to the polypeptide is made via a glycoside bond to the side chain of either serine, threonine, asparagine, or hydroxylysine (when it is present, as in collagen).

Oligosaccharide groups attached to the —OH group of serine and threonine are called *O-linked,* and those attached to the amide —NH_2 group of asparagine are called *N-linked.* The number of oligosaccharide groups per protein molecule is variable, but all groups in the molecule are usually identical.

The sugars that commonly occur in the oligosaccharide grouping include D-galactose, D-glucose, D-mannose, L-fucose, N-acetyl-D-glucosamine, N-acetyl-D-galactosamine, and sialic acid. In many cases the presence of sialic acid is especially important. The total individuality of the oligosaccharide grouping depends on (1) the composition of sugar residues, including the identification of the anomeric configuration of each residue, (2) the sequence of the residues, (3) the pattern of glycosidic linkages within the sequence, and (4) the nature of the linkage to the protein. A few examples of oligosaccharide groupings are shown in Table 10–3.

Assembly of N-Linked Oligosaccharides

The biosynthesis of O-linked oligosaccharides occurs in the Golgi body via the action of specific glycosyltransferases, as previously described for glycos-

TABLE 10–2 Listing of glycoproteins according to the type of function associated with a particular glycoprotein

Hormones

Antibodies

Enzymes[a]

Receptor proteins[a]

Transport proteins[a]

Cell adhesion proteins[a]

Growth control proteins[a]

Cell recognition proteins[a]

Proteins conferring blood group characteristics[a]

Proteins conferring structural stability to multimolecular aggregates

[a] Glycoproteins of these types are usually present in the surface coat and the plasma membrane of cells.

TABLE 10–3 Oligosaccharide groupings in some glycoproteins

1. A glycoprotein isolated from the membrane of the glomerulus in kidney:

$$\text{Glc} \xrightarrow{\alpha(1 \to 2)} \text{Gal} \xrightarrow{\beta} \text{OH of Hylys side chains}$$

2. One of the blood group–determining (A positive) proteins in pig:

$$\text{GalNAc} \xrightarrow{\alpha(1 \to 3)} \text{Gal} \xrightarrow{\beta(1 \to 3)} \text{GalNAc} \xrightarrow{\alpha} \text{OH of Ser and Thr side chains}$$

$$\uparrow \alpha(1 \to 2) \qquad \uparrow \alpha(2 \to 6)$$

Fuc   Nan   approximately 500 units per molecule

3. The major glycoprotein present in the membrane of human red blood cells:

$$\text{Nan} \xrightarrow{\alpha(2 \to 3)} \text{Gal} \xrightarrow{\beta(1 \to 3)} \text{GalNAc} \xrightarrow{\alpha} \text{OH of Ser and Thr side chains}$$

$$\uparrow \alpha(2 \to 6)$$

Nan

Exactly 16 units per molecule; this protein also contains a single copy of another oligosaccharide grouping attached to a specific asparagine residue.

4. In one of the human antibodies, namely, immunoglobulin G:

$$\text{Nan} \xrightarrow{\alpha(2 \to 6)} \text{Gal} \xrightarrow{\beta(1 \to 4)} \text{GlcNAc} \xrightarrow{\beta(1 \to 2)} \text{Man}$$

$$\alpha(1 \to 6)$$

$$\text{Man} \xrightarrow{\beta(1 \to 4)} \text{GlcNAc} \xrightarrow{\beta(1 \to 4)} \text{GlcNAc} \longrightarrow \text{NH}_2 \text{ of Asn side chains}$$

$$\alpha(1 \to 3)$$

$$\text{Nan} \xrightarrow{\alpha(2 \to 6)} \text{Gal} \xrightarrow{\beta(1 \to 4)} \text{GlcNAc} \xrightarrow{\beta(1 \to 2)} \text{Man}$$

$$\beta(1 \to 4) \qquad\qquad\qquad \alpha(1 \to 6)$$

GlcNAc   Fuc

Some glycoproteins contain oligosaccharide groups even more complex than this one.

5. A protein (antifreeze protein) isolated from the serum of Antarctic fishes; the intact protein has the amazing property of lowering the freezing point of water to the same extent as *an equal weight* of NaCl:

$$\text{Gal} \xrightarrow{\beta(1 \to 3)} \text{GalNAc} \xrightarrow{\alpha} \text{OH of Thr side chains} \qquad 31 \text{ units per molecule}$$

Note: Gal = D-galactose; Glc = D-glucose; Man = D-mannose; Fuc = L-fucose; GlcNAc = N-acetyl-D-2-glucosamine; GalNAc = N-acetyl-D-2-galactosamine; Nan = acetyl-neuraminic acid (sialic acid); Hylys = hydroxylysine; Asn = asparagine; Ser = serine; Thr = threonine.

NOTE 10 – 3
Dolichol phosphate is a polyisoprenoid lipid whose structure is shown in Figure 11–30.

aminoglycans (see Figure 10–33). However, research in the past few years has shown that the formation of N-linked oligosaccharides occurs in two phases, beginning in the endoplasmic reticulum and finishing in the Golgi body (see Figure 10–36). At some point during assembly or after entry into the lumen of endoplasmic reticulum, a polypeptide destined for conversion to an N-linked glycoprotein is selected (see below) for reaction with a lipid substance that carries an oligosaccharide group. The lipid carrier is called *dolichol phosphate* (Dol-P) (see Note 10–3). The major dolichol-linked oligosaccharide has 14 sugar units consisting of $\text{Glc}_3\text{Man}_9\text{GalNAc}_2$. The glycosyltransferases responsible for

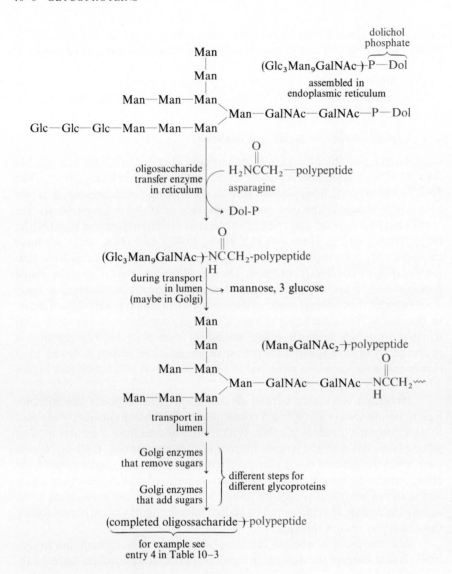

FIGURE 10–36 Biosynthesis of asparagine-linked oligosaccharide groups in glycoproteins. All oligosaccharides in this category originate in the endoplasmic reticulum, acquiring the same 14-residue oligo unit linked to dolichol phosphate as carrier. After transfer to asparagine residues is complete, further processing in both endoplasmic reticulum and Golgi body gives the completed oligosaccharide. Specific glycoside linkages are not shown.

the stepwise formation of the $Glc_3Man_9GalNAc_2$ linked to Dol-P are localized in the endoplasmic reticulum.

The reticulum enzyme responsible for transferring the lipid-linked oligosaccharide is highly specific for selecting the asparagine residues in the peptide. Recent studies indicate that the amino acid sequence of —asn—X—ser(thr)— (where X is any amino acid) is an absolute requirement and that the asparagine is preferably in an exposed tight β turn.

During transport in the lumen one of the terminal Man residues is removed (see Figure 10–36), the three Glc residues are removed, and a $Man_8GalNAc_2/$ polypeptide species is delivered to the Golgi body. (The Man and Glc_3 excisions may also occur in the Golgi body.) What happens to the glycoproteins in the Golgi body depends on their destination in the cell. The Golgi modifications include removing additional Man residues and adding other specific residues including sialic acid for those glycoproteins destined for secretion and assimilation into the plasma membrane. For the IgG protein shown in Table 10–3, five Man residues are removed and three N-acetylglucosamine residues, one fucose residue, and two sialic residues are added—all in specific positions and sequence.

Glycosylated Hemoglobin and Diabetes

Adult human blood contains three species of hemoglobin (Hb): hemoglobin A ($\alpha_2\beta_2$; 97%), hemoglobin A_2 ($\alpha_2\delta_2$; 2.5%), and hemoglobin F ($\alpha_2\gamma_2$; 0.5%). The HbA is the normal hemoglobin; HbF is fetal hemoglobin present in larger amounts in the fetus; the biological significance of HbA_2 is uncertain. Recent studies have established that there are at least five different forms of HbA: HbA_0 (96%), HbA_{Ia1} (0.2%), HbA_{Ia2} (0.2%), HbA_{Ib} (0.4%), and HbA_{Ic} (3%). All have the same $\alpha_2\beta_2$ composition, but the β chains in the four minor forms have simple sugar units covalently attached. They are thus referred to as *glycosylated hemoglobins*. All are apparently formed via spontaneous, nonenzymatic reactions between the carbonyl group of the sugar and an amino group of the β chain. In the case of HbA_{Ic} the sugar is free glucose (glc) and the amino group is contributed by the N terminal valine residue of each β chain. The reaction is proposed to involve a rapid formation of an aldimine adduct followed by a slower rearrangement to a more stable 1-deoxyfructosyl substituent (see Figure 10–37).

Uncertain is whether normal glycosylation of hemoglobin has any biological significance or if it is just a consequence of Hb being constantly exposed to an environment rich in glucose. Whatever, diabetics have a greater level of glycosylated Hb, with the largest increase (2–3 times normal) in HbA_{Ic}. Moreover, when the elevated blood glucose of a diabetic is brought under control, the HbA_{Ic} returns to normal levels. These correlations serve as the bases for a new and sensitive analytical procedure (HbA_{Ic} levels in blood are monitored) to diagnose the disease of diabetes and to evaluate the degree of success in controlling the disease by insulin therapy.

Also uncertain is whether the increased level of glycohemoglobin in diabetes causes some of the diabetic symptoms. However, researchers have estab-

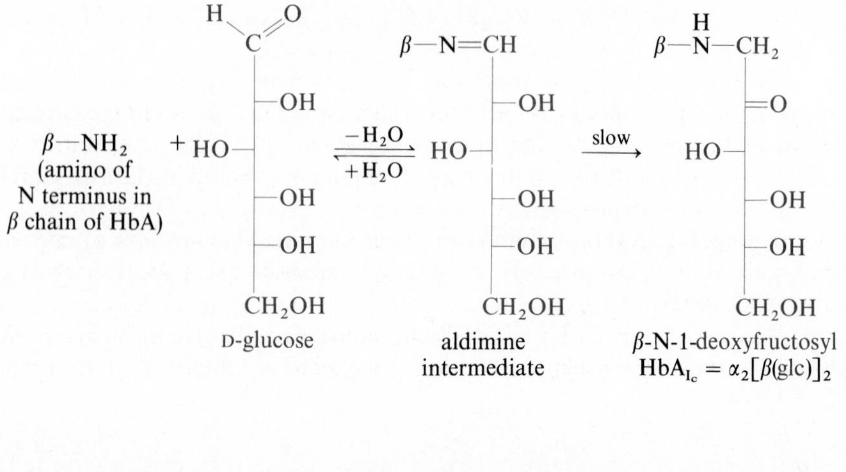

FIGURE 10 – 37 Nonenzymatic glycosylation of hemoglobin Other proteins also react in the same way when the blood glucose level increases.

lished that glycosylated hemoglobins have a lower affinity for oxygen. Thus elevated levels of these hemoglobins could critically diminish oxygen delivery to certain tissues. Some diabetic symptoms may also result from the glycosylation of other proteins in tissues, such as lens, kidney, and nerve, resulting in the formation of cataracts and impaired renal and nerve functions.

Lectins

Throughout the plant kingdom there exist proteins, called *lectins,* that bind very strongly to glycoproteins. The binding between lectin and glycoprotein involves specific binding sites on the lectin molecule that recognize specific carbohydrate groupings in the glycoprotein. In plants the lectins may serve as a defense mechanism against invasion of plant cells by foreign cells and/or viruses. Other evidence indicates that lectins may also explain how plants recognize certain "good" bacteria with which legume plants must associate for their very survival (the good bacteria being the nitrogen-fixing bacteria in the soil that grow symbiotically with the plant). Recent studies have uncovered the occurrence of lectins in several nonplant organisms, including mammals; here, too, they appear to function in recognition processes.

The most studied lectin is *concanavalin A* (conA) from the jack bean. The ligand specificity of conA is for D-glucose and D-mannose. The complete three-dimensional structure has been determined, and the binding-site events have been identified. For several years conA has proven to be a useful tool in probing the biochemistry of cell surfaces and in the study of cell agglutination (clumping), an event caused by the binding of conA. Modified forms of conA have been shown to be successful in restoring normal growth patterns to previously transformed cells; this result has generated a lot of interest in conA specifically and lectins in general. Can they be used in a therapeutic way to curb the growth of tumors? What can we learn about the relationship of cell surface biochemistry to the normal process of growth control? These and many other questions are under current study.

10–7 PROTEOGLYCAN AND PEPTIDOGLYCAN AGGREGATES

Cartilage Proteoglycan

Chondrocytes are specialized cells present in cartilage tissue that produce and excrete collagen and proteoglycan complexes that form the extracellular matrix of cartilage. This tissue mass of chondrocytes bathed in its extracellular matrix is not integrated with most body processes, since it is devoid of nerve cells, blood vessels, and a lymphathic system. Its functions, attributed to the extracellular matrix of collagen and proteoglycans, are to support, lubricate, and shape other tissues.

After many years of slow progress, the composition and the structure of the cartilage proteoglycan substance have now been defined. The carbo-

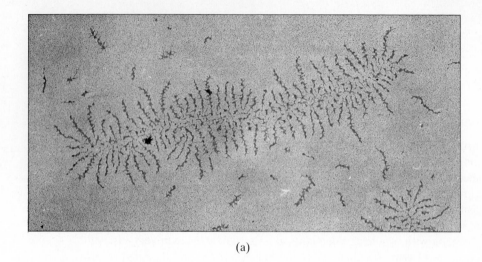

(a)

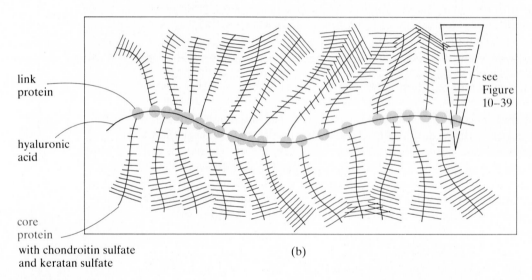

link
protein

hyaluronic
acid

see
Figure
10–39

core
protein
with chondroitin sulfate
and keratan sulfate

(b)

FIGURE 10–38 Proteoglycan aggregates. (a) Electron micrograph ($\approx 41,000 \times$) of the proteoglycan component of cartilage. (*Source:* Electron micrograph provided by Dr. David Pechak.) (b) Schematic representation of this bushy, brushlike structure.

hydrate components include hyaluronic acid, chondroitin sulfates, keratan sulfates, and short oligosaccharide units. There are two principal protein components, called a core protein and a link protein. The arrangement of these parts has been discovered by electron microscopy and extensive biochemical studies.

Microscopy of isolated proteoglycan aggregates in a reasonably intact state reveals a highly branched structure (see Figure 10–38). A good analogy is the structure of a test-tube brush. The backbone of the brush is a hyaluronic acid linear polymer. Numerous "bristles" radiate along the length of the hyaluronic acid (see Figure 10–39). Each bristle component is composed of a polypeptide backbone (the core protein) to which the other glycosaminoglycans and oligosaccharides are attached.

Another protein component (the link protein) acts as some "molecular glue" to bind an end of each core protein to the main hyaluronic acid backbone. The entire proteoglycan aggregate is very large, with a particle mass of about 50×10^6 daltons.

The expanded view of a single bristle element (Figure 10–39) indicates that the carbohydrates attached to the core protein are attached in a definite pattern: short oligosaccharides near the junction with hyaluronic acid, the keratan sulfates occupying the midregion, and the longer chondroitin sulfates at the outer region of the core protein. The size of one of these bristle proteoglycan aggregates is about $1.5–2.0 \times 10^6$ daltons.

The key to understanding the resiliency of this structure is the existence of thousands of $—OSO_3^-$ and $—COO^-$ groups contributed by hyaluronic acid, chondroitin sulfates, and keratan sulfates. The significance is that the proteoglycan aggregate can take on massive amounts of water and withstand the application of applied pressure by releasing the bound water. The viscous character of the hydrated aggregate also means that it has good lubricating properties.

Much current research is aimed at understanding the biological consequences of this structure. Of particular interest is the possibility that alterations in the normal proteoglycan structure of cartilage are related (maybe in a cause-effect manner) to certain diseases such as osteoarthritis and also to certain elements of the aging process.

Bacterial Cell Walls

All unicellular bacteria contain a rigid cell wall that serves as a protective barrier—its primary function but not the only one. The chemical structure of this cytological unit was first solved in 1965 for the cell wall material of *Staphylococcus aureus*. Since then, cell wall preparations of other organisms have been similarly examined. On a comparative basis, the picture that has emerged is that bacterial cell walls are both different and similar in composition. They are different in that the total composition of wall preparations differs from one genus to another, with varying levels of peptides, proteins, lipids, and carbohydrates being the major components. They are similar in that the peptide and carbohydrate parts are arranged in the same type of structural framework.

The common structural framework is a *gridlike network of* polysaccharide chains covalently cross-linked to each other via small peptide bridges.

Owing to this conjugation of peptide and carbohydrate, the material is termed a *peptidoglycan*. (Figure 10–40 should be referred to as you read the following descriptive material of the *S. aureus* cell wall structure.)

The polysaccharide moiety (see Note 10–4) is composed of a repeating disaccharide unit of N-acetyl-D-glucosamine and N-acetyl-muramic acid, with a $\beta(1 \rightarrow 4)$ linkage. Successive units are also attached via $\beta(1 \rightarrow 4)$ linkages. The peptide portion can be considered as consisting of two parts: (1) a *tetrapeptide unit* composed of both D- and L-amino acids and covalently attached

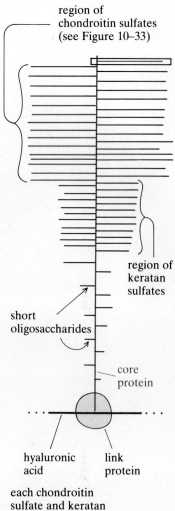

region of chondroitin sulfates (see Figure 10–33)

region of keratan sulfates

short oligosaccharides

core protein

hyaluronic acid   link protein

each chondroitin sulfate and keratan sulfate chain has many negative charges from $—COO^-$ and $—OSO_3^-$ groups; that is,

FIGURE 10–39 Enlarged view of the bristle of one core protein.

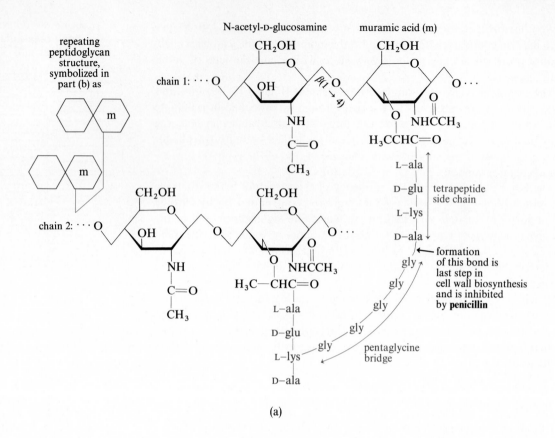

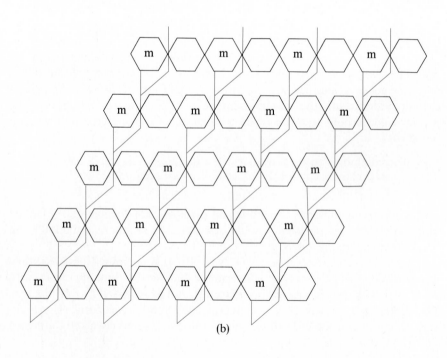

FIGURE 10–40 Peptidoglycan structure of the bacterial cell wall. (a) Chains 1 and 2. (b) Repetition of many repeating units through a cross-linking of several chains. This repetition yields a rigid, gridlike structure.

to the carboxyl group in the side chain of each muramic acid residue; and (2) *a pentaglycine unit* that acts as a bridge to the terminal alanine residue of the tetrapeptide unit of a neighboring polysaccharide chain.

Considering the extensive crisscross presence of covalent bonds, the structural rigidity of the cell wall should be evident. Thus we encounter another example of the structural fitness of a biological material relative to its natural function. (Note the reference to the mode of action of *penicillin* in Figure 10–40. Recall that the inhibitory action of penicillin was illustrated in Figure 5–24.) Other antibiotics that inhibit cell wall biosynthesis are *bacitracin, cycloserine, phosphonomycin,* and *vancomycin.*

NOTE 10–4
The enzyme *lysozyme* acts on the polysaccharide moiety of the cell wall structure, cleaving specific glycoside linkages and thus fragmenting the intact gridlike structure. The fragments dissociate from the cell surface, leaving the cell (now called a *protoplast*) in a deprotected state that is highly susceptible to rupture. Lysozyme is present in eye tears and the whites of eggs, serving as a defense against bacterial infection. Lysozyme preparations are also used in the laboratory for lysing bacterial cells to obtain cell-free systems for further fractionation (see Chapter 1, p. 31).

LITERATURE

ALBERSHEIM, P., and A. G. DARVILL. "Oligosaccharins." *Sci. Am.,* **253** (3), 58–64 (1985). Fragments of the plant cell wall have been discovered that serve as regulatory molecules.

BUNN, H. F., K. H. GABBY, and P. M. GALLOP. "The Glycosylation of Hemoglobin: Relevance to Diabetes Mellitus." *Science,* **200,** 21–27 (1978). A review article.

CAPLAN, A. I. "Cartilage." *Sci. Am.,* **251,** 84–97 (1984).

———. "Assembly of Asparagine-Linked Oligosaccharides." *Annu. Rev. Biochem.,* **54,** 631–664 (1985). A review article with a self-explanatory title.

KORNFELD, R., and S. KORNFELD. "Comparative Aspects of Glycoprotein Structure." *Annu. Rev. Biochem.,* **45,** 217–238 (1976). A review article focusing on the structure of the carbohydrate grouping of glycoproteins from varied sources.

PIGMAN, W., and D. HORTON, eds. *The Carbohydrates:* *Chemistry and Biochemistry.* 2nd ed., 3 vols. New York: Academic Press, 1972. An authoritative and complete reference source.

ROTHMAN, J. E. "The Compartmental Organization of the Golgi Apparatus." *Sci. Am.,* **253** (3), 74–89 (1985). A description of the substructure of the Golgi body and the participation of different Golgi compartments in the processing of oligosaccharides.

SHARON, N. "Glycoproteins." *Sci. Am.,* **230,** 78–86 (1974).

TIPPEN, P. J., and A. WRIGHT. "The Structure and Biosynthesis of Bacterial Cell Walls." In *The Bacteria,* edited by J. R. Sokatch and L. N. Ornsten, vol. VII, 291–426. New York: Academic Press, 1979. A thorough review article.

Textbooks of organic chemistry usually have one or two chapters devoted to the chemistry of carbohydrates.

EXERCISES

10–1. Transfer the Fischer projection formula for each of the following monosaccharides into its Haworth representation and then (when appropriate) into the chair conformation.

   (a) α-D-gulopyranose

   (b) β-D-allopyranose

   (c) α-D-galactopyranose-1-phosphate

   (d) α-L-lyxofuranose

   (e) the C4 epimer of (b)

10–2. In sugar X the orientations of the C1, C2, C3, and C4 hydroxyls are axial, equatorial, equatorial, and equatorial, respectively, and the CH_2OH orientation on C5 is axial. Draw the chair conformation. Then convert this formula to a Haworth representation, a Fischer projection, and a free carbonyl form. Now identify sugar X.

10–3. Draw the Haworth formula for each of the following carbohydrates.

(a) O-α-D-glucopyranosyl-(1 → 1)-α-D-glucopyranoside

(b) O-α-D-galactopyranosyl-(1 → 6)-β-D-glucopyranose

(c) O-α-D-glucopyranosyl-(1 → 1)-α-D-glucosaminopyranoside

(d) O-α-D-galactopyranosyl-(1 → 6)-O-α-D-glucopyranosyl-(1 → 2)-β-D-fructofuranoside

10–4. Label each of the following pairs of simple sugars as (A) enantiomers, (B) anomers, (C) epimers, (D) diastereomers, or (E) aldo-keto isomers.

(a) D-erythrose and D-threose
(b) D-xylose and D-arabinose
(c) α-D-idose and β-D-idose
(d) D-galactose and D-talose
(e) D-mannose and D-fructose
(f) α-D-glucose and α-L-glucose

10–5. Determine the maximum number of different disaccharides that could exist if they were composed only of D-glucose. Classify each as a reducing or nonreducing sugar.

10–6. How many different types of glucosyl residues are present in glycogen? Recall that the involvement in bonding distinguishes each residue.

10–7. Draw the complete structure of the carbohydrate moiety of examples 1, 2, and 4 in Table 10–3.

10–8. An important laboratory application of lectin proteins is the use of lectin-attached, solid-phase materials for the affinity chromatography separation of glycoproteins from nonglycoproteins. In addition, some lectins can be used to isolate one type of glycoprotein from another. Explain.

10–9. Define these terms: oligosaccharide, pyranose, furanose, NDP sugar, kinase, glycosyltransferase, glyconic acid, lactone, deoxy sugar, glycoside bond, glucan, amylases, glycoprotein, glycosaminoglycan, lectin.

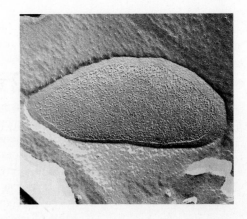

CHAPTER ELEVEN

LIPIDS, MEMBRANES, AND RECEPTORS

Lipid (from the Greek *lipos*, meaning "fat") refers to any naturally occurring, *nonpolar* substance that is nearly or totally insoluble in water but is soluble in other nonpolar solvents, such as chloroform, carbon disulfide, ether, and hot ethanol. Because of the structural and biofunctional diversity of the lipids, the statement is necessarily rather vague and very generalized.

In general terms, the major biological roles of lipids include serving as (1) components of membranes, (2) a major storage form of carbon and energy, (3) precursors of other important substances, (4) insulation barriers to avoid thermal, electrical and physical shock, (5) protective coatings to prevent infection and excessive loss or gain of water, and (6) some vitamins or hormones.

The chapter begins by examining *fatty acids,* which are common components of several lipids that confer the nonpolar character. Thereafter, the structures of various lipids are presented according to the following classification: (1) simple lipids, (2) compound lipids, and (3) derived lipids. The *simple lipids* include only those materials that are esters of fatty acids and an alcohol. The *compound lipids* include various materials that contain other substances in addition to an alcohol and fatty acids. The four major types of compound lipids are phosphoacylglycerols, sphingomyelins, cerebrosides, and gangliosides. Table 11–1 summarizes the composition of simple and compound lipids. The *derived lipids* include any lipids that cannot be neatly classified into either the simple or compound class. Steroids, prostaglandins, leukotrienes, and the lipid vitamins are just a few examples. In vivo many of these substances are produced (derived) from the carbons of fatty acids.

The rest of the chapter is devoted to the structure of membranes and a description of membrane-mediated activities, such as membrane transport and the transmission of regulatory signals for controlling intracellular processes.

TABLE 11–1 Comparison of simple and compound lipids in terms of their chemical composition, summarized here by the identification of the parts that are liberated on complete hydrolysis

	LIPID		COMPONENTS
Simple Lipids	1. Acylglycerols	$\xrightarrow[\text{H}_2\text{O}]{\text{hydrolysis}}$	Glycerol + fatty acid(s)
	2. Waxes	$\longrightarrow$	Alcohol + fatty acid (both long chain)
Compound Lipids	3. Phosphoacylglycerols	$\longrightarrow$	Glycerol + fatty acid(s) + HPO_4^{2-} + another HOR species
	4. Sphingomyelins	$\longrightarrow$	Sphingosine + fatty acid + HPO_4^{2-} + choline
	5. Cerebrosides	$\longrightarrow$	Sphingosine + fatty acid + simple sugar(s)
	6. Gangliosides	$\longrightarrow$	Sphingosine + fatty acid + 2–6 simple sugars, one of which is sialic acid

Note: Lipids 3 and 4 are also referred to as *phospholipids* because of the presence of phosphate. Lipids 4, 5, and 6 are also referred to as *sphingolipids* because of the presence of sphingosine. Lipids 5 and 6 are also referred to as *glycolipids* because of the presence of carbohydrate.

418

11-1 FATTY ACIDS

A *fatty acid* is a long-chain aliphatic carboxylic acid. (The term *fatty acids* originates from their isolation from fats.) Although there are a large number and variety of naturally occurring fatty acids, we can simplify matters with generalizations concerning those acids that occur most frequently in nature (see Table 11-2):

1. Most are *monocarboxylic acids* containing *linear* hydrocarbon chains with an *even* number of carbon atoms, generally in the range of C_{12}–C_{20}. Shorter- and longer-chain acids, branched- and cyclic-chain acids, and acids of odd-number carbon content do occur but at a much lower frequency.

TABLE 11-2 Naturally occurring fatty acids; a partial listing

NAME (CARBON CONTENT)	RCOOH STRUCTURE (DEPICTED PRIMARILY AS BOND-LINE FORMULAS)
Saturated Acids	
Lauric acid (C_{12})	$\overset{12}{C}H_3CH_2CH_2CH_2CH_2CH_2CH_2CH_2CH_2CH_2CH_2\overset{1}{C}OOH$ represented as ⋀⋁⋀⋁⋀⋁COOH 
Myristic acid (C_{14})	 ⋀⋁⋀⋁⋀⋁⋀COOH
Palmitic acid (C_{16})	 ⋀⋁⋀⋁⋀⋁⋀⋁COOH
Stearic acid (C_{18})	 ⋀⋁⋀⋁⋀⋁⋀⋁⋀COOH
Arachidic acid (C_{20})	 ⋀⋁⋀⋁⋀⋁⋀⋁⋀⋁COOH
Unsaturated Acids[a]	
Palmitoleic acid ($C_{16:1}$)Δ^9	$\overset{16}{C}H_3CH_2CH_2CH_2CH_2CH_2\overset{9}{C}H{=}CHCH_2CH_2CH_2CH_2CH_2CH_2CH_2\overset{1}{C}OOH$ represented as ⋀⋁⋀⋁═₉⋀⋁⋀⋁COOH 
Oleic acid ($C_{18:1}$)Δ^9	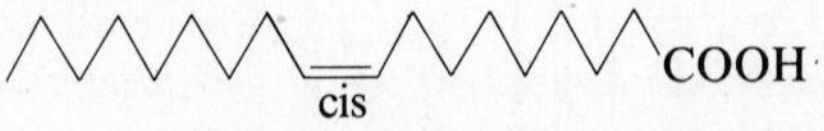 ⋀⋁⋀⋁═(cis)⋀⋁⋀⋁COOH
Linoleic acid ($C_{18:2}$)$\Delta^{9,12}$	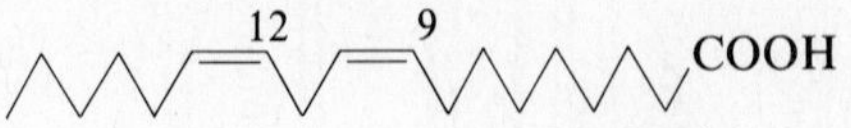 ⋀⋁═₁₂═₉⋀⋁COOH
Linolenic acid ($C_{18:3}$)$\Delta^{9,12,15}$	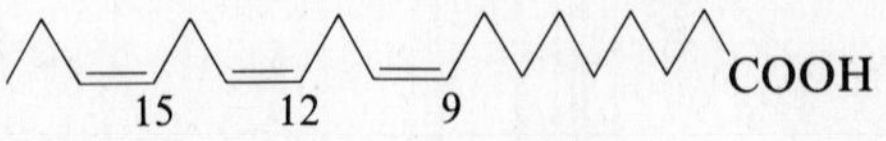 ═₁₅═₁₂═₉⋀⋁COOH
Arachidonic acid ($C_{20:4}$)$\Delta^{5,8,11,14}$	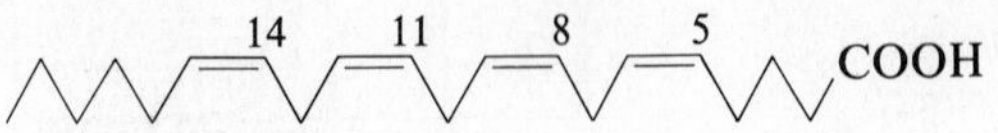 ⋀⋁═₁₄═₁₁═₈═₅⋀COOH

[a] In $C_{x:y}\Delta^z$, x is the carbon content, y is the number of double bonds, and z is the location of the double bonds; the carbon of the COOH group is number 1.

2. *Unsaturation* is common but largely confined to the C_{18} and C_{20} acids. When two or more double bonds exist they are almost always separated by a single methylene group; that is,

$$-CH=CH-CH_2-CH=CH-$$

and

$$-CH=CH-CH_2-CH=CH-CH_2-CH=CH-$$

3. In the unsaturated acids the double bonds are nearly always in the cis configuration.

The many C—C and C—H nonpolar bonds in the hydrocarbon chain confer considerable nonpolar character to the entire molecule even though there is one polar COOH group. Obviously, any substance composed in part of one or more fatty acids will also be largely nonpolar. In addition to explaining the insolubility of lipids in water, the apolar character of fatty acid chains is also basic to an appreciation of the assembly of lipids in biomembranes.

Laboratory Analysis

Mixtures of fatty acids are best analyzed by the technique of *gas-liquid column chromatography* (GLC). They are first converted to methyl esters; then a sample of the ester mixture is injected into a heated column constructed from a coiled tube. The tube is usually packed with a material coated with a liquid substance. On injection the sample is vaporized and mixes with an inert gas

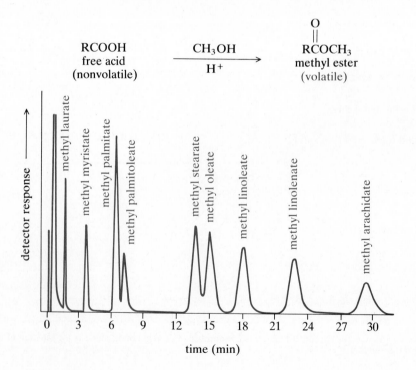

$$\underset{\substack{\text{free acid}\\\text{(nonvolatile)}}}{RCOOH} \quad \xrightarrow[H^+]{CH_3OH} \quad \underset{\substack{\text{methyl ester}\\\text{(volatile)}}}{R\overset{\displaystyle O}{\overset{\|}{C}}OCH_3}$$

FIGURE 11–1 Representation of a typical separation of a mixture of volatile methyl esters of fatty acids by gas-liquid chromatography. Note the rapid and very distinct separation of the sharp zones corresponding to homologous components. (*Source:* Elution pattern reproduced with permission of Applied Science Laboratories.)

such as argon. Under pressure the gas mixture then moves through the column, and substances are partitioned by virtue of their different solubilities in the stationary liquid phase.

Because the liquid phase is heated, the substances are continually revaporized to mix again with the moving gas phase. The gas emerges from the column into a device to detect and record the presence and the amount of substances in the individual zones (see Figure 11–1).

The only requirement to utilize GLC in the analysis of any type of non-polymeric material is volatility. If the substance is not naturally volatile, as in the case of fatty acids, it must be converted into another chemical form that can be vaporized. The GLC can be done rapidly, has a high degree of resolving power, and by interfacing with sophisticated detection components, has a high degree of sensitivity, as small as 10^{-12} mole.

Acyl Groups of Fatty Acids: Processed as Thioester Derivatives

Although the details of lipid metabolism are treated in Chapter 16, brief attention is given at this point to one aspect of the metabolism of fatty acids, namely, their conversion to a *thioester derivative*, the metabolically active form of the fatty acid. As shown in the following reaction,

$$\underset{\substack{\text{fatty acid}\\\text{(ionized}\\\text{at pH 7)}}}{\overset{\overset{\displaystyle O}{\parallel}}{R C O H}} + \underset{\text{coenzyme A}}{HSCoA} + ATP \xrightarrow{\text{thiokinase}} \underset{\substack{\text{acyl-SCoA}\\\text{(a thioester)}}}{R - \overset{\overset{\displaystyle O}{\parallel}}{C} - SCoA} + AMP + PP_i$$

the thioester species (*thio,* from the Greek *theion,* meaning "sulfur") is formed in an ATP-dependent reaction catalyzed by a *thiokinase enzyme*. Usually, the sulfhydryl group substrate is provided by the substance named *coenzyme A* (symbolized as CoASH). The acyl thioester linkage

$$\left(\overset{\overset{\displaystyle O}{\parallel}}{-C-S-} \right)$$

is most important and will be encountered repeatedly. Coenzyme A and the thioester linkage are discussed further in the next chapter.

Nonenzymatic Peroxidation of Fatty Acids

The double bonds of fatty acid chains are very susceptible to reaction with strong oxidizing agents such as hydrogen peroxide (H_2O_2), superoxide anion radical ($O_2 \cdot^-$), or the hydroxy radical ($\cdot OH$). These substances are toxic forms of oxygen that are discussed further in Chapter 15. The oxidation reaction converts the fatty acid to a hydroperoxide (ROOH) (see Figure 11–2).

The lipids most susceptible to hydroperoxide formation are the compound lipids present in biological membranes, which have an abundance of unsaturated

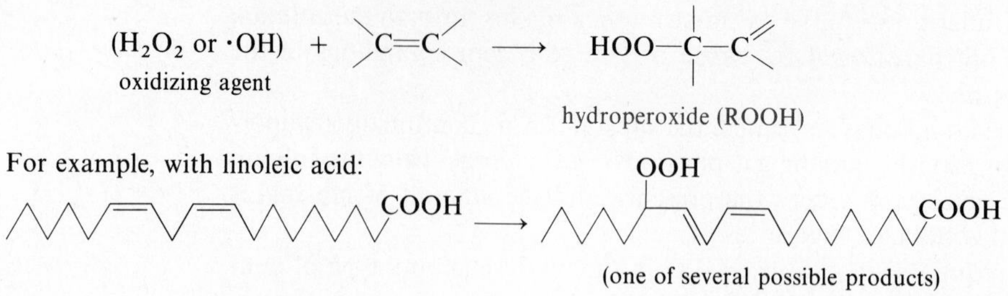

$$(H_2O_2 \text{ or } \cdot OH) + \quad \diagdown C = C \diagdown \quad \longrightarrow \quad HOO-\overset{|}{\underset{|}{C}}-C \diagdown$$

oxidizing agent

hydroperoxide (ROOH)

For example, with linoleic acid:

$$\text{(one of several possible products)}$$

FIGURE 11–2 Peroxidation of unsaturated fatty acids.

fatty acids. The peroxidation of membrane lipids can in turn cause oxidation of membrane proteins, an alteration that can have severe effects on the structure and the function of the membrane. The damage is attributed primarily to the conversion of the lipid hydroperoxides to peroxide radicals, which are highly re-active oxidizing agents. This action may be mediated by the presence of metal ions such as iron (see Figure 11–3). The rancidity and spoilage of food is also partially explained by these reactions.

Living cells are capable of protecting against this damage by detoxifying the original oxidizing agents or the lipid hydroperoxides. This detoxification is attributed to *glutathione, vitamin E* (see Section 11–4), and *ascorbic acid.* Each of these substances can function as an *antioxidant,* sacrificing their own structure to prevent damage to other sensitive biomolecules. Abnormalities caused by lipid hydroperoxides (or the normal process of aging) can be a result of a defect in the detoxification mechanism or an overproduction of lipid hydroperoxides in excesss of the level that can be efficiently handled.

ROOH

Fe^{2+}

Fe^{3+} OH$^-$

RO·

highly reactive
peroxy radical

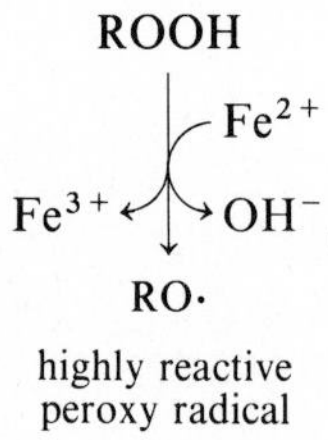

FIGURE 11–3 Hydroperoxide reduced to a peroxy radical.

Enzymatic Peroxidation: An Important Event

In view of these harmful effects of peroxidation, we note, interestingly, that some fatty acid peroxidation is in fact a normal, naturally occurring event catalyzed by specific enzymes. The substrates of these enzymes are polyunsatu-rated fatty acids, particularly arachidonic acid. These acids are converted to (hydro)peroxide intermediates used as precursors for the eventual biosynthesis of *prostaglandins* and *leukotrienes,* two very important classes of bioactive sub-stances in animals. Further consideration of this important aspect of lipid bio-chemistry is deferred until later in the chapter (see Section 11–4).

11–2 SIMPLE LIPIDS

The simple lipids consist of two groups: the neutral acylglycerols and the waxes.

Neutral Acylglycerols

An *acylglycerol* (also called a glyceride) is an ester of the trihydroxy alcohol *glycerol* and *fatty acid(s).* Depending on the number of acids in ester linkage, there are monoacylglycerols, diacylglycerols, and triacylglycerols. The triacyl species is the most abundant in nature. In any case, a simple acylglycerol does not contain any ionic functional groups and hence is said to be a *neutral lipid.*

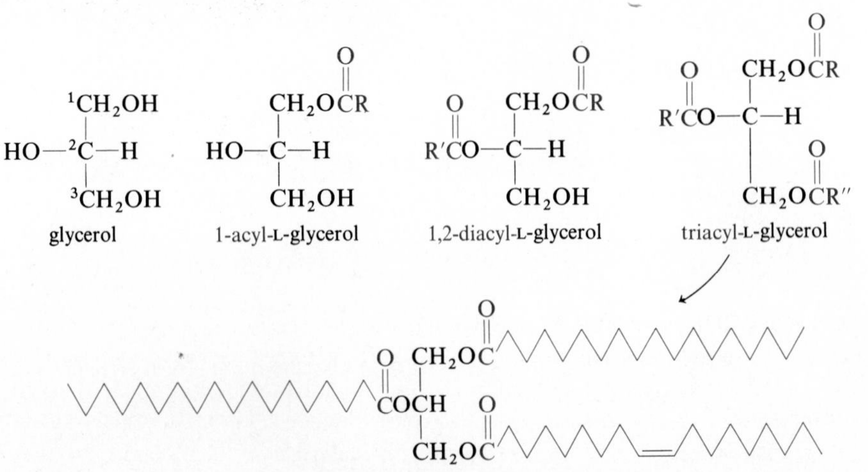

FIGURE 11–4 Acylglycerol structures. The naturally occurring acylglycerols are predominantly L isomers.

The structures of a mono-, a di-, and a triacylglycerol are shown in Figure 11–4. The R groups, which are usually different, designate the acyl side chains of the aliphatic fatty acids. Depending on their physical state at room temperature, triacylglycerols are termed neutral *fats* (solids) or neutral *oils* (liquids). Their water insolubility is common knowledge.

In animals the triacylglycerols fulfill three basic functions:

1. In adipose tissue they constitute the so-called fat depots, which are *storage forms* of carbon and energy. Since the bulk of the carbon is contributed by the acyl groups, the triacylglycerols can be considered as storage forms of the fatty acids. Although the fat depots are in dynamic equilibrium with all body components, we are all aware that excess carbon in the diet, primarily in the forms of carbohydrate and lipid, can cause large accumulation of triacylglycerols in adipose tissue and other organs.

2. In the form of lipoprotein particles such as *chylomicrons,* the triacylglycerols serve as the means by which ingested fatty acids are *transported* via the lymphatic system and blood for distribution within the animal body.

3. Triacylglycerols provide *physical protection* and *thermal insulation* for the various body organs.

There is only one relevant reaction of the neutral acylglycerols, namely, their hydrolysis to yield free glycerol and the fatty acids. In living cells, hydrolysis is achieved via enzymes termed *lipases.* Lipase-catalyzed hydrolysis is characterized by some degree of specificity, with different lipases acting preferentially on different ester linkages. Nonenzymatic degradation is readily accomplished by treatment with dilute acid or dilute alkali. When alkali is used, the process is termed *saponification* (see Figure 11–5). The result is a suspension of micelle particles commonly called a soap solution.

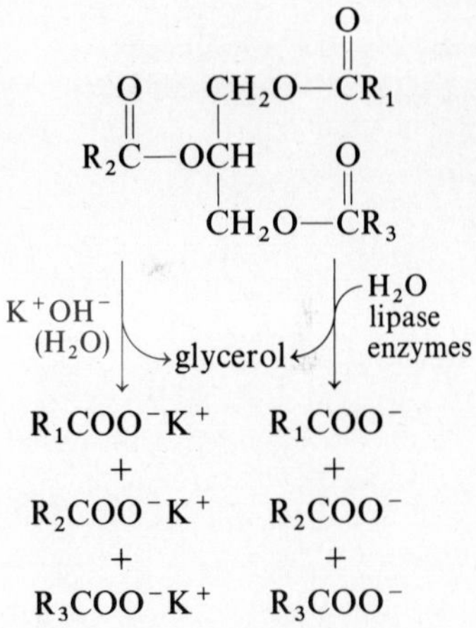

FIGURE 11–5 Hydrolysis of acylglycerols. The term *saponification* refers to the reaction (in color) of a glyceryl ester with potassium (or sodium) hydroxide to produce the potassium salts of the long-chain carboxylic acid, which is a soap.

Waxes

A *wax* is also an ester, but it is different in that the constituent alcohol and acid both contain *long hydrocarbon chains* (see Figure 11–6). Generally speaking, all waxes are totally insoluble in water. Commercial applications of

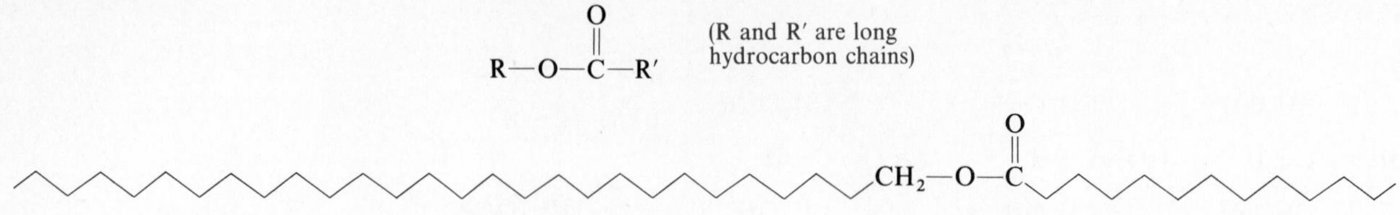

$$R-O-\overset{\overset{\displaystyle O}{\|}}{C}-R'$$

(R and R' are long hydrocarbon chains)

$$CH_2-O-\overset{\overset{\displaystyle O}{\|}}{C}$$

(a)

FIGURE 11–6 Wax. (a) General formula. (b) Example: myricyl palmitate, a primary component of beeswax.

$$CH_3(CH_2)_{28}CH_2-O-\overset{\overset{\displaystyle O}{\|}}{C}(CH_2)_{14}CH_3$$

myricyl palmitate

(b)

synthetic and naturally occurring waxes are widespread. In nature the waxes are generally metabolic end products, with the most important biological role being to serve as protective chemical coatings. The surface feathers and skin of animals are coated with a waxy covering that acts as a waterproofing agent. The waxy coating on the leaves and fruits of plants prevents loss of moisture and also minimizes the chance of infection.

11–3 COMPOUND LIPIDS

Because of their nonpolar character, compound lipids can be extracted from membranes with a nonpolar solvent such as a chloroform-methanol mixture. The lipid extract is then analyzed by thin-layer chromatography.

The lipid composition of a membrane (see Table 11–3) is characteristic of the type of membrane and the physiological age and state of the membrane. The fatty acids present in the membrane lipids may also be characteristic of some membranes.

Phosphoacylglycerols (Phosphoglycerides)

The most common phosphoacylglycerols are the following:

- phosphatidyl choline (PC) (also called lecithin)
- phosphatidyl ethanolamine (PE) (also called cephalin)
- phosphatidyl glycerol (PG)
- phosphatidyl serine (PS)
- diphosphatidyl glycerol (DPG) (also called cardiolipin)
- phosphatidyl inositol (PI) (also called phosphoinositide)
- phosphatidic acid (PA)

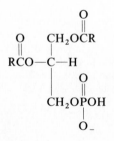

FIGURE 11–7 L-Phosphatidic acid.

Phosphatidic acid (see Figure 11–7) is the simplest structure and is a precursor in the biosynthesis of the other phosphoacylglycerols (see Section

TABLE 11–3 Lipid compositions of some membranes

	LIPID PERCENTAGE					
	PLASMA MEMBRANE					
LIPID	RED BLOOD CELL (RAT)	LIVER (RAT)	*E. coli*	MYELIN	ER	MITO
Phosphatidic acid	0.1	1.0	—	—	—	0.
Phosphatidyl choline	31.0	18.0	—	11.0	55.0	45.0
Phosphatidyl ethanolamine	15.0	11.0	80.0	14.0	16.0	25.0
Phosphatidyl serine	7.0	9.0	—	7.0	3.0	1.0
Phosphatidyl glycerol	—	—	15.0	—	—	2.0
Diphosphatidyl glycerol	—	—	5.0	—	—	18.0
Phosphatidyl inositol	2.2	4.0	—	—	8.0	6.0
Sphingomyelins	8.5	14.0	—	6.0	3.0	2.5
Glycosphingolipids	3.0	—	—	21.0	—	—
Cholesterol	24.0	30.0	—	22.0	6.0	3.0

Note: ER = rough endoplasmic reticulum; MITO = inner membrane of mitochondria. For a more complete compilation, see *Introduction to Biological Membranes* by M. K. Jain and R. C. Wagner (New York: Wiley, 1980).

17–3). It consists of a 1,2-diacylglycerol moiety in ester linkage with phosphoric acid at position 3. As illustrated in Figure 11–8, the other phosphoacylglycerols are distinguished by the identity of an additional HO—R group in phosphoester linkage to phosphatidic acid.

The fatty acid composition of a particular phosphoacylglycerol is not usually a unique characteristic. That is, phosphatidyl choline is phosphatidyl choline not because of the fatty acids that are present but because of the presence of choline. However, there are some generalities regarding the fatty acid components. For example, unsaturated fatty acids are often esterified at the C2 position of glycerol, and some phosphoacylglycerols appear to be primary carriers of certain fatty acids.

An example of the latter class is represented by a particular diacyl species of phosphatidyl choline having a special function. Lung tissue produces *1,2-dipalmitoylphosphatidyl choline,* which is then secreted into the lung chamber to act as a *lung surfactant coating.* This lipid coating prevents water from covering the alveolar walls, which can collapse the lungs. Infants that exhibit acute respiratory distress syndrome (and often die because of it) have lungs that are not yet producing adequate levels of the surfactant; its production begins in the last month of a normal gestation.

As another example, a large portion of arachidonic acid is carried by phosphatidyl inositol. Recent findings also implicate phosphatidyl inositol with an important regulatory function in eukaryotic cells. Described later in the chapter (see Section 11–8), this aspect of PI biochemistry involves two other forms of PI that contain additional phosphates: *phosphatidyl inositol-4-phosphate* (PIP) and *phosphatidyl inositol-4,5-bisphosphate* (PIP$_2$) [see Figure 11–8(b)].

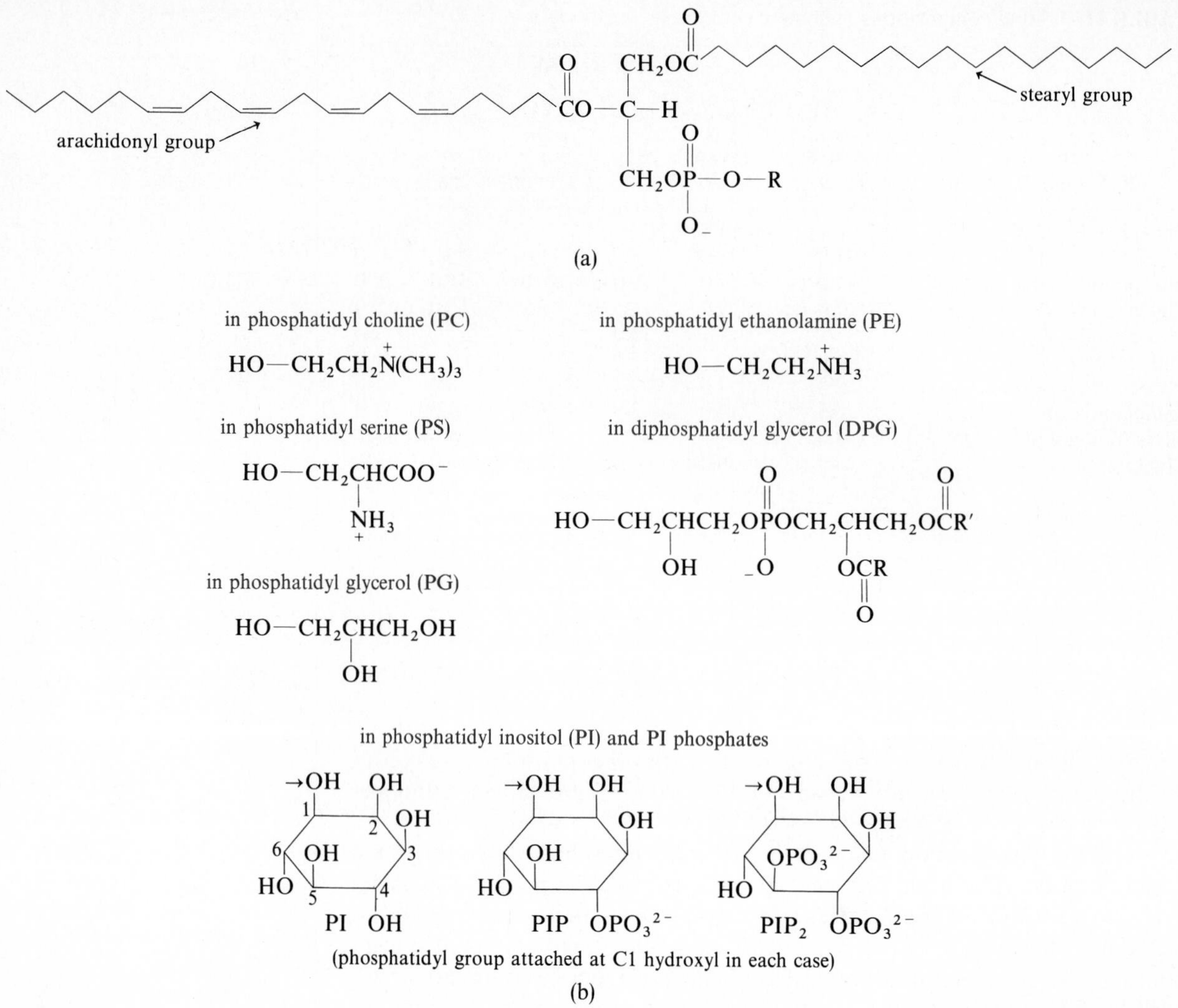

(a)

in phosphatidyl choline (PC)

$HO—CH_2CH_2\overset{+}{N}(CH_3)_3$

in phosphatidyl ethanolamine (PE)

$HO—CH_2CH_2\overset{+}{N}H_3$

in phosphatidyl serine (PS)

$HO—CH_2CHCOO^-$
$\qquad\qquad\underset{\underset{+}{NH_3}}{|}$

in diphosphatidyl glycerol (DPG)

$HO—CH_2CHCH_2OPOCH_2CHCH_2OCR'$

in phosphatidyl glycerol (PG)

$HO—CH_2CHCH_2OH$
$\qquad\qquad\underset{OH}{|}$

in phosphatidyl inositol (PI) and PI phosphates

(phosphatidyl group attached at C1 hydroxyl in each case)

(b)

FIGURE 11–8 Structures of the common phosphoacylglycerols. (a) Phosphatidyl esters. (b) Identity of HO—R source for —O—R grouping.

Variants of the Diacylglycerides

Instead of having two hydrocarbon chains attached to glycerol via ester linkages, some phosphoglycerides have one (monoalkyl) or both (dialkyl) chains attached by *ether bonds* (see Figure 11–9). In addition, many cells contain an enzyme catalyzing an α,β dehydrogenation (an oxidation) of the alkyl group in a monoalkyl ether phosphoglyceride to yield a phospholipid called a *plasmalogen*. Although some membranes (for example, the plasma membrane of heart cells) contain a considerable amount of these lipids, very little is known about their biological significance.

Apolar/Polar Character of Membrane Lipids

The formula and the space-filling model (see Figure 11–10) of the preferred conformation of phosphatidyl choline help us understand another structural

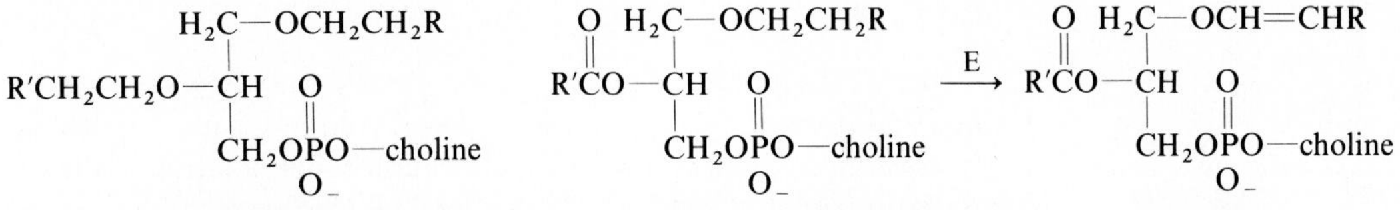

$$H_2C-OCH_2CH_2R$$
$$R'CH_2CH_2O-CH \quad O$$
$$CH_2OPO-choline$$
$$O_-$$

1,2-O-dialkyl ether

$$O \quad H_2C-OCH_2CH_2R$$
$$R'CO-CH \quad O$$
$$CH_2OPO-choline$$
$$O_-$$

I-O-monoalkyl ether

$$\xrightarrow{E}$$

$$O \quad H_2C-OCH=CHR$$
$$R'CO-CH \quad O$$
$$CH_2OPO-choline$$
$$O_-$$

plasmalogen

FIGURE 11–9 Structures of some variant phosphoglycerides.

FIGURE 11–10 Amphipathic structure of phosphoacylglycerols. (a) Structural formula. (b) Space-filling model.

feature of all phosphoglycerides and of the sphingomyelins and glycolipids as well. Note the presence of two distinct regions: a *nonpolar, hydrophobic "tail"* and a *polar, hydrophilic "head."* This dual nature is commonly called *amphipathic* (from the Greek *amphi,* "of both sides," and *pathos,* "feeling"). If the alcohol moiety is also charged at a physiological pH, as is the case with the quaternary amino grouping of choline, the polarity of the head portion is correspondingly greater. The same thing also applies to phosphatidyl ethanolamine and phosphatidyl serine. The discussion in Section 11–5 of the molecular ultrastructure of cell membranes will illustrate the biological significance of this aspect of lipid structure.

$$(CH_3)_3\overset{+}{N}CH_2CH_2O\overset{O}{\overset{\|}{P}}OCH_2-\overset{H}{\underset{|}{C}}-O\overset{O}{\overset{\|}{C}}CH_2CH_2CH_2CH_2CH_2CH_2CH_2CH=CHCH_2CH_2CH_2CH_2CH_2CH_2CH_2CH_3$$

underneath: O_- and $CH_2O\overset{O}{\overset{\|}{C}}CH_2CH_2CH_2CH_2CH_2CH_2CH_2CH_2CH_2CH_2CH_2CH_2CH_2CH_2CH_2CH_3$

(ionic) polar, hydrophilic "head"   nonpolar, hydrophobic "tail"

am*phipa*thic structure

symbolized as

(a)

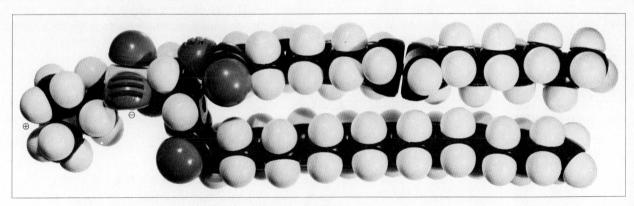

(b)

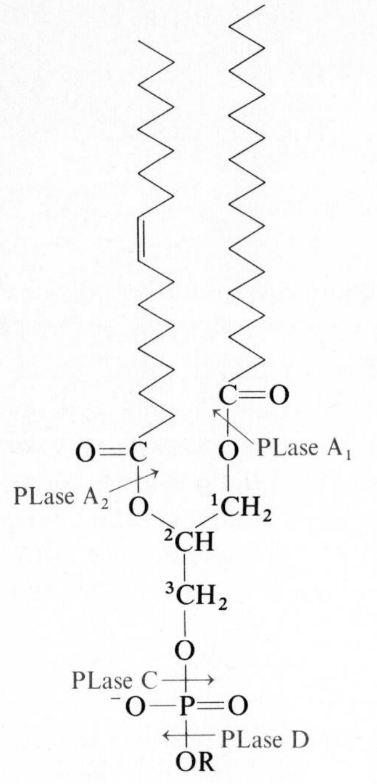

Phospholipases

Membrane lipids do undergo *turnover;* that is, they are constantly being formed and degraded. Biosynthesis is considered in Chapter 17. Because it will apply to a topic discussed later in the chapter, degradation is described now. Degradation involves the enzyme-catalyzed hydrolysis of the various ester bonds to release the constituents. These enzymes are called *phospholipases* (PLase), and they are found in all types of cells. Showing catalytic specificity for ester sites within a phosphoacylglycerol substrate, different phospholipases exist, some of which (A_1, A_2, C, and D) are indicated in Figure 11–11. After their release by hydrolysis, the constituents can enter other metabolic pathways or be used again for the resynthesis of membrane lipids.

Sphingomyelins: Nonglycosylated Phosphosphingolipids

The complete hydrolysis of a sphingomyelin yields one fatty acid, choline, phosphoric acid, and *sphingosine*—hence the name *sphingolipid.* No glycerol is present. The sphingosines are a family of *long-chain, unsaturated amino alcohols* varying in terms of carbon chain length (see Figure 11–12). Although two OH groups are present, the fatty acid moiety is linked via an amide bond to the lone NH_2 group of the sphingosine. This sphingolipid species is called a *ceramide.* The linkage of a phosphorylcholine moiety to the ceramide completes the *sphingomyelin* structure. Note that the sphingomyelins also have a distinct amphipathic character. Sphingomyelins are found in considerable quantities in the cellular membranes of nerve and brain tissue.

Glycolipids: Glycosylated Sphingolipids

Another route of lipid biosynthesis involves enzymes that convert ceramide precursors to *cerebrosides* and *gangliosides* by attaching a carbohydrate grouping to the acylsphingosine structure—hence the name *glycolipid* or *glycosphingolipid.* Cerebrosides and gangliosides are found primarily (though not exclusively) in the cell membranes of nerve and brain tissue, and of course they are also amphipathic.

The simplest cerebrosides contain a monosaccharide grouping in glycosidic linkage to the terminal —CH_2OH of the ceramide. Often the sugar is glucose, to give a glucocerebroside, or galactose, to give a galactocerebroside

FIGURE 11–11 Preferred sites of hydrolysis by different phospholipases (PLase).

FIGURE 11–12 Sphingosine and nonglycosylated sphingolipids. See Figures 11–13 and 11–14 for glycosylated sphingolipids.

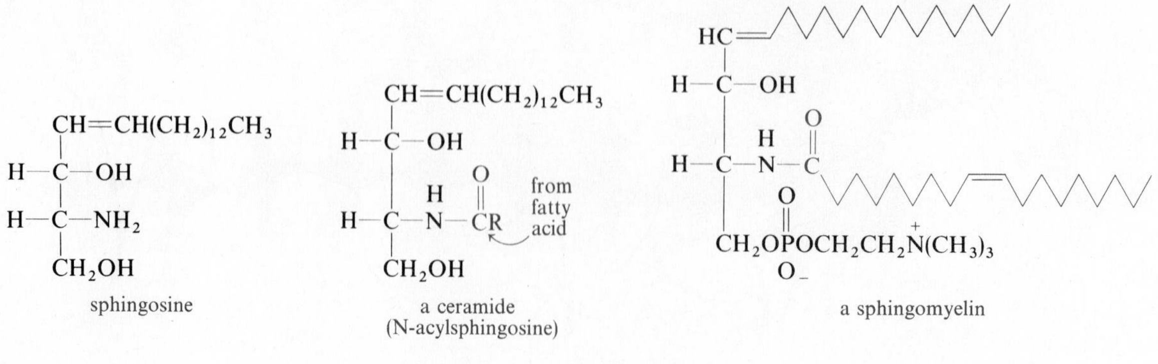

sphingosine

a ceramide
(N-acylsphingosine)

a sphingomyelin

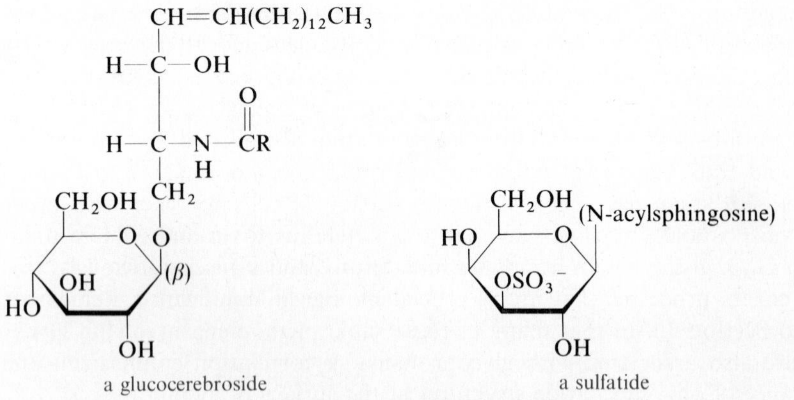

$$CH\!=\!CH(CH_2)_{12}CH_3$$

$$H\!-\!C\!-\!OH$$

$$H\!-\!C\!-\!N\!-\!CR$$
$$\qquad\quad H$$
$$\qquad O$$

CH_2OH   CH_2

(β)

OH

HO

OH

a glucocerebroside

CH_2OH   (N-acylsphingosine)

HO

OSO_3^-

OH

a sulfatide

FIGURE 11–13 Cerebroside structure. Galactocerebrosides are often sulfated at C3; the substance is called a *sulfatide*.

(see Figure 11–13). About 25% of brain cerebrosides consist of sulfated galacto-cerebrosides. Other cerebrosides contain larger oligosaccharide groupings (from 2–10 glycosyl residues) containing amino sugars, N-acetyl sugars, and fucose.

Gangliosides also contain an oligosaccharide grouping but are characterized by the presence of at least one *sialic acid* residue. The exact nature of the oligosaccharide component differentiates one ganglioside from another (and also one cerebroside from another). The structure in Figure 11–14 is that of G_{M_1}, one of the more common gangliosides. Others differ in the size, composition, and sequence of the oligosaccharide grouping and the number (1, 2, 3, or 4) of sialyl residues.

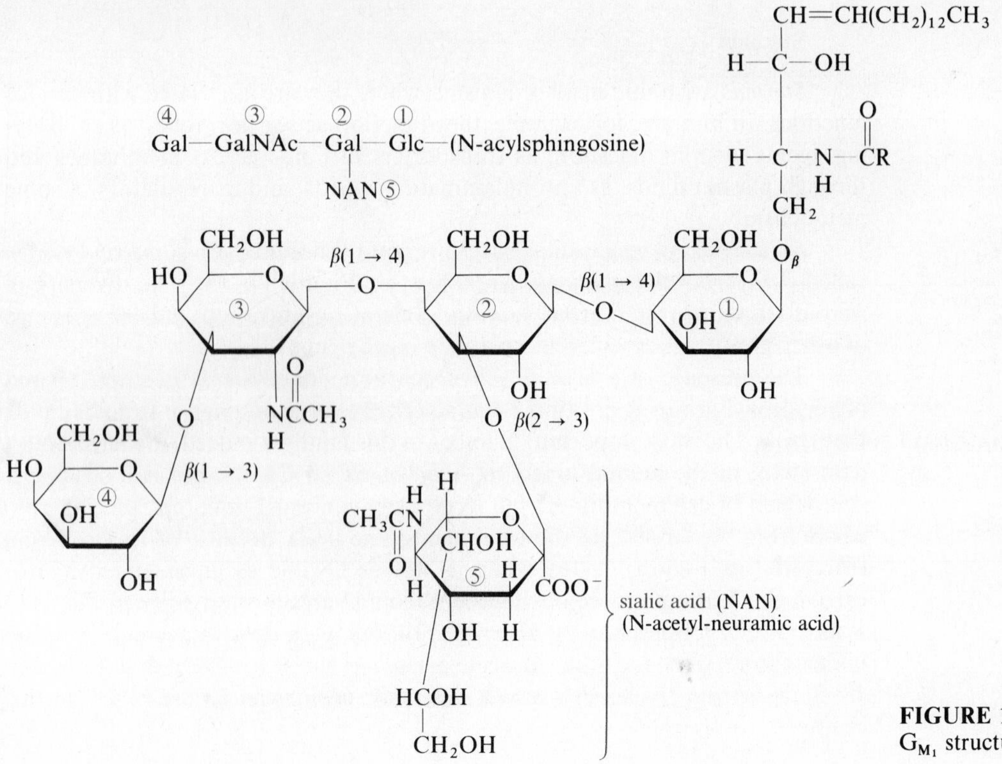

④   ③   ②   ①
Gal — GalNAc — Gal — Glc — (N-acylsphingosine)
                    │
                  NAN⑤

$$CH\!=\!CH(CH_2)_{12}CH_3$$

$$H\!-\!C\!-\!OH$$

$$H\!-\!C\!-\!N\!-\!CR$$
$$\qquad\quad H$$

CH_2OH β(1 → 4) CH_2OH β(1 → 4) CH_2OH O_β

HO ③   O   ② OH ①

O
NCCH_3
H

OH   O β(2 → 3)   OH

CH_2OH β(1 → 3)

HO ④ OH

OH

H H
CH_3CN   O
       CHOH
O   ⑤   H
H   COO^-
OH   H

sialic acid (NAN)
(N-acetyl-neuramic acid)

HCOH

CH_2OH

FIGURE 11–14 Ganglioside G_{M_1} structure.

In addition to contributing to the structural rigidity of membranes, glyco-lipids are implicated in many cellular functions mediated at the level of the cell surface. These functions include such important phenomena as (1) providing antigenic chemical markers to cells, (2) acting as chemical markers identifying the stages of cellular differentiation, (3) regulating the normal growth control of cells and, thus, being related to the transformation of a normal cell to a tumor cell, and (4) being responsible for the tendency of cells to react with other bioactive substances such as bacterial toxins (tetanus toxin and cholera toxin bind to G_{M_1}), glycoprotein hormones, interferons, and viruses. Obviously, this listing covers processes that are of crucial biological significance. Remember (refer to Section 10–6) that many of these same phenomena involving glyco-lipids are also associated with glycoproteins—a correlation emphasizing the importance of oligosaccharide structures at the surface of membranes.

11–4 DERIVED LIPIDS

As stated previously, in terms of both structure and function the *derived lipids* represent a very heterogeneous group, similar only in that their biosyn-thesis can be traced to carbon atoms derived from fatty acids. In the following pages we will consider a few important types: the steroids, the prostaglandins and leukotrienes, the carotenoids, and the lipid vitamins.

Steroids

Steroids are found in all organisms, where they are associated with various functions. In humans, for example, they function as sex hormones, as emulsify-ing agents in lipid digestion, as transporters of lipids across membranes and through plasma fluids, as anti-inflammatory agents, and as regulators of some metabolism.

All steroids have a similar basic structure consisting of a *fused ring system* called *perhydrocyclopentanophenanthrene* (see Figure 11–15). The diversity of steroid structure results from varying levels of unsaturation and the presence of other groupings at different positions on the ring.

The presence of a C_8—C_{10} hydrocarbon side chain at position 17 and of a hydroxyl group at position 3 characterizes a large number of steroids called the *sterols*. The most important member of this family—indeed, the most abun-dant sterol in the animal kingdom—is *cholesterol*. Cholesterol is a structural component of cell membranes, but it varies in concentration, representing any-where from 0% to 40% of the total membrane lipid. Because of its fused ring structure (see Figure 11–16), which is not as flexible as an extended hydro-carbon chain, the presence of cholesterol contributes more rigidity (stiffness) to a membrane. Cholesterol is also the primary metabolic precursor of other important steroids, including the bile acids and the sex hormones (see Section 17–3); in certain tissues it is also a precursor of vitamin D (see p. 437 in this chapter).

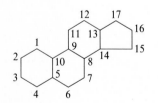

FIGURE 11–15 Perhydrocyclo-pentanophenanthrene.

CH$_3$
|
CHCH$_2$CH$_2$CH$_2$CH(CH$_3$)$_2$

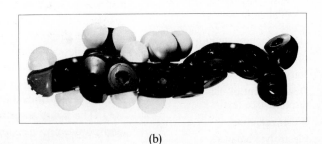

(a)

(b)

FIGURE 11–16 Cholesterol. (a) Structure. Eight asymmetric carbons are present. The indicated orientations (in color) of the two methyls and three hydrogens result from a specific pattern of ring fusion referred to as a **trans** fusion. Another consequence is that the overall structure of cholesterol is rather **flat**; see part (b). (b) Space-filling model. Most hydrogens are omitted to reveal the coplanarity of used rings.

Three steroids of particular interest are the male sex hormone, *testosterone*, and the female sex hormones, *estradiol* and *progesterone* (see Figure 11–17). In the male, testosterone regulates the development of nearly all sex characteristics, the maturation of the sperm, and the activity of the genital organs. In the female, estradiol and progesterone, both products of the ovary glands, are largely responsible for regulation of the menstrual cycle. Progesterone is produced only during a certain period of the cycle, most notably after the release of the ovum from the ruptured follicles, at which time it begins to regulate the preparation of the uterine mucosa for the deposition of the fertilized ovum. If fertilization occurs, the production of progesterone continues through pregnancy. If the egg is not fertilized, the level of progesterone drops, and production does not resume until the next cycle.

The biochemistry of the hormonal regulation of the menstrual cycle has been extensively studied, with a primary objective being the development of a safe and effective method of fertility control. The pioneering studies were performed in the late 1930s, and the first significant discovery was that daily injections of the natural progesterone inhibited ovulation. Since then, hundreds of steroid preparations, mostly synthetic, have been tested for inhibitory and/or regulatory properties affecting both ovulation and menstruation. Several of these materials are now available in oral pill form. Although these steroids have been declared safe, our knowledge of possible undesirable effects on a long-term basis is still not precise or complete.

Prostaglandins

Named because they were first detected in seminal plasma that originates in the prostate gland, prostaglandins (PG) are now known to be present in most mammalian tissues. In most cells they are present in extremely small quantities,

FIGURE 11–17 Sex hormones.

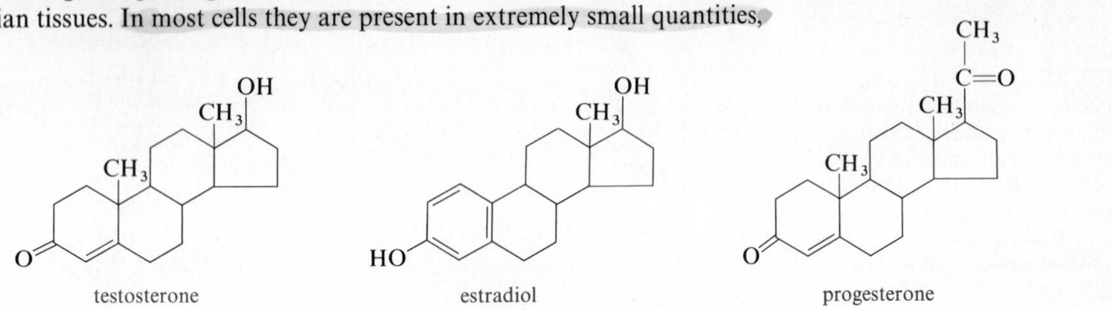

testosterone

estradiol

progesterone

TABLE 11–4 Physiological processes associated with prostaglandin control

Contraction of smooth muscle

Blood supply (blood pressure)

Nerve transmission

Development of the inflammatory response (see text)

Water retention

Electrolyte balances

Blood clotting (see text)

nanograms (10^{-9} g) or less. Many different cell processes are regulated by prostaglandins (see Table 11–4). Although in most cases precise explanations of prostaglandin effects are not yet available, two general features have been established: (1) many target cells controlled by prostaglandins contain specific membrane *receptor proteins* that bind with individual prostaglandins, and (2) in many cases the ultimate effect is mediated through *cyclic nucleotides,* with prostaglandins either increasing or decreasing (depending on the tissue) the cellular levels of cyclic AMP and/or cyclic GMP.

The basic structure of a prostaglandin is that of a C_{20} monocarboxylic acid containing an internal cyclopentane ring. Different prostaglandins have one or more double bonds in specific positions and are also oxygenated at specific positions. These features are illustrated in Figure 11–18 for PGE_1, PGE_2, and PGF_2.

Many of the prostaglandin effects involve their action on smooth muscle, with some PGs stimulating contraction and others stimulating relaxation. Moreover, two closely related PGs may exert completely opposite effects on the same tissue. For example, PGF_2 stimulates the contraction of venous smooth muscle, whereas PGE_1 stimulates the relaxation of venous smooth muscle.

Prostaglandin biosynthesis (see Figure 11–19) begins with the conversion of *arachidonic acid* to two endoperoxide intermediates, first PGG_2 and then PGH_2. The reactions are catalyzed by the same enzyme, *prostaglandin endoperoxide synthase,* which has two separate catalytic components, *cyclooxygenase* (O_2 incorporation and cyclopentane formation) and *peroxidase.* The PGG_2 species has an endoperoxide bridge (—O—O—) across the cyclopentane ring and a hydroperoxide (—OOH) group on a side chain. The PGH_2 retains the —O—O— function, but the —OOH group is reduced to —OH. As indicated in Figure 11–19, PGH_2 then serves as the precursor for the assembly of other prostaglandins.

In some cells PGH_2 is converted to two other products, *thromboxanes* (TXA) and *prostacyclin* (PGI_2). Thromboxane formation occurs in blood

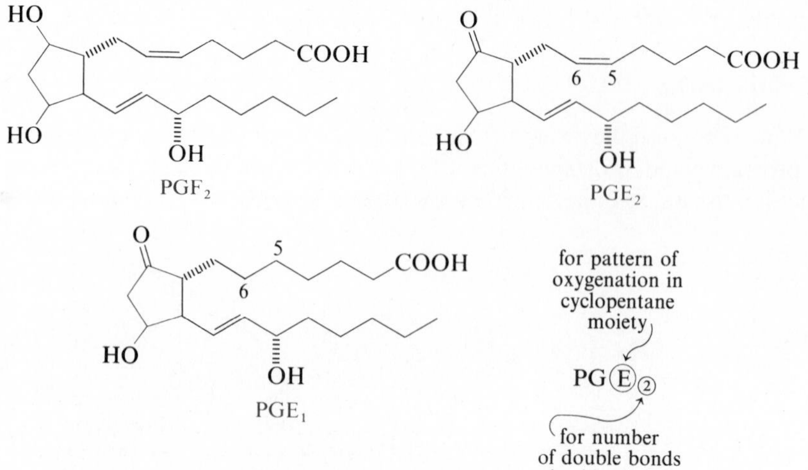

FIGURE 11–18 Structures of some prostaglandins. As indicated here, prostaglandins often have only slight differences in structure. For example, note the absence or presence of the double bond at C5 in PGE_1 and PGE_2, and the hydroxyl versus carbonyl in PGF_2 and PGE_2.

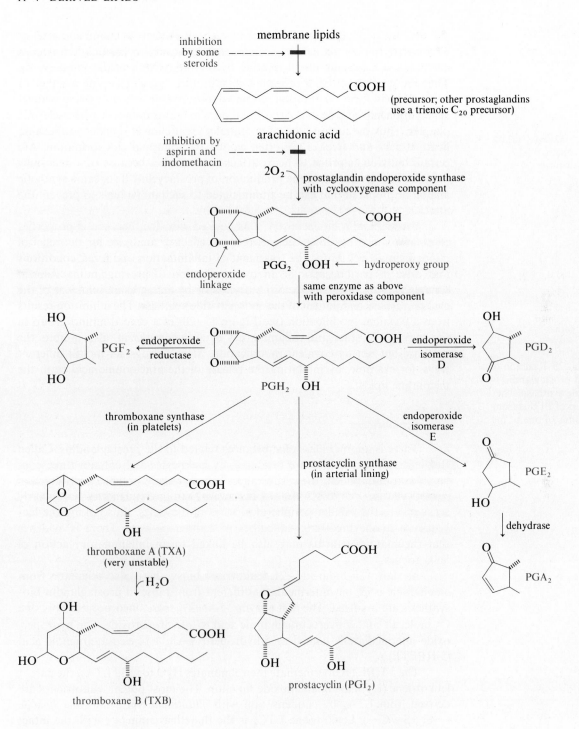

FIGURE 11–19 Summary of some important aspects of prostaglandin biosynthesis. The path from PGH$_2$ differs from tissue to tissue. Thromboxane B causes platelets to clump and causes constriction of the arteries. Prostacyclin inhibits the clumping of platelets and stimulates the arteries to dilate.

PES—NH$_2$

prostaglandin
endoperoxide
synthase

+

aspirin
(acetylsalicylic acid)

PES—N—CCH$_3$
　　 |
　　 H

inactive
acetylated
form of
enzyme

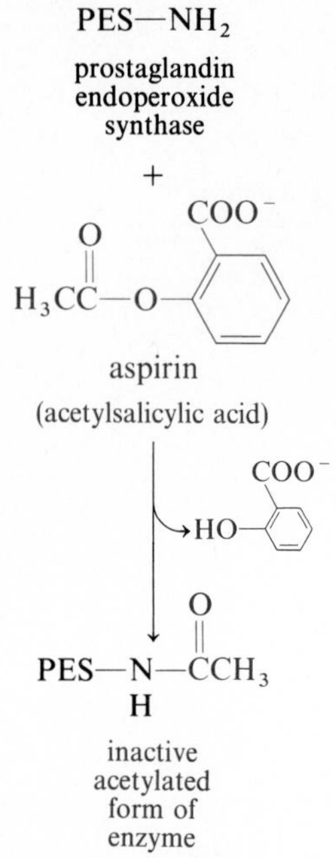

FIGURE 11–20 Reaction of aspirin with prostaglandin endoperoxide synthase. The amino group of PES is from the N terminus of one of the subunits.

platelets and promotes blood clotting by causing platelets to clump and arteries to constrict. In contrast, prostacyclin formation occurs in cells lining the arteries and veins and inhibits blood clotting by inhibiting both of the same events. Thus the relative activities of these two substances may determine whether or not a clot will form. In addition, an imbalance in their action favoring platelet aggregation may also be one of the initial steps in the formation of atherosclerotic plaques. Thus the findings are of potentially great clinical significance because heart attacks and strokes are often caused by abnormal clot formation. Are certain individuals prone to heart attacks and strokes because of a deficiency (inherited or developed) in the production of prostacyclin? If so, can a synthetic analog be prepared that can be administered to such individuals to prevent the attack?

Prostaglandin biochemistry is associated with the therapeutic properties of *aspirin,* which has long been used as an effective analgesic for the relief of minor pain as well as for the treatment of inflammation and fever, conditions that arise in part because of an overproduction of prostaglandins. Aspirin reduces the rate of prostaglandin biosynthesis by acting as an inhibitor of the cyclooxygenase component of the endoperoxide synthase. The inhibition results from a covalent modification (see Figure 11–20) of a crucial amino group in the enzyme. Another anti-inflammatory drug, *indomethacin,* also inhibits the synthase but not by covalent modification. *Steroid* drugs act in still different ways, for example, by inhibiting the release of the arachidonic acid from the membrane lipids.

Leukotrienes

There is another class of substances related to the prostaglandins. Called *leukotrienes* because they are produced by leukocytes and contain three conjugated double bonds, these substances are extremely potent constrictors of smooth muscle—20 to 400 times more active than prostaglandins. Particularly sensitive are the smaller peripheral airways of the lungs, thus linking the leukotrienes to the breathing difficulties of asthma patients. There is evidence that rheumatoid arthritis may also be linked to an inflammatory action of leukotrienes.

As shown in Figure 11–21, leukotriene biosynthesis also originates from arachidonic acid, but intermediates different from those in prostaglandin biosynthesis are involved. The first enzyme, *5-lipoxygenase,* incorporates only one O$_2$ molecule and converts arachidonic acid to a hydroperoxide. The hydroperoxide intermediate is named as 5-hydroperoxy-6,8,11,14-eicosatetraenoic acid (5-HPETE).

The 5-HPETE intermediate then eliminates H$_2$O to yield LTA$_4$, the parent leukotriene (LT) having an epoxide function. The most potent leukotrienes are derived from LTA$_4$ by condensation with *glutathione* via a *thioether linkage* (—C—S—C—). Leukotriene LTC$_4$ is the thioether conjugate with the intact tripeptide, with stepwise removal of glutamyl and glycinyl residues giving two additional thioether conjugates.

In addition to being converted to LTA$_4$, the 5-HPETE intermediate can also be converted to 5-hydroxyeicosatetraenoic acid (5-HETE). Moreover, other

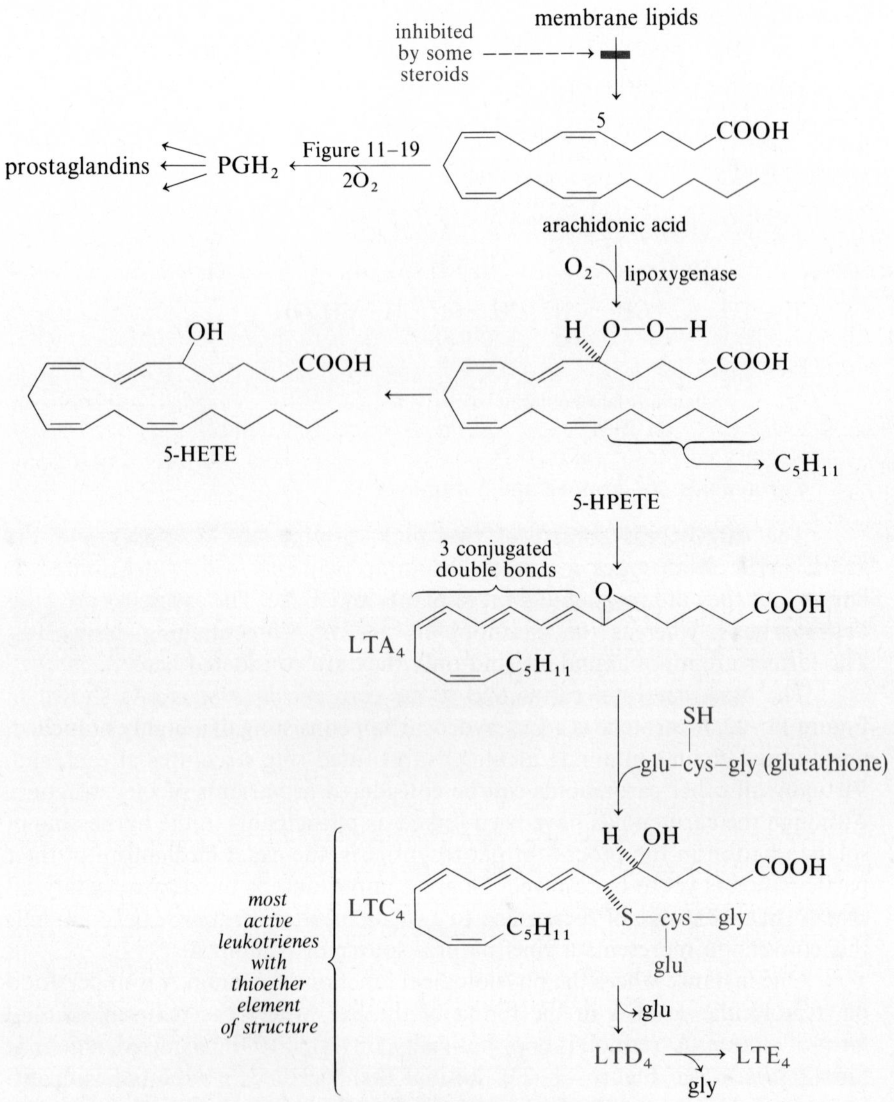

FIGURE 11–21 Some important aspects of leukotriene biosynthesis.

site-specific lipoxygenase enzymes have been detected recently that give rise to 11-HPETE, 12-HPETE, and 15-HPETE, which in turn can yield 11-HETE, 12-HETE, and 15-HETE. Some of these HPETE and HETE derivatives may be antagonists or protagonists of leukotriene and/or prostaglandin activity; some may be precursors of still other physiologically active substances not yet detected; and some may serve as *chemotactic agents*—substances that direct the migration of cells to specific locations. The biochemistry and physiology of these substances is currently under much investigation. Further study of the enzymes involved may result in the development of drugs to specifically inhibit leukotriene production and thus provide improved therapy for asthma and rheumatoid arthritis.

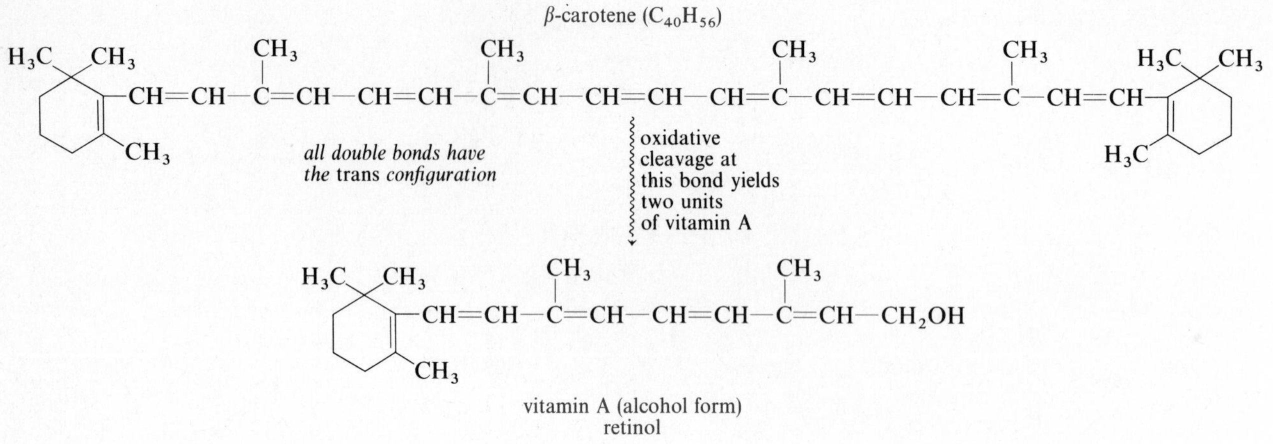

β-carotene ($C_{40}H_{56}$)

all double bonds have
the trans configuration

oxidative
cleavage at
this bond yields
two units
of vitamin A

vitamin A (alcohol form)
retinol

FIGURE 11–22 Conversion of
β-carotene to vitamin A.

Carotenoids: β-Carotene and Vitamin A

The carotenoids consist of two main groups, the *carotenes* and the *xanthophylls*. Both types are water-insoluble pigments widely distributed in nature, but they are most abundant in plants and algae. The carotenes are pure hydrocarbons, whereas the xanthophylls are oxygen-containing derivatives. The former are more abundant, and only they are considered here.

The most common carotenoid is the carotene *β-carotene*. As shown in Figure 11–22, β-carotene is a C_{40} hydrocarbon consisting of a highly branched, unsaturated chain containing identical substituted ring structures at each end. Virtually all other carotenoids can be considered as variants of this structure. Although the carotenoids have been linked as participants in the harnessing of solar radiation in the process of photosynthesis, the exact mechanism of their participation is yet to be resolved. Of great importance is the enzyme-catalyzed symmetrical cleavage of β-carotene to two molecules of *vitamin A*. In animals this conversion represents a chief natural source of vitamin A.

One instance where the physiological function of vitamin A is understood on a molecular level is in the retina of the eye, where the reduced, alcohol form of vitamin A (retinol) is enzymatically converted to the oxidized, aldehyde form (retinal) (see Figure 11–23). Retinal then becomes complexed with different retinal proteins, called *opsins*, and forms the active proteins that function in vision. The complexes of retinal and an opsin protein are the primary photoreceptors of incident light in the visual cells and transmit information to the nervous system. Most vertebrates contain two types of visual cells in the retina: (1) rod cells, which are dim-light receptors and do not perceive color, and (2) cone cells, which are bright-light receptors and are also responsible for color

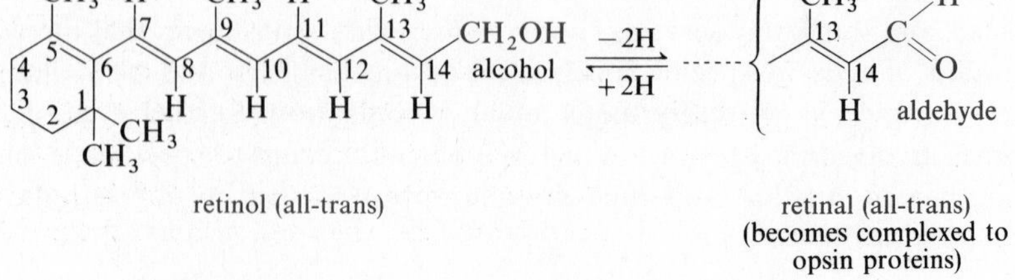

retinol (all-trans)

retinal (all-trans)
(becomes complexed to
opsin proteins)

FIGURE 11–23 Conversion of
retinol to retinal.

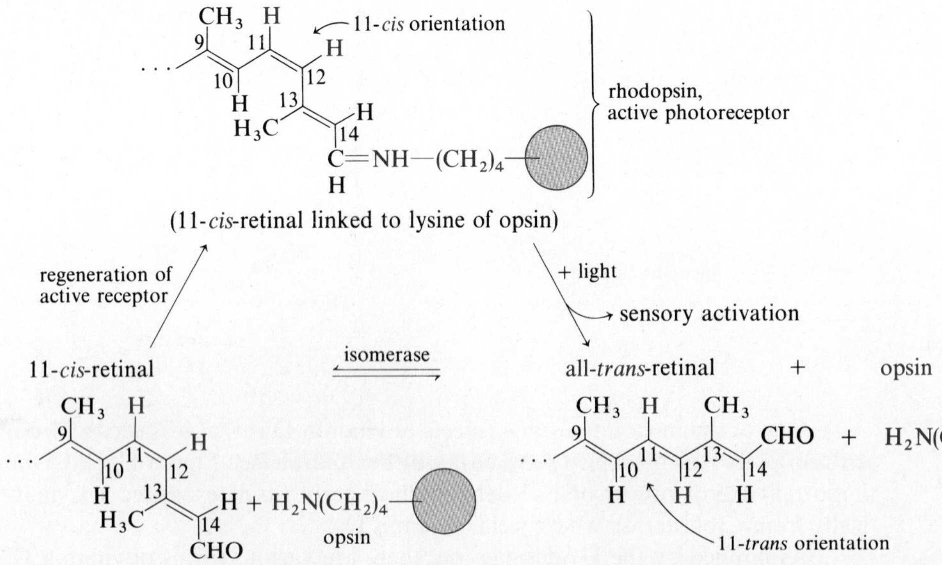

FIGURE 11–24 Cis-trans retinal isomerization cycle for rod pigments.

vision. In the rod cells there appears to be only one opsin, and the active lipid-protein receptor complex is called *rhodopsin*. In cone cells at least three different opsins are known to occur; these opsins are complexed to retinal to constitute a blue-sensitive receptor, a red-sensitive receptor, and a green-sensitive receptor. In the case of rhodopsin, retinal is covalently attached to the protein via the side-chain amino group of a lysine residue.

At one time only the all-trans form of retinal was known to be present in visual cells. However, in subsequent studies a second isomeric form was discovered, namely, 11-*cis*-retinal, in which one of the double bonds has a cis orientation. In fact, it is the 11-*cis* species that is complexed to the opsin protein. Another aspect of vision chemistry was uncovered with the discovery that 11-*cis*-retinal, when exposed to light, was converted to the all-trans isomer.

In accordance with these findings, the molecular events of vision are proposed to consist of a cis-trans isomerization cycle, as shown in Figure 11–24 for the rod pigment. The distinguishing photochemical act is the cleavage of the lipoprotein-rhodopsin complex, a complicated process accompanied by the isomerization of retinal. The conversion is not direct but involves many intermediates. Evidence suggests that one or more of these intermediate steps may be subsequently involved in generating extremely small electric potentials that activate the nervous system. The vision cycle, in terms of the fate of the visual pigment, is completed by the regeneration of 11-*cis*-retinal, which is required for the re-formation of the active rhodopsin pigment. One possible route is a direct enzymatic conversion catalyzed by an isomerase.

Vitamin D

Animals require *vitamin D* in the normal calcium and phosphorous metabolism necessary for healthy bone and tooth development. A deficiency leads to rickets, a disease in which the bones become soft and pliable, producing various deformities.

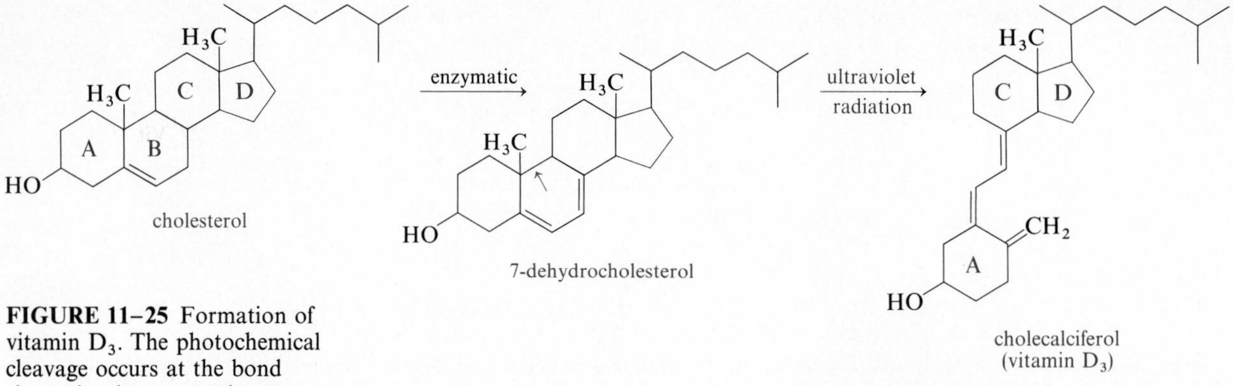

FIGURE 11–25 Formation of vitamin D$_3$. The photochemical cleavage occurs at the bond shown by the arrow; electron rearrangements after the cleavage yield the final product.

One of the most interesting aspects of vitamin D is that it is formed from a sterol precursor by exposure to ultraviolet radiation (see Figure 11–25). One important sterol precursor is 7-dehydrocholesterol (itself produced enzymatically from cholesterol), which yields vitamin D$_3$.

As implied by the D$_3$ designation, there are various forms of vitamin D. The D$_3$ species (cholecalciferol) is the form present in milk and fish liver oils, both of which are major dietary sources of the vitamin. The D$_3$ species can also be synthesized in the skin of animals. Because of this synthesis in skin tissue, adults receiving normal exposure to sunlight require less vitamin D in the diet than infants. Now established is that after D$_3$ intake or formation in the skin, the D$_3$ species (carried in plasma) is converted in the liver and the kidney to hydroxylated derivatives, which are even more active (see Figure 11–26). Indeed, there is strong evidence that the major active form in the body is the 1,25-dihydroxy derivative.

Vitamin E

FIGURE 11–26 Formation of 1,25-dihydroxycholecalciferol. This form is most active at stimulating the intestinal absorption of Ca^{2+}—and phosphate—and the mobilization of Ca^{2+} for bone development.

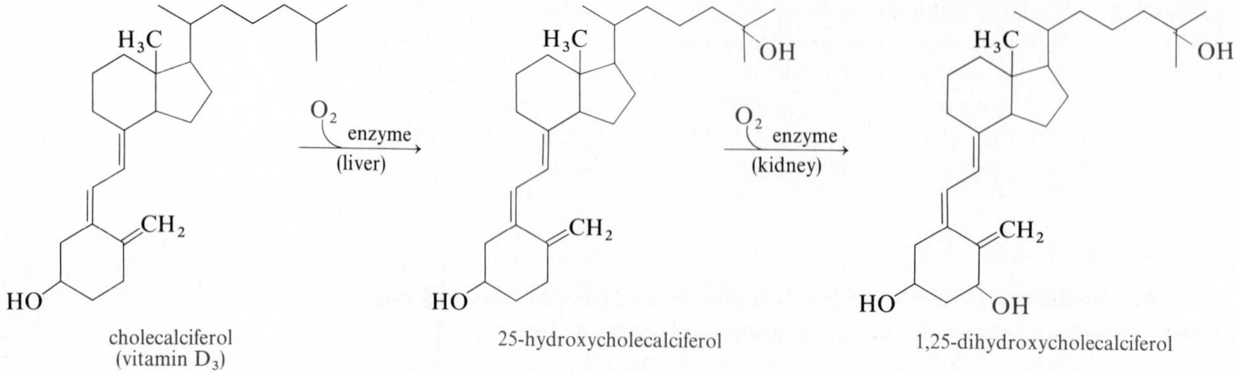

A substance required for reproduction in animals is *vitamin E*, the so-called antisterility or fertility vitamin. The basic structure of the E vitamins (there are different forms) is called *tocopherol* (from the Greek *tokos*, meaning "childbirth," and *pherein*, meaning "to carry"). The most potent form of vitamin E is α-tocopherol (see Figure 11–27).

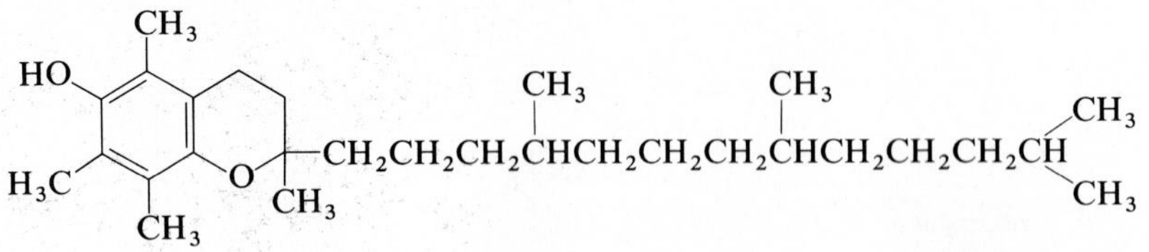

FIGURE 11–27 Vitamin E (α-tocopherol).

Although vitamin E has proven essential for reproduction in the laboratory rat, there is no conclusive evidence for the same relationship in humans. Of greater significance (in humans and various other animals) is the proposed effect vitamin E and close derivatives have in maintaining the normal chemical composition and function of muscle tissue. For example, in certain animals a vitamin E–deficient diet results in muscular dystrophy.

In the processing of food, vitamin E is commonly added because it acts as an antioxidant, preventing the spoilage of foods through oxidation. In living cells vitamin E may also function as an antioxidant, along with ascorbic acid and glutathione.

Vitamin K

A deficiency of *vitamin K* slows the rate of blood clotting. When first detected as being necessary for clotting, it was termed, in German, *Koagulations-vitamin*—hence the name *vitamin K*.

The basic structure of vitamin K is a *naphthoquinone* bicyclic ring system with a long hydrocarbon chain of variable length attached to the quinone ring (see Figure 11–28). The hydrocarbon chain in vitamin K_2 consists of repeating *isoprene* units. Animals depend on two sources of vitamin K: (1) diet, particularly one that includes green vegetables, tomatoes, and cheese, and (2) synthesis by bacteria in the intestinal tract.

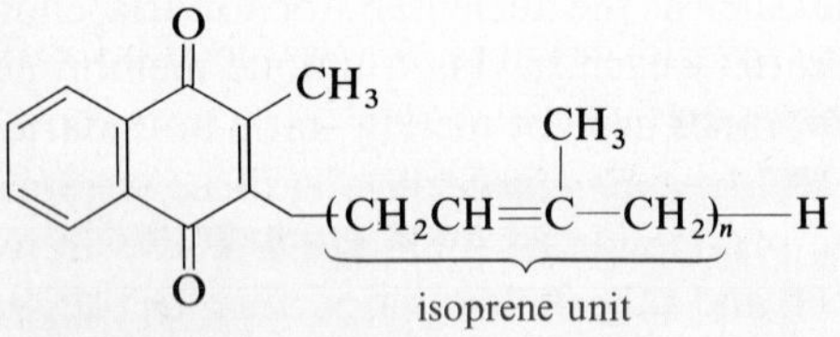

isoprene unit

FIGURE 11–28 Vitamin K (general structure); *n* is variable but usually less than 10.

The function of vitamin K in blood clotting is linked to the synthesis of a *complete prothrombin molecule*, the precursor of thrombin (see Figure 5–34). After the prothrombin molecule is assembled, vitamin K participates in a post-translational modification involving the carboxylation of several glutamic acid side chains to *γ-carboxyglutamic acid*. The insertion of the additional —COO⁻ sites is necessary for the optimum binding of Ca^{2+}, which activates the conversion of prothrombin to thrombin (see Figure 11–29).

Dolichol Phosphate

Many lipids in addition to vitamin K are assembled from the condensation of isoprene units. Cholesterol, carotenoids, and xanthophylls are examples from the previously described lipids. Another example is the polyisoprenoid *dolichol*

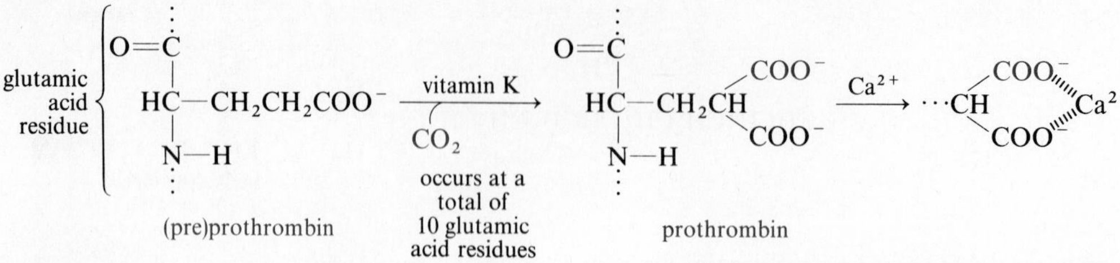

FIGURE 11–29 Vitamin K–dependent conversion of prothrombin to thrombin. Polypeptide backbone represented by $\cdots$

phosphate (see Figure 11–30), previously described in Chapter 10 as being the lipid carrier of the oligosaccharide group for the initial assembly of asparagine-linked glycoproteins in the endoplasmic reticulum. The metabolic pathway for forming the C_5 isoprene structure will be examined in Chapter 17 (see Figure 17–28).

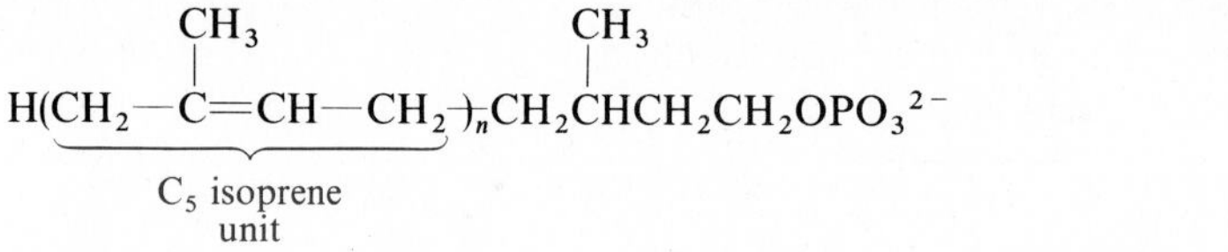

FIGURE 11–30 Dolichol phosphate, a polyisoprenoid. In vertebrates $n = 18-20$; in yeasts $n = 15-16$.

11–5 BIOMEMBRANES

In all living cells the cell interior is separated from the external environment by a boundary called the *cell membrane* or *plasma membrane* (see Figure 11–31). (In plants and bacteria the exterior face of the membrane is also associated with a cell wall structure.) Moreover, in eukaryotic cells there is also an elaborate membrane-bound compartmentalization within the cell, represented by the presence of organelles such as the nucleus, mitochondria, endoplasmic reticulum, and the Golgi apparatus—each having a unique membrane.

However, membranes are not merely static boundaries that segregate regions. In fact, they are *dynamic biochemical systems* responsible for many phenomena, such as the production of ATP; the selective transport of substances into and out of the cell and subcellular compartments; the conferring of the antigenic specificity of cell types; the binding of regulatory agents (such as hormones and growth factors); the binding of normal neurotransmitters and various pyschoactive drugs that mediate the transmission of electrical impulses in nerve tissue; the operation of various photoreceptor substances that harness light energy, as in photosynthetic membranes, or that transduce a light signal into specific chemical events, as in vision; the occurrence of many enzyme-catalyzed reactions; and the disease conditions that originate with the binding to the cell surface of viruses and several toxins. There are several other functions as well.

Not all membranes are identical, however, and specific examples of these functions are found in different types of membranes—sometimes, very specialized membrane systems. We have already encountered some examples, and others will be examined in this and later chapters. Before that, however, we will direct our attention to the molecular architecture of membranes.

Chemical Composition

Membranes are composed primarily of protein and lipid molecules. The relative amount of each varies considerably among different membranes, ranging from extremes of about 20%–80% protein and 80%–20% lipid (see Table 11–5). In the form of glycoproteins and glycolipids, carbohydrate material may account for 0.5%–10% of the mass of a membrane. The ultrastructural arrangement of this protein-lipid matrix was first proposed in 1935 by Danielli and Davson, followed by important modifications in the 1960s by S. J. Singer and others.

Lipid Alignment: Hydrophobically Associated Bilayers

Membrane lipids consist of phosphoacylglycerols, sphingomyelins, cerebrosides, gangliosides, and cholesterol. They can be isolated from the membrane by exhaustive extraction with nonpolar organic solvents and analyzed by standard chromatographic procedures. As previously displayed (see Table 11–3) the normal lipid composition of any given membrane is somewhat characteristic, although transient and permanent changes can occur as a result of physiological state, age, diet, temperature, and other factors.

As described in Chapter 2, when these amphipathic substances are dispersed in water, there is a natural *hydrophobic tendency* for the lipids to be arranged in the form of a *lipid bilayer*—a membrane. In each (mono)layer lipids are aligned so that the nonpolar, hydrophobic tail elements of structure—in an extended conformation—are in close contact with each other, and the polar, hydrophilic head elements of structure are in close contact with each other and with the polar, aqueous phase at each monolayer face. These interactions are exclusively noncovalent. The two monolayers are then aligned in a tail-to-tail orientation forming the bilayer structure (see Figure 11–32), with a distinct *nonpolar interior* and *two polar faces*. The thickness of the hydrophobic region of the bilayer is approximately 35–40 Å (3.5–4.0 nm). Taking into account the regions of the hydrophilic surfaces and associated proteins, the complete width of a lipid-protein bilayer matrix is about 70–90 Å (7–9 nm).

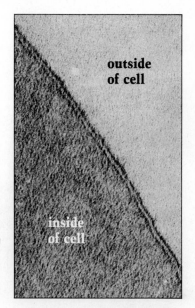

FIGURE 11–31 Electron micrograph of a portion of the plasma membrane of a red blood cell. The two electron-dense lines represent the inner and outer surfaces of the membrane. Magnification 240,000 ×. (*Source:* Photograph provided by J. David Robertson.)

TABLE 11–5 Chemical composition of some selected cell membranes

MEMBRANE SOURCE	% PROTEIN	% LIPID	% CARBOHYDRATE
Myelin membrane of nerve cells	18	79	3
Liver cells (mouse)	46	54	3
Liver cells (rat)	53	42	5–10
Red blood cells (human)	49	43	8
Nuclear membrane (rat liver cells)	59	35	2–4
Retinal rods (bovine)	51	49	4
Outer mitochondrial membrane	52	48	2–4
Inner mitochondrial membrane	76	24	1–2
Chloroplast lamellae (spinach)	70	30	6
Gram-positive bacteria	75	25	10
Mycoplasma bacteria	58	37	1.5

Source: Data taken from G. Guidotti, "Membrane Proteins," *Annu. Rev. Biochem.,* **41,** 731 (1972).

outside

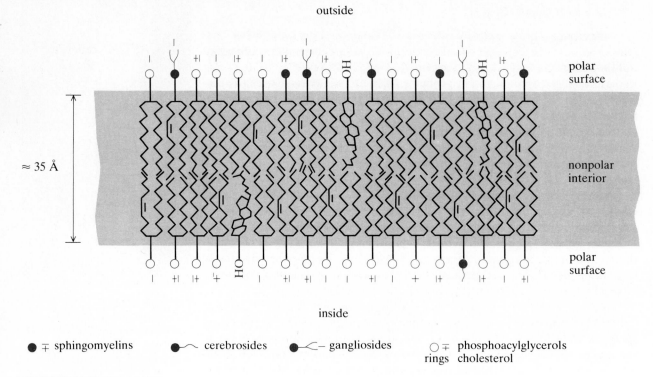

polar
surface

≈ 35 Å

nonpolar
interior

polar
surface

inside

● �⊤ sphingomyelins   ●⌒ cerebrosides   ●⌒ gangliosides   ○⊤ phosphoacylglycerols
                                                                      rings cholesterol

FIGURE 11-32 Schematic
illustration of the lipid bilayer
arrangement in membranes.

Although each monolayer has the same general arrangement of lipids, the lipid composition of each monolayer is usually different. For example, when they are present, cerebrosides and gangliosides are generally located in the outer monolayer (the external face of the membrane). The extent of this *monolayer asymmetry* of lipid composition varies among different membranes and can also vary as a cell engages in different activities and as a cell ages. There is also clear evidence that lipid molecules can move longitudinally within a monolayer and even laterally, changing positions from one monolayer to another. The biological significance of this lipid asymmetry and lipid migration is not yet fully understood.

The flexibility (stiffness) and fluidity of the lipid bilayer is determined by the type and the length of hydrocarbon groupings contributed by the fatty acid chains, the sphingosines, and cholesterol. Increased stiffness is associated with a high ratio of saturated/unsaturated chains and also with higher levels of cholesterol. These physical characteristics of membranes are also dependent on the type and the arrangement of proteins in association with the lipid bilayer.

The fluidity of the lipid bilayer matrix is governed by the extent to which the fatty acid hydrocarbon chains can be closely packed to each other. Able to assume fully extended conformations, saturated fatty acids can achieve maximal close-packing, a situation that increases stiffness (decreases fluidity) of the bilayer matrix. Because of the kinks in conformation produced by double bonds, unsaturated fatty acids diminish the extent of close-packing—an influence that decreases stiffness (increases fluidity) of the bilayer matrix. The bulky ring system of cholesterol also interferes with close-packing.

Membrane Proteins: Surface and Integral

ARRANGEMENT Membrane-localized proteins associate with the lipid bilayer in two ways (see Figure 11–33): (1) They are in association with the hydrophilic surfaces of the lipid bilayer, called *surface membrane proteins*, and (2) they are embedded into the hydrophobic region of the bilayer, called *integral membrane proteins*. Every membrane has both types, but different membranes have different proportions of each.

Surface proteins can associate with the hydrophilic faces of the lipid bilayer owing to noncovalent interactions between the polar head groups of lipids and the polar amino acid side chains on the protein surface. Only noncovalent associations were once thought to be involved. However, recent research has detected examples of proteins that are more strongly bound to membranes by a covalent association with lipid. The lipid site for covalent attachment appears to be quite specific, involving the inositol group of phosphatidyl inositol. Further research will determine how widespread the covalent attachment of membrane proteins may be. Since enzymes have been detected that will selectively cleave the protein-lipid covalent adduct, the phenomenon may represent a way in which certain stimuli can cause the release of certain proteins from certain membranes.

Integral proteins depend on hydrophobic interactions for association with membranes. This type of membrane protein would require nonpolar amino acid side chains to be exposed on its surface to associate with the nonpolar hydrocarbon chains that comprise the internal matrix of the bilayer. Depending on the extent of surface-exposed hydrophobic groups in the protein, an integral protein may penetrate the bilayer slightly, it may penetrate halfway, or it may span the entire width (*transmembrane protein*) of the bilayer.

Other distinguishing features of membrane proteins include the variety of functions they represent, the abundance of each protein present in the membrane, the pattern of *asymmetric* distribution of proteins in the two monolayers, and the particular pattern of protein-protein and protein-lipid interactions. The word *asymmetric* is highlighted again to emphasize this important aspect of membrane structure. To be discussed later in Chapter 15, one important consequence of membrane protein asymmetry is the establishment of proton (H^+) gradients as energy-rich states to sustain various cell processes such as the production of ATP in mitochondria and chloroplasts.

Amphipathic – having both nonpolar & polar character

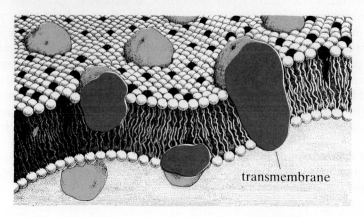

transmembrane

FIGURE 11–33 Membrane ultrastructure. The sketch represents what is sometimes called the **fluid-mosaic model.** Proteins are viewed as being predominately globular and amphipathic. Their hydrophilic ends protrude from the membrane, and their hydrophobic surfaces are embedded in the bilayer of lipids (grey) and cholesterol (black). The proteins make up the membrane's active sites; some are simply embedded on one or the other side, while others pass entirely through the bilayer. [*Source:* Reproduced with permission from S. J. Singer, "Architecture and Topography of Biologic Membranes," *Hosp. Pract.*, **8,** 81–90 (1973).]

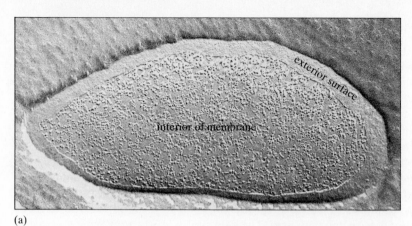

(a)

FIGURE 11–34 Membrane ultrastructure. (a) Electron microscopic image of a fracture plane through the erythrocyte (human) membrane, obtained by the technique of freeze-etching. The oblong body in the center of the field is a red blood cell with about 90% of the surface layer of its membrane removed. The surface layer itself, with a smooth appearance, is visible on the entire perimeter. The inner granular area is interpreted as representing the lipoid interior of the membrane, with many intramembrane protein-containing particles. Magnification 88,000 ×. (*Source:* Electron micrograph reproduced with permission from Daniel Branton.) (b) Drawing depicting how these intramembrane proteins are exposed when the membrane is cleaved so as to peel away the outside layer. [*Source:* Drawing reproduced with permission from S. J. Singer, "Architecture and Topography of Biologic Membranes," *Hosp. Pract.,* **8,** 81–90 (1973).]

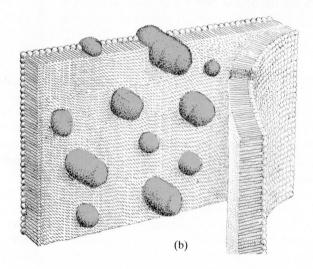

(b)

This knowledge of membrane *ultrastructure* has been convincingly established from many studies using electron microscopy, X-ray diffraction, nuclear magnetic resonance spectroscopy, electron spin resonance spectroscopy, and some other specialized techniques. Evidence from electron microscopy is presented in Figure 11–34. The photograph was obtained by the *freeze-etch technique,* revealing the molecular arrangement within the lipid bilayer. The outcome of this procedure is akin to peeling away one of the monolayers. Pictures suggest a static assembly, but there is much evidence that it is not. We have already referred to the movement of lipids. *Proteins can also change position.* The movement may be longitudinal in a monolayer; it may be rotational in a monolayer. Some proteins may even move or rotate to the other face of the bilayer.

In addition to changing location, a membrane protein is also capable of undergoing changes in its conformation while localized in the membrane.

This event can trigger other conformational changes in neighboring proteins and lipids in the immediate microenvironment.

ISOLATION   The isolation of a membrane protein first requires *solubilization,* releasing the protein from its membrane environment by disrupting its hydrophobic/hydrophobic and hydrophilic/hydrophilic interactions with membrane lipids and with other membrane proteins. Toward this end the tactics of *osmotic shock, ultrasonic treatment,* and *detergent treatment* are used, individually and in combination. Detergents (see Figure 11–35) include both ionic and nonionic types. Ionic detergents release membrane proteins involved in both hydrophobic/hydrophobic and hydrophilic/hydrophilic interactions. The nonionic detergents are applicable for proteins engaged primarily in hydrophobic/hydrophobic interactions.

After solubilization, classical procedures can be used to isolate a specific membrane protein from the mixture. The technique of affinity chromatography is particularly effective for isolating membrane proteins whose function is to bind with a specific ligand substance at the membrane; it works by attaching the ligand to the column support. For example, the insulin receptor protein has been successfully isolated by attaching insulin to a column support. Another effective strategy is to attach an antibody (specific for the membrane protein as an antigen) to the column support.

The solubilization and isolation of integral membrane proteins with full retention of bioactivity is especially difficult, because the active conformation of the protein may depend on very special protein-lipid and protein-protein associations in the natural microenvironment of the membrane. Thus any attempt to isolate a lipid-free, pure, fully functional protein may be doomed from the start. When success is achieved, the optimum "recipe" involving osmotic or ultrasonic shock and detergents is one that is arrived at by trial-and-error laboratory strategy, taking months or even years.

FUNCTIONS   Membrane proteins can be divided into two broad categories on the basis of their role in the membrane: (1) *structural proteins,* those that assist in holding the entire lipid-protein matrix together, and (2) *dynamic proteins,* those that actually participate in cellular processes that occur at the level of the membrane.

Structural proteins often have an elongated fibrous shape and reside on the hydrophilic surfaces, acting as a type of molecular tape. Examples are some of the proteins associated with the cytoskeletal network (refer to Section 1–3).

Three classes of dynamic proteins, having various sizes and shapes, are generally present in nearly all types of membranes: (1) *transport proteins* involved in the passage of substances in and out of the cell, (2) *enzymes* that catalyze reactions occurring at the membrane, and (3) *receptor proteins* that bind with specific substances on the exterior of the membrane, providing a signal that triggers changes in the membrane or inside the cell. Some specialized membranes contain proteins with other distinctive functions, such as photoreceptor proteins that respond to light signals.

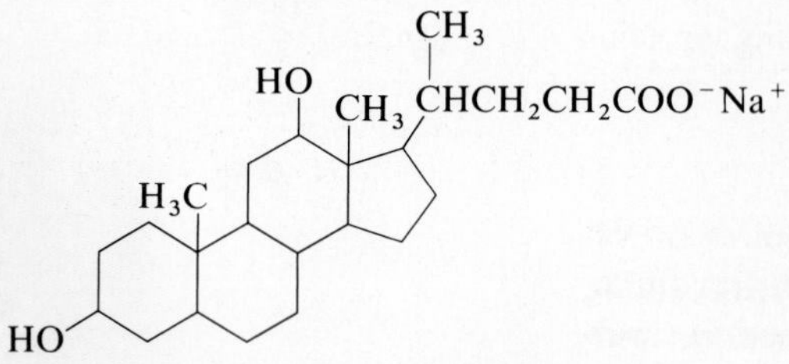

sodium deoxycholate
(bile salt detergent)

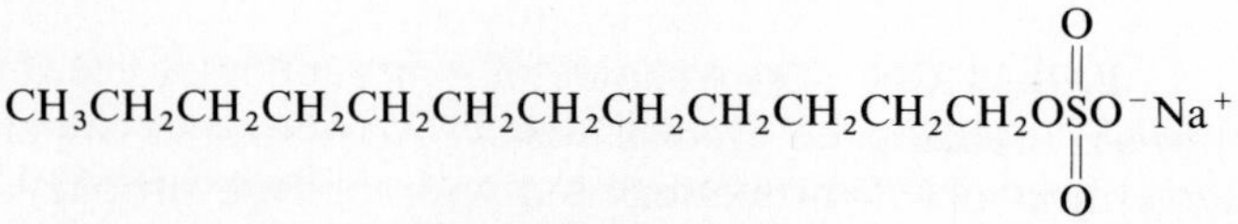

sodium dodecyl sulfate (SDS)
(lauryl sulfate)

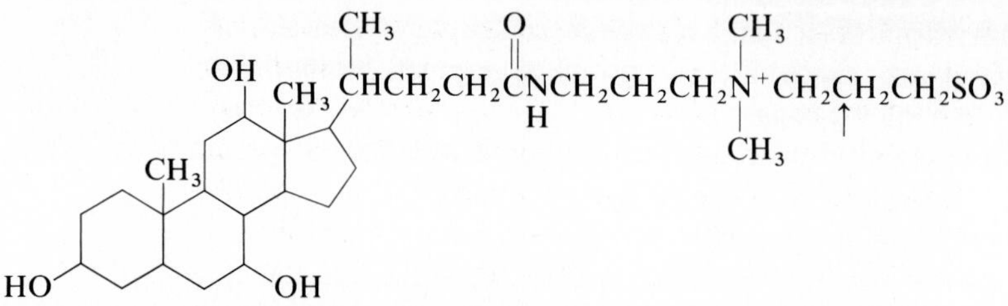

"CHAPS" and "CHAPSO" (HO at arrow)
3-[(3-cholamidopropyl)-1-dimethylammonio]-1-propanesulfonate

(a)

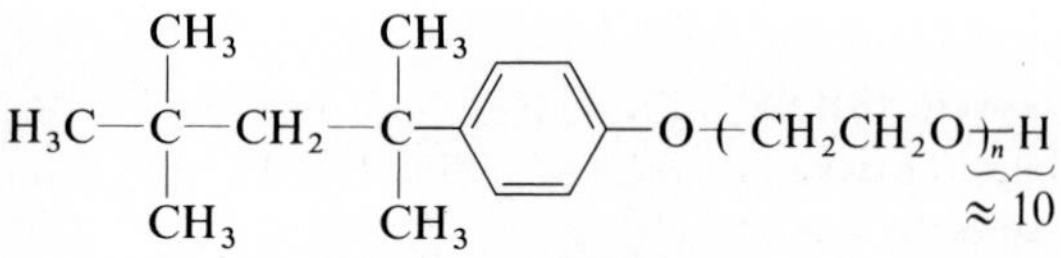

TRITON X–100

polyoxyethylene ethers
of p-(t)octylphenol

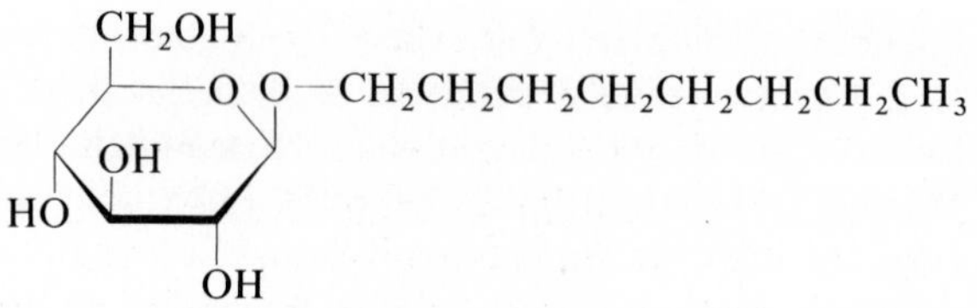

octyl glucoside

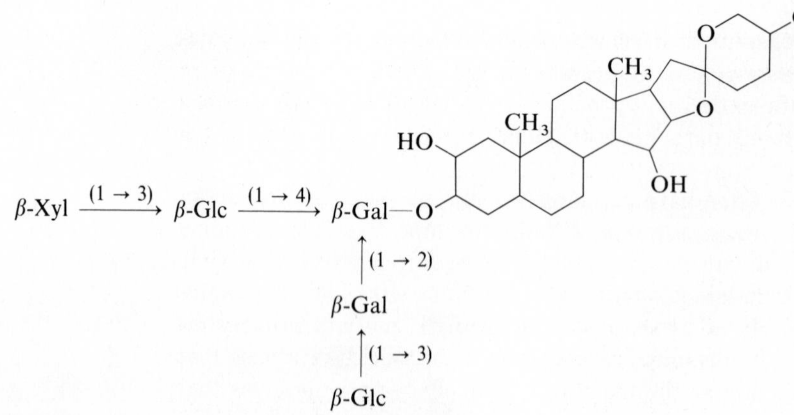

β-Xyl $\xrightarrow{(1 \rightarrow 3)}$ β-Glc $\xrightarrow{(1 \rightarrow 4)}$ β-Gal—O

$\uparrow$(1 $\rightarrow$ 2)

β-Gal

$\uparrow$(1 $\rightarrow$ 3)

β-Glc

digitonin

(b)

FIGURE 11–35 Structures of detergents for solubilizing membrane proteins. (a) Ionic. (b) Nonionic.

11–6 MEMBRANE TRANSPORT

General Principles

The passage (transport) of substances into and out of cells and also their transport between cytoplasm and the various subcellular organelles (mitochondria, nuclei, and so on) are determined by membranes. This role is a particularly important one of membranes. Obviously, if a biomembrane were a completely impermeable barrier, a cell would be totally isolated from its environment, and individual organelles inside a cell would be isolated from each other. In contrast, if membranes were completely permeable partitions, any and all substances would be free to move from one region to another.

Neither extreme applies. Rather, the transport properties of membranes are in between. They are *semipermeable* partitions: Some materials can move across, and others cannot. Depending on the substance and the membrane involved, direct transport may occur via one of two general processes: passive or active transport.

PASSIVE TRANSPORT When a substance moves across a membrane from a region where its concentration is high to a region of lower concentration, the movement is called *passive transport,* or *diffusion.* Because this transport is "down" a concentration gradient (high → low), *no energy* is expended by the cell to support it. There are two types of diffusion: simple diffusion and facilitated diffusion. Each type is described next.

Simple diffusion occurs without any direct interaction with a membrane protein. The movement occurs right through transient hydrophobic crevices in the lipid bilayer, through transient hydrophilic crevices that might form in regions where proteins are clustered, or through actual pores in the membrane (as in the nuclear membrane) formed by discontinuities in the lipid bilayer (see Figure 11–36). Water, some inorganic ions, several lipid-type substances, and the proteins and nucleic acids crossing the nuclear membrane are examples of substances that move in this way.

Facilitated diffusion involves the direct participation of one or more transport proteins located in the membrane. After an initial event of binding (often with specificity) the substance on one face of the membrane, the transport protein(s) may operate in one of three different ways (see Figure 11–37):

1. The transport protein, by a rotational movement in the membrane, delivers the substance to the other face, where it is released.

2. The transporting apparatus may consist of more than one protein component, with net transport due to a bucket brigade type of shuttle.

3. The transport protein apparatus may provide an interior channel specific for the passage of only a small number of different substances.

Channel-forming transport proteins are often associated with the transport of ions and are called *ionophores* (ion-carrying). An example of an important energy-independent Na^+/K^+ ionophore that responds to the presence of a neurotransmitter is described in the next section.

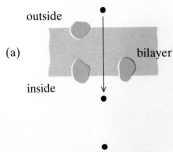
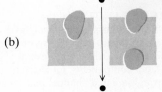
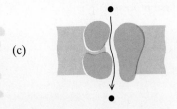

FIGURE 11–36 Simple diffusion across membranes. (a) Directly through a lipid bilayer. (b) Through a membrane pore. (c) Through a channel in a protein cluster.

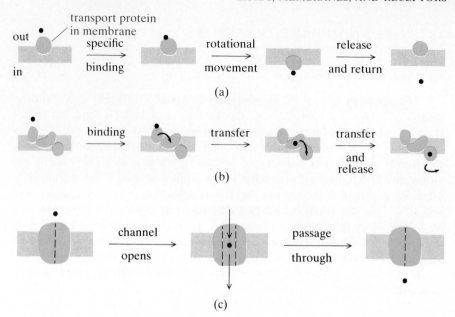

FIGURE 11-37 Facilitated diffusion across membranes. (a) Delivery by a single transport protein. (b) Delivery by several protein components. (c) Delivery through a gated channel in a transmembrane protein. For active transport, energy is required in any step of any of these processes.

ACTIVE TRANSPORT   When a facilitated mode of transport requires an expenditure of *energy* by the cell, the process is called *active transport*. Active transport will always be involved when the movement is against a concentration gradient (low → high). There is no single energy-dependent step in the process. Energy may be used to trigger conformational changes in a protein, thus promoting the movement of the protein, the opening of a channel in the protein, or the shuttle of a substance among the proteins, and/or energy may be used to modify the substance during the transport process. The energy requirement is often satisfied by a direct expenditure of ATP, but other more indirect energy-yielding processes are known. A direct ATP-dependent process is described shortly; it involves the Na^+/K^+-ATPase enzyme. An example of a more complex, indirect, energy-dependent process was previously described in explaining the operation of the γ-glutamyl cycle for amino acid transport (see Section 3-5).

Although useful to summarize general features, Figures 11-36 and 11-37 are oversimplified, particularly in terms of the involvement of transport proteins. In addition, many transport mechanisms also involve the participation of hormones and neurotransmitters that control the process.

There are two other strategies used to move *bulk quantities* of a substance in and out of cells: exocytosis and endocytosis. In *exocytosis* membrane-bound vesicles from the cytoplasm fuse with the plasma membrane and then open on the exterior face. In *endocytosis* a section of the plasma membrane (with bound substances on the external face) is pinched inward to form and release a vesicle in the cytoplasm. This vesicle then fuses with lysosomes, and the lysosome enzymes hydrolyze the internalized substance into fragments, which then are transported across the lysosome membrane into the cytoplasm. The secretion of packaged proteins from pancreas cells and the release of neurotransmitters by neuron cells are examples of exocytosis. The internalization of cholesterol-rich lipoproteins (see Section 17–4) is an example of endocytosis.

Transport ATPases: Focus on Na^+/K^+-ATPase

INTRODUCTION   Many essential body functions depend on the ATP-dependent transport of ions across a membrane. The membrane protein responsible is called a *transport adenosine triphosphatase* (ATPase), so-named because the active transport is accompanied by the overall hydrolysis of ATP to ADP and P_i, a reaction catalyzed by the transport protein itself. Different transport ATPases exist, showing almost absolute specificity for the ions they transport. The best-known and most widespread example is a *Na^+/K^+-ATPase,* responsible for the combined efflux of Na^+ and influx of K^+ across the plasma membrane. Some structure/function details of this protein are described in the following subsection. Other examples include a Ca^{2+}/Mg^{2+}-ATPase (found in various cells, with greatest activity in the sarcoplasmic reticulum of skeletal muscle), a K^+/H^+-ATPase (found so far only in the plasma membrane of specialized cells in the gastric mucosa), and a HCO_3^-/Cl^--ATPase (found in various cells in either the plasma membrane, the mitochondrial membrane, or the microsome membrane).

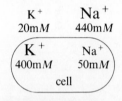

$$K^+ \qquad Na^+$$
$$20mM \qquad 440mM$$

K^+ Na^+
$400mM$ $50mM$
cell

FIGURE 11–38 Plasma levels of Na^+ and K^+.

Na^+/K^+-ATPase   Most cells maintain a cell interior that is high in K^+ and low in Na^+ by moving K^+ into the cell and Na^+ out of the cell. Since the plasma levels of K^+ and Na^+ are, respectively, lower and higher than cytoplasm levels (see Figure 11–38), both ion movements occur against concentration gradients, and thus energy is required. These ion gradients are important to the regulation of water content, to protein biosynthesis, and to the excitability of nerve and muscle cells. Estimates that nearly 25% of the total energy expenditure in the body is used for this purpose indicate how important these ion gradients are to a normal condition.

The Na^+/K^+-ATPase (also sometimes called an Na^+/K^+ pump) responsible for maintaining these gradients catalyzes an Na^+/K^+ *exchange* between the cytoplasm and the extracellular plasma. Based on extensive evidence, the stoichiometry of the exchange is $3Na^+$ for $2K^+$ per ATP. The overall reaction can be summarized as shown in Figure 11–39, where $E = Na^+/K^+$-ATPase.

The Na^+/K^+-ATPase is a single, oligomeric, transmembrane protein (MW $\approx$ 290,000). It has been extensively studied in whole cells, in membrane fragments, and also in detergent-solubilized preparations. Some distinctive features of structure/function include the following: (1) The oligomer is composed of two glycoprotein subunits (α and β); (2) of many subunit compositions proposed, recent evidence strongly supports an $\alpha_2\beta_2$ structure; (3) the ATPase active site appears to reside in the α subunit; (4) activity requires the presence of both Na^+ and K^+; (5) although K^+ can be replaced by some other monovalent cations, the specificity for Na^+ is absolute; (6) the enzyme can also catalyze a K^+-dependent phosphatase reaction; (7) the ATPase active site

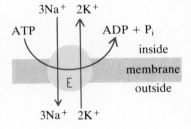

$3Na^+$ $2K^+$

ATP   ADP + P_i

inside

E   membrane

outside

$3Na^+$ $2K^+$

FIGURE 11–39 Na^+/K^+ exchange mediated by Na^+/K^+-ATPase

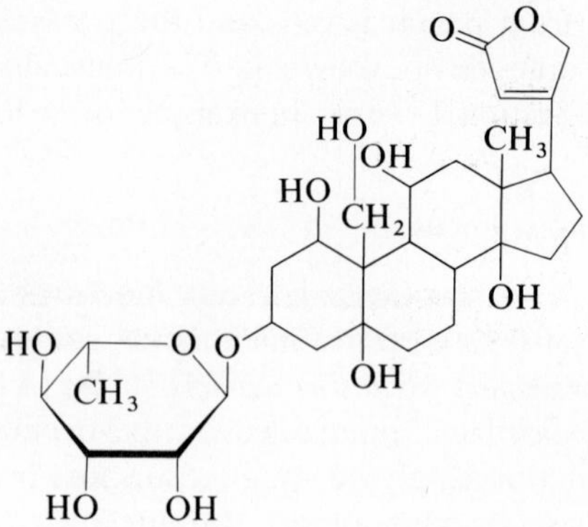

FIGURE 11–40 Ouabain.
Digitalis glycosides are used as
a heart stimulant in the treatment
of congestive heart failure.

involves tyrosine, arginine, cysteine, and histidine R groups; (8) optimum activity depends on some interaction with phospholipids carrying primarily unsaturated fatty acid chains; and (9) the enzyme is inhibited by various agents, including two digitalis glycosides, *ouabain* (pronounced wa-bane) (see Figure 11–40) and *digitoxin*, and the antibiotic *oligomycin*.

The stepwise operation of Na^+/K^+-ATPase is proposed to occur as shown in Figure 11–41. For simplicity, the diagram identifies only two domains of the intact protein, without specifying the arrangement of the α and β subunits. In step 1, ATP and Na^+ bind to the protein; ATP at the ATPase site and Na^+ in specific ion-binding locations probably located in a cavity accessible to the small Na^+ from the cytoplasmic side. In step 2 the ATPase activity initially operates as a kinase, catalyzing a self-phosphorylation modification, known to involve a specific aspartic acid group. The $-CH_2COO^- \rightarrow -CH_2COOPO_3^{2-}$ modification *triggers a profound conformational change* in the entire transport protein that closes the cavity leading to the cytoplasm and opens another on the extracellular face for release of Na^+.

In step 3, K^+ enters the cavity from the extracellular side. In step 4, activated by the bound K^+, the phosphatase activity of the same protein hydrolyzes the aspartyl phosphate, triggering *another conformational change* that restores the original orientation with an open cavity to the cytoplasm for K^+ entry. As indicated, the inhibition by ouabain and oligomycin is specifically targeted at the phosphatase activity in step 4.

Two other transport processes are described in the next section dealing with membrane receptors. One of these also involves the movement of Na^+ and K^+.

11–7 MEMBRANE RECEPTORS

The cellular response to various types of bioactive substances such as hormones, neurotransmitters, growth factors, pyschoactive drugs, plant toxins, bacterial toxins, and viruses begins at the exterior of the cell membrane with the binding of the bioactive substance to **specific receptor sites** in the membrane.

(Hereafter, the general term *ligand* will be used routinely in reference to a substance that binds to a receptor.) The receptor site is usually composed of a specific membrane protein, although some examples of lipid receptors are also known. One significant example of a lipid receptor is the G_{M_1} ganglioside that binds the cholera protein toxin.

The ligand/receptor-binding event provides an initial signal that controls one or more specific processes inside the cell. There are three phases in the process: (1) the identity and binding properties of the receptor, (2) the mechanism for transducing the extracellular binding signal into an intracellular signal, and (3) the specific molecules in the cell that are targeted for regulation. Each of these phases is considered in the material of this section and in Section 11–8.

Assay of Ligand Binding to a Receptor Protein

At the level of whole cells, purified membrane preparations, or a solubilized receptor protein, the interaction of ligand L with its receptor R can be assayed by *equilibrium-binding measurements* (refer to Section 4–6). Using a radioactively labeled form of the ligand allows for highly sensitive assays. After an $L + R \rightleftharpoons LR$ equilibrium is established, techniques of dialysis, ultrafiltration, or ultracentrifugation can be used to separate bound L (that is, LR) from free L, and the measured amounts are used to construct a *Scatchard plot*. A linear Scatchard plot suggests that L is binding to only one class of receptor having the same association constant K_a. Curvilinear plots suggest either cooperative interactions among the same class of receptors or the existence of different classes of receptors—high-affinity receptors versus low-affinity receptors. Figure 11–42(a) illustrates the format of a Scatchard plot, and Figure 11–42(b) shows actual results for the binding between insulin and its receptor protein in fat cells (adipocytes). The biphasic, curvilinear plot for the insulin receptor has been established to be due to the presence of high-affinity receptors ($K_a \approx 10^9 M^{-1}$) and low-affinity receptors ($K_a \approx 10^7 M^{-1}$), rather than negative cooperativity. The high-affinity sites appear to be associated with clustered insulin receptors and low-affinity sites with lone, nonclustered receptors. Additional information on the structure and the operation of the insulin receptor is discussed later.

Acetylcholine Receptor

In higher animals the nerve system—divided into the central nervous system (brain and spinal cord) and the peripheral nervous system—is composed of electrically excitable cell units called *neurons*. Highly specialized cells of various types, neurons can communicate, chemically and electrically, with other neurons, with cells of sense organs, and with other cells. The biochemistry and physiology of these processes are obviously of immense importance.

The stimulation of a neuron results in an influx of Na^+ and an accompanying efflux of K^+, changes that alter the voltage difference across the neuron membrane. The resting transmembrane potential is about -60 millivolts (mV), which in a local depolarization process is changed about 100 mV to yield a potential difference of about $+40$ mV. This local potential change is then rapidly conducted along the neuron. For communication with another resting neuron (or

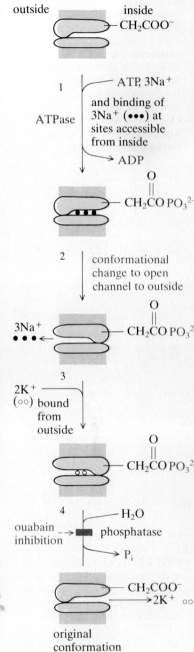

FIGURE 11–41 Schematic representation for the operation of Na^+/K^+-ATPase. Steps 1–4 correspond to the text description. For simplicity the $\alpha_2\beta_2$ oligomeric structure is not indicated.

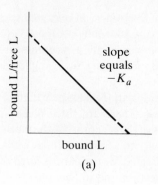
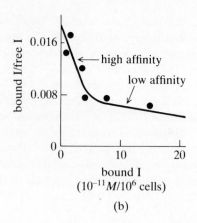

FIGURE 11–42 Scatchard plot. (a) General format. Linearity implies a single class of noncooperative receptors. The K_a values (association constants) are sometimes expressed as dissociation constants (K_d), where $K_a = 1/K_d$. (b) Binding of the insulin receptor to insulin (I), established to be due to the existence of at least two functionally different classes of the insulin receptor.

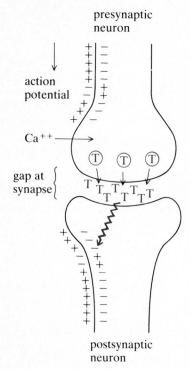

FIGURE 11–43 Chemical neurotransmission. Vesicles of neurotransmitters (T), formed in the presynaptic neuron, are released into the synapse on arrival of an action potential. The released transmitter then acts at the membrane of the postsynaptic neuron, resulting in the formation (wavy arrow) of a small action potential in the postsynaptic neuron. See Figure 11–44 for details at the synaptic gap of cholinergic neurons.

a muscle cell), this action potential must be transmitted to that cell. This transmission occurs primarily through the participation of *neurotransmitters* at the neuron-neuron junction or at the neuron-muscle junction. From here on we will limit consideration to the neuron-neuron junction, called the *synapse*.

The arrival of the action potential at the end of the presynaptic neuron causes the release of the neurotransmitter into the gap between the presynaptic neuron and the postsynaptic neuron (see Figure 11–43). The neurotransmitter communicates with the postsynaptic neuron by binding to a receptor site in the postsynaptic membrane, which initiates an action potential in the postsynaptic neuron. If enough of these excitatory transmissions—occurring at different synapses involving the same postsynaptic neuron—result in a strong signal, the entire neuron is stimulated.

Nerve transmission involving excitation by *acetylcholine* are called *cholinergic systems*. Acetylcholine's structure is

$$CH_3\overset{\overset{O}{\|}}{C}OCH_2CH_2\overset{+}{N}(CH_3)_3$$
acetylcholine

Another major class of transmitters includes adrenalin (also called epinephrine), noradrenalin (also called norepinephrine), and related compounds that control *adrenergic systems*—so-named because the natural source of these transmitters is the adrenal gland (medulla and cortex). The chemical transmission process is now known to be much more complex than envisioned years ago. For one thing, the list of known neurotransmitters numbers in the dozens and new ones are being discovered at a steady rate. Also, contrary to the old notion that one neurotransmitter is responsible for affecting only one type of neuron, we now know that many neurons can respond to different neurotransmitters, with both potentiating and inhibitory effects.

In cholinergic systems (see Figure 11–44) the arrival of the action potential at the end of the presynaptic neuron appears to initially stimulate the influx of Ca^{2+} across the presynaptic membrane. This action in turn stimulates an exo-

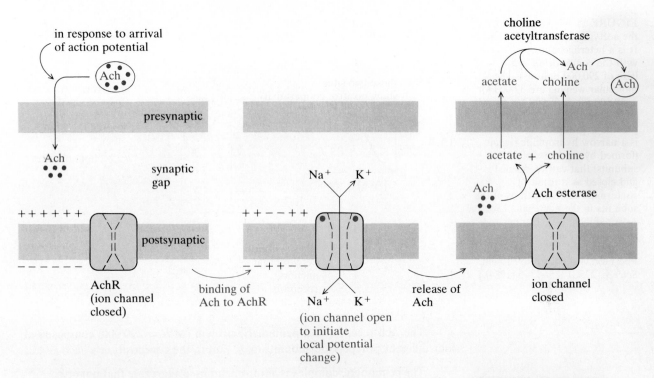

FIGURE 11–44 Stepwise summary of the events of transmission at a cholinergic synapse. Examine the diagram from left to right, and see text for description. See Figure 11–45 for additional details regarding the receptor.

cytosis-like release process of membrane-bound vesicles containing several acetyl-choline (Ach) molecules. In the synaptic gap Ach binds to a specific Ach receptor protein (AchR) in the postsynaptic membrane, causing a trans-membrane channel to open in the receptor complex and allowing the inward flow of Na^+ and some outward flow of K^+. This ion movement provides the basis for a local depolarization across the postsynaptic membrane, thus trans-mitting an action potential to the postsynaptic neuron. The transmembrane ion channel is closed by the dissociation of Ach from AchR, thus avoiding a permanent depolarization. The free Ach transmitter is degraded in the gap owing to the presence of the enzyme *acetylcholine esterase*.

Acetate and choline are transported back into the presynaptic neuron, wherein the enzyme *choline acetyltransferase* catalyzes the resynthesis of acetyl-choline. (The transport of choline back across the presynaptic membrane is crucial, because neurons are unable to synthesize choline.) The entire process occurs very, very quickly—in about a millisecond. As the action potential is conducted along the neuron, the original ion gradients are restored by the action of Na^+/K^+-ATPase.

The acetylcholine receptor has been under intense study for about thirty years. The first success in extracting a functional receptor from intact membranes of the electric organs of *Electrophorus* and *Torpedo* fish was achieved in 1970, and highly purified preparations were obtained in 1973. Since then, much has been learned about its structure and function and its relationship to some dis-eases and to various physiologically active substances. Recently, the genes cod-ing for the Ach receptor subunits have also been cloned.

FIGURE 11–45 Organization of the acetylcholine receptor AchR. It is a heterogeneous pentamer with the composition $\alpha_2\beta\gamma\delta$ (size of 290,000 daltons). The molecular weights are 39,000 for α, 48,000 for β, 58,000 for γ, and 64,000 for δ, plus each has glyco prosthetic groups. The ion channel is a narrow hydrophilic region (formed by segments of all subunits) that can be opened and closed as a result of conformational transitions of the subunits in the aggregate. [*Source:* Adapted from J. P. Changeux, A. Devillers-Thiery, and P. Chemouilli, "Acetylcholine Receptor: An Allosteric Protein," *Science,* **225,** 1335–1345 (1984).]

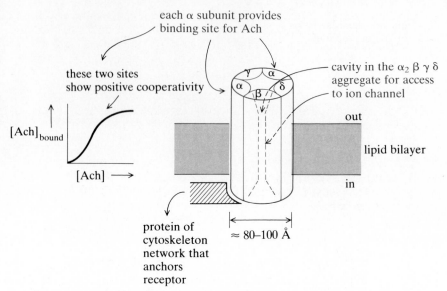

The Ach receptor is a pentameric protein (MW $\approx$ 290,000) composed of four different polypeptide subunits (α, β, γ, δ) in the composition $\alpha_2\beta\gamma\delta$.

The pentameric complex is a transmembrane aggregate that not only serves as the Ach receptor but also provides an internal transmembrane channel for ion movement.

The five subunits are associated in a rosette arrangement (see Figure 11–45), with each subunit having a transmembrane orientation. A protein component of the cytoskeleton network apparently interacts with the oligomer on the cytoplasmic face to anchor the receptor in the membrane. Disulfide bonding between the δ subunits of each of two neighboring receptors also reduces free movement of the receptor in the membrane.

The region of structure exposed on the synapse face of the membrane provides *two Ach binding sites*—one in each of the two α subunits—that exhibit strong positive cooperativity. The hydrophilic ion channel is formed in the interior of the oligomer, although it is not yet certain to what extent each subunit contributes to the formation of the channel. The channel entrance on the synapse side appears to reside at the bottom of a funnel-shaped crevice. When Ach is not bound, the conformation of the entire oligomer is such that this channel is constricted; that is, the channel is closed.

The binding of Ach to the α sites triggers local conformational changes, causing all subunits to shift positions and resulting in the opening of the channel.

(Additional details regarding the allosteric responses of AchR can be found in the article by J. P. Changeux, A. Devillers-Thiery, and P. Chemouilli cited in the Literature.)

The biochemistry of cholinergic systems is related to many other physiological phenomena:

1. The action of *local anesthetics* such as lidocaine, procaine, and novocaine involves inhibition of acetylcholine transmission by temporarily blocking the ion channel. One proposal is that these agents bind to AchR near the pit of the crevice. This binding may either physically obstruct access to the ion channel or cause local structural changes that prevent the channel from fully opening.

2. α-*Toxin polypeptides* from the venom of various poisonous snakes antagonize signal transmission by competitive binding with Ach at the α site of AchR.

3. The active ingredient of some *insecticides* (such as malathion) is a neurotoxin that deactivates the acetylcholine esterase. The consequence is that the postsynaptic membrane is kept in a depolarized state because the neurotransmitter is not turned over.

4. The rare but serious disease *myasthenia gravis* is an autoimmune condition wherein the body produces an antibody against its own Ach receptor.

5. Although a cause or effect correlation has not yet been determined, the activity of choline acetyltransferase is significantly reduced in the neurons of individuals known to have been afflicted with degenerative brain diseases associated with memory loss, as in *Alzheimer's disease*.

Insulin Receptor

The hormone insulin controls a variety of cellular processes, the most common being the stimulation of glucose transport into the cell. Other insulin-stimulated transport processes are the uptake of amino acids and Na^+. The molecular explanation begins with the participation of an insulin receptor protein, about which much has been learned in recent years. Some features of the insulin receptor that may apply to other receptors include the following: (1) There are at least two classes of insulin receptors, differing in their ability to bind the hormone [see Figure 11–42(b)], (2) insulin receptors display the ability to move in the membrane, and (3) the insulin receptor behaves differently in different tissues.

The solubilized insulin receptor from liver and fat cells is composed of two glycoprotein subunits, α (MW = 130,000) and β (MW = 95,000), connected by disulfide (—S—S—) bonds. The insulin-binding site is localized in the α subunit. Recent studies have established that an early event (perhaps the initial event) associated with the binding of insulin to its receptor is an ATP-dependent phosphorylation of the β subunit in the receptor. Moreover, the kinase activity responsible for this modification also appears to reside in the same β subunit of the receptor. In other words, the protein substrate and the protein kinase are one and the same. The process can be termed *autophosphorylation*.

Although changes in protein conformation are probably involved, what happens thereafter is not yet entirely certain. One suggestion is that the receptor phosphorylation may cause a futher conformational change in the receptor to permit —S—S—/—SH interaction between the receptor and other membrane proteins, perhaps a transport protein for glucose (see Figure 11–46). The

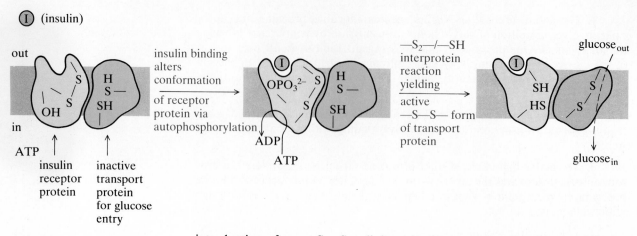

FIGURE 11–46 Proposal for insulin action. After insulin binding and the indicated insulin effects are completed, the release of insulin followed by dephosphorylation of the receptor and a reversal of the $—S_2—/—SH$ redox will restore the initial conditions.

introduction of an $—S—S—$ linkage in the transport protein would be the trigger for reorienting the conformation of the transport protein.

Other studies suggest that some insulin effects may result from the release into the cell of a polypeptide fragment from a membrane protein. This cleavage involves a proteolytic enzyme that gets activated by the insulin/receptor binding event. Thus there may be no single or simple explanation for the action of insulin. A final note: There is evidence that the dietary requirement of trace amounts of chromium (estimated at $50-200$ μg per day) is due to the ability of Cr^{3+} to potentiate the action of insulin in stimulating glucose uptake.

Cytoplasmic Receptors

The existence of receptors in membranes is complemented by *cytoplasmic receptors*, which are specific binding proteins found in the cell cytoplasm. In higher animals the most important type are *steroid-specific receptors*. Most steroid hormones are capable of freely passing back and forth across the lipid bilayer. When a particular steroid enters a cell containing a receptor protein for it, a steroid/receptor complex forms. A common function of the steroid/receptor complex is to control the expression of genes, thereby changing the amounts and/or types of proteins produced in the cell. Gene control involves the transfer of the steroid/receptor complex into the nucleus where binding occurs either to DNA or to DNA-associated proteins. Depending on the steroid and the target cell involved, the gene control may be highly specialized, affecting the synthesis of only certain proteins. Further discussion of details is beyond the scope of this book.

11–8 TRANSDUCTION OF SIGNALS ACROSS MEMBRANES

The control of intracellular processes in response to extracellular substances (hormones, neurotransmitters, and growth factors) is critical to the normal operation of many cells. In each instance the initial event is the binding of the bioactive ligand to its receptor. How, then, does this initial signal of binding cause other events to occur or not occur?

In the previous section two specific examples of signal-transducing mechanisms, involving two specific receptors in specific types of cells, were described: (1) the Ach/AchR interaction that triggers conformational changes in AchR to open an ion channel and (2) the insulin/insulin receptor interaction that triggers autophosphorylation of the receptor, stimulating the receptor to change conformation and interact with other membrane proteins. Although possibly each mechanism also applies to the operation of other specific systems—for example, the binding of the epidermal growth factor (EGF) is also known to involve the autophosphorylation of the EGF receptor—there are other signal-transducing mechanisms more commonly associated with different bioactive substances in many different cell types.

Receptor-Controlled Adenylate Cyclase System

The most widespread signal transduction system involves the role of *cyclic AMP as a secondary messenger.* The ligand/receptor-binding event on the extracellular face of the membrane activates (or inhibits) the enzyme *adenylate cyclase,* located on the intracellular face. Activation (see Figure 11–47) results in the formation of cyclic AMP from ATP, with the elevated cellular level of cyclic AMP acting as an internal signal of the extracellular signal. The immediate role of cyclic AMP is to *activate* the enzyme *cyclic AMP–dependent protein kinase* (cAMP–dPK) by binding to an inactive tetrameric form (R_2C_2) of the enzyme, causing its dissociation to release the active kinase subunit C. The active kinase then catalyzes the phosphorylation of one or more specific target proteins in the cell. In some cases these early events are part of an elaborate protein kinase cascade (refer to Section 9–6).

Although cyclase activation is more common, some ligand/receptor-binding events have an opposite effect, inhibiting the activity of adenylate cyclase. This inhibition counteracts the stimulatory effects of cyclic AMP by tuning off the production of this intracellular messenger.

How is adenylate cyclase activity controlled by the event of ligand/receptor-binding? Though not complete in every detail, general operation of the signal-transducing mechanism is now known. A key discovery was the determination that adenylate cyclase activity is associated with two components—one having a catalytic function, a second having a regulatory function. The catalytic component (C) carries the active site of cyclase for ATP → cAMP. The regulatory component (G) regulates the catalytic component and is designated G because it is a GTP/GDP-binding protein with GTPase activity. When GTP is bound (see Figure 11–48), the G protein has a conformation favoring a tight association with the C protein—an association that activates the cyclase site in C. When the GTPase activity of the G protein hydrolyzes the bound GTP to GDP and P_i, the G protein (now with bound GDP) assumes a different conformation incapable of maintaining a strong association with C. The G component dissociates, and the active site in C assumes an inactive conformation.

The participation of a ligand-responsive receptor in the GC $\rightleftharpoons$ G + C system is envisioned (see Figure 11–49) to involve an initial interaction between

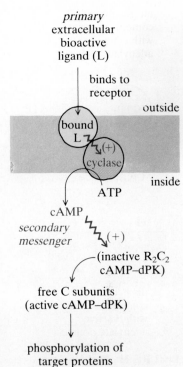

FIGURE 11–47 Cyclic AMP as a secondary messenger of primary extracellular signals. Proteins subsequently controlled by the phosphorylation activity of cAMP–dPK are identified in other chapters. The wavy arrow with (+) signifies conformational changes accompanied by activation of a protein.

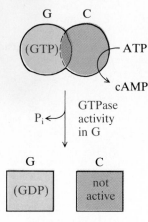

interacting GC
complex associated
with active
adenylate cyclase

dissociated GC
complex associated
with inactive
adenylate cyclase

FIGURE 11–48 Adenylate cyclase activity involves a GTP-binding regulatory subunit.

the receptor and the G component. In the absence of the bioactive ligand, the unoccupied receptor has contact with the G_{GDP} conformer, and the nonassociated cyclase component is inactive. The signal transduction mechanism associated with ligand/receptor binding is proposed to operate as follows: The binding of ligand L to its receptor triggers a structural change in the receptor, which transmits a structural change in the receptor-associated G_{GDP} species. The induced change in conformation of G_{GDP} causes the release of bound GDP and the binding of "fresh" GTP, which in turn causes another conformational change in G. The new G_{GTP} conformer then associates with C, activating the cyclase site. The subsequent activity of GTPase to re-form the G_{GDP} conformer triggers the release of G_{GDP} and relaxes the cyclase activity. The possible role of membrane lipids in this communication of receptor and the G and C proteins remains yet to be elucidated.

This proposal is based on much experimental work, with many studies making use of *hydrolysis-resistant, synthetic analogs of GTP*. One such analog is β,γ-imido-guanosine-5′-triphosphate [GPP(NH)P]:

$$Gua-rib-O-\overset{\overset{\displaystyle O}{\|}}{\underset{\underset{\displaystyle O_-}{|}}{P}}-O-\overset{\overset{\displaystyle O}{\|}}{\underset{\underset{\displaystyle \_O}{|}}{P}}-\underset{\underset{\displaystyle H}{|}}{N}-\overset{\overset{\displaystyle O}{\|}}{\underset{\underset{\displaystyle O_-}{|}}{P}}O^-$$

GPP(NH)P

When membrane preparations containing adenylate cyclase and, for example, an associated hormone receptor are assayed in the presence of GPP(NH)P rather than GTP, a sustained level of maximal cyclase activity can be observed, even in the absence of the hormone. Mimicking GTP, GPP(NH)P binds tightly

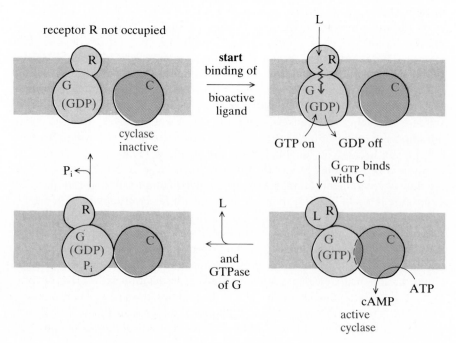

FIGURE 11–49 Diagrammatic summary of the receptor-mediated activation of adenylate cyclase to produce the cAMP secondary messenger.

to the G component and locks the GC complex into the active state, because GPP(NH)P cannot be hydrolyzed by the GTPase activity of G.

Ligand/receptor-binding events that deactivate cyclase activity are proposed to operate in a similar fashion but through a slightly different G protein. Yet uncertain, however, is whether this inhibitory-specific G protein deactivates cyclase activity by binding directly to the C protein or by binding to and short-circuiting the action of the activation-specific G protein (see Figure 11–50).

Receptor-Controlled Calcium Systems

Although cAMP is definitely established as the major secondary messenger for transducing regulatory signals across many types of membranes, it is not the only one. Messenger roles for *cyclic GMP* and for *calcium* are also established, with Ca^{2+}-sensitive processes (see Figure 11–51) being more widespread. The receptors involved are targeted at controlling the activity of guanylate cyclase and *increasing* the intracellular concentration of Ca^{2+}. Calcium levels may rise by receptor activation of a Ca^{2+} channel in the membrane or of the release of stored Ca^{2+} from reticulum systems in the cell.

Most calcium-sensitive processes in the cell are not directly mediated by free Ca^{2+} but indirectly mediated by a specific calcium-binding protein called *calmodulin* (CaM). Widely distributed in nature, calmodulin is composed of a single polypeptide chain (148 amino acid residues) containing four identical Ca^{2+} binding sites. (The amino acid sequence of the calmodulin polypeptide has been remarkably conserved throughout evolution of eukaryotic cells. Consistent with its cation-binding function, 27 of the residues are glutamate and 23 others are aspartate.)

The $CaM/(Ca^{2+})_4$ complex represents the regulatory species of Ca^{2+}, which by interacting with other specific proteins, modulates their activity. A

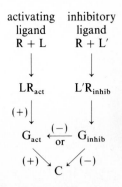

FIGURE 11–50 Summary of activation and inhibition of adenylate cyclase.

FIGURE 11–51 General summary of Ca^{2+} as a secondary messenger. The primary bioactive ligand binds to its receptor (R) to cause (wavy arrow) an increase in intracellular Ca^{2+}.

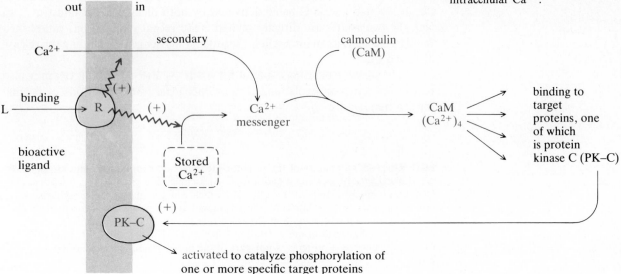

broad spectrum of about thirty different proteins from various cell types are linked to regulation by $CaM/(Ca^{2+})_4$. Many are enzymes from various areas of metabolism. Others are protein components of microfilaments and microtubules in the cytoskeleton network.

In some cells the action of the intracellular Ca^{2+} signal (via calmodulin) is associated with the phosphorylation of certain cellular proteins. The reason for this association wad discovered in 1980: Cells contain a specific Ca^{2+}-stimulated protein kinase. Called *protein kinase C* (or simply PK–C; C for calcium), this enzyme is localized in the cell membrane. For reasons that will soon be obvious, we will defer further discussion of protein kinase C.

Receptor-Controlled Phosphatidyl Inositol Systems

In the 1970s biochemists had observed that a wide variety of extracellular signals induce a rapid turnover of phosphatidyl inositol (PI) in the membrane of target cells. Recent studies have now confirmed that this turnover is not merely some remote consequence of cellular control. Rather:

The PI turnover is itself part of a primary signal-transducing mechanism.

Triggered by ligand/receptor binding, the key event is a conformational activation of the membrane enzyme *phospholipase C* (see Figure 11–52). Using the diphosphoinositide PIP_2 as substrate, phospholipase C releases IP_3 (*inositol-1,4,5-trisphosphate*) to the cytoplasm, and a *1,2-diacylglycerol* (DG) fragment is retained in the membrane, where it is proposed to function as a secondary messenger. The immediate protein target of the DG messenger is protein kinase C. The relationship of this signal mechanism to PI turnover is explained by the precursor pathway of $PI \rightarrow PIP \rightarrow PIP_2$ involving the participation of two specific enzymes, PI kinase and PIP kinase.

In addition to conferring a special importance to the presence of PI lipids in membranes, this discovery also suggests that some cells may use a single extracellular signal to generate two intracellular signals that act synergistically—that is, protein kinase C being activated by both diacylglycerol and Ca^{2+}. In fact, the two routes are directly related, since recent research has established that IP_3 also acts as an intracellular secondary messenger stimulating the release of stored Ca^{2+}.

The signal-generating route of $PI \rightarrow PIP \rightarrow PIP_2 \rightarrow DG + IP_3$ may also be of importance in explaining certain oncogenic transformations of normal cells to tumor cells.

FIGURE 11–52 Summary of the receptor-mediated activation of protein kinase C ▶ via a diacylglycerol messenger and a Ca^{2+} messenger. The activation can involve a response to ligand binding at receptor R_1, which will give rise to the diacylglycerol messenger; a separate response to another ligand at receptor R_2, which will elevate Ca^{2+} levels in the cell; or both at the same time. The CTP-dependent route of PI resynthesis is described in more detail in Chapter 17. The possible participation of *ras* and *src* oncogene proteins is not part of the normal signal transduction process.

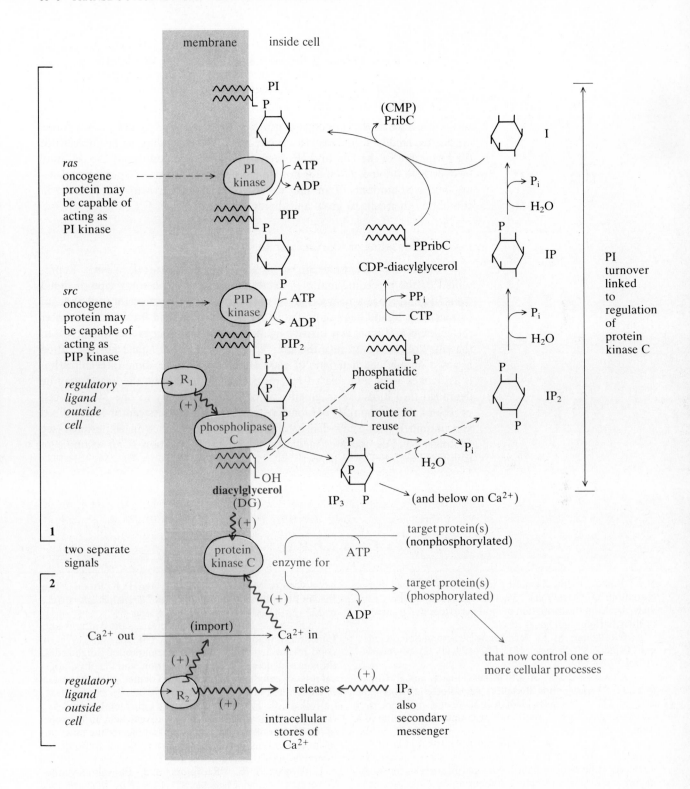

That is:

> The catalytic actions of the normal PI kinase and PIP kinase can be mimicked by two oncogene-derived protein kinases: the *ras* oncogene protein and the *src* oncogene protein, respectively (refer to Figure 11–52).

In the presence of these oncogene proteins the PI $\rightarrow$ PIP $\rightarrow$ PIP$_2$ conversions are oncogenically stimulated to elevate PIP$_2$ levels, leading to an increase in the formation of the DG and IP$_3$ signals. Under such conditions the cell may be overly stimulated, and if it is not transformed itself, it may respond by releasing increased amounts of another substance (such as a growth factor) that will stimulate other cells to grow and divide rapidly.

Internalization of Extracellular Ligands

Membrane receptors can be broadly classified into two classes. Represented by the preceding material, class I receptors rely on some type of signal transduction process to transmit a message to the inside of the cell that an extracellular ligand has bound to the receptor. The ligand itself does not enter the cell. Class II receptors involve ligand/receptor-binding events that result in the entry of the ligand into the cell. Note, however, that *ligand internalization* is not a unique characteristic of class II receptors, since some internalization is associated with some class I receptors. Once internalized, the ligand may be degraded into bioactive fragments or be utilized directly in one or more cell processes. Some examples of bioactive substances that are receptor-internalized are vitamin B$_{12}$ and low-density lipoprotein particles rich in cholesterol (see Chapter 17). Yes, receptor-mediated ligand internalization is an example of facilitated membrane transport.

LITERATURE

BLAND, J. "Biochemical Consequences of Lipid Peroxidation." *J. Chem. Educ.,* **55,** 151–155 (1978). A short review article on the formation of lipid peroxides and a survey of their harmful biological effects.

BRETSCHER, M. S. "Membrane Structure: Some General Principles." *Science,* **181,** 622–629 (1973). A review article.

CHANGEUX, J. P., A. DEVILLERS-THIERY, and P. CHEMOUILLI. "Acetylcholine Receptor: An Allosteric Protein." *Science,* **225,** 1335–1345 (1984). A short review article focusing on the structure of AchR and the cooperative binding of Ach.

CUATRECASAS, P. "Membrane Receptors." *Ann. Biochem.,* **43,** 169–214 (1974). A review of recent progress in the identification, isolation, and purification of various membrane-localized receptors. A focus on insulin receptors.

CZECH, M. P., J. MASSAQUE, and P. F. PILCH. "The Insulin Receptor: Structural Features." *Trends Biochem. Sci.,* **6,** 222–224 (1981). A short review article.

FISHMAN, P. H., and R. O. BRADY. "Biosynthesis and Function of Gangliosides." *Science,* **194,** 906–915 (1976). A good review article covering the formation of gangliosides, the consequences of impaired formation, and the physiological roles of gangliosides, such as that of mediating the action of cholera toxin.

FOX, C. F. "The Structure of Cell Membranes." *Sci. Am.,* **226,** 30–38 (1972). An informative article on the roles of proteins in membrane structure and membrane function; includes a discussion of the process of active transport of materials across membranes.

HOKIN, L. E. "Receptors and Phosphoinositide-Generated Second Messengers." *Annu. Rev. Biochem.,* **54,**

205–235 (1985). A review article on the secondary-messenger role of diacylglycerol (DG) and inositol trisphosphate (IP$_3$).

HUBBEL, W. L., and M. D. BOWNDS. "Visual Transduction in Vertebrate Photoreceptors." *Annu. Rev. Neurosci.,* **2,** 17–34 (1979). The role of cyclic GMP and Ca^{2+} in altering the permeability of the rod cell membrane in response to light absorption by rhodopsin is reviewed.

JACOBS, S., and P. CUATRECASAS. "The Mobile Receptor Hypothesis for Cell Membrane Receptor Action." *Trends Biochem. Sci.,* **2,** 280–282 (1977). A short review article.

JAIN, M. K., and R. C. WAGNER. *Introduction to Biological Membranes.* New York: Wiley, 1980. Structure, composition, isolation of components, fluidity, and formation are all discussed.

KABACK, H. R. "Transport Studies in Bacterial Membrane Vesicles." *Science,* **186,** 882–892 (1974). A review article of membrane transport, focusing on a few bacterial systems that have been characterized.

KLEE, C. B., T. H. CROUCH, and P. G. RICHMAN. "Calmodulin." *Annu. Rev. Biochem.,* **49,** 489–515 (1980). A thorough review article of calmodulin structure and calmodulin-controlled cellular processes.

MARCHESI, V. T. "The Structure and Function of a Membrane Protein." *Hosp. Pract.,* **8,** 76–84 (1973). Introductory-level discussion of a major glycoprotein isolated from the membrane of red blood cells. [Original research described in *Proc. Natl. Acad. Sci., USA,* **69,** 1445 (1972).]

MARCHESI, V. T., H. F. FURTHMAYR, and M. TOMITO. "The Red Cell Membrane." *Annu. Rev. Biochem.,* **45,** 667–698 (1976). A thorough review article.

MARX, J. L. "The Leukotrienes in Allergy and Inflammation." *Science,* **215,** 1380–1383 (1982). Brief *Research News* article on the possible biological functions.

MENGER, E. L., ed. *Acc. Chem. Res.,* **8,** no. 3, 81–112 (1975). A special issue containing articles devoted exclusively to various aspects on the chemistry of vision.

NELSON, N. A., R. C. KELLY, and R. A. JOHNSON. "Prostaglandins and the Arachidonic Acid Cascade." *Chem. Eng. News,* 30–44 (August 16, 1983). A short review article on the biosynthesis and degradation of prostaglandins

and related materials. Their use as therapeutic drugs is also covered.

NISHIZUKA, Y. "Turnover of Inositol Phospholipids and Signal Transduction." *Science,* **225,** 1365–1370 (1984). A short review article describing the messenger roles of diacylglycerol.

RAZIN, S., and S. ROTTEM. "Cholesterol in Membranes." *Trends Biochem. Sci.,* **3,** 51–55 (1978). A short review article on the role of cholesterol as a regulator of membrane fluidity.

SAMUELSSON, M. "Prostaglandins and Thromboxanes." *Annu. Rev. Biochem.,* **47,** 997–1029 (1978). Biosynthesis and interactions with cyclic nucleotides are reviewed.

SCHRAMM, M., and Z. SELINGER. "Message Transmission: Receptor-Controlled Adenylate Cyclase System." *Science,* **225,** 1350–1356 (1984). A short review article on the structure and the interaction of the G and C components of adenylate cyclase.

SINGER, S. J. "The Molecular Organization of Membranes." *Annu. Rev. Biochem.,* **43,** 805–833 (1974). A detailed review article focusing on the arrangement of proteins in membranes; some coverage of the movement (fluidity) of lipids and proteins in the membrane.

STENFLO, J., and J. W. SUTTIE. "Vitamin K-Dependent Formation of γ-Carboxyglutamic Acid." *Annu. Rev. Biochem.,* **46,** 157–172 (1977). A review article on the activation of prothrombin by vitamin K.

UNWIN, N., and R. HENDERSON. "The Structure of Proteins in Biological Membranes." *Sci. Am.,* **250,** 78–94 (1984). Description and illustrations of the three-dimensional structure and subunit arrangements of integral proteins with a particular focus on transmembrane proteins.

VAN OBBERGHEN, E., B. ROSSI, A. KOWLASKI, H. GAZZANO, and G. PONZIO. "Receptor-Mediated Phosphorylation of the Hepatic Insulin Receptor: Evidence That the M$_r$ 95,000 Receptor Subunit Is Its Own Kinase." *Proc. Natl. Acad. Sci., USA,* **80,** 945–949 (1983). A research article with a self-explanatory title.

YAMAKAWA, T., and Y. NAGAL. "Glycolipids at the Cell Surface and Their Biological Functions." *Trends Biochem. Sci.,* **3,** 128–131 (1978). A short review article.

EXERCISES

11–1. The melting points of a few fatty acids are listed in Table 11–6. On the basis of these data, explain the characteristic that distinguishes neutral fats and oils.

11–2. What prediction can be made regarding the value of the melting point temperature for (a) palmitoleic acid, (b) lauric acid, (c) arachidic acid, and (d) arachidonic acid? (Refer to the data in Table 11–6.)

11–3. Draw line formulas for (a) C$_{22}$, (b) (C$_{22:1}$)Δ^{10}(*cis*), and (c) (C$_{22:4}$)$\Delta^{a,b,c,d}$ (conjugated pattern of unsaturation beginning at carbon 7).

11–4. If a mixture consisting of a triglyceride and phosphatidyl choline were analyzed by thin-layer chromatography on silica gel in a chloroform-methanol-water developing solvent, you would observe complete separation, with the R$_f$

TABLE 11-6 Data for Exercise 11-1

ACID	MELTING POINT (°)
Linoleic acid	−5.0
Linolenic acid	−10.0
Myristic acid	53.9
Oleic acid	13.4
Palmitic acid	63.1
Stearic acid	69.6

of the triglyceride being approximately 1 and that of phosphatidyl choline approximately 0.4. Explain why the R_f values of these two lipids differ so widely.

11-5. Draw the structure of the intact lipid that corresponds to each of the following mixtures of products obtained from the complete hydrolysis of the lipid.

(a) glycerol, palmitic acid, stearic acid, inorganic phosphate
(b) glycerol, palmitoleic acid, oleic acid, ethanolamine, inorganic phosphate
(c) sphingosine, palmitic acid, inorganic phosphate
(d) sphingosine, glucose, oleic acid.

11-6. What will have a greater amphipathic character, a glucocerebroside or a sphingomyelin? Explain.

11-7. Explain why inorganic ions (K^+, Na^+, Ca^{2+}, and others) do not cross a biomembrane by simple diffusion.

11-8. When bacterial cells are grown at temperatures lower than normal, for example, at 15°C rather than at 37°C, the membrane lipids contain a larger proportion of unsaturated fatty acids than saturated fatty acids. When the temperature is restored to 37°C, the ratio of unsaturated fatty acids to saturated fatty acids is then observed to decrease. What is the possible significance of these observations?

11-9. Proteins that are covalently attached to membranes are linked to phosphatidyl inositol on the cytoplasmic face of the membrane. The covalent linkage occurs between the inositol group of PI and the C terminus residue of the polypeptide. (a) Illustrate the type of linkage that is involved. (b) How many possibilities are there for this linkage to form? (c) If the protein is released with inositol phosphate still attached, what class of phospholipase enzyme is involved in the release process?

11-10. What type of amino acid side chains would you expect to be clustered at the functional binding domains in the calmodulin molecule?

11-11. Define or describe these terms: thioester linkage, hydroperoxide formation of fatty acids, neutral lipid, lipase, saponification, lecithin, glycolipid, amphipathic, prostaglandin, leukotriene, ceramide, rhodopsin proteins, lipid bilayer, membrane fluidity, integral membrane protein, receptor, semipermeable, passive transport, active transport, transport channels, ion pumps, ionophores, synapse, neurotransmitter, secondary messenger, ATPase, GTPase, calmodulin, cAMP–dPK.

CHAPTER TWELVE

PRINCIPLES OF BIOENERGETICS

To be alive and grow, living cells depend on an intake of food to serve not only as sources of carbon, nitrogen, sulfur, phosphorus, and the other biologically essential elements but also as sources from which useful energy is extracted. Without energy, a cell is a nonfunctional machine.

For what purposes is energy used in living cells? In previous chapters we have already dealt with some examples: (1) the biosynthesis of proteins, DNA, and RNA; (2) the kinase-catalyzed phosphorylation of proteins to trigger conformational changes; (3) the supercoiling of DNA; (4) the conversion of simple sugars into sugar phosphates and NDP sugars and the conversion of fatty acids into thioesters; and (5) the operation of active transport systems. Many other examples will be encountered in future chapters dealing with specific reactions associated with the metabolism of carbohydrates, lipids, and amino acids.

In Chapter 6 we had our first encounter (see Section 6–2) with biochemical energetics in describing the general metabolic role of **adenosine triphosphate (ATP),** the preeminent energy material in a cell. To restate that role briefly:

> By forming ATP, cells conserve chemical energy released in energy-yielding degradative reactions; and then by degrading ATP, cells utilize this bioenergy to sustain the energy-requiring events of biosynthesis and other cell processes.

In this chapter we will expand on this theme, focus on ATP in more detail, and also examine other biomaterials that function in the transfer of energy.

Describing the energetics of any system—be it living or nonliving, organic or inorganic, chemical, physical, or biological—is the domain of a specialized field called *thermodynamics,* where emphasis is placed on energy changes as the system undergoes a transformation from one state to another. In the language of thermodynamics, energy changes can be described in a variety of ways, but the most useful description is given in terms of the *change in free energy.* For chemical systems the free-energy change is extremely useful, because under the commonly encountered conditions of constant temperature and constant pressure, it provides a valid method of predicting the feasibility of a reaction, as well as being representative of the maximum amount of chemical energy that is potentially available for doing useful work. For convenience, the first section of this chapter provides a minireview of some important principles of thermodynamics.

12–1 PRINCIPLES OF THERMODYNAMICS

Energetics of State Transitions

Any transformation is described by contrasting the physical and/or chemical properties of the initial and final states of the system. Included among these properties are pressure, temperature, volume, the physical states of the materials, the concentration of each material, and the chemical composition of each material. Two simple examples are given here, one a physical transformation and the other a chemical transformation. (For both we will impose conditions of constant temperature and pressure and, for reasons of simplicity, neglect any changes in volume.) In the first example the nature of the initial and final states is obvious, water is converted from a liquid to a vapor state. The chemical process is also simply described; gaseous propane (an organic alkane) and oxygen react to yield carbon dioxide and water in respective reacting mole proportions of $1 + 5 \rightarrow 3 + 4$. This reaction, of course, represents a complete oxidation (combustion) of propane corresponding to the use of natural gas as a fuel.

$$H_2O(\text{liq}) \xrightarrow{\text{vaporization}} H_2O(\text{vap}) \qquad (P \text{ and } T \text{ are constant})$$
$$\text{initial state} \qquad\qquad \text{final state}$$

$$CH_3CH_2CH_3(\text{vap}) + 5O_2(\text{vap}) \xrightarrow[\substack{\text{(complete} \\ \text{oxidation)}}]{\text{combustion}} 3CO_2(\text{vap}) + 4H_2O(\text{vap}) \qquad (P \text{ and } T \text{ are constant})$$
$$\text{initial state} \qquad\qquad\qquad \text{final state}$$

Common experience tells us these descriptions are incomplete. The major shortcoming is that we have neglected to indicate that one transformation requires an input of energy from the surroundings and the other results in an output of energy. That is, we have not included a *comparison of the energy levels of the initial and final states.* In the energy-requiring vaporization of water, the initial liquid state is at a lower energy level than the final vapor state, with the difference in energy levels primarily due to a greater amount of molecular motion in water molecules in the vapor state than in the liquid state. In the combustion of propane the difference in energy between the two states exists because there is a greater chemical bonding energy in the reactants than in the products. Thus in the course of the chemical change the conversion of the higher-energy reactants to lower-energy products is accompanied by a release of heat energy. A final point concerns the energetics of *forward* and *reverse processes,* namely, the two events are characterized by *equal but opposite energetics.* Thus the condensation of water vapor releases 9.7 kcal/mole, and the formation of propane and oxygen from carbon dioxide and water requires 531 kcal/mole of propane (see Note 12–1):

NOTE 12–1
In the SI system the unit of energy is the *joule*, J. The interconversion of thermochemical calories and joules is made according to these relationships:

1 calorie = 4.184 joules
1 kilocalorie = 4.184 kilojoules

$$H_2O(\text{liq}) + \text{energy} \longrightarrow H_2O(\text{vap}) \qquad CH_3CH_2CH_3 + 5O_2 \longrightarrow 3CO_2 + 4H_2O + \text{energy}$$
$$\substack{\text{energy-requiring process with} \\ \text{9.7 kcal required per} \\ \text{mole of water}} \qquad\qquad \substack{\text{energy-yielding process with} \\ \text{531 kcal produced per} \\ \text{mole of } C_3H_8}$$

The chemical reactions in living cells can be memorized. However, energy descriptions are useful in understanding the design and logic of these reactions—a topic we begin in the next chapter.

Thermodynamic State Functions

A description in thermodynamic terms of the transition from one state to another can be given in terms of *internal energy E, enthalpy H, entropy S,* and *free energy G.* The internal energy expresses the total energy of a system; the enthalpy expresses the heat content of a system (in units of energy, and not to be confused with temperature); the entropy expresses the degree of disorderliness of a system; and the free energy expresses the energy available for conversion to useful work.

Each of these thermodynamic properties is referred to as a *state function* because their values are determined only by the specific condition of the system, that is, the state of the system. In most instances the actual value of E, H, S, or G is difficult if not impossible to measure, and hence thermodynamics deals primarily with changes (Δ for finite change) in state functions: ΔE, ΔH, ΔS, and ΔG. For the general case of initial state → final state, the change is always expressed as *final state value minus initial state value* ($E_f - E_i$). For a chemical transformation of reactants → products, the difference is stated as *product state value minus reactant state value* ($H_p - H_r$).

Each of the expressions (ΔE, ΔH, ΔS, and ΔG) has a different meaning and application. For biochemical applications the free-energy change ΔG has greatest utility because it applies under conditions of constant pressure and constant temperature, the conditions under which cellular reactions occur. In addition, the value of ΔG for a chemical reaction is easily measured in various ways, and it has very useful interpretations. Although we could develop the concept of ΔG in terms of ΔE, ΔH, and ΔS (they are all interrelated; for example, $\Delta G = \Delta H - T\Delta S$), we will bypass this approach and proceed directly to an analysis of what ΔG means.

Free Energy

The function G was appropriately termed *free energy* because a decrease in its value from one state to another ($G_f - G_i$) is a measure of the *maximum amount of energy that is potentially available for useful work* when the change occurs under the conditions of constant T and P. There are other interpretations of the free-energy function. One of particular usefulness to chemical systems is that the free energy is *a thermodynamic property directly related to the total chemical energy of the system* and hence to the *chemical stability of the system.*

In this context a high free-energy value is representative of a potentially unstable system that under the proper conditions would spontaneously go to a lower level of free energy. In other words, a *negative ΔG value* ($-\Delta G$) corresponds to an *energy-yielding reaction* due to a change from an unstable state of high chemical energy content to a more stable state of lower chemical energy content. Such a reaction is termed an *exergonic reaction* and is said to be *thermodynamically favorable*. The amount of energy released can be used to do work. In contrast, a *positive ΔG value* ($+\Delta G$) corresponds to an *energy-requiring reaction* due to a change from a stable state of low chemical energy to a more unstable (less stable) state of higher chemical energy. Such a reaction is termed an *endergonic reaction* and is said to be *thermodynamically unfavorable* (see Note 12–2). Endergonic processes do not occur spontaneously unless energy is supplied. In other words, work must be done on the system.

If an exergonic reaction were to occur by itself, the output of chemical energy would be lost, primarily as heat energy. However, if it were to occur in the presence of an endergonic reaction, the output of chemical energy from the

NOTE 12–2

If the chemical energy of reactants (G_r) is less than that of products (G_p), the reaction is endergonic, that is, energy-requiring, and the value of ΔG is positive:

reactants $\longrightarrow$ products
(initial state)   (final state)
G_r G_p

$$\Delta G = G_p - G_r$$

If the chemical energy of reactants (G_r) is greater than that of products (G_p), the reaction is exergonic that is, energy-yielding, and the value of ΔG is negative.

exergonic reaction could serve as an input of chemical energy to drive the endergonic process. This type of behavior is termed *energy coupling* and represents the basic design of energy flow in living organisms. We will resume this subject later in the chapter.

The condition of $\Delta G = 0$ has a special meaning signifying that the forward and reverse processes of a reversible reaction system (one exergonic and the other endergonic) are occurring at the same rate and the system is at *equilibrium*, at which point there is no tendency to undergo any further net change.

Standard Free-Energy Change ($\Delta G°$)

Under conditions of constant temperature and pressure, for a general chemical transformation involving reactants A and B and products C and D,

$$aA + bB \rightleftharpoons cC + dD \qquad \text{(constant } T, P)$$

the value of the free-energy change is given by the relationship (see Note 12–3)

$$\Delta G = \Delta G° + RT \ln \frac{[C]^c[D]^d}{[A]^a[B]^b}$$

where

$[\ ]$ = molar concentration

$R = 2.0 \times 10^{-3}$ kcal/degree/mole

T = temperature (in degrees Kelvin)

NOTE 12–3
The abbreviation ln represents the natural base e logarithm, which is related to base 10 logarithms by

$$\ln(N) = 2.3 \log_{10}(N)$$

Here ΔG represents the free-energy change at any point in the course of the transformation, and $\Delta G°$ corresponds to the *standard* free-energy change, which applies only to a particular set of conditions, namely, when all participants are in their standard states. (Although the conditions of the standard state are purely arbitrary, there is a universal agreement; for solutes dissolved in solution, the accepted criterion is unit activity ($a = 1.0$), or approximately $1M$ for most materials.) Thus at a specific temperature and pressure, the value of ΔG will vary with changes in the existing concentrations of all participants, whereas $\Delta G°$ is constant for any reaction system and will change only if the temperature and/or pressure are altered. By using the standard-state condition to eliminate variations due to different concentrations of reactants and products, we can *compare the chemical energetics of different systems* at the same temperature and pressure.

A direct method of determining $\Delta G°$ is possible if we recognize that $\Delta G = 0$ when the system is at equilibrium, and hence

$$\Delta G° = -RT \ln \frac{[C]^c[D]^d}{[A]^a[B]^b}$$

But now since the system is at equilibrium, the concentration ratio is actually the equilibrium constant (K_{eq}) of the system, and we can write

$$\Delta G^\circ = -RT \ln K_{eq} = -2.3RT \log K_{eq}$$

This equation is one of the most useful relationships of thermodynamics. Its utility is obvious: at any temperature the ΔG° can be calculated directly from the equilibrium constant, which in most cases can be determined in the laboratory. Some problems based on the use of this equation are provided in the Exercises.

Many reactions, especially those that occur in nature, involve protons (H^+) as a product or reactant, and hence a quotation of their ΔG° necessarily means that the concentration of H^+ is approximately $1M$ (standard condition of unit activity). Since $1M$ H^+ corresponds to pH = 0, it is certainly unrealistic to quote the ΔG° of H^+-dependent reactions in a living cell where the pH is approximately 7. Accordingly, when the equilibrium constants of pH-dependent reactions are measured, the system is studied at a pH of 7, and the resultant standard free-energy changes are specified as being so calculated. To avoid confusion with a true ΔG°, the standard free-energy changes of H^+-dependent reactions calculated under a nonstandard-state condition of $[H^+]$ are symbolized differently, with a $\Delta G^{\circ\prime}$ notation being most common.

Note: ΔG° (or $\Delta G^{\circ\prime}$) values apply only when all reactants and products exist at standard concentration of $1M$ (and $10^{-7}M$ for H^+). In a cell, however, concentrations of materials are much smaller than $1M$ (the millimolar range is common for most), and the cellular concentrations can and do change. Thus to examine cellular reactions in terms of $\Delta G^{\circ\prime}$ values is not realistic; $\Delta G'$ values based on the actual concentrations are realistic. Nevertheless, $\Delta G^{\circ\prime}$ values are useful as *estimates* of the energy yield and energy need of cellular reactions. They are quite reliable when the actual concentrations of reactants and products are such that the expression of ($[C]^c[D]^d/[A]^a[B]^b$) is about 1. When it is 1, $\Delta G' = \Delta G^{\circ\prime} + RT \ln(1)$ and $\Delta G' = \Delta G^{\circ\prime}$. With this distinction in mind, the language previously introduced for $\Delta G'$ values—endergonic/energy-requiring and exergonic/energy-yielding—will also be applied to $\Delta G^{\circ\prime}$ values. The following principles are valid interpretations of $\Delta G^{\circ\prime}$ values:

$A + B \underset{\longrightarrow}{\longleftarrow} C + D$	$A + B \overset{\longleftarrow}{\longrightarrow} C + D$
equilibrium favors products	equilibrium favors reactants
$K'_{eq} >>> 1$	$K'_{eq} <<< 1$
$\Delta G^{\circ\prime} = -$ value	$\Delta G^{\circ\prime} = +$ value
stability$_{reactants}$ < stability$_{products}$	stability$_{reactants}$ > stability$_{products}$
conversion of reactants to products yields energy	conversion of reactants to products requires energy

Although the equation $\Delta G^\circ = -RT \ln K_{eq}$ has a general utility for all chemical reactions, one can prove that *for oxidation-reduction reactions* the standard free-energy change is given by

$$\Delta G^\circ = -n\mathscr{F}\mathscr{E}^\circ \quad \text{or} \quad \Delta G^{\circ\prime} = -n\mathscr{F}\mathscr{E}^{\circ\prime}$$

where n = number of moles of electrons transferred

$\mathscr{F}$ = Faraday's constant (≈ 23.0 kcal/V)

$\mathscr{E}^\circ$ = net standard oxidation-reduction potential (in volts)

$\mathscr{E}^{\circ\prime} = \mathscr{E}^\circ$ at pH 7

The basis of this relationship is that an oxidation-reduction system is capable of doing useful work owing to the transfer of electrons from that part of the system undergoing *oxidation (loss of electrons)* to that undergoing *reduction (gain of electrons)*. Such a system is called an *electrochemical cell,* and the net potential ($\mathscr{E}^\circ$) is merely a measure of the difference between each part of the system that undergoes oxidation and reduction. Although we could continue here with an analysis of the basic principles and thermodynamics of electro-chemical cells, the subject is more logically discussed in conjunction with the process of respiratory electron transfer, which is examined in Chapter 15.

12-2 HIGH-ENERGY BIOMOLECULES

The preceding discussion of the general principles of thermodynamics and of ΔG in particular is one of two areas needed to establish a foundation to study cellular energetics. The rest of the chapter will be devoted to the second area, wherein we will consider a special group of molecules that participate in the flow of cellular energy—with *adenosine triphosphate* (ATP) having a central importance. Once again, as summarized in Figure 12-1, we consider the general participation of ATP. Two important points are illustrated.

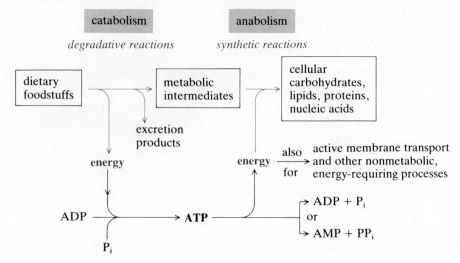

FIGURE 12-1 General features outlining the flow of chemical energy in living cells.

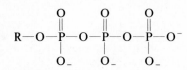

FIGURE 12–2 Highly reactive triphosphoanhydride function of nucleoside triphosphates.

1. Chemical energy is conserved via the formation of ATP in association with the energy-yielding degradative reactions of catabolism.

2. Chemical energy is then utilized via the cleavage of ATP for the energy-requiring synthetic reactions of anabolism and other energy-requiring processes such as triggering changes in protein conformations, active transport, and muscle contraction.

In this section our objective will be to establish why ATP is so well suited for this central role. Although the focus will be on ATP, the description applies to any *nucleoside triphosphate* due to the common presence of the *triphosphoanhydride* function (see Figure 12–2).

Other substances having crucial roles in cellular energetics include acyl phosphates, thioesters, enoyl phosphates, guanidinium phosphates, sulfonium ion species, certain oxyesters, and strong reducing agents. All have one thing in common: A large amount of energy is released when they undergo a chemical transformation (see Table 12–1). Accordingly, they are called *high-energy com-*

TABLE 12–1 Summary of high-energy compounds of biological importance

CLASS	EXAMPLE	$\Delta G^{\circ\prime}$ (pH 7) (kcal/mole)
Nucleoside triphosphates (and pyrophosphate)	$ATP + H_2O \longrightarrow ADP + P_i$	-8.0
	$ATP + H_2O \longrightarrow AMP + PP_i$	-8.0
	$PP_i + H_2O \longrightarrow 2P_i$	-8.0
Acyl phosphates	$RC{-}OPO_3^{2-} + H_2O \longrightarrow RCOO^- + P_i + H^+$ $\quad\ \ \overset{\|}{O}$	-11.8
Enoyl phosphates	$R\overset{H}{\underset{\overset{\|}{OPO_3^{2-}}}{C{=}CCOO^-}} + H_2O \longrightarrow RCH_2\overset{}{\underset{\overset{\|}{O}}{C}}COO^- + P_i$	-14.8
Acyl thioesters	$\overset{O}{\overset{\|}{RC}}{-}SR' + H_2O \longrightarrow RCOO^- + HSR' + H^+$	-7.4
Guanidinium phosphates	$RCH_2\overset{H\ \ H}{\underset{\overset{\|}{^+NH_2}}{NCNPO_3^{2-}}} + H_2O \longrightarrow RCH_2\overset{H}{\underset{\overset{\|}{^+NH_2}}{NCNH_2}} + P_i$	-10.3
Sulfonium ion compounds	$R{-}\overset{+}{\underset{\underset{R'}{\|}}{S}}{-}R + H_2O \longrightarrow R{-}S{-}R + R'OH$	-7.0
Certain acyl oxyesters	$\overset{O}{\overset{\|}{RC}}{-}OR' + H_2O \longrightarrow RCOO^- + HOR' + H^+$	-8.0
Strong reducing agents when oxidized by O_2	$NADH + O_2 \xrightarrow{H^+} NAD^+ + H_2O$	-53.0
	$FADH_2 + O_2 \longrightarrow FAD + H_2O$	-46.0
	$FMNH_2 + O_2 \longrightarrow FMN + H_2O$	-46.0

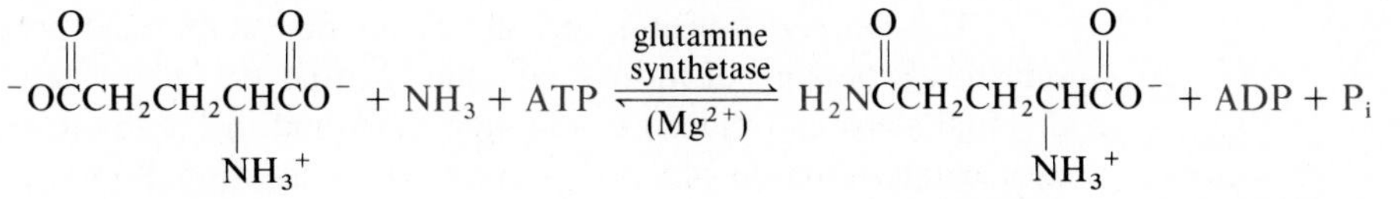

$$\underset{\substack{\text{glutamate}}}{{}^{-}\text{OCCH}_2\text{CH}_2\overset{\overset{\displaystyle O}{\parallel}}{\underset{\underset{\displaystyle \text{NH}_3^{+}}{|}}{\text{C}}}\text{HCO}^{-}} + \text{NH}_3 + \text{ATP} \underset{(\text{Mg}^{2+})}{\overset{\substack{\text{glutamine} \\ \text{synthetase}}}{\rightleftharpoons}} \underset{\substack{\text{glutamine}}}{\text{H}_2\text{NCCH}_2\text{CH}_2\text{CHCO}^{-}} + \text{ADP} + \text{P}_i$$

FIGURE 12–3 Enzymatic ATP-dependent conversion of glutamate and NH_3 to glutamine. (The Mg^{2+} is required as a cofactor for optimal activity.)

pounds. The term *high energy* is generally accepted to imply that the amount of energy involved is at least 7.0 kcal/mole. Being an arbitrary value, there is, of course, nothing sacred about it. It is merely an arbitrary standard that sets ATP and a few other compounds apart from other substances. In this group those compounds with a more general importance are the nucleoside triphosphates, the acyl phosphates, the thioesters, and the strong reductants like NADH. Each is described in the following pages.

Adenosine Triphosphate

Rather than proceed with a direct description of ATP energetics, let us develop the subject by analyzing a single reaction system that is representative of ATP participation in metabolism namely, the enzymatic conversion of glutamate and NH_3 to glutamine (see Figure 12–3). This reaction is particularly important in the area of nitrogen metabolism (see Section 18–3), but for the moment its metabolic function is not of interest. At this time we are primarily interested in analyzing the net reaction as a *model system* for learning something about ATP.

At pH 7 and 37°C (310°K),

$$K'_{eq} = 1.21 \times 10^3$$

Therefore

$$\begin{aligned} \Delta G^{\circ\prime} &= -2.3RT \log K'_{eq} \\ &= -2.3(2.0 \times 10^{-3})(310)[\log(1.21 \times 10^3)] \\ &= -4.4 \text{ kcal} \end{aligned}$$

From the negative value of $\Delta G^{\circ\prime}$ we conclude that the net process is exergonic, and hence the formation of glutamine from glutamate in the presence of ATP degradation is thermodynamically favorable. However, this conclusion does not in any way suggest what purpose is served by the conversion of ATP to ADP and P_i.

To resolve this question, we can look upon the net process as being composed of two separate reactions:

Step 1.	glutamate + $NH_3 \longrightarrow$ glutamine + H_2O	(12–1)
Step 2.	ATP + $H_2O \xrightarrow{Mg^{2+}}$ ADP + P_i	(12–2)
Net.	glutamate + NH_3 + ATP $\longrightarrow$ glutamine + ADP + P_i	(12–3)

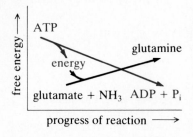

FIGURE 12–4 Energy transfer between exergonic (in color) and endergonic processes. See Figure 12–5 for specific details explaining how the transfer of energy occurs.

This maneuver is conceptually permissible since a net process can be considered as the sum of individual steps. Moreover, in this case both of the individual steps can indeed occur separately and in the absence of any enzyme. (As a catalyst, the enzyme only accelerates the reaction. Remember that although it reduces the energy of activation, the enzyme does not alter the overall thermodynamics of the reaction.)

In the absence of ATP, ADP, and P_i, equilibrium measurements on the glutamate-glutamine interconversion (at pH 7 and 37°C) show that $K'_{eq} = 3.13 \times 10^{-3}$. Thus

$$glutamate + NH_3 \rightarrow glutamine + H_2O$$
$$\Delta G^{\circ\prime} = -2.3RT \log(3.13 \times 10^{-3}) = +3.6 \text{ kcal}$$

Yes, the sign of the free-energy change is correct, and the formation of the amide from the free acid is actually an endergonic reaction. In the presence of ATP, however, we have already established that amide formation is exergonic (see Figure 12–4). With a little thought, a conclusion about the participation of ATP is inescapable:

The hydrolysis of ATP to ADP and P_i must be sufficiently exergonic to provide the chemical energy needed to mediate the endergonic formation of glutamine.

In other words, the complete reaction involving glutamate, NH_3, and ATP can be considered as a coupling of a thermodynamically unfavorable reaction to a thermodynamically favorable reaction, with the latter acting as an energetic driving force of the former.

But how does the exchange of energy occur? In this particular case the transfer of energy is mediated via the intermediate formation of phosphoglutamate, as shown in Figure 12–5. Prior to reaction with NH_3, the highly reactive triphosphoanhydride function of ATP is exploited to mediate the transfer of a phosphate group to the free acid —COO^- group in glutamate. The phosphorylated intermediate now has a highly reactive anhydride function wherein the carbon of $(\delta^+)C=O(\delta^-)$ is much more easily attacked by the NH_3 molecule. In other words, the energy of ATP has been used directly to provide a re-

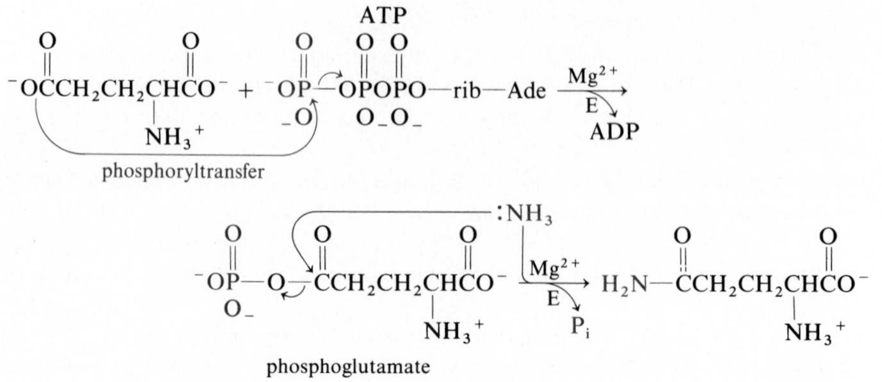

FIGURE 12–5 Glutamine synthetase (E) reaction. Phosphoglutamate is a highly reactive anhydride produced as a reaction intermediate.

action path in which the $C=O$ is activated before reaction with NH_3. The enzyme specificity is responsible for this overall sequence, including the selection of the side-chain carboxyl group over the alpha carboxyl group as the site for phosphoryltransfer.

To summarize: In our model system we have established that the ATP + $H_2O \rightarrow$ ADP + P_i conversion is a thermodynamic driving force for an otherwise unfavorable process. Although we have considered a specific chemical transformation of the living state, distinguished by specific thermodynamic values and a specific mechanism, the theme applies to *all ATP-dependent processes*. Many different examples will be encountered in subsequent chapters.

How much energy is available in ATP? To make this calculation, we need look no further than the glutamate-glutamine reaction system for which the $\Delta G^{\circ\prime}$ values for glutamine formation were given in the presence and the absence of ATP. Since changes in thermodynamic functions are dependent solely on the initial and final states, we can express the net, standard free-energy change as equal to the sum of the free-energy changes corresponding to any parts— an application of Hess's law of summation. Hence

$$\Delta G^{\circ\prime}_{net} = \Delta G^{\circ\prime}_1 + \Delta G^{\circ\prime}_2$$

where

$$\Delta G^{\circ\prime}_{net} = -4.4 \text{ kcal} \qquad \text{(reaction 12–3)}$$

$$\Delta G^{\circ\prime}_1 = +3.6 \text{ kcal} \qquad \text{(reaction 12–1)}$$

$$\Delta G^{\circ\prime}_2 = ? \qquad \text{(reaction 12–2)}$$

We then solve for $\Delta G^{\circ\prime}_2$:

$$\Delta G^{\circ\prime}_2 = \Delta G^{\circ\prime}_{net} - \Delta G^{\circ\prime}_1$$

$$= -4.4 - (+3.6) = -8.0 \text{ kcal}$$

a value at pH 7 and 37°C that applies to

$$-8.0 \text{ kcal for} \rightarrow$$

$$ATP + H_2O \underset{\xrightarrow{\hspace{1cm}}}{\overset{Mg^{2+}}{\rightleftharpoons}} ADP + P_i$$

$$+8.0 \text{ kcal for} \leftarrow$$

Over the years there has been considerable debate about the most accurate value of $\Delta G^{\circ\prime}$ for ATP hydrolysis, with reported values ranging from -7.0 to -9.0 kcal/mole. The chief difficulty is that the hydrolysis of ATP is quite sensitive to changes in temperature, pH, and Mg^{2+} concentration. The issue is even fuzzier when one considers what the value of $\Delta G^\prime$ (not $\Delta G^{\circ\prime}$) is in a living cell, where the environment does change, and the concentrations are anything but those that apply to standard conditions. One estimate is that in vivo a value of -11.0 kcal/mole or more may be more realistic. Whatever, we will not bicker over this point but, rather, make routine use of -8.0 kcal/mole and $+8.0$ kcal/mole in all future reference to ATP hydrolysis and formation, respectively.

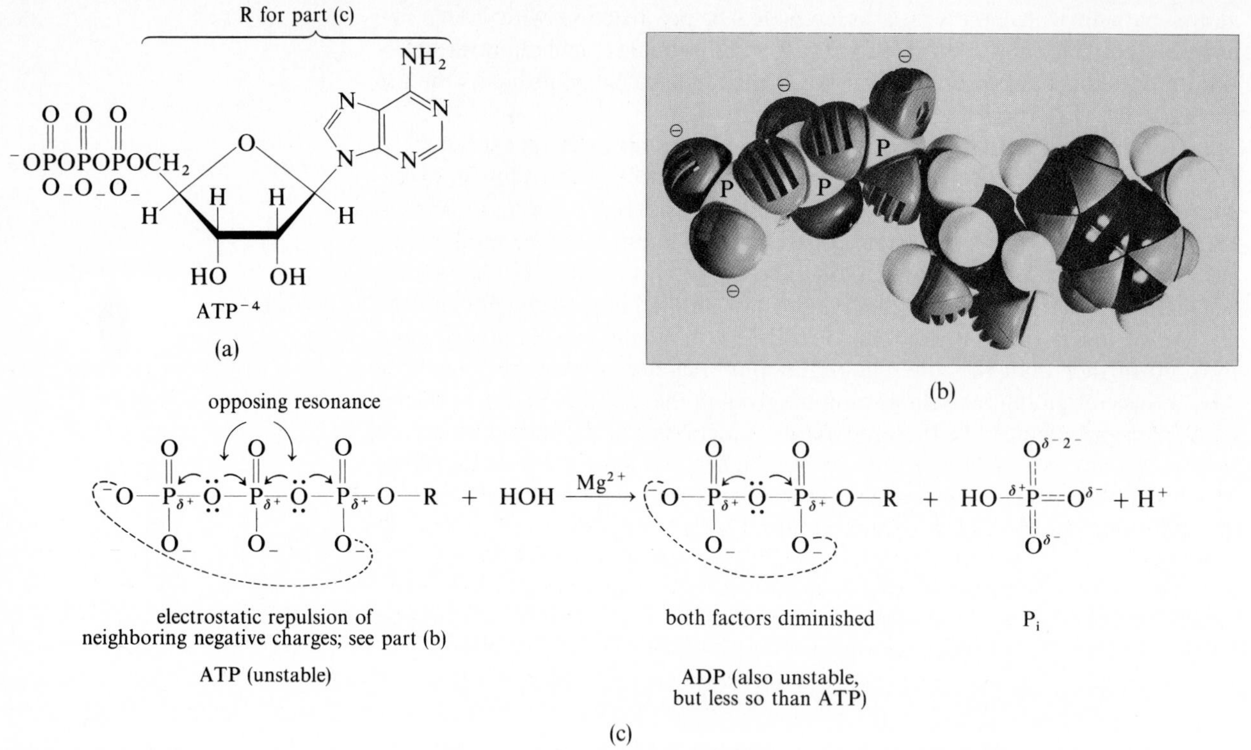

(a)

(b)

(c)

FIGURE 12–6 ATP.
(a) Tetraanion (ATP^{-4}) at pH 7.
(b) Space-filling model.
(c) Hydrolysis of the triphosphoanhydride group. In solution the Mg^{2+} is complexed to ATP and ADP at the sites of the negatively charged oxygens. The resultant P_i is shown here as a resonance-stabilized dianion, a factor that contributes to the greater stability of the product mixture.

And now let us address an obvious question: Why is ATP such a thermodynamically unstable compound? Or putting the question differently, why is a large amount of energy required to form ATP from ADP and P_i? Reasons involve two types of factors: greater destabilizing influences in ATP than in ADP and P_i and greater stabilizing influences in ADP and P_i than in ATP.

At physiological pH 7, where ATP would exist largely as a fully ionized tetraanion (see Figure 12–6), the primary factor of ATP instability is attributed to *electrostatic repulsion* among the clustered O$^-$ groups. A secondary (at pH 7) factor of ATP destabilization is attributed to a principle called *opposing resonance*. This principle argues that a potential competition will exist between successive (δ^+)P atoms for the unshared electrons of the sandwiched O atom in each of the P—O—P linkages. Thus the very existence of ATP is dependent on the presence of sufficient chemical energy within the molecule to overcome these physicochemical stresses. The energy is not confined to any particular bond but is distributed among several in the entire molecule. When the triphosphoanhydride function is cleaved, these stresses are lessened in ADP, and as a bonus, the product state is appreciably stabilized owing to resonance of the orthophosphate dianion (P_i^{2-}).

Finally, we note that the energetics of ATP hydrolysis are not dependent specifically on the presence of an adenine grouping. Thus all of the nucleoside triphosphates will have a very similar $\Delta G^{\circ\prime}$ value for their hydrolysis, so that thermodynamically speaking, ATP, GTP, UTP, and all of the others are considered to be energy-rich compounds. (Similar arguments apply to the hydrolysis of $ATP \rightarrow AMP + PP_i$ and $PP_i \rightarrow 2P_i$, both of which exhibit energy changes comparable to those for $ATP \rightarrow ADP + P_i$.) Indeed, so far we have encountered (a) several examples where specifically GTP rather than ATP is used as an energy source, (b) the frequent use of UTP in forming the activated UDP sugars, and (c) the role of NTPs and dNTPs as highly reactive substrates in RNA and DNA biosynthesis.

Energy Charge: Expressing the Concentration of Available Cellular Energy

A final look at the flow of energy in terms of ATP participation is given in Figure 12–7, highlighting again the interconversion of the three adenylate nucleotides. (Remember that use of ATP may involve its conversion to ADP or AMP; refer to Section 6–2.) Gradually, you will see that a living cell is capable of regulating its chemistry in response to changes in the intracellular concentrations of ATP, ADP, and AMP. D. Atkinson has termed this regulation *energy charge control*. Energy charge is an expression of the relative concentrations of the three adenylate-containing (AMP-containing) species. As Atkinson has defined it:

$$\text{energy charge} = \frac{[ATP] + \frac{1}{2}[ADP]}{[ATP] + [ADP] + [AMP]}$$

The maximum value of energy charge is 1.0, a state where [ADP] and [AMP] both equal zero and only ATP exists. A cell in such a state would be saturated with energy.

The minimum value of energy charge is 0, a state where [ADP] and [ATP] both equal zero and only AMP exists. A cell in such a state would be starved for energy.

In living cells these *extremes are avoided by the regulation of metabolism.*

The concept is that living cells are capable of maintaining an energy charge value within the narrow range of about 0.8–0.9. When the energy charge falls below 0.8, signaling a drop in the cellular ATP level due to an increased rate of energy consumption, the cell responds by increasing the rate of ATP-

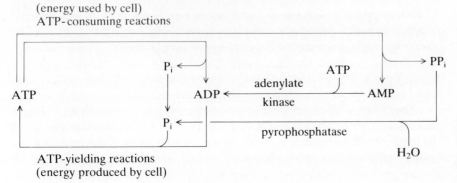

(energy used by cell)
ATP-consuming reactions

ATP-yielding reactions
(energy produced by cell)

FIGURE 12–7 Flow of energy in living cells and ATP participation. The two enzymes identified are important in recycling PP_i and AMP.

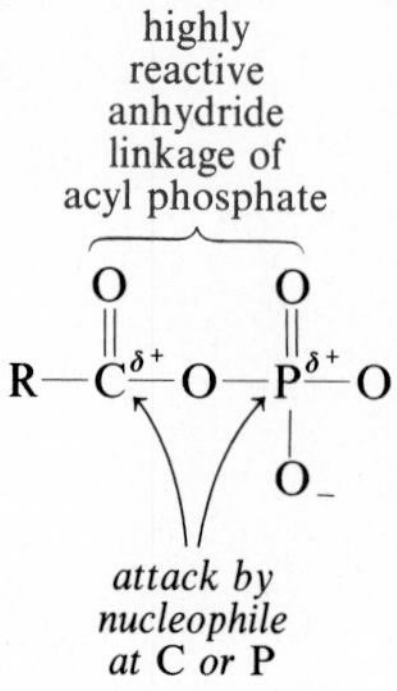

highly
reactive
anhydride
linkage of
acyl phosphate

attack by
nucleophile
at C or P

FIGURE 12–8 Acyl phosphate.

yielding metabolic reactions. In contrast, when the energy charge exceeds 0.9, signaling a rise in the cellular ATP level due to a diminished rate of energy consumption, the cell responds by decreasing the rate of ATP-yielding reactions. We will study the regulation of metabolism in these and other terms throughout future chapters.

Acyl Phosphates

Acyl phosphates (see Figure 12–8) have a *highly reactive anhydride linkage* wherein both the carbonyl carbon of the acyl group and the phosphorus atom are particularly susceptible to reaction with an attacking nucleophile. When the anhydride is hydrolyzed, the $\Delta G^{o\prime}$ value of about -12.0 kcal/mole is clear evidence of this instability. Again, a combination of factors explain the large $-\Delta G^{o\prime}$ value. The anhydride itself has inherent instability (opposing resonance), and the two products, P_i and $RCOO^-$, are resonance-stabilized (see Figure 12–9).

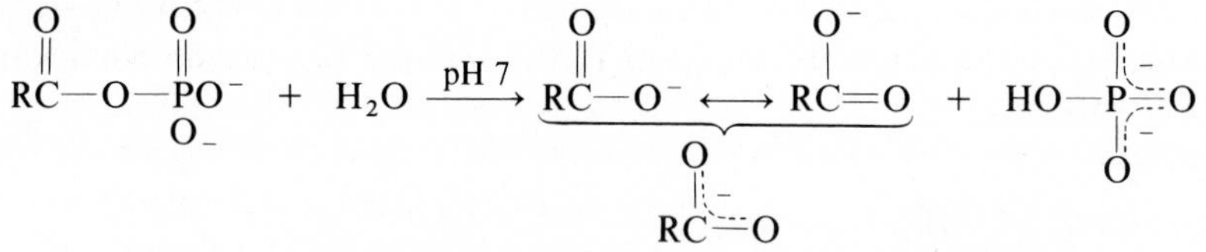

FIGURE 12–9 Hydrolysis of acyl phosphate. Both products are resonance-stabilized.

An example of biological importance illustrating reaction at the carbonyl group, shown here in Figure 12–10, was previously encountered in Chapter 9. It described the activation of amino acids via an aminoacyl-AMP anhydride intermediate prior to forming the aminoacyl-tRNA adducts used in translation.

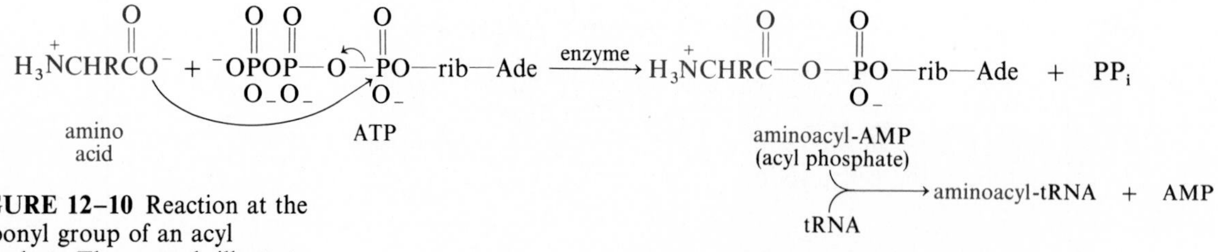

FIGURE 12–10 Reaction at the carbonyl group of an acyl phosphate. The example illustrates the use of ATP to form another high-energy molecule. The same enzyme, aminoacyl-tRNA synthetase, catalyzes both steps (see Section 9–2).

An example of reaction at the P atom (see Figure 12–11) involves the phosphoryltransfer between 1,3-bisphosphoglycerate and ADP, a key ATP-generating reaction during the degradation of simple sugars.

Thioesters

Many carboxylic acids exist in cells as free ionized acids ($RCOO^-$). Many others are activated by conversion (see Figure 12–12) to a *thioester derivative.* The major —SH-containing biomolecules are *coenzyme A, lipoic acid,* and the *cysteine side chains* of polypeptides. For now we will focus on coenzyme A (CoASH), whose structure (see Figure 12–13) consists in part of *pantothenic acid,* an essential vitamin in animals.

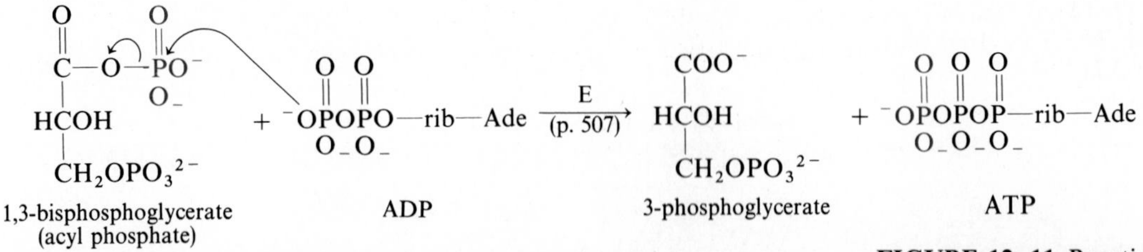

1,3-bisphosphoglycerate   ADP   3-phosphoglycerate   ATP
(acyl phosphate)

FIGURE 12–11 Reaction at the P atom in an acyl phosphate. In this case the phosphoryltransfer results in the formation of ATP.

The formation of thioesters is catalyzed by thiokinase enzymes in an ATP-dependent reaction (see Figure 12–14), indicating that the thioester represents a more reactive state (higher energy) than the free acid. The reaction consists of two phases: (1) the energy of ATP is used to first form an acyl-AMP, an activated anhydride; then (2) the —SH of CoASH attacks the carbonyl carbon. The thioester derivatives exhibit increased reactivity at both the carbonyl carbon and the alpha carbon atom. Examples of each will be encountered in later chapters.

Although the acyl group in an oxyester (RCOOR′) is also activated, the same group in a thioester (RCOSR′) is more reactive. The greater reactivity of thioesters is partly due to the larger atomic size of S over O, diminishing the extent to which a thioester linkage can exhibit resonance. However, one type of O-ester having comparable reactivity to an S-ester is the *carnitine ester* wherein a neighboring positively charged quaternary N in carnitine has a destabilizing influence on the acyl—O linkage (see Figure 12–15). Carnitine esters participate in the transport of fatty acids across the mitochondrial membrane (see Chapter 17).

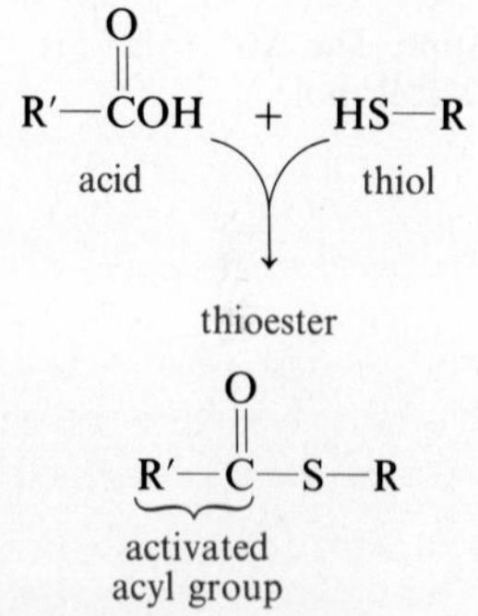

FIGURE 12–12 Thioester derivative.

Enoyl Phosphates

The single most important example of enoyl phosphates is *phosphoenolpyruvate* (see Figure 12–16). Like 1,3-bisphosphoglycerate, this substance is also produced during the degradation of simple sugars and then also serves as a high-energy donor of phosphate to ADP to yield ATP. The large $-\Delta G^{\circ\prime}$ associated with the loss of phosphate is attributed to two factors stabilizing the

FIGURE 12–13 Coenzyme A (CoASH).

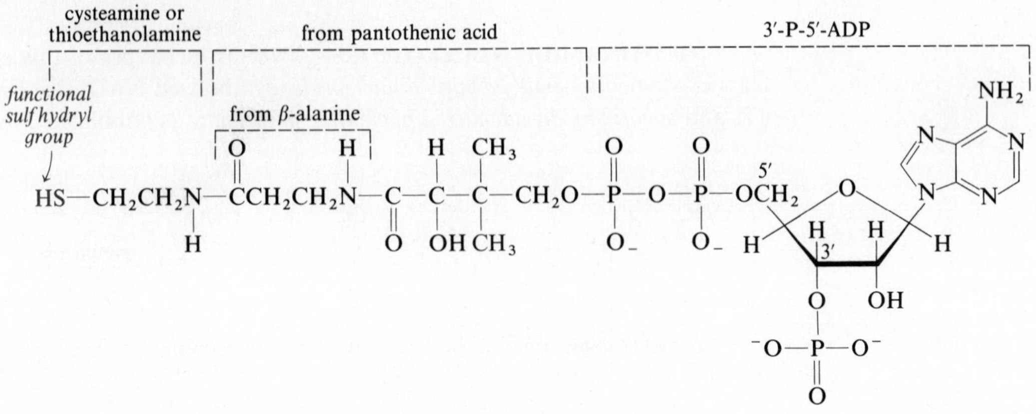

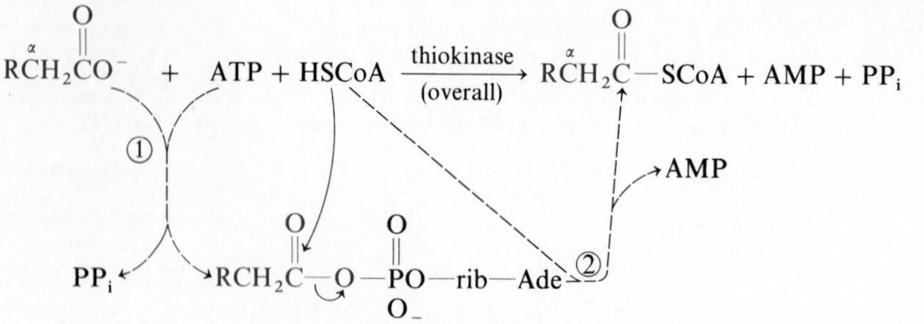

FIGURE 12–14 Formation of a thioester. Dashed lines identify the two steps of the overall reaction.

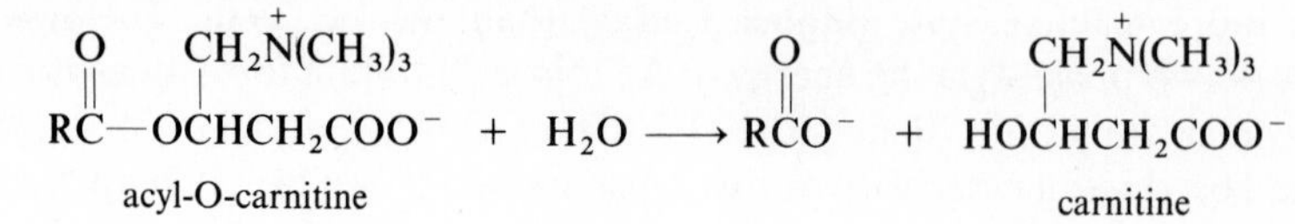

$$RCH_2\overset{O}{\overset{\|}{C}}O^- + ATP + HSCoA \xrightarrow[\text{(overall)}]{\text{thiokinase}} RCH_2\overset{\alpha}{C}-SCoA + AMP + PP_i$$

FIGURE 12–15 Oxyesters of carnitine. The $\Delta G^{\circ\prime}$ (pH 7) is -8.0 kcal/mole.

$$RC\overset{O}{\overset{\|}{}}-O\overset{CH_2\overset{+}{N}(CH_3)_3}{\underset{}{C}}HCH_2COO^- + H_2O \longrightarrow RC\overset{O}{\overset{\|}{}}O^- + HOCHCH_2COO^-$$

acyl-O-carnitine   carnitine

product state: (1) the usual resonance stabilization of the orthophosphate dianion and (2) the thermodynamically favorable tautomerism of the initial enol form of pyruvic acid (pyruvate) to the more stable keto form of pyruvate.

Strong Reducing Agents

Most nonphotosynthetic cells derive the chemical energy they need by degrading an organic source of reduced carbon (sugars, amino acids, fatty acids) to oxidized states. Using O_2, aerobic cells oxidize the carbon completely to CO_2, a conversion associated with a large amount of ATP formation. The bulk of the ATP produced is derived from specific oxidation steps in the transformation of reduced carbon $\rightarrow CO_2$.

However, most of these steps do not produce ATP directly. Rather, they produce the reduced forms of two classes of coenzyme compounds having the potential to support ATP formation in a separate process called *oxidative phosphorylation*. The coenzymes involved are nicotinamide nucleotides and flavin nucleotides. These reduced coenzymes are properly considered as high-energy biomolecules because when they are reoxidized by O_2, large amounts of energy are made available.

NICOTINAMIDE NUCLEOTIDES   Two distinct types of this coenzyme class exist: *nicotinamide adenine dinucleotide* (symbolized NAD) (see Figure 12–17) and *nicotinamide adenine dinucleotide phosphate* (symbolized NADP). As

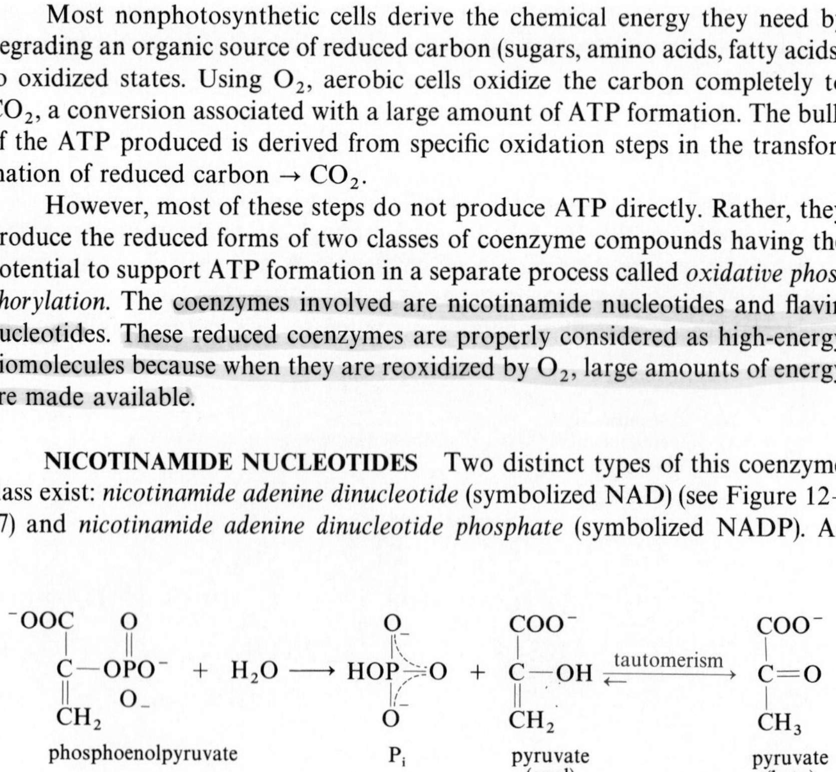

FIGURE 12–16 Hydrolysis of an enoyl phosphate. The tautomerism of the product state exerts a very strong influence in contributing to the negative ΔG value of this reaction.

$$\overset{^-OOC}{\underset{CH_2}{\overset{|}{C}}}-OPO^- + H_2O \longrightarrow HOP\!\!=\!\!O + \overset{COO^-}{\underset{CH_2}{\overset{|}{C}}}-OH \xrightarrow{\text{tautomerism}} \overset{COO^-}{\underset{CH_3}{\overset{|}{C}}}=O$$

phosphoenolpyruvate P_i pyruvate   pyruvate
                                                                (enol)   (keto)

shown in Figure 12–18, the difference between them in structure is slight, with NADP containing an extra phosphate group. In each the active domain of structure is the *nicotinamide* moiety, a substituted pyridine. The remainder is composed of adenine, β-D-ribose, and phosphate, all connected by now familiar linkages. In fact, as the name implies, the molecules can be considered as being composed of two nucleotide units linked by a phosphodiester bond—hence the name *dinucleotide*.

The nicotinamide component is important for two reasons. First, nicotinamide is a derivative of the vitamin *nicotinic acid,* sometimes called *niacin.* Since nicotinamide does not generally occur in the free state but rather in glycosidic linkage with ribose in NAD and NADP, the latter are assumed to be the metabolically active forms of the vitamin. The second important aspect is that the pyridine ring of nicotinamide comprises the active site of both NAD and NADP.

All reduction-oxidation (redox) reactions involve *electron transfer,* with the reactant undergoing reduction gaining electrons and the reactant undergoing oxidation losing electrons. When electron transfer is associated with the transfer of hydrogen atoms, the terms hydrogenation and dehydrogenation are used. In living cells many such reactions occur, catalyzed by enzymes called *dehydrogenases,* usually capable of catalyzing both forward and reverse reactions. According to systematic nomenclature (refer to Table 5–1), these enzymes are just one type of *oxidoreductases*—the largest enzyme class in nature.

All dehydrogenases depend on the participation of a coenzyme to serve as a cosubstrate in the electron transfer reaction. The majority use either NAD or NADP coenzymes for this purpose—some being NAD-specific and others being NADP-specific. When functioning as oxidizing agents, the oxidized forms of NAD^+ and $NADP^+$ (+ designates the charge on the N of the pyridine ring)

FIGURE 12–17 NAD^+. The view here is the same as the one drawn in Figure 12–18.

FIGURE 12-18 Nicotinamide adenine dinucleotides. NADP has OPO_3^{2-} at the arrow rather than OH.

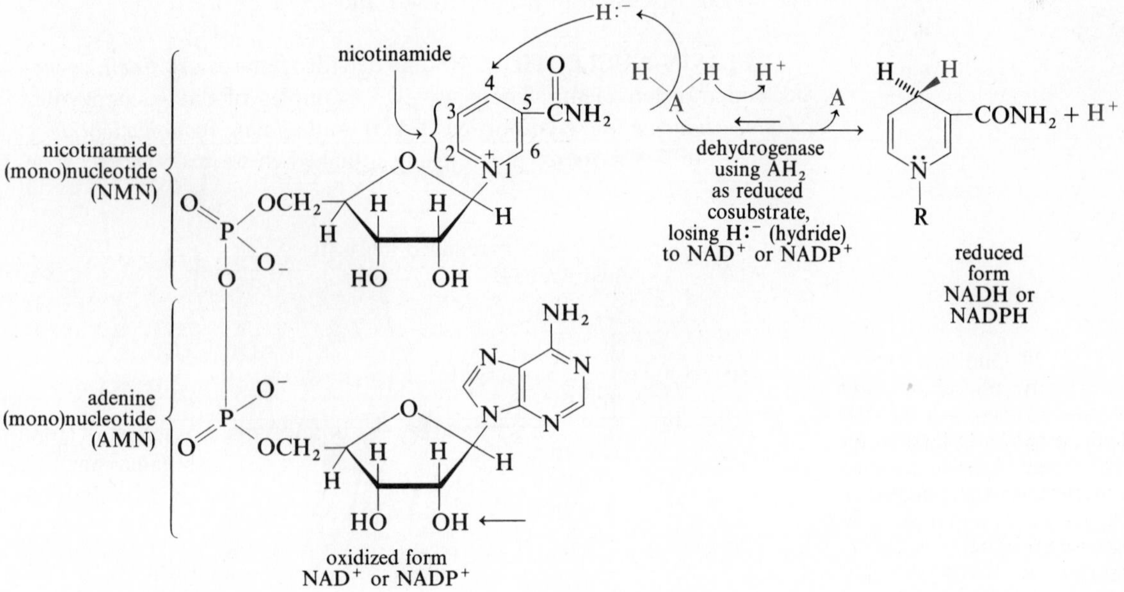

engage in a two-electron transfer by accepting a *hydride ion* H:$^-$ from the reduced substrate AH$_2$. Thus NAD$^+$ and NADP$^+$ are referred to as electron acceptors. (Note that the hydride ion is accepted at position 4 of the pyridine ring.) This reaction yields oxidized A, H$^+$, and the reduced forms of the coenzymes, symbolized as NADH or NADPH. In a reaction of the reverse type, NADH (or NADPH) functions as a reducing agent, acting as a two-electron donor (as hydride ion).

Once the reduced forms are produced in the cell, their primary metabolic fate, especially for NADH, is to be reoxidized as the first step in a series of consecutive oxidation-reduction reactions that terminate with the reduction of O$_2$. The complete process, termed *electron transport,* is most important because most organisms in the biosphere utilize this operation, or a variation of it (using a terminal electron acceptor other than O$_2$), as the main source of energy needed for the intracellular formation of ATP.

Transporting electrons from NADH to O$_2$ to yield H$_2$O represents a transition from a strong reducing agent (NADH) to a weak reducing agent (H$_2$O)—that is, from a state of high energy to a state of lower energy. This energy difference for the entire redox reaction is more technically described (see Section 15–1) as due to a combination of a reducing system and an oxidizing system separated by a *positive potential difference* ($\Delta\mathscr{E}_{net}^{o\prime}$ is a positive value). As evident from the relationship $\Delta G^{o\prime} = -n\mathscr{F}\mathscr{E}_{net}^{o\prime}$, a $+\mathscr{E}_{net}^{o\prime}$ value represents a $-\Delta G^{\circ}$ value, and so the process is exergonic. For the particular combination of NADH $\rightarrow$ NAD$^+$ + 2H$^+$ + 2e (the reducing system) and $\frac{1}{2}$O$_2$ + 2H$^+$ + 2e $\rightarrow$ H$_2$O (the oxidizing system), we can show (see p. 573) that $\mathscr{E}_{net}^{o\prime}$ = + 1.14 V. For this two-electron process, $\Delta G^{o\prime}$ is about -53 kcal (see Figure 12–19).

The steps of electron transport and the manner in which the available energy is conserved for forming ATP comprise one of the most fascinating biochemical processes in nature. The complete coupled process is termed *oxidative phosphorylation* (see Figure 12–20). Many of the details have been established and will be discussed in Sections 15–2 and 15–3.

FLAVIN NUCLEOTIDES   Some dehydrogenases use *flavin coenzymes* as electron transfer agents. There are two examples of flavin coenzymes: *flavin adenine dinucleotide* (symbolized FAD) and *flavin mononucleotide* (FMN). Usually, the flavin coenzyme is firmly attached to its protein via strong non-

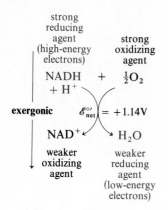

$$\Delta G_{net}^{o\prime} = -n\mathscr{F}\mathscr{E}_{net}^{o\prime}$$
$$= -(2)(23.1)(1.14)$$
$$= -52.7\,\text{kcal}$$

FIGURE 12–19 Overall representation of the large energy yield for the two-electron redox reaction between NADH and O$_2$. As indicated, the Faraday constant is 23.1 kcal/V.

FIGURE 12–20 Outline of the steps in oxidative phosphorylation. Step 1 is the formation of NADH by dehydrogenation of the reduced substrate. Step 2 is the oxidation of NADH by electron transport, with oxygen as the terminal electron acceptor. Step 3 is the phosphorylation of ADP to yield ATP.

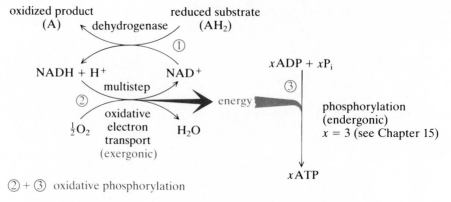

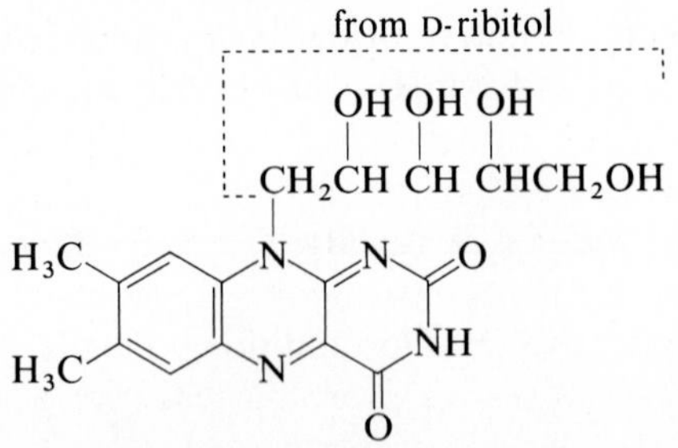

FIGURE 12–21 Riboflavin (vitamin B_2).

covalent forces or covalent bonding, and the complex is called a *flavoprotein* (fp). Both FAD and FMN represent the metabolically active forms of *riboflavin* (vitamin B_2) (see Figure 12–21).

The active portion of FAD (or FMN) is localized in the *isoalloxazine* fused ring system. In redox reactions the flavin structure participates in a two-electron transfer (see Figure 12–22) involving the equivalent of two hydrogen atoms ($2H = 2H^+ + 2e$). When reoxidized by O_2 in the process of electron transport, the energy difference corresponds to an $\mathscr{E}^{\circ\prime}_{net}$ of about $+1.0$ V, providing an energy yield ($-\Delta G^{\circ\prime}$) of about 46 kcal (see Figure 12–23). When fpH_2 reoxidation by O_2 is coupled to ADP phosphorylation, sufficient energy is conserved to account for the formation of 2ATP per fpH_2 reoxidized (see Chapter 15).

In two-electron transfer reactions the $NAD(P)^+/NAD(P)H$ coenzymes participate in a hydride ion ($H{:}^-$) transfer where both electrons are transferred simultaneously. Although flavin coenzymes sometimes participate in the same way, they usually engage in two sequential, one-electron transfer processes in-

FIGURE 12–22 Flavin coenzymes. The stepwise transfer of electrons involving free-radical forms of flavin is not shown. See text for some discussion.

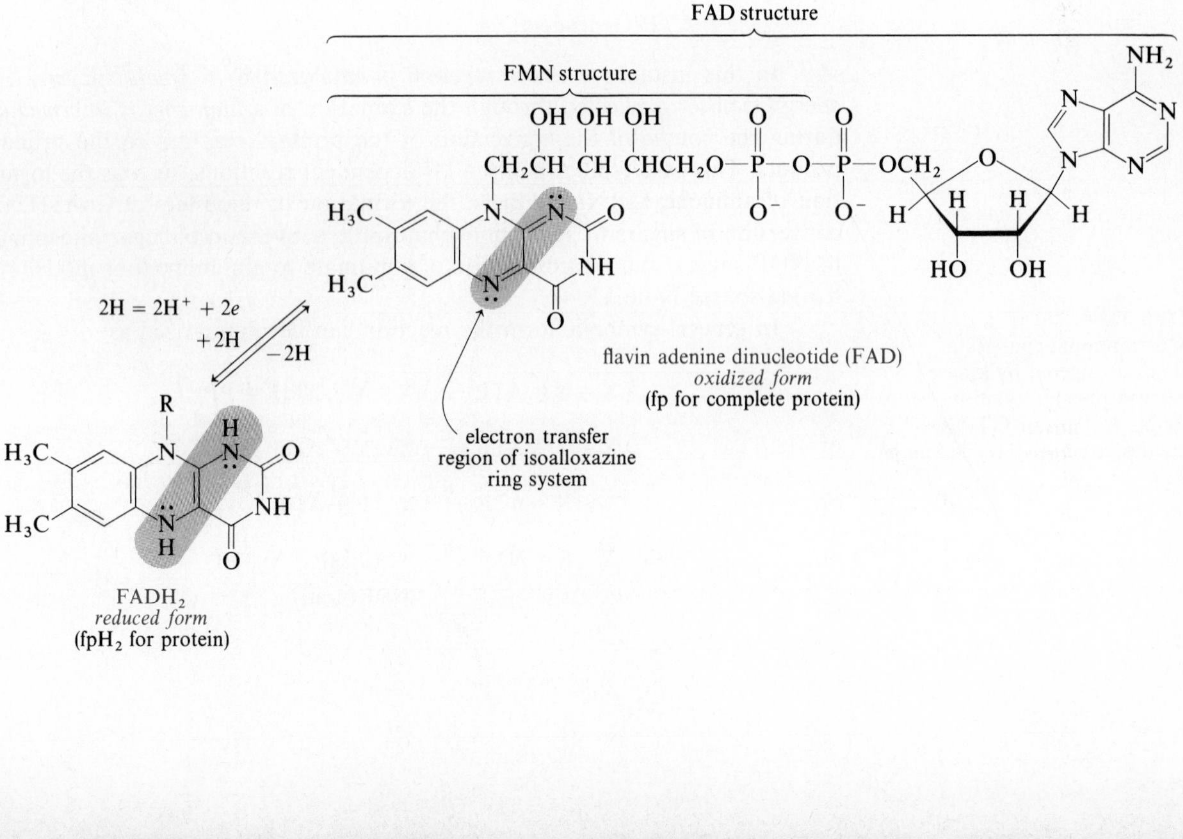

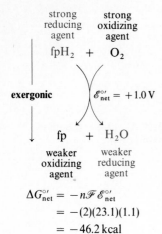

$$\Delta G^{\circ\prime}_{net} = -n\mathscr{F}\mathscr{E}^{\circ\prime}_{net}$$
$$= -(2)(23.1)(1.1)$$
$$= -46.2 \text{ kcal}$$

FIGURE 12–23 Overall representation of the large energy yield for the two-electron transfer between a reduced flavin and O_2.

NOTE 12–4
Conformational changes of proteins, triggered by kinase-catalyzed phosphorylations or by internal ATPase or GTPase activities, are also classified under case 1.

volving the intermediate formation of partially reduced (or partially oxidized) *radical forms,* symbolized as flavinH·. The overall interconversions are represented as

$$\text{flavin} \underset{-1H^+\ 1e}{\overset{+1H^+\ 1e}{\rightleftarrows}} \text{flavinH·} \underset{-1H^+\ 1e}{\overset{+1H^+\ 1e}{\rightleftarrows}} \text{flavinH}_2$$

Consult more comprehensive texts for additional details.

12–3   COUPLING PHENOMENON

As previously mentioned:

The coupling principle states that an energetically favorable reaction is linked to the formation of a product(s) that otherwise would be energetically unfavorable.

This principle is a theme through all of metabolism, explaining the efficient conservation and utilization of metabolic chemical energy and the efficient flow of metabolic intermediates through pathways composed of several consecutive reactions. Throughout the next few chapters we will encounter many specific examples of both points, but now we focus on the nature of coupled reactions. In so doing, we will briefly repeat some of the previous material and also develop other aspects not yet mentioned.

Energy transfer between coupled exergonic and endergonic reactions can occur in either of two ways: *direct* or *indirect*. Reactions involving a direct energy transfer occur by one of two mechanisms: case I and case II described in the following subsections.

Case I: ATP-Dependent

In this instance a *single reaction* is catalyzed by a *single enzyme*. The energy transfer is effected through the formation of a *high-energy intermediate* during the course of the conversion of the primary reactant to the primary product. This case typifies many ATP-dependent reactions, such as the formation of aminoacyl-tRNA adducts, the formation of thioesters of CoASH, the conversion of sugars to sugar phosphates, the conversion of sugar phosphates to NDP sugars, and the conversion of glutamate to glutamine (our model system) (also, see Note 12–4).

In general symbolic form the reaction can be summarized as

$$X + Y + ATP \xrightarrow{E} X{-}Y + AMP + PP_i$$

or

$$X + Y + ATP \xrightarrow{E} X{-}Y + ADP + P_i$$

or

$$X + ATP \xrightarrow{E} X{-}P + ADP$$

or

$$X{-}P + NTP \xrightarrow{E} NMP{-}P{-}X + PP_i$$
$$\text{(NDP sugar)}$$

Here the energetically unfavorable process of $X + Y \rightarrow X—Y$ with a $+\Delta G^{\circ\prime}$ is coupled to the energetically favorable process of ATP degradation (to AMP and PP_i or ADP and P_i), with a $-\Delta G^{\circ\prime}$ via a single reaction in two steps:

Step 1. $\qquad X + ATP \xrightarrow{E_1} [X—AMP] + PP_i$

Step 2. $\qquad [X—AMP] + Y \xrightarrow{E_1} X—Y + AMP$

$\qquad$ (where X—AMP is a high-energy reaction intermediate)

Net. $\qquad X + Y + ATP \xrightarrow{E_1} X—Y + AMP + PP_i$

EXAMPLE

$\qquad$ acetate + CoASH + ATP $\longrightarrow$ acetyl-SCoA + AMP + PP_i

Or via the following steps:

Step 1. $\qquad X + ATP \xrightarrow{E_1} [X—P] + ADP$

Step 2. $\qquad [X—P] + Y \xrightarrow{E_1} X—Y + P_i$

$\qquad$ (where X — P is a high-energy reaction intermediate)

Net. $\qquad X + Y + ATP \xrightarrow{E_1} X—Y + ADP + P_i$

EXAMPLES

$\qquad$ glutamate + NH_4^+ + ATP $\longrightarrow$ glutamine + ADP + P_i

$\qquad$ amino acid + tRNA + ATP $\longrightarrow$ aminoacyl-tRNA + AMP + PP_i

Case II: ATP-Dependent

A variation of case I occurs when the net reaction results from the consecutive occurrence of two (or more) *distinct reactions* catalyzed by two (or more) *separate enzymes*. Here the common intermediate is, in fact, the primary product of one reaction and the primary reactant of the second.

In general form,

$$X + Y + ATP \xrightarrow{E_1} \xrightarrow{E_2} X—Y + ADP + P_i$$

Here the energetically unfavorable process of $X + Y \rightarrow X—Y$ with a $+\Delta G^{\circ\prime}$ is coupled to the energetically favorable process of ATP degradation (to AMP and PP_i or ADP and P_i), with a $-\Delta G^{\circ\prime}$ via separate reactions:

Reaction 1. $\qquad X + ATP \xrightarrow{E_1} X—P + ADP$

Reaction 2. $\qquad X—P + Y \xrightarrow{E_2} X—Y + P_i$

Net. $\qquad X + Y + ATP \xrightarrow{E_1} \xrightarrow{E_2} X—Y + ADP + P_i$

EXAMPLE   (with three enzymes):

$$\text{glucose} + \text{ATP} \xrightarrow{E_1} \text{glucose-6-phosphate} + \text{ADP}$$

$$\text{glucose-6-phosphate} \xrightarrow{E_2} \text{glucose-1-phosphate}$$

$$\text{glucose-1-phosphate} + \text{fructose} \xrightarrow{E_3} \text{sucrose} + P_i$$

Net. $$\text{glucose} + \text{fructose} + \text{ATP} \xrightarrow{E_1} \xrightarrow{E_2} \xrightarrow{E_3} \text{sucrose} + \text{ADP} + P_i$$

Case III: ATP-Independent

In reactions where an actual energy transfer does not occur, the coupling is explained purely in terms of *equilibrium considerations;* ATP is not involved. However, as in case II, we are dealing with consecutive reactions catalyzed by different enzymes. As indicated, the unfavorable equilibrium of one reaction can be displaced if the product of that reactant then serves as a substrate (reactant) with a strong tendency to be converted to another material in a subsequent reaction.

In general form,

$$X \xrightarrow{E_1} Y \xrightarrow{E_2} Z \qquad \text{(no ATP)}$$

Here an energetically unfavorable process of $X \rightarrow Y$ with a $+\Delta G^{\circ\prime}$ is driven to completion by being coupled to the energetically favorable process of $Y \rightarrow Z$ with a $-\Delta G^{\circ\prime}$:

Reaction 1.	$X \underset{E_1}{\overset{}{\rightleftarrows}} Y$	$K_{eq_1} \ll 1$	$\Delta G_1^{\circ\prime} = +$
Reaction 2.	$Y \underset{E_2}{\overset{}{\rightleftarrows}} Z$	$K_{eq_2} \ggg 1$	$\Delta G_2^{\circ\prime} = -$
Net.	$X \xrightarrow{E_1} Y \xrightarrow{E_2} Z$	$K_{eq_{net}} > 1$	$\Delta G_{net}^{\circ\prime} = -$

EXAMPLE

$$\text{NAD}^+ + \text{malate} \xrightarrow{E_1} \text{oxaloacetate} + \text{NADH}(H^+) \qquad \Delta G^{\circ\prime} = +6.7 \text{ kcal}$$

$$\text{oxaloacetate} + \text{acetyl-SCoA} \xrightarrow{E_2} \text{citrate} + \text{HSCoA} \qquad \Delta G^{\circ\prime} = -9.0 \text{ kcal}$$

Net. $$\text{NAD}^+ + \text{malate} + \text{acetyl-SCoA} \xrightarrow{E_1} \xrightarrow{E_2}$$
$$\text{citrate} + \text{HSCoA} + \text{NADH}(H^+) \qquad \Delta G_{net}^{\circ\prime} = -2.3 \text{ kcal}$$

Chemiosmotic Coupling: A Special Case

In the process of *oxidative phosphorylation* wherein the exergonic process of electron transport ($\text{NADH} + H^+ + \frac{1}{2}O_2 \rightarrow \text{NAD}^+ + H_2O$) is used to provide energy for the endergonic process of phosphorylating ADP ($\text{ADP} + P_i \rightarrow \text{ATP}$), a special type of coupling mechanism is used, called *chemiosmotic coupling.* Based on an indirect mode of energy transfer, chemiosmotic coupling involves the establishment of *energy-rich concentration gradients* of ions—

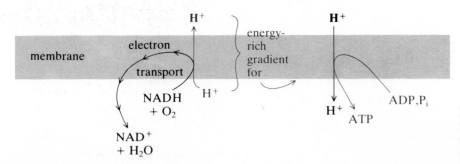

FIGURE 12–24 Simplified representation of chemiosmotic coupling. Chapter 15 gives more details.

notably H^+—across the mitochondrial membrane in conjunction with electron transport (see Figure 12–24). The thermodynamically favorable influx of ions back into the mitochondria provides the energy to support the process of ATP formation. Chemiosmotic coupling is described in more detail in Chapter 15.

LITERATURE

KLOTZ, I. M. *Energy Changes in Biochemical Reactions.* New York: Academic Press, 1967. A small monograph (available in paperback) on the fundamental concepts of thermodynamics as they apply to biological systems. Written by a chemist for the biologist.

LEHNINGER, A. L. *Bioenergetics.* 2nd ed. New York: W. A. Benjamin, 1971. A superb introductory book (available in paperback) emphasizing the energy relationships in living cells at the metabolic level.

MONTGOMERY, R., and C. A. SWENSON. *Quantitative Problems in the Biochemical Sciences.* 2nd ed. San Francisco: Freeman, 1976. Chapter 10 summarizes basic principles of biochemical energetics and applies them in problem solving.

TINOCO, I., K. SAVER, and J. C. WANG. *Physical Chemistry—Principles and Applications in Biological Sciences.* Englewood Cliffs, N.J.: Prentice-Hall, 1978. A textbook for life science students.

VAN HOLDE, K. E. *Physical Biochemistry.* Englewood Cliffs, N.J.: Prentice-Hall, 1971. A more sophisticated though not overpowering treatment of thermodynamics applied to biochemical systems. Available in paperback.

WILLIAMS, V. R., W. L. MATTICE, and H. B. WILLIAMS. *Basic Physical Chemistry for the Life Sciences.* 3rd ed. San Francisco: Freeman, 1978. Chapters 2 and 3 contain thorough presentations of the principles of thermodynamics and the free-energy concept as they apply to biochemical and biological systems.

EXERCISES

12–1. The standard free-energy change for the hydrolysis of glucose-1-phosphate at pH 7 and 37°C has been measured as -5.0 kcal/mole. Calculate the equilibrium constant for this reaction:

$$\text{glucose-1-phosphate} + H_2O \rightarrow \text{glucose} + P_i$$

12–2. On the basis of material presented in this chapter and the principle of the relationship between structure and function, do you predict the $\Delta G^{\circ\prime}$ for the reaction below to be approximately (a) $+3.6$ kcal/mole, (b) -3.6 kcal/mole, (c) $+6.8$ kcal/mole, (d) -6.8 kcal/mole, or (e) $+1.9$ kcal/mole?

$$\text{asparagine} + H_2O \rightarrow \text{aspartate} + NH_3$$

12–3. The standard free-energy change for the hydrolysis of glucose-6-phosphate at pH 7 and 25°C has been measured as -3.3 kcal/mole. Given this data and the information in Exercise 12–1, calculate the $\Delta G^{\circ\prime}$ for the following reaction at pH 7 and 37°C:

$$\text{glucose-1-phosphate} \rightarrow \text{glucose-6-phosphate}$$

12–4. Which of the reactions given in Table 12–2 would be likely to be coupled to the formation of ATP from ADP and P_i? (Assume a pH of 7 and a temperature of 37°C apply to both $\Delta G^{\circ\prime}$ and K_{eq}.)

TABLE 12–2 Reactions for Exercise 12–4

REACTION	$\Delta G^{\circ\prime}$ (kcal)	K_{eq}
(a) phosphoenolpyruvate + $H_2O \rightarrow$ pyruvate + P_i	—	2.5×10^{10}
(b) 3-phosphoglycerate $\rightarrow$ 2-phosphoglycerate	—	1.8×10^{-1}
(c) fructose-6-phosphate + $H_2O \rightarrow$ fructose + P_i	-3.3	—
(d) succinyl-SCoA + $H_2O \rightarrow$ succinate + HSCoA	-11.0	—

12–5. The overall standard free-energy change for the reaction

$$\text{pyruvate} + \text{ATP} + CO_2 \rightarrow \text{oxaloacetate} + P_i + \text{ADP}$$

is 1.1 kcal/mole. From this information, calculate the $\Delta G^{\circ\prime}$ for the reaction

$$\text{pyruvate} + CO_2 \rightarrow \text{oxaloacetate}$$

(Assume a pH of 7 and a temperature of 37°C for all $\Delta G^{\circ\prime}$ values.) Classify the coupled reaction involving ATP as case I, case II, or case III.

12–6. Dihydroxyacetone phosphate (DHAP) is one of the principal intermediates produced during the degradation of hexoses such as glucose. Under anaerobic conditions certain bacteria can produce glycerol from glucose due to the action of two enzymes that catalyze first the reduction of dihydroxyacetone phosphate to glycerol-1-phosphate and then the hydrolysis of glycerol phosphate to yield free glycerol and inorganic phosphate:

	$\Delta G^{\circ\prime}$ (kcal)
dihydroxyacetone phosphate + 2Hs $\rightarrow$ glycerol-1-phosphate	$+8.8$
glycerol-1-phosphate $\rightarrow$ glycerol + P_i	-2.4

Calculate whether the overall sequence from DHAP to glycerol is endergonic or exergonic. Is the overall sequence an example of coupling according to case I, case II, or case III?

12–7. In the cell the enzyme (a dehydrogenase) that catalyzes the reduction of DHAP to glycerol-1-phosphate utilizes NADH as the source of reducing power. The NADH is in turn oxidized to NAD^+. The complete reaction is

$$\text{DHAP} + \text{NADH} + H^+ \rightarrow \text{glycerol-1-phosphate} + NAD^+$$

Given that the $\Delta G^{\circ\prime}$ for the oxidation of NADH to NAD^+ is -14.8 kcal and the information in Exercise 12–6, calculate the $\Delta G^{\circ\prime}$ for the formation of glycerol phosphate according to this reaction. Compare this value with that given in Exercise 12–6, and explain the difference. What effect does the NADH-dependent reaction of DHAP have on the energetics of the overall conversion of DHAP to glycerol as it occurs in the cell?

12–8. The steady-state concentration of ATP in a red blood cell has been estimated to be approximately thirteen times greater than that of ADP. In addition, the concentration of inorganic phosphate has been estimated to be approximately eight times greater than that of ADP. Given this information, calculate the value of the free-energy change ($\Delta G'$) that applies to ATP hydrolysis in the red blood cell.

12–9. A buffered (pH 7) solution containing phosphoenolpyruvate (30 mmole), glucose (20 mmole), and small amounts of adenosine diphosphate, pyruvate kinase, and hexokinase was incubated at 37°C. Given the reactions below, (a) explain what will occur during the incubation process, particularly in the early stages; (b) write the net reaction that occurs; and (c) predict whether or not the net reaction will proceed to completion (that is, there will be little or none of the original substrates remaining).

$$\text{phosphoenolpyruvate} + \text{ADP} \xrightarrow{\text{pyruvate kinase}} \text{pyruvate} + \text{ATP}$$

$$\text{glucose} + \text{ATP} \xrightarrow{\text{hexokinase}} \text{glucose-6-phosphate} + \text{ADP}$$

12–10. Given the thermodynamic information in this chapter, calculate the theoretical yield of ATP from the oxidation of $NADH(H^+)$ during the coupled process of oxidative phosphorylation.

12–11. Propose an explanation for the difference between the $\Delta G^{\circ\prime}$ values for the hydrolysis of glucose-1-phosphate (-5.0 kcal/mole at pH 7 and 25°C) and glucose-6-phosphate (-3.3 kcal/mole, also at pH 7 and 25°C).

12–12. The $\Delta G^{\circ\prime}$ (at pH 7) for the hydrolysis of sucrose to glucose and fructose is -7.0 kcal/mole:

$$\text{sucrose} + H_2O \rightarrow \text{glucose} + \text{fructose}$$

Given this information, the information in Exercises 12–1 and 12–3, and the answer to Exercise 12–3, calculate the $\Delta G^{\circ\prime}$ for the formation of sucrose according to the example given for case II coupling in Section 12–3. Then do the same calculation in a simpler way.

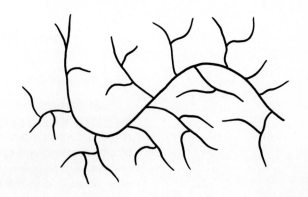

CHAPTER THIRTEEN

CARBOHYDRATE METABOLISM

13–1 PRINCIPLES OF METABOLISM

Introduction

Metabolism can be defined as the sum of all the chemical reactions that occur in a living organism. Thus metabolism is a vast subject. The number of reactions alone is staggering, with different organisms, depending on their complexity, characterized by several hundred to several thousand. Collectively, these reactions are responsible for sustaining the viability of the organism. Although each reaction does have individual importance, the operation of the whole organism is due to the integration of individual reactions into an intricate maze, *controlled by a host of sensitive regulatory checks and balances*. In general, the maintenance and the control of this design sustain normal metabolism, whereas its disruption contributes to abnormal metabolism.

The subject of metabolism is not only massive but is also complex. Yet because of the use of radioisotopes, of improved methods for isolating enzymes, and of the dedication of many past and current researchers, great strides have been made in unraveling many of the details. The growth of knowledge has also been aided by the nature of the subject itself, for despite the tremendous metabolic diversity represented by all living forms, there also is a very definite theme of *metabolic unity*. This unity should come as no surprise since we have already emphasized in previous chapters that all cells contain the same classes of biomolecules—proteins, nucleic acids, lipids, and carbohydrates. At this point we are saying that the general metabolism of all classes of biomolecules is, in principle, basically the same from one organism to another. Indeed, the metabolism of many substances is identical in all organisms.

Metabolism is customarily divided into *catabolism* and *anabolism*. Catabolism (from the Greek *cata,* meaning "down") refers to degradative-reaction sequences; anabolism (from the Greek *ana,* meaning "up") refers to synthetic-reaction sequences. The term *pathway* is commonly used to refer to a set of consecutive reactions accounting for a specific overall conversion. For example, the consecutive sequence of A → B → C → D → E → F → G has the net effect of A → G. Substances produced along the way (B, C, D, E, and F) are called *metabolic intermediates.*

The distinction of degradation versus synthesis is not the only manner in which we can differentiate the reactions occurring in a living cell. Table 13–1 lists additional comparisons in terms of whether the reactions involve oxidation or reduction, of the energetics of the reaction sequence, and of the nature of the starting materials and the end products.

Although catabolic and anabolic pathways are distinct in these respects, they are very much related to each other. Moreover, the relationships are in terms of the very characteristics that constitute their differences. That is:

1. *On the level of oxidation versus reduction:* Not every step of a catabolic pathway involves the oxidation of a metabolic intermediate; nor does every

TABLE 13–1 Comparison of the main features of catabolism and anabolism

CATABOLISM	ANABOLISM
Degradative	Synthetic
Oxidative in nature	Reductive in nature
Energy yielding	Energy requiring
Variety of starting materials with well-defined end products	Well-defined starting materials with a variety of end products

step of an anabolic pathway involve the reduction of an intermediate. However, in those reactions that are so characterized, the nicotinamide adenine dinucleotide coenzymes are common participants. More specifically, catabolism utilizes the oxidized forms (NAD^+ and $NADP^+$) and produces the reduced forms (NADH and NADPH), whereas anabolism requires the reduced forms and produces the oxidized forms. The one variation in this pattern is that anabolic reactions use primarily NADPH, producing $NADP^+$. Nevertheless, the general participation of nicotinamide adenine dinucleotides in both processes is a distinct common denominator. (Some redox reactions use the flavin coenzymes as electron transfer agents: $FAD \rightleftharpoons FADH_2$ and $FMN \rightleftharpoons FMNH_2$.)

2. *On the level of energetics:* Catabolism is exergonic (energy-yielding), with a *net* requirement for ADP and a *net* production of ATP. The ATP then serves as the source of energy for the endergonic (energy-requiring) reactions of anabolism, and ADP (and AMP) is produced.

3. *On the level of starting materials, end products, and intermediary metabolites:* The end products and intermediary metabolites that are generated in catabolism generally serve as the starting materials in anabolism. The reverse is also true.

Thus we conclude that catabolism and anabolism are *integrated, complementary processes*. As summarized in Figure 13–1, these relationships provide for an optimal level of metabolic efficiency in organisms and will serve to unify our analysis throughout the next several chapters.

Carbohydrate Metabolism

Our first consideration will be in the area of carbohydrate metabolism. To treat this fully would require more attention than the scope of this book allows. Rather, we will focus on three major reaction pathways: (1) *glycolysis* (carbohydrate breakdown), (2) *glycogenesis* (carbohydrate formation), also called *gluconeogenesis,* and (3) another sequence responsible for carbohydrate breakdown called by various names such as the *hexose monophosphate shunt* or the *pentose phosphate pathway.*

Depending on the organism and its growth conditions, the pathway of glycolysis fulfills many functions. In many microbes growing under *anaerobic* (oxygen absent) *conditions,* glycolysis serves as the main energy-yielding catabolic route for carbohydrate substrates, resulting in the production of specific metabolic end products such as ethanol, lactic acid (lactate), and glycerol. This

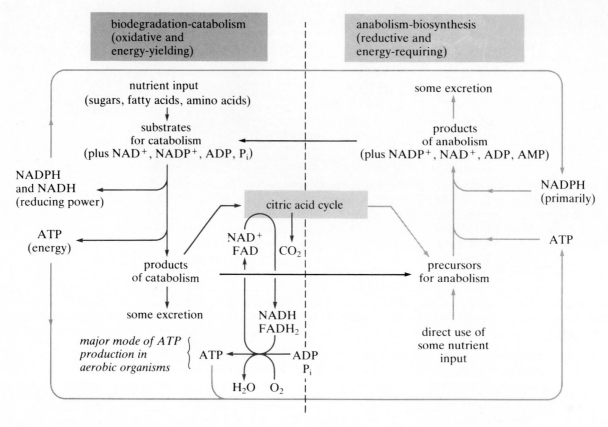

FIGURE 13–1 Diagrammatic summary of complementary relationships that provide for an integrated network of catabolism and anabolism. Attention is focused on the nutrient source of carbon because it is the major bioelement of organic biomolecules. In photosynthetic cells and a few classes of bacteria, the required form is CO_2. In all cases to which this diagram applies, a reduced, organic form of carbon, such as carbohydrates, is required. The central participation of the citric acid cycle and the participation of O_2 in the reoxidation of NADH associated with ATP production will be developed as we proceed. Organisms that require O_2 for this purpose are referred to as *aerobic*. This class includes all higher animals, most microorganisms, and nonphotosynthetic cells.

type of process is often referred to as a *fermentation*. There are numerous other fermentation routes, with each yielding a specific end product. However, many are either variations or offshoots of the glycolytic pathway. Fermentation is a much more general term, implying only that the initial substrate is degraded anaerobically. In animals the production of lactate via anaerobic glycolysis accounts for such important processes as supplying energy for muscle contraction in skeletal muscle when the oxygen supply is limited.

Under *aerobic conditions* (oxygen present), the reactions of glycolysis comprise the initial phase of carbohydrate degradation that is then linked to another extremely important set of reactions called the *citric acid cycle* (see Chapter 14). In this case glycolysis stops at the level of *pyruvate,* the immediate precursor of lactate (see Figure 13–2). Under aerobic conditions (respiration) the combined action of glycolysis and the citric acid cycle results in complete oxidation of the carbons in a hexose to CO_2. There is an accompanying release of large amounts of potentially useful metabolic energy, largely in the form of reduced nicotinamide adenine dinucleotides (NADH). The NADH is then reoxidized via the oxygen-dependent process of *oxidative phosphorylation* (see Chapter 15), resulting in the production of a large amount of ATP (see Note 13–1).

The consideration of glycogenesis will not only complement the coverage of glycolysis but will also provide us with the opportunity to become acquainted

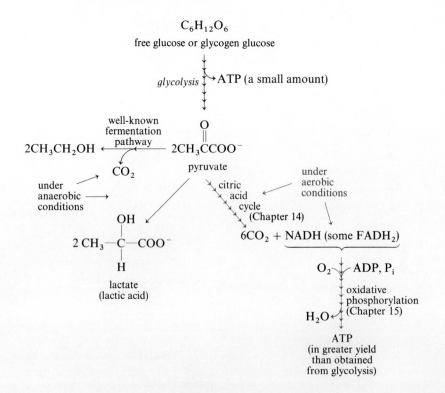

$$C_6H_{12}O_6$$
free glucose or glycogen glucose

glycolysis → ATP (a small amount)

FIGURE 13–2 Comparison of anaerobic and aerobic pathways for degrading glucose. The aerobic reactions account for the formation of 36ATP, compared with the formation in the glycolysis pathway of 2ATP (from glucose) or 3ATP (from glycogen glucose).

well-known fermentation pathway

$2CH_3CH_2OH \leftarrow$

CO_2

$2CH_3CCOO^-$
pyruvate

under anaerobic conditions

under aerobic conditions

citric acid cycle (Chapter 14)

OH
|
$2\ CH_3-C-COO^-$
|
H
lactate
(lactic acid)

$6CO_2 + NADH$ (some $FADH_2$)

O_2, ADP, P_i

oxidative phosphorylation (Chapter 15)

H_2O

ATP
(in greater yield than obtained from glycolysis)

with the remarkable network of regulatory features that control the supply-and-demand balance between events that yield ATP, NADH, and NADPH and events that consume these materials. The elucidation of this self-controlling capability of living cells is one of the most important developments of modern biochemistry and should stimulate your interest and also aid in your understanding of the subject of metabolism.

This rather lengthy introduction has dealt with the general principles of metabolism and of carbohydrate metabolism in particular. Do not be concerned if you feel somewhat confused with this initial exposure. The many principles will become more coherent chapter by chapter. In other words, there is no such thing as an instant appreciation of metabolism. On the contrary, it develops gradually. (Rote memorization, albeit certainly necessary to develop the basic language, is not the sole answer.)

13–2 GLYCOLYSIS: OVERALL VIEW

Introduction

The entire reaction sequence of glycolysis (see Note 13–2) is shown in Figure 13–3. Beginning with the initial substrates (glucose or glycogen are shown), degradation proceeds through a consecutive sequence of enzyme-catalyzed reactions resulting in the formation of specific products—lactate or

NOTE 13–1
In some cases glycolysis will operate anaerobically under aerobic conditions. That is, pyruvate will be converted to lactate when O_2 is present. Some important examples where this event occurs are in mature red blood cells, retina tissue, fetal tissues shortly after birth, and in the intestinal mucosa.

NOTE 13–2
The pathway of glycolysis has historical significance because it was the first complete biochemical sequence to be unraveled. The work spanned a period of nearly fifty years (1890s–1940s) and involved numerous individuals (the pioneers of biochemistry), each contributing important pieces to the puzzle. During this time the phosphoesters, ATP, and NAD^+ were discovered, their structures were determined, and their participation was defined. The sequence of conversions was identified, and many of the enzymes involved were isolated and characterized.

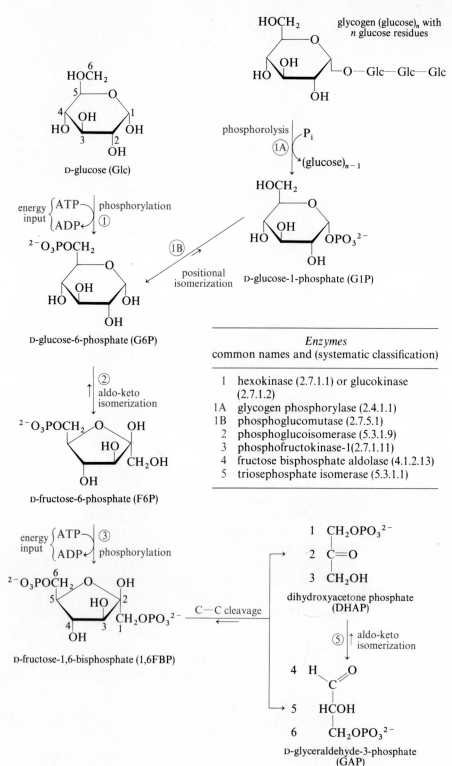

FIGURE 13–3 Pathway of glycolysis from free glucose or glycogen glucose to produce—under anaerobic conditions—lactate or ethyl alcohol. Under aerobic conditions the pathway stops at pyruvate formation. The ⇌ signifies that the enzyme specified will catalyze the reaction in both directions.
(*Figure continues to top of next page.*)

Enzymes
common names and (systematic classification)

1 hexokinase (2.7.1.1) or glucokinase (2.7.1.2)
1A glycogen phosphorylase (2.4.1.1)
1B phosphoglucomutase (2.7.5.1)
2 phosphoglucoisomerase (5.3.1.9)
3 phosphofructokinase-1(2.7.1.11)
4 fructose bisphosphate aldolase (4.1.2.13)
5 triosephosphate isomerase (5.3.1.1)

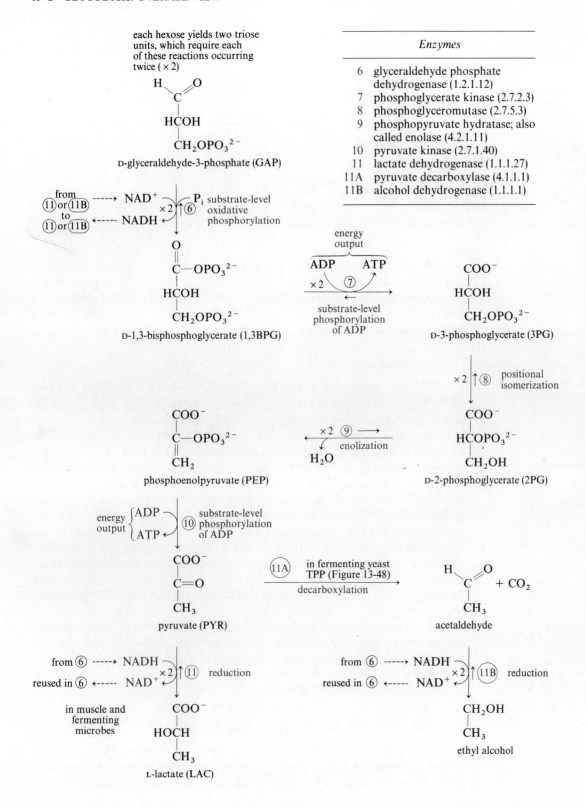

each hexose yields two triose units, which require each of these reactions occurring twice (× 2)

D-glyceraldehyde-3-phosphate (GAP)

from ⑪ or ⑪B to ⑪ or ⑪B

D-1,3-bisphosphoglycerate (1,3BPG)

energy output

substrate-level phosphorylation of ADP

D-3-phosphoglycerate (3PG)

× 2 ⑧ positional isomerization

phosphoenolpyruvate (PEP)

× 2 ⑨ enolization

D-2-phosphoglycerate (2PG)

energy output

substrate-level phosphorylation of ADP

pyruvate (PYR)

⑪A in fermenting yeast TPP (Figure 13-48) decarboxylation

acetaldehyde

from ⑥ → NADH reused in ⑥ ← NAD⁺ × 2 ⑪ reduction

in muscle and fermenting microbes

L-lactate (LAC)

from ⑥ → NADH reused in ⑥ ← NAD⁺ × 2 ⑪B reduction

ethyl alcohol

Enzymes

6 glyceraldehyde phosphate dehydrogenase (1.2.1.12)
7 phosphoglycerate kinase (2.7.2.3)
8 phosphoglyceromutase (2.7.5.3)
9 phosphopyruvate hydratase; also called enolase (4.2.1.11)
10 pyruvate kinase (2.7.1.40)
11 lactate dehydrogenase (1.1.1.27)
11A pyruvate decarboxylase (4.1.1.1)
11B alcohol dehydrogenase (1.1.1.1)

TABLE 13–2 Net reactions of glycolysis for lactate production

Lactate from Free Glucose

1.
$$\text{glucose} \xrightarrow[\text{enzymes}]{11} 2 \text{ lactate}$$
$$1C_6 \qquad\qquad 2C_3$$

2.
$$\text{glucose} + 2\text{ATP} + 4\text{ADP} + 2P_i \longrightarrow 2 \text{ lactate} + 2\text{ADP} + 4\text{ATP}$$

3. $\text{glucose} + 2\text{ATP} + 4\text{ADP} + 2P_i + 2\text{NAD}^+ \longrightarrow 2 \text{ lactate} + 2\text{ADP} + 4\text{ATP} + 2\text{NAD}^+$

Lactate from Glycogen Glucose

1.
$$(\text{glucose})_n \xrightarrow[\text{enzymes}]{12} 2 \text{ lactate} + (\text{glucose})_{n-1}$$

2.
$$(\text{glucose})_n + 1\text{ATP} + 4\text{ADP} + 3P_i \longrightarrow 2 \text{ lactate} + (\text{glucose})_{n-1} + 1\text{ADP} + 4\text{ATP}$$

3. $(\text{glucose})_n + 1\text{ATP} + 4\text{ADP} + 3P_i + 2\text{NAD}^+ \longrightarrow 2 \text{ lactate} + (\text{glucose})_{n-1} + 1\text{ADP} + 4\text{ATP} + 2\text{NAD}^+$

alcohol and CO_2. In the next several pages we will examine many facets of this pathway as a whole and also focus on most of the individual steps. When we have finished, the maze of structures and arrows should be better understood. One approach in deciphering a metabolic pathway of this type is to begin with an analysis of *net effects*.

Net Chemistry

The net reactions of glycolysis for lactate production are given in Table 13–2. For each process a family of equations is given, with each successive equation intended to be more descriptive than the previous ones.

Equation 1 merely defines the overall transformation in terms of the carbon skeletons of the reactants and products. Equation 2 includes the participation of inorganic phosphate (P_i) and the adenine nucleotides. The appearance of ATP on both sides of the equation indicates that *ATP is both required and produced* in the overall process. Equation 3, which is the most descriptive, includes the participation of the nicotinamide nucleotides. As with ATP, the involvement of NAD^+ is twofold: *NAD^+ is both required and produced*. In this case, however, note that there is neither a net requirement nor a net gain. Equation 3 is also the most accurate statement of what is needed in order for glycolysis to occur: enzymes, ATP, ADP, P_i, NAD^+, and, of course, glucose or glycogen. Put them all together in a reaction vessel (pH 7, 37°C) and lactate will be produced.

To become familiar with the specific chemical events of glycolysis, you should verify the validity of equation 3 in each set. To do so, you need recall only one point, namely, that a sequence of consecutive reactions such as A → B → C can be dissected into a pattern of A → B and B → C and then added to yield the overall effect of A → C.

Overall View of the Energetics of Glycolysis

The tally of net ATP production (and P_i consumption) is really quite simple. Reference to Figure 13–3 reveals that with free glucose as initial substrate there are two ATP-requiring reactions, the formation of glucose-6-

phosphate and that of fructose-1,6-bisphosphate, and two ATP-generating reactions, the formation of 3-phosphoglycerate and that of pyruvate. However, the ATP-generating reactions contribute a total of 4ATP since each hexose unit is cleaved to two three-carbon units. Overall, then, there is a requirement of 2ATP and a production of 4ATP, giving a *net gain* of 2ATP (see Figure 13–4). With glycogen as the initial substrate, ATP is required only in the formation of fructose bisphosphate, and thus a net gain of 3ATP is realized (see Figure 13–5). This type of analysis is relatively straightforward, and there should be no difficulty in understanding its significance: Although glycolysis requires an expenditure of cell energy to form 1,6FBP, a larger amount of energy is made available by the subsequent conversion of 1,6FBP to lactate—a *net gain* of ATP is achieved.

There is more to the energetics of glycolysis than merely counting ATPs. Some additional insight can be gained by focusing on the energetics of each step by examining the standard free-energy changes ($\Delta G^{\circ\prime}$) of each reaction (see Figure 13–6). Before using these data, we note that the free-energy changes given are only for the reactions involving the glycolytic intermediates and do not consider any coupling to ATP hydrolysis or ADP phosphorylation. For example, the value of $+3.3$ kcal corresponds to $Glc + P_i \rightarrow G6P + H_2O$ and not $Glc + ATP \rightarrow G6P + ADP$, and the value of -13.1 kcal corresponds to $PEP + H_2O \rightarrow PYR + P_i$ and not $PEP + ADP \rightarrow PYR + ATP$. (To compare $\Delta G^{\prime}$ values would be more realistic, but these values differ from cell to cell and change as the concentrations of intermediates change.)

Of what value are these data in understanding glycolysis? Given the free-energy change of each reaction, we can construct an *energy profile diagram* of the complete pathway that should depict the relative differences in free energy among all of the primary metabolic intermediates. Such a diagram is shown in Figure 13–7. A close study reveals that the tactic in constructing the diagram is really quite simple. Each intermediate, which is involved in at least two consecutive reactions, has been positioned on a free-energy scale in terms of its free-energy content relative to the free energy of both its immediate precursor and its immediate product. Note that there then emerges a pattern to the composite energetics of glycolysis in that we can clearly distinguish two distinct phases, one *endergonic* and the other *exergonic*.

Using free glucose as a starting point, we note that the endergonic phase, which consists of the initial part of the glycolytic pathway, is comprised of three *energy barriers*. In this context a rationale for an ATP requirement is now apparent. Two barriers, involving the formation of phosphoesters (G6P and 1,6FBP) are overcome by energy transfer using the highly reactive ATP as a donor of phosphate. With glycogen as initial substrate, the energy relationships among $(Glc)_n$, G1P, and G6P result in one less energy barrier. There is no need for energy transfer from ATP to mobilize a glucose residue as phosphoglucose. Inorganic P_i is used directly.

Still to be considered, however, is the final energy barrier from the level of 1,6FBP to 1,3BPG. Even though the magnitude of this barrier is considerable—being roughly equivalent to the first two combined—it is not coupled to ATP hydrolysis. Nevertheless, during glycolysis this barrier does not contribute to a metabolic blockage and an accumulation of 1,6FBP. The reason is provided

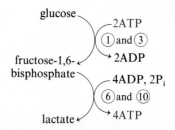

FIGURE 13–4 Consumption and production of ATP in glycolysis with glucose as the initial substrate.

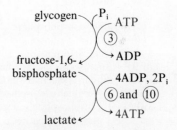

FIGURE 13–5 Consumption and production of ATP in glycolysis with glycogen as the initial substrate.

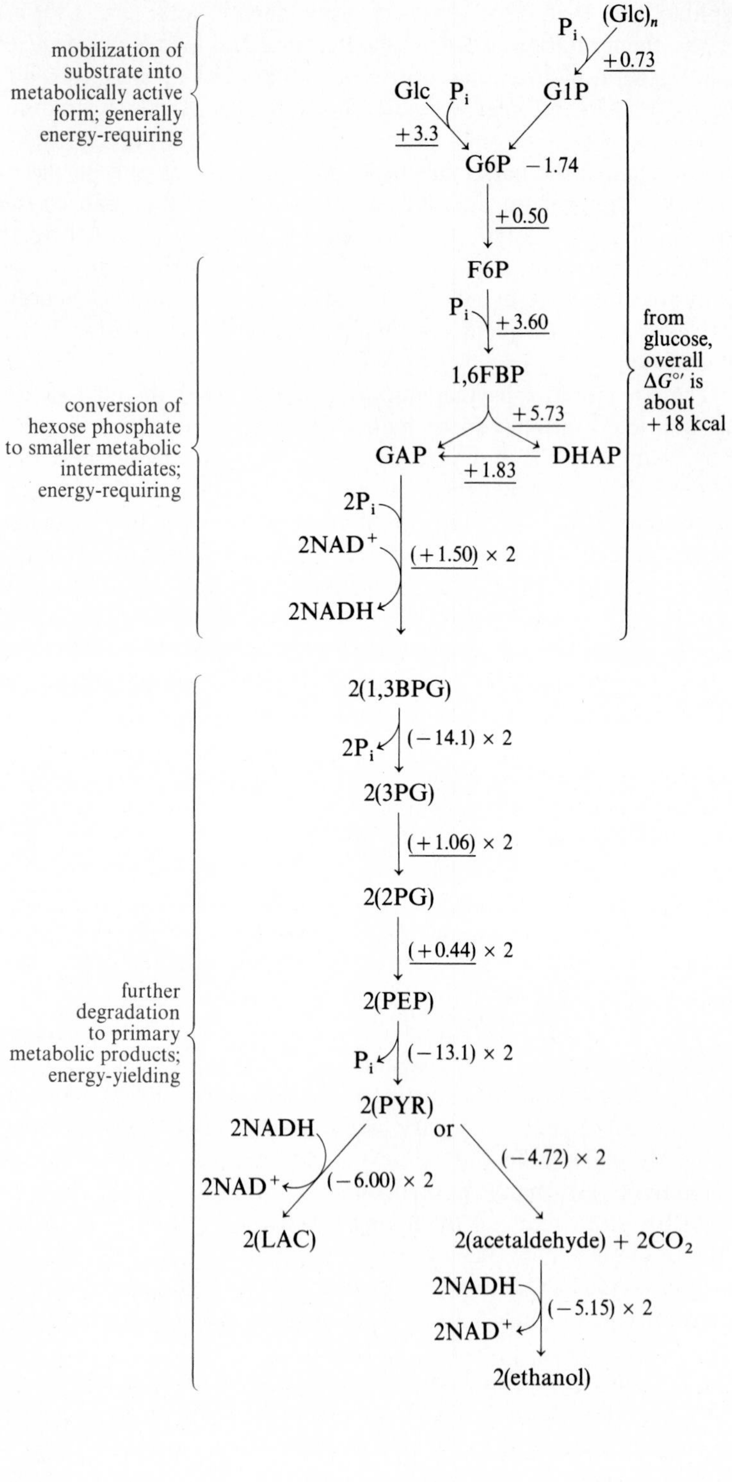

mobilization of
substrate into
metabolically active
form; generally
energy-requiring

P_i (Glc)$_n$

$\underline{+0.73}$

Glc P_i G1P

$\underline{+3.3}$

G6P -1.74

$\underline{+0.50}$

F6P

P_i $\underline{+3.60}$

1,6FBP

$\underline{+5.73}$

GAP $\underline{+1.83}$ DHAP

$2P_i$

$2NAD^+$ $(\underline{+1.50}) \times 2$

$2NADH$

conversion of
hexose phosphate
to smaller metabolic
intermediates;
energy-requiring

from
glucose,
overall
$\Delta G^{\circ\prime}$ is
about
$+18$ kcal

abbreviations as in Figure 13–3

(Glc)$_n$	glycogen
Glc	glucose
G1P	glucose-1-phosphate
G6P	glucose-6-phosphate
F6P	fructose-6-phosphate
1,6FBP	fructose-1,6-bisphosphate
DHAP	dihydroxyacetone phosphate
GAP	glyceraldehyde-3-phosphate
1,3BPG	1,3-bisphosphoglycerate
	3-phosphoglycerate
2PG	2-phosphoglycerate
PEP	phosphoenolpyruvate
PYR	pyruvate

(Although they are used in obvious ways,
some of the abbreviations used here and in
Chapter 14 are not conventional symbols
recommended by the International Committee of Biochemistry Nomenclature.)

2(1,3BPG)

$2P_i$ $(\underline{-14.1}) \times 2$

2(3PG)

$(\underline{+1.06}) \times 2$

2(2PG)

$(\underline{+0.44}) \times 2$

2(PEP)

P_i $(-13.1) \times 2$

further
degradation
to primary
metabolic products;
energy-yielding

2(PYR) or

$2NADH$

$2NAD^+$ $(-6.00) \times 2$ $(-4.72) \times 2$

2(LAC)   2(acetaldehyde) + 2CO$_2$

$2NADH$

$2NAD^+$ $(-5.15) \times 2$

2(ethanol)

FIGURE 13–6 Standard free-
energy changes (kilocalories per
mole) of the reactions of the
glycolytic pathway. Endergonic
(energy-requiring) reactions are
underlined.

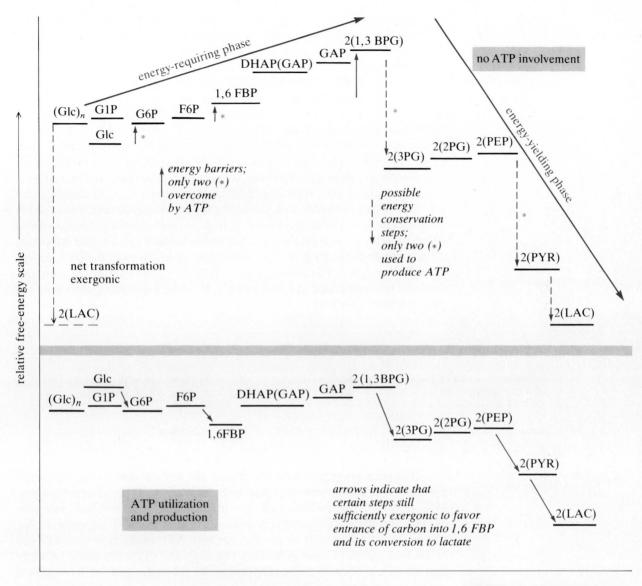

FIGURE 13–7 Skeleton view of the thermodynamic (free-energy) relationships among the metabolic intermediates of the glycolytic pathway. Horizontal lines indicate relative energy states of each intermediate. The upper part considers the reactions were they to occur in the absence of ATP involvement. The lower part considers the effect of ATP involvement in the two ATP-requiring and two ATP-yielding reactions.

by the thermodynamics of the very next reaction, which consists of the extremely favorable conversion of 1,3BPG to 3PG. (Recall that 1,3-bisphosphoglycerate, an acyl phosphate, is an unstable, high-energy compound; see Section 12–3.) Thus since the reactions are consecutive, the endergonic sequence of 1,6FBP through 1,3BPG is driven to completion by being coupled to the thermodynamically favorable removal of 1,3BPG as it is formed. In this sequence the coupling occurs without an actual transfer of energy (case III, Section 12–3).

The conversion of 2(1,3BPG) to 2(3PG) is especially important, because it is not only coupled to the degradation of 1,6FBP but also to the phosphorylation of 2ADP to yield 2ATP. The cell profits then in two ways: The flow of carbon through 1,6FBP is favored and ATP is formed. The PEP → PYR conversion is also of great metabolic significance. Here again, a high-energy phosphorylated compound (PEP is an enoyl phosphate) is used as a donor of phosphate for ATP formation. The formation of ATP in this manner—directly, during the course of a single enzyme-catalyzed reaction—is called *substrate-level phosphorylation*. Note that although the last reaction of glycolysis (PYR → LAC) has the capacity to be coupled to ATP formation, this does not occur, and the energy available is not conserved. You might argue that in view of the thermodynamic design already discussed, this nonconservation site is inconsistent with metabolic efficiency. However, just the opposite may be true. One suggestion is that the appearance of this type of reaction, occurring independently of an energy-requiring process and positioned at the end of a multistep sequence, may represent added insurance that preceding intermediates will be efficiently converted to the final product. The fact that several other metabolic pathways display a similar pattern supports such a suggestion. In so doing, the cell pays a price: Energy is not conserved as ATP. Another important metabolic role of the PYR → LAC conversion is described in the next section.

To summarize: The involvement of ATP in glycolysis (starting with glucose) can be summarized as follows (the -45.4-kcal value is obtained from the data in Figure 13–6):

1.   glucose + 4P$_i$ $\xrightarrow[\text{NADH}]{\text{NAD}^+}$ 2 lactate + 4P$_i$ $\Delta G^{\circ\prime} = -45.4$ kcal   (available)

2.   2ATP $\longrightarrow$ 2ADP + 2P$_i$ $\Delta G^{\circ\prime} = 2(-8) = -16$ kcal   (expended)

3.   4ADP + 4P$_i$ $\longrightarrow$ 4ATP $\Delta G^{\circ\prime} = 4(+8) = +32$ kcal   (conserved)

1 + 2 + 3.   glucose + 2ADP + 2P$_i$ $\xrightarrow[\text{NADH}]{\text{NAD}^+}$ 2 lactate + 2ATP $\Delta G^{\circ\prime} = -29.4$ kcal   (not conserved)

The numbers suggest that the overall efficiency of conserving the energy available from the conversion of glucose to lactate for ATP production is about 35%–40%. However, under cellular conditions (rather than standard-state conditions), the efficiency is probably higher. Without the ATP expenditure the efficiency would obviously be greater, but then ATP production would be impeded by a significant energy barrier. The evolution of the glycolytic pathway thus represents a compromise: Give some to get more, and get it without a serious impediment.

NAD$^+$ in Glycolysis

Operating anaerobically (be it for lactate or ethanol formation), glycolysis uses two different NAD-dependent dehydrogenases: One enzyme requires NAD$^+$ and produces NADH, whereas the other requires NADH and produces NAD$^+$ (see Figure 13–8). In other words, glyceraldehyde phosphate dehydrogenase and lactate dehydrogenase (or alcohol dehydrogenase) are complementary enzymes. This *internal coupling* means that the supply of NAD$^+$ is under constant replenishment, thus ensuring no interruption of the pathway at the level of the GAP → 1,3BPG reaction. The closing of the NAD$^+$ cycle is probably the major biochemical significance of the terminal dehydrogenase—more so than the energy flow argument mentioned in the preceding section.

Later in this chapter we will consider the regeneration of NAD^+ under aerobic conditions when the pyruvate $\rightarrow$ lactate reaction does not occur.

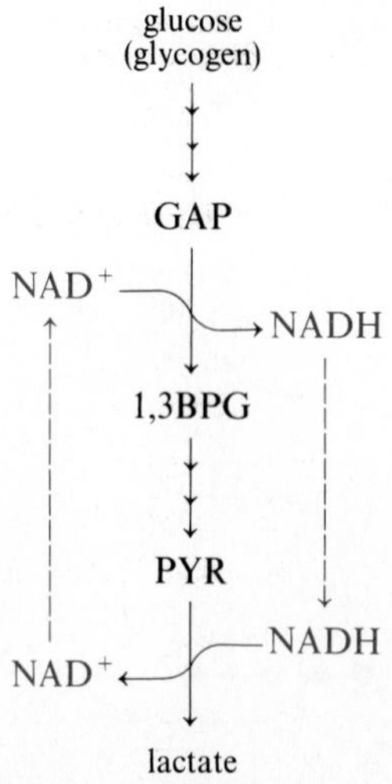

13-3 SOME INDIVIDUAL REACTIONS OF GLYCOLYSIS

A considerable part of biochemical research, past and present, is characterized by the detailed investigation of individual reactions. This type of research involves the isolation and purification of the enzyme, followed by extensive studies of its structure and catalytic properties. All the enzymes of glycolysis have been so studied, and accordingly, our current knowledge about this pathway is extensive and detailed. However, it is not our purpose to treat the subject in its entirety. This treatment is more appropriate to advanced courses. Alternatively, we will confine our attention to a few select reactions, with a focus on principles such as enzyme specificity, coenzyme participation, and, primarily, metabolic regulation.

Glycogen Phosphorylase

The first reaction in the catabolism of glycogen is the sequential removal of glucose residues, catalyzed by the widely distributed enzyme *glycogen phosphorylase*. The chemistry (see Figure 13-9) involves attacks by P_i on glucose residues at nonreducing ends of the glycogen structure. The chemistry occurs successively until an $\alpha(1 \rightarrow 6)$ branch site is approached to within about four residues. At this point another enzyme (*oligosaccharide transferase*) transfers an intact triglucose fragment to a neighboring branch stem, forming an $\alpha(1 \rightarrow 4)$ bond (see Figure 13-10). A third enzyme, an *$\alpha(1 \rightarrow 6)$ glucosidase*, then removes the remaining glucose unit involved in the $\alpha(1 \rightarrow 6)$ linkage and releases it as

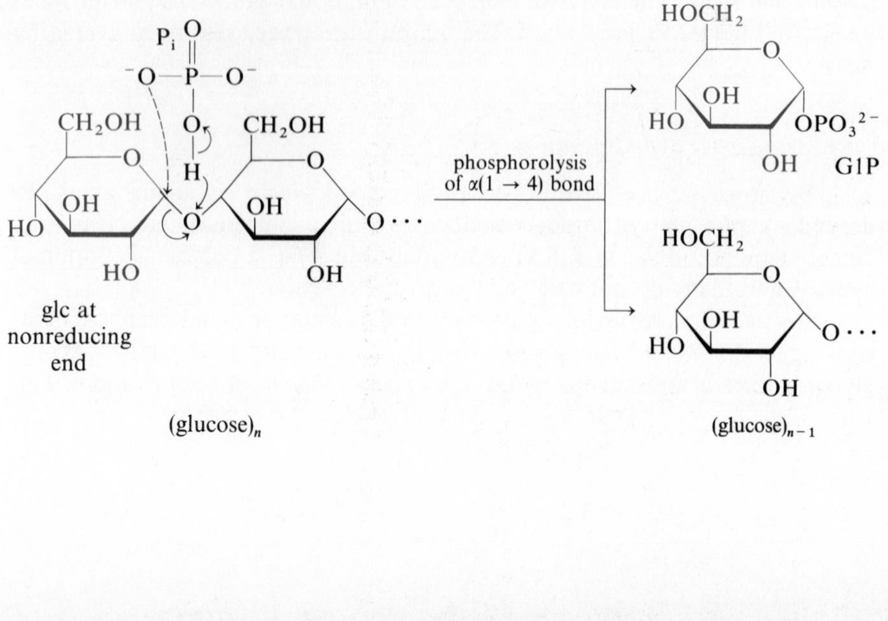

FIGURE 13-8 Internal coupling of NAD^+ in glycolysis under anaerobic conditions.

FIGURE 13-9 Reaction catalyzed by glycogen phosphorylase.

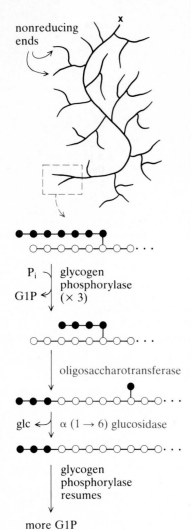

nonreducing ends

x

P$_i$ ⌐ glycogen
           phosphorylase
G1P ◁ (× 3)

oligosaccharotransferase

glc ◁ α (1 → 6) glucosidase

glycogen
phosphorylase
resumes

more G1P

FIGURE 13–10 Release of glucose residues at branch sites in glycogen, which requires two auxiliary enzymes (in color). The X at the top of the diagram marks the glucose residue at the single reducing end in glycogen.

free glucose, which is metabolized further after its ATP-dependent conversion to G6P. Then glycogen phosphorylase action resumes until another branch point is approached. The result is that glycogen glucose is mobilized as glucose-1-phosphate (primarily).

There are two forms of glycogen phosphorylase. Forty years ago C. Cori and G. Cori discovered the existence in the same cell of two different forms of glycogen phosphorylase—commonly designated as the *a* and *b forms* (see Figure 13–11). These forms are now understood to correspond to *phosphorylated* and *nonphosphorylated interconvertible forms, a* being phosphorylated and *b* being nonphosphorylated. Muscle phosphorylase is a homogeneous dimer (MW ≈ 200,000), composed of two identical polypeptide subunits (841 residues in each). (In some tissues the *a* form may exist as a tetramer.) Each subunit also contains the coenzyme *pyridoxal phosphate* (see Section 18–3), covalently attached to lysine residue 679. The location of each pyridoxal phosphate is near the active site in each subunit and appears to be intimately involved in the catalytic mechanism.

The phospho/dephospho converter enzymes are (1) *phosphorylase kinase,* a protein kinase of restricted specificity for phosphorylase and one other target protein, and (2) *protein phosphatase,* an enzyme that can operate on various phosphoproteins. The modification of phosphorylation *activates* the enzyme (*b → a*), and the modification of dephosphorylation *deactivates* the enzyme (*a →   b*). Later in this chapter we will describe how this interconversion is related to the hormonal control of phosphorylase activity, mediated by cyclic AMP and Ca^{2+}.

Although the *a* form is more active, the activity of the *b* form is increased in still another way, namely, allosteric activation by *AMP binding.* The AMP activation of phosphorylase *b* is not in response to an extracellular hormonal signal but, rather, is a regulatory response to an intracellular *metabolic signal.* The signal for an elevated concentration of AMP is an increased use (and hence a demand for more) of ATP. When energy demands are elevated, the rapid turnover of ATP will increase the cell levels of ADP and AMP, and the level of ATP will decrease. In other words, the energy charge (refer to p. 477) decreases. In this case, as the AMP concentration rises, AMP acts as an allosteric activator, binding to a regulatory site in each subunit of phosphorylase *b* and promoting a conformational change in *b* for enhanced formation of glucose-1-phosphate at the active site (see Figure 13–12). The *a* form is only very slightly activated by AMP.

Hexokinase and Glucokinase

By now you are conditioned to the word *kinase* signifying an ATP-dependent, phosphoryltransfer reaction to form a phosphoester. Nearly all kinases require a divalent cation (commonly Mg^{2+}) as a cofactor for optimal activity, and many do not catalyze the reverse reaction.

Hexokinases are obviously involved in the formation of hexose phosphates (see Figure 13–13). Although most display a high affinity ($K_m ≈ 10^{-5}M$) for glucose, other hexoses can serve as substrates. In some organisms more than

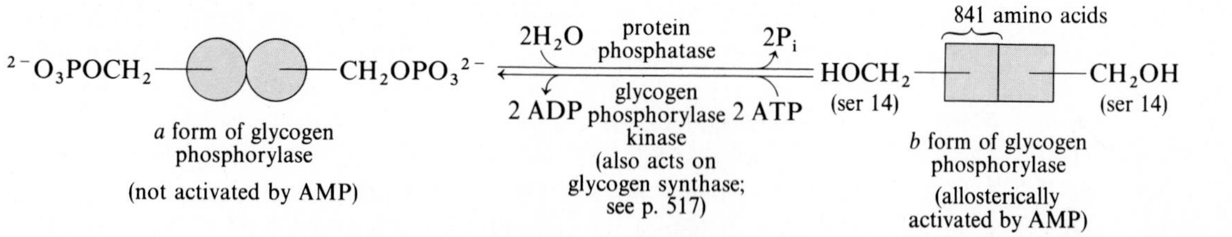

$^{2-}O_3POCH_2$—⬤⬤—$CH_2OPO_3^{2-}$

$$\xrightarrow[\substack{\text{glycogen} \\ \text{phosphorylase} \\ \text{kinase} \\ \text{(also acts on} \\ \text{glycogen synthase;} \\ \text{see p. 517)}}]{\substack{2H_2O \quad \text{protein} \\ \text{phosphatase} \\ \\ 2\,ADP \qquad\qquad 2\,ATP}}$$

2Pᵢ released / HOCH₂—▭▭—CH₂OH

a form of glycogen
phosphorylase

(not activated by AMP)

$\overbrace{\qquad}^{\text{841 amino acids}}$

HOCH₂ (ser 14) ... CH₂OH (ser 14)

b form of glycogen
phosphorylase
(allosterically
activated by AMP)

FIGURE 13–11 Interconvertible forms of glycogen phosphorylase. The sphere and the box are used to represent different conformations of the same polypeptide subunit resulting from the phosphorylated or nonphosphorylated serine residue.

one hexokinase is present. Mammals, for example, contain at least three hexokinases (isoenzymes), and in some tissues all three are present in the same cell. Some cell types also possess a *glucokinase,* which fulfills the same role as a hexokinase but is characterized by a much more restricted specificity, strongly favoring glucose as a substrate. In mammals it has been proposed that glucokinase serves as a metabolic safety valve in that it operates only when blood glucose levels are high. The basis of this suggestion is that the K_m value of glucokinase $(\approx 10^{-2}M)$ is about a thousand times greater than that for hexokinase, implying that glucokinase will become saturated only in the presence of high levels of glucose. By contributing to the prevention of excessively high glucose levels in blood, the glucokinase helps circumvent many physiological disorders.

Hexokinases also exhibit regulatory characteristics, with the regulators being ATP (it inhibits), ADP (it activates), and G6P (it inhibits) (see Figure 13–14). Notice that the regulators are in fact one of the substrates and two of the products of the reaction. All three signals complement each other in terms of the energy demands of a cell. When energy consumption is high (meaning again that ATP is turned over rapidly), the rise in ADP activates hexokinase and increases the rate of glucose mobilization to keep pace with the demand for more ATP. When energy consumption declines, the ATP level rises and the ADP level falls. For numerous reasons the intracellular G6P level will also rise in response to less energy demand. The elevation in ATP and G6P both inhibit the operation of hexokinase and decrease the rate of glucose mobilization.

$(\text{glucose})_n$

$\downarrow$ Pᵢ / AMP ⬇

b form

$(\text{glucose})_{n-1}$

glucose-1-phosphate
(G1P)

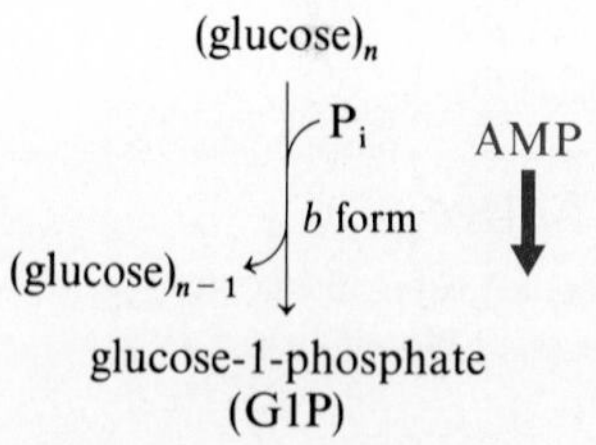

FIGURE 13–12 Metabolic control of the *b* form of glycogen phosphorylase by AMP. The heavy arrow parallel to the reaction arrow signifies an activating effect of AMP on the enzyme.

Phosphoglucomutase and Phosphoglyceromutase

Positional phosphate isomerization—the transfer of a phosphate group from one —OH position to another in a substrate—occurs twice in the glycolysis pathway: G1P → G6P and 3PG → 2PG. Although two different

FIGURE 13–13 Hexokinase-catalyzed phosphoryltransfer from ATP to D-glucose.

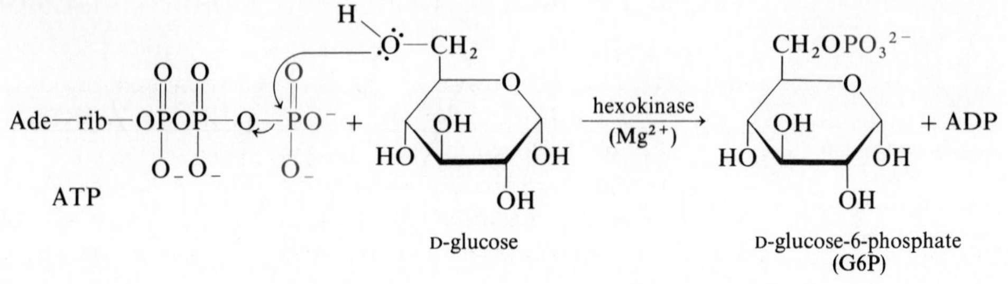

Ade—rib—OPOP—O—PO⁻ + [D-glucose structure] $\xrightarrow[(Mg^{2+})]{\text{hexokinase}}$ [D-glucose-6-phosphate structure] + ADP

ATP

D-glucose

D-glucose-6-phosphate
(G6P)

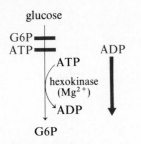

FIGURE 13–14 Metabolic regulation of hexokinase. The ▬ across the reaction arrow signifies inhibition; this symbol and the parallel arrow (for activation) will be used routinely.

G1P
glucose-1-phosphate

ATP ⌐
           │ phosphoglucokinase
ADP ⌐

glucose-1,6-bisphosphate
1,6GBP

1,3BPG
1,3-bisphosphoglycerate

3PG ⌐
           │ bisphosphoglyceromutase
3PG ⌐

2,3-bisphosphoglycerate
2,3BPG

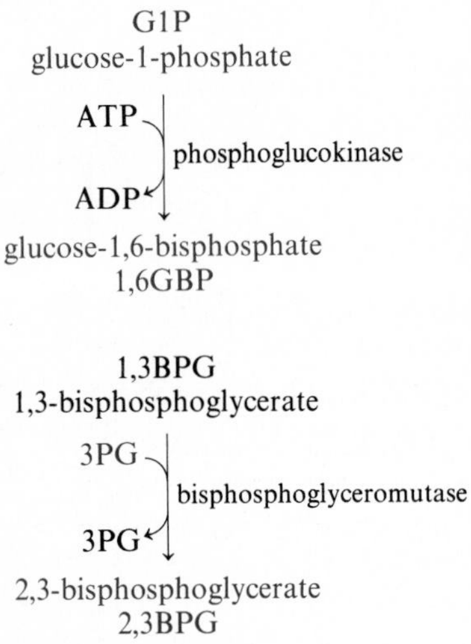

FIGURE 13–15 Formation of phosphodiester cofactors.

enzymes are involved, they both have a common reaction mechanism. The switch in phosphate positions is made indirectly through the intermediate participation of a required *phosphodiester cofactor, glucose-1,6-bisphosphate* (1,6GBP) for phosphoglucomutase and *2,3-bisphosphoglycerate* (2,3BPG) for phosphoglyceromutase. Only small amounts of each cofactor are necessary to initiate the action of each mutase, the cofactor being regenerated during the course of the reaction. Hence the cofactors are not considered as true metabolic intermediates of the glycolysis pathway.

The enzymes involved in their formation are identified in Figure 13–15. Whereas 1,6GBP is formed via an ATP-dependent kinase, the formation of 2,3BPG involves a phosphoryltransfer reaction between 1,3-bisphosphoglycerate and 3-phosphoglycerate. (*Note:* In red blood cells larger amounts of 2,3-bisphosphoglycerate are formed, where it has the specialized function of regulating the binding of O_2 to hemoglobin; refer to Figure 4–45.)

The phosphodiester cofactor and inactive enzyme first engage in a phosphoryltransfer reaction, yielding the active serine-phosphorylated form (E-OPO_3^{2-}) of the mutase and either G1P or G6P (see Figure 13–16). Thereafter, in a G1P → G6P conversion, the E-OPO_3^{2-} form of the enzyme can react with substrate G1P, re-forming 1,6GBP, which can then react again with E-OH to re-form E-OPO_3^{2-} and G6P. After the 1, 2, 3 sequence is complete, cycles of 2, 3/2, 3/2, 3/ and so on operate. The scheme is readily reversible. A similar sequence of events applies to the 3PG → 2PG interconversion involving the 2,3-bisphosphoglycerate cofactor.

Phosphofructokinase-1

After the aldo → keto isomerization of G6P → F6P (a readily reversible process) is complete, ATP is expended to convert fructose-6-phosphate to fructose-1,6-bisphosphate (1,6FBP) (see Figure 13–17). The enzyme involved, *phosphofructokinase-1*, is highly specific for the D isomer of F6P as substrate but otherwise has general features of operation typical of other ATP-dependent

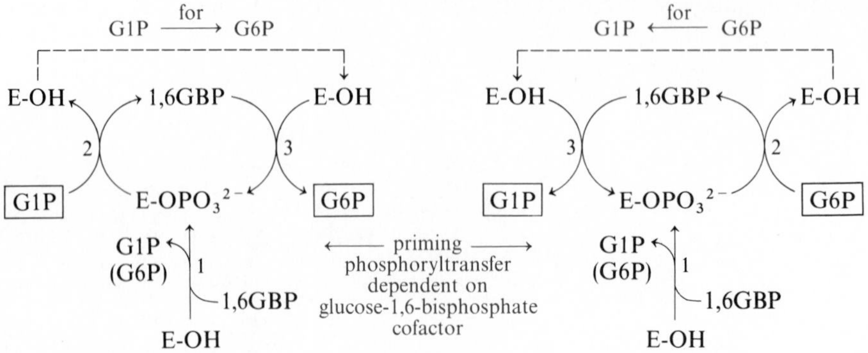

FIGURE 13–16 Role of glucose-1,6-bisphosphate as a cofactor for phosphoglucomutase (E-OH). The sequence at the left applies to the G1P → G6P reaction. The reverse reaction of G6P → G1P is shown at the right.

kinases. It is now called phosphofructokinase-1 to distinguish it from another enzyme, phosphofructokinase-2, that yields fructose-2,6-bisphosphate, an important regulator of glycolysis and glycogenesis (see Figure 13–36).

The F6P → 1,6FBP step is considered as the *committed step* of hexose catabolism by glycolysis. That is, as described later (see Figure 13–25), a variety of other reactions involve G1P, G6P, and F6P as substrates. But when F6P is converted to 1,6FBP, the 1,6FBP has only two subsequent metabolic fates: (1) cleavage by aldolase to prime the rest of glycolysis or (2) conversion back to F6P by a different enzyme (see Section 13–5).

Phosphofructokinase-1 activity is subject to many regulatory influences and is the *major regulatory enzyme* of the glycolysis pathway. This makes perfect sense, because being the committed step, the formation of 1,6FBP is an optimum control site on the flow of hexose carbon through glycolysis. The activity of phosphofructokinase-1 is stimulated by high levels of ADP and AMP and is depressed by high levels of ATP, NADH, and citrate. Later in the chapter we will examine the metabolic logic of these effects together with the complementary regulatory controls on the enzyme operating in the reverse reaction of 1,6FBP → F6P (see Figure 13–21).

Aldolase

Fructose bisphosphate aldolase (or just aldolase; see Note 13–3) catalyzes the only degradative step of glycolysis involving a C—C bond. This step results in the cleavage of one 6-carbon unit (a ketohexose) into two 3-carbon units (a ketotriose and an aldotriose) (see Figure 13–18). In conjunction with the triose isomerase, the overall effect is specifically the conversion of one unit of 1,6FBP to two units of glyceraldehyde-3-phosphate (GAP). Obviously, the isomerization is a key step in glycolysis, since it provides for the complete catabolism of the whole hexose unit rather than just half of it.

Aldolase preparations have been obtained from bacteria, plant, and animal sources, but the most studied is muscle aldolase. The predominant form (isoenzymes are known) of the muscle enzyme is a heterogeneous tetramer with an $\alpha_2\beta_2$ type of structure (MW $\approx$ 160,000). The α and β chains each contain 364 amino acids, a single disulfide bond, and in fact are almost identical in sequence. Although the reasons are not understood, the relative amounts of the five isoenzyme forms do vary with age and usually from tissue to tissue.

D-fructose-6-phosphate
F6P

ATP, NADH ——

ATP   (ADP, AMP)

phospho-
fructo-
kinase-1

ADP

D-fructose-1,6-bisphosphate
1,6FBP

FIGURE 13–17 Metabolic regulation of phosphofructokinase-1. The hexose unit is now committed to degradation.

NOTE 13–3
The name *aldolase* was coined in reference to the ability of the enzyme to also catalyze the reverse reaction (aldehyde + ketone → larger ketone), which in the language of organic chemistry is termed an *aldol condensation*.

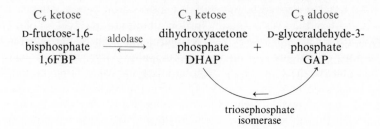

| C_6 ketose | | C_3 ketose | | C_3 aldose |
| D-fructose-1,6-bisphosphate 1,6FBP | aldolase | dihydroxyacetone phosphate DHAP | + | D-glyceraldehyde-3-phosphate GAP |

triosephosphate isomerase

FIGURE 13–18 Aldolase catalyzing the cleavage of a C_6 hexose into two C_3 trioses. See Figure 13–19 for a display of the mechanism of action.

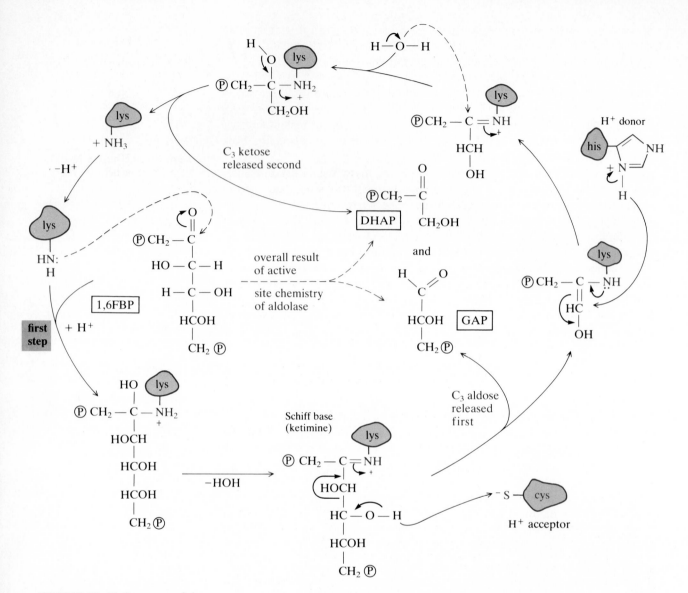

FIGURE 13–19 Summary of the catalytic mechanism of muscle aldolase. Although the sequence explains 1,6FBP → DHAP + GAP, all events are reversible to account for the condensation of the two triosephosphates to yield the hexose bisphosphate. Symbol Ⓟ is $-OPO_3^{2-}$.

The mechanism of action (see Figure 13–19) explaining how the C(3)—C(4) bond is cleaved begins with a condensation between the carbonyl carbon (C=O) of the ketose and an epsilon amino group (H_2N-) of a lysine side chain to yield a *Schiff base* ($>C=N$) intermediate. Thereafter, other active-site R groups of cysteine and histidine participate in a sequence of acid-base catalysis to cleave the C—C bond. The aldolase in yeasts, fungi, certain bacteria, and algae operates via a completely different mechanism. A Zn^{2+} cofactor is involved, and a covalent Schiff base intermediate is not formed. The point of mentioning this difference is to focus again on the concept that in different cells the same reaction may occur either (1) by an identical or very similar

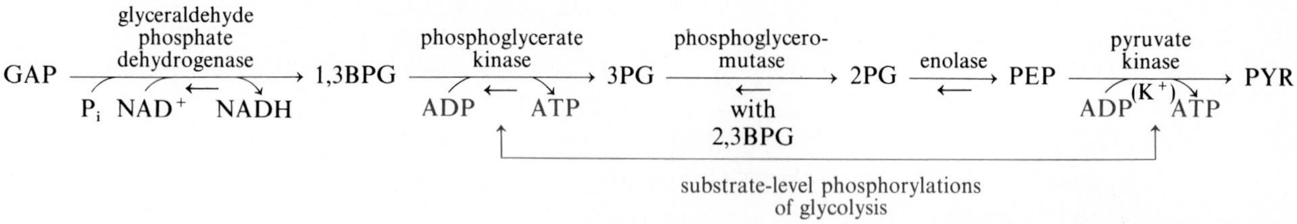

substrate-level phosphorylations
of glycolysis

FIGURE 13–20 Energy-yielding reactions in the glycolysis pathway. The reaction mechanism for the formation of 1,3-bisphosphoglycerate is shown in Figure 13–21.

process or (2) by a very different process. Thus an attitude of caution must prevail before drawing conclusions regarding the biochemical similarity of the same processes in different organisms.

Glyceraldehyde-3-Phosphate Dehydrogenase

The conversion of glyceraldehyde-3-phosphate (GAP) to 1,3-bisphosphoglycerate (1,3BPG) represents the distinct *oxidative step* of glycolysis. The most significant aspect of the reaction is that a high-energy acyl phosphate (bisphosphoglycerate) is formed as a product without a requirement for ATP. In fact, the reaction provides a basis to form ATP in the next reaction of 1,3BPG → 3PG (see Figure 13–20). The ATP formed here and in the later conversion of PEP → PYR are labeled as substrate-level phosphorylations, to distinguish them from the ATP production by the more elaborate process of oxidative phosphorylation.

Glyceraldehyde phosphate dehydrogenase is oligomeric, consisting of four identical chains—an α_4 type of structure. Regardless of specific structural differences in the enzyme from various sources, all preparations studied possess an active-site cysteine sulfhydryl (—SH) group absolutely required for activity. The active holoenzyme form of the enzyme contains bound NAD$^+$—one with each subunit.

The proposed mechanism (see Figure 13–21) begins with condensation of the active-site —SH and the carbonyl carbon of the aldotriose to yield a *thiohemiacetal* adduct. The bound NAD$^+$ then abstracts a hydride ion from the thiohemiacetal to yield an enzyme-bound, high-energy thioester linkage and bound NADH. Transfer of the acyl group of the thioester to the P$_i$ substrate yields 1,3-bisphoshoglycerate (1,3BPG), followed by oxidation of the bound NADH by substrate NAD$^+$. Overall, a high-energy phosphoryl donor (1,3BPG) has been formed without any expenditure of chemical energy as ATP.

Pyruvate Kinase

Catalyzing the second of two ATP-generating reactions in glycolysis, pyruvate kinase is unlike the enzyme involved in the first reaction (phosphoglycerate kinase) in several ways. For one thing, phosphoglycerate kinase is a reversible enzyme, capable of operating in either direction [(1,3BPG) + ADP $\rightleftharpoons$ 3PG + ATP], whereas pyruvate kinase operates only irreversibly for PEP + ADP → PYR + ATP (see Figure 13–22). In glycogenesis the irreversibility of pyruvate kinase is circumvented in ways described later in the chapter.

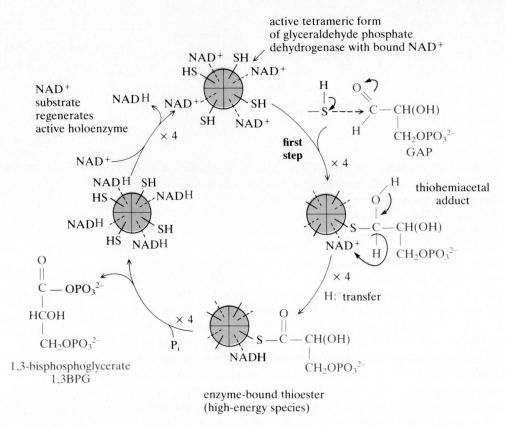

active tetrameric form
of glyceraldehyde phosphate
dehydrogenase with bound NAD$^+$

NAD$^+$ substrate regenerates active holoenzyme

$\times 4$

GAP

first step

$\times 4$

thiohemiacetal adduct

$\times 4$

H: transfer

enzyme-bound thioester
(high-energy species)

1,3-bisphosphoglycerate
1,3BPG

FIGURE 13–21 Catalytic mechanism of glyceraldehyde phosphate dehydrogenase. To minimize clutter in the diagram, the formation of an enzyme-bound thioester intermediate is shown on only one of the subunits of the tetramer.

Another difference is that pyruvate kinase is a regulated enzyme, whereas phosphoglycerate kinase is not. Energy charge regulatory signals of pyruvate kinase include activation by ADP and inhibition by ATP. Note that these ADP/ATP effects are in perfect harmony with those described earlier for other regulatory enzymes of glycolysis, notably phosphofructokinase-1. Elevated ADP levels (due to high-level energy use) enhance the rate of glycolysis to provide more ATP; elevated ATP levels (due to a reduction in energy use) diminish the rate of glycolysis. Moreover, the regulation of phosphofructokinase-1 and pyruvate kinase is also supplemented by a direct regulatory communication between the two steps. This communication is an example of a *feed-forward activation* due to an elevated level of fructose-1,6-bisphosphate acting as an activating effector of pyruvate kinase.

Although the ADP/ATP and 1,6FBP controls on pyruvate kinase are common to most cells, some cell types contain a pyruvate kinase that also responds to *hormonal signals*. (In animals at least three different forms are known to exist, and each appears to be a homogeneous tetramer, but each is composed of a different type of subunit.) Extracellular hormonal control applies to the type L form of pyruvate kinase in liver and kidney. This control is mediated by cyclic AMP as a secondary messenger via cyclic AMP–dependent protein kinase (cAMP–dPK; refer to Section 11–8), which catalyzes the phosphorylation of type L pyruvate kinase (see Figure 13–23). The phosphorylated form of pyruvate kinase has a K_m for the PEP substrate of about 0.9mM, whereas the nonphosphorylated form has a K_m of about 0.3mM; there is no difference in the V_{max}. In other words, phosphorylation *inhibits* the activity of pyruvate kinase. The physiological significance of phosphorylated pyruvate kinase in explaining the stimulation of gluconeogenesis in liver in response to the hormone glucagon is described later in the chapter.

Lactate Dehydrogenase

Under anaerobic conditions the significance of the last step of glycolysis was highlighted earlier: Through the reoxidization of NADH to NAD^+, the continued flux of carbon through the NAD^+-dependent step of glyceraldehyde-3-phosphate → 1,3-bisphosphoglycerate is maintained. Operating in reverse, however, lactate dehydrogenase (LDH) also serves as the first enzyme in the pathway of glycogenesis for re-forming glucose and glycogen from lactate.

Lactate dehydrogenase is a tetrameric enzyme existing in five isoenzyme forms (H_4, H_3M, H_2M_2, HM_3, M_4), with individual cell types having characteristic isoenzyme compositions. Although they all catalyze the same PYR ⇌ LAC reaction, subtle and distinct differences in kinetic characteristics do distinguish individual isoenzymes. These differences are often consistent with the metabolism of a given type of cell and/or of the various developmental stages of an organism. For example, the H_4 species of LDH—the major isoenzyme of heart muscle—is strongly inhibited by pyruvate; however, the M_4 species—the major form of skeletal muscle—is not inhibited by pyruvate.

This regulatory distinction between heart muscle (an aerobic tissue) and skeletal muscle (an anaerobic tissue) is useful to both tissues. Under conditions of excessive exercise skeletal muscle can engage in large bursts of anaerobic glycolysis. Being insensitive to pyruvate, the M_4 species of LDH will maintain the flux of carbon through pyruvate, removing all the pyruvate as lactate, most of which is released to blood. In contrast, heart can remove blood lactate as a source of carbon for its own aerobic metabolism. After using lactate (from blood) to produce pyruvate, the H_4 species responds to product inhibition by the pyruvate formed, providing an opportunity for the pyruvate to be metabolized in

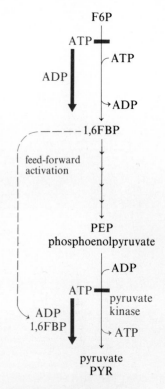

FIGURE 13–22 Metabolic regulation of pyruvate kinase. Note the complementary regulatory signals for phosphofructokinase-1 and pyruvate kinase and the regulatory connection involving 1,6FBP as an activator of pyruvate kinase.

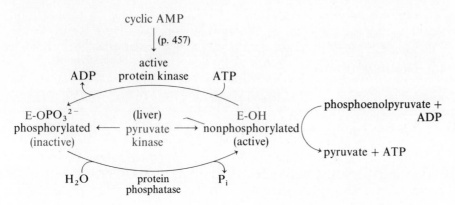

FIGURE 13–23 Hormonal regulation of pyruvate kinase. The formation of cyclic AMP, serving as secondary messenger, is stimulated by several hormones (see Figure 13–35).

heart mitochondria by the aerobic, citric acid cycle pathway. Recall (from Section 4–5) that elevated blood levels of the H_4–LDH can be used as a biochemical marker in the diagnosis of myocardial infarction.

13–4   OTHER REACTIONS ASSOCIATED WITH GLYCOLYSIS

Various microorganisms carry out glycolysis under anaerobic conditions by also producing lactate or some other end product. Whatever the end product, its formation always involves a step providing for the reoxidation of NADH to maintain a supply of NAD^+ to sustain the flow of carbon through the step of glyceraldehyde + NAD^+ + P_i → 1,3-bisphosphoglycerate + NADH.

Ethyl Alcohol

The successive action of *pyruvate decarboxylase* and *alcohol dehydrogenase* converts pyruvate to ethyl alcohol and CO_2 (see Figure 13–3, reactions 11A and 11B). The involvement of the coenzyme *thiamine pyrophosphate* (TPP) in the action of pyruvate decarboxylase is described later.

Glycerol

Via the action of *glycerol phosphate dehydrogenase* and *glycerophosphatase,* some microbes can accumulate free glycerol from dihydroxyacetone phosphate (see Figure 13–24). The same sequence can also occur in aerobic cells to provide L-glycerol phosphate for the biosynthesis of acylglycerols and phosphoacylglycerols, explaining a small part of converting carbohydrate to fat. The major part of the carbohydrate → fat conversion involves the sequence of hexose ⇶ pyruvate → acetyl-SCoA ⇶ fatty acids.

Aerobic Glycolysis

FIGURE 13–24 Use of glycolysis intermediates to form lipids and free glycerol.

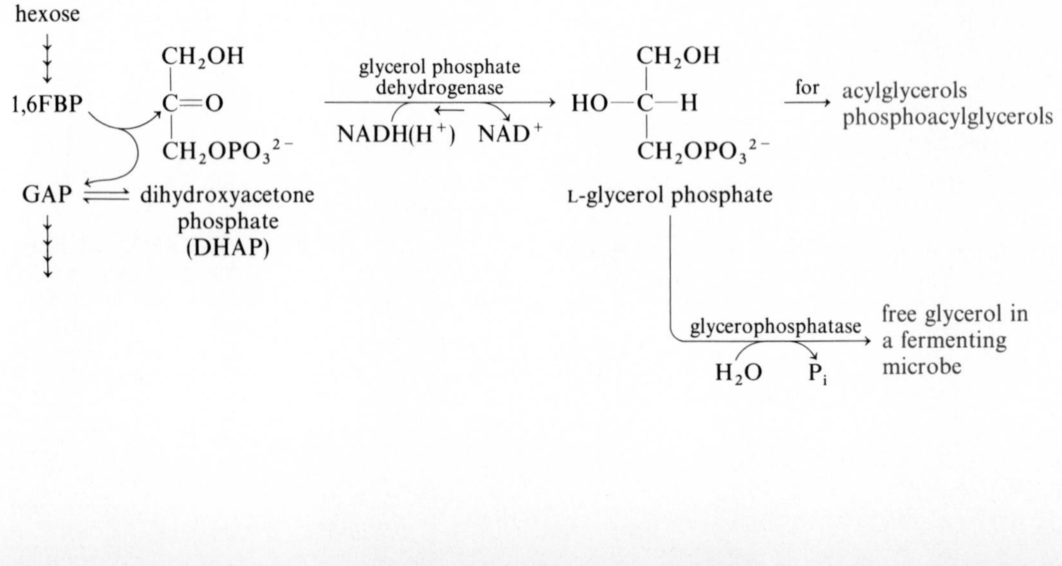

When lactate, an end product of anaerobic glycolysis, is also produced by cells growing in the presence of O_2, the process is called *aerobic glycolysis.* One example occurs in red blood cells that lack mitochondria. Unable to derive

ATP from the mitochondrial combustion of pyruvate $\rightarrow$ CO_2, red blood cells obtain ATP from glycolysis. Other examples of aerobic glycolysis include retina tissue, fetal tissues shortly after birth, intestinal mucosa, and several cancer cells.

13–5 OTHER FACETS OF HEXOSE METABOLISM

Glycolysis is not used exclusively for the degradation of glucose. Rather, the pathway will accommodate the catabolism of any hexose that can be converted to one of the hexose intermediates, G1P, G6P, or F6P. Some of these hexose connections to glycolysis are shown in Figure 13–25. The same display also summarizes some important precursor relationships that hexose phosphates have to many of the sugar derivatives we have encountered in previous chapters.

Fructose, Mannose, and Galactose Catabolism

Provided by the hydrolysis of ingested sucrose, fructose can be converted (panel A) to fructose-6-phosphate or fructose-1-phosphate by either hexokinase (giving F6P) or fructokinase (giving F1P). The hexokinase is the same used for glucose. The more specific fructokinase operates principally in liver, which also contains a specific *F1P-aldolase* converting F1P to dihydroxyacetone phosphate and free glyceraldehyde. Glyceraldehyde dehydrogenase and glycerate kinase convert the free glyceraldehyde to 2-phosphoglycerate, providing entry into glycolysis.

The same hexokinase used for glucose and fructose also converts mannose to mannose-6-phosphate (M6P; panel B). Similar to the operation of phosphoglucoisomerase, the action of *phosphomannoisomerase* converts M6P into fructose-6-phosphate.

Galactose catabolism (panel C) involves UDP sugars. First (step 1 in panel C) a specific *galactokinase* converts galactose to galactose-1-phosphate (Gal1P). Then *UDP-galactose pyrophosphorylase* operates (step 2) to convert Gal1P to UDP-Gal. Representing our first encounter of this type of enzyme, *UDP-glucose epimerase* (step 3) then catalyzes an inversion of configuration at C4, converting UDP-Gal to UDP-Glc (see Figure 13–26). The epimerization reaction uses enzyme-bound NAD^+ to abstract a hydride ion (H:$^-$) from the tetrahedral C4 position to generate a planar C4 carbonyl function. The NADH then donates H:$^-$ back to the C4 carbonyl position from the opposite face to complete the inversion. The action of an epimerase enzyme is readily reversible. Entry to glycolysis as glucose-1-phosphate (G1P) is achieved (step 4 in Figure 13–25) by *galactose phosphate uridyl transferase* catalyzing an exchange reaction between Gal1P and UDP-Glc to yield UDP-Gal and Glc1P.

Inborn errors of metabolism are associated with numerous hereditary diseases characterized by the presence of a defective enzyme or the failure to produce normal levels of an enzyme. One example in carbohydrate metabolism is *galactosemia,* characterized by high levels of galactose and galactose-1-phosphate in blood, in urine, and in cells. Galactose and its phosphoester may themselves be toxic or may give rise to toxic derivatives such as galactitol (see Figure 10–13). The elevated levels are explained by a deficiency of galactose

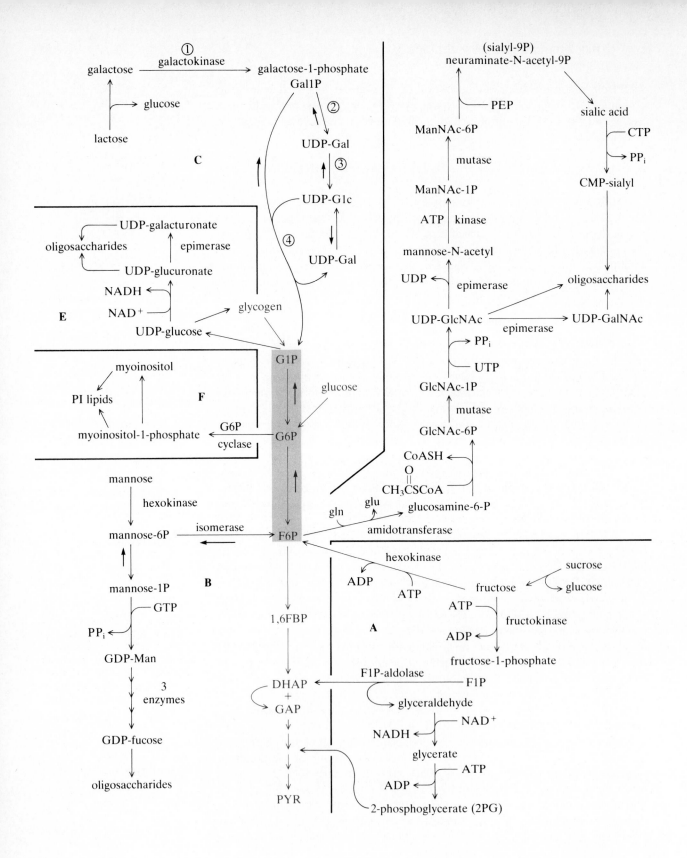

◀ **FIGURE 13–25** Summary of catabolism of hexoses and their conversion into sugar derivatives. The text gives descriptions to accompany each panel. The glycolysis pathway is shown in color. Some individual reactions are described further in Figures 13–26, 13–27, and 13–28.

phosphate uridyl transferase (step 4) due to a mutation in the uridyl transferase gene. The situation is serious in infants, where the major natural source of dietary carbohydrate is milk lactose, containing galactose. Severe cases of infant galactosemia result in cataract formation, cirrhosis of the liver and spleen, and in some cases irreversible mental retardation. Generally, however, all of these clinical manifestations can be avoided by simply eliminating galactose and galactose-containing substances from the diet. As the infant ages, less dietary control is required, since the adult galactosemic is capable of metabolizing galactose-1-phosphate by alternative routes.

Formation of Sugar Derivatives

In previous chapters we have encountered several examples of modified forms of simple sugars. Panels B, D, E, and F of Figure 13–25 summarize how they are formed.

Sialic acid, N-acetylated galactosamine (GalNAc), and *N-acetylated glucosamine* (GlcNAc) originate from fructose-6-phosphate (panel D). The sequence begins with F6P-amidotransferase catalyzing the transfer of an —NH$_2$ group from the amide group of amino acid glutamine to C2 of F6P, yielding glucosamine-6-phosphate (see Figure 13–27). A transfer of CH$_3$C=O from acetyl-SCoA and migration of phosphate yields N-acetyl-glucosamine-1-P, which is then converted to its UDP derivative, UDP-GlcNAc; it in turn undergoes an inversion in C4 configuration to yield UDP-GalNAc. Both UDP-GlcNAc and UDP-GalNAc can be used for biosynthesis of oligosaccharide groupings in various glycoproteins, glycosaminoglycans, and glycolipids. UDP-GlcNAc can also be converted to sialic acid by a sequence involving a condensation of a six-carbon hexose skeleton with the three-carbon skeleton of phosphoenolpyruvate (PEP). Sialic acid is incorporated into oligosaccharide structures after conversion to a CMP-sialyl derivative.

Glucuronate and *galacturonate* are formed (panel E) as UDP derivatives that can be traced back to glucose-1-phosphate via UDP-glucose. The key

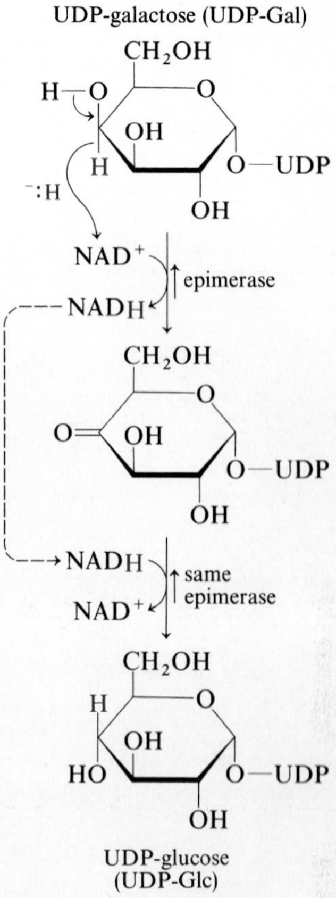

UDP-galactose (UDP-Gal)

FIGURE 13–26 Interconversion of glucose and galactose at the level of UDP derivatives.

FIGURE 13–27 Formation of glucosamine. Note that the enzyme uses the free carbonyl form of fructose-6-phosphate. See panel D in Figure 13–25 for the fate of glucosamine-6-phosphate.

fructose-6-phosphate

glucosamine-6-phosphate

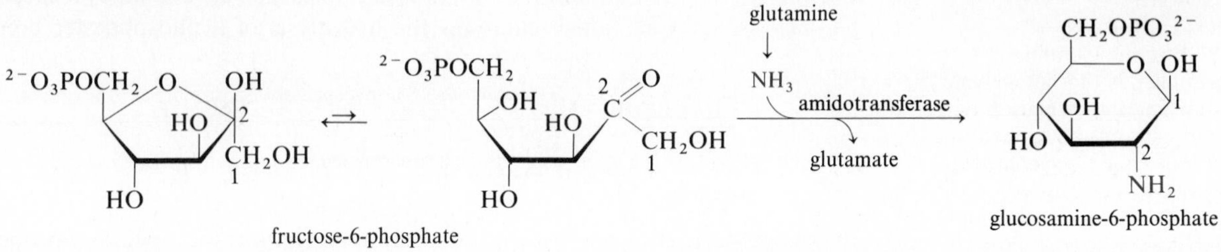

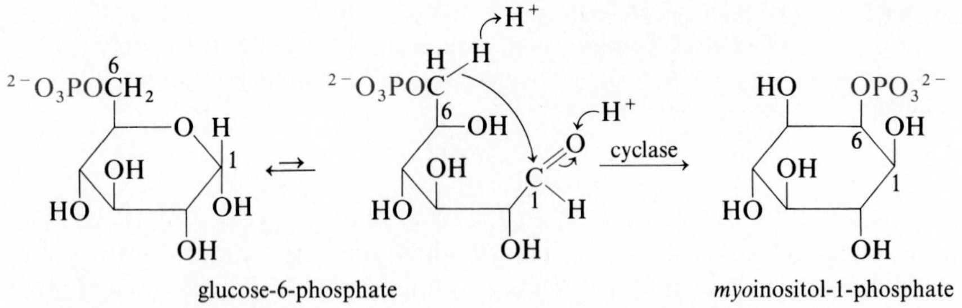

FIGURE 13–28 Formation of myoinositol by Cl–C6 cyclization of glucose.

glucose-6-phosphate   *myo*inositol-1-phosphate

oxidative step involves an NAD^+-dependent dehydrogenase forming UDP-glucuronate from UDP-glucose. An epimerase converts UDP-glucuronate to UDP-galacturonate.

The formation of *fucose* (L-6-deoxygalactose) can be traced (panel B) to mannose-6-phosphate, and the formation of *myoinositol* can be traced (panel F) to glucose-6-phosphate. The details of fucose formation are not shown. The formation of myoinositol involves the action of glucose-6-phosphate cyclase. Using the free carbonyl form of G6P as substrate, the cyclase catalyzes the intramolecular condensation of the C6 —OH and C1 carbonyl carbon to yield myoinositol-1-phosphate (see Figure 13–28).

13–6   GLYCOGENESIS

Glycogenesis, the assembly of glycogen and glucose from the smaller carbon fragments of pyruvate and lactate, occurs by the combined participation of several glycolysis enzymes capable of operating in reverse and different enzymes in place of those in glycolysis that operate irreversibly. The four irreversible steps in glycolysis that involve different enzymes in glycogenesis are as follows:

pyruvate (PYR) → phosphoenolpyruvate (PEP)

fructose-1,6-bisphosphate (1,6FBP) → fructose-6-phosphate (F6P)

glucose-6-phosphate (G6P) → glucose (Glc)

glucose-1-phosphate (G1P) → glycogen

For the 1,6FBP → F6P and G6P → Glc reactions, each is handled by a specific *phosphatase* enzyme, which catalyzes the hydrolysis of a phosphoester bond:

$$1,6FBP + H_2O \xrightarrow{\text{fructose-1,6-bisphosphatase}} F6P + P_i$$

$$G6P + H_2O \xrightarrow{\text{glucophosphatase}} Glc + P_i$$

In the next section we will have more to say about important regulatory properties of fructose-1,6-bisphosphatase.

NOTE 13–4
The Enzyme Commission recommends that an enzyme catalyzing the formation of a bond without the expenditure of ATP be called a *synthase* and classified as a transferase (class 2). If ATP is involved, the name *synthetase* is used, and it is classified as a ligase (class 6).

Pyruvate → Phosphoenolpyruvate

Depending on the cell, the PYR → PEP conversion occurs in one of three different ways (see Figures 13–29 and 13–30). In some bacteria a single ATP-dependent enzyme, *phosphoenolpyruvate synthetase,* catalyzes the reaction directly (route 1). Cells of higher plants and animals achieve the conversion indirectly involving two enzymes: *pyruvate carboxylase* and *phosphoenolpyruvate carboxykinase.* Both enzymes are usually located in the mitochondrion (route 2), but this location creates no problem since the mitochondrial membrane is permeable to the passage of pyruvate. The carboxylase catalyzes the ATP- and biotin-dependent carboxylation of pyruvate to yield *oxaloacetate* (OAA), a C_4 α-keto dicarboxylic acid. (The biochemistry of the coenzyme biotin will be deferred until Section 17–2.) The decarboxylation/phosphorylation reaction of OAA → PEP uses the carboxykinase for which GTP (not ATP) is the preferred phosphate donor. The exit of PEP from the mitochondrion to the cytoplasm (the membrane is also permeable to PEP) is necessary to resume glycogenesis.

Sometimes, the carboxykinase is located in the cytoplasm. This location creates a problem because the mitochondrial membrane is impermeable to the passage of oxaloacetate. The problem is circumvented (route 3) by two other enzymes: *malate dehydrogenase* in the mitochondrion and *malate dehydrogenase* in the cytoplasm—isoenzymes located in two different cell compartments. First, the mitochondrial dehydrogenase reduces OAA to the α-hydroxy dicarboxylic acid *malate* (MAL). Whereas the mitochondrial membrane is impermeable to oxaloacetate, remarkably the same membrane is permeable to malate. Then after exiting the mitochondrion, malate is reoxidized by the cytoplasmic enzyme, providing the OAA substrate for the cytoplasmic carboxykinase. In the next chapter we will encounter another metabolic function of this *malate shuttle.*

G1P → Glycogen

Glucose is assimilated (via UDP-glucose) into glycogen by the operation of two enzymes:

$$\text{glucose-1-phosphate} \atop \text{G1P} \quad \xrightarrow[\underset{\text{UTP} \qquad \text{PP}_i}{}]{\overset{\text{UDP-glucose}}{\text{pyrophosphorylase}}} \quad \text{UDP-glucose} \quad \xrightarrow[\underset{\text{(glucose)}_n \quad \text{UDP}}{}]{\overset{\text{glycogen}}{\text{synthase}}} \quad (\text{glucose})_{n+1}$$

Glucosyl addition to a nonreducing end of a glycogen primer is catalyzed by *glycogen synthase* (see Note 13–4). The muscle synthase is a homogeneous tetramer (MW ≈ 350,000) capable of existing in nonphosphorylated and phos-

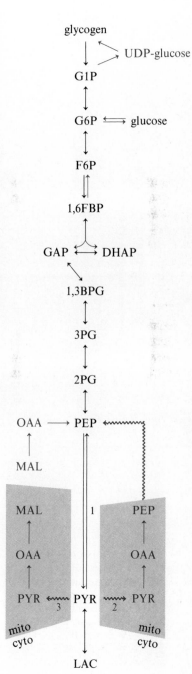

FIGURE 13–29 Outline of glycolysis and glycogenesis. The double barb arrow ↔ symbolizes a reversible enzyme that can operate in both directions. Color identifies the participation of different enzymes in glycogenesis to overcome irreversible enzymes of glycolysis. Wavy color arrows represent transport between cytoplasm and mitochondrion. See Figure 13–30 for additional details.

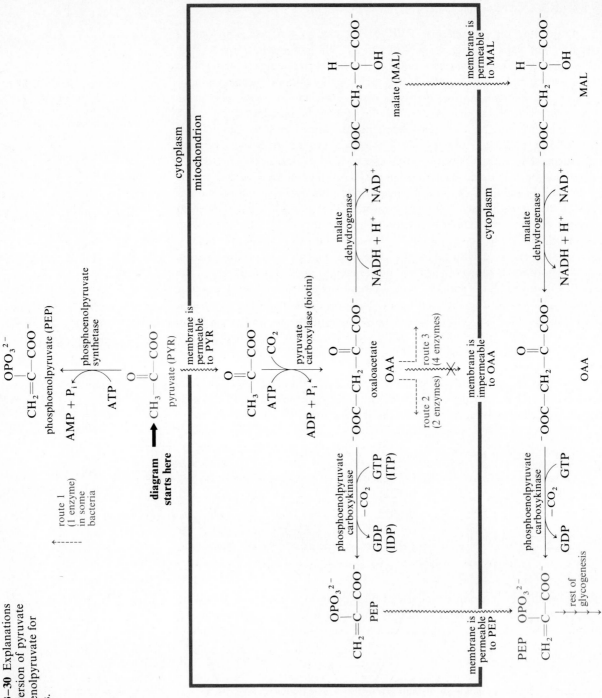

FIGURE 13–30 Explanations for the conversion of pyruvate to phosphoenolpyruvate for glycogenesis.

phorylated forms, interconvertible by a protein kinase and phosphoprotein phosphatase:

$$
\begin{array}{ccc}
\text{less active} & \overset{\text{H}_2\text{O}}{\underset{\text{ADP}}{\overset{\text{phosphoprotein phosphatase}}{\rightleftarrows}}} & \text{more active} \\
\text{phosphorylated} & & \text{nonphosphorylated} \\
\text{glycogen synthase} & \overset{\text{glycogen}}{\text{synthase/}} & \text{glycogen synthase} \\
\text{(dependent on G6P;} & \text{phosphorylase} & \text{(independent of G6P;} \\
\text{D form)} & \text{kinase} & \text{I form)}
\end{array}
$$

Both converter enzymes are the same enzymes used for interconverting the phospho/dephospho forms of glycogen phosphorylase. Thus we will use only one name for the phosphorylating kinase, namely, *glycogen synthase/phosphorylase kinase*. Because the phosphoprotein phosphatase is active toward several different phosphoproteins, its name does not identify a specific protein as target substrate.

Although the phospho/dephospho modifications apply to both glycogen synthase and glycogen phosphorylase, the correlation of modification and activity is opposite: Phosphorylation activates phosphorylase but deactivates synthase. In the next section we will integrate these complementary control features to the hormonal regulation of glycolysis/glycogenesis (see Figure 13–34). The two forms of glycogen synthase have still another distinction. The activity of the more active, nonphosphorylated form, called the *I form,* is independent of any allosteric enhancement effect by glucose-6-phosphate. However, the less active, phosphorylated form, called the *D form,* is dependent on glucose-6-phosphate for its activity.

13–7 OVERALL VIEW OF THE REGULATION OF GLYCOLYSIS AND GLYCOGENESIS

Having described the events of glycolysis and glycogenesis, we can now consider the intricate and fascinating features that *simultaneously regulate the flow of carbon in each pathway*—favoring glycolysis when a cell needs and uses a lot of ATP and favoring glycogenesis when the need and use of ATP is lessened. There are two types of control: *metabolic* (in all types of organisms) and *hormonal* (in higher animals).

Metabolic Control

When cells undergo a shift in metabolism, changes occur in the intracellular concentrations of metabolites. These concentration changes then serve as intracellular signals to regulate the activity of key enzymes in metabolic pathways in order to keep supply in pace with demand. Several substances can act as regulators, functioning as allosteric inhibitors or activators, noncompetitive inhibitors, and competitive inhibitors. For the moment we will focus attention on the regulatory influences of the adenylate species: AMP, ADP, and ATP. Remember, from earlier discussions, the energy charge concept: A cell shifting

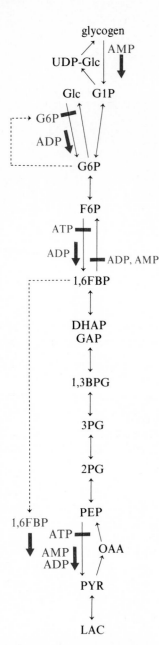

FIGURE 13–31 Coordinated metabolic control of glycolysis and glycogenesis via elevated levels of ATP, ADP, and AMP. The feedback inhibition by G6P and the feed-forward activation by 1,6FBP are also indicated (dashed lines). Other metabolic controls that supplement the ones shown here are presented in Figure 14–29.

into a state of high energy consumption will be characterized by increased levels of ADP and AMP and reduced levels of ATP; a shift back to a "resting" state will be characterized by reduced levels of ADP and AMP and increased levels of ATP.

The regulated enzymes of the interconnected and interdependent pathways of glycolysis and glycogenesis are identified in Figure 13–31. The complementary effects of all these controls are to increase the rate of carbohydrate breakdown and decrease the rate of hexose formation when the energy charge is decreased ([ATP] decreases and [AMP] and [ADP] increase) and to do just the reverse when the energy charge is increased ([ATP] increases and [AMP] and [ADP] decrease). The most pronounced complementary controls are exerted on the two enzymes involved in the F6P $\rightleftharpoons$ 1,6FBP interconversion, the major site of regulation. Shortly, we will discuss another metabolite that regulates F6P $\rightleftharpoons$ 1,6FBP, namely, fructose-2,6-bisphosphate. Figure 13–31 also identifies two control signals associated with the hexose phosphate intermediates: (1) the feedback inhibition by G6P on hexokinase and (2) the feed-forward activation by 1,6FBP on pyruvate kinase.

Hormonal Control

In several mammalian tissues the rates of glycolysis and glycogenesis are controlled by the action of hormones. Three major examples are *epinephrine* (also called adrenaline), *insulin,* and *glucagon.* Epinephrine (see Figure 13–32) is derived from the metabolism of tyrosine in the adrenal gland; insulin and glucagon are polypeptides (see Table 3–9) produced in the pancreas. The regulatory effects of each of these hormones begins with the binding of hormone to its specific *membrane receptor* to generate a signal that ultimately controls key enzymes in the cell.

Before examining each hormone, we will first consider some details about a topic mentioned before but never fully developed, namely, *protein phosphatases.* Unlike protein kinases, which often show a preference to phosphorylate only a small number of specific target proteins, individual protein phosphatase enzymes are less specific and can dephosphorylate several proteins. Although much less is known about protein phosphatases than protein kinases, it appears that many cells contain a major protein phosphatase capable of existing and operating as a monomer (MW $\approx$ 35,000) or an oligomer (MW $\approx$ 125,000). Furthermore, a hormone signal for the control of protein kinase activity can also regulate protein phosphatase activity.

Although different protein phosphatases may be regulated in different ways, the major phosphatase referred to above is directly controlled in the cell by a *protein phosphatase inhibitor* (PPI). The PPI substance is a cellular protein capable of binding to protein phosphatase (PP) to form a [PPI/PP] complex in which the phosphatase is inactive (see Figure 13–33). An interesting feature is that the PPI protein itself is controlled by phosphorylation, with phosphorylated PPI (that is, PPI–P) being the active form for binding to PP. The removal of phosphate from PPI–P in the complex triggers a conformational change in PPI favoring the dissociation of the complex to release the active protein phosphatase. Equally interesting, if not more so, is that the phosphatase catalyzing the dephosphorylation of PPI–P may in fact be protein phosphatase itself.

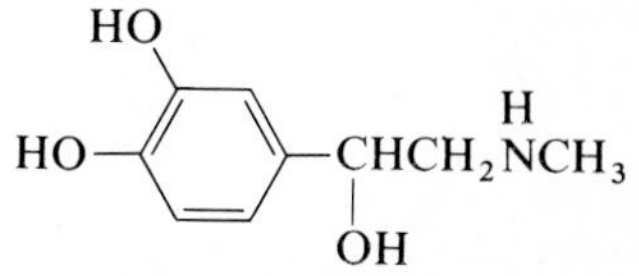

EPINEPHRINE   In response to various stimuli, such as low blood sugar or hypothalamus secretions, epinephrine is secreted from the adrenal gland (specifically, the adrenal medulla structure) and carried by the blood to target cells—primarily muscle. The effect on muscle is to increase the flow of hexose carbon through glycolysis to sustain the energy needs of muscle contraction. The biochemistry of epinephrine control occurs via a remarkable *regulatory cascade* involving 3′,5′-cyclic AMP acting as a secondary messenger.

The cascade (see Figure 13–34) begins at the exterior surface of the cell membrane with the binding of epinephrine to its specific receptor protein R. In turn, this event triggers the activation of adenylate cyclase, also localized in the membrane (refer to Figure 11–47). The activation of cyclase increases the production of cyclic AMP—an intracellular signal having two ultimate effects: (1) an increased rate of glycolysis and (2) a decreased rate of glycogenesis. The immediate effect of cyclic AMP is to activate the enzyme cAMP-dependent protein kinase (cAMP–dPK).

The active subunit of cAMP–dPK then catalyzes the phosphorylation of two target proteins: glycogen synthase/phosphorylase kinase (GSP kinase) and protein phosphatase inhibitor (PPI). In turn, the activated GSP kinase catalyzes the phosphorylation of its target proteins: glycogen synthase and glycogen phosphorylase. The complementary effects of these last two phosphorylations—deactivating synthase and activating phosphorylase—stimulate the unidirectional flow of carbon out of stored glycogen into glucose-1-phosphate. (The GSP kinase activity in muscle cells can also be stimulated by the release of stored Ca^{2+} from the reticulum compartment; see Note 13–5.)

The phosphorylation cascade is enforced by the other effect of cyclic AMP in triggering the activation of the protein phosphatase inhibitor. This effect

FIGURE 13–32 Epinephrine (adrenalin).

NOTE 13–5
The enzyme GSP kinase is a large oligomeric protein (MW $\approx 10^6$) having the structure $(\alpha\beta\gamma\delta)_4$. The β subunit is the target site for phosphorylation by cAMP–dPK. The δ subunit is identical to *calmodulin,* the Ca^{2+}-binding polypeptide we described earlier (see p. 459). Binding of Ca^{2+} to δ in $(\alpha\beta\gamma\delta)_4$ also activates the GSP kinase.

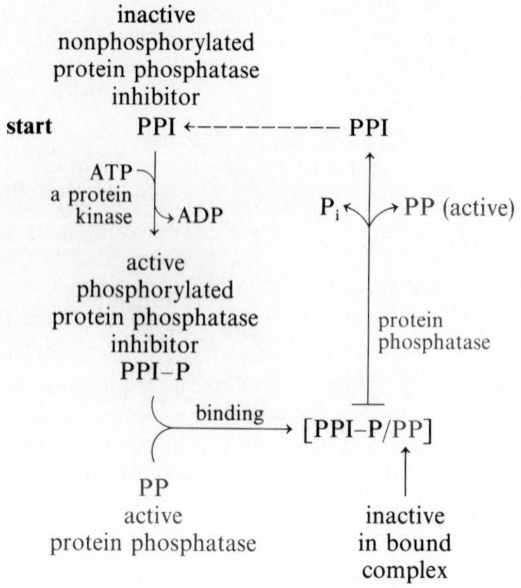

FIGURE 13–33 Control of protein phosphatase (PP) by protein phosphatase inhibitor (PPI). Although not shown here, the activity of the protein kinase is also regulated by another signal. When active PP is available, it will catalyze hydrolysis of various phosphoproteins, including PPI–P, as indicated.

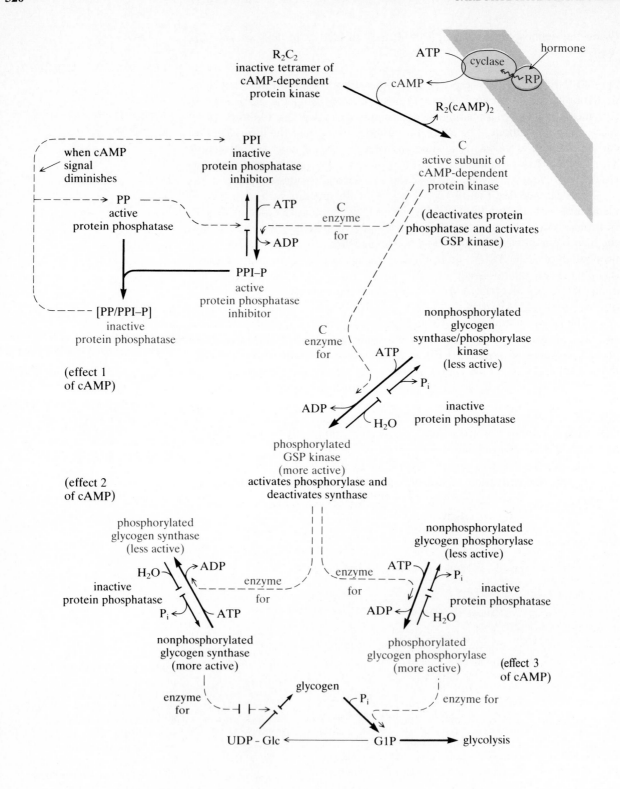

deactivates the protein phosphatase, thus preventing the reversal of the phosphorylation cascade.

With reduced levels of adrenal gland activity and the subsequent reduction of epinephrine blood levels, the hormone → receptor ↝ cyclase signal diminishes, and the rate of cAMP formation declines. In addition, the action of *cyclic phosphodiesterase* further reduces the cAMP level by hydrolyzing cAMP to AMP. The decline in the intracellular concentration of cAMP short-circuits the activation of cAMP–dPK, and all protein phosphorylations are diminished. Moreover, the emergence of the active protein phosphatase triggers the dephosphorylation of all previously phosphorylated proteins, returning synthase and phosphorylase to prehormonal nonphosphorylated forms.

INSULIN   In response to a rise in the level of glucose in blood, B-type islet cells of the pancreas produce and secrete the protein hormone insulin. In turn, insulin acts on various types of cells, signaling different cells to do different things. The primary target cells are those of muscle, liver, and adipose (fat) tissue. In each target cell the insulin effect begins with the binding of insulin to a specific receptor protein in the cell membrane. This binding stimulates glucose entry into the cell and, subsequently, also stimulates the use of the intracellular glucose for glycogen synthesis, protein synthesis, and lipid synthesis. The result is the reduction of blood glucose to a normal level—a signal for the pancreas to stop producing insulin.

The reasons for the effects of insulin are still not yet completely understood. There is more to an insulin effect than merely suggesting that various metabolic processes occur faster because more glucose is pumped into the cell. For example, the stimulation of glycogen synthesis may be attributed to an insulin-dependent activation of cAMP phosphodiesterase, which would lower the intracellular level of cAMP. This reduced level, in turn, would stimulate the conversion of the inactive glycogen synthase to the active glycogen synthase and also dilute the activation of glycogen phosphorylase. The net effect is the stimulation of G1P → glycogen.

GLUCAGON   In response to low blood glucose, the presence of epinephrine also stimulates the A-type islet cells of pancreas to secrete the polypeptide hormone glucagon. The primary target tissue of glucagon regulation is liver, where the effect is to *stimulate the production of glucose* for release to the blood and subsequent delivery to other tissues. The glucose in liver is derived in two ways: (1) the degradation of liver glycogen and (2) the resynthesis of glucose from pyruvate, the latter originating from lactate or from the amino acid alanine. (When alanine is the carbon source, the term *gluconeogenesis* is used in reference to pyruvate →→→→ glucose.) Both processes converge on

◀ **FIGURE 13–34** Hormonal control of glycogen ⇌ G1P via the secondary-messenger role of cyclic AMP. Note that cAMP enhances (bold arrows) the rate of glycogen degradation and diminishes (broken arrows) the rate of glycogen formation. The ─┤├──→ symbolism indicates the lack of PP activity due to formation of the inactive PP/PPI–P complex. Symbol RP represents the membrane receptor protein for the hormone.

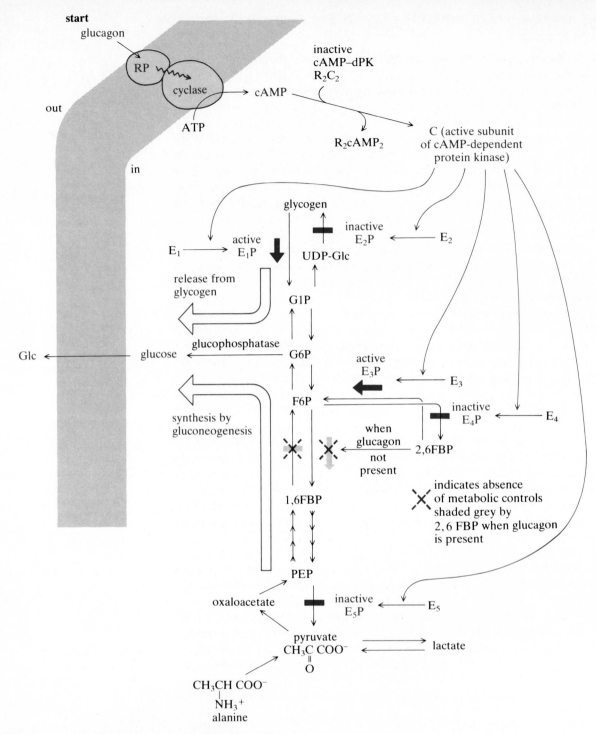

FIGURE 13–35 Explanation of the enhanced rate of glucose formation in a liver cell in response to glucagon stimulation. See Figure 13–34 for additional details of the glycogen → G1P interconversion. Symbols E_xP designate phosphorylated enzymes produced via the participation of cyclic AMP–dependent protein kinase. See text for an explanation of the alanine → pyruvate conversion. Symbol E_1 represents glycogen phosphorylase; E_2 is glycogen synthase; E_3 is fructose-2,6-bisphosphatase; E_4 is phosphofructokinase-2; E_5 is pyruvate kinase; RP is receptor protein for glucagon.

the formation of glucose-6-phosphate (G6P), which can then yield free glucose by the action of glucophosphatase. Although similar in part to the action of epinephrine on muscle cells, the glucagon regulation of liver cells does have two additional distinct features. One involves pyruvate kinase, and the second involves another regulatory substance, fructose-2,6-bisphosphate.

The glucagon cascade (see Figure 13–35) is initiated by glucagon binding to its specific membrane receptor protein. This binding triggers activation of adenylate cyclase, which elevates cyclic AMP, which activates cAMP-dependent protein kinase, which activates GSP kinase (and deactivates the protein phosphatase), which activates liver glycogen phosphorylase and deactivates liver glycogen synthase. In other words, we have the same scenario as depicted in Figure 13–34 for epinephrine stimulation of muscle. The result is an increase in the rate of formation of glucose-1-phosphate (G1P) and hence G6P. However, were these the only effects on liver by glucagon, liver would exhibit a high rate of glycolysis rather than a high rate of gluconeogenesis.

Enter the pyruvate kinase isoenzyme of liver. Recall (Figure 13–23) that this liver enzyme is also regulated by phosphorylation, being inactive when phosphorylated. The link to glucagon arises because the protein kinase catalyzing the phosphorylation of pyruvate kinase is cAMP-dependent protein kinase. Thus in liver cAMP not only increases the rate of glucose phosphate formation but also prevents its further catabolism by glycolysis.

This cAMP inhibition of F6P $\longrightarrow\rightarrow\longrightarrow$ pyruvate by deactivating pyruvate kinase is also supplemented by a cAMP activation of the reverse process (gluconeogenesis: pyruvate $\longrightarrow\rightarrow\longrightarrow$ F6P). This activation involves the biochemistry of *fructose-2,6-bisphosphate* (2,6FBP), recently discovered in 1980. It was previously undetected because it is present in extremely small quantities. In fact, under certain conditions the cellular concentration of 2,6FBP is virtually zero. Enzymes involved in the formation and degradation of 2,6FBP have also been isolated: Formation occurs via F6P + ATP → 2,6FBP + ADP, catalyzed by *phosphofructokinase-2;* degradation involves the action of *fructose-2,6-bisphosphatase* for 2,6FBP + H_2O → F6P + P_i (see Figure 13–36).

There is good evidence that the role of 2,6FBP is to control flux through the F6P $\rightleftharpoons$ 1,6FBP site of glycolysis/glycogenesis, with 2,6FBP acting as an *activator* of phosphofructokinase-1 and also as an *inhibitor* of fructose-1,6-bisphosphatase (see Figure 13–37). When glucose is abundant, combination of these effects by 2,6FBP represents a nonhormonal, metabolic control of carbohydrate metabolism that favors the degradation of hexose $\longrightarrow\rightarrow$ pyruvate.

What has this discussion to do with glucagon? Remarkably, it so happens that both enzymes of the F6P $\rightleftharpoons$ 2,6FBP interconversion are also controlled

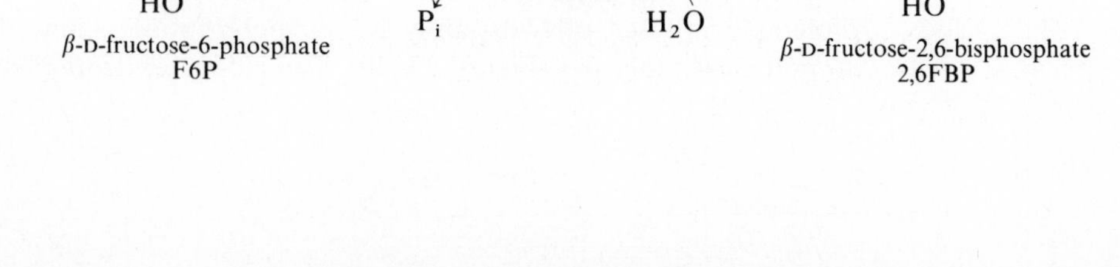

FIGURE 13–36 Formation and degradation of fructose-2,6-bisphosphate.

β-D-fructose-6-phosphate
F6P

phosphofructo-kinase-2

fructose-2,6-bisphosphatase

β-D-fructose-2,6-bisphosphate
2,6FBP

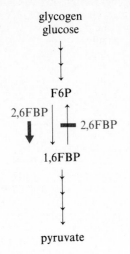

glycogen
glucose

F6P

2,6FBP

2,6FBP

1,6FBP

pyruvate

FIGURE 13–37 Control of carbohydrate metabolism by fructose-2,6-bisphosphate (2,6FBP). When 2,6FBP is present, it enhances the rate of glycolysis and diminishes the rate of glycogenesis. As shown in Figure 13–35, these 2,6FBP controls do not occur when liver cells respond to glucagon, which decreases the intracellular concentration of 2,6FBP.

in a complementary manner by phosphorylation, and the kinase enzyme involved is cAMP–dPK (see Figure 13–35). Phosphorylated phosphofructo-kinase-2 is deactivated, and phosphorylated fructose-2,6-bisphosphatase is activated. In other words, glucagon (via cAMP) *reduces* the intracellular concentration of the 2,6FBP regulator. Consequently, the 2,6FBP activation of glycolysis and inhibition of glycogenesis are removed.

To summarize:

- Glucagon activates adenylate cyclase for cAMP formation.
- Cyclic AMP activates cAMP-dependent protein kinase.
- The cAMP–dPK then catalyzes the phosphorylation of five different enzymes, thus controlling a total of five reactions:

Stimulating	*Inhibiting*
glycogen → G1P	UDP-Glc → glycogen
2,6FBP → F6P	PEP → pyruvate
	F6P → 2,6FBP

- All the reactions contribute to a high rate of glucose-6-phosphate formation.

13–8 CARBOHYDRATE CATABOLISM AND MUSCLE CONTRACTION

Major Proteins of Contraction

In 1954 A. F. Huxley and H. E. Huxley (not related) independently proposed a mechanism for muscle contraction involving the *sliding movement* of protein filaments within muscle tissue. Since then, many of the molecular details have been discovered. To describe the mechanism, we must first examine some of the details of the ultrastructure of muscle tissue.

As shown in Figure 13–38(a), muscle cells are bundles of longitudinal fibers called *myofibrils,* each composed of distinct repeating units called *sarcomeres* [see Figures 13–38(b) and 13–38(c)]. The ultrastructure of each sarcomere is due to horizontally aligned *myosin filaments* and *actin filaments,* containing the two contractile proteins actin and myosin.

Actin (see Figure 13–39) is a nearly spherical globular protein (MW ≈ 43,000), capable of extensive self-association under physiological conditions. In the actin filament several actin molecules associate in two strings of beads that intertwine to form a double-helix aggregate. In electron micrographs the thin actin filaments appear as faint wispy lines.

Two other important proteins are also associated with the actin filament: tropomyosin and troponin. *Tropomyosin* molecules (MW ≈ 66,000) are fibrous proteins composed of two purely helical polypeptide chains that also wrap around each other (see Figure 13–40). The elongated tropomyosin molecules

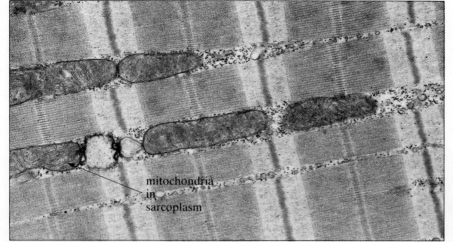

(a)

mitochondria in sarcoplasm

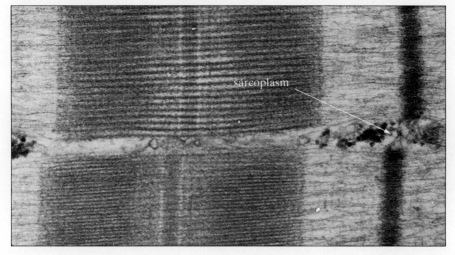

(b)

sarcoplasm

FIGURE 13–38 Ultrastructure of muscle. (a) Electron micrograph of myofibrils in a fiber of papillary muscle tissue of cat heart. The myofibrils run horizontally and are separated by the sarcoplasm, which contains mitochondria and some glycogen granules. The latter appear as black dots. Magnification 25,000 ×. (b) Electron micrograph of sarcomere unit. Magnification 77,800 ×. (c) Diagrammatic representation of sarcomere unit depicting the parallel alignment of myosin and actin filaments. Actin filaments occupy the light I band and penetrate some distance into the A band, where they can interact with projections of myosin filaments. The Z line corresponds to the joining of actin filaments from two adjacent sarcomeres. The M line corresponds to the joining of myosin filaments within the A band. [*Source:* Reproduced with permission from D. W. Fawcett, *An Atlas of Fine Structure* (Philadelphia: Saunders, 1966). Photographs in (a) and (b) provided by D. W. Fawcett.]

(c)

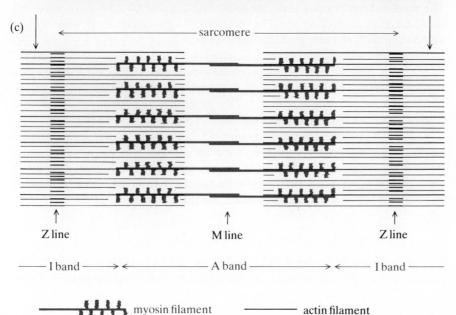

sarcomere

Z line   M line   Z line

I band   A band   I band

myosin filament   actin filament

525

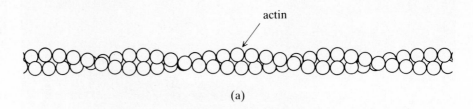

actin

(a)

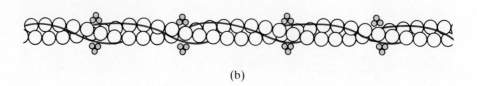

(b)

lie along the length of the actin filament, each tropomyosin molecule spanning
about seven actin monomers. The *troponin* protein complex of skeletal and
cardiac muscle contains three subunits: *troponin I* (I for inhibitor binding;
$MW \approx 22{,}000$), *troponin T* (T for tropomyosin binding; $MW \approx 37{,}000$), and
troponin C (C for Ca^{2+} binding; $MW \approx 18{,}000$). All three troponins contact
each other in the intact troponin complex, but only troponins T and I are in
contact with the actin filament—specifically bound to the tropomyosin molecule. As might be expected, troponin C is structurally very similar to calmodulin,
the major Ca^{2+}-binding protein in vertebrates.

Myosin ($MW \approx 460{,}000$) is a hexamer consisting of two heavy chains
($MW \approx 200{,}000$) and two pairs of light chains ($MW \approx 15{,}000–27{,}000$) and has
a very asymmetric structure (see Figure 13–41). The amino terminal portion of
each heavy chain gives rise to a folded, globular conformation (the myosin
"head"), while the remainder of each chain forms an extended α-helix fibril region
(the myosin "tail"), with the two helical segments intertwining with each other.
The myosin head regions carry an active site capable of *ATPase activity* and
also have an *actin-binding site*. The two pairs of light chains are in noncovalent
association with the two globular head regions.

In a myosin filament several hundred myosin hexamers are bundled together, with globular headpieces protruding on the surface of the filament in a
regular pattern. The overlapping of two such bundles via interactions of the
fibril tails from each bundle completes the molecular architecture of the myosin
filament, seen as thick electron-dense lines in an electron micrograph.

In each sarcomere [refer to Figures 13–38(b) and 13–38(c)], myosin filaments are located only in a region called the *A band*, whereas actin filaments
traverse the region called the *I band* and protrude into the A band. It is the A
band region where actin and myosin filaments are close enough to engage in
physical contact with each other during the process of contraction.

Calcium and ATP: Drivers of the Contraction Process

When muscle (skeletal and cardiac) is relaxed, ADP is bound to the
ATPase site on the myosin head, but the conformation of tropomyosin prevents

subunits of
troponin

Ca^{2+}

tropomyosin

C

I   T

actin   actin   actin

FIGURE 13–40 Close-up
representation of the troponin/
tropomyosin/actin interactions on
the actin filament.

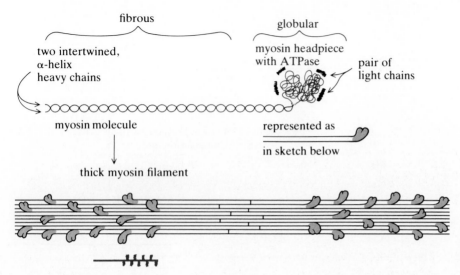

FIGURE 13–41 Myosin molecules and their association to form polymolecular myosin filaments.

an action contact between the actin and myosin filaments. Upon neural stimulation, stored Ca^{2+} is released into the sarcoplasm from the sarcoplasmic reticulum, with the sarcoplasm $[Ca^{2+}]$ rising from about 10^{-8}–$10^{-7}M$ to about $10^{-5}M$. The binding of Ca^{2+} to troponin C provides the signal to trigger a conformational change in the troponin complex, in turn triggering a change in tropomyosin, in turn triggering a change in the alignment of actin molecules. The final outcome is the formation of an *actomyosin complex* wherein actin monomers in the actin filament now make contact with myosin head regions of the myosin filament.

In the actomyosin complex a succession of cyclic events occur (see the next section) wherein the chemical energy of ATP is expended to support the mechanical process of contracting the sarcomeres of the myofibril. When the excitation subsides, Ca^{2+} is actively pumped out of the sarcoplasm back into the reticulum, a process involving the operation of a Ca^{2+}-dependent ATPase located in the membrane of the sarcoplasmic reticulum.

The Ca^{2+} activation of contraction via the $Ca^{2+} \rightarrow$ troponin C $\rightarrow$ tropomyosin $\rightarrow$ actin $\rightarrow$ actomyosin route of communication is not applicable to all muscles. In the muscle of a molluscs regulation by Ca^{2+} is achieved differently: by direct binding of Ca^{2+} to the myosin headpieces, resulting in the activation of the ATPase site. In vertebrate smooth muscle the process is different still: Ca^{2+} binds to calmodulin, which then activates a protein kinase that catalyzes the phosphorylation of one of the myosin light chains, and this phosphorylation triggers a conformation change in the headpiece, activating the latent ATPase site of nonphosphorylated myosin.

Sliding-Filament Model

When muscle contracts (see Figure 13–42), neither the actin nor myosin filaments change in length. Rather, the actin filaments are longitudinally displaced deeper into the A band, thus shortening the sarcomere length. The dis-

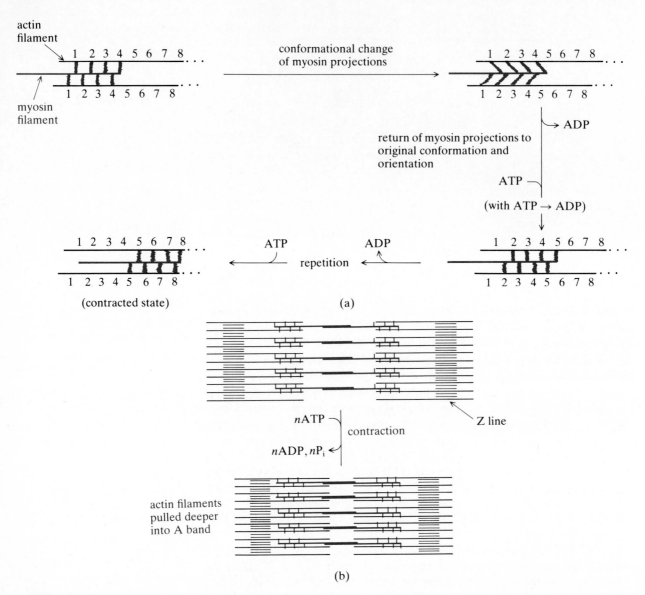

FIGURE 13–42 Mechanism of muscle contraction. (a) Power stroke events involving a single myosin filament (color). Begin your examination in the upper left. Although not shown, ADP is bound to each myosin head group. The numbers identify myosin contact sites to actin molecules in the actin filament. When stimulated to contract, a conformational change of the myosin projections causes them to shift position—a shift that provides the power stroke to displace the actin filament, pulling the actin filament to the left in this diagram. After dissociation of ADP occurs, the myosin projections undergo another conformational change that restores the original orientation but now gives a different set of contact sites with actin. The binding of ATP to the myosin projections and the subsequent ATPase action prepare the myosin filament for another power stroke. Repeating these events gives the contracted state. (b) Diagram representing the events in (a) occurring cooperatively throughout the entire sarcomere involving many myosin filaments.

placement force is provided by a reorientation of the myosin heads, which tilt, rotate, or pivot to the interior of the A band, pulling along the actin filaments with which they are in contact. One proposal is that this "power stroke" of the myosin filaments occurs after ADP dissociates from the ATPase site of the myosin heads.

What is ATP required for? After completion of a power stroke, ATP is proposed to bind at the vacated ATPase site to trigger the dissociation of the original actomyosin contact. This dissociation allows the myosin head to tilt, rotate, or pivot back to its original position—a process that may occur concurrently with the hydrolysis of ATP to ADP at the ATPase site. Once restored, the myosin head (with bound ADP again) makes a new actin contact, and the power stroke is repeated (several times), displacing the actin filament even deeper into the A band.

The ATP required is supplied almost exclusively by the catabolism of carbohydrates. In skeletal muscle, where the oxygen supply is low and the bulk of metabolism is therefore anaerobic, glycolysis supplies the ATP, with glycogen reserves acting as the main carbon substrate and lactate being the final product. In cardiac muscle, capable of more aerobic metabolism, the combined action of glycolysis through pyruvate formation and the citric acid cycle is more prevalent, with the bulk of the required ATP being supplied by mitochondrial oxidative phosphorylation. Consistent with this distinction between aerobic and anaerobic muscle tissues is the fact that cardiac muscle generally contains a much larger proportion of mitochondria than skeletal muscle does. Approximately 40% of the dry weight of heart represents mitochondria. Some heart cell mitochondria are visible in the micrograph displayed in Figure 13-38(a).

During prolonged periods of muscular activity, ATP is used directly after its formation. In resting periods the energy of ATP is stored in the form of the phosphoguanidine compound *creatine phosphate* (see Figure 13-43). During brief periods of contractile activity and in the initial phase of a prolonged stress, creatine phosphate serves as a high-energy (refer to Section 12-2) donor of phos-

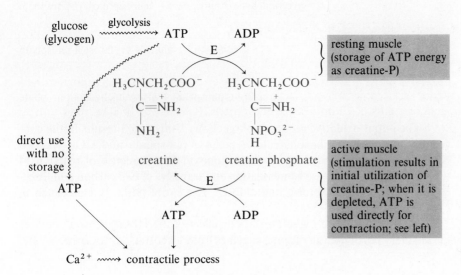

FIGURE 13-43 Creatine phosphate in muscle biochemistry. Symbol E represents creatine kinase (also called creatine phosphokinase).

phate to re-form ATP from ADP. Recall that an elevated blood level of an isoenzyme form of creatine kinase is used in the diagnosis of heart attack (see Section 4–5).

13–9   HEXOSE MONOPHOSPHATE SHUNT (PENTOSE PHOSPHATE PATHWAY)

Although most aerobic organisms utilize the combined reactions of glycolysis and the citric acid cycle as the main catabolic route for the complete oxidation of carbohydrates to CO_2, many also possess alternative pathways. First elucidated in the 1950s, one such pathway involves in part the conversion of glucose-6-phosphate to CO_2 and ribulose-5-phosphate, a pentose. Hence the entire sequence is called a *pentose phosphate pathway*. Because the pathway diverts G6P from metabolism via glycolysis, it is also called the *hexose monophosphate shunt* (HMS). We will use the HMS name. The HMS pathway is widely distributed (animals, plants, and bacteria) and is responsible for some important metabolic functions.

The following eight reactions (all occurring in the cytoplasm) comprise the HMS pathway:

oxidative (1 + 3)

1. $NADP^+ + \text{glucose-6-phosphate} \rightarrow \text{6-phosphogluconolactone} + NADPH + H^+$

2. $\text{6-phosphogluconolactone} + H_2O \rightarrow \text{6-phosphogluconate}$

3. $NADP^+ + \text{6-phosphogluconate} \rightarrow \text{ribulose-5-phosphate} + CO_2 + NADPH + H^+$

nonoxidative

4. $\text{ribulose-5-phosphate} \rightarrow \text{ribose-5-phosphate}$

5. $\text{ribulose-5-phosphate} \rightarrow \text{xylulose-5-phosphate}$

6. $\text{ribose-5-phosphate} + \text{xylulose-5-phosphate} \rightarrow$
   $\text{glyceraldehyde-3-phosphate} + \text{sedoheptulose-7-phosphate}$

7. $\text{sedoheptulose-7-phosphate} + \text{glyceraldehyde-3-phosphate} \rightarrow$
   $\text{fructose-6-phosphate} + \text{erythrose-4-phosphate}$

8. $\text{xylulose-5-phosphate} + \text{erythrose-4-phosphate} \rightarrow$
   $\text{fructose-6-phosphate} + \text{glyceraldehyde-3-phosphate}$

The initial oxidative sequence of reactions 1 through 3 results in the conversion of a hexose phosphate to a pentose phosphate and CO_2 via two $NADP^+$-dependent dehydrogenases. A nonoxidative sequence of reactions 4 through 8 involves pentose isomerization and transfers of two carbon and three-carbon units between ketoses and aldoses. The entire pathway is depicted in Figure 13–44 (pp. 532–533).

An analysis of the overall metabolic effect of the HMS pathway requires the auxiliary participation of some enzymes from glycolysis/glycogenesis. (*Note:*

Nearly all cells capable of pentose metabolism also carry out glycolysis. In other words, both pathways can operate simultaneously in the small cell. We will return to this point a little later.) Because of the branch in the pathway from ribulose-5-phosphate, we can conveniently examine net effects by considering the processing of more than one G6P molecule. As indicated in Figure 13–44 for 3G6P taken twice, the net chemical effect is the same as the combined operation of glycolysis and the citric acid cycle, namely, the equivalent of *one hexose is oxidized to CO_2*. Note, however, that each CO_2 originates from a different hexose.

Metabolic Functions

Owing to the operation of an enzyme that interconverts NADPH and NADH (see Figure 13–45), the HMS pathway under aerobic conditions can act as a catabolic, energy-yielding process. The NADH (from NADPH) can prime the process of oxidative phosphorylation to produce much ATP. However, the primary metabolic functions are believed to be (1) providing a *source of NADPH for anabolic pathways* that use enzymes that often require NADPH rather than NADH and (2) providing carbon intermediates for other areas of metabolism, notably *ribose-5-phosphate* for nucleic acid biosynthesis (see Section 18–7) and *erythrose-4-phosphate* for the biosynthesis of three amino acids, namely, phenylalanine, tyrosine, and tryptophan (see Section 18–5).

Extent of Operation and Regulation

In cells capable of degrading hexoses via both glycolysis and the HMS pathway, an obvious question concerns the distribution of metabolism. Intricate ^{14}C-tracing experiments have provided some understanding. In different animal cells the amount of glucose metabolism via the HMS pathway is quite variable. High levels of HMS activity (about 50%) have been measured in fat cells (adipose tissue) and mature red blood cells. Lower levels of activity have been measured in liver (5%–10%), skeletal and heart muscle (5%), brain (10%), thyroid (15%), and lung (15%).

In red blood cells (erythrocytes) the production of NADPH via pentose pathways is necessary to maintain high levels of *reduced glutathione* (GSH). The GSH in turn is required to sustain the structural integrity of the erythrocyte by preventing oxidative damage to cell proteins and to membrane lipids (refer to Figure 3–38 and Section 11–1). Severe anemia can result when such damage occurs. Unfortunately, individuals afflicted with the genetic metabolic disorder, called glucose-6-phosphate dehydrogenase syndrome, provide clinical evidence for these relationships. The problem is a deficiency of normal dehydrogenase activity, which obviously means the production of less-than-normal levels of NADPH.

Although we will not consider the details, the pentose pathway is also subject to metabolic regulation, with the key regulatory enzyme being glucose-6-phosphate dehydrogenase.

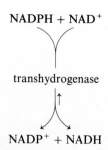

FIGURE 13–45 Transhydrogenation between NADP and NAD coenzymes. In cells lacking a metabolic pathway to produce NADPH, the enzyme operates in reverse to produce NADPH from NADH.

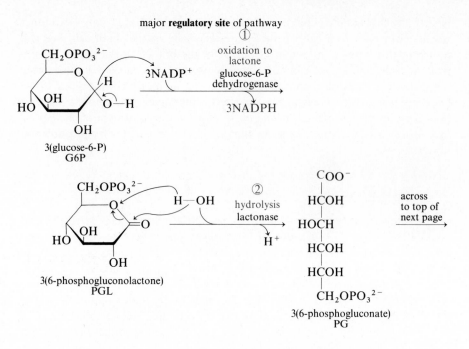

Reaction Summary

(1) $3G6P + 3NADP^+ \longrightarrow 3PGL + 3NADPH$
(2) $3PGL \longrightarrow 3PG$
(3) $3PG + 3NADP^+ \longrightarrow 3Ru5P + 3CO_2 + 3NADPH$
(4) $1Ru5P \longrightarrow 1R5P$
(5) $2Ru5P \longrightarrow 2Xu5P$
(6) $1R5P + 1Xu5P \longrightarrow 1GAP + 1S7P$
(7) $1GAP + 1S7P \longrightarrow 1F6P + 1E4P$
(8) $1E4P + 1Xu5P \longrightarrow 1F6P + 1GAP$

sum: $3G6P + 6NADP^+ \longrightarrow 2F6P + 3CO_2 + GAP + 6NADPH$

then from isomerase:

$2F6P \longrightarrow 2G6P$

sum: $3G6P + 6NADP^+ \longrightarrow 2G6P + 3CO_2 + GAP + 6NADPH$
or $6G6P + 12NADP^+ \longrightarrow 4G6P + 3CO_2 + 2GAP + 12NADPH$ **A**

then from reversal of glycolysis:

$GAP \longrightarrow DHAP$
$GAP + DHAP \rightarrowtail\!\!\!\rightarrow G6P$

net: $2GAP \longrightarrow G6P$ **B**

overall: $6G6P + 12NADP^+ \longrightarrow 5G6P + 6CO_2 + 12NADPH$
(A + B)   each from
                                           different molecule
                                              of G6P

FIGURE 13–44 Eight reactions of the hexose monophosphate shunt and auxiliary enzymes. The sequence starts above and runs over to the next page. All reactions occur in the cytoplasm. The flow of carbon atoms from three different molecules of ribulose-5-phosphate is indicated with black C symbols in regular and boldface and color C symbols.

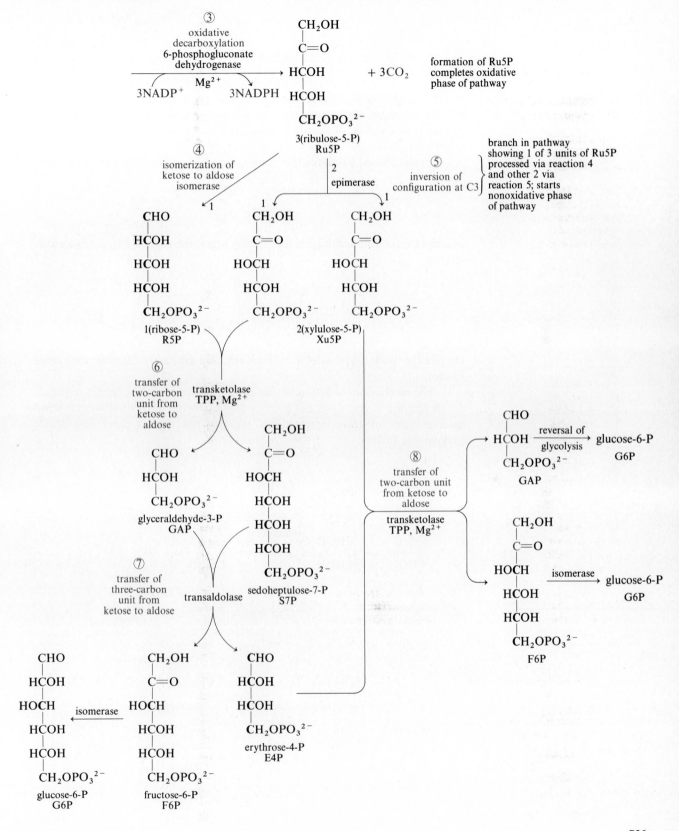

③ oxidative decarboxylation
6-phosphogluconate dehydrogenase

Mg^{2+}

$3NADP^+$ → $3NADPH$

CH_2OH
$C=O$
$HCOH$
$HCOH$
$CH_2OPO_3^{2-}$

$+ 3CO_2$

3(ribulose-5-P)
Ru5P

formation of Ru5P completes oxidative phase of pathway

④ isomerization of ketose to aldose isomerase

2 epimerase

⑤ inversion of configuration at C3

branch in pathway showing 1 of 3 units of Ru5P processed via reaction 4 and other 2 via reaction 5; starts nonoxidative phase of pathway

1

CHO
$HCOH$
$HCOH$
$HCOH$
$CH_2OPO_3^{2-}$

1(ribose-5-P)
R5P

1

CH_2OH
$C=O$
$HOCH$
$HCOH$
$CH_2OPO_3^{2-}$

1

CH_2OH
$C=O$
$HOCH$
$HCOH$
$CH_2OPO_3^{2-}$

2(xylulose-5-P)
Xu5P

⑥ transfer of two-carbon unit from ketose to aldose

transketolase
TPP, Mg^{2+}

CHO
$HCOH$
$CH_2OPO_3^{2-}$

glyceraldehyde-3-P
GAP

CH_2OH
$C=O$
$HOCH$
$HCOH$
$HCOH$
$HCOH$
$CH_2OPO_3^{2-}$

sedoheptulose-7-P
S7P

⑦ transfer of three-carbon unit from ketose to aldose

transaldolase

⑧ transfer of two-carbon unit from ketose to aldose

transketolase
TPP, Mg^{2+}

CHO
$HCOH$
$CH_2OPO_3^{2-}$

GAP

reversal of glycolysis → glucose-6-P
G6P

CH_2OH
$C=O$
$HOCH$
$HCOH$
$HCOH$
$CH_2OPO_3^{2-}$

F6P

isomerase → glucose-6-P
G6P

CHO
$HCOH$
$HOCH$
$HCOH$
$HCOH$
$CH_2OPO_3^{2-}$

glucose-6-P
G6P

isomerase

CH_2OH
$C=O$
$HOCH$
$HCOH$
$HCOH$
$CH_2OPO_3^{2-}$

fructose-6-P
F6P

CHO
$HCOH$
$HCOH$
$CH_2OPO_3^{2-}$

erythrose-4-P
E4P

Transaldolase and Transketolase

There are two distinctive enzymes of the HMS pathway: *transaldolase* and *transketolase*. Both enzymes catalyze the same general type of reaction: The transfer of a small carbon fragment from a donor ketose to an acceptor aldose to yield a new aldose of shorter chain length and a new ketose of longer chain length.

Transketolase catalyzes the transfer of a two-carbon glycoaldehyde unit:

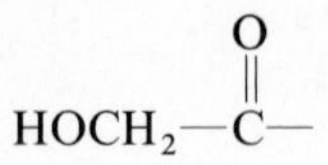

$$HOCH_2-\overset{\overset{\displaystyle O}{\|}}{C}-$$

whereas transaldolase catalyzes the transfer of a three-carbon dihydroxyacetone unit:

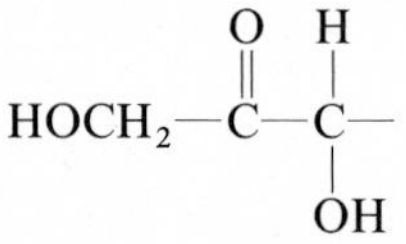

$$HOCH_2-\overset{\overset{\displaystyle O}{\|}}{C}-\underset{\underset{\displaystyle OH}{|}}{\overset{\overset{\displaystyle H}{|}}{C}}-$$

Despite the similarity, the details of each reaction are vastly different (see Figure 13–46).

The mechanism of action of transaldolase is very much like that of the aldolase enzyme of glycolysis and glycogenesis involving the formation of a Schiff base, enzyme-substrate intermediate. We will not discuss the transaldolase enzyme any further.

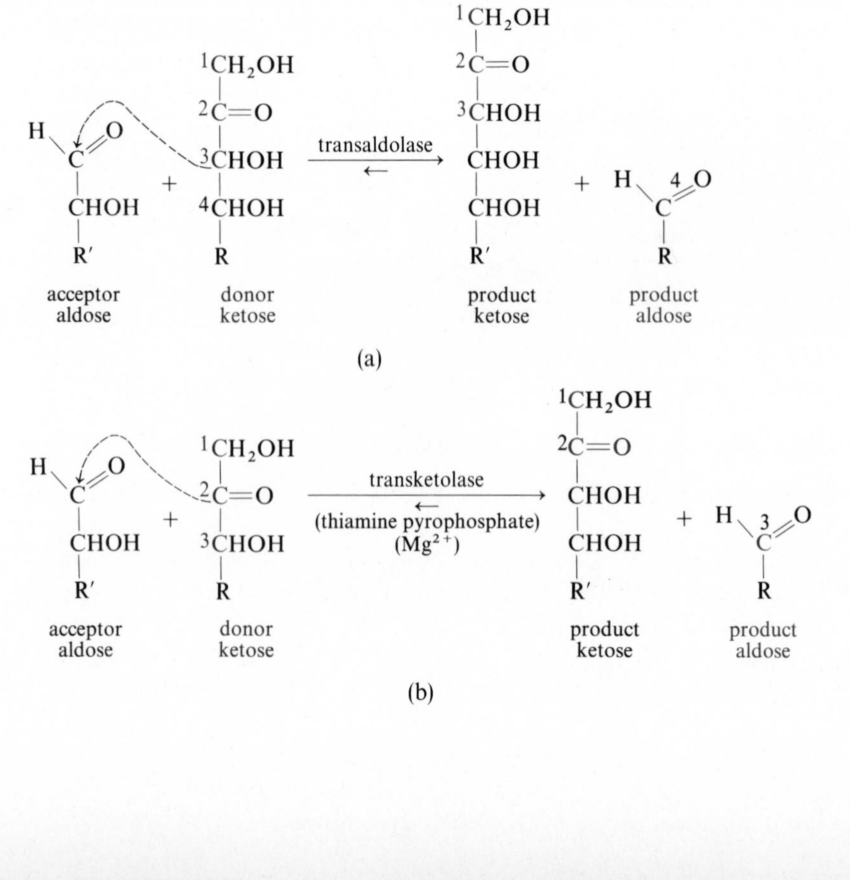

(a)

FIGURE 13–46 General ketose/aldose reactions. (a) Transfer of a C_3 fragment from ketose to aldose. (b) Transfer of a C_2 fragment from ketose to aldose.

(b)

Transketolase activity involves the obligatory participation of a metal ion cofactor (Mg^{2+} is optimal) and, more importantly, *thiamine pyrophosphate* (TPP) (see Figure 13–47), which is the metabolically active coenzyme form of thiamine (vitamin B_1). Because of a mechanistic relationship of the transketolase reaction to other TPP-dependent reactions, let us consider the latter topic as a unit. In so doing, we will not only explain the transketolase reaction but also describe events associated with the TPP-dependent conversions of pyruvate to acetaldehyde (the next-to-last reaction of alcohol fermentation) and the conversion of pyruvate to acetate (a very crucial reaction that links glycolysis to the citric acid cycle).

FIGURE 13–47 Thiamine pyrophosphate.

In Table 5–2 thiamine pyrophosphate was described as a coenzyme that participates in *acyl group,*

$$R-\overset{\overset{\displaystyle O}{\|}}{C}-$$

transfer reactions. More specifically, the TPP coenzyme participates in two types of reactions: (1) the decarboxylation of alpha keto acids and (2) the formation of alpha hydroxy carbonyl linkages, that is,

$$-\overset{\overset{\displaystyle O}{\|}}{C}-\overset{\overset{\displaystyle OH}{|}}{\underset{|}{C^{\alpha}}}-$$

The transketolase reaction is represented by type 2.

Certain features apply to each type of reaction (see Figure 13–48). The coenzyme is bound to the enzyme through several noncovalent interactions, which may include a bridged metal ion. This association with the enzyme enhances the reactivity of the coenzyme, which involves the ionization of $C-H$ in the thiazole moiety of TPP to yield a highly reactive *carbanion species*. The carbonyl carbon in the substrate of the reaction then forms a covalent adduct by condensing at the carbanion carbon. The subsequent fate of this ternary complex of enzyme-(Mg^{2+})TPP-substrate is then governed by the catalytic specificity of the enzyme, as shown in the three paths of the diagram.

The transketolase enzyme (path a in Figure 13–48) directs an electron rearrangement that results in the cleavage of a carbon-carbon bond in the original ketose to release the smaller aldose product, with the two-carbon fragment corresponding to the glycoaldehyde

$$HOCH_2\overset{\overset{\displaystyle O}{\|}}{C}-$$

terminus of the original ketose still attached. After an electron rearrangement to yield another active carbanion, a new carbon-carbon bond is formed with the carbonyl carbon of the aldose substrate. A final electron rearrangement releases the product ketose and the original state of the enzyme (coenzyme).

The paths catalyzed by pyruvate decarboxylase and pyruvate dehydrogenase represent alpha keto acid decarboxylation (path b in Figure 13–48). The events associated with acetaldehyde formation have been diagramed.

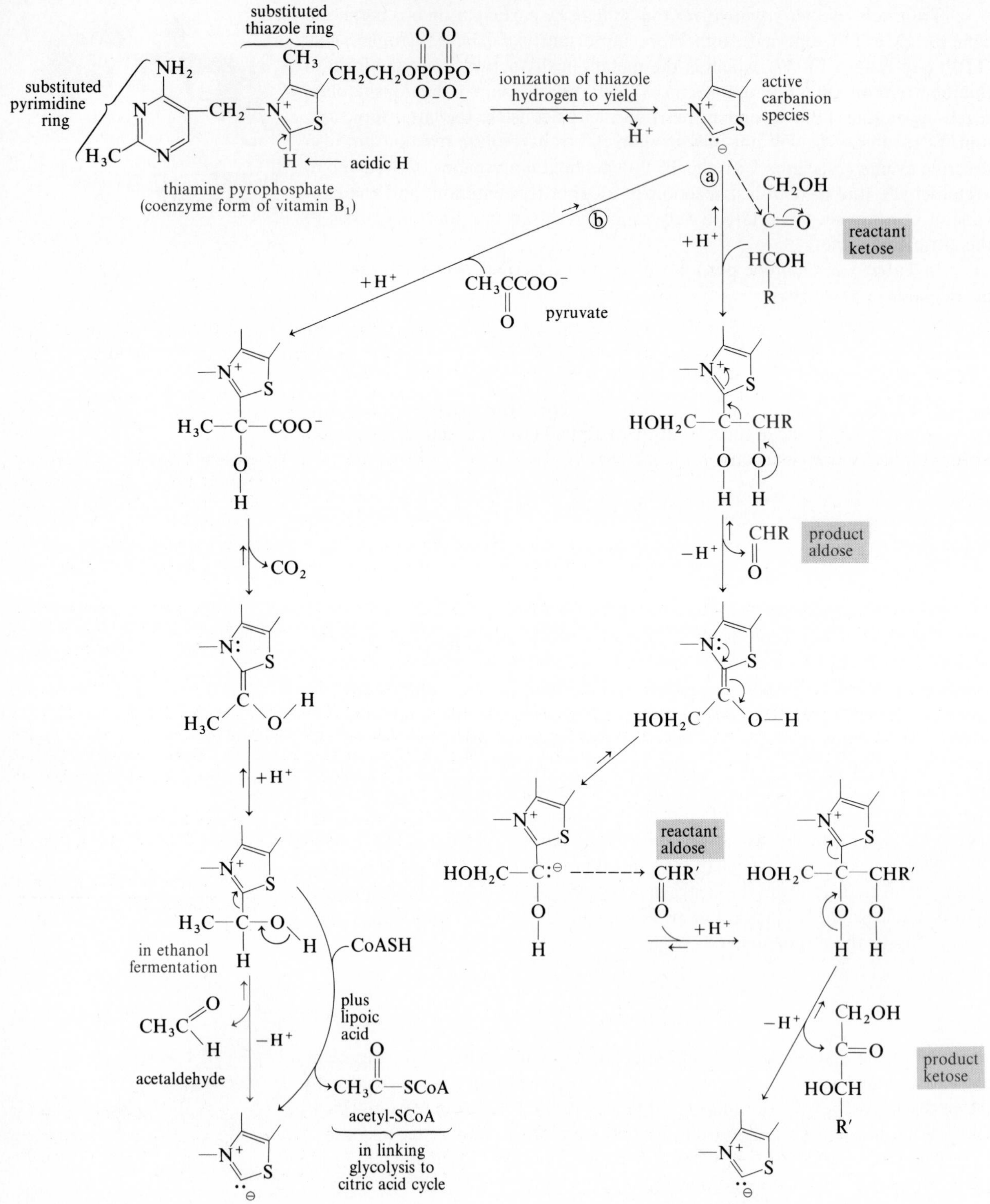
substituted thiazole ring
substituted pyrimidine ring
acidic H
thiamine pyrophosphate
(coenzyme form of vitamin B1)
ionization of thiazole hydrogen to yield
active carbanion species
reactant ketose
pyruvate
product aldose
reactant aldose
product ketose
in ethanol fermentation
acetaldehyde
CoASH
plus lipoic acid
acetyl-SCoA
in linking glycolysis to citric acid cycle

◄ **FIGURE 13–48** Chemistry of the coenzyme participation of thiamine pyrophosphate. Path a is an outline of TPP participation in the transketolase reaction. Path b applies to the decarboxylation of an α-keto acid (specifically, pyruvate). The production of acetaldehyde applies to the operation of pyruvate decarboxylase (see Figure 13–3). The production of acetyl-SCoA by a more complicated route involving lipoic acid applies to the operation of pyruvate dehydrogenase, which is discussed in Section 14–3 (see Figure 14–16).

LITERATURE

ADELSTEIN, R. S., and E. EISENBERG. "Regulation and Kinetics of the Actin-Myosin-ATP Interactions." *Annu. Rev. Biochem.,* **49,** 921–956 (1980). A detailed review article on the molecular events in muscle contraction, including regulation by calcium.

CHOCK, P. B., S. G. RHEE, and E. R. STADTMAN. "Interconvertible Enzyme Cascades in Cellular Regulation." *Annu. Rev. Biochem.,* **49,** 813–844 (1980). A review article with current coverage of the glycogen phosphorylase and glycogen synthase enzymes.

CLARKE, M., and J. A. SPUDICH. "Nonmuscle Contractile Proteins: The Role of Actin and Myosin in Cell Motility and Shape Determination." *Annu. Rev. Biochem.,* **46,** 797–822 (1977). Contractile proteins are present in most types of nonmuscle cells. This article reviews their occurrence, structure, and function.

CUNNINGHAM, E. B. *Biochemistry: Mechanisms of Metabolism.* New York: McGraw-Hill, 1978. A novel textbook highlighting the chemical details of enzyme action. The enzymes of glycolysis, glycogenesis, and pentose pathways are described in Chapter 9.

FLETTERICK, R. J., and P. C. HEINNICK. "The Structures and Related Function of Phosphorylase *a*." *Annu. Rev. Biochem.,* **49,** 31–62 (1980). A review article.

GOLDBERG, N. D. "Cyclic Nucleotides and Cell Function." *Hosp. Pract.,* **9,** 127–142 (May 1974). Good discussion of the central regulatory role of protein kinase, which is proposed to participate in several other processes besides glycogen phosphorylase activation.

HERS, H. G. "The Control of Glycogen Metabolism in the Liver." *Annu. Rev. Biochem.,* **45,** 167–189 (1976). A thorough review article.

HERS, H. G., L. HUE, and E. VAN SCHAFTINGEN. "Fructose-2,6-Biophosphate." *Trends Biochem. Sci.,* **7,** 329–331 (1982). A short review article.

HORECKER, B. L. "Transaldolase and Transketolase." In *Comprehensive Biochemistry,* edited by M. Florkin and E. H. Stotz, vol. 15. Amsterdam, New York: Elsevier, 1973. A review article on the mechanism of action of these two important enzymes in carbohydrate metabolism.

HOYLE, G. "How Is Muscle Turned On and Off?" *Sci. Am.,* **22,** 84–93 (1970). An article describing the role of calcium ion in the contraction and relaxation of muscle tissue.

KLEE, C B., T. H. CROUCH, and P. G. RICHMAN. "Calmodulin." *Annu. Rev. Biochem.,* **49,** 489–515 (1980). A review article of structure and cellular functions of this protein.

MURRAY, J. M., and A. WEBER. "The Cooperative Action of Muscle Proteins." *Sci. Am.,* **230,** 58–71 (1974). A good introductory discussion of current ideas regarding the molecular events of muscle contraction—the interactions of myosin, actin, troponin, and tropomyosin, controlled by calcium ions.

ROACH, P. J. "Functional Significance of Enzyme Cascade Systems." *Trends Biochem. Sci.,* **2,** 84–87 (1977). A short review article.

SEGAL, H. L. "Enzymatic Interconversion of Active and Inactive Forms of Enzymes." *Science,* **180,** 25–32 (1973). A good review article on this subject.

EXERCISES

13–1. What is meant by the statement that the reduced forms of nicotinamide adenine dinucleotides have complementary roles in catabolism and anabolism?

13–2. What is the biological significance of catabolic reactions that result in the formation of ATP?

13–3. If radioactive glucose, labeled in positions 3 and 4 with ^{14}C, was incubated with a cell-free liver homogenate under anaerobic conditions, what positions in the lactate produced would be labeled with ^{14}C?

13–4. Verify that $\Delta G^{\circ\prime} = -45.4$ kcal/mole for

$$\text{glucose} + 4P_i + 2NAD^+ \xrightarrow{\text{glycolysis}} 2\text{lactate} + 4P_i + 2NAD^+$$

13–5. If pure (nondenatured) samples of enzymes 1A, 1B, 2, 3, 4, 5, 6, 7, 8, 9, 10, and 11 (see the code in Figure 13–3) were incubated at 37°C in a buffered system (pH 7) containing glycogen, ATP, Mg^{2+}, and NAD^+, very little lactate would be produced. Explain.

13–6. How many oxidoreductases, transferases, hydrolases, lyases, isomerases, and ligases participate in the conversion of glucose to ethanol?

13–7. The metabolism of sucrose first involves the action of sucrose phosphorylase to yield glucose-1-phosphate and fructose. Assuming that both glucose-1-phosphate and fructose are further metabolized to lactate, (a) how many ATPs would be required, and (b) how many ATPs would be produced?

13–8. Under a growth condition resulting in the production of ethyl alcohol via the glycolytic pathway in fermenting bacteria, what internal coupling mechanism involving NAD will parallel that in muscle cells actively degrading carbohydrate under the same condition? What type of condition is in effect?

13–9. If 0.001 mole of 3-phosphoglyceraldehyde is incubated with 0.005 mole of P_i, 0.0001 mole of NAD^+, and 3-phosphoglyceraldehyde dehydrogenase, a reaction will occur that will reach equilibrium rather quickly. This equilibrium is characterized by the presence of a considerable amount of the 3-phosphoglyceraldehyde originally added. Predict what will happen to this equilibrium mixture if 0.005 mole of pyruvate and lactate dehydrogenase is added.

13–10. Write a complete balanced equation that best describes the net catabolism of mannose to ethanol.

13–11. Arrange the following proteins in the order in which they participate (beginning with the action of adrenalin) in the conversion of glycogen to glucose-1-phosphate: glycogen synthase/phosphorylase kinase (active and inactive forms), phosphorylase b, adenylate cyclase (active and inactive forms), phosphorylase a, and cyclic AMP–dependent protein kinase (active and inactive forms).

13–12. Explain the following statement: Phosphofructokinase catalyzes the committed step of glycolysis.

13–13. Explain the meaning of the following term: complementary metabolic regulation of the interconversion of two substances.

13–14. Write a complete net reaction for the enzymatic conversion of two lactate units into a glucose unit of glycogen. Assume that the sequence will occur in a eukaryotic cell and that phosphoenolpyruvate carboxykinase is located in the cytoplasm.

13–15. Does the concept of internal coupling (as described for NAD in glycolysis in Section 13–2) apply to the pathway of glycogenesis? Examine the pathway you worked out for Exercise 13–14.

13–16. The active form of cyclic AMP–dependent protein kinase increases the rate of glycolysis but decreases the rate of glycogenesis. To some students this dual role is baffling—how can the active form of the same enzyme activate one process but inhibit another? What is your explanation?

13–17. Is the net requirement of metabolic energy as ATP for the production of glucose-1-phosphate from lactate greater than, less than, or the same as the net production of ATP from the catabolism of glucose-1-phosphate to lactate? (In solving this problem, remember that all of the purine and pyrimidine nucleotides are thermodynamically equivalent. In addition, assume that the reactions are those that would occur in eukaryotic cells. Have you done 13–14 yet?)

13–18. In thermodynamic terms, what is the significance of the reactions between 1,3-bisphosphoglycerate and fructose-6-phosphate during glycogenesis?

13–19. If ribose-5-phosphate, uniformly labeled with radioactive carbon, was incubated in a suitably buffered solution containing xylulose-5-phosphate (no ^{14}C), thiamine pyrophosphate, Mg^{2+}, and transketolase, what two new carbohydrates would be produced, and what would be the labeling pattern of carbon 14 in each one?

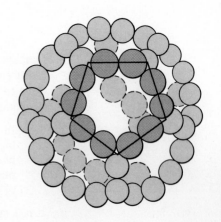

CHAPTER FOURTEEN
CITRIC ACID CYCLE

In the preceding chapter we referred to the citric acid cycle as the set of reactions responsible for the complete, aerobic degradation to carbon dioxide of the pyruvate carbons produced in glycolysis. In that context the citric acid cycle can be considered as a pathway of carbohydrate metabolism. However, its metabolic involvement is much broader in scope:

In nearly all organisms it serves as (1) the central pathway integrating the metabolic flow of carbon among all of the main classes of biomolecules and, in conjunction with the process of oxidative phosphorylation, as (2) the major source of metabolic energy in the form of ATP.

It is a metabolic pathway of immense significance.

The pathway is also called the tricarboxylic acid cycle (TCA cycle) or the Krebs cycle [after the metabolic studies of Hans Krebs (Nobel Prize, 1953) who elucidated the reaction sequence].

NOTE 14–1
The following abbreviations are used in Table 14–1:

OAA = oxaloacetate
CIT = citrate
ISOCIT = isocitrate
αKG = α-ketoglutarate
SUC-SCoA = succinyl-SCoA
SUC = succinate
FUM = fumarate
MAL = malate

14–1 OVERALL VIEW OF THE CITRIC ACID CYCLE

The citric acid cycle consists of eight steps, beginning with the condensation of oxaloacetate and the acetyl unit of acetyl-SCoA to yield citrate and ending with the reformation of oxaloacetate (see Figure 14–1). Symbolic representations of each reaction are shown in Table 14–1 (see Note 14–1 for abbreviations used in the table), along with a listing of the standard free-energy changes. The overall chemical effect is clearly summarized by the net equation, which states

TABLE 14–1 Reactions of the citric acid cycle

TRANSFORMATIONS OF THE CITRIC ACID CYCLE		$\Delta G^{\circ\prime}$ (kcal)
(1) C_2—SCoA(acetyl-SCoA) + C_4(OAA) + H_2O	$\xrightarrow{E_1} C_6$(CIT) + CoASH	−9.08
(2) C_6(CIT)	$\xrightarrow{E_2} C_6$(ISOCIT)	+1.59
(3) C_6(ISOCIT) + NAD^+	$\xrightarrow{E_3} C_5$(αKG) + CO_2 + NADH + H^+	−1.70
(4) C_5(αKG) + NAD^+ + CoASH	$\xrightarrow{E_4} C_4$(SUC-SCoA) + CO_2 + NADH + H^+	−9.32
(5) C_4(SUC-SCoA) + GDP + P_i	$\xrightarrow{E_5} C_4$(SUC) + GTP + CoASH	−2.12
(6) C_4(SUC) + FAD	$\xrightarrow{E_6} C_4$(FUM) + $FADH_2$	0
(7) C_4(FUM) + H_2O	$\xrightarrow{E_7} C_4$(MAL)	−0.88
(8) C_4(MAL) + NAD^+	$\xrightarrow{E_8} C_4$(OAA) + NADH + H^+	+6.69
Net: (Sum of 1 through 8) $CH_3\overset{\overset{\displaystyle O}{\|}}{C}SCoA + 3NAD^+ + FAD + 2H_2O + GDP + P_i \xrightarrow[\text{enzymes}]{8} 2CO_2 + 3NADH + 3H^+ + FADH_2 + GTP + CoASH$		−14.82

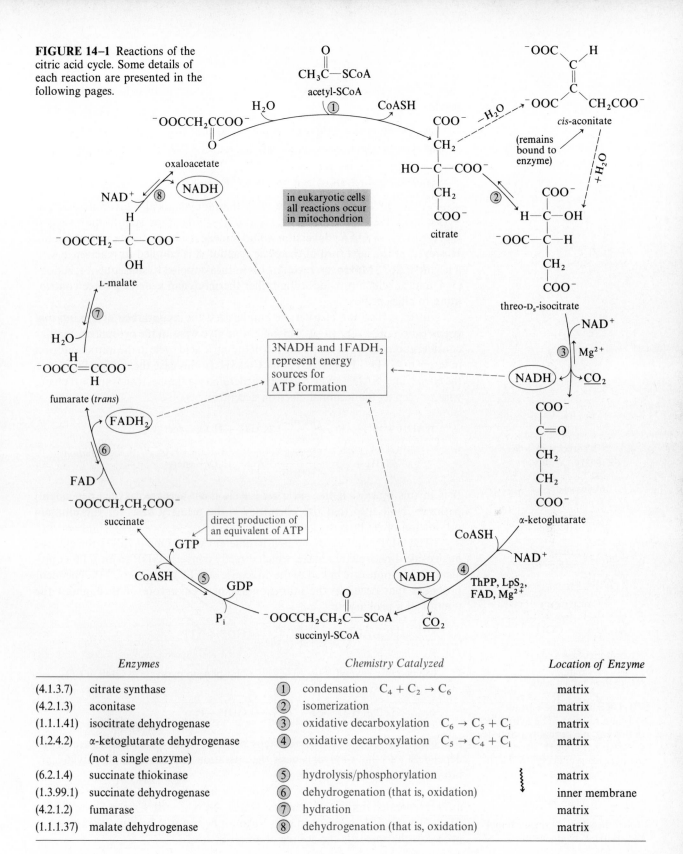

FIGURE 14–1 Reactions of the citric acid cycle. Some details of each reaction are presented in the following pages.

in eukaryotic cells all reactions occur in mitochondrion

3NADH and 1FADH$_2$ represent energy sources for ATP formation

direct production of an equivalent of ATP

Enzymes		Chemistry Catalyzed	Location of Enzyme
(4.1.3.7)	citrate synthase	① condensation $C_4 + C_2 \rightarrow C_6$	matrix
(4.2.1.3)	aconitase	② isomerization	matrix
(1.1.1.41)	isocitrate dehydrogenase	③ oxidative decarboxylation $C_6 \rightarrow C_5 + C_i$	matrix
(1.2.4.2)	α-ketoglutarate dehydrogenase (not a single enzyme)	④ oxidative decarboxylation $C_5 \rightarrow C_4 + C_i$	matrix
(6.2.1.4)	succinate thiokinase	⑤ hydrolysis/phosphorylation	matrix
(1.3.99.1)	succinate dehydrogenase	⑥ dehydrogenation (that is, oxidation)	inner membrane
(4.2.1.2)	fumarase	⑦ hydration	matrix
(1.1.1.37)	malate dehydrogenase	⑧ dehydrogenation (that is, oxidation)	matrix

541

that the operation of the cycle through one complete turn has the following results:

- The complete oxidation of the acetyl unit to two units of CO_2.
- The production of three units of reduced NADH and one unit of reduced $FADH_2$.
- The production of one unit of GTP.

Although the overall pathway is thermodynamically favored ($\Delta G_{net}^{\circ\prime} = -14.8$ kcal), the $\Delta G^{\circ\prime}$ of the last reaction ($+6.7$ kcal for step 8) may suggest that the MAL → OAA conversion would impede the flow of other intermediates. However, in the next turn of the cycle reaction 8 is coupled to reaction 1 with a negative $\Delta G^{\circ\prime}$. Moreover, reaction 1 is in turn coupled to subsequent reactions (3, 4, and 5), which provide still another thermodynamic driving force contributing to efficient flow.

In the preceding chapter we emphasized the importance of the internal regeneration of NAD^+ to sustain anaerobic glycolysis in the cytoplasm. Similar considerations apply to the operation of the citric acid cycle in the mitochondrion in terms of NAD^+, FAD, GDP, and CoASH. In this case the oxidized forms of the coenzymes NAD^+ and FAD are regenerated via the electron transport chain, with oxygen as the terminal electron acceptor:

$$NADH + H^+ + \tfrac{1}{2}O_2 \xrightarrow[\text{transport}]{\text{electron}} NAD^+ + H_2O + \text{energy}$$

$$FADH_2 + \tfrac{1}{2}O_2 \xrightarrow[\text{transport}]{\text{electron}} FAD + H_2O + \text{energy}$$

for ATP formation

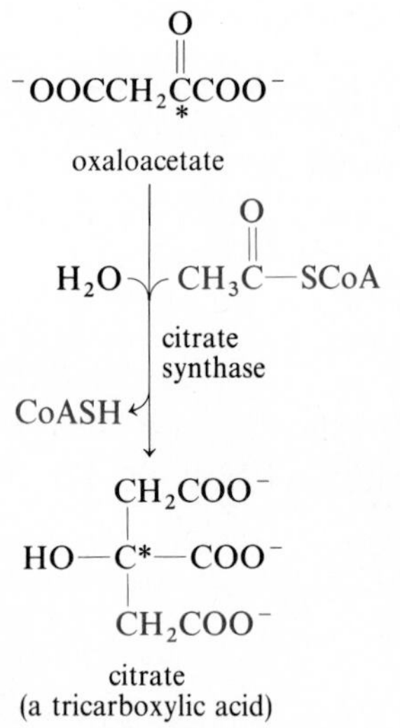

It is in this sense that the citric acid cycle is an aerobic (oxygen-dependent) pathway. Note also that this chemistry is the primary basis for the ultimate production of ATP energy.

The GDP can be regenerated in different ways. One involves the enzyme nucleoside diphosphate kinase, which directly converts GDP to an ATP equivalent. Other routes are linked to the direct use of GTP in specific GTP-dependent processes that occur in the mitochondria of a eukaryote or throughout the interior of a prokaryote:

$$GTP + ADP \xrightarrow[\text{diphosphate kinase}]{\text{nucleoside}} GDP + ATP$$

$$\text{oxaloacetate} + GTP \xrightarrow[\text{carboxykinase}]{\text{PEP}} \text{phosphoenolpyruvate} + GDP + CO_2$$

$$GTP + H_2O \xrightarrow[\text{enzymes}]{\text{GTPase}} GDP + P_i$$

For example, recall that several steps of polypeptide biosynthesis are GTP-dependent and that in glycogenesis the operation of phosphoenolpyruvate carboxykinase utilizes GTP.

Coenzyme A, an essential reactant consumed in αKG → SUC-SCoA (step 4), is regenerated in the next reaction of SUC-SCoA → SUC (step 5), see Figure 14–1. (The CoASH released in step 1 cannot be counted because it is reused in processes that provide the source of acetyl-SCoA.)

$$\overset{O}{\overset{\|}{^-OOCCH_2\overset{*}{C}COO^-}}$$

oxaloacetate

$$H_2O \quad CH_3\overset{O}{\overset{\|}{C}}{-}SCoA$$

citrate
synthase

CoASH

$$\begin{array}{c} CH_2COO^- \\ | \\ HO{-}C^*{-}COO^- \\ | \\ CH_2COO^- \end{array}$$

citrate
(a tricarboxylic acid)

FIGURE 14–2 The citrate synthase reaction. The active site of this enzyme probably promotes the abstraction of H^+ from the methyl group of acetyl-SCoA to yield the carbanion

$$:\overset{\ominus}{C}H_2COSCoA$$

which then attacks the carbonyl carbon (∗) of oxaloacetate.

14–2 INDIVIDUAL REACTIONS OF THE CITRIC ACID CYCLE

Our attention now turns to some details of the individual reactions of the cycle, some of which are very remarkable. In addition to having significance to the operation of the citric acid cycle, several of the details represent important principles in the operation of related enzymes and of enzymes in general.

Citrate Synthase

Despite the net degradative effect of the cycle, the first reaction involves the formation of a C—C bond (see Figure 14–2). Moreover, the reaction does not require ATP. The energy needed is provided by the hydrolysis of the thioester linkage, which in this case is also responsible for enhancing the reactivity of the alpha carbon in the acetyl group.

Typical of mammalian sources, the citrate synthase enzyme of pig heart is oligomeric (MW = 98,000), composed of two identical subunits. Catalytically, the enzyme is quite specific for acetyl-SCoA and oxaloacetate as substrates, with fluoroacetyl-SCoA and fluorooxaloacetate being the only known alternative substrates. (*Fluoroacetate* is found in certain locoweeds and is used as a coyote poison, with the poisonous effect due to its conversion to *fluorocitrate*, which then is a potent competitive inhibitor of aconitase, the next enzyme of the citric acid cycle.) There is evidence that citrate synthase is an important regulatory enzyme of the citric acid cycle. Some particularly significant effects include a strong competitive inhibition by ATP (competitive toward acetyl-SCoA) and separate allosteric inhibitions by NADH, succinyl-SCoA, and palmityl-SCoA (see Figure 14–3). The significance of these kinetic controls to the regulation of metabolism will be explored later.

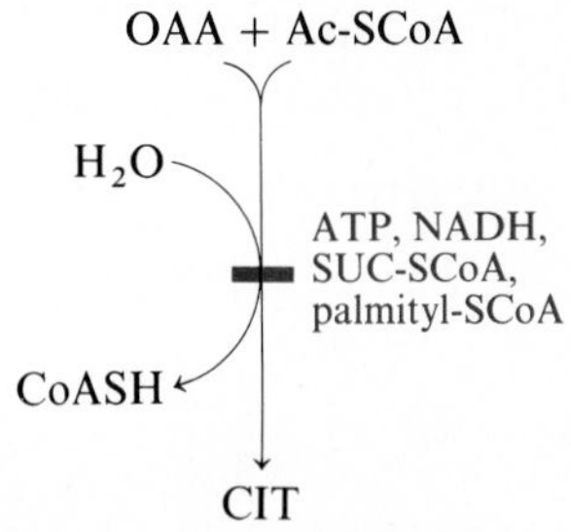

Aconitase

Proceeding through *cis*-aconitate, an enzyme-bound intermediate, the —OH of citrate is moved to one of the —CH_2— carbons via a sequence of dehydration and rehydration. Both processes exhibit stereospecific operation, and only one stereoisomer of isocitrate is produced, namely, threo-D_s-isocitrate (see Note 14–2 and Figure 14–4).

FIGURE 14–3 Metabolic controls on citrate synthase.

NOTE 14–2

The *threo* and *erythro* designations refer to the spatial orientations of the —OH and —COO^- groups relative to each other, using the structures of the sugars threose and erythrose as frames of reference. Threo implies a trans orientation and erythro a cis orientation. The D_s notation refers to the specific configuration of the —OH carbon, using the structure of D-serine as a frame of reference.

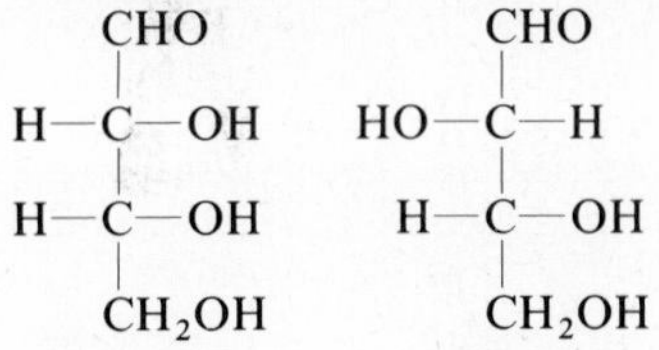

FIGURE 14–4 Stereoisomers of isocitrate. The aconitase enzyme produces only one (color).

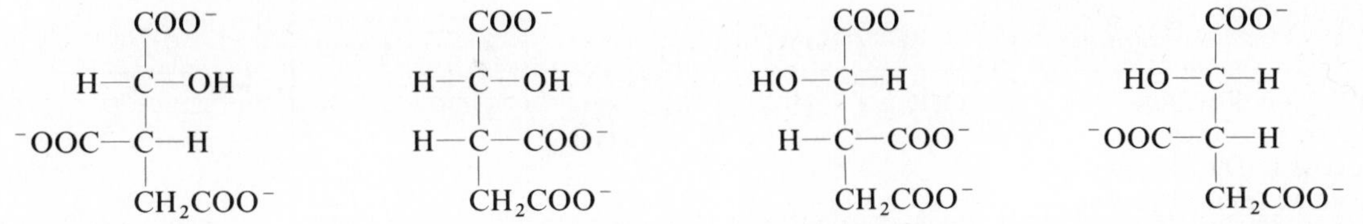

threo-D_s-isocitrate (only product)

erythro-D_s-isocitrate

threo-L_s-isocitrate

erythro-L_s-isocitrate

(none of these three isomers are produced)

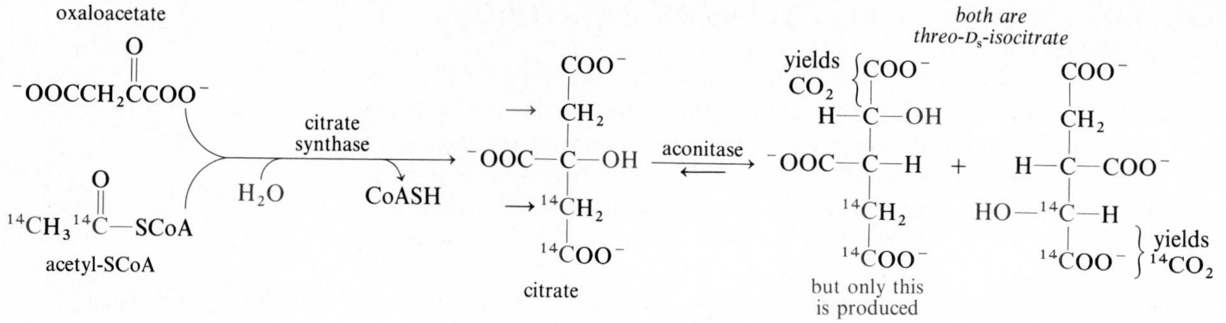

FIGURE 14–5 Evidence by carbon-14 radiolabeling for asymmetric catalysis by aconitase. When isocitrate dehydrogenase is present, no radioactive $^{14}CO_2$ is detected (see Figure 14–6). This result means that the —OH group is not transferred to the carbon atom that originated from acetyl-SCoA. The two small arrows on citrate identify identical ⟩CH$_2$ groups in citrate, from which the aconitase enzyme always selects the one marked by the color arrow.

Aconitase has still another remarkable element of specificity: the ability to select between identical carbons in the symmetrical citrate molecule (refer to Section 5–7). Proof of this ability to discriminate can be provided by tracing the CO_2 produced in the isocitrate → α-ketoglutarate conversion back to its origin in either acetyl-SCoA or oxaloacetate. By using radiolabeled substrates, one observes the following patterns of CO_2 production (see Figure 14–5): If unlabeled oxaloacetate and acetyl(^{14}C-labeled)-SCoA are incubated in the presence of citrate synthase, aconitase, and isocitrate dehydrogenase with some added NAD^+, only unlabeled (nonradioactive) CO_2 is produced. The production of $^{14}CO_2$ is detected only in the reciprocal experiment, starting with ^{14}C-labeled oxaloacetate and unlabeled acetyl-SCoA. Such results can only be explained by the action of aconitase eliminating H_2O across only one of two equivalent C—C bonds in citrate, in this case the C—C bond involving carbons that are originally contributed by oxaloacetate. If no such discrimination occurred, either of two different ^{14}C-labeling patterns would be generated in threo-D_s-isocitrate, one giving rise to $^{12}CO_2$ and the other to $^{14}CO_2$.

The mechanism of aconitase action depends on bound Fe^{2+}, complexed in iron-sulfur reaction clusters (see Chapter 15). Aside from the inhibitory effect of fluorocitrate mentioned in the description of citrate synthase, there are no known physiological activators or inhibitors, suggesting that aconitase is not a regulatory enzyme.

Isocitrate Dehydrogenase

Having a substrate specificity consistent with the action of aconitase, the isocitrate dehydrogenase in the mitochondrion catalyzes the oxidative decar-

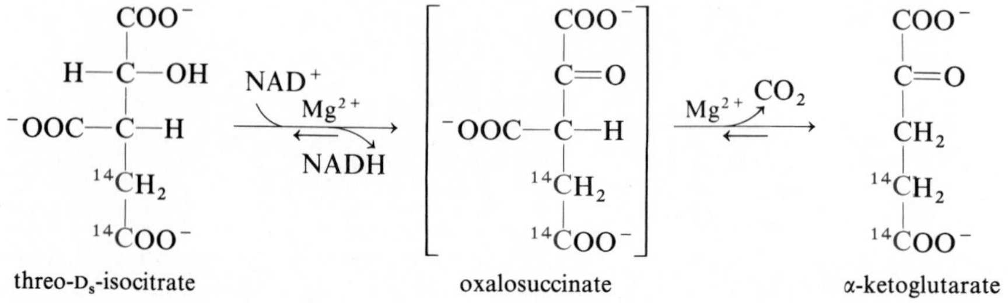

FIGURE 14–6 Reaction catalyzed by isocitrate dehydrogenase. The ^{14}C-labeling pattern shows results by using ^{14}C-labeled acetyl-SCoA as in Figure 14–5.

boxylation of threo-D_s-isocitrate (ISOCIT) to the C_5 α-ketodicarboxylic acid, α-ketoglutarate (αKG) (see Figure 14–6). This reaction is the first of three NADH-producing reactions in the cycle. As indicated in Figure 14–6, the chemistry of decarboxylation occurs subsequent to oxidation, involving the intermediate formation of an enzyme-bound oxalosuccinate intermediate.

The activity of isocitrate dehydrogenase, an oligomeric protein (MW ≈ 150,000) consisting of three different subunits, is regulated by several complementary effectors: ADP and NAD^+ are activators; ATP and NADH are inhibitors (see Figure 14–7). Indeed, isocitrate dehydrogenase is proposed to be the major *metabolic control site* in the cycle. More on this point will be given later in the chapter.

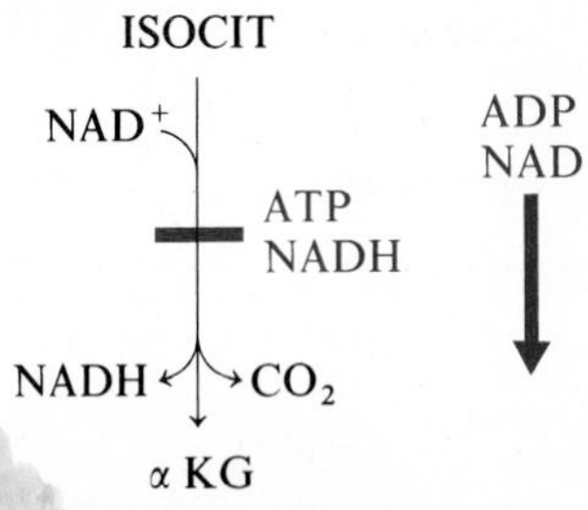

Isocitrate dehydrogenase is also found in the cytoplasm (1.1.1.42) and is also responsible for an ISOCIT → αKG conversion. However, the cytoplasmic enzyme uses $NADP^+$ as its oxidizing agent. The role of this enzyme is to provide a source of α-ketoglutarate in the cytoplasm for subsequent use in amino acid biosynthesis and a source of NADPH for subsequent use in anabolic reactions. The integration of the citric acid cycle to anabolic processes is developed more fully later in the chapter.

α-Ketoglutarate Dehydrogenase: A Multienzyme Complex

The oxidative decarboxylation of α-ketoglutarate to succinyl-SCoA (SUC-SCoA) is similar to the previous reaction only in that both require NAD^+. However, the αKG → SUC-SCoA conversion is more complex, requiring not only NAD^+ and CoASH as coenzymes but also requiring *three additional coenzymes:* thiamine pyrophosphate (TPP), flavin adenine dinucleotide (FAD), and lipoic acid (LpS_2) (see Figure 14–8).

An obvious question is why such a large number of different coenzymes are required. The fact is that this reaction is actually catalyzed by an *aggregate of three different enzymes,* each present in multiple copies. In this *multienzyme complex* each enzyme participates in distinct stages of the overall reaction. We will defer further analysis to a later discussion of the pyruvate dehydrogenase complex, also a multienzyme complex very similar in structure and mode of action to the α-ketoglutarate dehydrogenase complex. This similarity is not surprising, since both enzyme systems catalyze the same general type of reaction: the oxidative decarboxylation of an α-keto acid to an acyl thioester.

FIGURE 14–7 Metabolic controls on isocitrate dehydrogenase.

FIGURE 14–8 Reactions catalyzed by multienzyme complexes. Additional details are examined later in this chapter (see Section 14–3). The pyruvate → acetyl-SCoA reaction is of great importance.

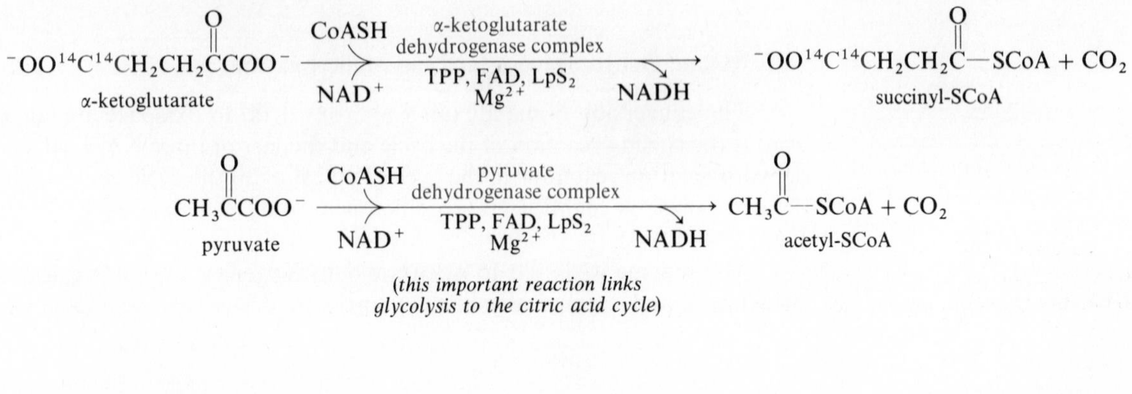

(*this important reaction links glycolysis to the citric acid cycle*)

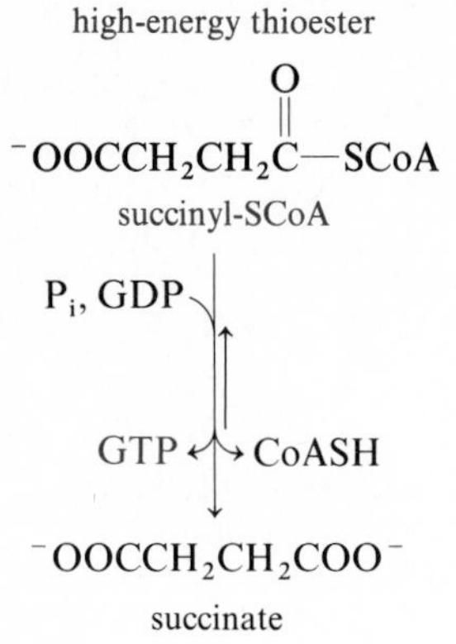

high-energy thioester

$$\begin{array}{c} \overset{O}{\underset{\|}{}} \\ ^-OOCCH_2CH_2C{-}SCoA \end{array}$$

succinyl-SCoA

P_i, GDP

GTP ⤨ CoASH

$^-OOCCH_2CH_2COO^-$

succinate

FIGURE 14–9 Substrate-level phosphorylation in the citric acid cycle.

FIGURE 14–10 Stereospecific catalysis by succinate dehydrogenase.

maleate   →no reaction (cis elimination)   succinate   →succinate dehydrogenase (trans elimination)→   fumarate

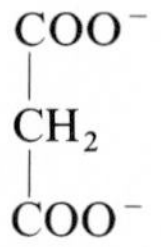

COO⁻
|
CH₂
|
COO⁻

FIGURE 14–11 Malonate.

Succinyl Thiokinase

The fate of the high-energy succinyl-SCoA generated from α-ketoglutarate is hydrolysis to the free acid (see Figure 14–9), with the energy released being conserved through the coupled formation of a nucleoside triphosphate, specifically, GTP. Remember that reactions of this type (such as 1,3BPG + ADP → 3PG + ATP and PEP + ADP → PYR + ATP in glycolysis) are referred to as *substrate-level phosphorylations* to distinguish them from the production of ATP coupled to electron transport. This step is the only step of the cycle that generates metabolic energy directly.

Succinate Dehydrogenase

Succinate → fumarate, the third oxidation of the citric acid cycle, is catalyzed by succinate dehydrogenase, an FAD-containing *flavoprotein*. (Review the FAD and FMN coenzyme species in Chapter 12.) The enzyme exhibits strict stereochemical specificity, removing hydrogen exclusively via a *trans elimination* (see Figure 14–10). No significant in vivo regulatory features are known for this step. The enzyme is subject to potent competitive inhibition by *malonate* (see Figure 14–11), providing a way to inhibit citric acid cycle activity in laboratory systems. Whereas all other citric acid cycle enzymes are soluble proteins in the mitochondrial matrix, succinate dehydrogenase is unique in that it is membrane-associated, specifically localized in the inner mitochondrial membrane.

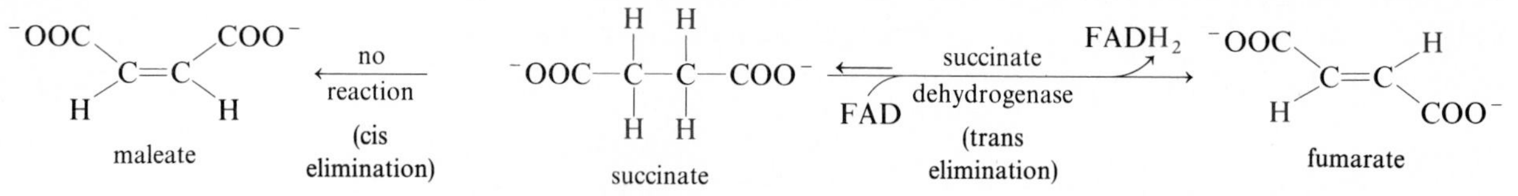

Fumarase

The hydration of fumarate to malate, catalyzed by fumarase, is still another example of enzyme stereospecificity. Fumarate is the only active substrate, and L-malate (not D-malate) is the product. No significant in vivo regulatory features have been detected.

Malate Dehydrogenase (and the Malic Enzyme)

The conversion of malate (an α-hydroxy acid) to oxaloacetate (an α-keto acid) is the closing reaction of the cycle and the last of three NAD⁺-dependent dehydrogenation reactions in the cycle. Typical of all the cycle enzymes, malate dehydrogenase is highly stereospecific, using only L-malate as substrate (see Figure 14–12).

The enzyme is similar to isocitrate dehydrogenase in that it exists in two structurally and catalytically distinct forms in the cells of higher organisms.

One form is located in the mitochondrion, and the other is in the cytoplasm. However, unlike the two isocitrate dehydrogenases that require either NAD^+ or $NADP^+$, both malate dehydrogenases are NAD^+-dependent. Of course, the mitochondrial enzyme is the one that operates in the citric acid cycle.

One function of the cytoplasmic malate dehydrogenase was previously described in Section 13–6 as part of the malate shuttle, which in eukaryotes provides one explanation for the pyruvate → phosphoenolpyruvate conversion in glycogenesis. However, some of the same shuttle events operating in reverse may be more important, accounting for the entry into the mitochondrion of NADH that is originally formed in the cytoplasm. The latter process is discussed further in Section 15–4. The cytoplasmic enzyme also provides a link of the citric acid cycle to the biosynthesis of pyrimidines and of several amino acids (discussed later in the chapter).

Although not a participant of the citric acid cycle, another important enzyme is involved in the metabolism of malate. Called *malic enzyme* and located in the cytoplasm of eukaryotes, it catalyzes the oxidative decarboxylation of malate to pyruvate:

$$\underset{\text{malate}}{^-OOCCH_2\overset{\overset{\text{OH}}{|}}{C}HCOO^-} \xrightarrow[\underset{NADP^+ \quad NADPH}{}]{\text{malic} \atop \text{enzyme}} CO_2 + \underset{\text{pyruvate}}{CH_3\overset{\overset{O}{||}}{C}COO^-}$$

Malic enzyme uses $NADP^+$ specifically and thus represents a source of reduced NADPH to be used in various anabolic reactions that specifically use NADPH rather than NADH.

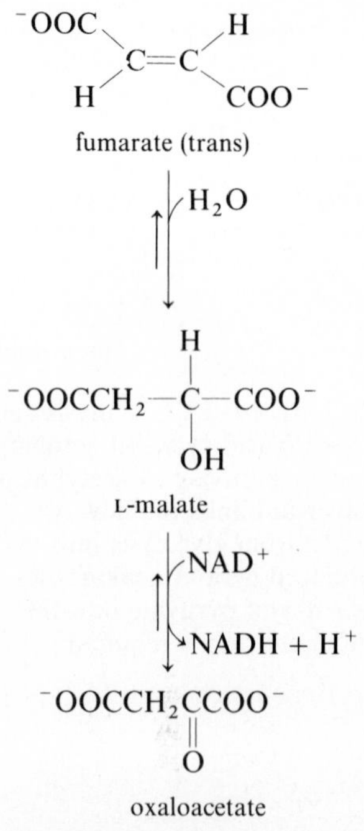

FIGURE 14–12 Stereochemical features of the fumarase and malate dehydrogenase reactions.

14–3 ROLE OF THE CITRIC ACID CYCLE IN METABOLISM

The material in this section is divided into two parts, providing an analysis of the role of the citric acid cycle in both the catabolism and the anabolism of the main classes of biomolecules. We emphasize again, however, that the cycle is not a pathway common to either phase of metabolism or to the metabolism of any specific class of molecules. Rather:

> The citric acid cycle is a central pathway that integrates and unifies the whole of metabolism.

It is unfortunate that we must fragment this subject in order to discuss it.

Role of the Citric Acid Cycle in Catabolism

CARBOHYDRATE CATABOLISM Most aerobic organisms utilize carbohydrate materials as the major source of carbon and energy via the combined metabolic processes of glycolysis and the citric acid cycle (see Figure 14–13). The connection between the two pathways occurs at the level of pyruvate →

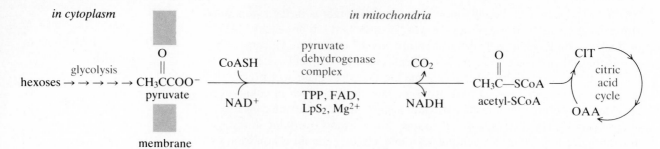

FIGURE 14–13 Glycolysis and the citric acid cycle. In aerobic cells the pyruvate → acetyl-SCoA conversion links the flow of carbon from glycolysis into the citric acid cycle. In eukaryotes the entry of pyruvate into the mitochondrion is required.

acetyl-SCoA, a conversion catalyzed in the matrix of the mitochondrion by the *pyruvate dehydrogenase complex* (PD complex).

Identical in form to the conversion of α-ketoglutarate → succinyl-SCoA in the citric acid cycle, the reaction involves a thioester-forming, oxidative decarboxylation of an α-keto acid—without any expenditure of ATP energy. The overall reaction is quite thermodynamically favorable ($\Delta G^{\circ\prime}$ is approximately -8.0 kcal/mole) and is enzymatically irreversible. Both characteristics contribute to a unidirectional flow of carbon from pyruvate to acetyl-SCoA. Moreover, when one considers the pyruvate as originating from phosphoenolpyruvate and then the acetyl-SCoA entering the citric acid cycle, the net thermodynamic effect is even more one-sided (see Figure 14–14). That is, all three of the reactions are exergonic and thus strongly favor the flow of carbon from glycolysis into the citric acid cycle.

Unlike a typical cellular reaction requiring a single enzyme, the pyruvate → acetyl-SCoA transformation requires three different enzymes—*pyruvate dehydrogenase, dihydrolipoyl transacetylase,* and *dihydrolipoyl dehydrogenase*—and five different coenzymes—NAD^+, CoASH, FAD, thiamine pyrophosphate, and lipoic acid. The PD complex is composed of several copies of each of these enzymes, clustered through noncovalent interactions.

Before examining the stepwise chemistry of the overall reaction, we need to describe the structure and function of *lipoic acid* (LpS_2). This substance is a low–molecular weight, sulfur-containing, carboxylic acid and is most always covalently attached to a polypeptide backbone via an amide linkage to the epsilon amino group of a lysine side chain—an adduct called *lysyl-lipoamide* (see Figure 14–15).

The active function group in the lipoyl structure is contributed by the two sulfur atoms, capable of existing as a disulfide (oxidized lipoyl) or as two sulfhydryls (reduced lipoyl), and the two forms are interconvertible. Enzymes de-

FIGURE 14–14 Flow of carbon from glycolysis into citric acid cycle, a flow that is thermodynamically favored.

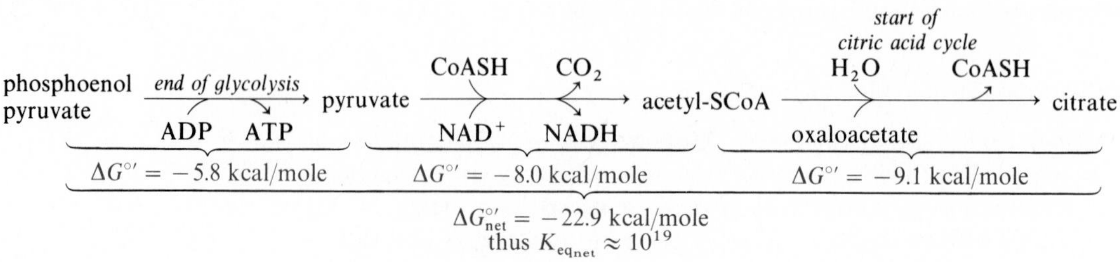

phosphoenol pyruvate $\xrightarrow[\text{ADP} \quad \text{ATP}]{\textit{end of glycolysis}}$ pyruvate $\xrightarrow[\text{NAD}^+ \quad \text{NADH}]{\text{CoASH} \quad CO_2}$ acetyl-SCoA $\xrightarrow[\text{oxaloacetate}]{\overset{\textit{start of}}{\underset{}{\textit{citric acid cycle}}} \; H_2O \qquad \text{CoASH}}$ citrate

$\Delta G^{\circ\prime} = -5.8$ kcal/mole $\Delta G^{\circ\prime} = -8.0$ kcal/mole $\Delta G^{\circ\prime} = -9.1$ kcal/mole

$$\Delta G^{\circ\prime}_{\text{net}} = -22.9 \text{ kcal/mole}$$
$$\text{thus } K_{\text{eq}_{\text{net}}} \approx 10^{19}$$

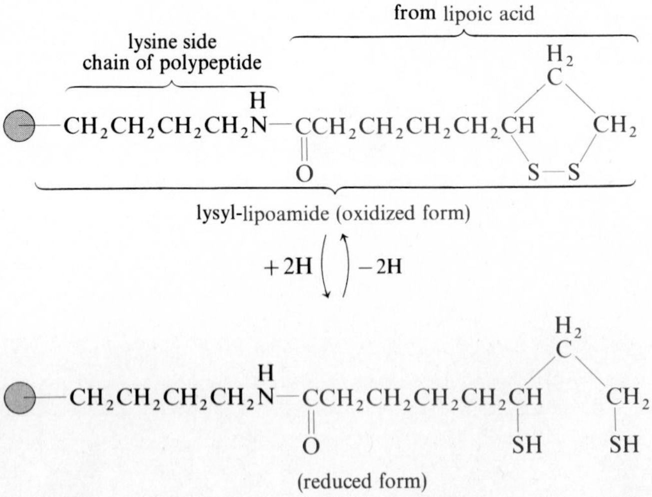

from lipoic acid

lysine side
chain of polypeptide

lysyl-lipoamide (oxidized form)

$+2H \left(\updownarrow \right) -2H$

(reduced form)

FIGURE 14–15 Covalent adduct between lipoic acid and a specific lysine side chain in a polypeptide.

pendent on lipoic acid, such as dihydrolipoyl transacetylase, catalyze *acyl group transfer reactions,* wherein the lipoyl cofactor acts as an intermediate carrier of the acyl $\left(\begin{array}{c} O \\ \parallel \\ RC- \end{array} \right)$ group. The details for dihydrolipoyl transacetylase are described shortly.

The operation of enzymes in the pyruvate dehydrogenase complex, involving five sequential steps, is illustrated in Figure 14–16. Pyruvate dehydrogenase (E_{PDH}), dependent on thiamine pyrophosphate (TPP), catalyzes (step 1) the decarboxylation of the α-keto acid, with the C_2 unit becoming attached to TPP as a hydroxymethyl adduct. (Refer to Section 13–9 for a review of TPP chemistry.) In step 2 the C_2 unit is transferred to the internal S of the oxidized lipoyl function of the transacetylase (E_{TA}). The acetylated transacetylase is then attacked (step 3) by CoASH to produce acetyl-SCoA, leaving the lipoyl function

FIGURE 14–16 Events associated with the three enzymes of the pyruvate dehydrogenase complex. Steps 1 through 5 are described in the text.

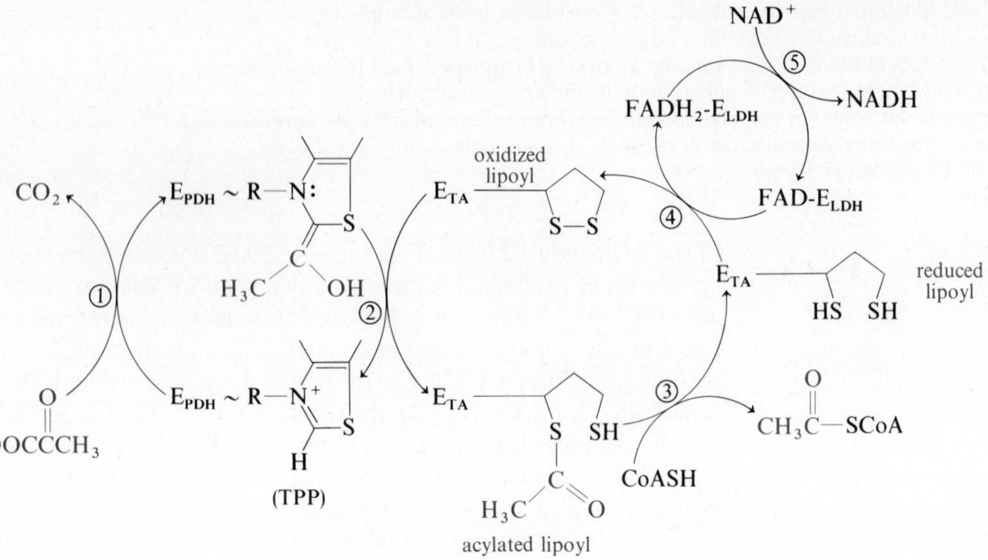

In E. coli

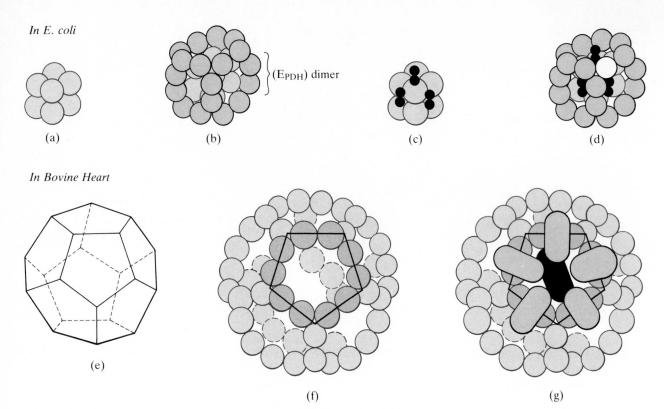

(a)   (b)   (c)   (d)

In Bovine Heart

(e)   (f)   (g)

FIGURE 14–17 Comparison of the proposed molecular anatomy of the pyruvate dehydrogenase complex (PDC) from *E. coli* and bovine heart mitochondria. *In E. coli:* (a) Cubic cluster of 24 transacetylase (E_{TA}) molecules (color), possibly arranged as eight trimers. (b) Twelve pyruvate dehydrogenase (E_{PDH}) dimers (grey spheres) on the edges of a transacetylase core. (c) Six dimers (black spheres) of dihydrolipoyl dehydrogenase (E_{LDH}) on each face of a transacetylase core. Three dimers are shown here. (d) Complete pyruvate dehydrogenase aggregate. *In bovine heart:* (e) Pentagonal dodecahedron, illustrated here for reference. (f) Core cluster of 60 subunits of dihydrolipoyl transacetylase (color). Each edge of a pentagonal face is occupied by two subunits. The edges of the front face (heavier color) are highlighted. (g) Five molecules of pyruvate dehydrogenase (grey shading) and one molecule of dihydrolipoyl dehydrogenase (solid black) in association with one face of the transacetylase core. Not shown are the other 25 molecules of pyruvate dehydrogenase and the other 5 molecules of dihydrolipoyl dehydrogenase.

of E_{TA} in its reduced form. Thus steps 1, 2, and 3 account for the complete conversion of the primary substrates (pyruvate and CoASH) to products (CO_2 and acetyl-SCoA). The last two steps (4 and 5) are required to regenerate the oxidized lipoyl unit for another round of reaction. The immediate oxidant (step 4) of the reduced lipoyl diol is FAD, carried by the flavoprotein dihydrolipoyl dehydrogenase (E_{LDH}). Finally (step 5), the oxidized form of the flavoprotein is regenerated by the participation of NAD^+ as the terminal oxidizing agent.

And now, what about the structure of the intact PD complex? Although the most simple composition would be a trimer with a single copy of each enzyme, nature has evolved a much more elaborate assembly, with *multiple copies*

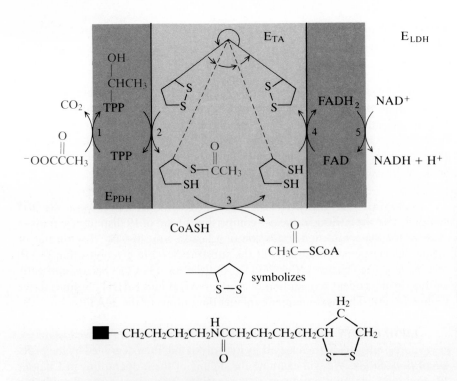

FIGURE 14–18 Diagrammatic representation of the extended lysyl-lipoamide function of the dihydrolipoyl transacetylase component in establishing communication among the active sites in the pyruvate dehydrogenase complex. The lysyl-lipoamide extended arm of E_{TA} can move (different positions of the dashed line) from one site to another. Steps 1 through 5 are the same as in Figure 14–17.

of each enzyme (see Figure 14–17). In *E. coli* (typical of prokaryotes) the PD complex, with a total particle weight of about 4×10^6, contains 12 dimeric molecules of TPP-E_{PDH}, 24 molecules of LpS_2-E_{TA}, and 6 dimeric molecules of FAD-E_{TA}. The total parts are clustered in a preferred cubic geometrical arrangement, using the multiple copies of the transacetylase as an internal core. The PD complex of bovine heart (representing animal eukaryotes) is even larger (about 8.5×10^6), with 30 tetrameric molecules of TPP-E_{PDH}, 60 molecules of LpS_2-E_{TA}, and 6–12 dimeric molecules of FAD-E_{LDH}. The geometry of the internal E_{TA} cluster is also different, namely, a pentagonal dodecahedron.

Although the precise operational advantage of a multienzyme complex is debatable, we can reasonably suggest that it would be an extremely efficient metabolic system, since all of the required participants are located in one position. Because multiple copies of each enzyme are present, the same reaction can occur simultaneously at several different locations on the aggregate. At each location where the reaction occurs, the interaction among the three enzymes involves the lysyl-lipoamide function as an *extended swinging arm* (see Figure 14–18), sweeping from one active-site region to another.

Whatever the advantage, multienzyme systems generally exhibit strict *regulatory controls*. The PD complex, for example, is inhibited by ATP and NADH and activated by ADP. There are several other inhibitors and activators (see Section 14–4)—a situation typical of a critical step in metabolism. Although most of these controls involve allosteric effects, the ATP inhibition in eukaryotes is part of a phospho/dephospho interconversion (see Figure 14–19). The kinase and phosphatase converter enzymes act specifically on the E_{PDH} component of the complex, and both are also in association with the intact complex. Formed

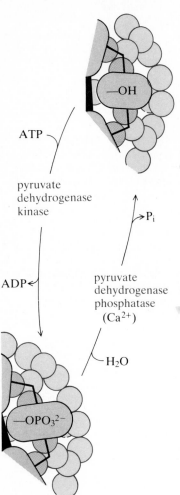

active E_PDH
in the complex

ATP

pyruvate
dehydrogenase
kinase

ADP

pyruvate
dehydrogenase
phosphatase
(Ca^{2+})

P_i

H_2O

inactive E_PDH
in the complex

FIGURE 14–19 Control of the PD complex (in eukaryotes) by phosphorylation/dephosphorylation. Only one E_PDH in one section of the intact complex is shown, for simplicity. The binding of kinase and phosphatase enzymes to the PD complex is not shown.

by the action of pyruvate dehydrogenase kinase, the phosphorylated E_PDH component is inactive; the action of pyruvate dehydrogenase phosphatase, itself requiring Ca^{2+} for optimal activity, generates the active, nonphosphorylated species.

The PD complex is but one of several multienzyme aggregates known to exist in nature. Mention has already been made in the citric acid cycle of the α-ketoglutarate complex, a close relative of the pyruvate dehydrogenase complex on both a structural and a functional level. The *fatty acid synthetase complex*, the *respiratory electron transport chain*, and the *fatty acid desaturase* system are some other examples to be highlighted in later chapters.

SUMMARY OF CARBOHYDRATE CATABOLISM Glycolysis, the PD complex, and the citric acid cycle—comprising a total of 19 distinct reactions—account for the complete degradation of a hexose unit to CO_2. The output of metabolic energy occurs directly at the substrate level in glycolysis (for 2ATP) and in the citric acid cycle (for 2GTP), and the rest (34ATP) occurs indirectly via the O_2-dependent reoxidation of all the NADH and $FADH_2$ produced (see Figure 14–20). The next chapter explains the output of the 34ATP.

LIPID (FATTY ACID) CATABOLISM When lipids are utilized as an energy source, fatty acids are released by hydrolysis and then degraded by the process of *β-oxidation*. We will examine the details of these operations in Chapter 17. For now, we only point out that repetitive operation of the β-oxidation pathway results in the fragmentation of a fatty acid molecule into several molecules of acetyl-SCoA:

$$CH_3(CH_2)_{16}COO^- \xrightarrow[\text{CoASH}]{\text{enzymes of } \beta\text{-oxidation}} 9CH_3\overset{O}{\overset{||}{C}}-SCoA \xrightarrow[\text{cycle}]{\text{enzymes of citric acid}} 18CO_2 + \text{energy}$$

fatty acid
(such as stearic acid)   acetyl-SCoA

Obviously, the acetyl-SCoA is then available to enter into the citric acid cycle for complete oxidation to $2CO_2$.

PROTEIN (AMINO ACID) CATABOLISM Figure 14–21 summarizes the catabolic relationships of the citric acid cycle to the energy-yielding oxidation of carbon substrates that can be traced to carbohydrates, fatty acids, and amino acids. The link to carbohydrates and fatty acids involves a single metabolite in each case, pyruvate and acetyl-SCoA, respectively. However, for amino acids there are several possible links. All of the specific reactions are not important at this time, but some introduction is appropriate.

When amino acids are utilized as an energy source, the first phase of degradation gives rise to products such as pyruvate, acetyl-SCoA, α-ketoglutarate, fumarate, succinyl-SCoA, and oxaloacetate—the specific one dependent on the amino acid involved. For some the entire amino acid carbon skeleton is present in the product. Included in this group are alanine, aspartate, and glutam-

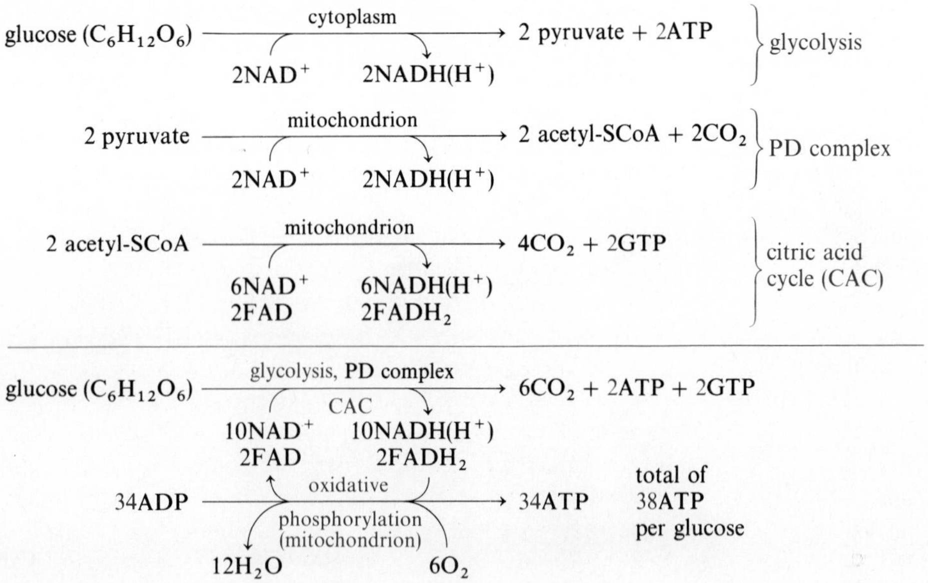

$$\text{glucose } (C_6H_{12}O_6) \xrightarrow[\underset{2NAD^+ \qquad 2NADH(H^+)}{}]{\text{cytoplasm}} 2 \text{ pyruvate} + 2ATP \quad \Big\} \text{glycolysis}$$

$$2 \text{ pyruvate} \xrightarrow[\underset{2NAD^+ \qquad 2NADH(H^+)}{}]{\text{mitochondrion}} 2 \text{ acetyl-SCoA} + 2CO_2 \quad \Big\} \text{PD complex}$$

$$2 \text{ acetyl-SCoA} \xrightarrow[\underset{\substack{6NAD^+ \qquad 6NADH(H^+) \\ 2FAD \qquad 2FADH_2}}{}]{\text{mitochondrion}} 4CO_2 + 2GTP \quad \Big\} \begin{array}{l}\text{citric acid} \\ \text{cycle (CAC)}\end{array}$$

$$\text{glucose } (C_6H_{12}O_6) \xrightarrow[\underset{\substack{10NAD^+ \qquad 10NADH(H^+) \\ 2FAD \qquad 2FADH_2}}{\text{CAC}}]{\text{glycolysis, PD complex}} 6CO_2 + 2ATP + 2GTP$$

$$34ADP \xrightarrow[\underset{\substack{\text{phosphorylation} \\ \text{(mitochondrion)} \\ 12H_2O \qquad\qquad 6O_2}}{}]{\text{oxidative}} 34ATP \qquad \begin{array}{l}\text{total of} \\ \text{38ATP} \\ \text{per glucose}\end{array}$$

FIGURE 14–20 Summary of energy (ATP) production from glucose catabolism in aerobic cells. Although not shown in all reactions, ADP, GDP, and P_i are involved.

ate, which yield pyruvate, oxaloacetate, and α-ketoglutarate—the respective α-keto acids of these amino acids.

The major type of reaction responsible for the α-amino acid → α-keto acid conversion is *transamination,* wherein the amino group of a donor amino acid is transferred to an acceptor α-keto acid, resulting in the formation of a new α-keto acid (derived from the original amino acid) and a new amino acid (derived from the original α-keto acid). The enzymes involved, called *transaminases* or *aminotransferases,* occur in both the cytoplasm and mitochondrion. The reaction is outlined in Figure 14–22 and is described in more detail in Chapter 18, as are some of the reactions associated with the formation of intermediates arising from just part of an amino acid carbon skeleton.

Role of the Citric Acid Cycle in Anabolism

In the preceding section we have described how various metabolites originating from different sources can be shunted into the cycle for further degradation, resulting in the production of NADH and FADH$_2$ and ultimately the production of ATP. By the same token, intermediates of the cycle can be bled off at various points for use as precursors in the biosynthesis of different materials. However, the removal of intermediates must occur with the continued catabolic operation of the cycle for the purpose of supplying the ATP that is also needed for anabolism. In other words, the cycle must simultaneously fulfill two roles. Such is not the case when the cycle is linked to catabolic reactions, because here the production of energy is, in fact, a natural consequence of the exergonic and oxidative operation of the cycle and its coupling to oxidative phosphorylation. The following material focuses on some of the details of the anabolic relationships, which are summarized in Figure 14–23 (p. 556).

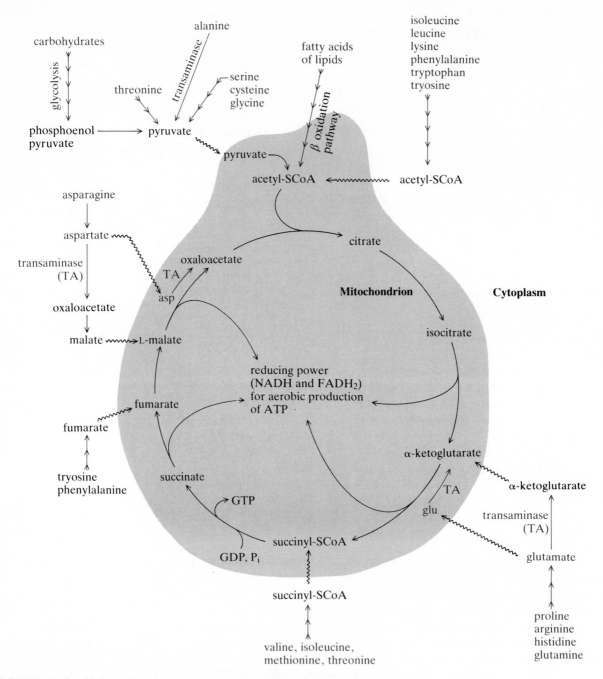

FIGURE 14–21 Citric acid cycle in catabolism. Degradative pathways involving more than one enzymatic step are indicated by multiple arrows ($\rightarrow\rightarrow\rightarrow\rightarrow$). The shaded area corresponds to intramitochondrial processes. Wavy lines represent transport across the mitochondrial membrane.

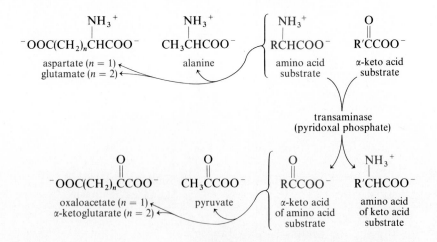

FIGURE 14–22 Transamination reaction. The general reaction is shown at the right. The fate of three specific amino acid substrates to yield α-keto acids that enter the main stream of metabolism are shown at the left.

PROTEIN (AMINO ACID) ANABOLISM   Virtually all organisms are capable of synthesizing amino acids, although there are differences in the number of acids synthesized by any one type of organism. For example, most plants and bacteria can synthesize all of the 20 amino acids required for proteins; but animals (including humans) are capable of synthesizing only certain ones, relying on a dietary intake for the rest. Whatever the organism, however, some of the intermediates of the citric acid cycle serve as important precursors for amino acids. Two particularly important compounds are the α-keto acids, oxaloacetate (OAA), and α-ketoglutarate (αKG). A third important precursor is pyruvate (PYR), also an α-keto acid and, of course, closely linked to the citric acid cycle.

Since the mitochondrial membrane is impermeable to the movement of OAA and αKG (more so to OAA), these intermediates are bled off (see Figure 14–24) in the form of their precursors—malate (MAL) and isocitrate (ISOCIT), respectively—which can be transported across the membrane. The action of dehydrogenases in the cytoplasm is to produce OAA and αKG in the cytoplasm. Then OAA and αKG can be acted on by cytoplasmic transaminases to yield aspartate (ASP) and glutamate (GLU), respectively. (In some cells transaminase enzymes may be present in the mitochondrion, thus permitting a more direct route of ASP and GLU formation.) Pyruvate can be formed from malate in one of two routes: indirectly, via oxaloacetate, or directly, via the action of malic enzyme.

The removal of intermediates for amino acid biosynthesis represents a serious problem to the continued operation of the cycle. That is, if an intermediate is diverted for use in synthesis, the cycle is interrupted, and OAA production will fall off. Since OAA is needed to condense with acetyl-SCoA, a process that keeps carbon coming into the cycle, any reduction in OAA production will reduce the activity of the whole cycle and any process related to it. In fact, since the cycle is at the center of all metabolism, the entire metabolic activity of the cell will become sluggish.

As a counteraction for this possibility, there is a remarkable set of safety valves, called *anaplerotic reactions* (from Greek meaning "to fill up"). The basic

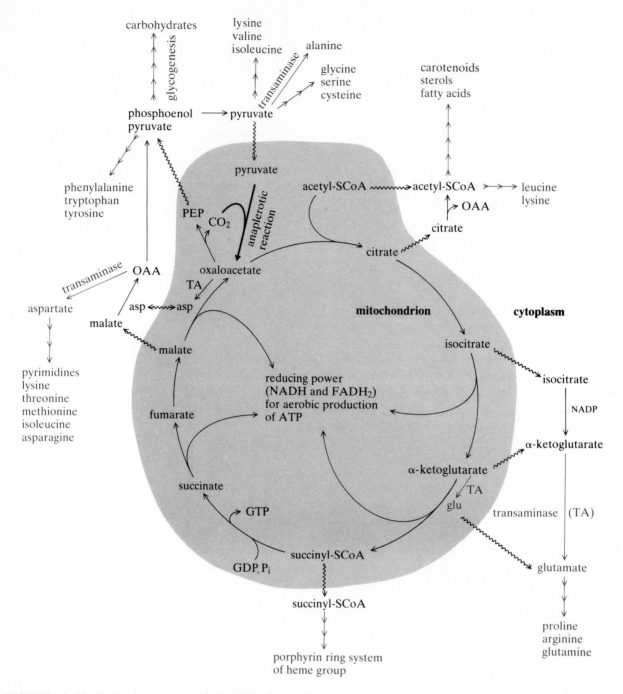

FIGURE 14–23 Citric acid cycle in anabolism. Synthetic pathways involving more than one enzymatic step are indicated by multiple arrows ($\rightarrow \rightarrow \rightarrow \rightarrow$). The shaded areas correspond to intramitochondrial processes. The heavy arrow indicates a major anaplerotic reaction. Wavy lines represent transport across the mitochondrial membrane.

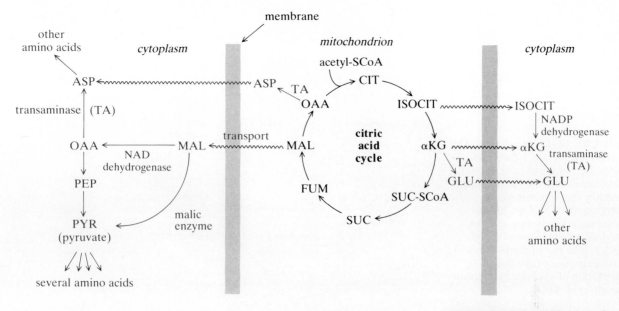

FIGURE 14–24 Utilization of citric acid cycle intermediates for amino acid biosynthesis.

function of these reactions is to maintain an adequate supply of a crucial intermediate that participates in a central pathway such as the citric acid cycle when that intermediate is being bled off into other metabolic pathways. Relative to our current discussion, the most important reaction of this type—particularly in mammals—is the conversion of pyruvate to oxaloacetate catalyzed by pyruvate carboxylase:

$$\underset{\text{pyruvate}}{\text{CH}_3\overset{\overset{\textstyle O}{\|}}{\text{C}}\text{COO}^-} + \text{CO}_2 \xrightarrow[\underset{\text{ATP}}{\text{carboxylase (biotin)}}]{\overset{\text{ADP} + \text{P}_i}{\underset{\text{pyruvate}}{}}} \underset{\text{oxaloacetate (OAA)}}{{}^-\text{OOCCH}_2\overset{\overset{\textstyle O}{\|}}{\text{C}}\text{COO}^-}$$

Recall that this enzyme, which is also localized in the mitochondrion, has already been discussed as the catalyst for the first step in glycogenesis (see Section 13–6). Thus we see how one enzyme can have different metabolic roles.

Because pyruvate carboxylase is such a crucial enzyme, you might suspect that its catalytic activity would be under strict metabolic control. Indeed, this control has been confirmed by studies that establish acetyl-SCoA as a strong allosteric activator. The activation by acetyl-SCoA fits a pattern of metabolic logic. When a cycle intermediate is diverted (for example, when ISOCIT or αKG is used for glutamate synthesis), the failure to regenerate OAA contributes to an accumulation of acetyl-SCoA, because it requires OAA to form citrate (see Figure 14–25). This buildup of acetyl-SCoA serves as a metabolic signal to activate the anaplerotic enzyme. More OAA is produced, ensuring the formation of citrate. Hence the cycle can continue to supply αKG for glutamate biosynthesis and also permit some of the αKG to flow through the cycle so that the entire sequence is not totally interrupted.

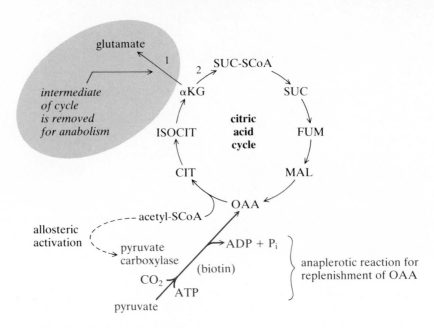

FIGURE 14-25 Anaplerosis. The occurrence of step 1 competes with the continuation of normal cycle operation, step 2, resulting initially in a diminished production of SUC-SCoA and eventually a reduction of OAA and an accumulation of acetyl-SCoA. The rise of acetyl-SCoA concentration signals the activation of pyruvate carboxylase to replenish OAA.

The same principle and the same anaplerotic reaction operate when any intermediate is diverted from the cycle.

LIPID ANABOLISM In Chapter 17 we will discover that the biosynthesis of most lipids originates primarily with acetyl-SCoA. In fact, acetyl-SCoA is the metabolic source of all the carbon atoms in the synthesis of fatty acids, carotenoids, and steroids (see Figure 14–26). The anabolic utilization of acetyl-SCoA does not, however, create an operational strain on the normal functioning of the citric acid cycle, as does the removal of one of the internal intermediates.

The anabolic utilization of acetyl-SCoA does create one problem: Most of the biosynthetic reactions, including those for lipids, occur in the cytoplasm, whereas the bulk of acetyl-SCoA production occurs within the mitochondria. While some acetyl-SCoA may exit from the mitochondrion directly, the major exit route is an indirect one mediated by a special enzyme found in the cytoplasm. The enzyme, called *citrate lyase,* catalyzes the breakdown of citrate to acetyl-SCoA and oxaloacetate. Thus citrate moves across the mitochondrial membrane and is cleaved in the cytoplasm. The return of OAA to the mitochondrion as MAL sustains the cycle and allows continued operation. It is worth repeating that continued operation is needed not only to process acetyl-SCoA but also to supply ATP necessary for the anabolic utilization of acetyl-SCoA in the cytoplasm.

CARBOHYDRATE ANABOLISM The connection of the cycle to the biosynthesis of carbohydrates is made via *phosphoenolpyruvate carboxykinase,* a mitochondrial enzyme catalyzing the GTP-dependent conversion of oxaloacetate to phosphoenolpyruvate (see Figure 14–27). This enzyme is the same one previously discussed (see Section 13–6) as being one of two enzymes respon-

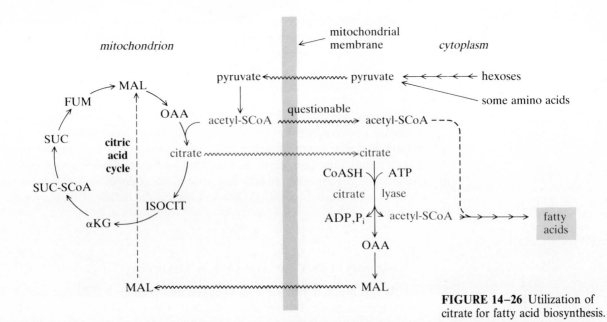

FIGURE 14–26 Utilization of citrate for fatty acid biosynthesis.

sible for the conversion of pyruvate to phosphoenolpyruvate in glycogenesis. Our current discussion does not change any of that. But now we are seeing exactly how the enzyme is integrated into the whole of metabolism. Clearly, in this integrated framework the o aloacetate precursor for phosphoenolpyruvate is not necessarily formed only from pyruvate. Rather, the OAA can originate from pyruvate or any metabolite that can be converted to one of the inter- mediates of the citric acid cycle, each of which can then be converted to oxaloacetate.

FIGURE 14–27 Utilization of cycle intermediates for the biosynthesis of hexoses. As described in the text, OAA can be removed via PEP, and it can also exit via aspartate or as the malate precursor. The $\oplus$ indicates the allosteric activation by acetyl-SCoA on pyruvate carboxylase.

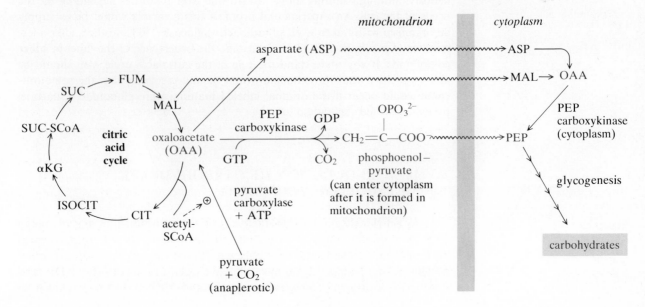

When the carboxykinase is located in the cytoplasm, the link to carbohydrate biosynthesis involves the exit of malate or aspartate from the mitochondrion to the cytoplasm. In either case, the anaplerotic role of pyruvate carboxylase is necessary to sustain the level of OAA required for uninterrupted metabolism through the cycle.

OTHER ANABOLIC RELATIONSHIPS   The involvement of the cycle intermediates in biosynthetic pathways is not confined to carbohydrates, lipids, and amino acids. Other critical relationships exist in the biosynthesis of nucleotides and heme groupings. As indicated in Figure 14–23, *pyrimidine biosynthesis* utilizes aspartate (which can arise from oxaloacetate), and *heme biosynthesis* requires succinyl-SCoA. We will inspect the details of these processes in Chapter 18.

Summary of the Citric Acid Cycle in Metabolism

At this point, if you feel overwhelmed and perhaps a little confused, it is quite understandable. The chapter is entitled "Citric Acid Cycle," but we have digressed into many other areas. You can take solace in one respect, however; our digressions could have been even more extensive.

Regardless, any uncertainty in your mind about what the metabolic role of the cycle is can now be removed if you think back over the past several pages and realize that all of our separate digressions were analyzed in terms of the participation of the citric acid cycle. Thus we have not fragmented and separated metabolism; rather, we have *unified* and *integrated* it. Therein lies the most important function of the cycle. The whole of cellular metabolism is interconnected—and hence interrelated and interdependent. Moreover, the events of catabolism (Figure 14–21) and of anabolism (Figure 14–23) occur simultaneously, although shifting more toward one over the other depending on the needs of the cell. An experimental proof of this principle would be to supply an organism with tracer levels of radioactive glucose (^{14}C) and then, after a few minutes of incubation at 37°C, determine the occurrence of the label in other compounds. If you understand the role of the citric acid cycle, you should be able to predict what the outcome would be and recognize that the same outcome would occur if the original labeled material were glutamate, aspartate, pyruvate, or acetate, and so on.

14–4   REGULATION OF CARBOHYDRATE METABOLISM: A SUMMARY

As previously emphasized, the rates of energy production and of energy utilization are not constant throughout the lifetime of a cell. Rather, the production of energy changes to keep pace with a changing need of energy. Earlier, we focused on changes in the intracellular concentrations of ATP, ADP, and AMP as control signals: a high level of ADP and AMP and a low level of ATP

cells in resting metabolic state $\rightleftharpoons$ cells in very active metabolic state

(energy utilization
is less than that of cells
in very active metabolic state)

(energy utilization
is greater than that of cells
in resting metabolic state)

$$
\left.\begin{array}{c}
\text{high ATP} \\
\text{low ADP + AMP} \\
\text{high}\left(\dfrac{\text{ATP}}{\text{ADP + AMP}}\right)
\end{array}\right\} \rightleftharpoons \left\{\begin{array}{c}
\text{low ATP} \\
\text{high ADP + AMP} \\
\text{low}\left(\dfrac{\text{ATP}}{\text{ADP + AMP}}\right)
\end{array}\right.
$$

$$
\left.\begin{array}{c}
\text{high NADH} \\
\text{low NAD}^+ \\
\text{high}\left(\dfrac{\text{NADH}}{\text{NAD}^+}\right)
\end{array}\right\} \rightleftharpoons \left\{\begin{array}{c}
\text{low NADH} \\
\text{high NAD}^+ \\
\text{low}\left(\dfrac{\text{NADH}}{\text{NAD}^+}\right)
\end{array}\right.
$$

FIGURE 14–28 Summary of metabolic changes in concentrations of ATP, ADP, AMP, NADH, and NAD$^+$ that provide intracellular signals to regulate metabolism.

signaling a very active metabolic state; and a shift of concentrations in the reverse direction signaling a return to a less active resting state (see Figure 14–28).

However, the ATP–ADP–AMP signals are only part of the scenario. Another set of signals is composed of the concentrations of NADH and NAD$^+$, which in aerobic organisms represent the primary source of ATP energy. In a very active metabolic state the high ADP and AMP and the low ATP will also be accompanied by a high level of NAD$^+$ and a low level of NADH—because the increased rate of NADH formation will be coupled to an increased rate of oxidation back to NAD$^+$. Conversely, a shift back to a resting state will involve a decrease in NAD$^+$ and an increase in NADH.

Figure 14–29(a) summarizes how the changing ratios of ATP/(ADP + AMP) and NADH/NAD$^+$ combine in a most remarkable manner to control the flow of carbon through glycolysis, glycogenesis, and the citric acid cycle. The diagram is a collection of all of the regulatory characteristics of enzymes that were previously mentioned in Chapter 13 and in this chapter (see Note 14–3). In Figure 14–29 the ➡ symbol for activation is replaced by (+) and the ⊣ symbol for inhibition is replaced by (−). Close study reveals that increases in the cellular levels of ADP, AMP, and NAD$^+$ are all signals that enhance the rate of catabolism *and* diminish the rate of anabolism. In other words, the "turning on" of a few key enzymes of carbohydrate catabolism and the complementary "turning off" of a key anabolic enzyme (FBP → F6P) results in the "turning on" of ATP production. Note that the term *turning on* does not mean that these reactions were not occurring prior to activation. The point is that after activation the net catabolism of carbohydrates and ATP production occur at a greater rate in order to keep pace with the demand. In contrast, ATP and NADH are signals that diminish the rate of catabolism.

If the beauty of the ATP, ADP, AMP, NADH, and NAD$^+$ effects aren't enough to dazzle you, consider then Figure 14–29(b), which summarizes a host of other metabolic signals involving the pathway intermediates themselves. All of the feedback and feed-forward effects are mutually consistent with each other and with those involving ATP, ADP, AMP, NADH, and NAD$^+$. In fact, the

NOTE 14–3
The regulatory characteristics of enzymes are established by in vitro kinetic assays on purified enzymes. The biochemist then argues, of course, that the behavior of enzymes in vitro also applies to the natural in vivo environment. However, the interpretation of in vivo significance is not necessarily a certainty. Some in vitro observations may not apply in vivo. Even more noteworthy is the possibility that many aspects of metabolic regulation that do operate in vivo are not observed in vitro.

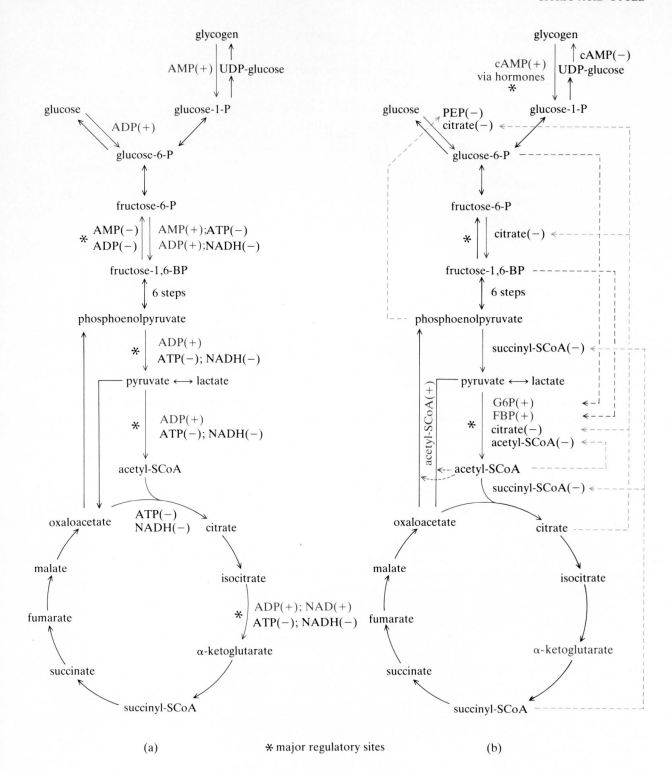

(a)   * major regulatory sites   (b)

metabolite signals would arise from the latter. For example, high ADP and AMP would initially increase the rate of production of G6P and 1,6FBP and the elevated G6P and 1,6FBP could then increase the rate of conversion of PYR → acetyl-SCoA. Thus one set of regulatory signals generates a second set of signals that operates in a complementary fashion. A sort of "all-alert" warning system is operating.

To indicate a sequence for all of these regulatory effects is difficult. However complicated the network may appear (patience and close study will render it less complicated), one thing should be obvious:

> A living organism is not a conglomerate of chemical processes that operate randomly, but an efficiently coordinated symphony of integrated reactions subject to an exquisite pattern of regulatory checks and balances.

Remember that all of these controls are exerted at the level of enzymes and that many involve allosteric effects.

LITERATURE

ATKINSON, D. *Cellular Energy Metabolism and Its Regulation.* New York: Academic Press, 1977. A discussion of the interplay between thermodynamic and kinetic factors in maintaining the metabolic stability that underlies life. Atkinson explains his hypothesis of energy charge control.

CUNNINGHAM, E. B. *Biochemistry: Mechanisms of Metabolism.* New York: McGraw-Hill, 1978. Chapter 10 deals with CAC enzymes.

GINSBURG, A., and E. R. STADTMAN. "Multienzyme Systems." *Annu. Rev. Biochem.,* **39** (1970). A review article on the biochemistry of multienzyme complexes including α-keto acid dehydrogenase complexes.

KREBS, H. A. "The History of the Tricarboxylic Acid Cycle." *Perspect. Biol. Med.,* **13,** 154 (1970).

LOWENSTEIN, J. M. "The Tricarboxylic Acid Cycle." In *Metabolic Pathways,* 3rd ed., edited by D. Greenberg. New York: Academic Press, 1967. A review article giving a thorough analysis of this reaction pathway.

MAHLER, H. R., and E. H. CORDES. *Biological Chemistry.* 2nd ed. New York: Harper & Row, 1971. Chapter 14 of this textbook contains an excellent discussion of the individual enzymes of the cycle, the stereospecificity of the cycle, and the integration of the cycle with other areas of metabolism.

◄ **FIGURE 14–29** Summary of regulatory patterns operating in carbohydrate metabolism. Regulatory enzymes are symbolized by colored arrows. The * identifies major regulatory enzymes. (a) Control sites involving ATP, ADP, AMP, NADH, and NAD^+. The use of color and (+) represents activation; the (−) represents inhibition. Elevated levels of ADP, AMP, and NAD^+ favor the stimulation of carbohydrate catabolism, whereas elevated levels of ATP and NADH favor the depression of carbohydrate catabolism. (b) Control sites involving feedback inhibition and feed-forward stimulation (dashed lines for both) by metabolic intermediates and (in animals) also involving hormonal regulation by cyclic AMP. The AMP, ADP, ATP, NAD^+, and NADH effects are supplemented by a host of other feedback inhibitions, feed-forward activations, and hormonal effects. (*Note:* Although the text discussion deals primarily with metabolism under aerobic conditions, which would then include the operation of the citric acid cycle, portions of these diagrams also apply to anaerobic conditions, in which catabolism terminates with lactate production.) Compartmentalization is not indicated, for simplicity.

EXERCISES

14–1. Write all of the reactions in the citric acid cycle that comprise the oxidative transformations of the pathway.

14–2. Relative to the other reactions of the citric acid cycle, what is unique about the transformations catalyzed by (a) succinyl thiokinase and (b) α-ketoglutarate dehydrogenase?

14–3. If a molecule of uniformly labeled (^{14}C) pyruvate were oxidatively degraded to acetyl-SCoA and the latter then entered the citric acid cycle, which of the label patterns below would correspond to the α-ketoglutarate that would be produced in the first turn of the cycle?

$$^-OO^{14}C^{14}CH_2CH_2\overset{\overset{\displaystyle O}{\|}}{C}COO^-$$

or

$$^-OOCCH_2CH_2{}^{14}\overset{\overset{\displaystyle O}{\|}}{C}{}^{14}COO^-$$

14–4. To continue 14–3, what percentage of the radioactivity in the original molecule of labeled acetyl-SCoA would be produced as $^{14}CO_2$ after three complete turns of the cycle? (*Note:* Assume that the acetyl-SCoA molecules that would function in the second and third condensations with oxaloacetate are not labeled. *Hint:* The enzymes converting succinate to L-malate do not differentiate between identical carboxyl groupings in succinate or fumarate.)

14–5. The citric acid cycle is frequently described as the major pathway of aerobic metabolism, which means that it is an oxygen-dependent, degradative process. Yet none of the reactions of the cycle directly involves oxygen as a reactant. Why then is the pathway oxygen-dependent (aerobic) rather than oxygen-independent (anaerobic)?

14–6. Which of the following equations best describes the net aerobic catabolism of one molecule of pyruvate to α-ketoglutarate?

(a) $PYR + OAA + 2NAD^+ + HSCoA \rightarrow$
$$\alpha KG + 2CO_2 + 2NADH + 2H^+$$

(b) $PYR + 2NAD^+ + HSCoA \rightarrow$
$$\alpha KG + 2CO_2 + 2NADH + 2H^+$$

(c) $PYR + OAA + 2NAD^+ \rightarrow$
$$\alpha KG + 2CO_2 + 2NADH + 2H^+$$

(d) $PYR + OAA + O_2 \rightarrow \alpha KG + 2CO_2 + 2H_2O$

(e) $PYR + 2\text{acetyl-SCoA} + O_2 \rightarrow$
$$\alpha KG + 2CO_2 + 2H_2O$$

(f) $PYR + OAA + \frac{1}{2}O_2 \rightarrow \alpha KG + 2CO_2 + H_2O$

14–7. When fluoroacetyl-SCoA ($FCH_2COSCoA$) condenses with oxaloacetate, fluorocitrate is produced. If the fluorocitrate were acted on by aconitase according to the citrate $\rightarrow$ isocitrate conversion, would the F and OH in fluoroisocitrate be on the same carbon or on different carbons?

14–8. If the catalytic residues in citrate synthase promoted the formation of the carbanion species of oxaloacetate shown below, what product (show its structure) would likely be formed on subsequent condensation with acetyl-SCoA?

$$^-OOC-\underset{\ominus}{\overset{..}{C}H}-\overset{\overset{\displaystyle O}{\|}}{C}-COO^-$$

14–9. If radioactive CO_2 could be detected when acetyl-SCoA (^{14}C-labeled) and oxaloacetate (no ^{14}C label) were incubated in a suitably buffered solution containing NAD^+ and pure samples of citrate synthase, aconitase, and isocitrate dehydrogenase, it would be an observation contradictory to our understanding of the stereospecificity that applies to this section of the citric acid cycle. Explain.

14–10. If uniformly labeled tyrosine (^{14}C) were incubated with a cell-free liver extract, some of the tyrosine molecules would be metabolized in the following manner:

$$HO-\underset{}{\text{⬡}}-CH_2\overset{\overset{\displaystyle NH_3^+}{|}}{C}HCOO^- \rightarrow\rightarrow\rightarrow\rightarrow\rightarrow$$

(all carbons are ^{14}C)

$$^-OO^{14}C^{14}\overset{\overset{\displaystyle H}{|}}{C}={}^{14}C^{14}\underset{\underset{\displaystyle H}{|}}{}COO^-$$

Show all of the necessary reactions that would account for the subsequent appearance in fructose-1,6-bisphosphate of the carbons arising from fumarate. Indicate the distribution of radioactive carbons—and nonradioactive carbons, if there will be any—in the hexose.

14–11. If the citric acid cycle were to be primed with intermediates originating from aspartate (oxaloacetate) and glutamate (α-ketoglutarate), what compound would have to be generated by some other source in order for the enzymes of the cycle to continually and efficiently process these intermediates? Explain.

14–12. Write a net equation for the conversion of a molecule of isocitrate to oxaloacetate via the enzymes of the citric acid cycle. Write the equation for the same conversion when it is linked to the mitochondrial electron transport chain.

14-13. If a sample of a freshly prepared muscle homogenate were added to a buffered solution containing pyruvate and oxaloacetate, the amount of CO_2 produced would be greater than if pyruvate were added alone. Propose a sequence of reactions that would explain the stimulation of oxaloacetate without the participation of extramitochondrial NADH.

14-14. How would you compare the ratios of both $NADH/NAD^+$ and ATP/ADP in heart muscle during periods of sleep and handball playing?

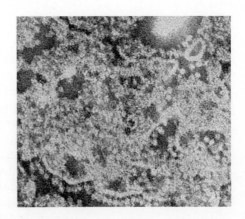

CHAPTER FIFTEEN
OXIDATIVE PHOSPHORYLATION

In order to exist, all living organisms must have a supply of energy from the surrounding environment. Photosynthetic organisms depend on sunlight as a source of *radiant energy*. With the exception of a few microbes that can utilize inorganic substances as energy sources, nonphotosynthetic organisms depend on an input of *reduced organic compounds*. Although carbohydrate is the primary organic energy source, lipid and protein are also utilized.

Given an external source of energy, each type of organism is capable of providing itself with an intracellular supply of metabolically useful chemical energy in the form of ATP. In Chapters 13 and 14 we have seen how some ATP can be produced directly by substrate-level phosphorylations. However, in most cells the major ATP-forming process is **oxidative phosphorylation.** The basic design of oxidative phosphorylation is as follows (see Figure 15–1):

> The formation of ATP from ADP and P_i is coupled to a process of electron transport involving the transfer of electrons from a high-energy reducing agent via intermediate electron carriers to a terminal electron acceptor.

This type of chemistry is common to all organisms and is a major unifying biochemical principle of our biosphere.

This chapter deals with the formation of ATP in nonphotosynthetic organisms, with a focus on *aerobic organisms* where molecular *oxygen* (O_2) *is the terminal electron acceptor*. The prerequisite for oxidative phosphorylation in (most) nonphotosynthetic cells is the production of high-energy electron donors such as NADH and $FADH_2$ during catabolism—a process to which we have referred on several occasions. The same design of oxidative phosphorylation also operates in photosynthesizing cells, but there the process involves the formation of

FIGURE 15–1 Summary of ATP formation via oxidative phosphorylation. The electron acceptor substances that participate in photosynthetic organisms are discussed in Chapter 16.

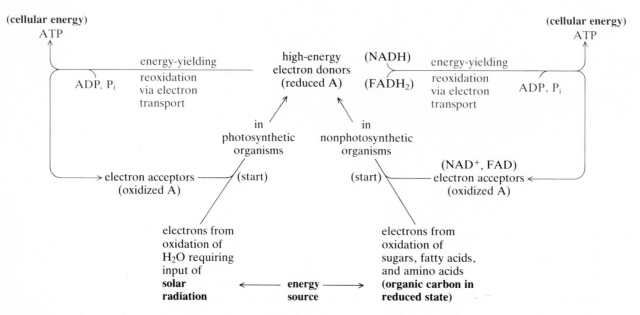

high-energy electron donors other than NADH, and their formation is a consequence of the absorption of solar radiation. Because of the dependence on external radiant energy, the latter process is termed *photophosphorylation*. We will consider it in the next chapter.

15–1 ELECTRON TRANSFER SYSTEMS: SOME BASIC PRINCIPLES

Oxidative phosphorylation is a term that collectively refers to two chemical processes: (1) the exergonic oxidation of reduced species such as NADH and FADH$_2$ via a sequence of electron transfer reactions, and (2) the endergonic phosphorylation of ADP to yield ATP. Although distinct, both are *energetically coupled processes*, with the energy released in electron transport being used for ATP formation. In addition, the two processes are also *spatially coupled*, meaning that the molecules involved in each event are located in the same compartment in a close-packed and highly ordered arrangement. In eukaryotes the molecular apparatus is located in the *inner mitochondrial membrane* (see Figure 15–2); in prokaryotes the location is the cell membrane.

A discussion and an appreciation of the design of the energy-yielding electron transport process and the mechanism of energy coupling require an understanding of certain physicochemical principles. Depending on your previous traning, you will find that the following material may or may not be largely a review.

Electrochemical Reduction Potentials

One of the basic concepts learned in any general chemistry course is that all oxidation-reduction reactions involve the transfer of electrons. Electrons are lost from the component undergoing oxidation (the reducing agent) and gained by the component undergoing reduction (the oxidizing agent). The overall process is called a *redox reaction*. The theme of this type of process is summarized in the following equations, particularly in the representation of the two *half-reactions*, which together comprise the complete system:

oxidized A + reduced B → reduced A + oxidized B

from left to right [→]:

A undergoes reduction (electrons gained are donated by B)

B undergoes oxidation (electrons lost are donated to A)

composed of two half-reactions (*couples*):

reduced B $\rightleftharpoons$ oxidized B + ne *oxidation*

oxidized A + ne $\rightleftharpoons$ reduced A   *reduction*

Half-reactions are frequently called *couples*, signifying the pair of oxidized and reduced forms of a given substance. Two principles are implicit in a redox reaction: (1) the reacting participants differ in terms of their affinity for electrons,

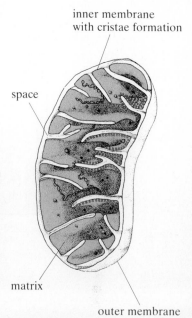

Outer membrane
enzymes such as
    amino acyl-SCoA synthetase
    monoamine oxidase
    nucleoside diphospho kinase

Space
adenylate kinase

Inner membrane
oxidative phosphorylation
succinate dehydrogenase
carnitine-acyl transferase
ATP transport system

Matrix:
pyruvate dehydrogenase
citric acid cycle enzymes except
    succinate dehydrogenase
β-oxidation enzymes
    phosphoenolpyruvate carboxy-
    kinase.

FIGURE 15–2 Compartments of a mitochondrion. The locations of some enzymes are listed. Note that the process of oxidative phosphorylation occurs in the inner mitochondrial membrane.

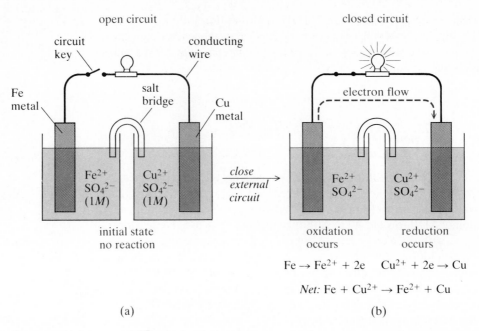

open circuit

closed circuit

(a)

initial state
no reaction

oxidation
occurs

reduction
occurs

$$Fe \rightarrow Fe^{2+} + 2e \quad Cu^{2+} + 2e \rightarrow Cu$$

Net: $Fe + Cu^{2+} \rightarrow Fe^{2+} + Cu$

(b)

FIGURE 15–3 Electrochemical cell. (a) Circuit open; no reaction. (b) Circuit closed. Electrons will spontaneously flow from iron couple to copper couple.

and (2) an event of oxidation must be accompanied by an event of reduction. Our major concern here will be addressed to the first principle. Finally, we note that like any other chemical reaction, an oxidation-reduction process can be described in terms of its overall energetics and whether or not it can occur spontaneously. The point to be established in this section is as follows:

> Both factors, energetics and spontaneity, are determined by the relative differences in the tendency of the reduced reactant of one couple to lose electrons and be oxidized, and of the oxidized reactant of the other couple to gain electrons and be reduced.

Let us further develop these principles by considering the system illustrated in Figure 15–3. The diagram is a conventional representation of a simple *electrochemical cell,* which by definition is a chemical system capable of generating a flow of current, that is, electricity. The assembly has four parts: (1) and (2) two containers, each consisting of a slab of pure metal in contact with a unit molar solution of its ions; for example, Fe metal with $Fe^{2+}SO_4^{2-}$ in one and Cu metal with $Cu^{2+}SO_4^{2-}$ in the other; (3) a salt bridge acting as an essential internal circuit connecting the two systems without permitting mixing; and (4) an external circuit consisting of a conducting wire.

After assembly but before the circuit key is closed, the bulb will be off, indicating the absence of any current flow. In chemical terms, neither chemical couple is undergoing reaction, and no equations are appropriate. When the key is closed, however, the bulb will light, signifying a spontaneous flow of current through the external circuit. Putting it another way, the system is *generating useful energy.*

The reason for the current flow is a net oxidation-reduction reaction involving each compartment, with the iron couple undergoing oxidation and the copper couple undergoing reduction. Thus the concentration of Cu^{2+} in the right compartment will decrease, whereas the concentration of Fe^{2+} in the left compartment will increase. Reflecting on the spontaneity of the process, we can conclude that on a relative basis the copper couple (Cu^{2+}, Cu) has a greater tendency to accept electrons and undergo reduction than does the iron couple (Fe^{2+}, Fe).

In physicochemical terms, the relative tendency of any couple to undergo reduction is expressed in terms of a *standard electrochemical reduction potential,* symbolized as $\mathscr{E}^{\circ}$ [with volts (V) as units]. The value of each potential is measured against the standard hydrogen couple ($2H^{+}$, H_2), a universally accepted frame of reference. Under strict standard conditions of unit pressure and unit molar concentration ($[H^{+}] = 1M$, or pH $= 0$), the value of the reduction potential for the hydrogen couple at 25°C is arbitrarily defined as 0.00 V. Under conditions more appropriate for biochemical use (namely, $[H^{+}] = 10^{-7}M$, or pH $= 7$), the value of the adjusted standard reduction potential ($\mathscr{E}^{\circ\prime}$) is -0.42 V. (We will not examine how the pH 7 correction is calculated.) Thus:

$$2H^{+} \,(1M, \text{pH} = 0) + 2e \rightleftharpoons H_2 \,(1 \text{ atm}) \qquad \mathscr{E}^{\circ} = 0.00 \text{ V}$$

$$2H^{+} \,(10^{-7}M, \text{pH} = 7) + 2e \rightleftharpoons H_2 \,(1 \text{ atm}) \qquad \mathscr{E}^{\circ\prime} = -0.42 \text{ V}$$

A direct measurement of the $\mathscr{E}^{\circ}$ for other couples is made by utilizing a setup similar to that shown in Figure 15–3, with one compartment being the (H^{+}, H_2) couple and the other being the couple to be measured. Both are initially at standard conditions. The external circuit contains a voltage-measuring device such as voltmeter or a potentiometer, and the observed voltage will be a measure of the capacity for current flow in the complete system. In other words, the voltage will reflect the net potential difference $\mathscr{E}^{\circ}_{\text{net}}$ between the reduction potential of the two couples. Since the reduction potential of the hydrogen system is arbitrarily set at 0, the observed net voltage will correspond to the reduction potential of the other couple. In general, for the *redox process* that can occur *spontaneously:*

$$\frac{\text{net potential}}{\text{complete system}} = \frac{\text{positive difference between the}}{\text{reduction potentials of each couple}}$$

That is,

$$\mathscr{E}^{\circ}_{\text{net}} = \mathscr{E}^{\circ}_{\substack{\text{couple} \\ \text{undergoing} \\ \text{reduction}}} - \mathscr{E}^{\circ}_{\substack{\text{couple} \\ \text{undergoing} \\ \text{oxidation}}}$$

By convention, a positive ($+$) sign accompanies the reduction potential of any couple that in fact has a greater tendency to undergo reduction relative to the hydrogen system. A negative ($-$) sign accompanies the reduction potential of any couple that has a lesser tendency to undergo reduction relative to the hydrogen system and, in fact, undergoes an oxidation instead. If the couple is pH-dependent, appropriate calculations are made to adjust the $\mathscr{E}^{\circ}$ value to a condition applicable to pH ($\mathscr{E}^{\circ\prime}$). The conventional standard is pH 7. The reduction potentials of certain couples, including some of biological importance, are listed in Table 15–1.

TABLE 15–1 Listing of some standard reduction potentials, in volts (for $\mathscr{E}^{\circ\prime}$: pH 7; T of 20°–30°C)

HALF-REACTION COUPLE	$\mathscr{E}^{\circ}$	$\mathscr{E}^{\circ\prime}$
$\frac{1}{2}O_2 + 2H^+ + 2e \rightleftharpoons H_2O$	—	+0.82
$Fe^{3+} + 1e \rightleftharpoons Fe^{2+}$	+0.77	—
$Cu^+ + 1e \rightleftharpoons Cu$	+0.52	—
$Cu^{2+} + 2e \rightleftharpoons Cu$	+0.34	—
$2H^+ + 2e \rightleftharpoons H_2$	0.00	−0.42
$Fe^{2+} + 2e \rightleftharpoons Fe$	−0.44	—
cytochrome $a_3 \cdot Fe^{3+} + 1e \rightleftharpoons$ cytochrome $a_3 \cdot Fe^{2+}$	—	+0.3 → 0.5[a]
cytochrome $f \cdot Fe^{3+} + 1e \rightleftharpoons$ cytochrome $f \cdot Fe^{2+}$	—	+0.37
cytochrome $a \cdot Fe^{3+} + 1e \rightleftharpoons$ cytochrome $a \cdot Fe^{2+}$	—	+0.29
cytochrome $c \cdot Fe^{3+} + 1e \rightleftharpoons$ cytochrome $c \cdot Fe^{2+}$	—	+0.25
cytochrome $c_1 \cdot Fe^{3+} + 1e \rightleftharpoons$ cytochrome $c_1 \cdot Fe^{2+}$	—	+0.22
coenzyme Q + $2H^+ + 2e \rightleftharpoons$ coenzyme QH_2 (in ethanol)	—	+0.10
cytochrome $b \cdot Fe^{3+} + 1e \rightleftharpoons$ cytochrome $b \cdot Fe^{2+}$	—	+0.04
$FAD + 2H^+ + 2e \rightleftharpoons FADH_2$	—	−0.12[b]
$NADP^+ + 2H^+ + 2e \rightleftharpoons NADPH[H^+]$	—	−0.32
$NAD^+ + 2H^+ + 2e \rightleftharpoons NADH[H^+]$	—	−0.32
oxidized ferredoxin + $1e \rightleftharpoons$ reduced ferredoxin	—	−0.43

[a] Very doubtful value; in vivo cyt a_3 is complexed with cyt a; the $\mathscr{E}^{\circ\prime}$ of the (a, a_3) complex, called *cytochrome oxidase,* is approximately +0.29.

[b] In vivo FAD exists in a firmly bound state to its protein. The potential for FAD in each of these individual flavoproteins may be more or less negative than −0.12 owing to the uniqueness of the association between FAD and the protein. The value of −0.12 given here is a representative average.

Since all reduction potentials are determined in the same manner and against the same reference system, the relative tendency of any two couples to participate in an electron transfer reaction can be readily predicted:

> Given any two couples, under appropriate conditions the couple with the more positive reduction potential will spontaneously tend to gain electrons and undergo reduction.

Consider again, for example, the system of Figure 15–3, consisting of the iron and copper couples:

Reduction Half-Reactions	*Reduction Potentials*
$Fe^{2+} + 2e \rightleftharpoons Fe$	$\mathscr{E}^{\circ} = -0.44$ V
$Cu^{2+} + 2e \rightleftharpoons Cu$	$\mathscr{E}^{\circ} = +0.34$ V

A comparison of $\mathscr{E}^{\circ}$ values predicts that Cu^{2+} will be reduced (its $\mathscr{E}^{\circ}$ is more positive), that Fe will be oxidized (its $\mathscr{E}^{\circ}$ is more negative), and that the net electrochemical redox potential is

$$\mathscr{E}^{\circ}_{net} = \mathscr{E}^{\circ}_{Cu^{2+}, Cu} - \mathscr{E}^{\circ}_{Fe^{2+}, Fe} = +0.78 \text{ V}$$

Note that this prediction is consistent with our earlier description of what indeed occurs in an iron-copper electrochemical cell.

Finally, let us consider an example of biological importance, namely, a reaction involving the NAD and oxygen couples. Applying the same principle reveals that the transfer of electrons from NADH to molecular oxygen will be the thermodynamically favored process:

Reduction Half-Reactions	*Reduction Potentials (pH 7)*
$NAD^+ + 2H^+ + 2e \rightleftharpoons NADH + H^+$	$\mathscr{E}^{\circ\prime} = -0.32$ V
$\frac{1}{2}O_2 + 2H^+ + 2e \rightleftharpoons H_2O$	$\mathscr{E}^{\circ\prime} = +0.82$ V

In other words, a comparison of $\mathscr{E}^{\circ\prime}$ values predicts that O_2 will be reduced and NADH will be oxidized, and that

$$\mathscr{E}^{\circ\prime}_{net} = \mathscr{E}^{\circ\prime}_{O_2, H_2O} - \mathscr{E}^{\circ\prime}_{NAD^+, NADH} = +1.14 \text{ V}$$

As previously indicated, the separate elements of a redox reaction can also be represented in terms of separate oxidation and reduction half-reactions. An oxidation half-reaction is merely the reverse of a reduction half-reaction, and the electrochemical oxidation potential $\mathscr{E}^\circ_{ox}$ has the same value but opposite sign of the reduction potential. When expressed this way, the net potential is calculated as $\mathscr{E}^\circ_{net} = \mathscr{E}^\circ_{ox} + \mathscr{E}^\circ_{red}$. For example, in $NADH + \frac{1}{2}O_2 \rightarrow NAD^+ + H_2O$:

oxidation:	$NADH + H^+ \rightarrow NAD^+ + 2e + 2H^+$	$\mathscr{E}^{\circ\prime}_{ox} = +0.32$ V
reduction:	$\frac{1}{2}O_2 + 2H^+ + 2e \rightarrow H_2O$	$\mathscr{E}^{\circ\prime}_{red} = +0.82$ V
		$\mathscr{E}^{\circ\prime}_{net} = +1.14$ V

Energetics of Electron Transfer

By recalling the principles of Chapter 12, you should recognize that the overall energetics of electron transfer reactions, embodied by $\Delta G^{\circ\prime}$, ought to be related to the net potential difference $\mathscr{E}^{\circ\prime}_{net}$. Neglecting any development of this relation, we simply state that such is the case. The exact form of the mathematical relationship has already been given in Chapter 12 but should be briefly reviewed. It is

$$\Delta G^{\circ\prime} = -n \mathscr{F} \mathscr{E}^{\circ\prime}_{net}$$

where n is the number of electrons transferred in the overall process and $\mathscr{F}$ is a proportionality constant equal to 23,060 cal/V. The point to note is that whenever the potential difference $\mathscr{E}^{\circ\prime}_{net}$ has a positive value, the $\Delta G^{\circ\prime}$ will be negative, signifying that the process is not only capable of occurring spontaneously but also is energy-yielding.

Applying this equation to the reduction of NADH by molecular O_2 reveals, for example, that the potential output of energy is 52.6 kcal, a sizable energy yield indeed. That is,

$$NADH + H^+ + \tfrac{1}{2}O_2 \rightleftharpoons NAD^+ + H_2O \qquad \mathscr{E}^{\circ\prime} = +1.14 \text{ V}$$

Since $n = 2$ electrons,

$$\Delta G^{\circ\prime} = -(2)(23,060)(1.14) = -52.6 \text{ kcal}$$

Electrochemical Potentials Due to Ion Gradients

A voltage difference—that is, an electrochemical potential—between two phases can result from situations other than the occurrence of redox chemistry. However they arise, all electrochemical potentials can be expressed in terms of a free energy (ΔG) equivalent as previously identified for a net redox potential. A nonredox electrochemical potential of immense biological significance involves the *unequal distribution of ions* across a membrane. The potential arises from two factors: (1) the difference in the actual concentration of the ion on the two sides of the membrane, a true *concentration gradient*, and (2) the difference in electric charge between the two regions on each side of the membrane, a *charge gradient*.

In Section 11–6 we highlighted the unequal distribution of K^+, Na^+, and Cl^- across membranes, particularly in nerve and muscle cells. In this chapter we will encounter another example, namely, the unequal distribution of *protons* H^+ in mitochondria, chloroplasts, and prokaryote bacteria. It is referred to as an *electrochemical H^+ potential* and is symbolized as $\Delta\mu_{H^+}$. Its formation serves as the basis of transducing the energy available from electron transport into the energy required for phosphorylation of ADP. The total value of $\Delta\mu_{H^+}$ is expressed as

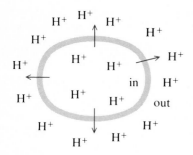

vesicle with a membrane

$$\Delta\mu_{H^+} = \Delta\Psi + (-Z\,\Delta pH) = \Delta\Psi - Z\,\Delta pH$$

where $\Delta\Psi$ is the membrane potential (in volts), reflecting a region on one side of the membrane being more negatively charged than the region on the other side; and ΔpH is the actual concentration gradient of H^+ across the membrane, expressed in terms of the pH values on each side of the membrane. The symbol Z is a factor used to convert the ΔpH value into units of volts ($Z = 0.06$ at 30°C).

Because ΔpH is expressed as ($pH_{out} - pH_{in}$), and because electron transport pumps H^+ out of a membrane-bound compartment so that pH_{out} is less than pH_{in}, the value of ΔpH is negative (see Figure 15–4). Thus multiplying $Z\,\Delta pH$ by a minus sign gives a positive value.

We will return to this material later in the chapter, focusing on the formation of the H^+ gradient, its magnitude, and how it is proposed to drive ATP formation.

FIGURE 15–4 Electrochemical proton potential ($\Delta\mu_{H^+}$), with $[H^+]_{out} > [H^+]_{in}$, or $pH_{out} <$ pH_{in}. The inside compartment is less positive (or more negative) than the outside. Colored arrows identify the vectorial displacement of H^+ out of the vesicle.

15–2 RESPIRATORY CHAIN OF ELECTRON TRANSPORT

Oxygen as a Terminal Electron Acceptor

The various ways in which O_2 participates as an oxidizing agent in biochemical reactions and also the matter of oxygen toxicity will be surveyed later in the chapter. For now, our focus is on the major role of O_2 as an essential nutrient of aerobic organisms. That role is represented by the following equation involving a type of enzyme identified for now merely as an *oxidase:*

$$SH_2 \; + \; \tfrac{1}{2}O_2 \; \xrightarrow{\text{oxidase}} \; S \; + \; H_2O$$

electron   electron
donor   acceptor

The generalized form of this oxidase reaction is somewhat deceptive regarding the role of oxygen in respiration, because it suggests that the oxidation of a reduced substrate (SH_2; for example, malate) proceeds via a direct dehydrogenation, with oxygen serving as an immediate electron acceptor. Although there are examples of this type of reaction in cells, the participation of oxygen in electron transport is of a completely different nature, with more than one catalyst involved.

The pioneering work in this area was done independently by H. Wieland, T. Thunberg, O. Warburg, and D. Keilin in the 1920s and 1930s. They proposed that O_2-dependent dehydrogenation reactions involve *intermediate electron carriers* that intervene in the flow of electrons between the initial electron donor SH_2 and the *terminal electron acceptor* O_2. The entire process consists of a chain of redox reactions involving the *successive interaction* of carriers (see Figure 15–5). Each intermediate carrier first participates in its oxidized state (C_{ox}^1) as an acceptor of electrons and is converted to its reduced state (C_{red}^1). In the reduced state the carrier then transfers electrons to the next carrier in its oxidized state (C_{ox}^2) to yield C_{red}^2, and in so doing, C_{red}^1 is reconverted to the original oxidized state (C_{ox}^1). The final carrier transfers electrons to O_2, which is reduced to water. The operation of a successive series of oxidation-reduction reactions is called an *electron transport chain.* In aerobic cells the process is called a *respiratory chain.*

The existence of electron transport chains was a key discovery in biochemistry, because this pattern of chemistry occurs in all types of cells. After nearly sixty years of study, many of the details of operation have been discovered, but some gaps in our knowledge still remain. We will now consider some of the details.

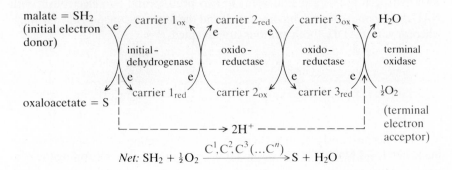

FIGURE 15–5 Transport chain involving intermediate electron carriers. Symbol *e* represents electron flow.

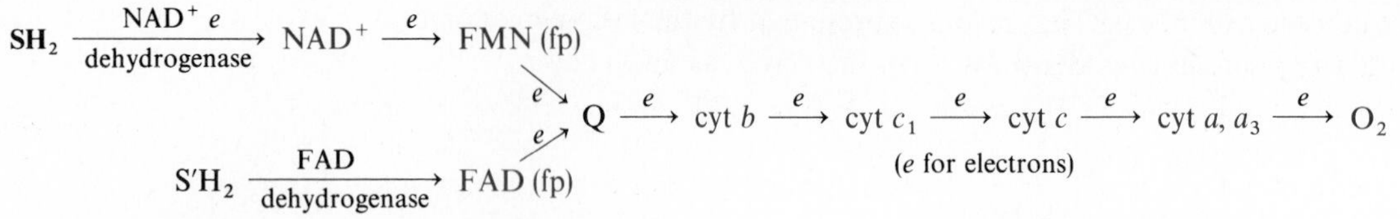

$$SH_2 \xrightarrow[\text{dehydrogenase}]{NAD^+ \; e} NAD^+ \xrightarrow{e} FMN \;(fp)$$

$$\xrightarrow{e} \; Q \xrightarrow{e} cyt \; b \xrightarrow{e} cyt \; c_1 \xrightarrow{e} cyt \; c \xrightarrow{e} cyt \; a, a_3 \xrightarrow{e} O_2$$

$$S'H_2 \xrightarrow[\text{dehydrogenase}]{FAD} FAD \;(fp)$$

(e for electrons)

FIGURE 15–6 Basic design of the electron transport chain in respiring mitochondria. The oxidation of SH_2 substrates such as pyruvate, isocitrate, α-ketoglutarate, and malate primes the chain with NADH. The oxidation of other substrates by a flavin-dependent dehydrogenase primes the chain with fpH_2.

Electron Carriers

In higher organisms the known carriers of the respiratory chain are nicotinamide adenine dinucleotide (NAD), flavin nucleotides (FAD and FMN), coenzyme Q quinones (Q), a family of cytochrome (cyt) proteins (designated as b, c_1, c, a, a_3), and iron-sulfur proteins (designated as FeS). Their sequence of interaction is shown in Figure 15–6. Iron-sulfur proteins are not indicated in this representation because they are proposed to intervene at several positions rather than just one.

The process begins with a transfer of electrons from a reduced substrate (SH_2 or $S'H_2$) to either NAD^+ or FAD, depending on whether the initial dehydrogenase involved is NAD^+-dependent or FAD-dependent. When the process begins at the level of NAD^+, note that the next carrier is FMN. Whatever the initial substrate, the electrons from the flavin carrier are then transferred to coenzyme Q, and from there they go through a specific sequence of the cytochrome carriers and ultimately are delivered to O_2, the terminal acceptor.

NAD^+, FAD, AND FMN   The structure and oxidation-reduction chemistry of the NAD and flavin (FAD and FMN) coenzymes have been examined in Section 12–2. Now we focus on the fact that the reduced state of each coenzyme (NADH, $FADH_2$, and $FMNH_2$) represents the initial high-energy state (high energy relative to the terminal reaction of $O_2 \rightarrow H_2O$) that primes the respiratory chain. Ordinarily, NAD^+ is a free coenzyme that binds to a dehydrogenase enzyme only during the course of the reaction catalyzed by the dehydrogenase. In contrast, the flavin function is generally firmly bound to its dehydrogenase, and the complete protein is called a *flavoprotein* (fp).

Although in our previous discussions of NAD^+ and FAD, we referred to each as a hydrogen acceptor coenzyme, the term *electron carrier* is wholly equivalent, since each coenzyme gains the equivalent of two electrons. (Remember: Two hydrogen atoms are equivalent to two protons and two electrons.) With NADH only one proton is incorporated, and the other is gained by the medium; whereas with $FADH_2$ both protons are accepted:

$$SH_2 + NAD^+ \rightarrow S + NADH + H^+$$

$$\frac{S'H_2 + FAD \rightarrow S' + FADH_2}{-2Hs\;(2H^+, 2e)}$$

In referring to the reduced state of NAD throughout this chapter, we use the NADH and NADH + H^+ designations interchangeably. Dropping the proton is merely a matter of convenience. Later, however, we will have reason to highlight the involvement of H^+, emphasizing the difference between carriers of *only electrons* and carriers of *both protons and electrons*.

Although unrelated to their biological function, the reduced and oxidized forms of NAD have distinct light-absorbing properties (see Figure 15–7 and Appendix I). Although both forms display an absorption maximum at 260 nm, the reduced form also absorbs at 340 nm, whereas the oxidized form does not. Thus at 340 nm one can detect NADH in the presence of NAD^+ and correlate any change in absorption at 340 nm to an interconversion of the two forms. Similar light absorption differences also typify the oxidized and reduced forms of the flavin and cytochrome carriers. Some of the cytochrome details are presented later in this section. These differences can be exploited in the laboratory to follow the progress of a reaction involving these substances as electron carriers, including the progress of electron flow through the entire respiratory chain.

QUINONES The coenzyme Q designation refers to a family of compounds, sometimes called *ubiquinones* (Q for quinone), because of their ubiquitous occurrence in nature. Coenzyme Q molecules differ from source to source in the length of the hydrocarbon chain, with the value of n ranging from 6 (in microbes) to 10 (in mammals). As an electron and proton carrier, the quinone grouping is capable of undergoing a reversible conversion to the hydroquinone grouping (see Figure 15–8). As shown in Figure 15–8, the overall $Q \rightleftharpoons QH_2$ interconversion can occur in two stages, each being a one-electron (and one-proton, H^+) transfer reaction, involving the intermediate formation of a *semiquinone radical* species.

In support of its participation in the respiratory chain is the fact that the bulk of the lipidlike, ubiquinone is found in the inner mitochondrial membrane (the subcellular site of electron transport), from which it can be readily extracted by chloroform and other liquid solvents. In the lipid bilayer individual Q molecules can move laterally and transversely. Whether or not the in vivo activity of coenzyme Q is dependent on its being complexed to a protein within the membrane is unknown.

Certain bacteria utilize vitamin K naphthoquinones (see Section 11–4) as intermediate carriers. Whether the vitamin K type of compounds play a corresponding role in electron transport in higher organisms remains unresolved. Vitamin E may also participate.

CYTOCHROMES First detected in the late nineteenth century, *cytochromes* ("cellular pigments") are present in all types of cells. They are localized

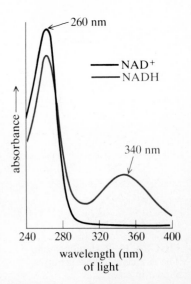

FIGURE 15–7 Spectrum of light absorption for NAD^+ (black) and NADH (color).

FIGURE 15–8 Redox chemistry of coenzyme Q via one-electron and one-proton transfer reactions.

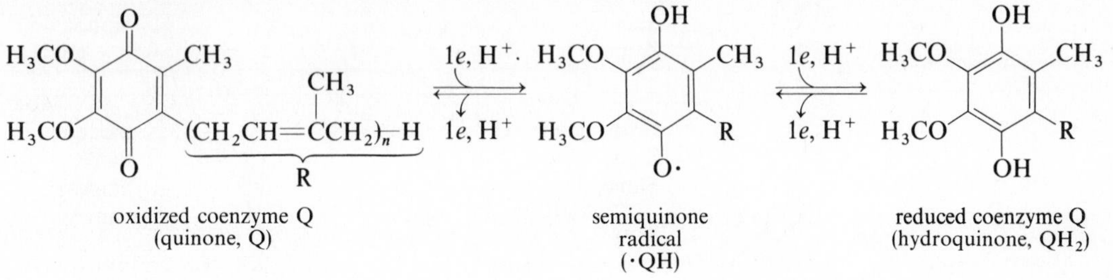

oxidized coenzyme Q
(quinone, Q)

semiquinone
radical
(·QH)

reduced coenzyme Q
(hydroquinone, QH_2)

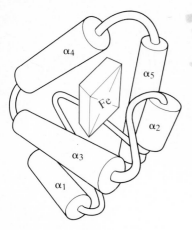

FIGURE 15–9 Hemoprotein. The associated heme group is shown in color.

primarily in membranes and, in eukaryotic cells, specifically in the mitochondrial membrane and the endoplasmic reticulum. Every cytochrome is a *hemoprotein*— a protein composed of an Fe-containing, planar heme group in association with a polypeptide (see Figure 15–9).

Our only previous consideration of hemoproteins dealt with myoglobin and hemoglobin, both of which require the reduced form of iron (Fe^{2+}) to bind molecular oxygen. The heme iron is also the active center of cytochromes but in a different way.

Cytochrome proteins participate in redox reactions with the heme iron capable of undergoing a reversible oxidation-reduction via the exchange of one electron per heme group.

$$Fe^{2+} \underset{+1\ electron}{\overset{-1\ electron}{\rightleftharpoons}} Fe^{3+}$$
$$\text{(reduced} \qquad\qquad \text{(oxidized}$$
$$\text{state)} \qquad\qquad\quad \text{state)}$$

This reaction is the most simplified statement of the biochemical role of the cytochromes.

Approximately 25–30 different cytochromes are known to exist throughout nature. All types are classified on the basis of distinctive light-absorbing properties, with those of similar properties segregated into groups designated by lowercase letters (*a*, *b*, *c*, and so on). Within each group, individual cytochromes with unique spectral properties are then designated by numerical subscripts (b, b_1, b_2, b_3). The structural factors accounting for the many cytochromes are (1) variations in the side-chain substituents of the tetrapyrrole moiety of the heme group (see Table 15–2 and Figure 15–10), (2) variations in the structure of the polypeptide component, and (3) variations in the way the polypeptide is bound to the heme unit.

TABLE 15–2 Comparison of heme in Figure 15–10 with other hemes

POSITION	a CYTOCHROMES	c CYTOCHROMES
1	Same	Same
2 (in a)	$-CHCH_2CH(CH_2)_3CH(CH_2)_3CH(CH_3)_2$ 　　|　　　　|　　　　| 　OH　　CH$_3$　　CH$_3$	
2 (in c)		$-CHCH_3$ （S—protein (covalent attachment)
3	Same	Same
4	Same	$-CHCH_3$ | S—protein
5	Hydrogen (—H)	Same
6	Same	Same
7	Same	Same
8	$-C{=}O$ (formyl group) 　H	Same

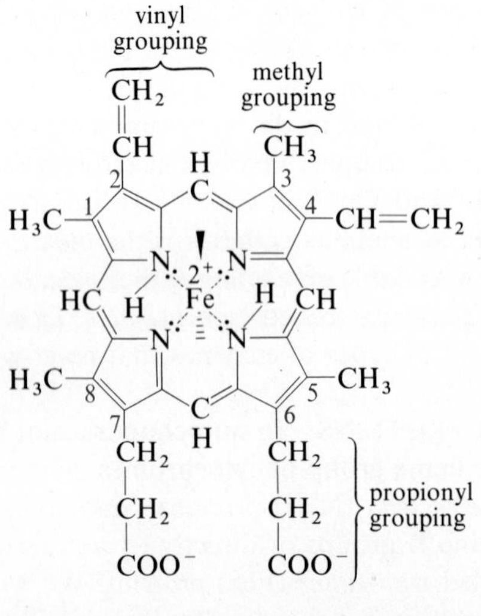

vinyl
grouping

methyl
grouping

propionyl
grouping

FIGURE 15–10 Heme of all b cytochromes and also in hemoglobin, myoglobin, and catalase. Wedges represent the fifth and sixth coordinate bonds of the heme iron to the polypeptide.

The graph in Figure 15–11 and the data of Table 15–3 summarize the general and specific absorption properties of some cytochromes. Note that the absorption spectrum for cytochromes in the reduced state usually contains three distinct maxima. The precise values of these absorption maxima (called the α, β, and γ bands) are used to differentiate among the individual cytochromes. Since the spectrum of the oxidized form of each cytochrome usually lacks these three maxima, one can also distinguish between the oxidized and the reduced forms and even determine the relative concentration of the two when they are present.

In the context of their biological role as intermediate electron carriers in the respiratory chain, a more significant distinction among the cytochromes is the following:

The variations in cytochrome structure are responsible for differences in the ability of individual cytochromes to participate in redox reactions.

This distinction is reflected in the specific $\mathscr{E}^{\circ\prime}$ values (see Table 15–1).

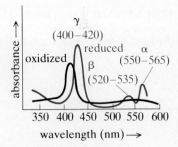

FIGURE 15–11 Typical absorption spectrum of cytochromes. The usual ranges of the α, β, and γ bands in the spectrum of a reduced cytochrome are identified.

TABLE 15–3 Light-absorbing characteristics of some cytochromes

CYTOCHROME (SOURCE)	MOLECULAR WEIGHT	ABSORPTION BANDS		
		α	β	γ
b	25,000	563	532	429
c	12,500	550	521	415
c_1 (mitochondria of animals, plants, yeast, and fungi)	37,000	554	524	418
a	$\approx 600,000$ for a, a_3 aggregate	600	Absent	439
a_3		604	Absent	443
b_1 (E. coli)	500,000	558	528	425
b_2 (yeast)	170,000	557	528	424
b_5 (microsomes)	25,000	556	526	423
f (chloroplasts)	?	555	525	423

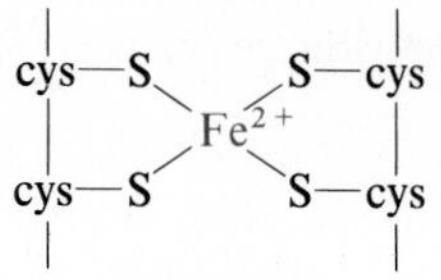

Cytochromes a and a_3 are very difficult to isolate as separate substances. In the membrane they exist and function as a tightly associated aggregate, referred to as the a, a_3 complex. This complex is further distinguished by the presence of *copper* (Cu), bound to the a_3 component, which is essential for the activity. In fact, the a, a_3 complex involves electron transfer, with both events of $Fe^{+2} \rightleftharpoons Fe^{+3} + 1e$ and $Cu^{+1} \rightleftharpoons Cu^{+2} + 1e$ occurring. The unique functional feature of the a, a_3 complex is that it is the only cytochrome component of the respiratory chain capable of reacting with molecular oxygen. Accordingly, the a, a_3 complex is frequently called *cytochrome oxidase*. Later in the chapter we will survey the various types of enzymes that react with O_2.

IRON-SULFUR PROTEINS   In mitochondria not all of the iron present is associated with the heme group of cytochromes. Some is complexed to other proteins called *nonheme iron* (NHI) *proteins*. Since the Fe is bound through sulfur atoms of cysteine R groups or directly to complexed sulfide (S^{-2}), these proteins are also called *iron-sulfur* (FeS) *proteins*. We will use the latter term. Examples of FeS_4 and Fe_2S_6 clusters are diagramed in Figure 15–12. Although different FeS proteins participate as carriers in the respiratory chain, there is still uncertainty about their exact locations in the carrier sequence and the exact manner in which the iron-sulfur cluster undergoes reversible redox.

In addition to the FeS proteins that participate in the electron transport of the respiratory chain, there exist throughout nature many other FeS proteins involved in various processes. The protein ferredoxin is of particular importance in the process of photosynthesis (see Chapter 16).

Design for the Respiratory Chain

A more detailed representation of the respiratory chain is given in Figure 15–13. Once again note that the process can be primed by electrons donated from NADH (originating from the oxidation of a substrate via an NAD-dependent dehydrogenase) or from $FADH_2$ (originating from an FAD-dependent dehydrogenase).

The reason that the NAD → flavin → CoQ → cytochromes → O_2 sequence has evolved is evident by examining the standard reduction potentials. Note that with the exception of the CoQ couple, the value becomes progressively more positive from NAD to oxygen. In other words, the carriers operate in order of an *increasing tendency to undergo reduction*. On an energy scale, then, each carrier in its reduced state is of higher energy than the reduced form of the next carrier in the chain. Thus *electrons proceed to a lower energy level*.

For example, when the chain is primed with NADH, the transfer of electrons in the first redox step—the reduced NADH donor is oxidized and the oxidized FMN acceptor is reduced—is an energetically favorable process, with a positive $\mathscr{E}^{\circ\prime}_{net}$ and thus a negative $\Delta G^{\circ\prime}_{net}$:

FeS$_4$ cluster

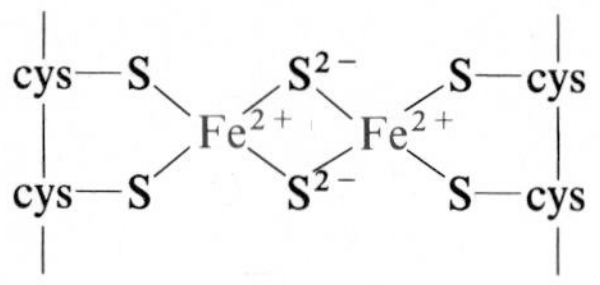

Fe$_2$S$_6$ cluster

FIGURE 15–12 Examples of iron-sulfur clusters in FeS proteins. During redox reactions $Fe^{2+} \rightleftharpoons Fe^{3+}$ occurs.

$$
\begin{array}{ll}
NADH + H^+ \rightarrow NAD^+ + 2H^+ + 2e & \mathscr{E}^{\circ\prime}_{ox} = -(-0.32) \\
FMN + 2H^+ + 2e \rightarrow FMNH_2 & \mathscr{E}^{\circ\prime}_{red} = -0.12 \\
\hline
NADH + H^+ + FMN \rightarrow NAD^+ + FMNH_2 & \mathscr{E}^{\circ\prime}_{net} = +0.20 = \mathscr{E}^{\circ\prime}_{ox} + \mathscr{E}^{\circ\prime}_{red}
\end{array}
$$

$$\Delta G^{\circ\prime}_{net} = -n\mathscr{F}\mathscr{E}^{\circ\prime}_{net} = -(2)(23,060)(0.20) = -9.2 \text{ kcal}$$

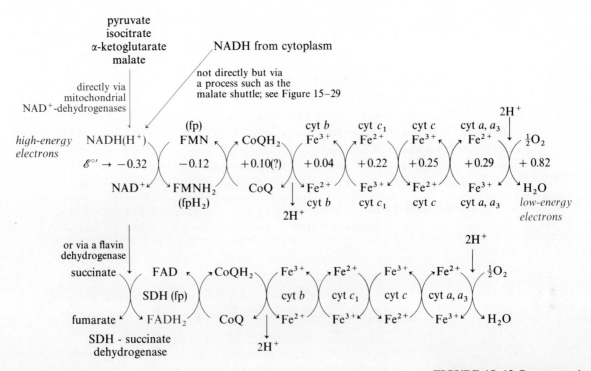

FIGURE 15–13 Representation of the respiratory chain linked to the oxidation of a substrate by an NAD-dependent dehydrogenase (*top*) and by a flavin-dependent dehydrogenase (*bottom*). The reduction potential values are as listed in Table 15–1. Iron-sulfur protein carriers are not shown here.

The same thing is true of the $FMNH_2$ and oxidized CoQ involved in the next step.

Since this type of pattern is continuous, the thermodynamic design is clearly evident:

The electron carriers are so arranged that each transfer has the capacity to proceed spontaneously and exergonically.

Note: The nonfit of the CoQ couple should be considered in the context of the conditions under which the $\mathscr{E}^{\circ\prime}$ has been measured, namely, in 95% ethanol, a condition not exactly representative of the natural environment of a biological membrane. The physiological in vivo function of CoQ as an electron carrier may be more closely approximated by a value of 0.0 V for $\mathscr{E}^{\circ\prime}$.

Although two electrons are transferred in the overall process, there is no proof whether the cytochrome sequence in particular involves one cytochrome molecule functioning twice or two separate molecules transferring one electron each. The $(cyt)_2$ designation is employed in Table 15–4 only for bookkeeping purposes, to account for the net transfer of $2e$.

The $\Delta G^{\circ\prime}$ analysis of each redox step was once used as a basis to evaluate which specific steps would be candidates for providing sufficient energy to form ATP for which the $\Delta G^{\circ\prime}$ of formation is about 8 kcal. However, the modern chemiosmotic theory of explaining the coupling of ADP phosphorylation and electron transport is not based on the premise of specific redox steps having the necessary $\Delta G^{\circ\prime}$ value. Rather, the chemiosmotic theory proposes that certain steps (sites) of the transport chain are associated with contributing to the

TABLE 15–4 Energetics of individual steps in respiratory chain

RESPIRATORY CHAIN REACTIONS	$\Delta \mathscr{E}^{\circ\prime}$ (V)	$\Delta G^{\circ\prime}$ (kcal)
$NADH(H^+) + fp \rightarrow NAD^+ + fpH_2$	+0.20	−9.2
$fpH_2 + CoQ(ox) \rightarrow fp + CoQH_2(red)$	+0.22	−10.2
$CoQH_2 + (cyt\ b \cdot Fe^{3+})_2 \rightarrow CoQ + (cyt\ b \cdot Fe^{2+})_2 + 2H^+$	−0.06	+2.8
$(cyt\ b \cdot Fe^{2+})_2 + (cyt\ c_1 \cdot Fe^{3+})_2 \rightarrow (cyt\ b \cdot Fe^{3+})_2 + (cyt\ c_1 \cdot Fe^{2+})_2$	+0.18	−8.3
$(cyt\ c_1 \cdot Fe^{2+})_2 + (cyt\ c \cdot Fe^{3+})_2 \rightarrow (cyt\ c_1 \cdot Fe^{3+})_2 + (cyt\ c \cdot Fe^{2+})_2$	+0.03	−1.4
$(cyt\ c \cdot Fe^{2+})_2 + (cyt\ a, a_3 \cdot Fe^{3+})_2 \rightarrow (cyt\ c \cdot Fe^{3+})_2 + (cyt\ a, a_3 \cdot Fe^{2+})_2$	+0.04	−1.8
$(cyt\ a, a_3 \cdot Fe^{2+})_2 + \frac{1}{2}O_2 + 2H^+ \rightarrow (cyt\ a, a_3 \cdot Fe^{3+})_2 + H_2O$	+0.53	−24.4
Net: $NADH(H^+) + \frac{1}{2}O_2 \rightarrow NAD^+ + H_2O$	+1.14	−52.6
or $\quad fpH_2 + \frac{1}{2}O_2 \rightarrow fp + H_2O$	+0.94	−43.4

establishment of an energy-rich electrochemical proton potential $(\Delta \mu_{H+})$ by pumping H^+ out of the mitochondrion. Shortly, we will examine this feature in greater detail.

Establishing the Sequence of Electron Carriers

Direct experimental proof of the carrier sequence was obtained in different ways. One involved the use of *respiratory inhibitors,* substances that interrupt the flow of electrons at specific locations along the chain. For example, *barbituates,* such as *amytal,* are potent inhibitors of electron transfer in the flavin → CoQ (cyt *b*) region. The antibiotic *antimycin A* specifically inhibits electron transfer at the cyt *b* → *c*₁ site. *Cyanides* (CN^-), *azides* (N_3^-), and *carbon monoxide* (CO) are potent inhibitors of cytochrome oxidase.

By incubating mitochondria with an oxidizable substrate such as malate in the presence of a respiratory inhibitor, one can measure the amount of the various carriers that exist in the reduced state and compare these results with levels of reduced carriers in a control system lacking the inhibitor.

The presence of the respiratory inhibitor will decrease the levels of the reduced form of the carriers after the site of inhibition and increase the levels of reduced carriers that exist after the site of inhibition (see Figure 15–14).

For example, in the presence of antimycin, inhibiting the cyt *b* → cyt *c*₁ transfer, one will observe that the amounts of cytochromes c_1, c, a, and a_3 in the reduced state will decrease and the amounts of cyt *b*, fpH_2, and NADH will increase. One can detect and measure such changes with sensitive spectroscopic techniques because each of the carriers displays a different absorption spectrum in its oxidized and reduced states. For example, reduced cyt *c* shows a strong absorption maximum at approximately 550 nm (the alpha band of reduced cyt *c*), whereas oxidized cyt *c* shows less absorption at the same wavelength (see Table 15–3). Thus in a comparison of the inhibited and control systems, a measure

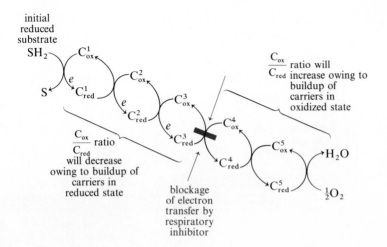

FIGURE 15–14 Diagram showing how the presence of a respiratory inhibitor will affect the amount of a carrier in its two states.

of the decrease in absorption at 550 nm will reflect a decrease in the amount of reduced cyt c and an increase in the amount of oxidized cyt c.

Further proof of the NAD $\rightarrow$ flavin $\rightarrow$ CoQ $\rightarrow$ b $\rightarrow$ c_1 $\rightarrow$ c $\rightarrow$ a, a_3 $\rightarrow$ O_2 sequence came in the 1960s with the development of gentle laboratory procedures to disrupt intact mitochondria into different types of functional fragments. Osmotic shock, exposure to ultrasonic frequency, and detergents can be used to accomplish the following stages of mitochondrial fractionation (see Figure 15–15):

1. Disrupting only the outer membrane, leaving a vesicle bounded only by a single membrane, which originally was the inner mitochondrial membrane.

2. Forming smaller vesicles (submitochondrial particles, SMP) from the cristae folds due to fracture planes occurring at the heel of the crista, followed by a fusion of the two fracture faces. (Note that the membrane surfaces of the SMP vesicles have the reverse orientation than in the stripped mitochondria from which they were formed, as if the crista has been turned inside out.)

3. Dislodging the protein components (color) that are in association with the external face of the SMP, to yield vesicles called *respiratory particles,* capable only of electron transport but not phosphorylation.

4. Disrupting the respiratory particles into membrane fragments (*respiratory complexes*), different ones capable of catalyzing only a part of the overall NADH $\rightarrow$ O_2 process.

Before considering the catalytic properties of the respiratory complexes, we should highlight what else can be concluded from such results: (1) The processes of both electron transport and phosphorylation are localized only in the inner mitochondrial membrane; (2) the coupling of electron transport to ATP formation occurs in association with closed vesicles; and (3) the protein component in association with the matrix surface of the inner mitochondrial membrane is necessary for ATP formation but not required for electron transport.

As indicated in Figure 15–15, four different respiratory complexes (I, II, III, IV) can be isolated from the inner membrane of mitochondria. Complex I

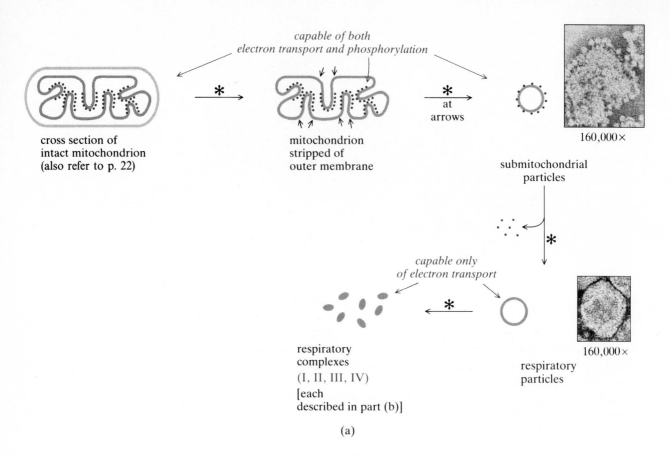

(a)

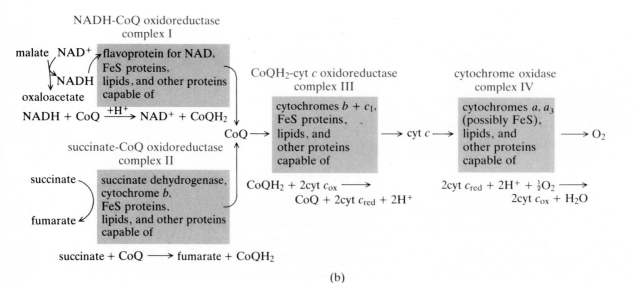

(b)

(NADH-CoQ oxidoreductase) catalyzes the transfer of electrons from NADH to CoQ. Complex II (succinate-CoQ oxidoreductase) catalyzes the transfer of electrons from succinate to CoQ. Complex III (CoQH$_2$-cytochrome c oxidoreductase) catalyzes the transfer of electrons from CoQH$_2$ to cyt c. Complex IV (cytochrome oxidase) catalyzes the transfer of electrons from cyt c to oxygen. Notice that the specific catalytic properties of each complex are perfectly consistent with the sequence of NADH $\rightarrow$ flavin $\rightarrow$ CoQ $\rightarrow$ b $\rightarrow$ c_1 $\rightarrow$ c $\rightarrow$ a, a_3 $\rightarrow$ O$_2$. Furthermore, the activities of each complex are additive, meaning that two complexes together will result in redox chemistry corresponding to the sum of the two separate complexes:

$$\text{NADH(H}^+\text{)} + (\text{cyt } c\cdot\text{Fe}^{3+})_2 \xrightarrow[\text{CoQ}]{\text{I + III}} \text{NAD}^+ + (\text{cyt } c\cdot\text{Fe}^{2+})_2 + 2\text{H}^+$$

$$\text{succinate} + (\text{cyt } c\cdot\text{Fe}^{3+})_2 \xrightarrow[\text{CoQ}]{\text{II + III}} \text{fumarate} + (\text{cyt } c\cdot\text{Fe}^{2+})_2 + 2\text{H}^+$$

$$\text{CoQH}_2 + \tfrac{1}{2}\text{O}_2 + 2\text{H}^+ \xrightarrow{\text{III + IV}} \text{CoQ} + \text{H}_2\text{O}$$

Note that measurable activity of each complex requires the addition of the substrate of each complex. In other words, the complexes themselves do not contain appreciable levels of NAD, CoQ, succinate, and cytochrome c. The fact that CoQ and cytochrome c must be added to demonstrate activity is the result of their being extracted during the fractionation procedure, indicating that at least these two substances are less tightly packaged in the mitochondrial membrane than are the other components. Both NAD and succinate are soluble substances to begin with.

Topology of Electron Carriers Within the Membrane

Although we cannot yet describe how all the electron carriers and ATP-forming proteins are arranged in the inner mitochondrial membrane, progress is being made. For example, there is firm evidence that the ultrastructural topology of the membrane (see Figure 15–16) is *asymmetric,* with some of the carrier proteins located on one side of the membrane and some located on the other side. In addition, some carriers span the entire width of the membrane.

Note that two cytochrome b species are identified in complex III. There is evidence that these two species represent a dimeric state of the cyt b carrier, with each subunit existing in a different microenvironment in the bilayer. The

◀ **FIGURE 15–15** Stepwise disruption of mitochondria. (a) Controlled conditions of osmotic shock, exposure to ultrasonic frequency, and/or exposure to mild detergents (identified by ∗). Actual electron micrographs of submitochondrial particles and respiratory particles are shown. (b) Composition and function of individual respiratory complexes (identified in the shaded areas). Each rectangle represents an active respiratory complex. Primary constituents of each are identified. The designation of lipids and other proteins refers to the lipoprotein matrix in which the functional constituents are embedded. In other words, these multimolecular aggregates are fragments of the inner mitochondrial membrane.

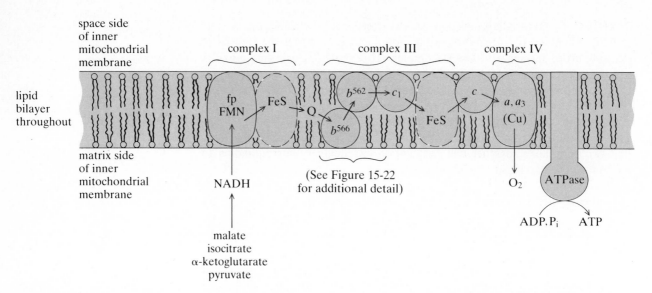

FIGURE 15–16 Arrangement and orientation of electron carriers and ATPase component in inner mitochondrial membrane. Not all FeS proteins are shown here. For example, one or more FeS proteins operate in complex III between Q and cyt c_1. The projection on the matrix side of the membrane represents the molecular apparatus involved in ATP formation; it is not necessarily localized only at the end of the chain.

evidence comes from spectroscopic measurements that detect one b species absorbing at 562 nm and another absorbing at 566 nm—hence the designations in Figure 15–16 of b^{562} and b^{566}. Later in the chapter we will examine the proposed involvement of these b subunits in the transfer of electrons through complex III.

Figure 15–16 also highlights the knoblike particles in association with the matrix side of the inner membrane. Called H^+-$ATPase$ (see Figure 15–17), this transmembrane protein complex has a *direct role in the formation of ATP*. The ATPase protein is more thoroughly described later in the chapter. Numerous copies of the carriers and the ATPase are present throughout the inner membrane.

15–3 COUPLING OF THE RESPIRATORY CHAIN TO ATP FORMATION

Yield of ATP from Electron Transport Chain

Using $\Delta G^{\circ\prime}$ values to compare the energy yield of electron transport ($FADH_2 \rightarrow O_2$ or $NADH \rightarrow O_2$) and the energy requirement of phosphorylation ($ADP + P_i \rightarrow ATP$), we see that the formation of about five or six ATP molecules could be energetically coupled to the transport of two electrons ($2e$). See Table 15–5.

However, $\Delta G^{\circ\prime}$ values do not reflect true in vivo conditions, and so such an evaluation is, at best, only a rough approximation. Indeed, actual measured values of the ATP yield in active mitochondria are significantly different: 2ATP for $FADH_2$ oxidation and 3ATP for NADH oxidation.

TABLE 15–5 Comparison of energy yields and energy need

REACTION	$\Delta G^{\circ\prime}$ (kcal/mole)	ATP Yield ESTIMATED	ATP Yield ACTUAL
$NADH + H^+ + \frac{1}{2}O_2 \xrightarrow{2e} NAD^+ + H_2O$	-52.6		3
$ADP + P_i \xrightarrow{2e} ATP + H_2O$	$+8.0$	≈ 6	
$FADH_2 + \frac{1}{2}O_2 \xrightarrow{2e} FAD + H_2O$	-43.4	≈ 5	2

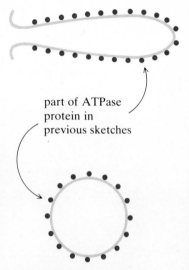

FIGURE 15–17 Membrane-localized H^+-ATPase component (color).

How are such measurements made? A typical experimental design is shown in Figure 15–18. Aside from the usual control of pH, temperature, and ionic strength, the basic requirements are that (1) purified mitochondria be used; (2) the reduced substrate added be oxidized by a mitochondrial dehydrogenase; (3) NAD^+ be added if the substrate is to be oxidized by an NAD^+-dependent dehydrogenase; (4) known amounts of ADP and P_i be added; and (5) an accurate measure be made of the amount of P_i consumed or ATP produced and of the amount of O_2 or reduced substrate consumed. Results are conventionally expressed as a *P/O ratio* (or, more generally, as a *P/2e ratio*), which corresponds to the number of moles of P_i consumed per gram atom of oxygen consumed. Alternative meanings are the number of moles of ATP produced per gram atom of oxygen consumed, or the number of moles of ATP produced per mole of substrate consumed, or the number of molecules of ATP produced for each pair of electrons transported along the respiratory chain.

Typical data obtained in this way are shown in Table 15–6. As stated earlier, when an oxidizable substrate primes the respiratory chain with electrons at the level of NADH, there is (P/O = 3) a maximum yield of 3ATP per *2e* transported to oxygen. And when $FADH_2$ primes the chain, there is (P/O = 2) a maximum yield of 2ATP.

The next logical question is, What regions of the respiratory chain are coupled to ATP formation? We have already implicated one, namely, the NAD → flavin region. The basis for this conclusion is simple: The flavin-dependent oxidation of a reduced substrate is indeed accompanied by a P/O

TABLE 15–6 Oxidative phosphorylation with intact mitochondria; P/O values expressed as nearest integer.

SUBSTRATE	DEHYDROGENASE	INHIBITOR	P/O RATIO
Pyruvate	NAD	—	3
Isocitrate	NAD	—	3
Malate	NAD	—	3
Succinate	FAD (flavin)	—	2
Isocitrate	NAD	antimycin	1
Isocitrate	NAD	CN^-	2
Succinate	FAD (flavin)	antimycin	0
Succinate	FAD (flavin)	CN^-	1

whole cells

↓ rupture

homogenate

↓ fractionate

purified mitochondria

O_2 ⟍⟋ P_i

NAD^+ ⟍⟋ ADP (buffer; 37°C)

malate ⟍

↓

measure amount of P_i and O_2 consumed and express as a P/O ratio; in this case P/O is 3 with malate

FIGURE 15–18 Performing quantitative measurements for oxidative phosphorylation. See Table 15–6.

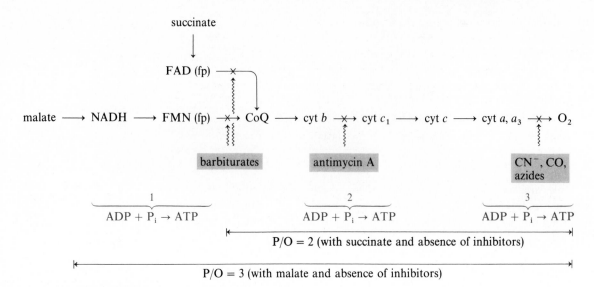

FIGURE 15–19 Phosphorylation regions. Brackets identify segments of the electron transport chain that provide energy used in ATP formation. However, the ATP is not actually formed during these steps. The FeS (NHI) carriers are not shown; X = sites of action by respiratory inhibitors.

value one unit smaller than that observed for an NAD^+-dependent oxidation of a reduced substrate. Thus we can assign the NAD → flavin transfer as phosphorylation region 1.

The remaining two regions have been established by measurement of P/O values in the presence of respiratory inhibitors with different substrates (see Table 15–6 and Figure 15–19). For example, the incubation of mitochondria with NAD^+-dependent substrates in the presence of antimycin A results in a P/O value of about 1. Since the specific site of antimycin A inhibition is the cyt b → cyt c_1 transfer, a P/O of 1 suggests that the region of the flavin → CoQ transfer is not linked to phosphorylation. If it were, we would predict a P/O of 2, since the NAD → flavin region alone would account for a P/O of 1. Corroborating this conclusion is the observation that very little P_i is consumed when mitochondria are incubated in the presence of antimycin A and succinate.

In addition to ruling out the flavin → CoQ region, the antimycin effect suggests then, that regions 2 and 3 are associated with the redox steps of cyt b → cyt c_1 and cyt a, a_3 → O_2, respectively. The former is, of course, implicated directly by the antimycin effect, since it is the specific site of antimycin inhibition. The cyt a, a_3 → O_2 region is directly confirmed by the addition of respiratory inhibitors of cytochrome oxidase—such as cyanide ion (CN^-), carbon monoxide (CO), or azides (N_3^-)—which yield P/O values of about 2 when the respiratory chain is primed with NADH and about 1 when the chain is primed with $FADH_2$.

Chemiosmotic Mechanism of Coupling:
The $\Delta\mu_{H+}$ Electrochemical Potential

Despite years of research, we do not yet know in complete detail how the available energy of electron transport is used for ATP formation. For a long time biochemists assumed that the coupling mechanism operated in a way similar to certain classical substrate-level phosphorylation reactions. An example is the reaction catalyzed by glyceraldehyde phosphate dehydrogenase (see Figure 13–21), where a high-energy thioester protein intermediate is formed prior to reaction with P_i. In oxidative phosphorylation the old hypothesis proposed by analogy that during specific carrier $\rightarrow$ carrier redox reactions, a high-energy phosphorylated intermediate would be formed that could then serve as a reactive donor of phosphate to ADP. However, all attempts to prove that this type of direct chemical coupling operates during oxidative phosphorylation have failed.

In 1961 P. Mitchell (Nobel Prize, 1978) proposed a radically different explanation. Originally greeted with skeptical interest, his novel proposal, called *chemiosmotic coupling,* is now accepted as fundamentally correct, although uncertainty exists regarding the precise molecular details of operation. A key observation for Mitchell's suggestion was that hydrogen ions (H^+, protons) are released from the matrix of mitochondria when they are in an active state of respiration. Assuming that the mitochondrial membrane is impermeable to the free passage of H^+, Mitchell suggested that the process of electron transport operates as an *H^+ pump* (a proton pump), translocating H^+ from the mitochondrial matrix across the inner mitochondrial membrane into the mitochondrial intermembrane space (see Figure 15–20). The result would be the formation of an *energy-rich H^+ gradient,* which would then be used to drive the formation of ATP.

As noted earlier in Section 15–1, this energized state of mitochondria is commonly described in term of an *electrochemical proton potential,* $\Delta\mu_{H^+}$, the magnitude of which is determined by the gradient of H^+ concentration (ΔpH) and by the gradient of charge ($\Delta\Psi$) due to the matrix compartment being more negative than the space compartment. That is,

$$\Delta\mu_{H^+} = \Delta\Psi - Z(\Delta pH)$$

Measured by various techniques, values for the contribution of the $\Delta\Psi$ membrane potential and ΔpH gradient for mitochondria, chloroplasts, and prokaryote bacteria show wide variation, but their sum gives a value for $\Delta\mu_{H^+}$ of about 0.2 V (200 mV). In mitochondria 15%–30% of this total potential difference may be due to the ΔpH gradient, with the rest contributed by $\Delta\Psi$.

Mechanisms for Proton Translocation

An obvious question is, How does the translocation of H^+ occur? Although different suggestions have been made, nothing has yet been proven. The simplest proposal is based on the asymmetric orientation of carriers and an alternating sequence of the two types of carriers—those that are carriers of H^+ and e and those engaged only in e transfer. An $H^+ + e$ carrier—so oriented as to acquire H^+ from the matrix—when interacting with the active site of the next e carrier, will transfer the electrons but release the H^+ for passage to the space compartment. These ideas are illustrated in Figure 15–21 for H^+ translocation associated with respiratory complex I.

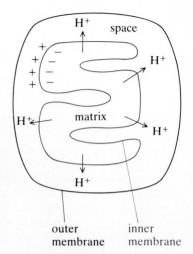

FIGURE 15–20 Proton displacement in respiring mitochondria. This high-energy condition occurs as a result of electron transport and drives the formation of ATP. The intermembrane space becomes more acidic and the matrix more basic. That is, pH of space < pH of matrix. In addition, the loss of H^+ renders the matrix compartment more negative than the space compartment.

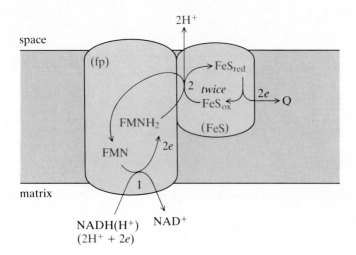

FIGURE 15–21 Proposed scheme for the translocation of H^+ in respiratory complex I. The initial redox step (1) consumes NADH(H^+) from the matrix to form $FMNH_2$. Then (2) the interaction of active sites between the reduced flavoprotein and the FeS protein occurs in such a way that when electrons are transferred to FeS, two protons are released with an exit path to the intermembrane space. A similar explanation would also apply to H^+ translocation in the operation of respiratory complex II primed by the oxidation of succinate, except that a different flavoprotein and FeS protein would be involved.

One hypothesis for the translocation of H^+ associated with respiratory complex III is shown in Figure 15–22. Called the *quinone cycle* or *Q cycle,* the process involves the formation of the semiquinone $\cdot QH$ radical and the participation of two different subunits of the cyt b carrier. The individual steps of one Q cycle are as follows:

1. One electron is transferred from an FeS protein of complex I to Q with a proton consumed from the matrix to give the semiquinone radical $\cdot QH$.

2. $\cdot QH$ acquires a second electron from the reduced state of one of the cyt b subunits and a second proton from the matrix to form the fully reduced quinone QH_2.

3. In a step releasing a proton to the space compartment, the mobile QH_2 then reacts in a one-electron transfer reaction with an FeS protein oriented at the outer surface of the membrane. The $\cdot QH$ radical and a reduced FeS protein are formed, with the latter then transferring an electron to cyt c_1 (step 6).

4. A second proton is then released to the intermembrane space in a one-electron transfer between the $\cdot QH$ radical and the oxidized form of the other subunit of cyt b, also oriented at the outer surface of the membrane. This redox step also provides the basis to recycle the cyt b subunit carriers.

5. A one-electron transfer occurs between the reduced cyt b subunit formed in step 4 and the oxidized cyt b subunit formed in step 2 to recycle each cyt b subunit to participate again in 2 and 4.

The net result is that $2H^+$'s are translocated for each electron transferred from the FeS protein of complex I (or complex II) to cyt c_1 or $4H^+$'s for an overall two-electron transfer through complex III.

Various mechanisms have been proposed to explain how additional protons are translocated in association with electron transfer through complex IV. Owing to their very speculative and complex nature, we will not examine them here.

NOTE 15–1

$$\frac{(H^+/O)}{(P/O)} = \frac{H^+}{P} \text{ ratio}$$

An H^+/P ratio expresses the number of protons required to drive the formation of 1ATP. With H^+/O and P/O values for priming the respiratory chain with NADH electrons, an approximation of H^+/P is

$$\frac{6\text{–}12\ H^+/O}{3\ P/O} = 2\text{–}4\ \frac{H^+}{P}$$

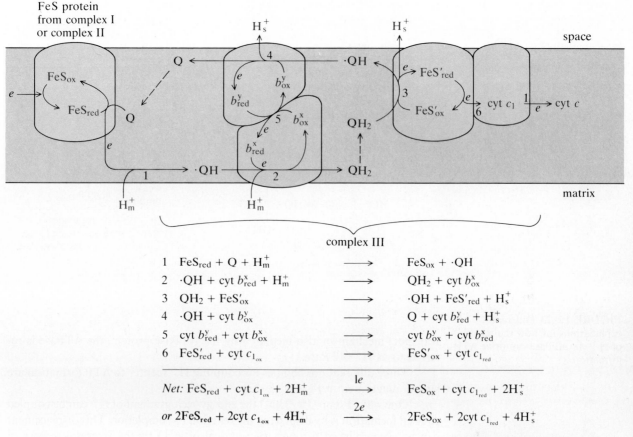

$$1 \quad FeS_{red} + Q + H_m^+ \longrightarrow FeS_{ox} + \cdot QH$$

$$2 \quad \cdot QH + cyt\ b_{red}^x + H_m^+ \longrightarrow QH_2 + cyt\ b_{ox}^x$$

$$3 \quad QH_2 + FeS'_{ox} \longrightarrow \cdot QH + FeS'_{red} + H_s^+$$

$$4 \quad \cdot QH + cyt\ b_{ox}^y \longrightarrow Q + cyt\ b_{red}^y + H_s^+$$

$$5 \quad cyt\ b_{red}^y + cyt\ b_{ox}^x \longrightarrow cyt\ b_{ox}^y + cyt\ b_{red}^x$$

$$6 \quad FeS'_{red} + cyt\ c_{1_{ox}} \longrightarrow FeS'_{ox} + cyt\ c_{1_{red}}$$

$$Net:\ FeS_{red} + cyt\ c_{1_{ox}} + 2H_m^+ \xrightarrow{\ 1e\ } FeS_{ox} + cyt\ c_{1_{red}} + 2H_s^+$$

$$or\ 2FeS_{red} + 2cyt\ c_{1_{ox}} + 4H_m^+ \xrightarrow{\ 2e\ } 2FeS_{ox} + 2cyt\ c_{1_{red}} + 4H_s^+$$

FIGURE 15–22 Proposed Q cycle for H^+ translocation during electron transport in complex III. The semiquinone radical species is represented by $\cdot QH$. The two cytochrome b subunits are designated b^x and b^y. Steps 1 through 5 comprise the Q cycle. Step 6 transfers electrons to cyt c_1, the last carrier in complex III, which then transfers electrons to cyt c for subsequent transfer to cyt a, a_3 (complex IV).

Stoichiometry of Electron Transport and Proton Translocation

There is still uncertainty about the stoichiometry of the H^+ pump, expressed as an H^+/O ratio, for the number of protons translocated per gram atom of oxygen consumed. Different laboratories have reported 8–12 H^+/O for priming the chain with NADH and 4–8 H^+/O for priming with $FADH_2$. In conjunction with P/O values, these H^+/O values suggest (see Note 15–1) that the translocation of about 2–4 protons is required to drive the formation of ATP.

Reentry of Protons Coupled to ATP Formation

Once formed, how is the $\Delta\mu_{H^+}$ potential used to drive ATP formation? Although the actual mechanism is still unknown, the major membrane protein whose operation is affected by the $\Delta\mu_{H^+}$ potential is the massive ATPase (also called ATP synthase) that spans the entire inner mitochondrial membrane and protrudes from the matrix face. A similar protein is found in chloroplasts and the cytoplasmic membrane of prokaryotes. Because the action of this ATPase is dependent on the reentry of H^+, a reentry that involves some sort of H^+

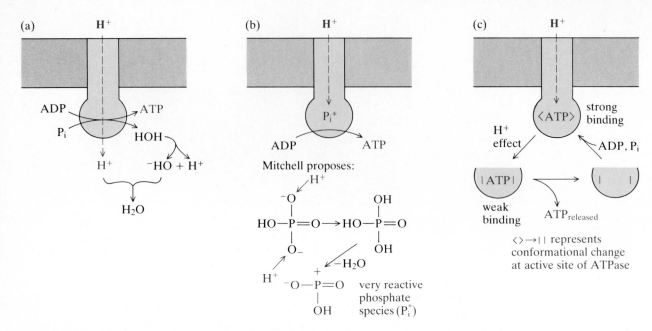

FIGURE 15–23 Different explanations for how the reentry of H^+ via ATPase can drive ATP formation.

transport mechanism also mediated by the ATPase protein, the ATPase is referred to as *H^+-ATPase*.

Three different possibilities for coupling H^+ reentry to ATP formation are simply diagrammed in Figure 15–23.

One way [Figure 15–23(a)] the energy-rich gradient of H^+ can be coupled to ATP formation is by displacing the reaction to completion. This displacement can be accomplished through the removal of H_2O by the vectorial reentry of H^+ from the space compartment through a "channel" in the ATPase component into the low-H^+ area inside the mitochondrion, where it will consume OH^-. Another proposal [Figure 15–23(b)] is that the reentry of H^+ through the ATPase channel may be funneled to the active site of ATPase, where it will promote the formation of a highly reactive form of phosphate (specifically, an ylid form), which then reacts rapidly with ADP, also bound at the active site. Another argument [Figure 15–23(c)] is that the H^+ reentry promotes a change in the conformation of the ATPase, resulting in the release of ATP, which was already formed at the active site but firmly bound.

H^+-ATPase of Oxidative Phosphorylation

Owing to its immense importance in cell metabolism, the H^+-ATPase has and continues to be under intense study. Since the late 1970s about two thousand research papers have been published on various aspects of solubilization, isolation, structure, and function. The oligomeric protein is large (MW $\approx$ 450,000) and is one of the most complex enzyme systems known. The intact H^+-ATPase complex is readily dissociated into two oligomeric components, designated F_0 and F_1 (hence, H^+-ATPase = F_0F_1), and each component has a different location, structure, and function (see Figure 15–24).

The F_0 segment is embedded in the lipid bilayer of the membrane. Because of its appreciable hydrophobic character and the tight association with the membrane bilayer, the F_0 component is difficult to solubilize—a situation that has impaired progress on structure/function studies. The F_0 component appears to be composed of at least three different polypeptides, but the exact subunit composition is unknown at present. The important function associated with the F_0 sector is that it provides the molecular apparatus for the reentry of H^+, although the way in which the H^+ reentry occurs is not yet clear. An interior hydrophilic channel, small enough to allow only the passage of protons, may be involved.

Although associated with the F_0 sector, the F_1 sector is not embedded in the membrane. In mitochondria it protrudes from the matrix face of the inner membrane. As previously indicated (see Figure 15–15), the F_0/F_1 association can be disrupted to release F_1 as a soluble component and thus is more easily isolated. Having a molecular weight of about 350,000, the F_1 component is composed of five different polypeptide subunits ($\alpha, \beta, \gamma, \delta, \varepsilon$) with most evidence supporting a composition of $\alpha_3\beta_3\gamma\delta\varepsilon$. Amino acid sequences of all subunits have been solved, and low-resolution diffraction studies of single F_1 crystals suggest that the aggregate is arranged in two equivalent halves. The single most important function associated with the F_1 sector is that it carries the *ATPase active site* (see Figure 15–25).

The H^+-ATPases also contain a third functional domain loosely associated with the F_1 sector. It is designated as the I protein, and its function is to regulate the F_1 ATPase site. When bound to F_1, the regulatory protein inhibits the ATPase site from catalyzing the hydrolysis of the newly formed ATP.

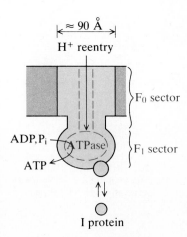

FIGURE 15–24 H^+-ATPase.

FIGURE 15–25 Chemiosmotic theory. Communicating a complex concept of science is sometimes made easier by using unorthodox tactics. One such tactic, popularized in recent years by the publishers of the journal *Trends in Biochemical Sciences* (TIBS), is the use of science cartoons. This sample may help you appreciate the meaning of the formal description of the chemiosmotic theory given in the text. (The cartoon is reproduced with permission of the TIBS publisher.)

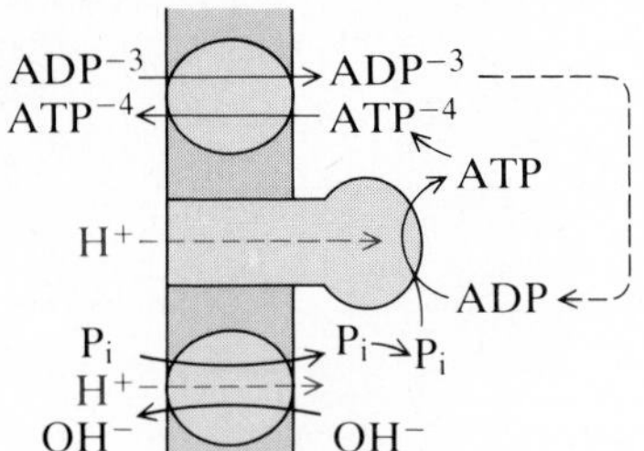

FIGURE 15–26 Two antiport transport proteins associated with the operation of H$^+$-ATPase.

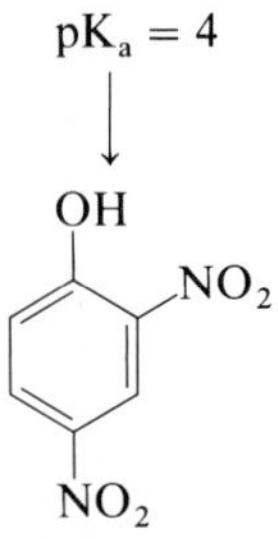

FIGURE 15–27
2,4-Dinitrophenol.

Entrance of Inorganic Phosphate and ADP and Exit of ATP

Since (1) the complete process of oxidative phosphorylation is intramitochondrial, (2) the bulk of ATP-requiring processes are extramitochondrial, and (3) the mitochondrial membrane is impermeable to the free diffusion of ADP, P$_i$, and ATP, an obvious question is, How do these substances enter and leave the mitochondrion? As you might expect, the membrane contains specific transport proteins that participate. There are two transport systems, one for ADP and ATP and another for P$_i$. The ADP/ATP transport system operates by exchanging an incoming ADP for an outgoing ATP. Such a system is referred to as an *antiport mechanism* (see Figure 15–26). The driving force of the ADP/ATP antiporter is the transmembrane potential ($\Delta\Psi$) component of $\Delta\mu_{H^+}$. Under physiological conditions, ATP is ATP^{-4} and ADP is ADP^{-3} (see Figure 12–6). Since the matrix compartment is more negative, the exchange of a -4 species for a -3 species is an energetically favorable process. A unique feature of the ADP/ATP antiporter is its strong inhibition by *atracyloside* (a plant toxin) and *bongkrekic acid* (a fungal antibiotic).

The entrance of P$_i$ occurs by a different protein system, but one that also operates as an *antiporter involving OH$^-$*. For each P$_i^{2-}$ that enters, an OH$^-$ is transported out. The P$_i^{2-}$/OH$^-$ antiporter is driven by the ΔpH gradient of the $\Delta\mu_{H^+}$ potential, meaning that it is also driven by the reentry of H$^+$.

Uncouplers of Oxidative Phosphorylation

By inhibiting the normal transfer of electrons between carriers and thus affecting the magnitude of the $\Delta\mu_{H^+}$ produced, respiratory inhibitors like barbituates, antimycin, CO, and CN$^-$ also diminish the amount of ATP produced. Other inhibitors, called *uncouplers*, diminish ATP formation without inhibiting electron transport. In general terms, the mode of action of an uncoupler involves disrupting the $\Delta\mu_{H^+}$ potential *before* it can be used to drive ATP formation. A few specific examples include *2,4-dinitrophenol* (see Figure 15–27), *valinomycin*, and *gramicidin*.

The specific mode of action of 2,4-dinitrophenol (DNPH) can be explained by its ability to destroy the ΔpH gradient. At physiological pH, DNPH ionizes to DNP$^-$, the latter then capable of functioning as a strong H$^+$ acceptor. Any DNP$^-$ entering the intermembrane space of mitochondria will scavenge protons translocated by electron transport and then, as DPNH, pass through either the inner or the outer membrane. In either case the ΔpH gradient will be destroyed, and ATP formation will not occur.

Valinomycin and gramicidin (both peptides) are specific examples of a class of uncouplers called *ionophores* (ion carriers). Valinomycin (Vin) is a doughnut-shaped cyclic peptide. The inner ring of valinomycin provides a small hydrophilic cavity that can specifically accommodate the binding of a potassium ion (K$^+$). The outer surface of valinomycin is sufficiently hydrophobic to penetrate and pass through a lipid bilayer. After binding to K$^+$ outside the mitochondrion, the Vin-K$^+$ complex can pass through the mitochondrial membrane into the matrix, and there it dissociates to release free K$^+$. Because the transmembrane potential involves a matrix more negative than the space side, the entry of K$^+$ (positive charge) into the matrix will act to neutralize the negative charge and thus diminish the $\Delta\Psi$ potential. Among other possible effects, a diminished $\Delta\Psi$ potential would reduce the operation of the ADP/ATP antiporter.

Gramicidin operates by embedding itself in the lipid bilayer of the inner membrane in such a manner as to provide hydrophilic channels (gates or pores) through which small positive ions (Na^+, K^+, and H^+) can pass. Not only will the entry of these positive ions diminish the transmembrane charge gradient ($\Delta\Psi$), but also the entry of H^+ will specifically dilute the ΔpH gradient.

Oligomycin: A Specific Inhibitor of H^+-ATPase

Respiratory inhibitors block the entire process of oxidative phosphorylation by inhibition of electron transfer. *Oligomycin*, a cyclic peptide antibiotic, also inhibits the process of oxidative phosphorylation, but it does so in a different way. Oligomycin inhibition is linked to a protein of the F_1 sector of H^+-ATPase, located in the vicinity of the F_0/F_1 association. When oligomycin binds to this protein, the reentry of H^+ through the H^+-ATPase is impaired. This impairment will inhibit ATP formation, and because the ΔpH gradient is not turned over, it will also back up the flow of electrons through the respiratory chain.

15-4 ATP YIELD FROM CARBOHYDRATE METABOLISM: REVIEW AND SUMMARY

Given the quantitative relationships of oxidative phosphorylation concerning the uptake of P_i and O_2, one can now *estimate* the efficiency of the combined operations of glycolysis and the citric acid cycle in terms of energy conservation (see Figure 15-28). Since the oxidation of one NADH unit yields 3ATP, and of one $FADH_2$ unit yields 2ATP, the yield of the combined route is 34ATP. Then consider (1) that for every molecule of glucose processed through glycolysis there is a requirement of 2ATP and a production of 4ATP via substrate-level phosphorylations, and (2) that the complete oxidation of 2 units of acetyl-SCoA in the citric acid cycle is accompained by the production of 2GTP (also by a substrate-level phosphorylation), which are thermodynamically and metabolically equivalent to 2ATP. Hence we can say that the total net gain of high-energy triphosphonucleotides (expressed as ATP) accompanying the complete oxidation of glucose to $6CO_2$ via glycolysis and the citric acid cycle is 38 ($34 - 2 + 4 + 2$). On the basis again of a $\Delta G^{\circ\prime}$ value of $+8.0$ kcal/mole for ATP formation, this result means that the cellular efficiency of energy conversion in a metabolically useful form is approximately 44%. Inasmuch as the energy required for in vivo ATP formation is probably greater than 8 kcal, the 44% figure is a conservative estimate. Thus:

	$\Delta G^{\circ\prime}$ (kcal)	
$38ADP + 38P_i \rightarrow 38ATP$	$+304$	conserved
$glucose + 6O_2 \rightarrow 6CO_2 + 6H_2O$	-686	available

$glucose + 6O_2 + 38ADP + 38P_i \rightarrow 6CO_2 + 6H_2O + 38ATP$	-382

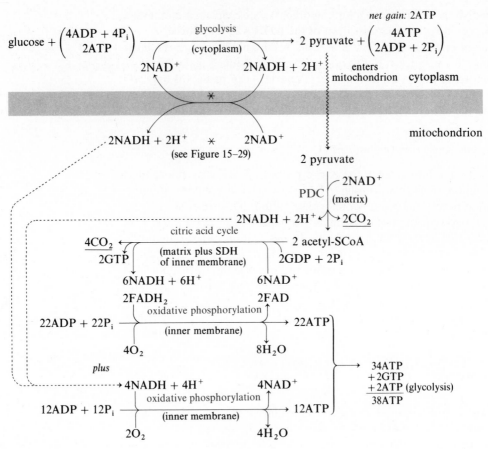

FIGURE 15–28 ATP yield for complete oxidation of glucose to $6CO_2$ via glycolysis, pyruvate dehydrogenase complex (PDC), and citric acid cycle.

There is one item in this analysis that merits a brief explanation. In the tally made, the net production of 38ATP includes 6ATP that would result from the utilization in electron transport of 2NADH produced during glycolysis. Since glycolysis occurs in the cytoplasm and, furthermore, since the mitochondrial membrane is impermeable to the direct transport of NADH, you might ask, How can we legitimately include the NADH generated by glycolysis in the mitochondrial production of ATP? The explanation is that the NADH produced in the cytoplasm can enter the mitochondrion by processes termed *shuttle pathways,* which provide for an indirect movement of substances across a membrane.

In the case of NADH one of two shuttle pathways is shown in Figure 15–29. Called the *malate/aspartate shuttle,* the key participants are (1) two isoenzyme forms of malate dehydrogenase, one located in the cytoplasm and the other in the mitochondrial matrix; (2) transaminase enzymes in each compartment to convert oxaloacetate into aspartate, a conversion that circumvents the membrane impermeability to oxaloacetate; and (3) two antiport transport proteins that participate in the exchange passage of malate and α-ketoglutarate and of glutamate and aspartate. Because the shuttle is cyclic, it is self-sustaining.

In animal organisms different shuttle pathways operate in different tissues.

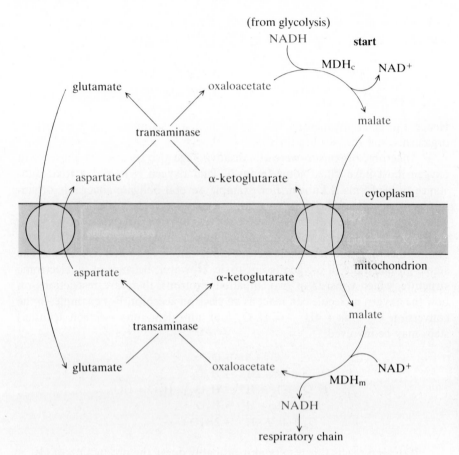

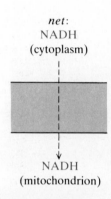

FIGURE 15–29 Malate/aspartate shuttle for the indirect transport of NADH from the cytoplasm into the mitochondrion. The abbreviation MDH_c represents cytoplasmic malate dehydrogenase, and MDH_m is mitochondrial malate dehydrogenase.

The malate/aspartate shuttle is important in liver and heart tissues. Many other tissues use a shuttle involving glycerol phosphate. You can learn something about the glycerol phosphate shuttle by solving Exercise 15–11.

15–5 OXYGEN TOXICITY AND SUPEROXIDE DISMUTASE

There is considerable evidence that oxygen was not present in the atmosphere of the earth at the time of its formation 4.5–4.8 billion years ago. In fact, some argue that oxygen became available only about 2 billion years ago as a result of the evolution of the photosynthetic organisms, the first being the blue-green algae. Until then, the only organisms existing were anaerobic cells. The gradual accumulation of oxygen accompanying the evolution of photosynthetic organisms was of great significance, because it provided for the subsequent evolution of aerobic organisms (first appearing about 1.5 billion years ago). By using oxygen as a terminal oxidizing agent, aerobic cells could extract more energy from reduced food substrates such as glucose, because the substrate can be completely oxidized to CO_2:

$$\text{glucose} \xrightarrow{\text{anaerobic}} \text{2 lactate} + 56 \text{ kcal/mole glucose}$$

$$\text{glucose} + 6O_2 \xrightarrow{\text{aerobic}} 6CO_2 + 6H_2O + 686 \text{ kcal/mole glucose}$$

Hence the emerging aerobic life forms had an advantage over anaerobic organisms, and the aerobes thrived.

Anaerobic organisms were at a disadvantage also because the presence of oxygen must have killed some (many?). That oxygen is indeed a toxic substance to life forms is known. For example, several obligate anaerobic organisms (soil organisms) thrive only in the absence of oxygen and die in its presence. The toxicity of oxygen to organisms raises two obvious questions: (1) Why is it toxic, and (2) why are aerobic organisms able to thrive in an atmosphere containing this toxic substance? These questions can be answered as follows: The stable ground state of oxygen is not toxic. However, because of its electronic structure, which consists of two unpaired electrons, there are restrictions on how the oxygen molecule can react as an electron acceptor. For example, in the conversion $O_2 + 4e + 4H^+ \rightarrow 2H_2O$, four univalent (one electron transfer) steps may be involved:

$$O_2 + 1e \rightarrow O_2^- \cdot$$
$$O_2^- \cdot + 1e + 2H^+ \rightarrow H_2O_2$$
$$H_2O_2 + 1e + H^+ \rightarrow H_3O_2 \rightarrow H_2O + HO \cdot$$
$$\underline{HO \cdot + 1e + H^+ \rightarrow H_2O}$$
$$net: \quad O_2 + 4e + 4H^+ \rightarrow 2H_2O$$

If oxygen reacts this way (and it probably does), the production of $O_2^- \cdot$, H_2O_2 (hydrogen peroxide), and $\cdot OH$ as intermediates presents a serious problem to life, because they are all potent oxidizing agents. The $O_2^- \cdot$ species is called the *superoxide anion radical,* or simply superoxide. The $\cdot OH$ species is called the *hydroxy radical* and is the strongest oxidizing agent known. The hydroxy radical can be formed not only by the scheme shown above but also as follows:

$$H_2O_2 + O_2^- \cdot \rightarrow O_2 + OH^- + \cdot OH$$

Each species represents a potential threat to living cells because of the damage each can cause to all the classes of biomolecules, notably the proteins and lipids. In other words, the toxicity of oxygen is due to the toxicity of the species ($HO \cdot$, $O_2^- \cdot$, and H_2O_2) that can be formed from it.

In addition to the possible occurrence of the univalent reaction sequence for $O_2 \rightarrow 2H_2O$ in living cells, H_2O_2 and $O_2^- \cdot$ are also known to be produced in several enzyme-catalyzed reactions that utilize O_2 as one of the substrates. Thus the reaction of H_2O_2 and $O_2^- \cdot$ to yield $\cdot OH$ is a real threat to cells. Fortunately, self-defense mechanisms serve to scavenge the H_2O_2 and $O_2^- \cdot$, and thus the $\cdot OH$ is unable to form and the danger of H_2O_2 and $O_2^- \cdot$ existing as powerful oxidizing agents in their own right is also removed.

Both H_2O_2 and $O_2^-\cdot$ can be detoxified (in part) through the intervention of naturally occurring *antioxidants* such as ascorbic acid, vitamin E, and glutathione. The H_2O_2 is also enzymatically detoxified by *catalase* and *peroxidase* enzymes:

$$2H_2O_2 \xrightarrow{\text{catalases}} O_2 + 2H_2O$$

$$H_2O_2 + DH_2 \xrightarrow{\text{peroxidases}} D + 2H_2O$$

where DH_2 is a reduced organic substance, of which there are several, that functions as a hydrogen donor.

In 1969 I. Fridovich and J. McCord established that the primary mode of detoxifying $O_2^-\cdot$ is also enzymatic and is a result of the action of *superoxide dismutase* (see Note 15–2):

$$O_2^-\cdot + O_2^-\cdot + 2H^+ \xrightarrow[\text{dismutase}]{\text{superoxide}} H_2O_2 + O_2$$

Although H_2O_2 is produced, it can be disposed of by antioxidants, catalases, and peroxidases. In view of the fact that $O_2^-\cdot$ (superoxide) is a powerful oxidizing agent, it is remarkable that there is an enzyme that can use it as a substrate without being damaged by it. (The same can be said of the catalases and peroxidases.)

Superoxide dismutase has since been detected in all types of prokaryotic and eukaryotic aerobic cells, thus supporting its significance as a self-defense mechanism against oxygen toxicity. If this assigned role of dismutase action is correct, what do you predict about its occurrence in obligate anaerobes? Yes, you are right—the enzyme is not present in obligate anaerobes, and apparently this is the primary reason why O_2 is lethal to such organisms. There are several other recent experimental findings that support this theory of oxygen defense (see the Fridovich article listed in the Literature).

There has been increased interest in superoxide dismutase in recent years because of evidence that granulocytes (a type of white blood cell) liberate large amounts of $O_2^-\cdot$ during a surge of respiratory metabolism accompanying their development into phagocytic cells that ingest and destroy foreign particles, bacteria, and other cells. In addition to participating in the killing of bacteria, $O_2^-\cdot$ may also damage other tissues, contributing to an inflammation of the tissue. One suggestion for combating this inflammation is that superoxide dismutase be used (by injection) as an anti-inflammatory drug. Some successes for this potential clinical application have already been reported, but further research is necessary.

NOTE 15–2
A *dismutation reaction* is one in which a reactant undergoes both oxidation and reduction.

15–6 SURVEY OF OXYGEN IN METABOLISM

Introduction

There are about two hundred different enzymes known that utilize O_2 (dioxygen) as one of their substrates. These enzymes are divided into two main categories on the basis of whether O_2 is or is not incorporated into the other substrate (S, SH, or SH_2) during the reaction: (1) *oxidases,* wherein O_2 is not incorporated but used only as an acceptor of electrons, and (2) *oxygenases,* wherein O_2 is incorporated.

There are two types of oxidase enzymes. One type is *H_2O-forming oxidases:*

$$SH_2 + \tfrac{1}{2}O_2 \xrightarrow{\text{oxidase}} S + H_2O$$

or
$$S_{red} + 2H^+ \rightarrow \tfrac{1}{2}O_2 \xrightarrow{\text{oxidase}} S_{ox} + H_2O$$

An example from the respiratory chain is

$$2Fe^{2+} + 2H^+ + \tfrac{1}{2}O_2 \xrightarrow[\text{oxidase}]{\text{cyt } a,\, a_3} 2Fe^{3+} + H_2O$$

The second type is *H_2O_2-forming oxidases:*

$$SH_2 + O_2 \xrightarrow{\text{oxidase}} S + H_2O_2$$

An example is

$$\underset{\text{amine}}{RCH_2NH_2} + O_2 \xrightarrow[\substack{\text{oxidase}\\ \text{(see p. 677)}}]{\text{monoamine}} \overset{\text{H}}{\underset{\text{aldehyde}}{RC}}{=}O + NH_3 + H_2O_2$$

There are also two types of oxygenase enzymes. One type is *monoxygenase* (also called *hydroxylases*), which incorporate only one atom of O_2. This type of enzyme usually requires a co-reducing agent (DH_2):

$$SH + DH_2 + O_2 \xrightarrow[\text{(hydroxylase)}]{\text{monoxygenase}} S{-}OH + H_2O + D$$

An example is

$$\text{phenylalanine} + DH_2 + O_2 \xrightarrow[\substack{\text{hydroxylase}\\ \text{(see p. 682)}}]{\text{phenylalanine}} \text{tyrosine} + D + H_2O$$

Other examples are provided by cytochrome P–450 enzymes, which are described in the next section.

The second type of oxygenase enzyme is *dioxygenase,* which incorporates both atoms of O_2 to yield a diol, a hydroperoxide, or a peroxide:

$$SH_2 + O_2 \xrightarrow{\text{dioxygenase}} \underset{\text{diol}}{S(OH)_2} \text{ or } \underset{\text{hydroperoxide}}{SOOH} \text{ or } \underset{\text{endoperoxide}}{\overset{O{-}\!\!-\!\!-O}{S}}$$

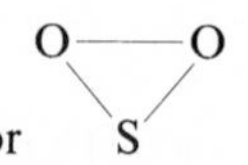

Some examples are

$$\text{arachidonic acid} + 2O_2 \xrightarrow{\text{E}} PGG_2 \text{ (see p. 433)}$$

where E is prostaglandin endoperoxide synthase, and

$$\text{arachidonic acid} + O_2 \xrightarrow{\text{E}} \underset{\substack{\text{precursor of} \\ \text{leukotrienes}}}{\text{hydroperoxide}}$$

where E is lipoxygenase.

Cytochrome P–450 Hydroxylases

An important enzyme function in animals is represented by a group of hydroxylases collectively referred to as *cytochrome P–450*. Usually localized in reticulum membranes, P–450 activity is found in many tissues but is particularly abundant in liver, where it represents about 14% of the total microsomal protein and about 50% of the total liver heme. Presently, eight different P–450 enzymes have been detected in liver. The active subunit of each P–450 is a hemoprotein (MW ≈ 50,000) in close association with other electron transfer proteins in the endoplasmic reticulum membrane. This O_2-activating, electron transfer protein is called cytochrome P–450 because the reduced form exhibits an intense absorption band at 450 nm when complexed to carbon monoxide (CO).

The P–450 enzymes are much studied because they are responsible for (1) processing numerous foreign compounds that enter the body (drugs, pesticides, carcinogens, and mutagens) and for (2) catalyzing some key reactions in normal cellular metabolism, particularly in the biosynthesis of cholesterol in liver and cholesterol-derived steroids in the adrenal gland. A P–450 conversion of a foreign compound may have different effects: A toxic foreign compound is detoxified and prepared for excretion, or a harmless, nontoxic foreign compound is transformed into a toxic derivative.

Two remarkable aspects of P–450 enzymes are that they act on a variety of structurally different substrates and can catalyze many types of chemical modifications. Some specific and general examples of P–450 reactions are given in Figure 15–30. Despite the obvious catalytic versatility, however, note that all of these reactions involve O_2 as substrate in a hydroxylase type of reaction—a co-reducing agent is usually required and only one O atom is incorporated into the primary substrate.

The chemistry of P–450 catalysis involves a mini–electron transport system composed of cytochrome P–450 reductase, sometimes an FeS non–heme iron carrier, and cytochrome P–450 itself. The reaction begins (see Figure 15–31) with the chain being primed by the co-reducing agent (DH_2), which is usually NADPH; then transfer to the reductase carrier; then, when it exists—as in adrenal tissue—transfer to the FeS carrier; then transfer to P–450. The reduced form of P–450 then reacts with O_2, converting it to a highly activated form, believed to be the hydroxyl radical ·OH. The oxygen activation is also accompanied by an activation of the SH substrate, proposed to involve the formation of a radical form (·S) of the substrate. The last step in the proposed mechanism involves the combination of the ·OH and ·S radicals to yield SOH.

hydroxylation

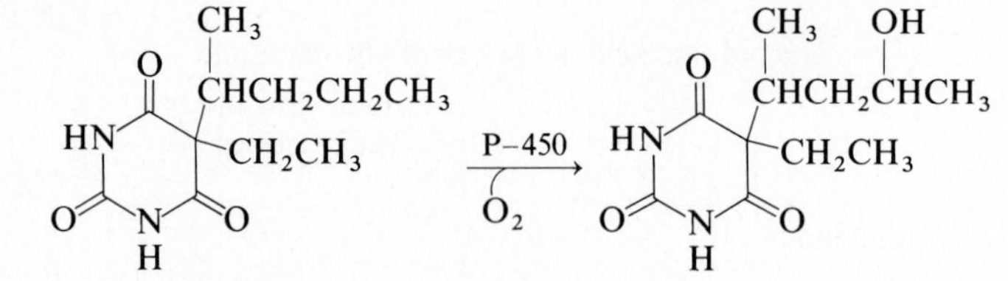

pentobarbital
(sedative; general anesthetic)

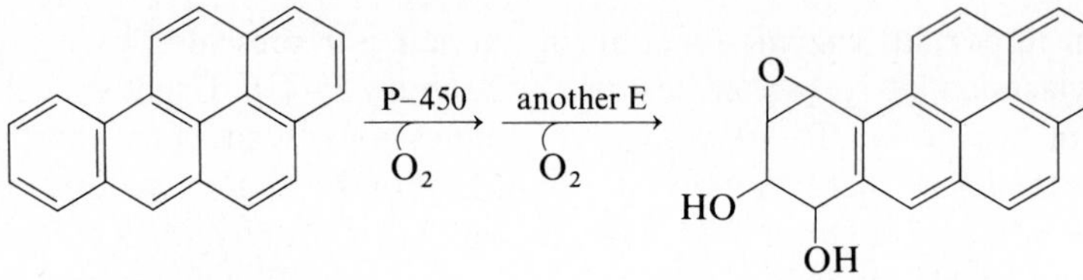

benzo[a]pyrene

diol epoxide
(DNA mutagen and
protein modifier)

oxidative deamination

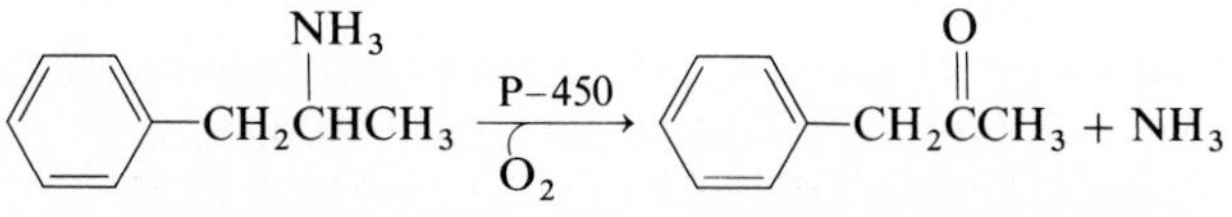

amphetamine

N-dealkylation

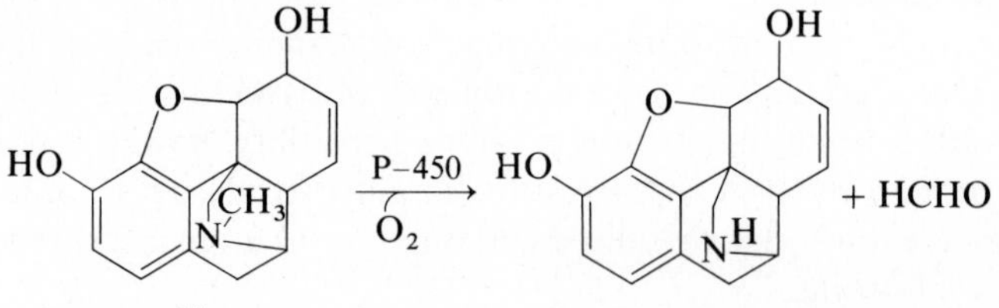

morphine

O-dealkylation

$$R-O-CH_3 \xrightarrow[O_2]{P-450} ROH + HCHO$$

S-dealkylation

$$R-S-CH_3 \xrightarrow[O_2]{P-450} RSH + HCHO$$

N-oxidation

$$R_3N \xrightarrow[O_2]{P-450} R_3N=O$$

S-oxidation

$$R-S-R' \xrightarrow[O_2]{P-450} R-\overset{\overset{\displaystyle O}{\|}}{S}-R'$$

FIGURE 15–30 Some specific and general examples of reactions catalyzed by cytochrome P–450.

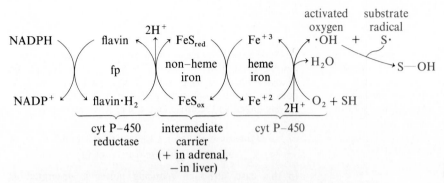

FIGURE 15–31 Electron transport associated with cytochrome P–450 catalysis.

net: $NADPH + H^+ + O_2 + SH \rightarrow NADP^+ + H_2O + SOH$

There is much yet to be learned about the biology and the chemistry of P–450. Are individuals with elevated P–450 levels more susceptible to cancer? If so, what genetic factors are responsible for the increased activity, and can they be controlled? Are P–450 levels affected by hormonal factors, nutrition, age, and/or other diseases? What is the mechanism whereby foreign compounds cause induction of the P–450 gene? Such questions (and others) obviously have direct relevance to human biochemistry.

LITERATURE

BOYER, P. B., B. CHANCE, L. ERNSTER, P. MITCHELL, E. RACKER, and E. C. SLATER. "Oxidative Phosphorylation and Photophosphorylation." *Annu. Rev. Biochem.*, **46,** 955–1026 (1977). Each author reviews an aspect of energy-coupled phosphorylation in membranes, particularly in mitochondria and chloroplasts. All models of energy transduction are covered.

DICKERSON, R. E. "The Structure and History of an Ancient Protein." *Sci. Am.*, **226,** 58–72 (1972). A concise discussion of the structure of cytochrome *c* and its evolution as a molecule over 1.2 billion years. The amino acid sequence from 38 different species is examined.

ERNSTER, L., ed. *Bioenergetics.* Volume 9 of *New Comprehensive Biochemistry,* edited by A. Neuberger and L. L. M. van Deenen. New York: Elsevier, 1985. A current overview of energy transduction mechanisms in mitochondria, chloroplasts, and bacteria.

FRIDOVICH, I. "The Biology of Oxygen Radicals." *Science,* **201,** 875–880 (1978). A review article on the superoxide radical and superoxide dismutase.

GREEN, D. E., and H. J. H. YOUNG. "Energy Transduction in Membrane Systems." *Am. Sci.,* **59,** 92 (1971). A review article.

HINKLE, P. C., and R. E. McCARTY. "How Cells Make ATP." *Sci. Am.,* **238,** 104–123 (1978). Excellent article emphasizing the chemiosmotic coupling mechanism and the mode of action of ionophore uncouplers.

KASHKET, E. R. "The Proton Motive Force in Bacteria: A Critical Assessment of Methods." *Annu. Rev. Microbiol.,* **39,** 219–242 (1985). A thorough review article describing laboratory methods currently in use for measuring $\Delta\mu_{H^+}$.

MITCHELL, P. "Protonmotive Chemiosmotic Mechanisms in Oxidative and Photosynthetic Phosphorylation." *Trends Biochem. Sci.,* **3,** N58–N61 (1978). A brief description of the chemiosmotic hypothesis by the original proponent.

NICHOLLS, D. G. *Bioenergetics: An Introduction to Chemiosmotic Theory.* New York: Academic Press, 1982. Available in paperback; 190 pages. Good treatment of a complex topic.

RACKER, E. "Inner Mitochondrial Membranes: Basic and Applied Aspects." *Hosp. Pract.,* **9,** 87 (1974); and "The Membrane of the Mitochondrion." *Sci. Am.* February 1968. Excellent introductory review articles of structure and evidence for the chemiosmotic hypothesis of coupling.

———. *A New Look at Mechanisms in Bioenergetics.* New York: Academic Press, 1976. A short book for student reading on mitochondrial ATP production and related processes. Available in paperback.

E X E R C I S E S

15–1. Calculate the net oxidation-reduction potential and the standard free-energy change for each of the following reactions as written from left to right, and indicate whether or not the reaction will tend to occur spontaneously given the proper conditions. (Assume that pH and temperature are as specified in Table 15–1.)

(a) $NADH + H^+ + CoQ \rightarrow NAD^+ + CoQH_2$

(b) $CoQH_2 + 2[cyt\ b\ (oxidized)] \rightarrow$
$$CoQ + 2[cyt\ b\ (reduced)] + 2H^+$$

(c) 2 (reduced ferredoxin) $+ NADP^+ + 2H^+ \rightarrow$
$$2\ (oxidized\ ferredoxin) + NADPH + H^+$$

15–2. The standard reduction potential (pH 7) of the lactate-pyruvate couple is -0.19 V, which means then that the reaction

$$pyruvate + 2H^+ + 2e \rightarrow lactate \qquad \mathscr{E}^{o\prime}_{red} = -0.19\ V$$

is endergonic. Yet in our analysis of the glycolytic pathway in Chapter 13, we have seen that the reduction of pyruvate to lactate, catalyzed by lactate dehydrogenase, is strongly exergonic. Explain. The reaction is

$$pyruvate + NADH + H^+ \xrightarrow[\text{dehydrogenase}]{\text{lactate}} lactate + NAD^+$$

15–3. If a respiratory inhibitor functioned at the site of electron transfer between cytochromes c_1 and c, what approximate value for the P/O ratio would be obtained on incubation (aerobically) of intact mitochondria in the presence of malate, ADP, P_i, NAD^+, the inhibitor, and a buffer?

15–4. The incubation of mitochondria in the presence of succinate and malonate will result in less oxygen consumption than in the presence of succinate alone, but the P/O ratios will not be significantly different. Explain.

15–5. If respiratory complexes II and IV were incubated together under aerobic conditions in the presence of added succinate, coenzyme Q, and cytochrome c, what overall oxidation-reduction reaction would occur?

15–6. In the system described in Exercise 15–5, would you expect to detect the formation of any significant amount of reduced cytochrome c_1 during the reaction? Explain.

15–7. What will be the net yield of ATP from the complete aerobic catabolism of glucose to $6CO_2$ according to the hexose monophosphate shunt (HMS) pathway, and how does the number compare with the ATP yield from glycolysis linked to the citric acid cycle? The reaction is

$$6(\text{glucose-6-P}) \xrightarrow{\text{HMS}} 5(\text{glucose-6-P}) + 6CO_2$$
$$net: \qquad 1(\text{glucose-6-P}) \rightarrow 6CO_2$$

In this case, although reducing power is generated as NADPH and in the cytoplasm, appropriate mechanisms exist in most cells to allow for the transfer of reducing power from NADPH to NADH in the cytoplasm and then the shuttling of the cytoplasmic NADH into the mitochondrion. (Remember that the calculation is to be made on the basis of the catabolism of free glucose.)

15–8. If purified mitochondria are incubated in a buffered solution containing an oxidizable substrate X, cyt c, ADP, and P_i, product Y is produced. In addition, P_i is consumed and O_2 is produced. If representative data for a 30-min incubation period are 31.3 μmole of P_i consumed and 15.2 μg atoms of oxygen produced, what can you conclude about the biochemistry of the $X \rightarrow Y$ conversion?

15–9. If a suspension of mitochondria was incubated in the presence of malate under anaerobic conditions, what prediction would you make about the ratios of electron carriers in their oxidized and reduced states?

15–10. When purified mitochondria are isolated and then assayed for oxidative phosphorylation activity, the buffered reaction system (containing the oxidizable substrate, ADP, and P_i) is usually supplemented with a small amount of pure cytochrome c obtained from almost any source. Why do you suppose the cytochrome c addition is necessary, and why does it not have to be obtained from the same source as the mitochondria?

15–11. Some eukaryotic cells contain an NAD-dependent glycerol phosphate dehydrogenase in the cytoplasm as well as an FAD-dependent glycerol phosphate dehydrogenase in the mitochondrion. Both enzymes catalyze the interconversion of dihydroxyacetone phosphate and L-glycerol phosphate, metabolites that can both move across the mitochondrial membrane. What metabolic significance might this compartmentalization of enzymes have?

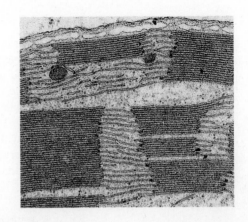

CHAPTER SIXTEEN
PHOTOSYNTHESIS

The range of solar radiation reaching the surface of earth is commonly called visible or white light, with a low wavelength limit of about 400 nm (or 4000 Å) and a high wavelength limit of about 700 nm (or 7000 Å). The light is the *energy of all life*. Photosynthetic organisms perform the supremely important role of harnessing this radiant energy and converting it into useful chemical energy to sustain not only their own existence but also the existence of all aerobic nonphotosynthetic organisms.

The crucial chemistry of photosynthesis is summarized by the following net equation:

$$6CO_2 \;+\; 6H_2O \;+\; \textbf{solar energy} \xrightarrow{\text{photosynthesis}} C_6H_{12}O_6 \;+\; 6O_2$$

low-energy oxidized carbon	weak reducing agent		higher-energy reduced carbon	strong oxidizing agent

The results are (1) light energy is used to convert a low-energy oxidized form of inorganic carbon (CO_2) into a higher-energy reduced form of organic carbon (carbohydrate), using water as a reducing agent; and (2) molecular oxygen is produced—the only chemical source of oxygen on this planet. These products of photosynthetic organisms are then used to support the life of aerobic, nonphotosynthetic organisms, which extract the chemical energy of reduced organic carbon (originally solar energy) by oxidizing it back to CO_2. This reciprocal chemistry in our biosphere is commonly referred to as the *carbon-oxygen cycle* (or simply the *carbon cycle*) (see Figure 16–1). We emphasize the supreme importance of photosynthesis because it begins the cycle, and hence without it, life as we know it would soon cease to exist.

The specialized organelle unique to photosynthetic cells where this chemistry occurs is the *chloroplast*. The chemical events occur in two distinct phases, termed the *light reaction* and the *dark reaction*. As the name implies, the light reaction is dependent on an input of radiant energy. The complete process involves many substances, the most important being *chlorophyll*. As indicated in Figure 16–2, the light reaction involves three events: the oxidative cleavage of water, the formation of NADPH, and the formation of ATP. The dark re-

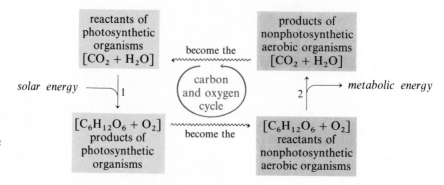

FIGURE 16–1 Summary of the interdependence of photosynthetic and nonphotosynthetic organisms.

action involves the enzymatic assimilation and conversion of CO_2 into carbohydrate. The NADPH and ATP formed in the light reaction are used in the dark reaction as sources of reducing power and energy, respectively. Each of these processes is examined in this chapter.

Note: Some of you may need a brief review of the modern quantum theory of light as energy, first formulated by Max Planck (1900) and Albert Einstein (1905). According to classical wave theory, the energy of a light wave depends only on its amplitude. Planck's modification was to propose that the precise energy E of a specific light wave is directly dependent on the frequency v of the wave:

$$E \propto v \qquad \text{or} \qquad E = hv$$

where h is Planck's constant, equal to 1.585×10^{-34} cal·s. Einstein proposed that the energy of a wave is emitted and absorbed only as whole-number multiples of hv—$1hv$, $2hv$, $3hv$—with each hv amount called a *quantum of energy* or a *photon of light*. Equivalent statements of light energy quanta can be written by using the reciprocal relationship between the frequency of a wave and the wavelength (λ) of a wave:

$$v = \frac{c}{\lambda}$$

where c is the speed of light, equal to 3×10^{10} cm/s. Therefore

$$E_{\text{photon}} = hv = \frac{hc}{\lambda}$$

or

$$E_{n\text{ photons}} = n\frac{(hc)}{\lambda}$$

Thus the shorter the wavelength, the higher is the energy.

16-1 CHLOROPLASTS: CELLULAR SITE OF PHOTOSYNTHESIS

General Comments

The specialized chloroplast organelle is found in the green leaf cells of higher plants and in the cells of photosynthetic algae, except the blue-green algae. Photosynthetic bacteria and blue-green algae lack the elaborate chloroplast structure but do have other light-trapping systems bound to the plasma membrane and also packaged into granules.

In the 1790s Ingen-Housz identified that light is essential for plants to produce O_2 and fix CO_2, and established that these processes only occur with the green parts of plants. The discovery and first description of chloroplasts in

---

results of light reaction

(directly dependent on radiant energy)

(1) $H_2O \rightarrow (2H^+ + 2e) + \frac{1}{2}O_2$

(2) $NADP^+ + 2H^+ + 2e \rightarrow$
$$NADPH + H^+$$

(3) $ADP + P_i \rightarrow ATP$

results of dark reaction

(indirectly dependent on radiant energy)

$$CO_2 \xrightarrow{\frac{NADPH}{ATP}} (CH_2O)$$

FIGURE 16-2 Summary of the major chemical events in photosynthesis.

the green parts is attributed to A. Meyer in 1883. In 1938 R. Hill was the first to demonstrate that isolated chloroplasts could produce O_2 in response to light. D. Arnon and co-workers were the first, in 1954, to show that isolated chloroplasts could carry out the complete process of photosynthesis—produce O_2 and fix CO_2. Chloroplasts have since been under study in regard to their structures, origin, replication, and differentiation.

Chloroplast Ultrastructure

The size, shape, and number of chloroplasts vary widely among photosynthesizing cells. In higher plants they are generally cylindrical (see Figure 16–3), ranging anywhere from 5–10 μm in length and 0.5–2 μm in diameter. A comparison of these dimensions with those of mitochondria reveals that the chloroplasts are much larger (by 2–5 times). Indeed, among the defined organelles of higher cells, only the nucleus is larger. The most conspicuous feature is that the chloroplast, like the mitochondrion, possesses a characteristic and highly organized fine level of internal ultrastructure. The ordered array of several electron-dense bodies throughout the interior is interpreted as corresponding to stacks of flattened membranous bodies, called *lamellae* or *thylakoid disks*. The stacked piles are called *grana* and number 40–80 per chloroplast. As indicated in Figure 16–3, the number of stacked thylakoid disks per granum is quite variable. As with the cristae of mitochondria, the thylakoid disks result from folded protrusions of an inner chloroplast membrane into the core of the chloroplast. Each thylakoid disk contains all of the necessary molecules (photosensitive pigments, electron carriers, and accessory components) for the crucial light reaction of photosynthesis. The bulk of the enzymes and coenzymes responsible for the assimilation of CO_2 into organic material are found in the soluble portion (*lumen*) of the chloroplast.

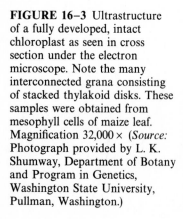
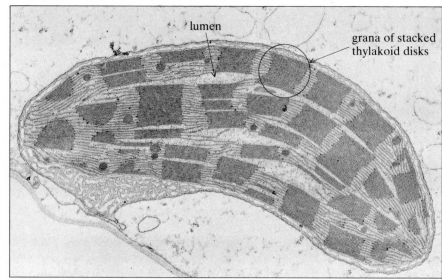

FIGURE 16–3 Ultrastructure of a fully developed, intact chloroplast as seen in cross section under the electron microscope. Note the many interconnected grana consisting of stacked thylakoid disks. These samples were obtained from mesophyll cells of maize leaf. Magnification 32,000× (*Source:* Photograph provided by L. K. Shumway, Department of Botany and Program in Genetics, Washington State University, Pullman, Washington.)

Despite the fact that each thylakoid disk contains all the materials required for the light reaction, it is not the basic unit of photosynthesis. High-resolution electron microscopic studies of isolated and partially fragmented thylakoids suggest that the basic unit (termed a *quantosome*) may be a small, somewhat spherical unit, several of which are embedded in the phospholipoprotein matrix of the thylakoid membrane (see Figure 16–4). What the exact composition and arrangement of the quantosomes may be and whether or not all quantosomes are identical are questions yet to be answered. Regardless of these uncertainties, the ultrastructure of the chloroplast is another example of the fascinating and, indeed, truly remarkable constructions of nature. Even more fascinating, however, is what goes on inside the chloroplast.

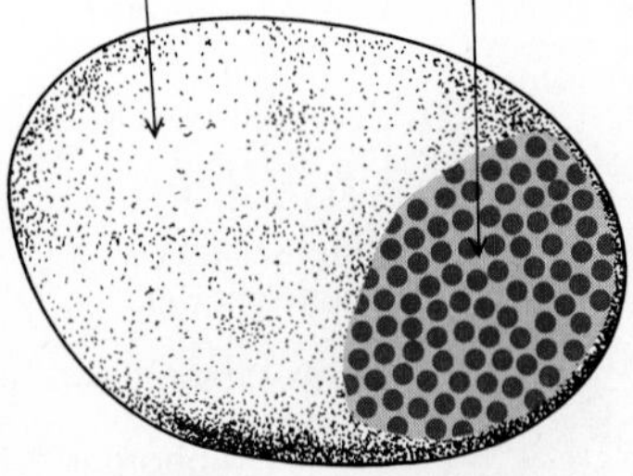

FIGURE 16–4 Idealized drawing of a thylakoid disk. The sketch depicts a top view of a thylakoid, with a portion of the surface membrane peeled off to reveal the presence of an array of multimolecular aggregates proposed to represent the elementary photochemical units, termed quantosomes.

16–2   LIGHT REACTION OF PHOTOSYNTHESIS

The combined effect of the light-dependent ($nh\nu$ is required) phase in all photosynthetic green plants and several algae can be given by the following equation (see Note 16–1):

$$2NADP^+ + mADP + mP_i + 2H_2O \xrightarrow{\text{chloroplasts}} 2NADPH + 2H^+ + mATP + O_2$$

It is a net equation summarizing the four main events of the light reaction; (1) the *photochemical excitation* of chlorophyll; (2) the oxidative cleavage of water, called *photooxidation;* (3) the reduction of $NADP^+$, called *photoreduction;* and (4) the formation of ATP, called *photophosphorylation*.

The previous equation is written to show the evolution of one unit of molecular oxygen. So that we avoid unnecessary confusion, this stoichiometry will be used throughout this chapter. Since the production of $1O_2$ involves the transfer of 4 electrons from H_2O to $NADP^+$, the equation also depicts an idealized 2 + 2 stoichiometry involving H_2O and $NADP^+$. Still uncertain are the number m of ATP molecules produced per molecule of oxygen evolved and the number n of light quanta required per molecule of oxygen evolved. For ATP formation several investigators have suggested a value of 4 for m. Others propose that $m = 2$. For the amount of light energy there are also two suggested values: $n = 4$ and $n = 8$. According to Einstein's law of photochemical equivalence stating that 1 quantum is required to excite 1 electron, the proposal that $n = 4$ represents 100% quantum efficiency (that is, 4 quanta for a 4-electron process), whereas $n = 8$ represents 50% efficiency. The controversy over these different values has existed since 1922 when O. Warburg first reported, and defended vigorously, that $n = 4$. The issue is still not resolved.

NOTE 16–1
A similar equation can be written for the photosynthetic bacteria. In these organisms, however, the source of reducing power is not H_2O but generally a special inorganic donor, such as H_2 or H_2S, or certain reduced organic acids. This variation in the photosynthetic bacteria will not be discussed further.

Primary Photochemical Act

The primary photochemical act is the absorption of photons of light by chlorophyll molecules, resulting in their excitation to higher energy states.

Thereafter, electrons from the excited chlorophyll molecules are transferred to

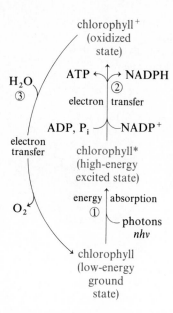

FIGURE 16–5 Brief summary of the role of chlorophyll in photosynthesis.

FIGURE 16–6 Chlorophyll structure. Note:

$Y = -CH_3$ in chl a
$Y = -CHO$ in chl b

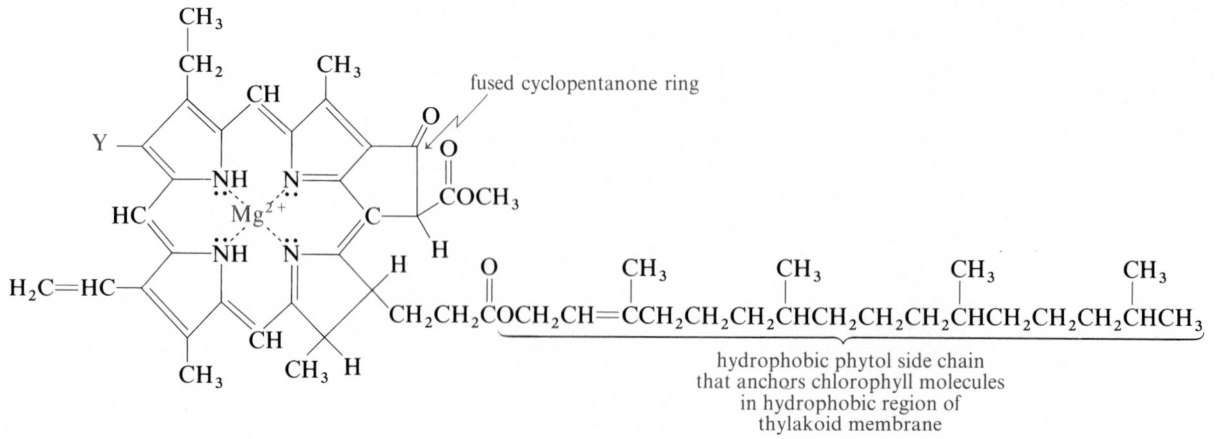

specialized acceptor molecules and ultimately to $NADP^+$. This process of electron transfer is also accompanied by ATP formation (see Figure 16–5). The role of water is that of electron donor (reducing agent) to return the electron-deficient (oxidized) chlorophyll molecules to the ground state—a process accompanied by the release of molecular oxygen.

Chlorophyll

With the exception of the photosynthetic bacteria and a few algae, almost all photosynthetic cells contain identical molecular species of chlorophyll, one designated *chlorophyll a* and the other *chlorophyll b*. (Hereafter, we will use chl a, and chl b.) Green plants and green algae contain both chl a and chl b. Other algae contain only chl a. Photosynthetic bacteria also contain only a single chlorophyll that is different from both chl a and chl b; it is called *bacteriochlorophyll*.

The basic structure of all chlorophylls is that of a planar, metal-containing tetrapyrrole unit similar to the heme prosthetic group of myoglobin and the cytochromes. The chlorophyll molecule, however, is distinctly different from the heme prosthetic group in three respects: (1) In chlorophyll the metal coordinated to the tetrapyrrole moiety is *magnesium* (as Mg^{2+}) rather than iron; (2) chlorophylls are *not* dependent on attachment to a protein grouping for activity; and (3) the chlorophylls contain a characteristic set of pyrrole side-chain groupings, the most notable of which are (a) a large, nonpolar, alcohol grouping (*phytol*) in ester linkage to a propionic acid side-chain residue and (b) a fused cyclopentanone ring. The nonpolar phytol grouping is particularly noteworthy, since it provides the structural basis for integration of chlorophyll molecules into the lipoprotein bilayer matrix of the thylakoid membrane. All of these features are depicted in Figure 16–6. As indicated, the structural difference between chl a and chl b is merely a variation in one pyrrole ring substituent.

The physicochemical property of the chlorophyll molecule that is biologically important is, of course, its capacity to absorb light energy from the visible region of the electromagnetic spectrum. In all likelihood, the electrons excited in this process are those in the conjugated double-bond system of the tetrapyrrole unit. In any event, because of the difference in structure between

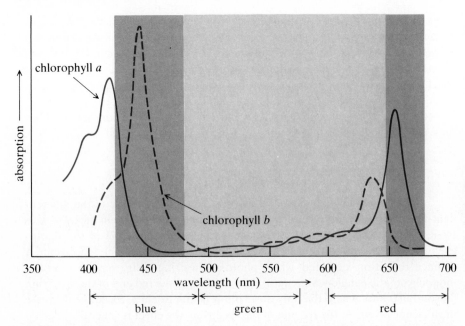

FIGURE 16–7 Absorption characteristics of chlorophyll in the visible region of the electromagnetic spectrum. The shaded areas represent the regions of photon absorption that result in chloroplast activity; the darker region represents greater chloroplast activity. Note the overlap of these regions with the absorption maxima for chlorophyll. Chloroplasts exhibit high levels of O_2 production when photons of light are absorbed in the two dark grey areas. Smaller but still significant levels of chloroplast activity are observed in the light grey region, where some of the accessory pigments absorb.

chl *a* and chl *b*, each displays a unique absorption (excitation) spectrum (see Figure 16–7). The composite result of this pattern of absorption is the green color of chlorophyll. Of greater significance, however, is the extensive evidence that the maximum activity of chloroplasts in producing oxygen is at wavelengths in the vicinity of the absorption maxima of chlorophyll (see Figure 16–7). Observations of this type have been made with chloroplasts from many different sources and, of course, have been interpreted to mean that chlorophyll is the primary photosensitive pigment in the photosynthetic apparatus. Chl *a* is regarded as the *primary* pigment because it is present in larger amounts than chl *b*.

Within the confines of the thylakoid membrane there are even differences in absorption among chl *a* molecules and perhaps also among chl *b* molecules. These differences arise because of a particular microenvironment in the lipoprotein matrix of the membrane where the chlorophyll molecules reside. Owing to a unique location and arrangement, these molecules can participate in unique interactions with other components of the membrane that produce unique light excitation properties in one or more chlorophyll molecules. For example, whereas most chl *a* molecules show a maximal excitation by photons from a range of about 665–670 nm, there is good evidence supporting the existence of two specialized chlorophyll molecules that absorb maximally at 683 and 700 nm. Labeling these molecules as specialized chlorophylls means that they occupy a particular location in the quantosome aggregate, where they act as terminal participants in the capture of radiant energy absorbed by other chlorophyll molecules (and possibly other accessory pigments; see the next paragraph) and then as the immediate donors of electrons to oxidized carriers.

In addition to containing chlorophyll, chloroplasts also contain other substances that absorb in the 400–700 nm range. They are called *accessory pigments* because their absorption of light supplements the action of chlorophyll. These pigments include materials from two classes of compounds, the *carotenoids* (β-carotene is most important in higher plants; see Section 11–4) and the *phycobilins,* which are open-chain tetrapyrrole compounds. Although their exact function has not yet been identified, evidence suggests that by acting as secondary absorbers, these accessory pigments assist in the transfer of energy to the specialized chlorophylls within the quantosome reaction center. They may also protect chlorophyll molecules from getting bleached by solar rays.

Emerson Effect and the Dual-Photosystem Hypothesis

One of the most significant discoveries regarding the role of the different light-absorbing chlorophyll species in photosynthesis was made in the late 1950s by R. Emerson. He found that the efficiency (see Note 16–2) of the photochemical phase was not constant throughout the entire visible spectrum. A distinct reduction occurred when photosynthesizing cells were exposed to monochromatic (single-wavelength) light sources at the red end of the spectrum (beyond 680 nm), where the only absorption is due to chl *a*. Since the decrease in efficiency occurs at the red end of the spectrum, the effect is generally referred to as the *red-drop phenomenon.* Emerson subsequently demonstrated that full efficiency could be restored if the cells were simultaneously exposed to another light source with a wavelength of 650 nm, a point of maximum absorption in chl *b*. In other words, when light absorption by chl *a* was accompanied by light absorption from chl *b*, the cells displayed an enhancement (*Emerson effect*) in the photoevolution of oxygen as compared with separate light absorption by either chl *a* or chl *b*.

Emerson suggested then that the light-dependent phase of photosynthesis includes *two separate photosystems,* both of which must be activated for maximum efficiency of the light-dependent reaction. One (photosystem I, or PSI) contains largely chl *a* (including the specialized chl *a* absorbing at 683); the other (photosystem II, or PSII) contains both chl *a* (including the specialized chl *a* absorbing at 700) and chl *b*. The hypothesis of two separate photosystems has since been supported by several different lines of evidence and currently governs much of the modern thinking concerning the photochemical apparatus. The present viewpoint is that both systems are separately localized within each quantosome of the thylakoid membrane and that each system participates in a separate phase of the overall light reaction.

Electron Transport in Photosynthesis

At this point a logical question is, What is the function of each photosystem in the overall light reaction? To pursue this subject, we must first recognize the basic type of chemistry that is involved. To analyze it, let us return to the equation given earlier for the net effect of the light reaction:

$$2NADP^+ + mADP + mP_i + 2H_2O \xrightarrow[nhv]{\text{chloroplasts}} 2NADPH + 2H^+ + mATP + O_2$$

NOTE 16–2
Photosynthetic efficiency is usually measured as the amount of oxygen produced per quantum of radiant energy absorbed.

Close examination reveals that the basic chemistry is an oxidation-reduction reaction involving a transfer of two hydrogen atoms ($2H^+$ and 2 electrons) from H_2O (the reduced donor) to $NADP^+$ (the oxidized acceptor). Recalling our discussion in the previous chapter concerning the difference between the $NAD^+/NADH$ and O_2/H_2O couples, we should also note that the direction of electron transfer involves a transition from a low energy state (H_2O) to a higher energy state (NADPH). It is this unfavorable energy barrier that is overcome by the absorption of light energy, with the accompanying formation of ATP representing a conservation of part of the absorbed energy:

$$NADP^+ + 2H^+ + 2e \rightarrow NADPH + H^+ \qquad \mathscr{E}^{\circ\prime}_{red} = -0.32 \text{ V}$$
$$H_2O \rightarrow \tfrac{1}{2}O_2 + 2H^+ + 2e \qquad \mathscr{E}^{\circ\prime}_{ox} = -0.82 \text{ V}$$
$$\overline{NADP^+ + H_2O \rightarrow NADPH + H^+ + \tfrac{1}{2}O_2 \qquad \mathscr{E}^{\circ\prime}_{net} = -1.14 \text{ V}}$$

Here $\mathscr{E}^{\circ\prime}_{net}$ is negative; therefore the reaction from left to right is endergonic. Specifically, $\Delta G^{\circ\prime} = +52.6$ kcal.

In the pages to follow, our objective will be to elaborate on how these events occur and to explain the suggested participation of photosystems I and II. You will discover shortly that the modes of electron transport and phosphorylation of ADP are *identical in general design* to the events of oxidative phosphorylation. In view of our frequent referral to the theme of biochemical unity, you should find this relationship to be a logical one. Our discussion of the whole subject will be somewhat brief, in part because of the restricted limits of this book but also because the complete details of this process have yet to be determined exactly.

Photoreduction (PSI) and Photooxidation (PSII)

Despite the unknown nature of many phases of the light reaction, there is considerable evidence to suggest that both photosystems fulfill separate but complementary roles (see Figure 16–8). These proposed functions are as follows

FIGURE 16–8 Outline of the interaction between two photosystems, each activated by light absorption at different wavelengths. (See Figure 16–9 for additional details.)

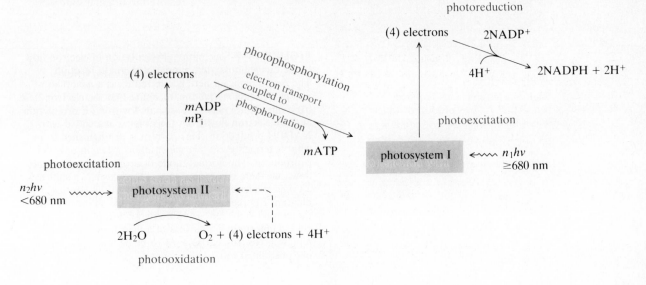

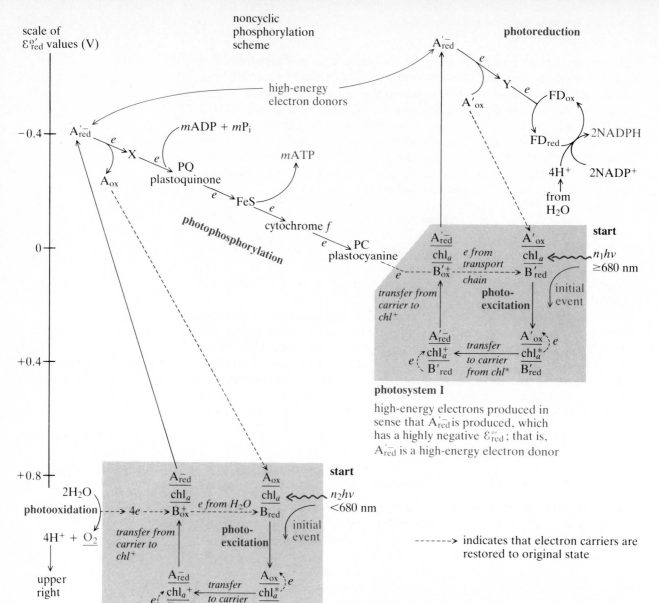

FIGURE 16–9 Schematic representation of electron flow in the dual-photosystem model of the light reaction. The model depicted here is referred to as a *noncyclic phosphorylation mechanism,* meaning that the electron flow from one photosystem is associated with ATP production and the electron flow from the other is associated with NADPH production. This pattern is distinguished from what is termed *cyclic phosphorylation,* wherein only ATP is produced. The normal mode of operation is noncyclic flow, but there is evidence that chloroplasts can shift to a cyclic mechanism under certain conditions. In this diagram FD represents ferredoxin; chl is chlorophyll in the ground state; chl* is an excited chlorophyll molecule; chl$^+$ is chlorophyll after loss of an electron; X and Y represent electron carriers yet to be definitely identified but postulated to be FeS proteins.

(we will not review the substance of the experimental proof): The activation of photosystem I at wavelengths equal to or greater than 680 nm results in a photoreduction ($2NADP^+ + 4H^+ + 4e \rightarrow 2NADPH + 2H^+$), whereas the activation of photosystem II at wavelengths less than 680 nm results in a photo-oxidation ($2H_2O \rightarrow O_2 + 4H^+ + 4e$). As implied by the two half-reactions, the complementary relationship between PSI and PSII involves the gain and the loss of electrons. However, the flow of electrons from H_2O to $NADP^+$ is indirect. The electrons required for NADPH formation arise directly from the photoexcitation of PSI. What, then, is the fate of the electrons generated in the photolytic cleavage of H_2O in PSII?

According to current interpretations, the electrons generated from the photooxidation of H_2O via PSII are returned to the same photosystem, which becomes electron-deficient after undergoing excitation (loss of electrons to a higher energy level) by light absorption. The balance between the two photosystems is restored by the transfer of electrons liberated in the excitation of PSII to the electron-deficient PSI. Furthermore, this linkage of *electron transport* from PSII to PSI is the exergonic sequence that is *coupled to the phosphorylation of ADP*.

A more detailed summary of the most widely supported proposal for the events occurring in the light reaction is diagramed in Figure 16–9. It is sometimes called the Z scheme. A key to understanding the message of this diagram resides in the A, B, and A′, B′ symbolism in the shaded rectangles. These shaded areas correspond to the two active photosystems, each containing the primary photosensitive chlorophyll molecules and other accessory pigments (all of course embedded in a lipoprotein matrix). The materials symbolized as A, B, A′, and B′ are proposed to be highly specialized *electron carriers* unique to each photosystem and required for the processes of photooxidation, photoreduction, and photophosphorylation. Each photosystem is believed to contain two such materials (see Figure 16–10), with one acting as an oxidized electron acceptor (A_{ox} in PSII and A'_{ox} in PSI) and the other as a reduced electron donor (B_{red} in PSII and B'_{red} in PSI). A and A′ may be FeS proteins; B appears to be a manganese-containing protein.

Also hypothesized is that A_{ox} and A'_{ox} undergo reduction when each receives an electron from a specialized (see the next paragraph) chlorophyll *a* molecule in each photosystem that has been excited (chl_a^*) by the absorption of light energy. On the other hand, B_{red} and B'_{red} are hypothesized to act as immediate electron donors, returning the electron-deficient chlorophyll molecule (chl_a^+) to its ground state. The resultant formation of the reduced carriers (A_{red}^- and $A_{red}'^-$) represents the raising of electrons to higher energy levels, from which they are then transferred to still other specialized acceptors. The electrons from A_{red}^- generated in PSII are ultimately transferred "downhill" (that is, to a lower energy level) to PSI, with the accompanying formation of ATP. The electrons from $A_{red}'^-$ generated in PSI are used in the reduction of $NADP^+$. The resultant formation of B_{ox}^+ and $B_{ox}'^+$ represents the formation of low-energy electron acceptors. The B_{ox}^+ can act as a suitable electron acceptor to drive the oxidation of H_2O, which is linked to PSII. In PSI, $B_{ox}'^+$ can act a suitable terminal acceptor in the transport of electrons from the A_{red}^- coming from PSII.

$$(chl^*)A_{ox} \xrightarrow[\substack{\text{electron from} \\ \text{excited} \\ \text{chlorophyll}}]{\substack{\text{substance A} \\ \text{accepts}}} (chl^+)A_{red}^-$$

(same for A′ in PSI)

$$B_{red}(chl^+) \xrightarrow[\substack{\text{electron} \\ \text{to deficient} \\ \text{chlorophyll}}]{\substack{\text{substance B} \\ \text{donates}}} B_{ox}^+ (chl)$$

(same for B′ in PSI)

FIGURE 16–10 Participation of the immediate electron carriers in close association with chlorophyll.

The reference to a specialized chl *a* concerns one particular molecule of chl *a* that is in a particular position in the quantosome; this molecule becomes ionized by virtue of being the recipient of energy from a "bucket brigade type" of transfer mechanism involving other molecules of chl *a*, chl *b*, and possibly the accessory pigments. In other words, the specialized chl *a* molecule does not absorb light directly. The energy transfer mechanism is probably the least understood aspect of photosynthesis and certainly the most remarkable. For example, estimates are that it occurs in about 10^{-12} to 10^{-15} s. The close association with the electron carriers (A, A′, B, and B′) that permits direct reactions with them is, of course, another unique feature of these specialized molecules.

Some of the electron carriers have been identified. A *plastoquinone* (PQ) molecule is the first acceptor of electrons from A_{red}^-, whose identity is still unknown. It may be a chlorophyll molecule. Others have reported evidence that B′ is a specialized cytochrome carrier whose immediate electron donor in the electron transport chain is *plastocyanine* (PC), a copper-containing protein, which in turn acts as an acceptor of electrons from *cytochrome f*. Another specialized FeS protein is suggested as a carrier between plastoquinone and cytochrome *f*.

The scheme in Figure 16–9 and its specific features constitute, of course, only a working model of the light reaction. This scheme is widely accepted as representative of what is occurring in vivo, because it agrees with most experimental studies made with illuminated chloroplasts and thylakoid preparations. As research continues, the picture will become more definitive and perhaps even modified.

Regardless of the many uncertainties that still exist, such as the nature of A, B, A′, and B′, the location and number of ATP-generating sites, and the possible existence of other electron carriers, there is universal agreement that electron transfer among specific carriers is an integral part of the light reaction. Moreover, in organisms characterized by the operation of two separate photosystems, and presumably this category includes virtually all photosynthesizing cells except the photosynthetic bacteria, it should occur to you that if the scheme in Figure 16–9 is correct, the electron transfer system between A_{red}^- (PSII) and $B_{ox}'^+$ (PSI) is a crucial step of the light reaction, since it constitutes the link between photooxidation and photoreduction in addition to serving as the driving force for the formation of ATP, that is, photophosphorylation.

To close our discussion of the dual-photosystem model, let us take note of the participation of the substance symbolized as FD in Figure 16–9, a substance proposed to be the immediate electron acceptor of $A_{red}'^-$ and the immediate electron donor of $NADP^+$ in the photoreduction phase. The FD notation represents *ferredoxin,* an FeS nonheme protein known to be present in all types of photosynthetic organisms, including the photosynthetic bacteria. Regardless of the source, ferredoxin has two distinguishing properties: (1) It can undergo a reversible oxidation-reduction via electron transfer, and (2) the FD_{ox}–FD_{red} couple has a reduction potential even more negative than that of the $NADP^+$–NADPH couple (see Figure 16–11 and Table 15–1). The latter property is one of the strong arguments for the proposed role of FD_{red} as the

$$FD_{ox} + 1e \rightarrow FD_{red}$$

oxidized   reduced
ferredoxin   ferredoxin

$$\mathscr{E}_{red}^{\circ\prime} = -0.43 \text{ V}$$

$$NADP^+ + 2H^+ + 2e \rightarrow$$
$$NADPH + H^+$$

$$\mathscr{E}_{red}^{\circ\prime} = -0.32 \text{ V}$$

FIGURE 16–11 Ferredoxin, a strong reducing agent, capable of reducing $NADP^+$.

immediate donor of electrons to $NADP^+$, since the transfer would be energetically favorable.

Coupling Mechanism of Photophosphorylation

In chloroplasts the energy-yielding electron transfer sequence is coupled to ATP formation in the same way as it is in mitochondria—via an *electrochemical proton potential* ($\Delta\mu_{H+}$). Electron carriers pump H^+ from the chloroplast matrix (stroma) into the thylakoid space, and the resultant $\Delta\mu_{H+}$ is used to drive the formation of the chloroplast H^+-ATPase, which protrudes from the thylakoid membrane into the matrix. Yes, the H^+-ATPases from chloroplasts and mitochondria are similar in structure and function. However, because of ultrastructure differences, the disruption of mitochondria and chloroplasts gives topologically distinct suborganelle particles. Whereas the submitochondrial particles are inverted with the H^+-ATPase on the inside of the vesicle, the subchloroplast particles are intact, noninverted thylakoid disks, with the H^+-ATPase on the outside of the vesicle (see Figure 16–12).

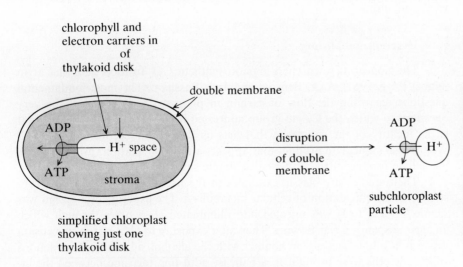

simplified chloroplast
showing just one
thylakoid disk

FIGURE 16–12 Origin of $\Delta\mu_{H+}$ in chloroplasts. Because of electron transport occurring in the thylakoid membrane (color) between photosystems I and II, protons are pumped from the stroma into the thylakoid space. The exit of H^+ via the H^+-ATPase drives ATP formation. Compare the reverse topology of the submitochondrial particle in Figure 15–20 with that of the subchloroplast particle.

16–3   DARK REACTION OF PHOTOSYNTHESIS

Introduction

A brief and convenient statement of what happens in the dark reaction (see Note 16–3) is given by the following equation:

$$6CO_2 \xrightarrow[\text{NADPH ATP}]{\substack{\text{chloroplast} \\ \text{enzymes}}} C_6H_{12}O_6$$

The process occurs in the chloroplast, involves many enzyme-catalyzed

NOTE 16–3
Although we customarily refer to the fixation of CO_2 as the dark reaction, this use does not mean that the reactions involved do not occur in the presence of light—they do. When illumination ceases, CO_2 fixation will continue to occur for a brief period until the levels of ATP and NADPH become limiting.

reactions, requires NADPH and ATP, and results in the formation of carbo-hydrate material ($C_6H_{12}O_6$) from inorganic CO_2. Although the equation is use-ful in summarizing the overall effect, it can be misleading, since it implies that (1) the carbon atoms of each of six separate molecules of CO_2 become part of the same hexose molecule, and (2) the dark reaction is limited to this one type of metabolic activity. Actually, neither of these implications is correct. Regarding the former, we will shortly discover that the relationship of $6CO_2 \rightarrow C_6H_{12}O_6$ represents only a *net* conversion rather than an actual synthesis of a single hexose molecule from six molecules of CO_2. As for the latter, it should occur to you that the complete enzymatic metabolism of a plant cell must consist of more than a fixation and conversion of CO_2 to carbohydrates. While a signifi-cant portion of the assimilated carbon will be stored as sucrose and utilized for the biosynthesis of cellulose, the remainder, by being channeled into central metabolic pathways such as the citric acid cycle, will serve as a chemical source of energy and carbon for the anabolism of other carbohydrates, amino acids, proteins, fatty acids, lipids, purine and pyrimidine nucleotides, nucleic acids, and even the tetrapyrrole moiety for chlorophyll itself. In other words, the original carbon of CO_2 is ultimately incorporated into the entire metabolism of the whole organism.

Experimental History

The pioneering researchers in photosynthetic CO_2 fixation were M. Calvin (Nobel Prize, 1961), A. A. Benson, and J. A. Bassham. The most fundamental questions regarding the flow of carbon in photosynthesis—indeed, the very questions to which the Calvin group addressed itself—are as follows: (1) What is the identity of the substance that acts as the initial acceptor of CO_2? (2) What is the immediate product that appears after CO_2 fixation? (3) How is the immediate product then converted to simple sugars?

Their studies are classic examples of the use of radioactive tracer sub-stances in the elucidation of cellular metabolism. The experimental design was simple. First, $^{14}CO_2$ was injected into illuminated glass tubes through which an algae suspension was flowing. Then after exposure to $^{14}CO_2$, the suspension was run into hot alcohol. On contact with the alcohol, all enzymatic reactions within the cells were brought to a halt. By adjusting the time between the in-jection of $^{14}CO_2$ and the final mixing with alcohol, the researchers were able to limit exposure of the cells to the carbon source to any desired interval, from a few minutes to a fraction of a second. Afterward, samples of the alcohol solu-tion, which contained dissolved compounds extracted from the cell, were analyzed chromatographically for the appearance of ^{14}C-labeled compounds.

As you might expect, with prolonged exposure times (10 min), the extract contained a large assortment of ^{14}C-labeled compounds including many simple carbohydrates (mostly phosphorylated sugars), several amino acids, all of the major nucleotides, and others. Shorter exposure intervals (30 s) considerably reduced the number of labeled materials, with virtually all of the ^{14}C being found in a restricted number of phosphorylated carbohydrates. Those iden-tified were trioses (dihydroxyacetone phosphate, glyceraldehyde-3-phosphate, and 3-phosphoglycerate); a tetrose (erythrose-4-phosphate); pentoses (ribose-5-phosphate, ribulose-5-phosphate, ribulose-1,5-bisphosphate, and xylulose-

5-phosphate); hexoses fructose-1,6-bisphosphate, fructose-6-phosphate, and glucose-6-phosphate); and heptoses (sedoheptulose-7-phosphate, and sedoheptulose-1,7-bisphosphate).

When the exposure period was reduced to only a fraction of a second, the extract was found to contain only a single compound with any appreciable ^{14}C content: *3-phosphoglycerate*. Although the initial studies were done with algae, there is now considerable evidence that the initial formation of 3-phosphoglycerate and its subsequent metabolism constitute the primary metabolic pathway utilized by most green plants as well. Plants utilizing this pathway are called *three-carbon plants,* or simply C_3 plants, because the immediate product of CO_2 fixation consists of three carbons. Recent studies have uncovered the existence of at least one major alternate pathway in some plants, which we will examine later in this section. Photosynthetic bacteria use alternate routes as well.

Although the appearance of 3-phosphoglycerate as the primary product of CO_2 assimilation suggested that the initial acceptor was probably a C_2 compound ($C_2 + CO_2 \rightarrow C_3$), Calvin and co-workers subsequently demonstrated that the acceptor molecule was *ribulose-1,5-bisphosphate,* a C_5 compound. In other words, the initial flow of carbon was $C_5 + CO_2 \rightarrow 2C_3$. Moreover, the flow of carbon during periods of light and dark indicated that the acceptor → product relationship was *cyclic,* with cellular levels of labeled bisphosphate decreasing in the dark and levels of phosphoglycerate increasing. The label pattern during dark periods also suggested that the subsequent metabolism of 3-phosphoglycerate, particularly the regeneration of ribulose-1,5-bisphosphate, was limited by factor(s) supplied only during periods of illumination. These factors are now known to be NADPH and ATP. After several years of study, the Calvin group eventually proposed a scheme identifying all the steps in getting from 3-phosphoglycerate to ribulose-1,5-bisphosphate. The task was extremely difficult, because the constituent reactions occur in a highly branched pathway. For traditional reasons, the sequence is frequently called the Calvin-Benson-Bassham cycle or the Calvin cycle (see Figure 16–13).

FIGURE 16–13 Brief summary of the Calvin cycle. A complete reaction scheme is shown in Figure 16–14. Whether in the light or the dark, the flow of carbon depends on a supply of NADPH and ATP from the light reaction.

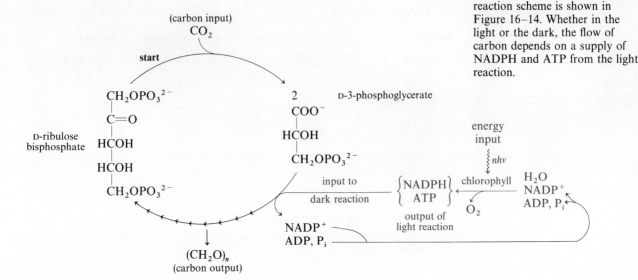

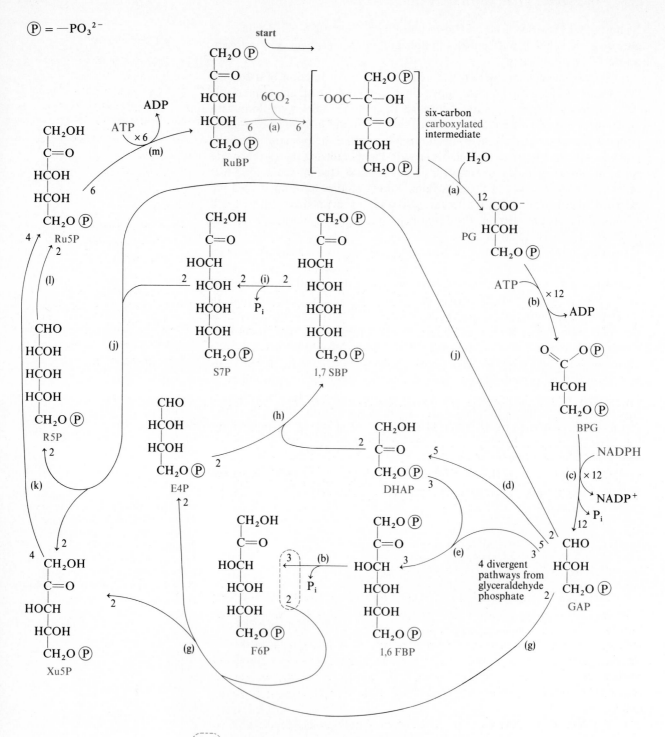

identifies the net gain of one hexose unit—
3F6P is formed and only 2F6P is consumed during
the course of fixing 6CO₂

Enzymes

(a) ribulose-1,5-bisphosphate carboxylase

(b) phosphoglycerate kinase

(c) glyceraldehyde phosphate dehydrogenase

(d) triose phosphate isomerase

(e) aldolase

(f) bisphosphatase

(g) transketolase (thiamine PP_i)

(h) aldolase

(i) diphosphatase

(j) transketolase (thiamine PP_i)

(k) epimerase

(l) isomerase

(m) phosphoribulo kinase

$$6RuBP + 6CO_2 \xrightarrow{(a)} 6[intermediate] \xrightarrow{(a)} 12PG$$
$$12PG + 12ATP \xrightarrow{(b)} 12BPG + 12ADP$$
$$12BPG + 12NADPH(H^+) \xrightarrow{(c)} 12GAP + 12NADP^+ + 12P_i$$
$$5GAP \xrightarrow{(d)} 5DHAP$$
$$3DHAP + 3GAP \xrightarrow{(e)} 3FBP$$
$$3FBP \xrightarrow{(f)} 3F6P + 3P_i$$
$$2F6P + 2GAP \xrightarrow{(g)} 2Xu5P + 2E4P$$
$$2E4P + 2DHAP \xrightarrow{(h)} 2SBP$$
$$2SBP \xrightarrow{(i)} 2S7P + 2P_i$$
$$2S7P + 2GAP \xrightarrow{(j)} 2Xu5P + 2R5P$$
$$4Xu5P \xrightarrow{(k)} 4Ru5P$$
$$2R5P \xrightarrow{(l)} 2Ru5P$$
$$6Ru5P + 6ATP \xrightarrow{(m)} 6RuBP + 6ADP$$

net: $6CO_2 + 12NADPH(H^+) + 18ATP \xrightarrow[steps]{13}$ fructose-6-phosphate(F6P) $+ 12NADP^+ + 18ADP + 17P_i$

subsequently:

fructose-6-P $\longrightarrow$ glucose-6-P $\longrightarrow\longrightarrow$ sucrose, cellulose, and general metabolism via glycolysis and citric acid cycle

◀ **FIGURE 16–14** Flow of carbon in photosynthesis (opposite page) according to the Calvin-Benson-Bassham scheme, a three-carbon (C_3) pathway. The net chemistry of the pathway for the assimilation of $6CO_2$ is summarized here. Abbreviations used are RuBP, ribulose-1,5-bisphosphate; PG, 3-phosphoglycerate; BPG, 1,3-bisphosphoglycerate; GAP, glyceraldehyde-3-phosphate; DHAP, dihydroxyacetone phosphate; FBP, fructose-1,6-bisphosphate; F6P, fructose-6-phosphate; Xu5P, xylulose-5-phosphate; E4P, erythrose-4-phosphate; R5P, ribose-5-phosphate; SBP, sedoheptulose-1,7-bisphosphate; S7P, sedoheptulose-7-phosphate.

Primary Path of Carbon in Photosynthesis

The Calvin-Benson-Bassham scheme is shown in Figure 16–14. The first reaction, catalyzed by *ribulose-1,5-bisphosphate carboxylase,* is the carboxylation of ribulose-1,5-bisphosphate (RuBP) to yield an unstable six-carbon intermediate, which is then cleaved to give two units of 3-phosphoglycerate [PG; reaction (a)]. The crucial nature of this reaction is evidenced by the fact that the carboxylase enzyme is known to occur in very large amounts within the chloroplasts, accounting for roughly a sixth of all the soluble protein in the lumen.

The immediate metabolic fate of 3-phosphoglycerate (in the chloroplast) is a two-step conversion to glyceraldehyde-3-phosphate (GAP), involving two enzymes already encountered in glycolysis (phosphoglycerate kinase and triose-phosphate dehydrogenase), which function here in reverse. The former requires ATP and the latter NADPH, both of which are supplied from the light reaction.

The subsequent metabolism of glyceraldehyde-3-phosphate diverges through four distinct *branches* [reactions (d), (e), (g), (j)]. Note that the chemistry of these and all subsequent steps involves only the modification and interconversion of various phosphorylated sugars. These reactions include isomerizations, an epimerization, transfers between aldoses and ketoses, dephosphorylations, and a phosphorylation—all transformations previously encountered in our earlier discussions of glycolysis and the pentose phosphate pathways. Note the two reactions catalyzed by *aldolase* and the two by *transketolase*.

Although each of the four branches of glyceraldehyde-3-phosphate metabolism is important, the isomerization to [reaction (d)] and condensation with [reaction (e)] dihydroxyacetone phosphate are especially crucial, since they account for the incorporation of the C of the original CO_2 into hexose units. Ultimately, all of these conversions [reactions (c) through (k)] converge on the production of ribulose-5-phosphate. The ribulose-5-phosphate is then phosphorylated in an ATP-dependent, kinase-catalyzed reaction to yield the original bisphosphate, thus closing the cycle. As in the formation of glyceraldehyde-3-phosphate, the ATP required in this step is supplied from the light reaction.

From a net analysis (see the caption of Figure 16–14) with $6CO_2$, reactions (a) through (c) yield 12GAP. Then from a distribution of the subsequent metabolism of the 12GAP through the four divergent routes indicated in Figure 16–14, the overall chemical effect is evident. The equivalent of one hexose molecule is produced from six molecules of CO_2, with 2NADPH and 3ATP being required for every CO_2 molecule that is assimilated.

Ribulose-1,5-bisphosphate Carboxylase

The most distinctive enzyme of the Calvin cycle is ribulose-1,5-bisphosphate carboxylase (RuBPCase). Present in large quantities in the matrix of chloroplasts, representing about 25% of the total green leaf protein, it is probably the most abundant enzyme in our biosphere.

STRUCTURE AND MECHANISM OF ACTION   The carboxylase molecule (MW = 560,000) is composed of eight large subunits (L, with MW about 55,000) and eight small subunits (S, with MW about 12,500) for a formula of L_8S_8. The X-ray diffraction studies show the molecule to be double-layered, each layer composed of L_4S_4 (see Figure 16–15). The two layers appear to be nearly eclipsed.

The proposed mechanism of action for RuBPCase (see Figure 16–16) suggests that active-site R groups promote the ionization of the enol form of ribulose-1,5-bisphosphate. Resonance of the resulting enolate ion explains how C2 of the ketopentose can acquire nucleophilic (δ^-) character for condensation with the electrophilic (δ^+) carbon of CO_2. After condensation, hydrolysis of the six-carbon intermediate yields two 3-phosphoglycerate fragments.

Optimum activity requires a divalent metal ion (M^{2+}). The M^{2+} effect involves an unusual aspect of RuBPCase activity, namely, a CO_2 enhancement. This role of CO_2 is distinctly different and separate from the use of CO_2 as substrate. The M^{2+}-mediated enhancement by CO_2 involves the formation of a modified form of the enzyme. The CO_2 reacts with the epsilon NH_2 function

(a)

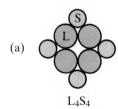

L_4S_4

(b)

L_4S_4

(c)

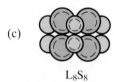

L_8S_8

FIGURE 16–15 Views of RuBPCase. (a) Top view of L_4S_4 layer. (b) Side view of L_4S_4 layer. (c) Side view of L_8S_8 two-layered molecule.

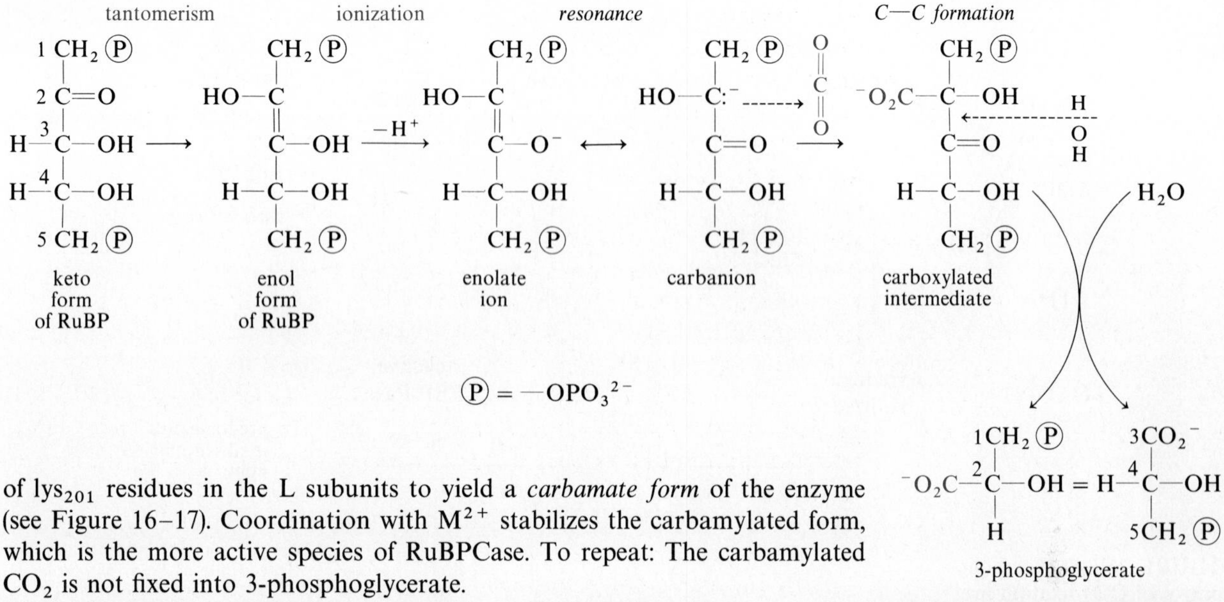

tantomerism   ionization   *resonance*   *C—C formation*

$$P = -OPO_3^{2-}$$

of lys_{201} residues in the L subunits to yield a *carbamate form* of the enzyme (see Figure 16–17). Coordination with M^{2+} stabilizes the carbamylated form, which is the more active species of RuBPCase. To repeat: The carbamylated CO_2 is not fixed into 3-phosphoglycerate.

METABOLIC REGULATION Of several metabolite signals that may affect RuBPCase activity, inhibition by fructose-1,6-bisphosphate (1,6-FBP) and activation by fructose-6-phosphate (F6P) are particularly significant. Both effects are related to the *activation of the Calvin cycle by light*. The control by light (*hv*) is mediated via a complex cascade system involving the participation of several proteins: ferredoxin, thioredoxin, ferredoxin-thioredoxin reductase, and fructose-1,6-bisphosphate phosphatase. In the absence of light, when there is no active photosynthesis, a high concentration of 1,6-FBP in the matrix serves to inhibit the carboxylase. When light is present, the phosphatase is activated, thus converting the carboxylase inhibitor 1,6-FBP to the carboxylase activator F6P.

As shown in Figure 16–18, the link between light and the activation of the converter phosphatase enzyme begins with the light-dependent formation of reduced ferredoxin. The enzyme ferredoxin-thioredoxin reductase then catalyzes a protein/protein redox reaction between reduced ferredoxin and the disulfide oxidized form of thioredoxin. [Thioredoxin (TR) is a small (MW = 12,000), heat-stable protein containing two —SH groups that can be reversibly converted to an intrachain disulfide bond.] The reduced thioredoxin then reacts with the phosphatase (FBP-Pase) in a disulfide/sulfhydryl redox reaction, converting the inactive, oxidized form of phosphatase to its active, reduced form.

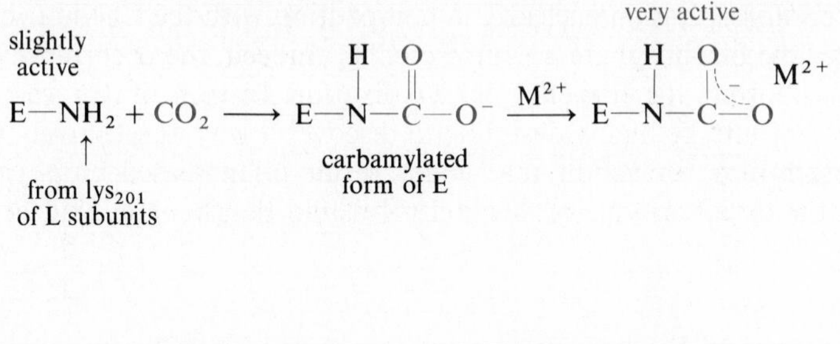

FIGURE 16–16 Proposed mechanism for the carboxylation of ribulose-1,5-bisphosphate via RuBP carboxylase. Active-site R groups are not shown.

FIGURE 16–17 Activation of ribulose-1,5-bisphosphate carboxylase (E) by CO_2.

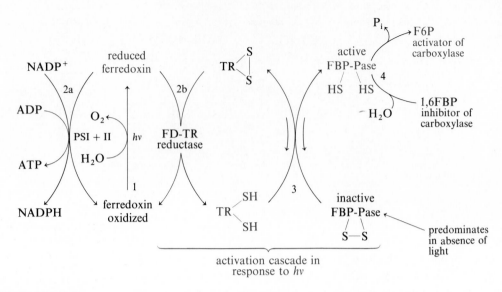

FIGURE 16–18 Metabolic control of CO_2 fixation in response to the presence or absence of light. Here TR is thioredoxin; FD-TR reductase is ferredoxin-thioredoxin reductase; FBP-Pase is fructose bisphosphate phosphatase.

When light is removed, the reverse of the thioredoxin/phosphatase reaction will restore the inactive phosphatase, thus allowing the level of the inhibitor FBP to increase.

There is evidence that fructose-1,6-bisphosphate phosphatase is not the only enzyme controlled by the thioredoxin protein. Perhaps at least three other Calvin cycle enzymes and two other enzymes involved in subsequent plant metabolism are affected. In addition, a thioredoxin protein plays an important role in the conversion of ribose to deoxyribose (see Figure 18–39) in all types of organisms.

Photorespiration

For reasons not yet fully understood, most plants are also capable of consuming O_2 in periods of illumination. The process is termed *photorespiration* and differs from mitochondrial respiration, which also occurs in plants. The O_2 consumed in photorespiration involves two enzymes (see Figure 16–19). The first is *RuBP carboxylase,* which also displays an *oxygenase activity.* Ribulose-1,5-bisphosphate is oxygenated to yield a C_5 hydroperoxide intermediate, which is then cleaved to yield 3-phosphoglycerate (C_3) and phosphoglycolate (C_2). The second enzyme is *glycolate oxidase,* which converts glycolic acid (derived from the hydrolysis of phosphoglycolate) to glyoxylic acid. This sequence of chemistry does not appear to be of any significant benefit to plants. In fact, photorespiration is clearly in competition with the Calvin cycle since it scavenges the bisphosphate acceptor of CO_2. Indeed, the occurrence of photorespiration lowers the efficiency of CO_2 fixation. In view of this wasteful competition, it would be highly desirable to discover a way of selectively inhibiting photorespiration, an action that could result in increased crop yields. Obviously, the target enzyme of this control would be glycolate oxidase.

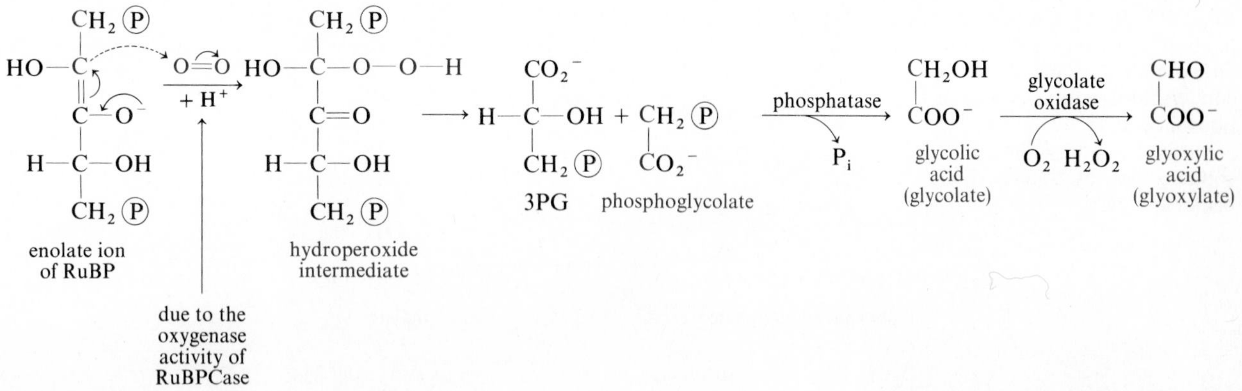

$$CH_2 \, \textcircled{P}$$
$$HO-C \quad O \overset{\frown}{=} O \quad HO-C-O-O-H$$
$$\qquad + H^+$$
$$C-O \qquad\qquad C=O$$
$$H-C-OH \qquad H-C-OH$$
$$CH_2 \, \textcircled{P} \qquad\quad CH_2 \, \textcircled{P}$$

enolate ion of RuBP

due to the oxygenase activity of RuBPCase

hydroperoxide intermediate

$$\longrightarrow \quad \begin{array}{c} CO_2^- \\ | \\ H-C-OH \\ | \\ CH_2 \, \textcircled{P} \end{array} + \begin{array}{c} CH_2 \, \textcircled{P} \\ | \\ CO_2^- \end{array}$$

3PG   phosphoglycolate

$$\xrightarrow{\text{phosphatase}} \begin{array}{c} CH_2OH \\ | \\ COO^- \end{array}$$
$$P_i$$

glycolic acid (glycolate)

$$\xrightarrow[O_2 \; H_2O_2]{\text{glycolate oxidase}} \begin{array}{c} CHO \\ | \\ COO^- \end{array}$$

glyoxylic acid (glyoxylate)

FIGURE 16–19 Photorespiration in green plants, due to the CO_2-fixing RuBP carboxylase, which is also capable of functioning as an O_2-fixing oxygenase.

Alternate Pathway of CO_2 Fixation

Although the Calvin-Benson-Bassham sequence is definitely the major pathway of CO_2 fixation, it is not the only pathway. In 1970 M. Hatch and C. Slack discovered that certain plants are also capable of fixing CO_2 by a different route, as indicated by the rapid appearance of ^{14}C from $^{14}CO_2$ in four-carbon products such as malate and aspartate. Some ^{14}C is eventually detected in phosphoglycerate and phosphorylated sugars but only after a longer time of incubation. This new route is called the *Hatch-Slack pathway* or the *four-carbon* (C_4) *pathway*.

The explanation of this alternate pathway involves a new CO_2 fixation step, the key feature of which is that the primary acceptor of CO_2 is *phosphoenolpyruvate* (PEP) rather than ribulose-1,5-bisphosphate, and that the immediate product is *oxaloacetate* (OAA) rather than 3-phosphoglycerate (see Figure 16–20). Strong evidence in support of this alternate carboxylation step was obtained with the isolation from the chloroplasts of these plants of *phosphoenolpyruvate carboxylase,* the enzyme that catalyzes the reaction. This reaction of course, accounts for the early appearance of both malate and aspartate, both of which can be produced directly from oxaloacetate.

Depending on the type of C_4 plant, the immediate fate of oxaloacetate is varied. We will consider just one type (see Figure 16–21) wherein the OAA is first reduced to malate (MAL) by an NADP-dependent dehydrogenase and the malate is then decarboxylated to yield pyruvate and CO_2. The enzyme pyruvate phosphate dikinase catalyzes the ATP-dependent conversion of pyruvate to phosphoenolpyruvate, thus closing the cycle. The CO_2 is then refixed by the RuBP carboxylase for entry into the C_3 pathway.

The four-carbon pathway operates in about a hundred different plants and grasses that have one thing in common—they thrive in hot arid environments. Crabgrass, Bermuda grass, and many tropical plants such as sugarcane are just a few examples of these plants. The adaptability of plants to such conditions is related to the four-carbon route because the *four-carbon pathway increases the efficiency of CO_2 fixation* (in the plants capable of it) in regions of maximum solar radiation, high temperatures, and a limited supply of water.

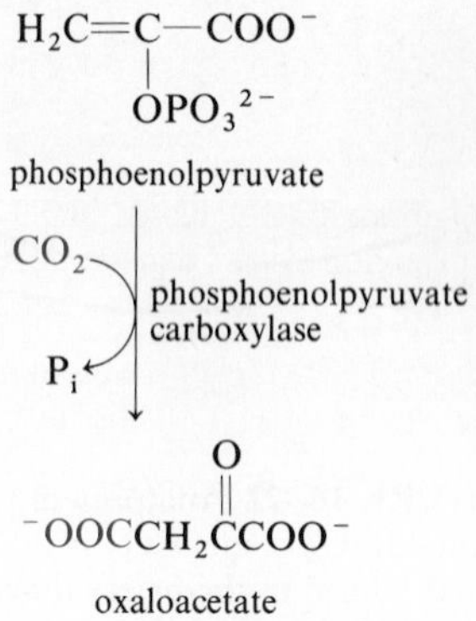

$$H_2C=C-COO^-$$
$$\qquad\quad | $$
$$\qquad\quad OPO_3^{2-}$$

phosphoenolpyruvate

$$CO_2 \rightharpoonup$$
phosphoenolpyruvate carboxylase
$$P_i \leftharpoonup$$

$$\begin{array}{c} \quad\; O \\ \quad\; || \\ {}^-OOCCH_2CCOO^- \end{array}$$

oxaloacetate

FIGURE 16–20 CO_2 fixation reaction in the four-carbon pathway.

FIGURE 16–21 Carbon dioxide fixation via a four-carbon (C_4) pathway. After initial condensation with phosphoenolpyruvate, the CO_2 is released for subsequent condensation by ribulose-1,5-bisphosphate.

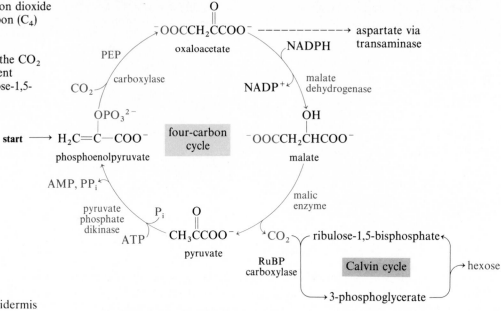

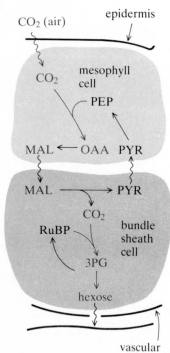

FIGURE 16–22 Anatomy of C_4 plants. In C_4 plants CO_2 is initially fixed in the external layer of mesophyll cells and then transported to the internal bundle sheath cells for entry into the C_3 pathway.

In full sunlight the leaves of C_4 plants generally fix CO_2 at about twice the rate of plants that use only the C_3 pathway.

There are three reasons for this increased efficiency of CO_2 fixation in C_4 plants. One involves the anatomy of the green tissue of C_4 plants, which have a double layer of cells between the epidermis and the vascular tissue (see Figure 16–22). *Mesophyll cells* comprise the outer layer, and *bundle sheath cells* comprise the inner layer. Moreover, each cell type is characterized by a different mode of CO_2 fixation. The PEP carboxylase for C_4 fixation is located in the chloroplasts of the outer mesophyll cells, whereas RuBP carboxylase for C_3 fixation is located in the chloroplasts of the interior bundle sheath cells. Because the PEP carboxylase is able to react with lower levels of CO_2 than required by the RuBP carboxylase, the pores of the leaf through which the CO_2 enters the plant need open just enough to maintain the low levels of CO_2. The passage of malate from the mesophyll cells to the interior bundle sheath cells serves to deliver and *concentrate* CO_2 to the less efficient RuBP carboxylase.

A second reason for increased efficiency follows from the first. Because the surface pores of C_4 plants need open only slightly to prime PEP carboxylase, *less water vapor leaves the leaf*. In other words, the four-carbon plant can make better use of its water, being able to fix CO_2 without being dehydrated when the water supply is low. A third reason for increased efficiency in C_4 plants is that they do *not* exhibit photorespiration.

Because of these special attributes that four-carbon plants have, researchers are attempting to hybridize them with three-carbon plants (Calvin cycle only) in order to increase the growth efficiency of the latter in climates where the three-carbon plants are difficult and expensive to grow—that is, where there is lots of sun and a limited water supply. Recombinant DNA technology may aid in developing such plants. Successes would have a revolutionary effect on agricultural practices and also contribute significantly to solving the world food problem.

16–4 PHOTOSYNTHESIS: A SUMMARY

Because of the complex biochemistry of photosynthesis, let us close our discussion of this subject by briefly reviewing the major events (see Figure 16–23). Photosynthesis consists of two separate but related processes, the so-called light reaction and the dark reaction, with the latter being absolutely dependent on the former. In the light reaction solar light energy is harnessed by chlorophyll-containing photosystems compartmentalized within the thylakoid membranes, resulting in the formation of metabolically useful reducing power (NADPH) and metabolically useful energy (ATP). The ultimate source of the reducing power is H_2O. In the dark reaction, occurring in the lumen of the chloroplasts, the ATP and the NADPH are utilized in the enzymatic assimilation of CO_2 into organic material, with the major acceptor molecule being ribulose-1,5-bisphosphate. The subsequent oxidation of the photosynthetically produced carbohydrates by aerobic, nonphotosynthetic organisms completes the major biochemical relationship in our biosphere.

Regarding the origin of life, one proposal is that photosynthetic organisms evolved much earlier than nonphotosynthetic organisms, the emergence of the latter being linked to the O_2 enrichment of the atmosphere by photosynthesizing systems.

FIGURE 16–23 Summary of photosynthesis and the interdependency to respiring cells.

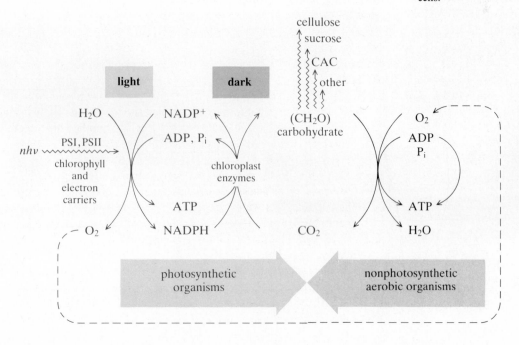

LITERATURE

BJORKMAN, O., and J. BERRY. "High-Efficiency Photosynthesis." *Sci. Am.,* **229,** 80 (1973). The four-carbon pathway is described.

CHOLLET, R. "The Biochemistry of Photorespiration." *Trends Biochem. Sci.,* **2,** 155–159 (1977). A short review article.

GOVINDJEE. *Photosynthesis.* Two volumes. New York: Academic Press, 1982, 1983. An authoritative and current source for many known details of the light and the dark reactions.

GOVINDJEE, and R. GOVINDJEE. "The Primary Events of Photosynthesis." *Sci. Am.,* **231,** 68 (1974). The absorption of radiation by chlorophyll and accessory pigments is discussed.

HATCH, M. D. "C_4 Pathway Photosynthesis: Mechanism and Physiological Function." *Trends Biochem. Sci.,* **2,** 199–202 (1977). A short review article.

HILL, R. "The Biochemist's Green Mansion: The Photosynthetic Electron Transport Chain in Plants." In *Essays in Biochemistry,* edited by P. N. Campbell and G. D. Greville, vol. 1, 121–152. New York: Academic Press, 1965. A review article with historical perspective of the light reaction by one of the pioneering researchers in the field.

LEVINE, R. P. "The Mechanism of Photosynthesis." *Sci. Am.,* **221,** 58–70 (1969). A good summary of the important features of the light reaction of photosynthesis, some of which are not treated in this chapter.

MACHLIS, L., ed. *Annual Review of Plant Physiology.* Palo Alto: Annual Reviews, Inc. An annual publication containing review articles of current developments concerning various aspects of plant physiology and biochemistry.

RABINOWITCH, E. I., and GOVINDJEE. "The Role of Chlorophyll in Photosynthesis." *Sci. Am.,* **213,** 74–83 (1965). A description of experiments that resulted in the suggestion of two separate photosystems.

CHAPTER SEVENTEEN

LIPID METABOLISM

Lipid metabolism is an extensive subject. There are a large number of lipid classes, each having unique anabolic and catabolic pathways. We will explore only certain facets, with particular emphasis given to major pathways of *fatty acid metabolism.* Since fatty acids are the chief structural components of the simple and compound lipids, the emphasis is a logical one. Other aspects of lipid metabolism covered in this chapter include the *biosynthesis of compound lipids,* the *biosynthesis of cholesterol* and other derived lipids, and the ability of many organisms to *convert lipids to carbohydrates.*

There is more to be gained from this study than knowledge of a new set of pathways. Lipid metabolism also provides (1) a clear example of the general principles that differentiate catabolic and anabolic pathways; (2) further examples of multienzyme complexes, enzyme stereospecificity, compartmentalized biochemistry, metabolic regulation, and receptor biochemistry; (3) the opportunity to examine the coenzyme participation of *biotin;* and (4) some examples of *multifunctional proteins.*

17–1 CATABOLISM OF FATTY ACIDS

Cellular Source of Fatty Acids and Thioester Activation

RELEASE As described in Chapter 11, fatty acids are found, for the most part, in ester linkage, primarily as acylglycerols, phosphoacylglycerols, and sphingolipids. Fatty acids are released from these lipids by the action of hydrolase enzymes called *lipases* (see Figure 17–1). Recall our previous description (Section 11–3) of the family of site-specific *phospholipases,* acting on different bonds in a phosphoacylglycerol.

FIGURE 17–1 Enzyme-catalyzed hydrolysis to release fatty acids.

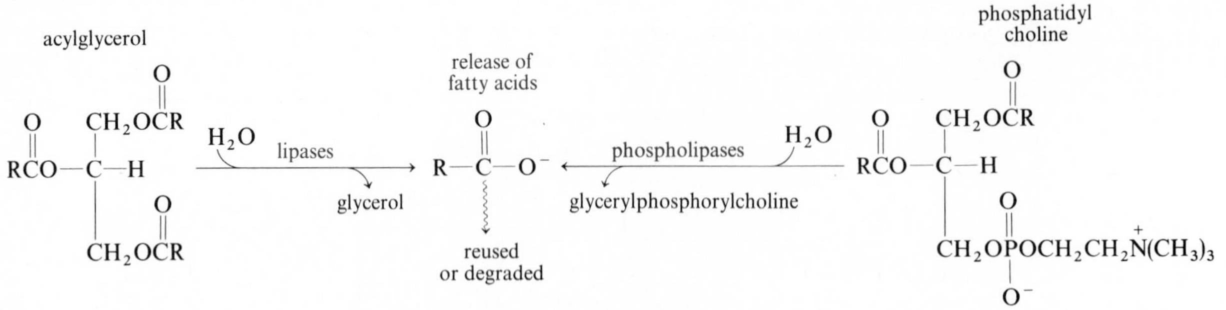

Once released from ester linkage, the free fatty acids can be further degraded to provide carbon and/or energy or be reutilized in the biosynthesis of various lipids, including the class of lipids from which they were released. Our focus now will be on their catabolism to CO_2. Certain long-chain, polyunsaturated fatty acids, such as arachidonic acid, can be used for the production of prostaglandins and leukotrienes (see Chapter 11).

ACTIVATION   Be it for reuse or further degradation, free acids are first converted to *thioesters* of coenzyme A (CoASH)—the high-energy activated forms of the acyl group. For the thioester

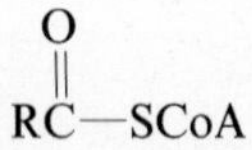

$$\underset{\displaystyle RC-SCoA}{\overset{\displaystyle O}{\overset{\displaystyle \|}{}}}$$

species an increase in acyl group activity is observed in reactions involving a condensation at the carbonyl carbon and in elimination or addition reactions involving the α and β carbons of the acyl group (see Figure 17–2).

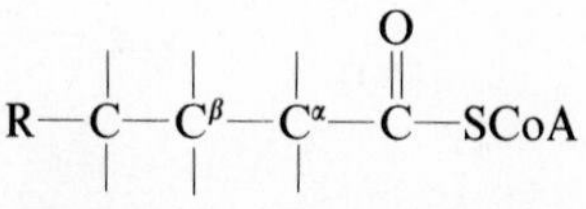

The most widely distributed activation process involves an ATP-dependent *fatty acid thiokinase* (see Figure 17–3). Varying in substrate specificity, at least three different thiokinases are known to exist in nature. One is highly specific for acetate (C_2), a second for acids of medium chain length (C_4–C_{12}), and the third for long-chain acids (C_{14}–C_{22}). The latter two act on both saturated and unsaturated acids. Regardless of type, the activating thiokinases are known to be particulate enzymes localized in cellular membranes. In eukaryotes they are found in the outer mitochondrial membrane (E_1 in Figure 17–4).

Entry of Acyl Group into Mitochondrion: Carnitine as a Carrier

The enzymes participating in the degradation of the acyl group are located in the mitochondrion—a compartmentalization that requires the passage of acyl-SCoA from the cytoplasm into the mitochondrion. Although capable of crossing the outer mitochondrial membrane, the acyl-SCoA species cannot cross the inner membrane. This impermeability barrier is overcome by the participation of *carnitine* (see Figure 17–4), which serves as an *acyl group carrier*. Two

$$\underset{\displaystyle R-C-C^{\beta}-C^{\alpha}-C-SCoA}{\overset{\displaystyle O}{\overset{\displaystyle \|}{}}}$$

FIGURE 17–2 Acyl-SCoA thioester.

FIGURE 17–3 Fatty acid thiokinase reaction.

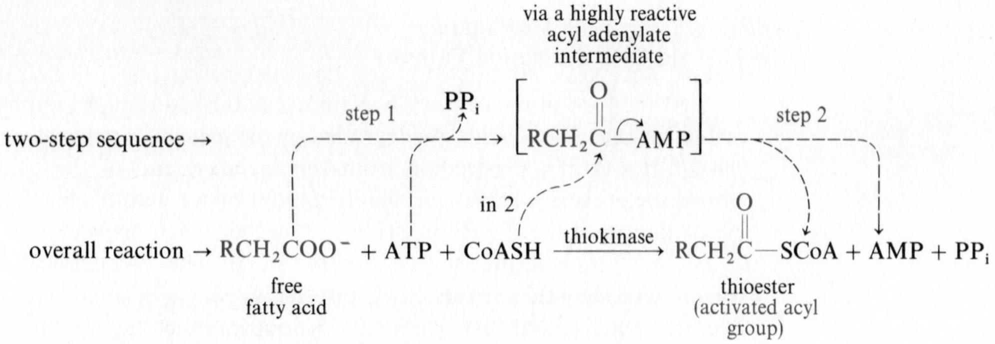

via a highly reactive
acyl adenylate
intermediate

two-step sequence →   step 1 PP_i $\left[\underset{\displaystyle RCH_2C \frown AMP}{\overset{\displaystyle O}{\overset{\displaystyle \|}{}}} \right]$ step 2

in 2

overall reaction → $RCH_2COO^- + ATP + CoASH \xrightarrow{\text{thiokinase}} RCH_2C{-}SCoA + AMP + PP_i$

free
fatty acid

thioester
(activated acyl
group)

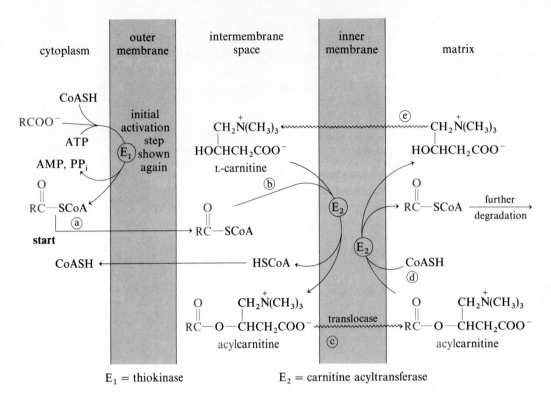

FIGURE 17–4 Role of L-carnitine in transporting fatty acids across the inner mitochondrial membrane.

proteins are involved, both localized in the inner mitochondrial membrane: carnitine acyltransferase (an enzyme) and carnitine:acylcarnitine translocase (a transport protein).

In the intermembrane space the acyltransferase (E_2) catalyzes the formation of an acylcarnitine oxyester. The translocase then facilitates the passage of the acylcarnitine ester to the other side of the membrane, releasing it into the matrix. Acyltransferase action on the matrix side of the membrane (the enzyme appears to be present on both sides) re-forms acyl-SCoA and free carnitine. Completing the transport cycle, the return of free carnitine to the intermembrane space may be facilitated by the same translocase protein.

Acyl Group Degradation via the β-Oxidation Pathway

At the turn of the century F. Knoop established that the aliphatic hydrocarbon chains of fatty acids are degraded by the *sequential removal of two-carbon units* (CH_3COO^-), proceeding from the carboxyl end of the chain. Knoop termed the process *β-oxidation,* signifying that each round of chemistry involves the oxidation of the β carbon prior to bond cleavage between C^β—C^α at the C^β—C^α—COO^- terminus (see Figure 17–5). The biochemical details, involving intermediates, sequence of conversion, enzymes, coenzymes, and so on, were unraveled about fifty years after Knoop's pioneering studies.

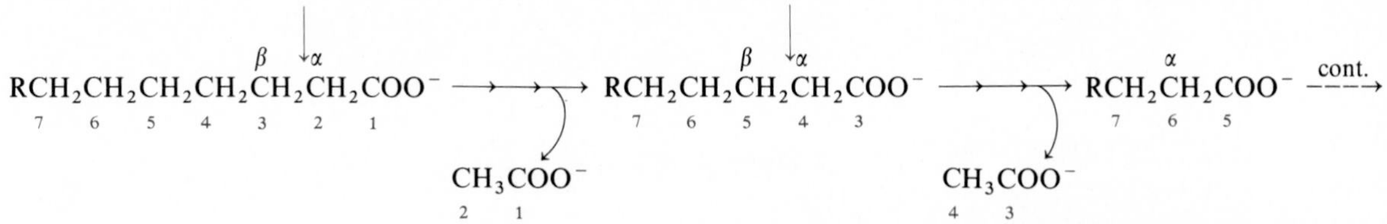

$$
\begin{array}{c}
\overset{\displaystyle\downarrow}{\underset{\substack{7\ \ \ 6\ \ \ \ 5\ \ \ 4\ \ \ 3\ \ \ 2\ \ \ 1}}{\underset{\beta\ \ \ \ \alpha}{RCH_2CH_2CH_2CH_2CH_2CH_2COO^-}}} \longrightarrow
\end{array}
$$

FIGURE 17–5 Pattern of sequential removal of two-carbon fragments in β-oxidation.

SEQUENCE OF CHEMICAL EVENTS   After entry into the matrix of mitochondria, acyl-SCoA species are degraded through the sequential action of four enzymes in successive cycles. With stearyl-SCoA (C_{18}-SCoA) as initial substrate, the chemistry is illustrated in Figure 17–6. The individual steps are as follows:

A. An FAD-dependent *dehydrogenation* to yield an α,β unsaturated acyl-SCoA.

B. *Hydration* to yield a β-hydroxyacyl-SCoA.

C. An NAD^+-dependent *dehydrogenation* to yield a β-ketoacyl-SCoA.

D. *Thiolytic cleavage* to yield acetyl-SCoA and a second acyl-SCoA now shortened by a two-carbon unit.

E. Recycling of the shortened acyl-SCoA through steps A and D, again and again.

Successive cycles of β-oxidation continue until the formation of the four-carbon β-keto metabolite,

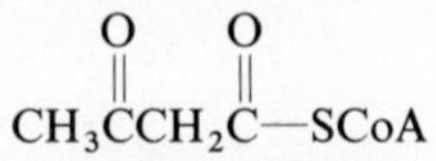

$$
\underset{\text{acetoacetyl-SCoA}}{CH_3\overset{\displaystyle O}{\overset{\|}{C}}CH_2\overset{\displaystyle O}{\overset{\|}{C}}-SCoA}
$$

The last thiolytic cleavage of acetoacetyl-SCoA yields two units of acetyl-SCoA and thus complete the process: 1 stearyl-SCoA → 9 acetyl-SCoA. All of the enzymes have been isolated in pure form. Note the stereospecific actions of the enzymes in A, B, and C.

ENERGETICS OF β-OXIDATION   An ideal analysis of the bioenergetics of fatty acid catabolism requires an assumption that the fate of acetyl-SCoA in the mitochondrion would be to enter the citric acid cycle for complete oxidation to CO_2. The assumption is not unrealistic. Indeed, such would be the case when the physiological state of the organism and/or dietary factors dictate that lipids rather than carbohydrates be utilized as the primary energy source. Consistent with Figure 17–6, the following analysis uses stearyl-SCoA as the initial substrate.

The pertinent equations follow. (Note that no distinction is made between the metabolic sources of $FADH_2$ and NADH in the equation for their coupled reoxidation to ATP formation. Regardless of source, they are metaboli-

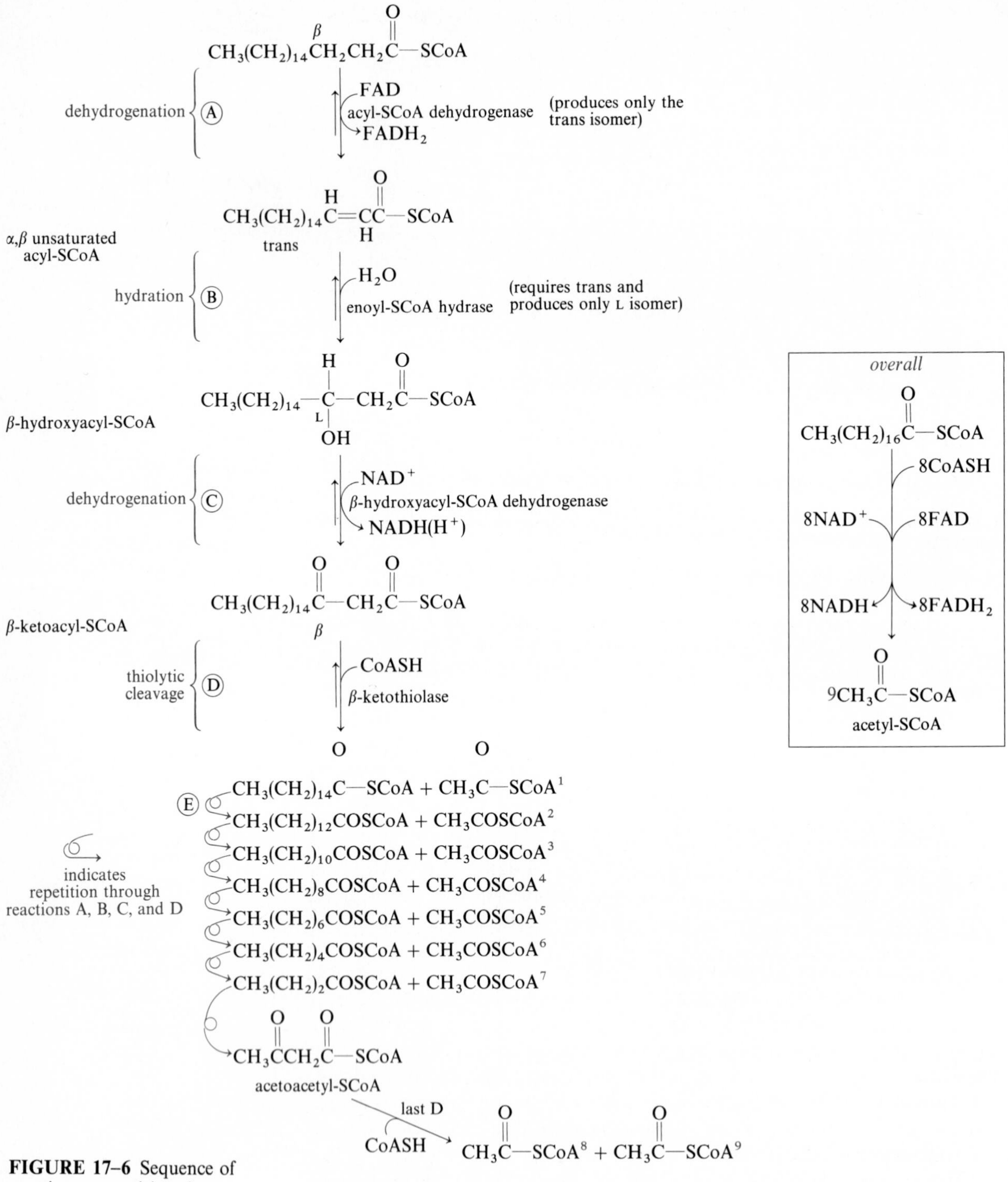

FIGURE 17–6 Sequence of reactions comprising the β-oxidation pathway for the degradation of fatty acids.

cally equivalent. That is, the P/O ratios are always 2 for $FADH_2$ and 3 for $NADH + H^+$.)

A. Balanced equation for β-oxidation (8 cycles for C_{18}-SCoA):

$$CH_3(CH_2)_{16}\overset{\overset{\displaystyle O}{\|}}{C}\!\!-\!\!SCoA + 8FAD + 8NAD^+ + 8H_2O + 8CoASH \longrightarrow$$

$$9CH_3\overset{\overset{\displaystyle O}{\|}}{C}SCoA + 8FADH_2 + 8NADH + 8H^+$$

B. Balanced equation for citric acid cycle (9 cycles):

$$9CH_3\overset{\overset{\displaystyle O}{\|}}{C}SCoA + 9FAD + 27NAD^+ + 9GDP + 9P_i + 27H_2O \longrightarrow$$

$$18CO_2 + 9CoASH + 9FADH_2 + 27NADH + 27H^+ + 9GTP$$

C. Balanced equations for oxidative phosphorylation:

$$17FADH_2 + 8.5O_2 + 34ADP + 34P_i \rightarrow 17FAD + 17H_2O + 34ATP$$

$$35NADH + 35H^+ + 17.5O_2 + 105ADP + 105P_i \rightarrow 35NAD^+ + 35H_2O + 105ATP$$

A + B + C. Net balanced equation (assuming GDP = ADP and GTP = ATP):

$$CH_3(CH_2)_{16}\overset{\overset{\displaystyle O}{\|}}{C}\!\!-\!\!SCoA + 26O_2 + 148ADP + 148P_i \longrightarrow$$

$$18CO_2 + 17H_2O + 148ATP + CoASH$$

After we subtract one ATP used for the initial thioester activation, the net tally is 147ATP gained from the complete catabolism of one stearic acid (18 carbons). By comparison, the complete oxidation of three glucose molecules (18 carbons $\rightarrow$ $18CO_2$) will only yield $3 \times 36 = 108ATP$. This greater ATP yield per carbon reflects the difference in the state of reduction of fatty acid carbons (mostly $-CH_2-$) versus carbohydrate carbons (mostly $>CHOH$).

Fatty acid carbons are in a higher state of reduction and therefore can yield more energy on oxidation.

Auxiliary Enzymes to β-Oxidation

ODD-NUMBERED ACIDS   Cycles of β-oxidation on a fatty acid having an odd number of carbon atoms yield several acetyl-SCoA fragments until a C_5 β-ketoacyl-SCoA is produced. Then in the last thiolytic cleavage the five-carbon ketoacyl thioester formed will yield acetyl-SCoA and *propionyl-SCoA*

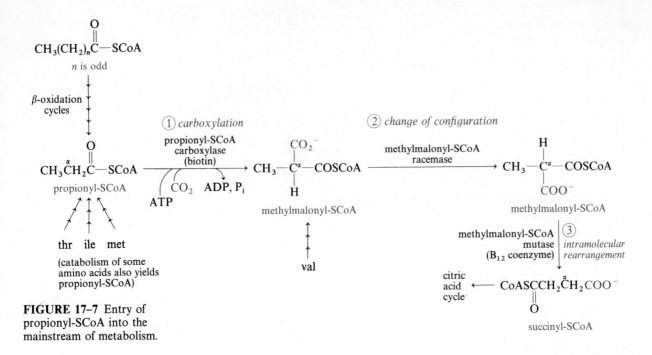

FIGURE 17–7 Entry of propionyl-SCoA into the mainstream of metabolism.

(see Figure 17–7). The subsequent metabolic fate of the propionyl-SCoA is entrance into the citric acid cycle in the form of succinyl-SCoA via the following set of reactions: (1) a biotin-dependent carboxylation converting propionyl-SCoA to methylmalonyl-SCoA; (2) an epimerization, inverting the configuration of methylmalonyl; and (3) a vitamin B_{12}–dependent intramolecular migration. The role of biotin in carboxylation reactions is examined later in this chapter (see Section 17–2). We will postpone vitamin B_{12}–dependent reactions until the next chapter (Section 18–5).

Odd-numbered fatty acids are, however, not that prevalent in nature. A more significant aspect of the propionyl-SCoA → succinyl-SCoA conversion relates to the catabolism of four amino acids: threonine, isoleucine, methionine, and valine. When used as carbon and energy sources, these amino acids are degraded in part to either propionyl-SCoA or methylmalonyl-SCoA. The subject of amino acid metabolism is dealt with more fully in Chapter 18.

UNSATURATED ACIDS The degradation of unsaturated fatty acids by β-oxidation also requires some additional steps because of the original position and geometry of the double bond(s) in the common unsaturated acids. For example (see Figure 17–8), with linoleic acid the first three cycles of β-oxidation produce a 12-carbon unsaturated chain with cis double bonds at positions 3 and 6. This substance is an unsuitable substrate for either the dehydrogenase or the hydrase of the β-oxidation sequence. The hydrase in particular requires a trans double bond between C2 and C3. As shown in the figure, this requirement is satisfied by the participation of an auxiliary *cis-trans isomerase enzyme,* which catalyzes a double isomerization—both the position and the geometry of the double bond are changed. Thereafter, β-oxidation can resume.

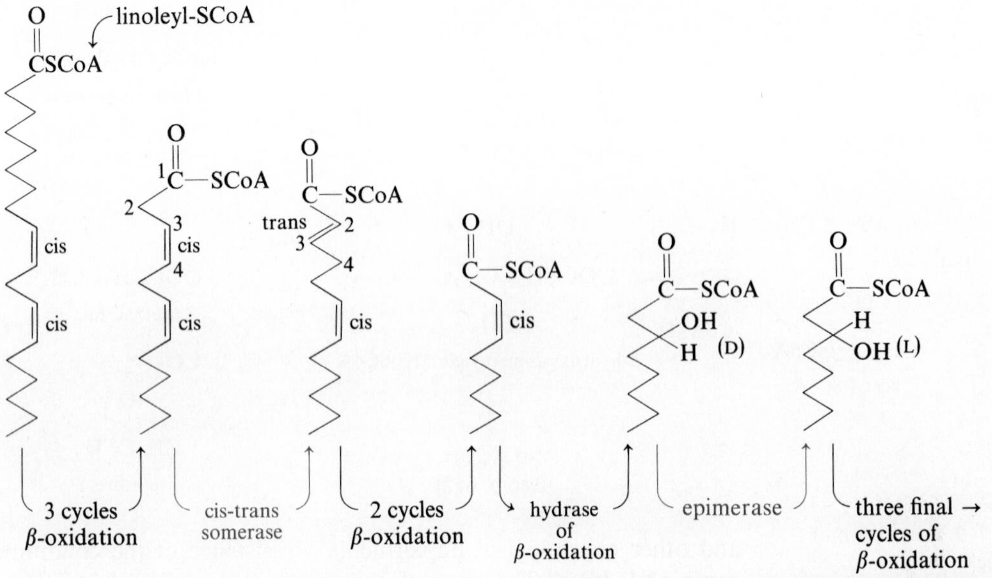

FIGURE 17–8 Degradation of unsaturated fatty acids. Sometimes the degradation requires the participation of one or more auxiliary enzymes to those of β-oxidation. The degradation of linoleic acid requires two auxiliary enzymes, identified here with an asterisk.

However, after two further cycles of β-oxidation a shortened and unsaturated acyl chain is produced with a C2/C3 double bond but with a cis geometry. Although different from the trans geometry produced by the dehydrogenase of β-oxidation, the hydrase can act on this cis isomer but forms the D isomer of the β-hydroxyacyl product rather than the L isomer (refer to Figure 17–6). But the D isomer is inconsistent with the absolute specificity of the NAD-dependent dehydrogenase of β-oxidation, which requires the L isomer as substrate. This obstacle is overcome by a second auxiliary enzyme, an *epimerase*, which catalyzes a D → L conversion. Thereafter, β-oxidation continues without any further need for auxiliary enzymes.

Ketone Bodies

In many organisms another fate of the acetoacetyl-SCoA derived from fatty acid catabolism is condensation with acetyl-SCoA to form *β-hydroxy-β-methylglutaryl-SCoA* (HMG-SCoA) (see Figure 17–9). Later in the chapter (p. 631) we will see that HMG-SCoA is an important precursor in the biosynthesis of cholesterol in animals and various other lipid types of materials in plants and bacteria. Under normal conditions in animals a small amount of HMG-SCoA is also converted in liver to acetyl-SCoA and the free acid *acetoacetate*, which in turn can be enzymatically reduced to *β-hydroxybutyrate*.

Under certain conditions, such as diabetes, starvation, and a lipid-rich diet, there is a shift to excessive utilization of fatty acids. The result is a production of acetyl-SCoA in excess of normal need. This excess acetyl-SCoA is funneled into HMG-SCoA, leading to an increased production of acetoacetate and β-hydroxybutyrate, also called *ketone bodies*. Abnormally elevated blood levels of these materials result in a condition called *acidosis* in the early stages and *ketosis* in the later stages. The consequences are a lowering of the blood pH

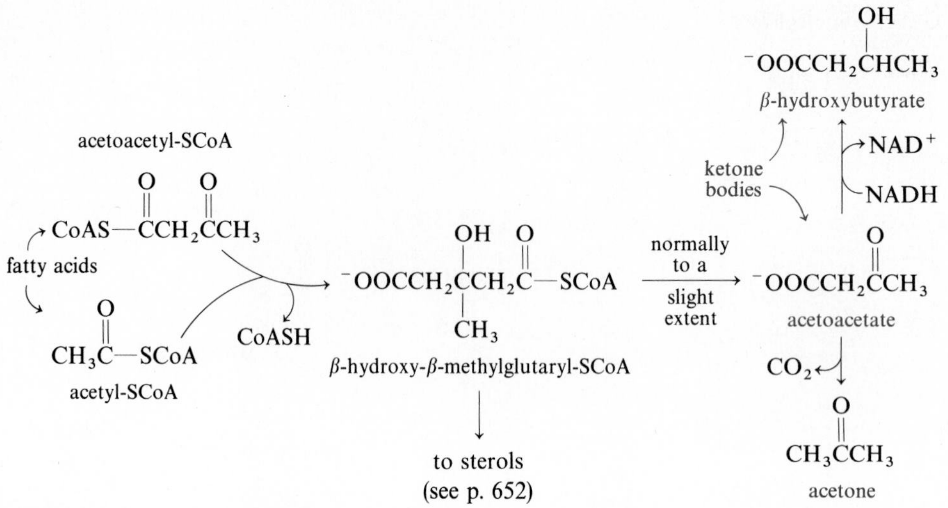

FIGURE 17–9 Formation of ketone bodies.

and other physiological impairments. Persistence of the condition can cause coma and death. The odor of acetone in the breath of individuals in acute acidosis is due to the nonenzymatic decarboxylation of acetoacetate.

Under such abnormal conditions the ketone bodies do have a useful function by serving as lipid-derived substitutes for blood glucose, the normal energy source for the brain. For example, during starvation the function of brain (and other organs) can be sustained for several weeks by β-hydroxybutyrate produced in liver. In brain cells the β-hydroxybutyrate can be converted to acetyl-SCoA for entry into the citric acid cycle. After oxidation of β-hydroxybutyrate to acetoacetate, the latter is converted to acetoacetyl-SCoA via an acyl-thioester exchange reaction involving succinyl-SCoA (see Figure 17–10). Acetoacetyl-SCoA formation by a direct reaction of acetoacetate with CoASH does not occur, because brain has very little ATP-dependent thiokinase activity.

Glyoxylate Cycle

Plants and some bacteria are capable of using acetyl-SCoA arising from lipid catabolism (or any other source) not only as a source of energy but also

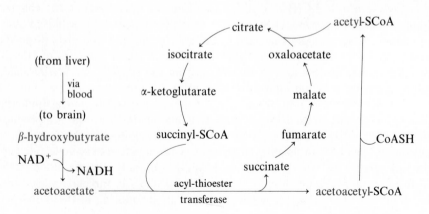

FIGURE 17–10 Utilization of β-hydroxybutyrate as a carbon source when the blood glucose is low.

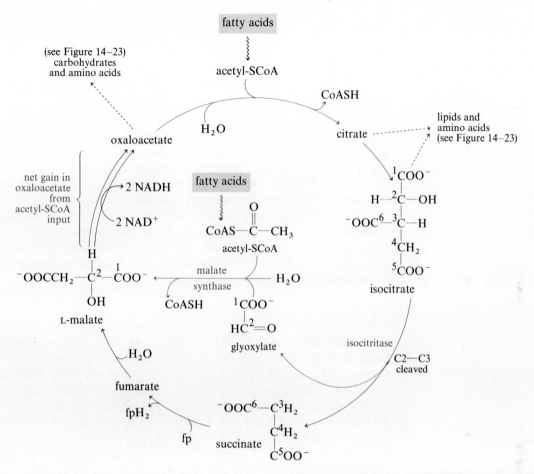

FIGURE 17–11 Glyoxylate cycle in plants and bacteria.

as a supply of carbon to produce most other classes of compounds. Animals are almost as versatile but are particularly inefficient at using lipid carbon to produce carbohydrates. The reason is that plants and bacteria can produce two auxiliary enzymes, *isocitritase* (or isocitrate lyase) and *malate synthase,* but neither one is produced in animals. In conjunction with some of the enzymes of the citric acid cycle, these additional enzymes contribute to a pathway called the *glyoxylate cycle.* In the fat-storing seeds of plants the glyoxylate cycle occurs in a separate organelle, namely, the *glyoxosome* (see Chapter 1, p. 24). The net result of the glyoxylate cycle is

$$2 \text{ acetyl-SCoA} + \text{fp} + 2\text{NAD}^+ + 3\text{H}_2\text{O} \rightarrow \text{oxaloacetate} + 2\text{CoASH} + \text{fpH}_2 + 2\text{NADH}$$

As indicated in Figure 17–11, the glyoxylate pathway provides for a twofold input of acetyl-SCoA: once by condensation with oxaloacetate and once by condensation with glyoxylate. This input not only generates intermediates for use in amino acid biosynthesis but also results in the net production of oxaloacetate (OAA). Thus if one of the intermediates prior to OAA is removed, the glyoxylate cycle has a self-contained anaplerotic reaction. In addition, the OAA itself can be diverted into amino acids or carbohydrates.

17–2 ANABOLISM OF FATTY ACIDS

Introduction

Fatty acids are assembled from acetyl-SCoA, the same substance to which they are degraded in β-oxidation. In fact, after an initial condensation reaction, the sequence of fatty acid biosynthesis proceeds through the same acyl group intermediates as in β-oxidation but in reverse: β-ketoacyl $\rightarrow$ β-hydroxyacyl $\rightarrow$ α,β unsaturated acyl $\rightarrow$ saturated acyl. There are, however, several differences between the catabolic and anabolic pathways. In addition, there are different phases of biosynthesis, and in eukaryotes these phases occur in different cellular locations: in the cytoplasm, with the endoplasmic reticulum, and in the mitochondrion. The cytoplasmic pathway is responsible for the assembly of saturated chain lengths up to C_{16} (palmitate); the mitochondrial system catalyzes the further extension of the chain length; and the reticulum contributes enzymes responsible for the conversion of saturated to unsaturated fatty acids and also for the further extension of chain length.

Soluble Anabolic Pathway

SOURCE OF ACETYL-SCoA The other major source of acetyl-SCoA besides β-oxidation is *glycolysis,* to yield pyruvate and then pyruvate $\rightarrow$ acetyl-SCoA + CO_2. But since the decarboxylation of pyruvate and β-oxidation occur in the mitochondria, since the enzymes for fatty acid biosynthesis are in the cytoplasm, and since the mitochondrial membrane is relatively impermeable to the free diffusion of acetyl-SCoA, there is a problem. How does acetyl-SCoA get across the membrane to participate in fatty acid anabolism? As shown in Figure 17–12, the exit of acetyl-SCoA from the mitochondrion is indirect, involving the passage of citrate into the cytoplasm where the enzyme *ATP citrate lyase* (also called citrate cleavage enzyme) cleaves citrate to re-form oxaloacetate and acetyl-SCoA. The oxaloacetate could reenter the mitochondrion as malate.

MALONYL-SCoA AND BIOTIN The most distinctive step in the assembly of palmitate is the initial conversion of acetyl-SCoA into *malonyl-SCoA* via

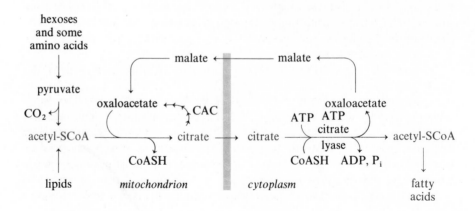

FIGURE 17–12 Transport of acetyl-SCoA to cytoplasm.

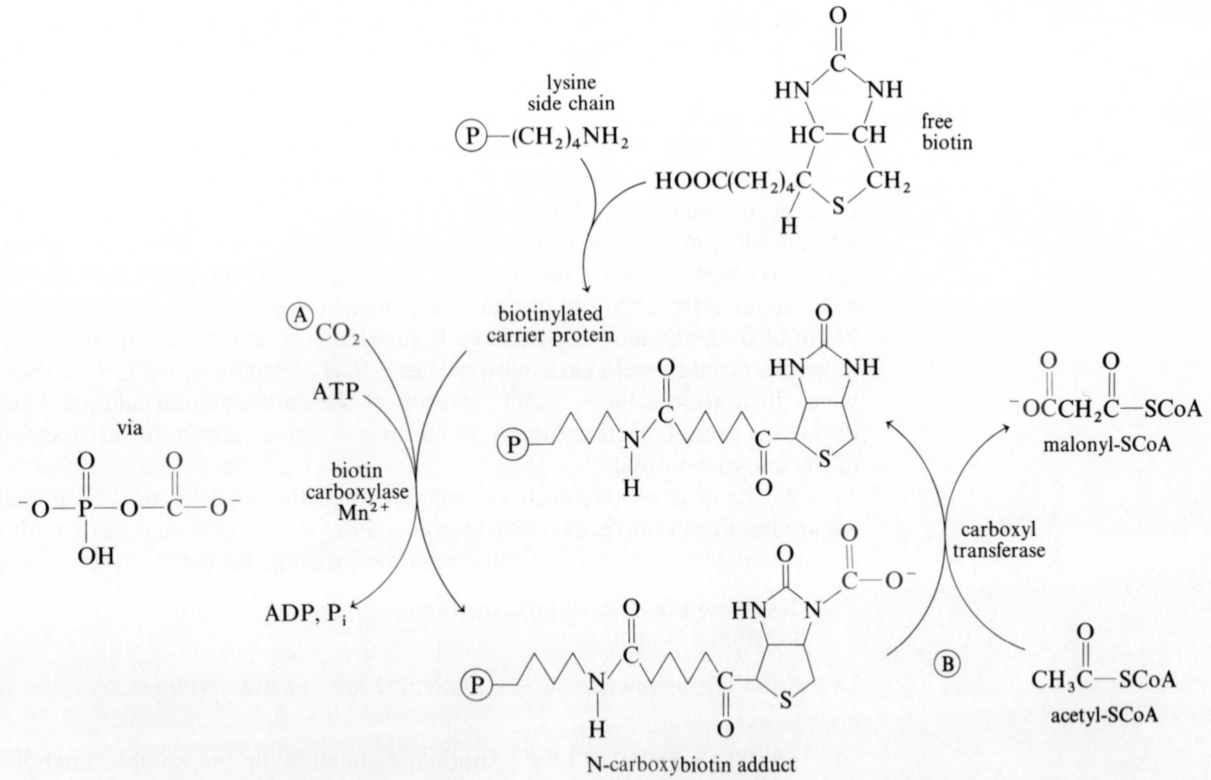

FIGURE 17–13 Structure of the coenzyme biotin and its role as a transfer agent of CO_2. The specific proteins identified here represent those that comprise acetyl-SCoA carboxylase.

an ATP-dependent carboxylation. The enzyme *acetyl-SCoA carboxylase* requires Mn^{2+} for optimal activity and also uses *biotin* as an essential coenzyme:

$$CO_2 + CH_3\overset{O}{\overset{\|}{C}}-SCoA + ATP \xrightarrow[Mn^{2+},\ biotin]{acetyl\text{-}SCoA\ carboxylase} {}^-OOCCH_2\overset{O}{\overset{\|}{C}}-SCoA + ADP + P_i$$

acetyl-SCoA   malonyl-SCoA

By producing malonyl-SCoA, the methyl (CH_3) carbon of the acetyl unit is converted to a more reactive methylene (CH_2) carbon atom—an activation that enhances the condensation reaction beginning each round of chain elongation. Before describing these matters, we will first examine the biochemistry of biotin (see Figure 17–13).

Excluding the CO_2 fixation reactions of photosynthesis, the small number of other naturally occurring carboxylation reactions involve enzymes that require the coenzyme participation of biotin. Examples of such reactions, in addition to acetyl-SCoA carboxylase (our current focus), include pyruvate carboxylase (p. 515) and propionyl-SCoA carboxylase (p. 636). To higher animals, incapable of synthesizing biotin, it is an essential vitamin. A sensitive and specific test for a biotin-dependent reaction is based on inhibition by the egg white protein *avidin,* a strong binder of biotin.

Whatever the particular carboxylase, biotin occurs covalently attached to the enzyme via an amide linkage to the epsilon amino group of a lysine side chain. In prokaryotes the acetyl-SCoA carboxylase is a multiprotein complex consisting of three protein subunits. Only one of these subunits, the *biotin carrier protein,* contains the biotin. The other two proteins serve as enzymes in the two stages of the carboxylation reaction (see Figure 17–13).

Stage A, in which a *biotin carboxylase* component participates, involves the energy-dependent carboxylation of the protein-bound biotin. In this reaction the substrate form of CO_2 is bicarbonate (HCO_3^-) ion. The carboxybiotin adduct is a highly reactive species of the one-carbon unit and is termed *activated CO_2.* As indicated, the site of CO_2 attachment to biotin occurs at one of the N atoms of the ureido ring. In stage B a *carboxyl transferase* component promotes the formation of a carbanion species ($^-:CH_2—$) of the acetyl CH_3 carbon, which then attacks the $—COO^-$ carbon of the carboxybiotin adduct. Thus during the overall reaction the role of biotin is to act as a carrier (transfer agent) of the one-carbon unit.

An identical process occurs in eukaryotes, with one distinction. In animal eukaryotes (for example, liver cells) the acetyl-SCoA carboxylase is not a multienzyme complex composed of three separate proteins. Rather:

> It is a multifunctional protein, a single protein having three separate functional domains.

Discussion of this phenomenon is presented later in this section.

PROTEIN THIOESTERS   Another distinction of the soluble anabolic pathway is that after the carboxylase step is complete, the remaining acyl group reactions proceed via a different thioester involving a substance called *acyl carrier protein,* symbolized here as ACP–SH. The functional sulfhydryl (SH) group of this protein is not contributed by the R group of cysteine but rather by a *phosphopantetheine* grouping, which is covalently attached to the polypeptide chain via a serine residue (see Figure 17–14). This same grouping, you may recall, is also found in the structure of coenzyme A (see Figure 12–13).

The initial involvement of ACP–SH involves a transacylation reaction with malonyl-SCoA to give malonyl-S–ACP. The enzyme is *malonyl transacylase,* and it operates at the beginning of each condensation cycle. Another enzyme, *acetyl transacylase,* catalyzes a similar reaction, but it is specific for acetyl-SCoA, which is used only in the first condensation cycle:

$$\text{ACP-SH} + {}^-\text{OOCCH}_2\overset{\displaystyle\overset{O}{\|}}{\text{C}}\text{—SCoA} \xrightarrow[\text{transacylase}]{\text{malonyl}} \text{ACP-S—}\overset{\displaystyle\overset{O}{\|}}{\text{C}}\text{CH}_2\text{COO}^- + \text{CoASH}$$
$$\text{malonyl-SCoA} \qquad\qquad\qquad\qquad \text{malonyl-S–ACP}$$

REACTION SEQUENCE   In a reaction catalyzed by *β-ketoacyl-S–ACP synthase,* the immediate fate of malonyl-S–ACP is condensation with an acetyl unit to yield acetoacetyl-S–ACP. Although it originates from acetyl-SCoA, the immediate source of the acetyl unit is an acetyl-S(cysteinyl) thioester. The latter is formed from acetyl-SCoA and a reactive cysteine-SH side chain of the synthase in a transacylation catalyzed by acetyl transacylase (see Figure 17–15).

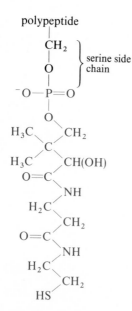

FIGURE 17–14 Acyl carrier protein, ACP–SH. Color identifies the 4′-phosphopantetheine group.

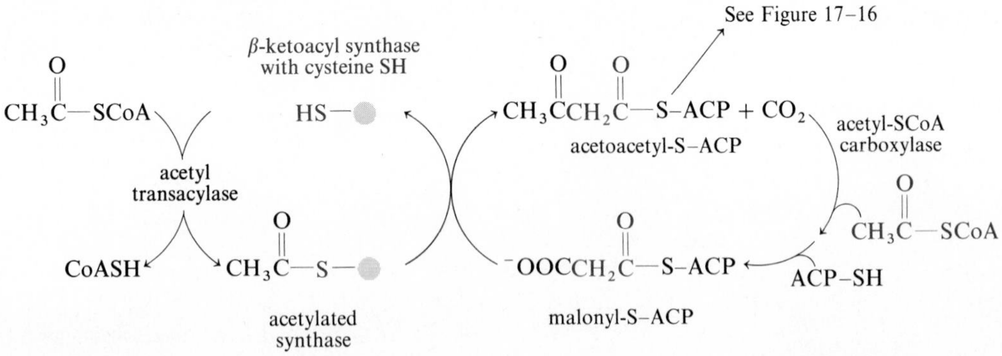

FIGURE 17–15 Initial condensation step of fatty acid biosynthesis. The β-ketoacyl synthase forms a covalent thioester adduct prior to condensation with malonyl-S–ACP.

Note that the CO_2 released during the condensation can be reused in the acetyl-SCoA carboxylase reaction to convert additional acetyl-SCoA to malonyl-SCoA (and then malonyl-S–ACP) for use in the next round of condensation.

After the β-ketoacyl condensation product is formed, a three-step reaction sequence (see Figure 17–16) occurs, each step still involving acyl-S–ACP thioester intermediates: (1) an NADPH-dependent reduction to yield a β-hydroxyacyl group; (2) a dehydration to an α,β-enoyl group; and (3) another NADPH-dependent reduction to form the fully saturated acyl group. Although this pathway is just the reverse of the chemical transformations in the β-oxidation pathway, different enzymes are involved in the two pathways.

To start a second cycle of chain extension, the β-ketoacyl synthase enzyme condenses a second malonyl unit with the C_4 butyryl function to give a C_6 β-ketoacyl-S–ACP. The same sequence of steps (1, 2, 3) then yields a C_6-saturated acyl-S–ACP. This pattern of chain extension and modification continues until palmityl-S–ACP is formed (a total of seven rounds). During each condensation reaction the newly formed acyl-S–ACP species first undergoes a transacylation reaction involving the same SH function of the synthase described earlier. This transacylation produces an acyl-S(cysteinyl) thioester form of the synthase and free ACP–SH, the latter now available to accept another malonyl unit from malonyl-SCoA.

Although all of the carbon atoms ultimately originate from eight units of acetyl-SCoA, note that there is a direct entry of only the first one, with the other seven entering via malonyl-SCoA. The final step involves hydrolysis of the palmityl thioester by a *thioesterase enzyme*. The free palmitate can then be converted to palmityl-SCoA by an ATP-dependent thiokinase.

Owing to the nature of the condensation reaction, you can see that the acyl chain grows in two-carbon increments from the methyl end to the carboxyl end (see Figure 17–17). Recall that in β-oxidation the acyl chain is shortened by two-carbon increments in the opposite direction.

FATE OF PALMITYL-SCoA   In the cytoplasm the palmityl-SCoA can be utilized directly in the assembly of any of the simple and compound lipids. Alternatively, the palmityl-SCoA can enter the mitochondrion (via the carnitine transport system), where the enzymes of β-oxidation can participate in reverse

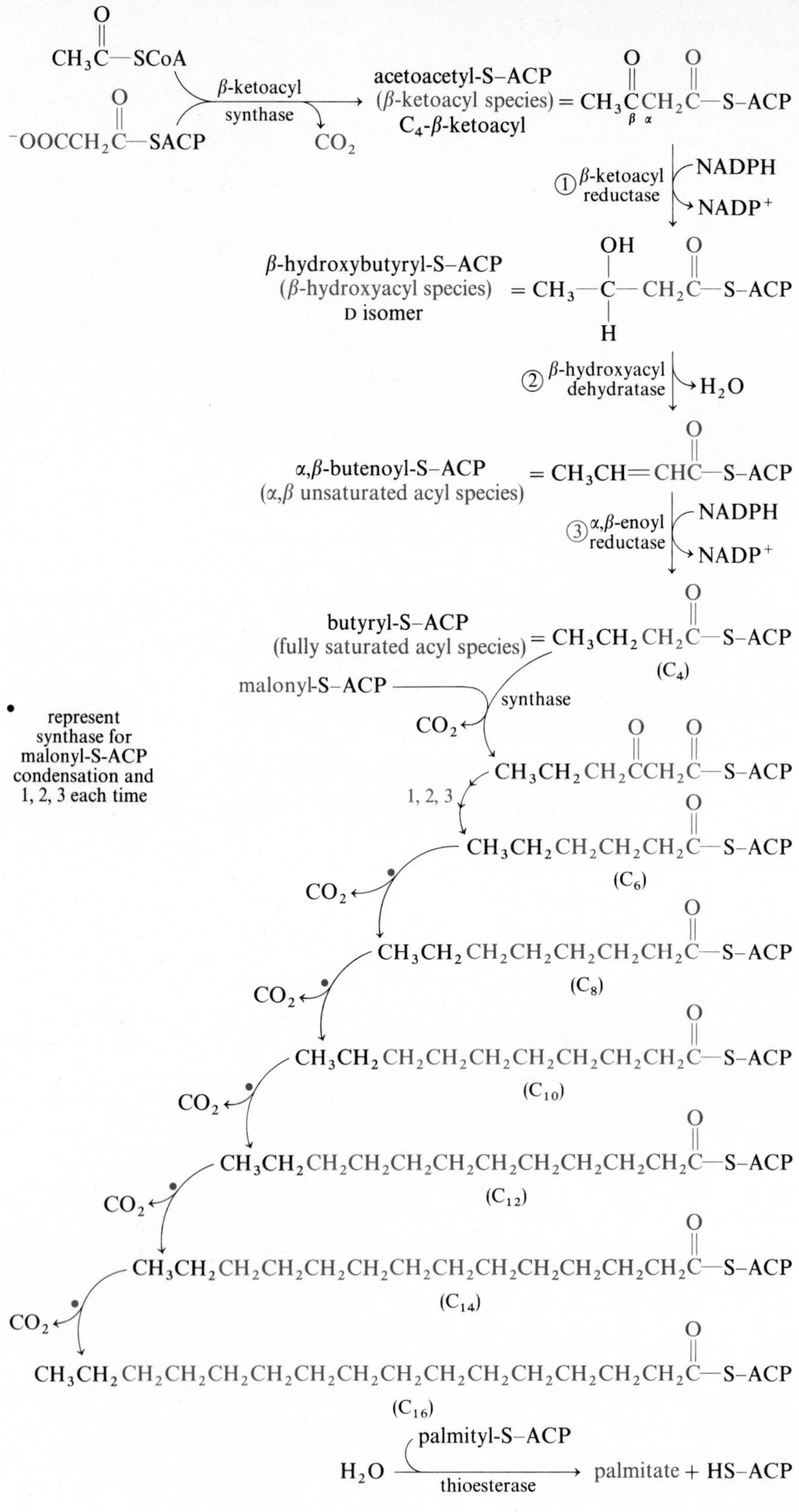

$$CH_3\overset{\overset{\displaystyle O}{\|}}{C}-SCoA$$

$$^-OOCCH_2\overset{\overset{\displaystyle O}{\|}}{C}-SACP$$

β-ketoacyl synthase → CO_2

acetoacetyl-S–ACP
(β-ketoacyl species) = $CH_3\overset{\overset{\displaystyle O}{\|}}{\underset{\beta}{C}}CH_2\overset{\overset{\displaystyle O}{\|}}{\underset{\alpha}{C}}-S-ACP$
C_4-β-ketoacyl

① β-ketoacyl reductase — NADPH → NADP$^+$

β-hydroxybutyryl-S–ACP
(β-hydroxyacyl species) = $CH_3-\overset{\overset{\displaystyle OH}{|}}{\underset{\underset{\displaystyle H}{|}}{C}}-CH_2\overset{\overset{\displaystyle O}{\|}}{C}-S-ACP$
D isomer

② β-hydroxyacyl dehydratase → H_2O

α,β-butenoyl-S–ACP
(α,β unsaturated acyl species) = $CH_3CH=CHC\overset{\overset{\displaystyle O}{\|}}{}-S-ACP$

③ α,β-enoyl reductase — NADPH → NADP$^+$

butyryl-S–ACP
(fully saturated acyl species) = $CH_3CH_2CH_2\overset{\overset{\displaystyle O}{\|}}{C}-S-ACP$
(C_4)

malonyl-S–ACP ——→
CO_2 ← synthase

$CH_3CH_2CH_2\overset{\overset{\displaystyle O}{\|}}{C}CH_2\overset{\overset{\displaystyle O}{\|}}{C}-S-ACP$

1, 2, 3 ↓

$CH_3CH_2CH_2CH_2CH_2\overset{\overset{\displaystyle O}{\|}}{C}-S-ACP$
(C_6)

CO_2 ←•

$CH_3CH_2CH_2CH_2CH_2CH_2CH_2\overset{\overset{\displaystyle O}{\|}}{C}-S-ACP$
(C_8)

CO_2 ←•

$CH_3CH_2CH_2CH_2CH_2CH_2CH_2CH_2CH_2\overset{\overset{\displaystyle O}{\|}}{C}-S-ACP$
(C_{10})

CO_2 ←•

$CH_3CH_2CH_2CH_2CH_2CH_2CH_2CH_2CH_2CH_2CH_2\overset{\overset{\displaystyle O}{\|}}{C}-S-ACP$
(C_{12})

CO_2 ←•

$CH_3CH_2CH_2CH_2CH_2CH_2CH_2CH_2CH_2CH_2CH_2CH_2CH_2\overset{\overset{\displaystyle O}{\|}}{C}-S-ACP$
(C_{14})

CO_2 ←•

$CH_3CH_2CH_2CH_2CH_2CH_2CH_2CH_2CH_2CH_2CH_2CH_2CH_2CH_2CH_2\overset{\overset{\displaystyle O}{\|}}{C}-S-ACP$
(C_{16})

palmityl-S–ACP

H_2O ——→ palmitate + HS–ACP
thioesterase

• represent synthase for malonyl-S-ACP condensation and 1, 2, 3 each time

FIGURE 17–16 Sequence of enzyme-catalyzed reactions that comprise the malonyl-SCoA pathway for the assembly of fatty acid chains. Also see Figure 17–19 for the operation of the multifunctional protein called fatty acid synthetase.

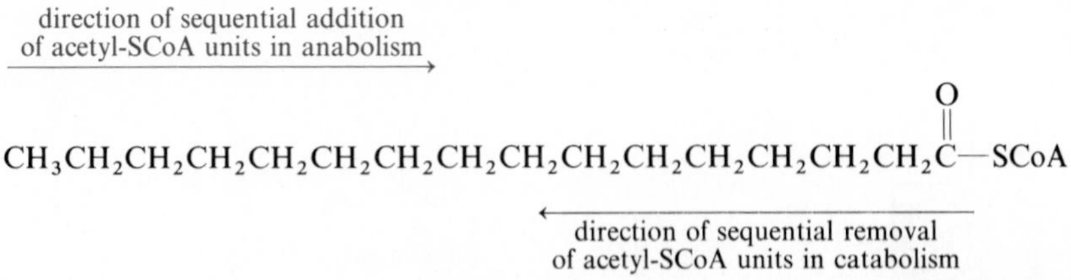

direction of sequential addition
of acetyl-SCoA units in anabolism

$$CH_3CH_2CH_2CH_2CH_2CH_2CH_2CH_2CH_2CH_2CH_2CH_2CH_2CH_2CH_2\overset{\displaystyle O}{\overset{\|}{C}}\!\!-\!\!SCoA$$

direction of sequential removal
of acetyl-SCoA units in catabolism

FIGURE 17–17 Assembly versus degradation of fatty acid chains.

to extend the C_{16} chain even further (see Figure 17–18). The only different step is that the $-CH=CH- \rightarrow -CH_2CH_2-$ conversion is catalyzed by an NADPH-dependent dehydrogenase rather than the $FADH_2$-dependent dehydrogenase. Elongation of palmityl-SCoA may also occur in association with the endoplasmic reticulum (also called microsomes). Enzymes involved in the production of unsaturated fatty acids are also localized in microsomes. We will discuss this behavior in a later section.

Fatty Acid Synthetase: A Multifunctional Protein

For about fifty years a dogma of biochemistry was that each enzyme has one type of active site capable of catalyzing only one type of chemical reaction. Then multifunctional proteins were discovered.

Multifunctional proteins are protein molecules having two or more different active sites that catalyze different reactions.

The most remarkable example of a multifunction protein is *fatty acid synthetase* (FAS). In animal eukaryotes FAS is an α_2 dimer, with each α chain (MW $\approx$ 250,000) composed of about 2300 amino acids. Remarkably, each α chain contains eight distinct functional sites. One site contributes the phosphopantetheine grouping, and the other seven are catalytic sites representing all of the enzyme functions just described as responsible for the conversion of malonyl units to palmitate.

The proposed architecture and function of the intact α_2 FAS molecule is illustrated in Figure 17–19. As shown, the current suggestion is that the two α chains are oriented in an antiparallel alignment and that half of each α subunit operates in conjunction with the opposite half of the other α subunit, allowing the dimer to function simultaneously in two rounds of palmitate assembly. In each complete α/α domain of function, the phosphopantetheine group pivots and moves the attached acyl chain from site to site. After a complete round of acyl group modification is done, the saturated acyl group is transferred to the cysteine SH at the ketoacyl synthase site and the free ACP–SH group acquires another malonyl unit.

Regulation of Fatty Acid Biosynthesis

Both the acetyl-SCoA $\rightarrow$ malonyl-SCoA and malonyl-SCoA $\rightarrow$ palmitate phases of fatty acid biosynthesis are regulated in vivo. Adaptive controls occur in response to both diet changes and hormones. For example, in a way yet unknown, insulin stimulates fatty acid synthetase activity. Examples of some im-

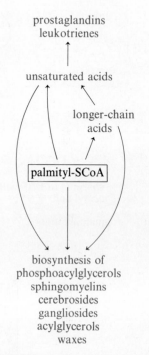

prostaglandins
leukotrienes

unsaturated acids

longer-chain acids

palmityl-SCoA

biosynthesis of
phosphoacylglycerols
sphingomyelins
cerebrosides
gangliosides
acylglycerols
waxes

FIGURE 17–18 Metabolic fates of palmityl-SCoA.

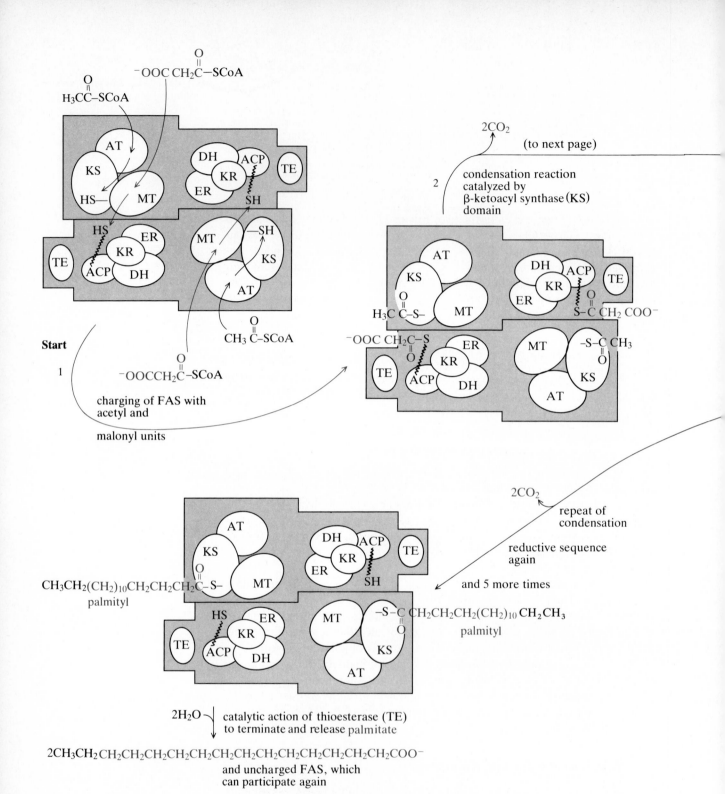

FIGURE 17–19 Architecture and operation of the multifunctional fatty acid synthetase molecule. Individual functional domains (unshaded regions) are as follows: KS = ketoacyl synthase; AT = acetyl transferase; MT = malonyl transferase; KR = ketoacyl reductase; DH = hydroxyacyl dehydratase; ER = enoyl reductase; ACP = acyl carrier protein; TE = thioesterase.

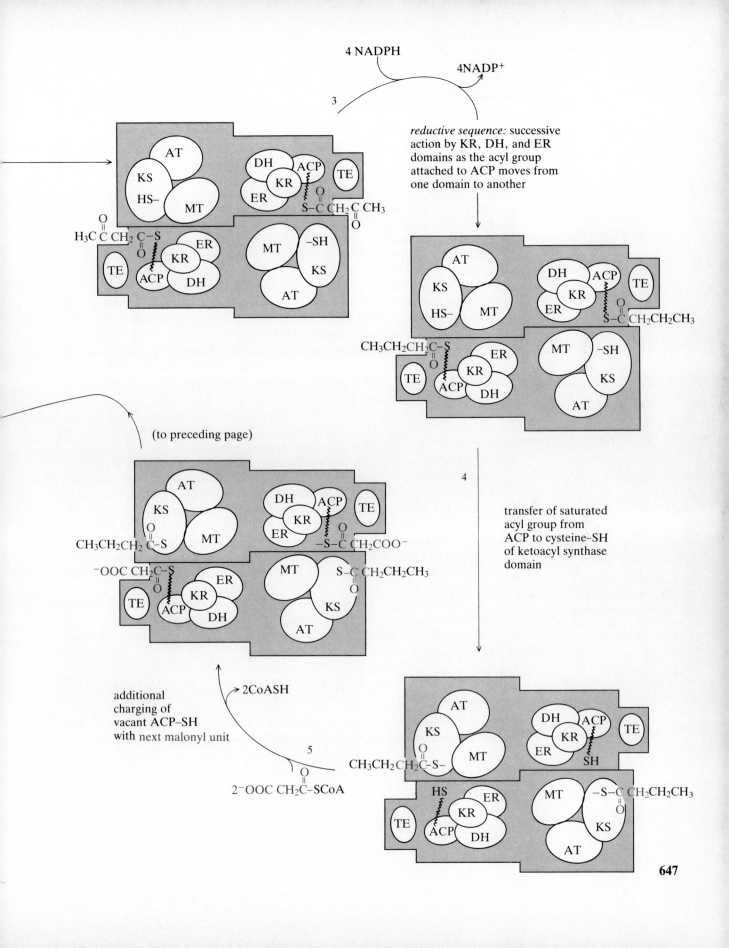

4 NADPH

4NADP⁺

3

reductive sequence: successive action by KR, DH, and ER domains as the acyl group attached to ACP moves from one domain to another

4

transfer of saturated acyl group from ACP to cysteine-SH of ketoacyl synthase domain

(to preceding page)

2CoASH

additional charging of vacant ACP–SH with next malonyl unit

5

2⁻OOC CH₂C–SCoA

647

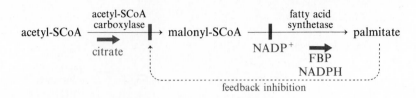

FIGURE 17–20 Metabolic controls for fatty acid biosynthesis.

portant metabolic signals (see Figure 17–20) include (1) activation of FAS by fructose-1,6-bisphosphate (FBP) and NADPH; (2) inhibition of FAS by $NADP^+$; (3) activation of acetyl-SCoA carboxylase by citrate; and (4) feedback inhibition of carboxylase by palmityl-SCoA and other long-chain, acyl-SCoA species.

The controls on carboxylase by citrate and palmityl-SCoA have an interesting explanation: Both regulate the association and dissociation of carboxylase protomer subunits (see Figure 17–21). The active state of carboxylase, stimulated by citrate, is an aggregate (MW ≈ 4–8×10^6) of 20–40 carboxylase protomers. The inactive state, responsive to the inhibition by palmityl-SCoA, is the nonassociated protomer. There is some evidence that the protomer $\rightleftharpoons$ aggregate interconversion is also controlled by a phosphorylation/dephosphorylation mechanism.

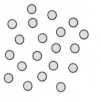

(inactive)
protomers of
acetyl-SCoA
carboxylase

citrate   palmityl-SCoA
effect   effect

associated state
of acetyl-SCoA
carboxylase
(active)

FIGURE 17–21 Explanation of the citrate and palmityl-SCoA controls on acetyl-SCoA carboxylase.

Biosynthesis of Unsaturated Fatty Acids

The microsomes of eukaryotic cells contain a multimolecular enzyme system, *fatty acid desaturase,* that catalyzes the elimination of hydrogen from long-chain acyl groups to yield unsaturated acyl groups. In effect, the overall process is a miniature electron transport system where molecular oxygen acts as a hydrogen acceptor for two hydrogen donors, the fatty acid and NADPH. The process is similar to monoxygenase activity of the cytochrome P–450 system described in Chapter 15. Proteins comprising the fatty acid desaturase system are cytochrome b_5, cytochrome b_5 reductase, and a monoxygenase (see Figure 17–22).

The site of double-bond formation depends on the specific fatty acid that participates as substrate. For example, linoleic acid (designated as 9,12-$C_{18:2}$, indicating a chain length of 18 carbons with two double bonds at carbons 9 and 12) is produced from stearic acid as follows:

$$C_{18} \xrightarrow[\text{desaturase}]{-2H} 9\text{-}C_{18:1} \xrightarrow[\text{desaturase}]{-2H} 9,12\text{-}C_{18:2}$$

stearyl-SCoA   oleyl-SCoA   linoleyl-SCoA

FIGURE 17–22 Formation of double bonds in unsaturated fatty acids.

The desaturase system of mammals is incapable of converting oleic acid to linoleic acid, and so mammals depend on a dietary supply (from plants) of

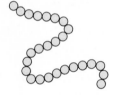

microsomal
desaturase system

linoleic acid. For mammals linoleic acid is an essential fatty acid because it is a precursor for arachidonic acid, which in turn serves as the precursor of the hydroperoxide intermediates in the formation of prostaglandins and leukotrienes (see Section 11–4), as shown here:

$$9,12\text{-}C_{18:2} \xrightarrow[\text{desaturase}]{-2H} 6,9,12\text{-}C_{18:3} \xrightarrow[\substack{\text{chain} \\ \text{extension}}]{+2\,\text{carbons}} 8,11,14\text{-}C_{20:3} \xrightarrow[\text{desaturase}]{-2H}$$

linoleic acid

$$5,8,11,14\text{-}C_{20:4} \longrightarrow \substack{\text{prostaglandins} \\ \text{and} \\ \text{leukotrienes}}$$

arachidonic acid

17–3 BIOSYNTHESIS OF OTHER LIPIDS

Acylglycerols

In fat cells (adipocytes) the triacylglycerols of fat granules are formed from L-glycerol phosphate by the sequence shown in Figure 17–23. As we all know, excess carbon from a diet rich in lipid or carbohydrate is efficiently channeled into this sequence. Under normal conditions metabolic and hormonal signals regulate the lipase-catalyzed degradation of the stored fat in adipocytes for delivery of fatty acids to cells of other organs.

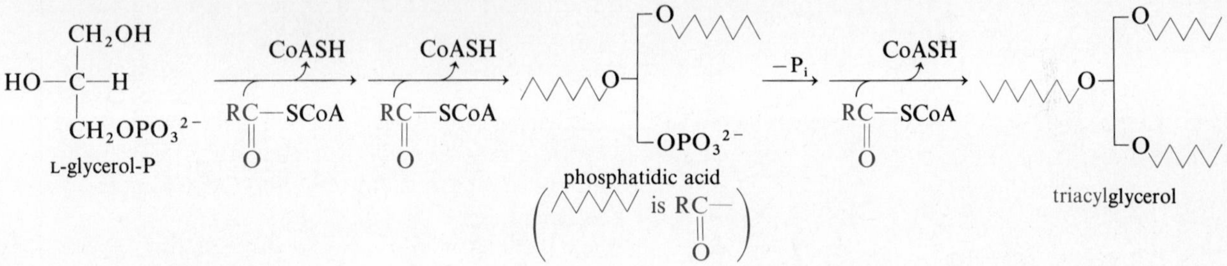

FIGURE 17–23 Biosynthesis of triacylglycerols. Phosphatidic acid is also a precursor of phosphoacylglycerols (see Figure 17–24).

Phosphoacylglycerols

In other cells phosphatidic acid also provides the backbone for the synthesis of the various phosphoacylglycerols (see Figure 17–24) used in the assembly of biomembranes. An interesting feature of these conversions is the specific involvement of CTP to produce activated intermediates for condensation with either phosphatidic acid or its dephosphorylated product, diacylglycerol. Phosphatidyl choline and phosphatidyl ethanolamine are formed by the condensation of a diacylglycerol and *CDP-choline* or *CDP-ethanolamine,* respectively. (In the next chapter we will see how phosphatidyl choline can also be formed directly from phosphatidyl ethanolamine.) Similarly, phosphatidyl inositol and phosphatidyl glycerol are formed by the condensation of a *CDP-diglyceride* with inositol or glycerol, respectively.

FIGURE 17–24 Biosynthesis of phosphoacylglycerols.

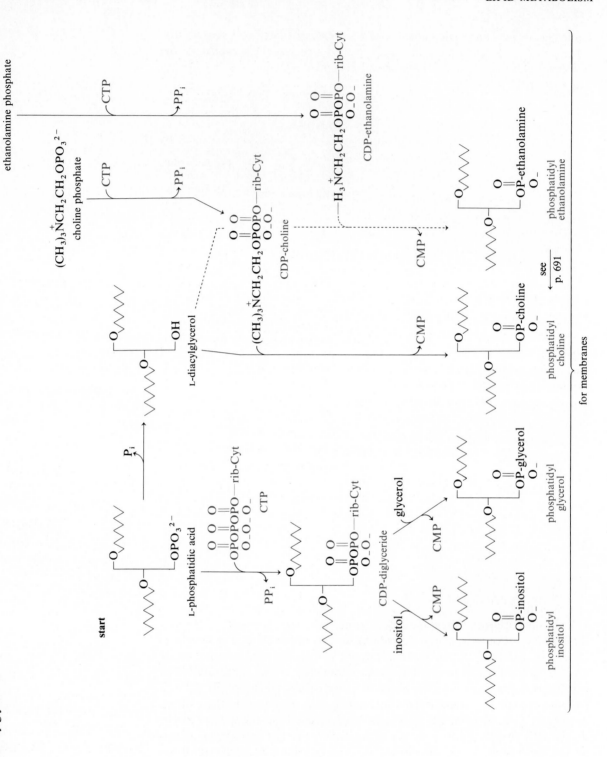

see p. 691

Sphingomyelins, Cerebrosides, and Gangliosides

Sphingosine, the common constituent of the three types of sphingolipids, is assembled from palmityl-SCoA and serine, as shown in Figure 17–25. The subsequent acylation of the NH_2 group of sphingosine yields a *ceramide,* the precursor of the sphingomyelins, cerebrosides, and gangliosides. Though not shown here, assembly of the cerebrosides and gangliosides is completed by attachment of glycosyl and sialyl units, previously activated for transfer by conversion to nucleoside diphospho derivatives (see Chapter 10).

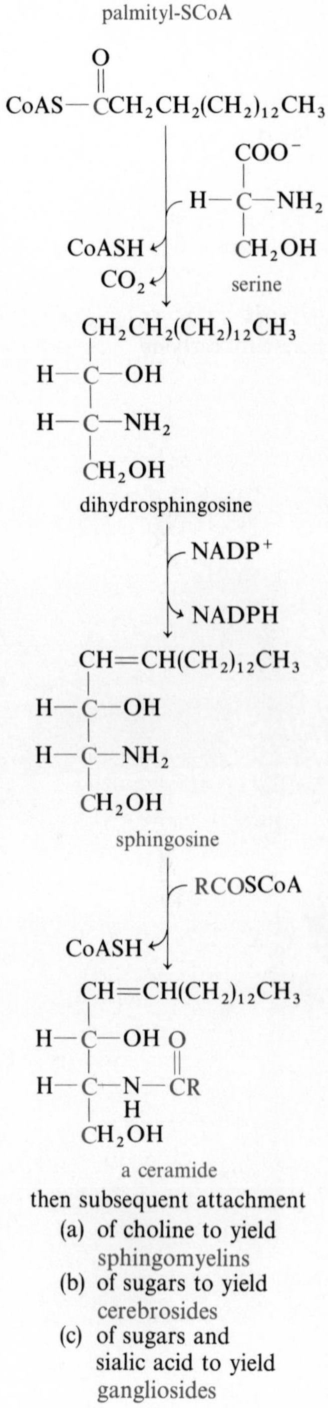

Cholesterol, Steroids, Terpenes, and Other Lipids

One of the most remarkable reaction sequences in all of nature is that responsible for the biosynthesis of cholesterol. The overall process results in the utilization of 18 acetyl-SCoA molecules to form one cholesterol molecule. Of the 36 acetyl group carbons, 27 are incorporated into cholesterol. Elegant studies established the following pattern of synthesis: 12 specific carbons in cholesterol originate from the carbonyl carbon (c) of the acetyl group and the other 15 from the methyl (m) carbon (see Figure 17–26).

The overall reaction sequence from acetyl-SCoA → cholesterol is outlined here in three phases:

acetyl-SCoA $\xrightarrow{\text{phase A}}$ mevalonate $\xrightarrow{\text{phase B}}$ squalene $\xrightarrow{\text{phase C}}$ cholesterol

(C_6 branched hydroxy acid)   (C_{30} unsaturated hydrocarbon)

The complete steps in phases A and B have been known for some time. However, despite about thirty years of study, some events of phase C still remain unsolved.

In phase A (see Figure 17–27) three units of acetyl-SCoA condense to give mevalonic acid (mevalonate). The process involves the intermediate formation of acetoacetyl-SCoA and then *β-hydroxy-β-methylglutaryl-SCoA* (HMG-SCoA), a substituted dicarboxylic acid. The next step, catalyzed by *β-hydroxy-β-methylglutaryl-SCoA reductase* (HMG-SCoA reductase), reduces HMG-SCoA to mevalonate as a free acid. The HMG-SCoA reductase is the *key regulatory enzyme* of cholesterol assembly. This regulation is described later in the chapter. The reduction of the $>\!C{=}O$ acyl group to $-CH_2OH$ represents a gain of four electrons—hence the requirement for 2NADPH. (Recall that we mentioned earlier in this chapter that HMG-SCoA is also converted to the ketone bodies.)

In phase B (see Figure 17–28) mevalonate is first activated by a trio of ATP-dependent kinases to yield 3-phospho-5-pyrophosphomevalonate. The latter is extremely unstable and undergoes a decarboxylation and a dephosphorylation to yield *isopentenyl pyrophosphate* (I-PP; C_5), which is then isomerized to give *dimethylallylpyrophosphate* (also C_5). The condensation of these two C_5 intermediates yields *geranyl pyrophosphate* (C_{10}), which can then condense with another molecule of I-PP to yield *farnesyl pyrophosphate* (C_{15}). After an isomerization, the two C_{15} isomers condense to yield *squalene* (C_{30}).

FIGURE 17–25 Biosynthesis of sphingolipids.

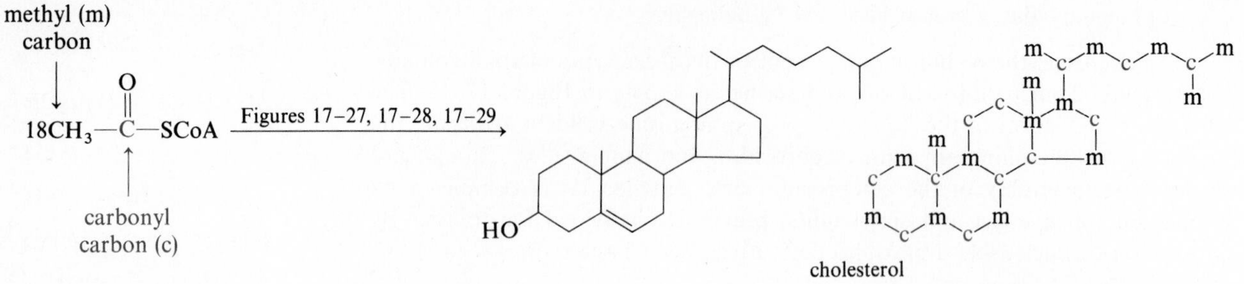

FIGURE 17–26 Origins of cholesterol carbons.

Although our focus is on cholesterol assembly, these same reactions are significant for other important reasons. In all types of organisms the C_5 isopentenyl unit serves as an intermediate in the assembly of branched hydrocarbon chain structures in various substances such as *vitamins A, K,* and *E, coenzyme Q,* the *phytol chain* of chlorophyll, *carotenoids,* and many *terpenes.* The term *terpene* refers to a hydrocarbon structure derived from the isopentenyl unit and having a minimum carbon content of C_{10}—a monoterpene. Other terpenes are C_{15}, C_{20}, C_{25}, C_{30}, and C_{40}. Squalene itself is considered a C_{30} terpene, and β-carotene is a C_{40} terpene. Some common C_{10} terpenes are camphor, menthol, α-pinene (turpentine ingredient), and α-terpineol (juniper oil). Other larger terpenes include pheromones (insect sex hormones), gibberellins and abscisic acid (plant hormones), and tetrahydrocannabinol (the active psychotropic component of marijuana).

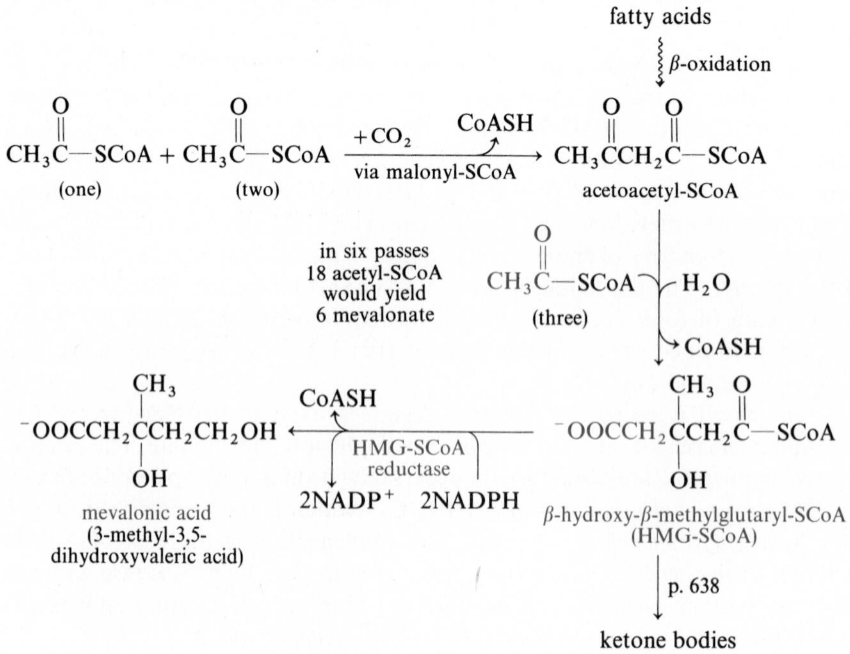

FIGURE 17–27 Phase A of cholesterol biosynthesis: formation of mevalonic acid.

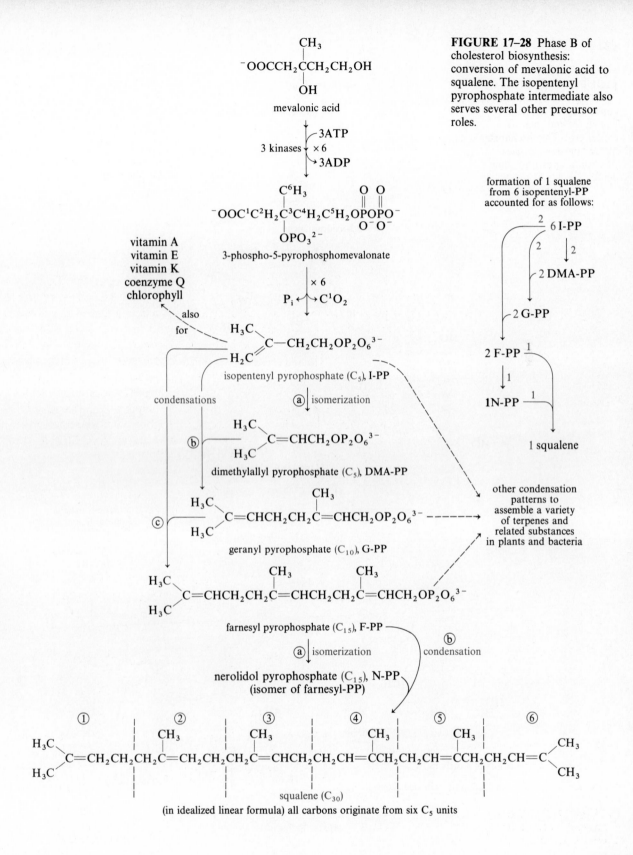

FIGURE 17–28 Phase B of cholesterol biosynthesis: conversion of mevalonic acid to squalene. The isopentenyl pyrophosphate intermediate also serves several other precursor roles.

mevalonic acid

3 kinases $\times$ 6, 3ATP → 3ADP

3-phospho-5-pyrophosphomevalonate

$\times$ 6, P_i ← C^1O_2

vitamin A
vitamin E
vitamin K
coenzyme Q
chlorophyll

also for

condensations

isopentenyl pyrophosphate (C_5), I-PP

(a) isomerization

dimethylallyl pyrophosphate (C_5), DMA-PP

geranyl pyrophosphate (C_{10}), G-PP

farnesyl pyrophosphate (C_{15}), F-PP

(a) isomerization   (b) condensation

nerolidol pyrophosphate (C_{15}), N-PP
(isomer of farnesyl-PP)

formation of 1 squalene from 6 isopentenyl-PP accounted for as follows:

6 I-PP
2 DMA-PP
2 G-PP
2 F-PP
1N-PP
1 squalene

other condensation patterns to assemble a variety of terpenes and related substances in plants and bacteria

squalene (C_{30})
(in idealized linear formula) all carbons originate from six C_5 units

653

FIGURE 17–29 Phase C of cholesterol biosynthesis: conversion of squalene to lanosterol and then to cholesterol. The mechanism of the lanosterol cyclase reaction is shown in some detail.

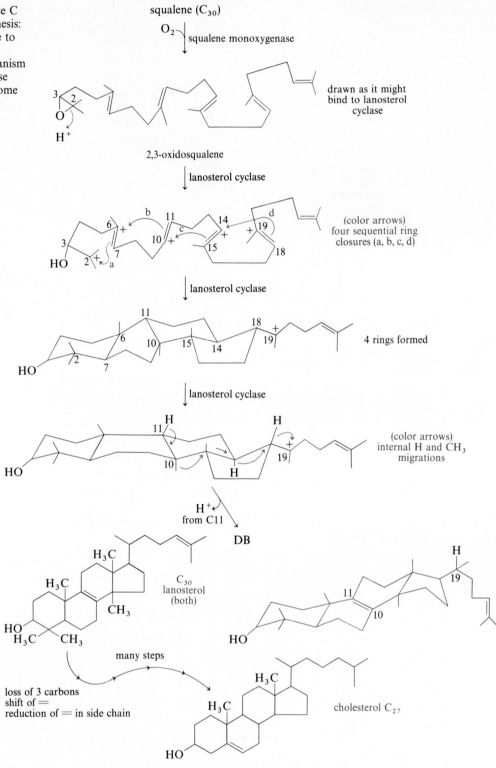

Now let us return to cholesterol biosynthesis. The conversion of the noncyclic squalene hydrocarbon to the polycyclic cholesterol structure (see Figure 17–29) is by far the most complex and still incompletely understood part in the entire pathway. It begins by the action of *squalene monoxygenase,* catalyzing a specific O incorporation at one end of the squalene chain to form an epoxide product, *2,3-oxidosqualene.* Then in a most remarkable reaction, *lanosterol cyclase* catalyzes the transformation of 2,3-oxidosqualene into the polycyclic product *lanosterol.* The polycyclization activity of lanosterol cyclase begins with a protonation of the epoxide O, which opens the epoxide ring to produce a tertiary carbonium ion (a carbocation; C^+) intermediate.

Subsequent binding and active-site events then direct two successive phases of interesting chemistry. The first phase involves a concerted sequence of intramolecular condensations to form three six-membered rings and one five-membered ring. Each condensation involves a pi (π) electron pair from a double bond condensing at a C^+ center and leaving behind another C^+. First (arrow a in Figure 17–29), $C2^+$ condenses with C7, forming a C_6 ring and a $C6^+$; second (arrow b), $C6^+$ condenses with C11, forming another C_6 ring and leaving $C10^+$; third (arrow c), $C10^+$ condenses with C15, forming another C_6 ring and leaving $C14^+$; and finally (d), $C14^+$ condenses with C18, forming a C_5 ring and leaving $C19^+$. The intramolecular condensations are almost entirely consistent with what one would expect on the basis of the Markovnikov rule, which predicts that each new C^+ will be a more stable tertiary carbonium ion rather than a secondary carbonium ion. The only inconsistency occurs in the third condensation (c), where the Markovnikov rule would predict a $C10^+/C_{14}$ condensation forming a tertiary $C15^+$ rather than the $C10^+/C_{15}$ condensation forming a secondary $C14^+$.

In the second phase of the cyclase reaction, the enzyme then directs a series of internal H and CH_3 migrations (1,2 shifts) that eliminate the $C19^+$ center and leave behind a double bond at C10/C11. Although several stereoisomers are possible (there are seven asymmetric carbons in lanosterol), all of the cyclization and migration events occur in such a manner as to form only one stereoisomer of lanosterol.

The chemical sequence and the enzymes involved in the subsequent multistep conversion of lanosterol → cholesterol are not yet completely understood. The overall process involves the removal of three ring-attached CH_3 groups, the migration of the C10/C11 double bond to C7/C8, and the reduction of the double bond in the cyclopentane side chain.

Regulation of Cholesterol Biosynthesis

The key regulatory enzyme in cholesterol biosynthesis is HMG-CoA reductase. The regulatory properties are complex and involve both long-term and short-term effects. Long-term regulation involves factors that affect the synthesis and/or degradation of the enzyme. Although how is still unclear, the synthesis of reductase is reduced by cholesterol itself as part of a feedback inhibition mechanism. Inhibition by cholesterol may also have a short-term explanation by inhibiting the activity of the HMG-CoA reductase already present.

Discovered in the late 1970s, the existence of phosphorylated and non-phosphorylated forms of reductase is at the center of regulation in the short term.

Phosphorylated reductase is inactive; nonphosphorylated reductase is active.

The active $\rightleftharpoons$ inactive interconversion is proposed to involve a protein kinase cascade (see Figure 17–30), part of which is ultimately responsive to hormones that operate via cyclic AMP (cAMP) as secondary messenger that inhibits cholesterol biosynthesis. The immediate HMG-CoA reductase converter enzymes

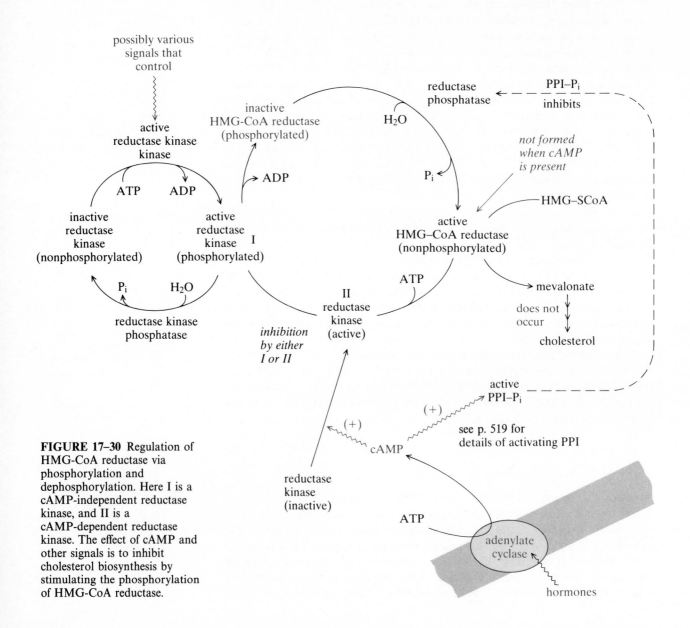

FIGURE 17–30 Regulation of HMG-CoA reductase via phosphorylation and dephosphorylation. Here I is a cAMP-independent reductase kinase, and II is a cAMP-dependent reductase kinase. The effect of cAMP and other signals is to inhibit cholesterol biosynthesis by stimulating the phosphorylation of HMG-CoA reductase.

are (1) either of two *reductase kinases* (for active nonphosphorylated reductase →
inactive phosphorylated) and (2) *reductase phosphatase* (for inactive phosphorylated reductase → active nonphosphorylated).

One reductase kinase is cAMP-independent, and the other is cAMP-
dependent. The cAMP-independent reductase kinase is itself also controlled by
phosphorylation/dephosphorylation involving two additional converter enzymes: *reductase kinase kinase,* which yields an active phosphorylated reductase
kinase, and *reductase phosphatase,* which yields an inactive nonphosphorylated
reductase kinase.

The cAMP-dependent reductase kinase appears similar to the R_2C_2
cAMP-dependent protein kinase described in earlier chapters (see Section 11–8).
Because of the cAMP involvement, this mode of control on HMG-CoA reductase can be traced to any hormone signal that would regulate the activity of
adenylate cyclase in the cell membrane. Note that the cAMP connection is
proposed to also involve the phosphoprotein phosphatase inhibitor (PPI; see
Figure 13–33 for a review), which would diminish the activity of the two steps
involving a protein phosphatase. Thus signals that enhance the formation of
the inactive, phosphorylated form of HMG-CoA reductase also diminish the
formation of the active, nonphosphorylated form of the reductase.

Not yet known is at what level cholesterol might intervene to inhibit the
formation of the active reductase. Also unknown are the signals regulating the
activity of the reductase kinase kinase.

Cholesterol Conversion to Bile Acids and Steroids

In mammals most (about 90%) of the cholesterol synthesis takes place in
the liver. Most (about 75%) of this liver-produced cholesterol is utilized to form
the so-called *bile acids* (see Figure 17–31). Bile acids assist in the digestion of
dietary lipids in the intestine by making lipid droplets smaller and water-soluble,
and hence more susceptible to the action of lipase enzymes. The major bile acid
is *cholic acid.* It, like other bile acids, is present in bile in conjugation with
glycine (*glycocholic acid*) or taurine (*taurocholic acid*).

In addition to being a precursor of bile acids and a component of membranes, cholesterol is also the *precursor* of all other important steroids (see
Figure 17–31), many of which function as hormones. *Pregnenolone* is the first
cholesterol-derived steroid, produced in a three-step reaction, with the first two
reactions catalyzed by the monoxygenase activity of *cytochrome P–450* (see
Chapter 15). Both reactions result in the hydroxylation of the cyclopentane side
chain. The third reaction then cleaves a six-carbon fragment from the hydroxylated side chain. The subsequent fate of pregnenolone, via its conversion to
progesterone, varies from tissue to tissue. The adrenal gland, for example, produces *aldosterone* and *cortisone.* The former influences electrolyte and water
metabolism; the latter regulates the metabolism of carbohydrates, fatty acids,
and proteins. The ovaries and testes produce, respectively, the estrogens (*estrone*) and androgens (*testosterone* and *androsterone*), both classes being responsible for the development of sex characteristics. The corpus luteum of the
ovaries is the chief site of progesterone production in the female.

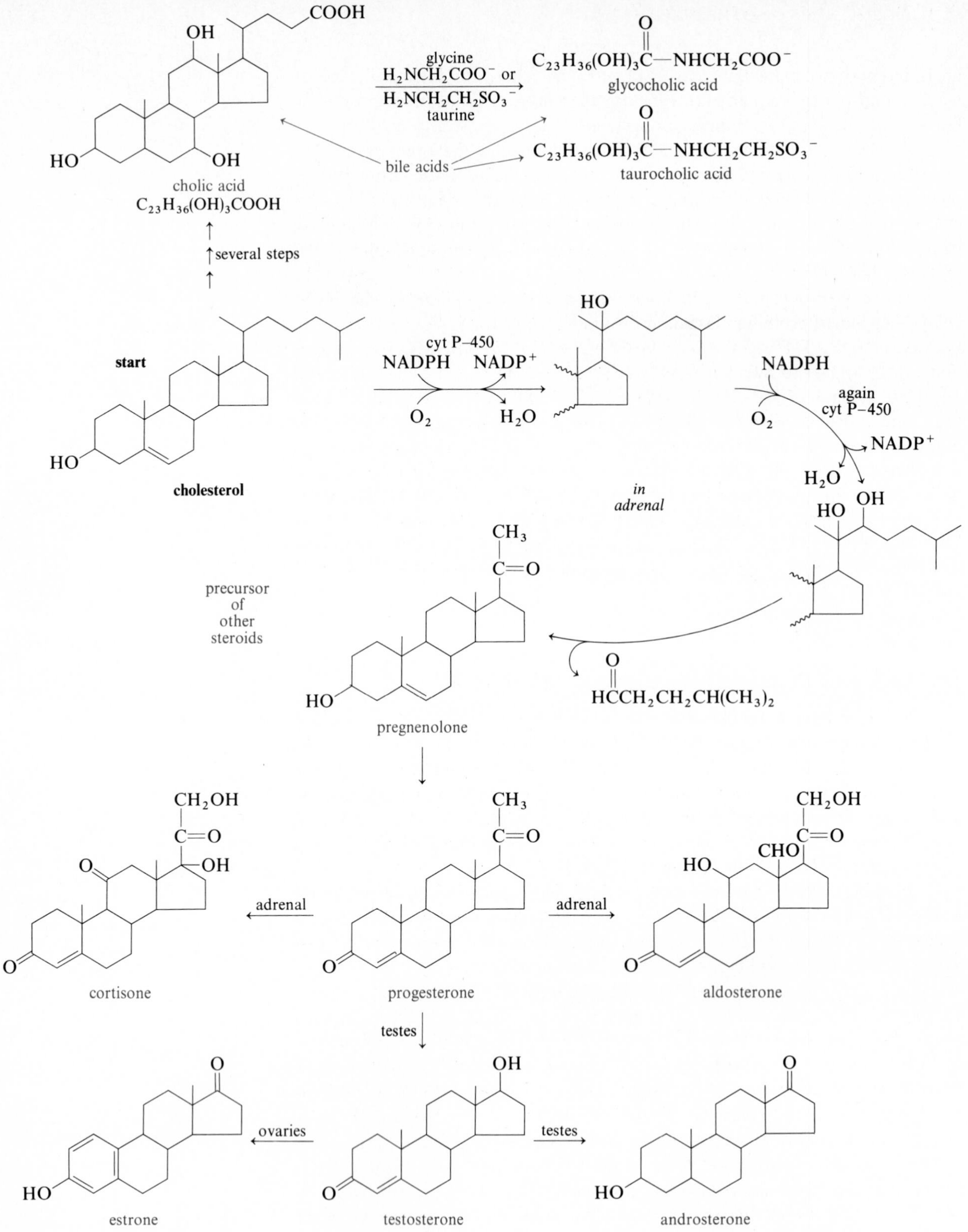

cholic acid
$C_{23}H_{36}(OH)_3COOH$

$\xrightarrow[\text{taurine}]{\substack{\text{glycine} \\ H_2NCH_2COO^- \text{ or} \\ H_2NCH_2CH_2SO_3^-}}$

$C_{23}H_{36}(OH)_3\overset{\overset{\textstyle O}{\|}}{C}-NHCH_2COO^-$
glycocholic acid

$C_{23}H_{36}(OH)_3\overset{\overset{\textstyle O}{\|}}{C}-NHCH_2CH_2SO_3^-$
taurocholic acid

bile acids

several steps

start

cholesterol

cyt P–450

NADPH   NADP$^+$

O_2 H_2O

in adrenal

NADPH

again
cyt P–450

O_2

NADP$^+$

H_2O

precursor
of
other
steroids

pregnenolone

$\overset{O}{\underset{}{\|}}$
$HCCH_2CH_2CH(CH_3)_2$

cortisone   adrenal ←   progesterone   → adrenal   aldosterone

testes

estrone   ← ovaries   testosterone   testes →   androsterone

17–4 CHOLESTEROL AND LDL RECEPTORS

Atherosclerosis

The normal level of cholesterol in blood ranges from 150 to 200 mg/100 mL. The persistence of abnormally high levels can cause plaques to occur in the aorta and in lesser arteries, a disease condition known as *atherosclerosis,* which can contribute to heart failure and strokes. Estimates are that upwards to 50% of the annual deaths in the United States are related to atherosclerosis. For many years a low-cholesterol diet has been recommended as a preventative measure against the disease, and recent results of an extensive clinical study now support the correlation. Diet control is also a standard part of disease management in addition to the use of drugs designed to reduce the body's ability to synthesize cholesterol. Many of these drugs operate as enzyme inhibitors specifically targeted at HMG-CoA reductase. Yet for many individuals these treatments are not wholly effective.

The design of new treatments is now possible because of advances made in understanding other aspects of cholesterol biochemistry. Of particular significance was the discovery that the entry of blood cholesterol into cells involved the operation of specific receptor proteins in the cell membrane. To further describe these receptors, we first need to describe the biochemistry of cholesterol transport in blood.

Receptor-Mediated Endocytosis

After its assembly (primarily in liver), free cholesterol is esterified at its C3—OH with long-chain fatty acids (usually linoleic acid). These cholesterol esters are then packaged inside lipid vesicles having an outer monolayer sheath composed of phospholipids and unesterified cholesterol. The hydrophobic tails in the lipid monolayer are oriented to the oily interior of cholesterol esters. The polar hydrophilic heads of the monolayer lipids are exposed on the surface of the particle, allowing it to be solvated by water. When proteins associate with such a vesicle by embedding in the outer lipid sheath, the resultant structure is called a *lipoprotein complex*. Depending on the ratio of lipid to protein and the nature of the packaged lipid, complexes of different density are formed: very low-density lipoproteins (VLDL), low-density lipoproteins (LDL), and high-density lipoproteins (HDL).

Cholesterol is packaged in LDL (primarily) and HDL complexes.

The LDL particles (see Figure 17–32; about 22 nm in diameter) contain about 1500 molecules of cholesterol esters surrounded by a lipid sheath having about 800 molecules of phospholipids and 500 molecules of unesterified cholesterol and one large protein molecule called *apoprotein B–100* (apo-*B*100). Cells that

◀ **FIGURE 17–31** Conversion of cholesterol to bile acids (upper part) and other steroids (lower part). For a description of cytochrome P–450 operation, see Chapter 15.

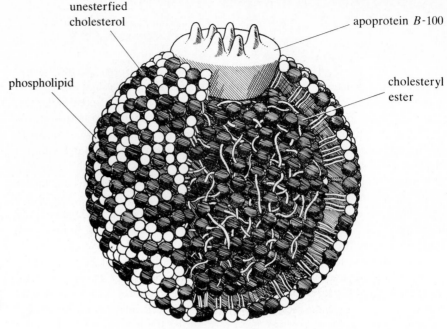

unesterified
cholesterol

apoprotein *B*-100

phospholipid

cholesteryl
ester

FIGURE 17–32 Diagrammatic representation of LDL complex. (*Source:* From "How LDL Receptors Influence Cholesterol and Atherosclerosis," by Michael S. Brown and Joseph L. Goldstein. *Scientific American,* November 1984, p. 60. Copyright © 1984 by Scientific American, Inc. All rights reserved.)

need cholesterol remove these cholesterol-rich LDL particles from the circulating blood by a process called *receptor-mediated endocytosis* (M. S. Brown and J. L. Goldstein, Nobel Prize 1985).

The following description applies to the diagram in Figure 17–33. The process begins with LDL particles being bound to the plasma membrane via specific *LDL receptors,* which recognize the apo-*B*100 protein on the surface of the LDL complex. The LDL receptors are clustered in depressions on the cell membrane called *coated pits.* Why and how the LDL receptor protein migrates to coated-pit regions are questions still unanswered. The cytoplasmic surface of a coated pit is completely engulfed by a layer of protein called *clathrin,* which self-associates on the inner membrane surface to form a polyhedral lattice structure (*clathri,* meaning lattice) having a bristlelike appearance in electron micrographs.

After the LDL complex binds to the LDL receptors, the entire coated-pit region is internalized as a pinched-off vesicle. Inside the cell the clathrin-coated vesicles then shed the clathrin coat and apparently fuse with one another to form larger vesicles called *endosomes.* An ATP-driven proton pump in the endosome membrane is proposed to transport H$^+$ into the endosome, lowering the internal pH, a condition that promotes the dissociation of the LDL particle from the LDL receptors. The region of the endosome containing the clustered LDL receptors forms a bud and is pinched off from the main endosome body, giving rise to LDL receptor vesicles that re-fuse with the main plasma membrane. Thus these events provide for the *recycling* of the LDL receptors to be used again. The remnant of the endosome vesicle still containing the LDL particle then fuses with lysosomes. Enzymes in the lysosome degrade the cholesterol esters to free cholesterol, which is released into the cell.

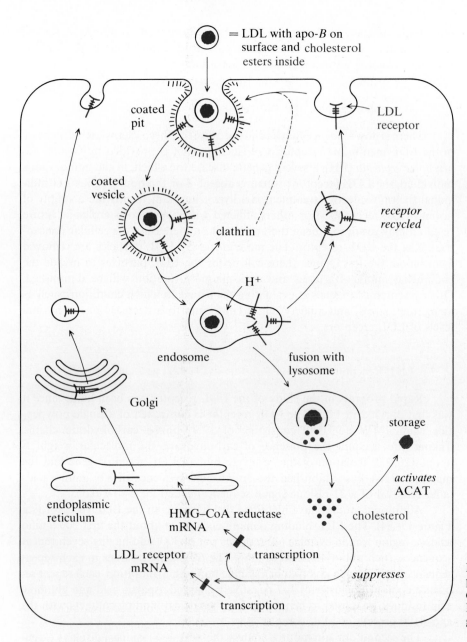

FIGURE 17–33 LDL receptor pathway for internalization of cholesterol-rich LDL particles. Descriptive explanations are given in the text.

The LDL-delivered cholesterol in the cell acts to regulate three events that serve to maintain a constant level of cholesterol within the cell (see Figure 17–33). First, cholesterol suppresses the transcription of mRNA for HMG-CoA reductase, thus assuring that cells will preferentially use LDL-derived cholesterol and will not overproduce their own cholesterol. Second, cholesterol activates the cytoplasmic enzyme acyl-CoA:cholesterol acyltransferase (ACAT), which re-forms cholesterol esters that then are stored as lipid droplets. Third, and most important, the cholesterol suppresses the transcription of mRNA for the LDL receptor, causing the cell to reduce its production of the LDL receptor and thus avoiding an increase in cholesterol intake.

Atherosclerosis and LDL Receptors

There is now evidence that a primary cause of atherosclerosis is a *deficiency* of the LDL membrane receptor. Convincing proof is provided by *familial hypercholesterolemia* (FH), a severe genetic disease for which, in the homozygous individual, the LDL receptor protein is absent. For the homozygous FH individual the only effective treatment is a liver transplant to provide a supply of normal LDL receptors. For others afflicted with hypercholesterolemia, future research may result in a drug therapy based on increasing the cellular concentration of the LDL receptor. The more efficiently LDL particles are removed from blood, the less chance there will be for the LDL particles to invade the endothelial lining of arteries and thus plaque formation will be diminished. Other research is directed at evaluating the extent to which conditions such as aging, diet, stress, and anxiety might contribute to a decrease in the normal levels of LDL receptors.

LDL Receptor Protein

Rapid progress in the study of the LDL receptor has been made since it was first purified in 1982. The LDL receptor is composed of a single polypeptide chain (MW $\approx$ 93,000) consisting of 17 O-linked carbohydrate chains. Taking into account the presence of carbohydrate, the molecular weight is about 120,000. Isolation of the complete human LDL receptor gene and the mature mRNA has established the amino acid sequence of 839 amino acids (after removal of a 21-residue signal sequence) coded for by 18 exons.

Additional studies have identified five domains in the LDL receptor (see Figure 17–34). The LDL-binding domain consists of about the first 300 amino acids, a region rich in cysteine (42 residues out of 300) and having seven repeat sequences, each about 40 residues long. The 6 cysteine residues in each repeat sequence appear to be involved in disulfide bonding. In addition, each repeat sequence contains a cluster of 4–7 negatively charged aspartic acid and glutamic acid residues, proposed to be involved in the critical binding contacts with the apo-B100 protein of LDL complexes.

The second domain (approximately the next 400 residues) displays extensive sequence homology to the precursor polypeptide for the epidermal growth factor. The significance of this domain is not known. The third domain (the next 55–60 residues) contains a total of 18 serine and threonine residues to which are attached O-linked carbohydrate chains. These clustered carbohydrate chains may act as struts to keep the receptors extended from the membrane surface to maximize their binding with the LDL complexes.

The fourth region consists of a stretch of 22 hydrophobic amino acids that anchor the receptor in the lipid bilayer and allow it to span the membrane. The fifth domain (about the last 50 amino acid residues at the carboxyl end)

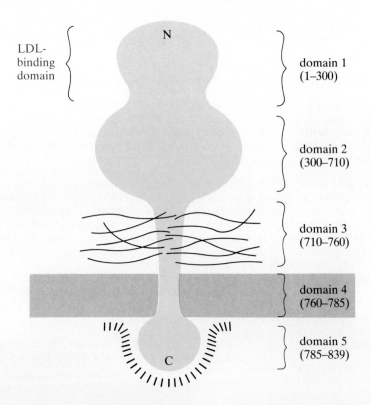

LDL-binding domain {

N

domain 1
(1–300)

domain 2
(300–710)

domain 3
(710–760)

domain 4
(760–785)

C

domain 5
(785–839)

FIGURE 17–34 Proposed molecular anatomy of the LDL receptor protein. The five domains are described in the text. Amino acid residue numbers for each domain are approximate sizes. The thin wavy lines in domain 3 represent O-linked oligosaccharide chains. Short thin lines in association with domain 5 represent clathrin. Letters N and C identify the amino and carboxyl ends of the polypeptide chain.

protrude into the cytoplasm. In addition to having possible associations with clathrin, the clustered LDL receptors may interact with each other through contacts in this region to stabilize the cluster.

Receptor-Mediated Endocytosis: Application to Other Transport Processes

Although first identified for LDL internalization, the phenomenon of receptor-mediated endocytosis has subsequently been shown to apply to a variety of other substances. Some of the ligand/receptor combinations known to operate in this manner include insulin/insulin receptor, iron/transferrin, epidermal growth factor (EGF)/EGF receptor, and vitamin B_{12}/transcobalamin II. Some protein toxins and viruses also enter cells by receptor-mediated endocytosis. The uptake of some plasma globular proteins and immune complexes also occurs by this mechanism. As of 1985 about thirty specific receptors had been established to participate in receptor-mediated endocytosis.

LITERATURE

BREMER, J. "Carnitine and Its Role in Fatty Acid Metabolism." *Trends Biochem. Sci.,* **2,** 207–209 (1977). A short review article.

BROWN, M. S., and J. L. GOLDSTEIN. "How LDL Receptors Influence Cholesterol and Atherosclerosis." *Sci. Am.,* **251,** 58–66 (1984). An excellent introductory article explaining cholesterol transport and the correlation of diminished LDL receptors to hypercholesterolemia.

DEMPSEY, M. E. "Regulation of Steroid Biosynthesis." *Annu. Rev. Biochem.,* **43,** 967–990 (1974). A very good review article describing current knowledge regarding cholesterol biosynthesis. The squalene carrier protein (SCP) is described.

FISHMAN, P. H., and R. O. BRADY. "Biosynthesis and Function of Gangliosides." *Science,* **194,** 906–915 (1976). A good review article covering the formation of gangliosides, the consequences of impaired formation, and the physiological roles of gangliosides, such as mediating the action of cholera toxin.

FULCO, A. J. "Metabolic Alterations of Fatty Acids." *Annu. Rev. Biochem.,* **43,** 215 (1974). A review article focusing on the biosynthesis of unsaturated fatty acids and hydroxyl-containing fatty acids, complemented by the Volpe and Vagelos review cited below.

GATT, S., and Y. BARENHOLZ. "Enzymes of Complex Lipid Metabolism." *Annu. Rev. Biochem.,* **42,** 61–90 (1973). Review article devoted to phosphoacylglycerols and sphingolipids.

GOLDSTEIN, J. L., M. S. BROWN, R. G. W. ANDERSON, D. W. RUSSELL, and W. J. SCHNEIDER. "Receptor-Mediated Endocytosis." *Annu. Rev. Cell Biol.,* **1,** 1–40 (1985). A detailed review article of this phenomenon with emphasis on the LDL receptor.

MATTICK, J. S., Y. TSUKAMOTO, J. NICKLESS, and S. J. WAKIL. "The Architecture of Animal Fatty Acid Synthetase." *J. Biol. Chem.,* **258,** 15,291–15,322 (1983). A series of four articles describing this multifunctional protein.

VOLPE, J. J., and P. R. VAGELOS. "Saturated Fatty Acid Biosynthesis and Its Regulation." *Annu. Rev. Biochem.,* **42,** 21–60 (1973). A good review article, complemented by the Fulco review cited above.

WAKIL, S. J., J. K. STOOPS, and V. C. JOSHI. "Fatty Acid Synthesis and Its Regulation." *Annu. Rev. Biochem.,* **52,** 537–579 (1983). A thorough review article on acetyl-SCoA carboxylase and fatty acid synthetase structure, function, and regulation.

WOOD, H. G., and R. E. BARDEN. "Biotin Enzymes." *Annu. Rev. Biochem.,* **46,** 385–414 (1977). A review article.

EXERCISES

17–1. If only the first of eight molecules of acetyl-SCoA used in the biosynthesis of palmitic acid were labeled with ^{14}C, which of the following would represent the location of the label in palmitic acid?

$$^-OO^{14}C^{14}CH_2CH_2CH_2CH_2CH_2CH_2CH_2-$$
$$CH_2CH_2CH_2CH_2CH_2CH_2CH_2CH_3$$

or

$$^-OOCCH_2CH_2CH_2CH_2CH_2CH_2CH_2CH_2-$$
$$CH_2CH_2CH_2CH_2CH_2{}^{14}CH_2{}^{14}CH_3$$

17–2. Determine the *net yield* of ATP molecules from the complete aerobic oxidation of one molecule of myristyl-SCoA to CO_2. Also determine the number of ATPs *required* in this process.

17–3. If the palmitic acid produced in Exercise 17–1 were degraded to acetyl-SCoA via β-oxidation, would the labeled acetyl-SCoA unit be produced from the thiolytic cleavage of β-ketomyristyl-SCoA or the thiolytic cleavage of acetoacetyl-SCoA?

17–4. Outline (as shown in Figure 17–8) the complete catabolism of oleic acid, linolenic acid, and arachidonic acid to acetyl-SCoA. (See also Table 11–1.)

17–5. Refer to Equation A on p. 635. Why does the conversion of stearic acid to nine units of acetyl-SCoA require nine rather than eight units of coenzyme A?

17–6. Why is the ATP yield per six carbons of fatty acid greater than the ATP yield per six carbons of a hexose? (Consider, of course, that the catabolism in each case would proceed all the way to CO_2.)

17–7. The metabolism of aspartate is linked to the citric acid cycle by the following reversible transamination reaction:

$$\text{aspartate} + \alpha\text{-keto acid} \xrightarrow[\text{aminase}]{\text{transaminase}}$$

$$\text{oxaloacetate} + \text{amino acid}$$

Using this information and referring to Chapter 14, suggest a reaction sequence that will account for the fact that some of the carbons of aspartate are eventually converted to a metabolite, which can then be used in the biosynthesis of *either* fatty acids or carbohydrates.

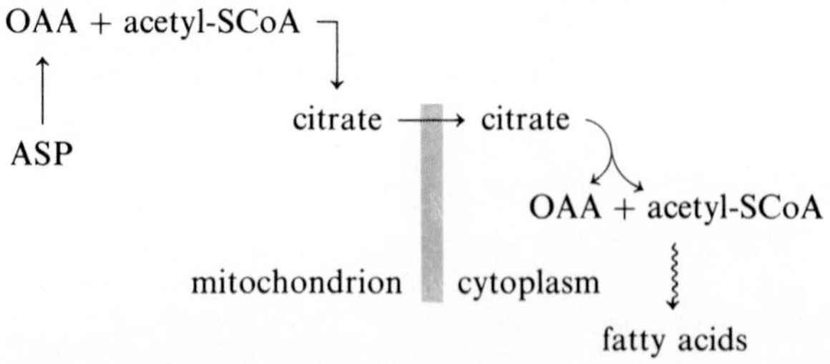

OAA + acetyl-SCoA ⎤

↑

ASP   citrate ⟶ citrate

                                          OAA + acetyl-SCoA

    mitochondrion   cytoplasm

                                          fatty acids

FIGURE 17–35. Reaction
sequence for Exercise 17–8.

17–8. In the situation posed in Exercise 17–7, what is the
only possible explanation for the conversion of carbons in
aspartate to carbons in fatty acids, according to the reaction
sequence in Figure 17–35?

17–9. How do the following reactions differ?

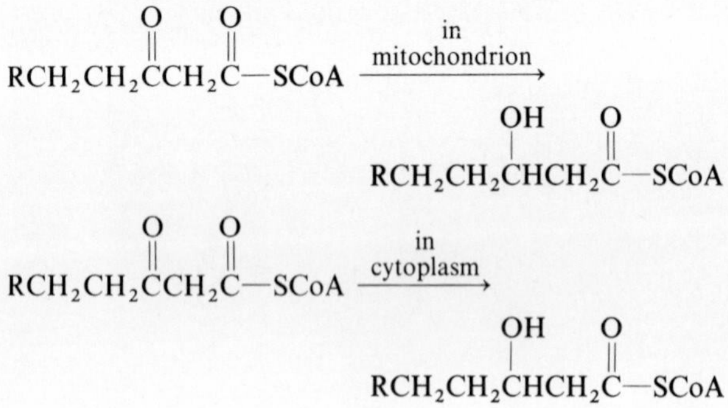

$$RCH_2CH_2\overset{O}{\underset{\|}{C}}CH_2\overset{O}{\underset{\|}{C}}\!-\!SCoA \xrightarrow[\text{mitochondrion}]{\text{in}}$$

$$RCH_2CH_2\overset{OH}{\underset{|}{C}}HCH_2\overset{O}{\underset{\|}{C}}\!-\!SCoA$$

$$RCH_2CH_2\overset{O}{\underset{\|}{C}}CH_2\overset{O}{\underset{\|}{C}}\!-\!SCoA \xrightarrow[\text{cytoplasm}]{\text{in}}$$

$$RCH_2CH_2\overset{OH}{\underset{|}{C}}HCH_2\overset{O}{\underset{\|}{C}}\!-\!SCoA$$

17–10. The active site of malate synthase probably promotes
the formation of an active carbanion species of acetyl-SCoA,
which then condenses with glyoxylate to form malate. Illustrate the mechanism.

17–11. Refer to Exercise 17–10. Compare the ATP yield of
the glyoxylate cycle to that of the citric acid cycle. Consider
both pathways linked to oxidative phosphorylation.

17–12. If a molecule of uniformly labeled acetyl-SCoA (^{14}C)
condensed with acetoacetyl-SCoA to form β-hydroxy-β-
methylglutaryl-SCoA, which was then converted to mevalonate, what would be the pattern of the ^{14}C label in
mevalonate?

17–13. According to the outline of the anabolic pathway
given in this chapter, estimate the amount of energy required
to produce one molecule of cholesterol from acetyl-SCoA.
Make the estimate in terms of the number of high-energy
compounds that are required, expressing each in terms of
its ATP equivalent. Assume that an acyl-SCoA compound
is thermodynamically equivalent to one ATP, and that one
molecule of NADH or NADPH is potentially equivalent to
three ATPs. Of course, ATP is ATP.

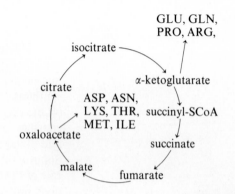

CHAPTER EIGHTEEN

AMINO ACID AND NUCLEOTIDE METABOLISM

Having previously emphasized (Chapters 7, 8, and 9) the use of amino acids and purine and pyrimidine nucleotides in the assembly of proteins and nucleic acids, we will now examine other aspects of their metabolism: (1) biosynthesis, (2) degradation, (3) transformations to other bioactive substances, (4) the excretion of excess nitrogen, (5) the biochemistry of three additional coenzymes, pyridoxal phosphate, tetrahydrofolic acid, and vitamin B_{12}, (6) some additional principles of nutrition, and (7) the flow of nitrogen in the biosphere. Some of this material was introduced in Chapter 14 in describing the amphibolic role of the citric acid cycle. Although the information contained in this chapter does not entirely cover the subject, particularly with regard to amino acid metabolism, it does cover the highlights and representative details. The first topic examined focuses on the natural beginning of nitrogen metabolism—the fixation of atmospheric nitrogen.

18–1 BIOLOGICAL NITROGEN FIXATION

Introduction

Excluding the use of synthetic fertilizers, living organisms in our biosphere ultimately depend on a supply of nitrogen from the atmosphere: 78%, by volume, of atmospheric material is N_2, and the earth is covered by about 4 billion tons of nitrogen. Each year about 30–60 million tons of atmospheric N_2 are fixed by a variety of bacteria. Plants and animals lack this ability. (Nonbiological fixation of atmospheric N_2 resulting from chemical reactions occurring during electrical storms is minimal.) Barring an ecological disaster, the nitrogen supply in the atmosphere will remain stable because of other microorganisms that convert nitrate (NO_3^-) to nitrite (NO_2^-) and nitrite to N_2 for return to the atmosphere.

Nitrogenase

The biological fixation of nitrogen involves the reduction of N_2 to NH_3:

$$:N{\equiv}N: \xrightarrow[\substack{6\ \text{electrons} \\ 6\ \text{protons}\ (6H^+)}]{\text{nitrogenase}} 2\left(\begin{array}{c} H \\ | \\ :N-H \\ | \\ H \end{array}\right)$$
$$(N_2)$$

The process occurs in various microbes—including certain anaerobic and aerobic bacteria capable of growing independently or in symbiosis with plants—photosynthetic bacteria, and blue-green algae. The enzyme system responsible

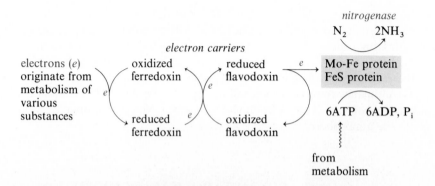

FIGURE 18–1 Electron carriers delivering electrons to nitrogenase.

for the chemistry of N_2 fixation is called *nitrogenase*. Since 1965, when it was first isolated from the anaerobic bacterium *Clostridium pasteurianum*, researchers have identified many features of structure and operation. Summarizing the main features (see Figure 18–1): (1) At least two distinct proteins are present; (2) molybdenum (Mo) and non–heme iron (FeS) are present; (3) electrons are delivered to the nitrogenase proteins by at least two different types of electron carrier proteins, ferredoxin (FeS protein) and flavodoxin (FMN-containing); and (4) the process is ATP-dependent, apparently 1ATP for each electron transferred.

Although of primary importance, N_2 fixation via nitrogenase is only one part of an integrated set of processes in our biosphere that comprise the *nitrogen cycle* (see Figure 18–2). The NH_3 product of microbial nitrogen fixation can be utilized directly by plants for the production of all nitrogen-containing organic compounds. Plants are then ingested by animals. Ammonia is regenerated by animal excretions and by the death and decay of both plants and animals. Certain soil bacteria can also utilize NH_3 as a source of energy by

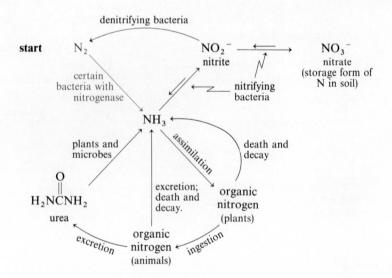

FIGURE 18–2 Integrated processes of the nitrogen cycle involving animals, plants, bacteria, and physical phenomena.

oxidizing it to NO_2^- and to NO_3^-. These oxidative processes of *nitrification* are complex and not completely understood. Because NH_3 is a toxic substance, the nitrifying bacteria play an important role when they convert it to nontoxic NO_3^-, the storage form of nitrogen in the soil. Many microbes are capable of the reverse of nitrification (that is, the reductive conversion of NO_3^- and NO_2^- ultimately to NH_3), which will resupply the pool of NH_3. Other microbes are also capable of *denitrification* ($NO_2^- \rightarrow N_2$), which will, of course, return N_2 to the atmosphere.

18–2   AMINO ACID METABOLISM: AN OVERVIEW

In Chapter 14 (see Figures 14–21 and 14–23) we introduced the catabolic and anabolic relationships of the citric acid cycle to amino acids. Those relationships are summarized again in Figure 18–3 showing (1) an input to the citric acid cycle of a portion of (or the entire carbon skeleton of) nearly all the common amino acids during their catabolism and (2) an output of carbon from the citric acid cycle via cycle intermediates that serve as precursors of the carbon skeletons of many amino acids during their anabolism.

Because of the individuality of the pathways involved, a multitude of reactions would be necessary to define each type of link for all the amino acids. While it is not our intent to do that, a few words are in order concerning the general metabolic significance of these relationships. As for the several catabolic links to the citric acid cycle, these relationships account for the fact stated earlier that amino acids can be effective intracellular sources of both metabolic energy

FIGURE 18–3 Summary of the degradation and biosynthesis of amino acids in terms of being integrated into the mainstream of metabolism. The utilization of amino acid carbons for the assembly of carbohydrates and/or lipids is the basis for the classification given in Table 18–1.

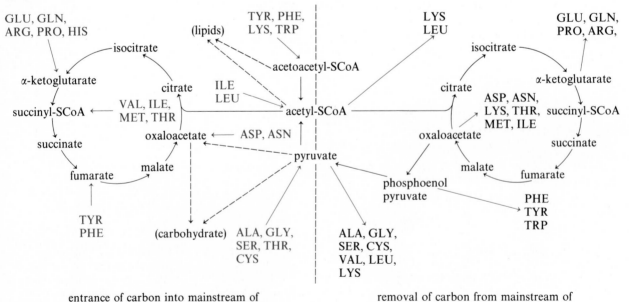

entrance of carbon into mainstream of
metabolism from amino acid degradation

removal of carbon from mainstream of
metabolism for amino acid formation

TABLE 18–1 Classification of amino acids based on their metabolic conversions to carbohydrates and/or lipids

GLYCOGENIC	KETOGENIC	BOTH GLYCOGENIC AND KETOGENIC
Glycine	Leucine	Phenylalanine
Alanine		Isoleucine
Cysteine		Lysine
Serine		Tryptophan
Threonine		Tyrosine
Aspartic acid		
Asparagine		
Glutamic acid		
Glutamine		
Arginine		
Methionine		
Valine		
Histidine		
Proline		

(ATP) and carbon needed for the formation of carbohydrates and lipids. The anabolic link, of course, accounts for the production of amino acids from carbohydrates and lipids. And yes, the carbons of one amino acid can be used for the production of another amino acid.

An old distinction—but still a helpful aid in learning—is based on tracing the degradation of an amino acid to the eventual formation of glucose (a *glycogenic* amino acid) or to the formation of lipids and ketone bodies (a *ketogenic* amino acid) (see Table 18–1). To be glycogenic, the amino acid must yield *pyruvate* or *oxaloacetate* (directly or as a cycle precursor), both of which can yield phosphoenolpyruvate for glycogenesis. This requirement is met by nearly two-thirds of the common amino acids. Ketogenic amino acids are those that directly yield *acetyl-SCoA* or *acetoacetyl-SCoA*. Several amino acids are only glycogenic, some are both glycogenic and ketogenic, and leucine is the only amino acid that is only ketogenic.

18–3 TRANSAMINATION, DECARBOXYLATION, AND DEAMINATION

The most prevalent and important type of reactions involving the amino acids are transamination, decarboxylation, and oxidative deamination. Transamination and decarboxylation are both dependent on the coenzyme participation of pyridoxal phosphate.

Transamination and Pyridoxal Phosphate

TRANSAMINATION Frequently, the first chemical event occurring in amino acid degradation and the last step in the synthesis of amino acids is a *transamination:*

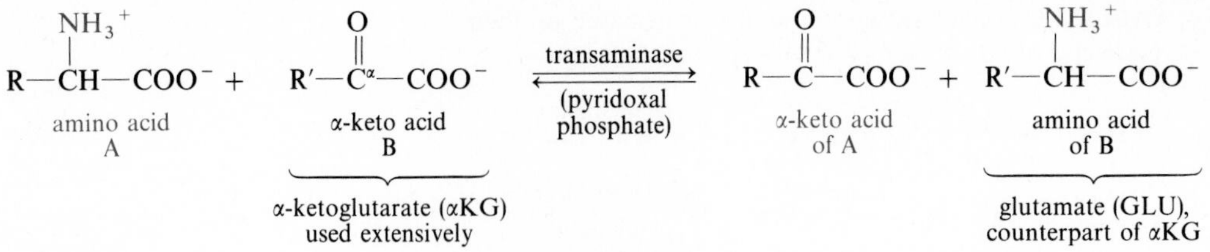

$$\underset{\substack{\text{amino acid} \\ A}}{R-\overset{\overset{\displaystyle NH_3^+}{|}}{C}H-COO^-} + \underset{\substack{\text{α-keto acid} \\ B}}{R'-\overset{\overset{\displaystyle O}{\|}}{C^\alpha}-COO^-} \underset{\substack{\text{(pyridoxal} \\ \text{phosphate)}}}{\overset{\text{transaminase}}{\rightleftarrows}} \underset{\substack{\text{α-keto acid} \\ \text{of A}}}{R-\overset{\overset{\displaystyle O}{\|}}{C}-COO^-} + \underset{\substack{\text{amino acid} \\ \text{of B}}}{R'-\overset{\overset{\displaystyle NH_3^+}{|}}{C}H-COO^-}$$

α-ketoglutarate (αKG) used extensively

glutamate (GLU), counterpart of αKG

FIGURE 18–4 Amino transfer reaction catalyzed by a transaminase. See Figures 18–6 and 18–7 for additional details.

In transamination an amino grouping is transferred from a donor amino acid to an acceptor α-keto acid to yield the α-keto acid of the donor amino acid and the amino acid of the original α-keto acid acceptor.

The reaction (see Figure 18–4) is catalyzed by a pyridoxal phosphate–dependent enzyme called *transaminase* or *aminotransferase*. As you might expect, the reaction is readily reversible.

Most cells contain several transaminases, and in eukaryotes they are found in both the cytoplasm and the mitochondria. Each enzyme usually has two elements of specificity: (1) having a strong preference for a particular amino acid/keto acid couple A but (2) using the same α-ketoglutarate/glutamate combination for the other couple B. The common specificity involving the αKG/GLU couple has been interpreted to mean that these two metabolites are centrally involved (paths 1 and 2 in Figure 18–5) in the metabolic flow of amino acid nitrogen.

In this flow α-ketoglutarate is the major nitrogen acceptor in amino acid catabolism, and glutamate is the major nitrogen dispenser in amino acid anabolism.

The cyclic flow of the αKG/GLU couple in transamination is also linked to other reactions (see Figure 18–5), some of which involve *glutamine* (GLN), a third metabolite having the same carbon skeleton as α-ketoglutarate and glutamate. In all types of cells the formation of glutamine (path 3) from glutamate provides a nitrogen source used in the biosynthesis of purines and pyrimidines. In mammals glutamine formation also serves as a way of transporting, in blood, some of the excess NH_3 of other tissues to kidney, where the glutamine is then hydrolyzed (path 5) to release the NH_3 for excretion in urine. A high blood level of NH_3 is toxic; a high blood level of glutamine is not. In higher animals the bulk of the excess glutamine in blood enters liver, where it releases the NH_3 for conversion to urea. This process is described later in Section 18–4. From liver the urea is transported in blood to kidney for excretion in urine. (Path 6 identifies an exchange reaction of an amino function between GLN and αKG to yield 2GLU, an important process in many bacteria.)

Another widely distributed and important reaction involving the αKG/GLU/GLN troika is the reversible redox interconversion (path 4) between α-ketoglutarate and glutamate. Further description of this reaction is given later in this section.

The display in Figure 18–5 merits two additional notes: (1) the precursor roles of aspartate and glutamate for the biosynthesis of several other amino

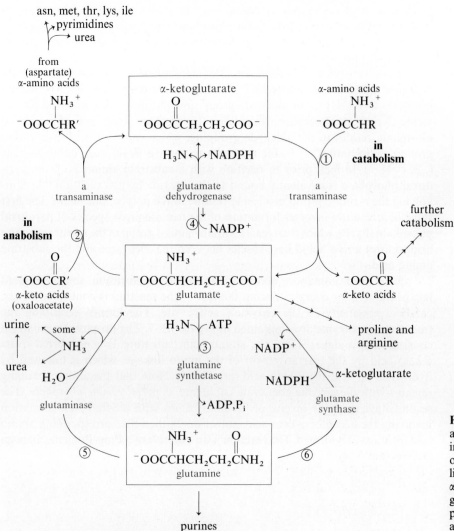

FIGURE 18–5 Summary of some amino acid reactions of importance to the metabolic flow of nitrogen in most types of living cells. Reactions involving α-ketoglutarate, glutamate, and glutamine occupy a central position. (*Note:* Some NH_3 is also utilized for pyrimidine biosynthesis.)

acids and (2) the precursor role of aspartate for the biosynthesis of pyrimidines. Several other metabolic fates of glutamate and aspartate are not shown in Figure 18–5 to avoid overkill at this point. We will encounter some of these reactions later in the chapter.

Beyond what has already been said, it is difficult to discuss further the significance of these reactions in different cells, particularly in mammals, because of tissue specificity regarding the interplay and regulation of these reactions under various physiological conditions.

PYRIDOXAL PHOSPHATE The distinctive feature of transaminase catalysis is the coenzyme participation of *pyridoxal phosphate,* sometimes called the amino acid coenzyme. Although our immediate concern is its role in transamination reactions, pyridoxal phosphate also functions in other types of amino acid transformations, such as decarboxylation and racemization (L-amino acid $\rightleftharpoons$ D-amino acid).

Pyridoxal phosphate represents the metabolically active form of vitamin B_6 called *pyridoxal,* a substituted pyridine. Oxidation of one hydroxymethyl group ($—CH_2OH$) to an aldehyde group and phosphorylation of the second yields pyridoxal phosphate (see Figure 18–6). The active site of pyridoxal phosphate resides in the aldehyde grouping, which can react with free amino groupings (in amino acids, for example) to form a *Schiff base* imino linkage ($\diagdown C{=}N{—}$). In fact, prior to reaction with a substrate amino acid, the pyridoxal phosphate is covalently bound to the enzyme by just such a linkage, involving the ε-amino group of a lysine residue in the polypeptide chain. The first stage of a reaction involves formation of the free aldehyde species of pyridoxal phosphate (PyrP), which then reacts with the amino group of the substrate amino acid to form a new Schiff base species between the coenzyme and the substrate amino acid.

Be it transamination, decarboxylation, or racemization, the subsequent fate of the ternary complex during the rest of the reaction is controlled by the catalytic specificity of the enzyme's active site. The events describing the transaminase enzyme are diagramed in Figure 18–7. The proposed mechanism consists of two stages. First, the amino acid (substrate 1) is converted to its α-keto acid by the rearrangement of the imino linkage, which is followed by hydrolysis to yield the α-keto acid (product 1). Note that the amino grouping remains with the enzyme-coenzyme conjugate as *pyridoxamine phosphate.* The second stage (just the reverse of the first) begins with a Schiff base formation involving the acceptor α-keto acid (substrate 2); then the corresponding amino acid (product 2) is formed. The reaction is an example of a Ping-Pong mechanism (see Section 5–4).

FIGURE 18–6 Pyridoxal phosphate structure, association with proteins, and reaction with an amino acid.

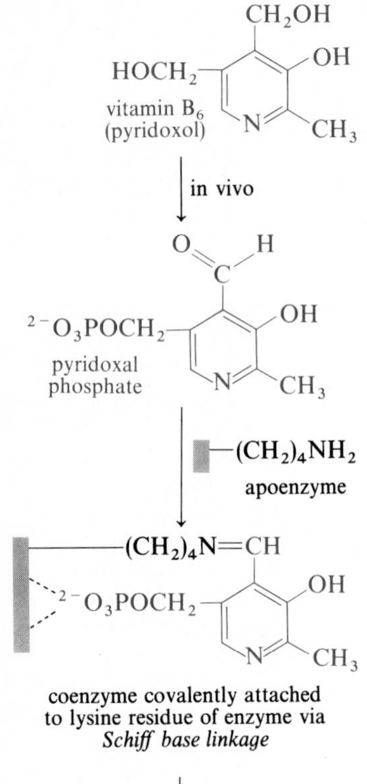

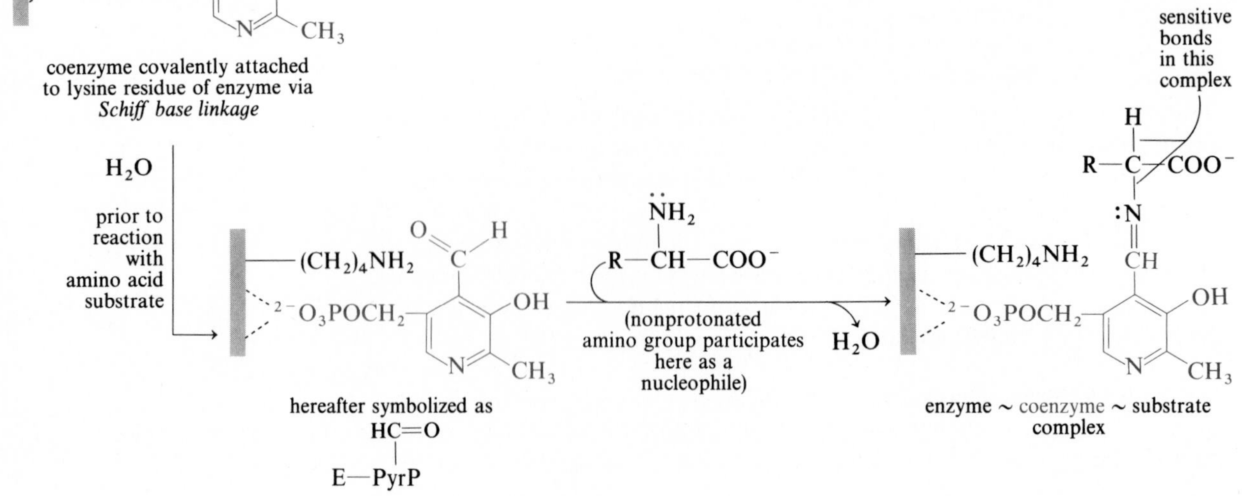

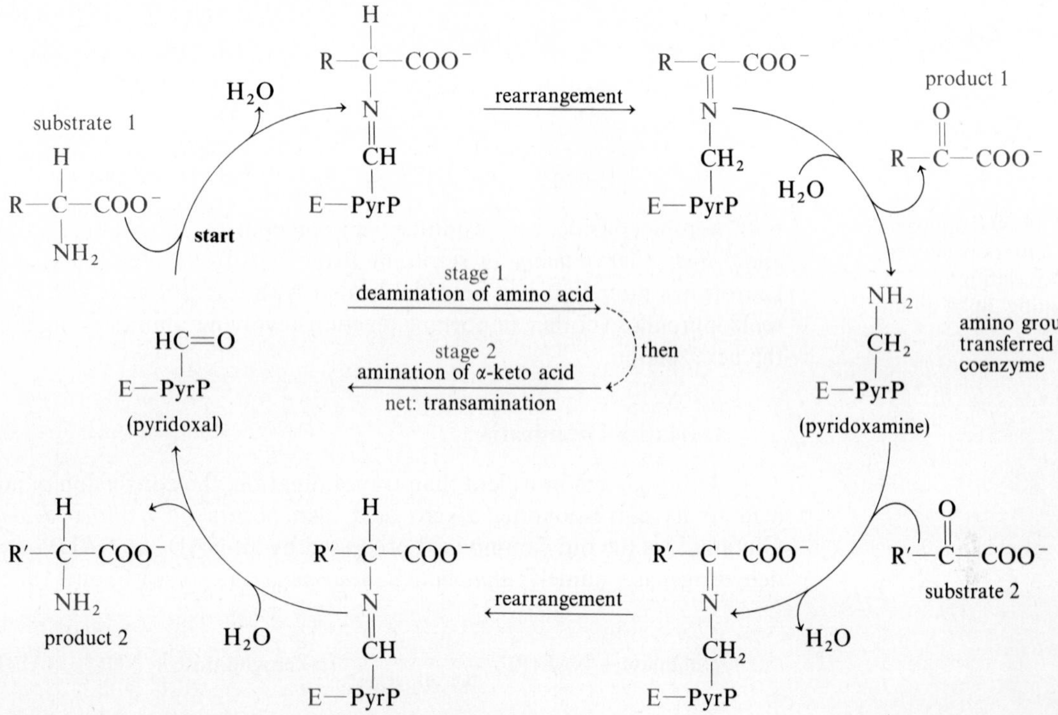

FIGURE 18–7 Sequence of events catalyzed by a transaminase enzyme, highlighting the involvement of the pyridoxal phosphate coenzyme as an amino group transfer agent.

Decarboxylation

The production of *amines* by removal of the α-carboxyl group of amino acids via pyridoxal phosphate-dependent decarboxylases occurs in all types of organisms:

$$^-OOC-CH(R)-\overset{+}{N}H_3 \xrightarrow[\text{(PyrP)}]{\text{amino acid decarboxylase}} RCH_2NH_2 + CO_2$$

amino acid — amine

After formation of the same type of E-PyrP-acid complex as in transamination, the catalytic action of the decarboxylase cleaves the C^α—COO$^-$ bond and releases the amine.

In mammals some amines derived from amino acid decarboxylation have important biological functions. For example, in brain cells *γ-amino butyric acid* (GABA) is produced from glutamate (see Figure 18–8). Both substances are classified as neurotransmitters, that is, chemical substances that affect the flow of electric potentials at synapses of neuron-neuron and of the neuromuscular junction. *GABA inhibits* nerve transmission, whereas *glutamate activates*.

Other examples (see Figure 18–9) include *histamine* (from histidine), large amounts of which are released from mast cells in response to the presence of an allergen, causing the allergic reaction; *dopamine* [from 3,4-dihydroxyphenylalanine (DOPA)], itself a neurotransmitter but also the precursor of

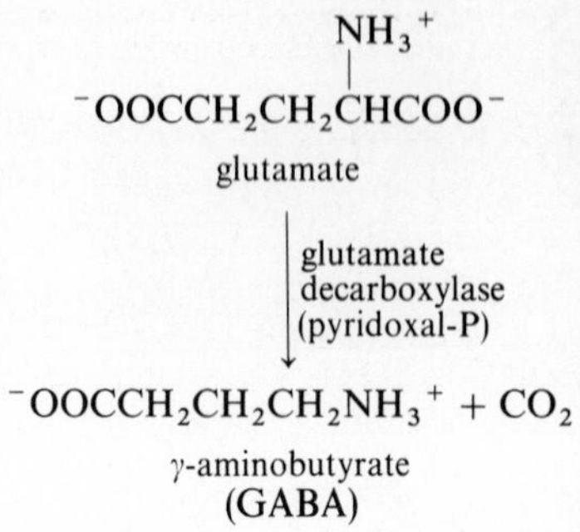

$$\overset{\overset{\displaystyle NH_3^+}{|}}{^-OOCCH_2CH_2CHCOO^-}$$

glutamate

↓ glutamate decarboxylase (pyridoxal-P)

$$^-OOCCH_2CH_2CH_2NH_3^+ + CO_2$$

γ-aminobutyrate (GABA)

FIGURE 18–8 Formation of γ-aminobutyrate.

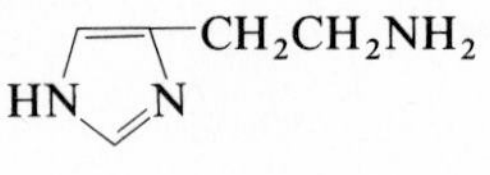

histamine

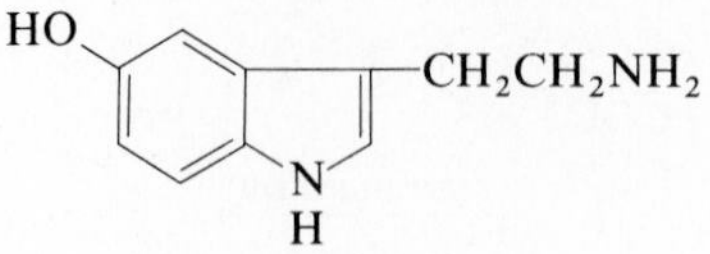

5-hydroxytryptamine
(serotonin)

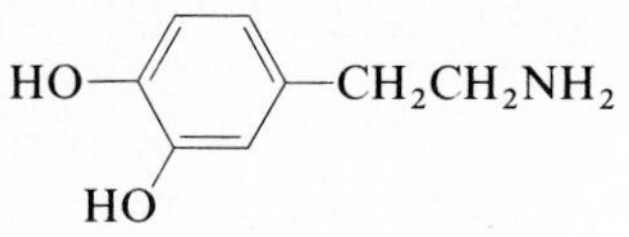

3,4-dihydroxyphenethylamine
(dopamine)

FIGURE 18–9 Examples of bioactive amines produced by decarboxylation of corresponding alpha amino acids.

both norepinephrine, a transmitter, and epinephrine (adrenaline), a hormone; and 5-*hydroxytryptamine* or serotonin (from 5-hydroxytryptophan), a possible neurotransmitter that has been implicated with alterations in behavior, as in schizophrenia. Another important reaction involving amines is highlighted in the next section.

Oxidative Deamination

Although less prevalent than transamination, the conversion of an amino acid to its corresponding α-keto acid also occurs by *oxiative deamination*. Glutamate is the only amino acid processed by an NAD- or NADP-dependent dehydrogenase, namely, *glutamate dehydrogenase* (also see Figure 18–5):

$$\text{L-glutamate} + \text{NAD(P)}^+ \xrightleftharpoons[\text{dehydrogenase}]{\text{glutamate}} \alpha\text{-ketoglutarate} + \text{NH}_3 + \text{NAD(P)H}$$

Other amino acids are processed by flavin-dependent oxidases, which will be described shortly.

When bacteria depend on extracellular NH_4^+ as the major source of nitrogen, the glutamate dehydrogenase reaction operates as an N-fixing reaction in the direction of glutamate formation. In other instances, such as in liver cells charged with a high intracellular level of glutamine and glutamate, the enzyme operates in the direction of deamination. Depending on the cell source, glutamate dehydrogenase prefers NAD or NADP or can use either coenzyme—hence the NAD(P) symbolism in the reaction display. Possibly, some cells use the NADP-dependent enzyme to provide some intracellular NADPH for anabolic reactions.

Two different types of flavin-dependent amino acid oxidases exist. One utilizes FAD as the electron acceptor and the other uses FMN. The FAD oxidase acts on D-amino acids, whereas the FMN-dependent oxidase acts on L-amino acids. In each instance the reduced flavoprotein can react directly with oxygen to produce hydrogen peroxide, a toxic substance. The enzyme *catalase* acts as a safety valve by catalyzing the decomposition of hydrogen peroxide to water and oxygen. The reaction is

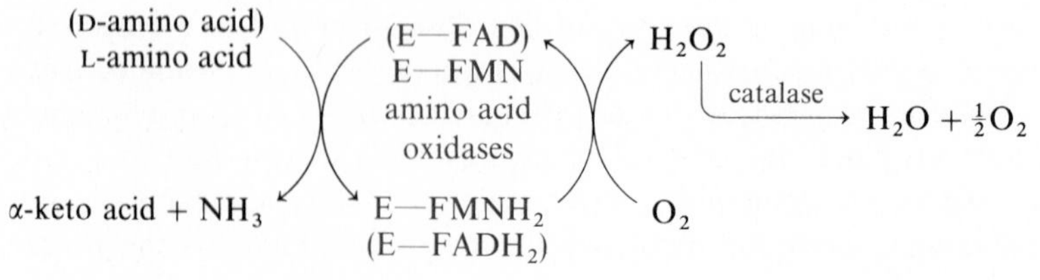

An enzyme widespread in the animal kingdom and important to normal brain metabolism is *monoamine oxidase*. This enzyme, which is related in action but not in structure to the amino acid oxidases, catalyzes the oxidative deamination of primary amines to the corresponding aldehydes:

$$RCH_2NH_2 + O_2 \xrightarrow[\text{oxidase}]{\text{monoamine}} RCHO + NH_3 + H_2O_2$$
$$\text{(amine)} \qquad\qquad\qquad \text{(aldehyde)}$$

The significance of this reaction is further discussed in Section 18–5 in conjunction with the metabolism of phenylalanine and tyrosine.

18–4 UREA CYCLE

Excess or unused nitrogen resulting from amino acid intake and degradation is excreted from an organism in the form of *ammonia, urea,* and *uric acid.* Each is a toxic substance when present in large concentrations. A major source of ammonia is deamination of glutamate by glutamate dehydrogenase. Urea is formed by a special group of enzymes whose combined operation constitutes the *urea cycle,* which will be examined here. Uric acid originates from the degradation of purines (see Section 18–7). Although the major end product varies from one type of organism to another, certain generalizations are possible. For example, most bacteria, plants, and fish excrete ammonia, whereas birds and most invertebrates excrete uric acid. In most mammals, including humans, the primary product is urea, owing to the operation of the enzymes of the urea cycle. In mammals these enzymes are localized in the liver. From the liver the urea is passed into the circulating blood, which is eventually dialyzed in the kidneys, resulting in loss of the low–molecular weight urea to the urine.

The urea cycle is composed of five reactions, each catalyzed by a different enzyme. As indicated in the following net reaction, (1) the immediate source of urea nitrogen is twofold, namely, ammonia and the amino group of aspartate; (2) the urea carbon originates from CO_2, which thus can be thought of as the nitrogen acceptor; and (3) the process requires 3ATP per molecule of urea produced:

$$CO_2 + NH_3 + H_2O + \underset{\text{(N)}}{\text{aspartate}} \xrightarrow[\underset{\text{3ATP}}{\text{5 enzymes}}]{\overset{\text{AMP, PP}_i}{\underset{\text{2ADP, 2P}_i}{}}} \underset{\text{urea}}{H_2N-\overset{\overset{\text{O}}{\|}}{C}-NH_2} + \text{fumarate}$$

As illustrated in Figure 18–10, the reaction sequence involves enzymes located in two different compartments, mitochondrion and cytoplasm. The interplay of these enzymes is possible by the transport of two metabolites, ornithine and citrulline, across the mitochondrial membrane. The process begins in the mitochondrion with the formation of *carbamyl phosphate* from CO_2, NH_3, and ATP, catalyzed by carbamyl phosphate synthetase (see Note 18–1).

NOTE 18–1
There are two forms of carbamyl phosphate synthetase (CPS) in eukaryotes, and liver cells have both. The CPS–I enzyme for the urea cycle is in the mitochondrion. The CPS–II enzyme is in the cytoplasm and functions in pyrimidine biosynthesis (see Figure 18–34).

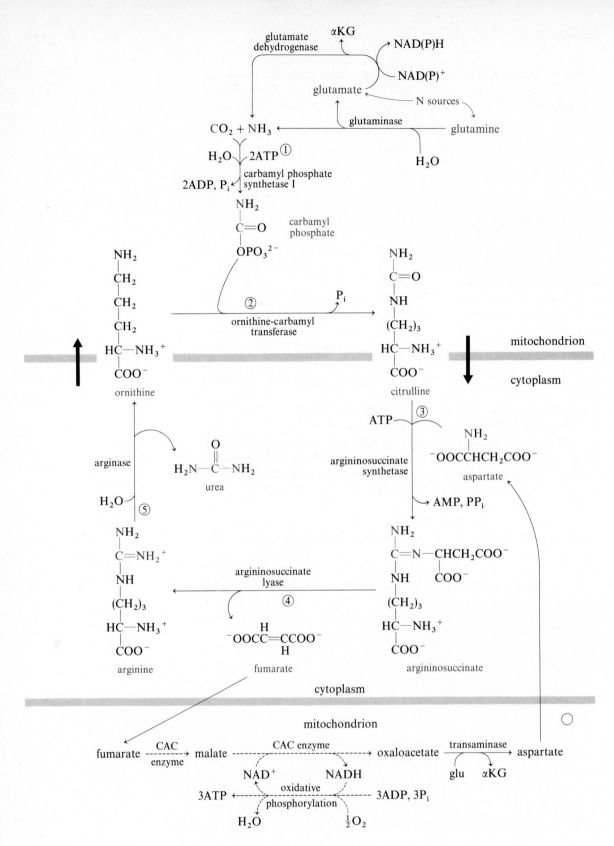

The involvement of two ATP molecules is explained by the following sequence:

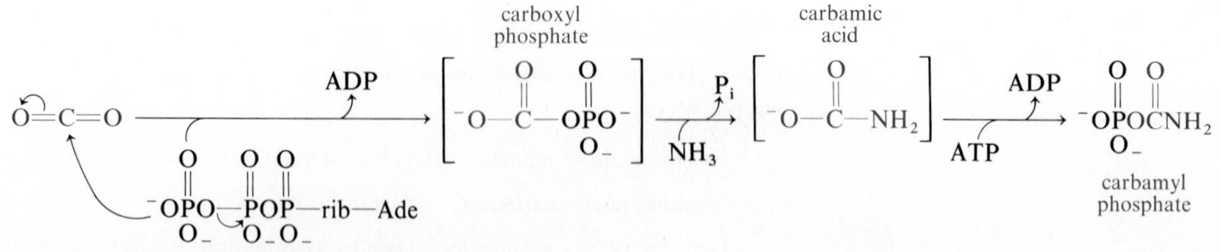

Carbamyl phosphate then condenses with *ornithine* to form *citrulline*. Both substances are basic α-amino acids found in most organisms but not known to occur in any of the cellular proteins. After exiting the mitochondrion, citrulline is condensed with aspartate (the second N source) to give *argininosuccinate*. In turn, the argininosuccinate is cleaved to yield *arginine* and *fumarate*. The unique and last step in the urea cycle is catalyzed by arginase, which converts arginine to urea and ornithine, the latter entering the mitochondrion to function again in another cycle.

Although ammonia and aspartate are the immediate nitrogen sources, note that both can be traced back to glutamate and glutamine. The NH_3 can be provided from glutamate by glutamate dehydrogenase or from glutamine by glutaminase. Aspartate can acquire its amino nitrogen from glutamate via transamination. By considering the transamination as involving oxaloacetate as the acceptor keto acid, we can indicate how the fumarate product of the urea cycle can be recycled to provide the carbon skeleton for aspartate. In addition, the recycling of fumarate also sustains the ATP requirement of the urea cycle because of the malate → oxaloacetate step in conjunction with oxidative phosphorylation.

For humans the average daily excretion of urea is about 30 g, an amount representing about 80%–90% of the total urine N. This amount is roughly 20 times the combined amount of ammonia and uric acid and 150 times the amount of free amino acids excreted in urine in the same period. An obvious question is, What happens to all this urea? Eventually, it is used as a nitrogen source by plants and bacteria, which possess the enzyme *urease* for converting urea to ammonia and carbon dioxide (see Figure 18–11). Mammals do not do this conversion because they lack the urease enzyme.

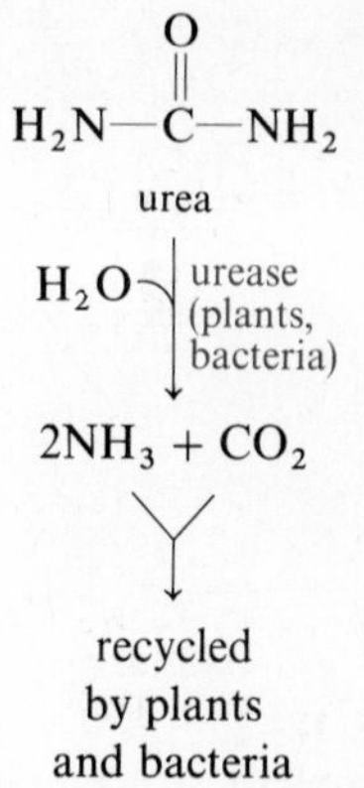

FIGURE 18–11 Urea degradation.

◄ **FIGURE 18–10** Urea cycle, reactions 1 through 5. Mitochondrion and cytoplasm compartments are separated by shaded regions. The two bold arrows represent the transport of citrulline and ornithine across the mitochondrial membranes. The reaction at the top traces the N back to glutamate and glutamine, and the reaction at the bottom identifies the recycling of fumarate via enzymes of the citric acid cycle (CAC).

18–5 METABOLISM OF SOME SPECIFIC AMINO ACIDS

Phenylalanine and Tyrosine

BIOSYNTHESIS (including tryptophan)   Some amino acids are assembled in very simple ways:

$$pyruvate \rightarrow alanine \quad \text{(via transamination)}$$

$$oxaloacetate \rightarrow aspartate \quad \text{(via transamination)}$$

$$\alpha\text{-ketoglutarate} \rightarrow glutamate \quad \begin{array}{l}\text{(via transamination} \\ \text{and glutamate dehydrogenase)}\end{array}$$

$$glutamate \rightarrow glutamine \quad \text{(via glutamine synthetase)}$$

$$aspartate \rightarrow asparagine \quad \text{(via asparagine synthetase)}$$

Others involve more elaborate chemistry, such as the synthesis of the aromatic amino acids *phenylalanine* and *tyrosine* (see Figure 18–12). Both are derived from *chorismic acid,* the carbons of which can be traced back to erythrose-4-phosphate and phosphoenolpyruvate. Participating enzymes are identified merely as $E_1, E_2, \ldots, E_n$. Although the details are not presented here, note that chorismic acid is also the precursor of several other important substances, namely, *p*-aminobenzoic acid (used in the biosynthesis of folic acid), tryptophan, the CoQ ubiquinones, vitamin K, vitamin E, and plastoquinone.

Most higher animals lack the ability (enzymes absent) to perform these transformations, nor can they produce chorismic acid. Thus they depend on a dietary supply of the substances derived from chorismic acid. There is one exception, tyrosine, which is not required because animals do contain *phenylalanine hydroxylase,* a monoxygenase enzyme that converts phenylalanine to tyrosine. In part because a genetic deficiency of the hydroxylase causes the disease of *phenylketonuria* (PKU) in infants, the enzyme has been studied extensively. It is a very specific enzyme, catalyzing hydroxylation only at the para position of the phenyl ring and having a strong preference for *tetrahydrobiopterin* as a cooxidizable substrate. The reducing power ultimately originates from NADPH (see Note 18–2 and Figure 18–13).

The absence of hydroxylase activity in PKU infants elevates the blood (and urine) levels of phenylalanine and three other metabolites derived from phenylalanine: phenylpyruvate (a phenylketone), phenyllactate, and phenylacetate (see Figure 18–14). Undetected and unchecked, the PKU child will rapidly exhibit extensive mental retardation. Fortunately, detection is easy, and treatment is also easy (feeding the child a formula with a low phenylalanine content). In adult life there is less need for dietary control because the toxic agents are excreted with greater efficiency. The primary toxic agent in PKU may be phenylpyruvate. There is some evidence that phenylpyruvate inhibits the action of pyruvate dehydrogenase in brain cells. The inhibition of the pyruvate $\rightarrow$ acetyl-SCoA conversion would severely affect the normal metabolism, development, and function of the brain. Disease PKU is but one of several hereditary diseases associated with disorders of amino acid metabolism. In fact, there are at least three other

NOTE 18–2
The biopterin structure is that of a substituted *pteridine,* the parent heterobicyclic material shown here. *Tetrahydrofolic acid,* another pteridine-derived coenzyme of immense importance in the metabolism of amino acids, purines, and pyrimidines, will be discussed later.

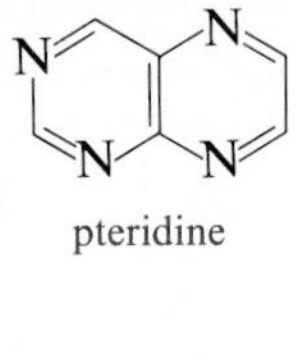

pteridine

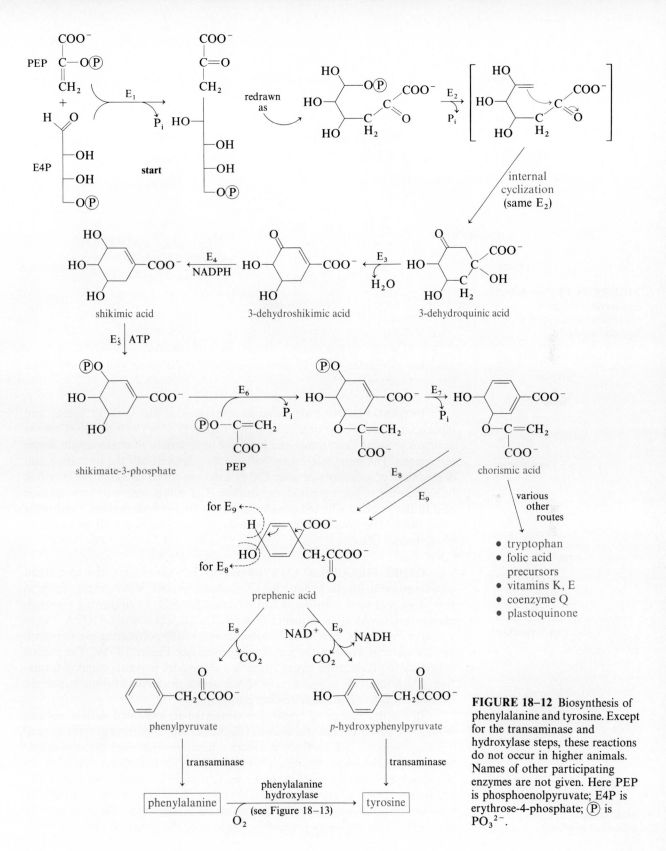

FIGURE 18–12 Biosynthesis of phenylalanine and tyrosine. Except for the transaminase and hydroxylase steps, these reactions do not occur in higher animals. Names of other participating enzymes are not given. Here PEP is phosphoenolpyruvate; E4P is erythrose-4-phosphate; Ⓟ is PO_3^{2-}.

681

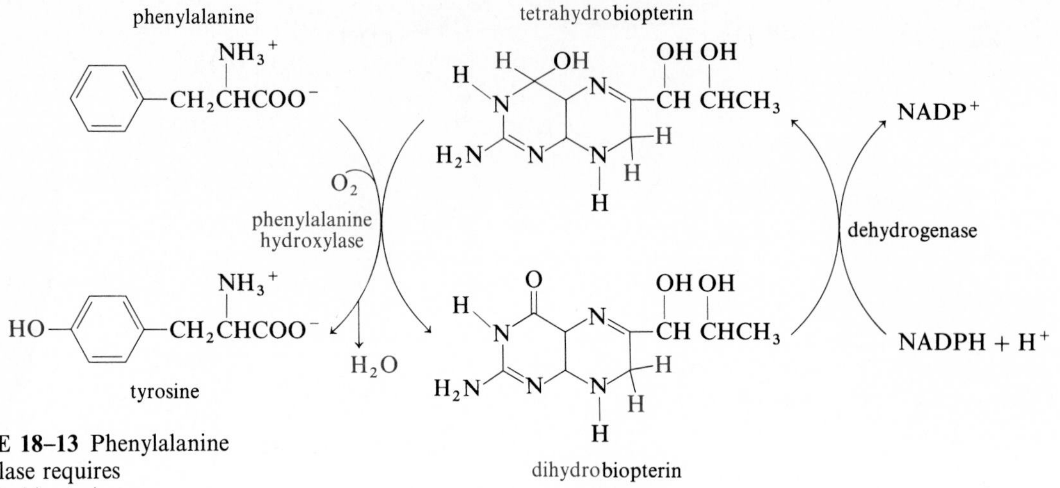

FIGURE 18–13 Phenylalanine hydroxylase requires tetrahydrobiopterin as a co-reducing agent.

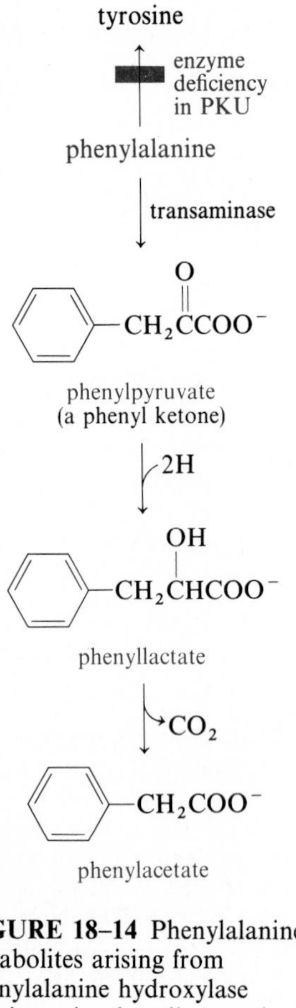

FIGURE 18–14 Phenylalanine metabolites arising from phenylalanine hydroxylase deficiency in phenylketonuria.

disorders associated with the metabolism of phenylalanine and tyrosine alone (see next sections).

DEGRADATION Phenylalanine and tyrosine are both ketogenic and glycogenic because their degradation (see Figure 18–15) yields acetyl-SCoA via acetoacetyl-SCoA (ketogenic) and fumarate (a precursor of oxaloacetate, hence glycogenic). The most interesting parts of the sequence are the formation and degradation of *homogentisic acid*. The rare disease called *alkaptonuria* is caused by a deficiency of homogentisic acid oxidase. The accumulation of homogentisic acid in the urine of afflicted persons has an alarming consequence—the urine turns black. The condition is not known to be associated with any adverse physiological effects.

OTHER METABOLIC CONVERSIONS In *melanocytes,* the specialized cells responsible for the production of *melanin pigments,* which impart color to skin, eyes, and hair, tyrosine is hydroxylated to yield *3,4-dihydroxy-L-phenyl-alanine* (L-DOPA). By a sequence of reactions still unknown, L-DOPA is converted to substances such as 5,6-indolequinone, which polymerize, possibly with protein material, to form the melanin pigments (see Figure 18–16). The genetic deficiency to produce the hydroxylase in melanocytes prevents melanin formation, which results in *albinism.* L-DOPA is also a precursor of some important neurotransmitters (see the following paragraphs).

In the thyroid gland tyrosine is converted to iodinated derivatives and conjugates called *thyroid hormones.* The substances 3-monoiodotyrosine and 3,5-diiodotyrosine are formed via direct iodination and yield the conjugated products 3,5,3'-triiodothyronine (T3) and 3,5,3',5'-tetraiodothyronine (T4, or *thyroxine*), both of which are the more active thyroid hormones (see Figure 18–17). In the thyroid T3 and T4 are present almost exclusively in peptide linkage in a protein called *thyroglobulin.* When the thyroid is stimulated, the thyroglobulin molecule is hydrolyzed by proteolytic enzymes in the gland, and the hormones

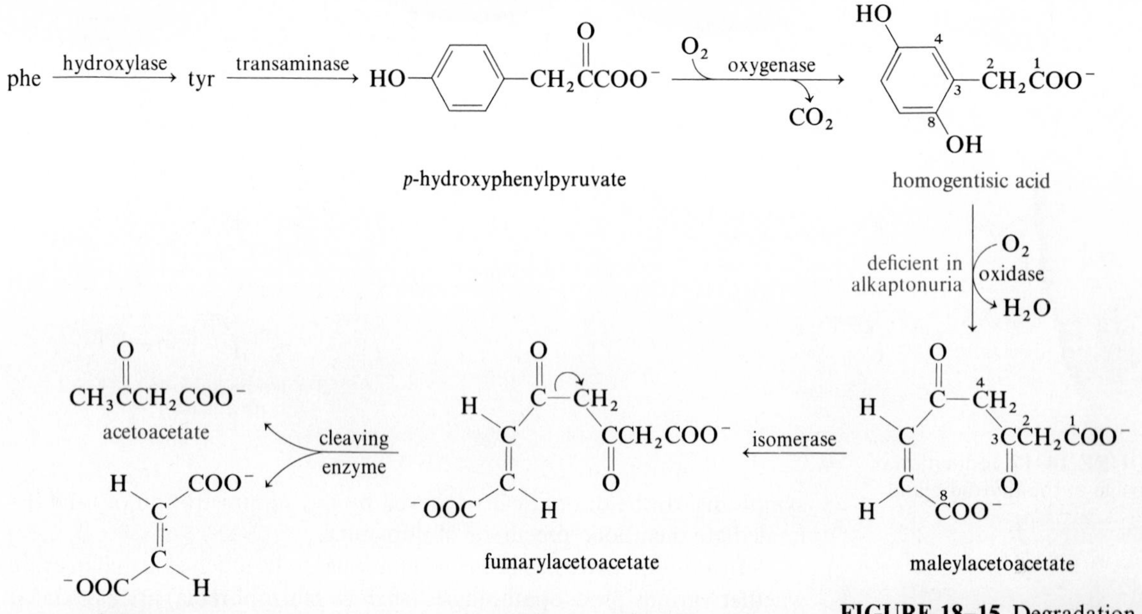

p-hydroxyphenylpyruvate

homogentisic acid

acetoacetate

fumarate

fumarylacetoacetate

maleylacetoacetate

FIGURE 18–15 Degradation of phenylalanine and tyrosine to provide carbon sources and/or ATP energy.

are secreted into the circulating blood and carried to other tissues. By processes yet largely unknown, the thyroid gland profoundly influences a broad spectrum of metabolic and physiological processes. Actually, there are few phases of the general growth and development of humans that are not regulated to some extent by these hormones.

In cells of the adrenal gland and brain (and some other tissues), tyrosine is converted to *dopamine, norepinephrine,* and *epinephrine*—all called *catecholamines.* The sequence begins (see Figure 18–18) with the hydroxylation of tyrosine to L-DOPA. (The hydroxylase is a different type than in melanocytes, but the overall result is the same.) Then L-DOPA is decarboxylated to dopamine, which is then side-chain-hydroxylated to norepinephrine, which in turn is N-methylated to yield epinephrine. The hormone biochemistry of epinephrine (also called adrenaline) as a regulator of glycolysis was discussed earlier in Chapter 13. Dopamine and norepinephrine are neurotransmitters.

Dopamine is the transmitter for a region of brain associated with the central control of movement. There is strong evidence that a *dopamine deficiency* is responsible for the symptoms of *Parkinson's disease.* For some individuals these

FIGURE 18–16 Tyrosine hydroxylation to yield L-DOPA and subsequent formation of melanin pigments. See Figure 18–18 for another conversion from L-DOPA.

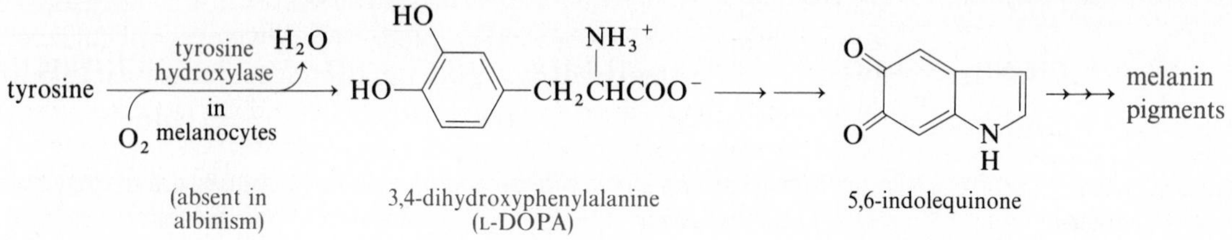

tyrosine

3,4-dihydroxyphenylalanine
(L-DOPA)

(absent in albinism)

5,6-indolequinone

melanin pigments

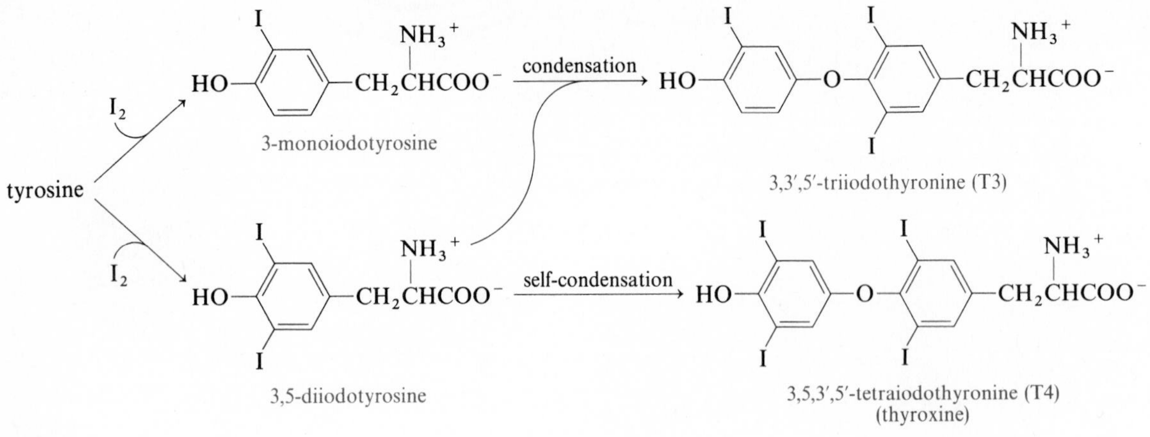

FIGURE 18–17 Iodination of tyrosine in the thyroid gland.

FIGURE 18–18 Conversion of L-DOPA to bioactive catecholamines.

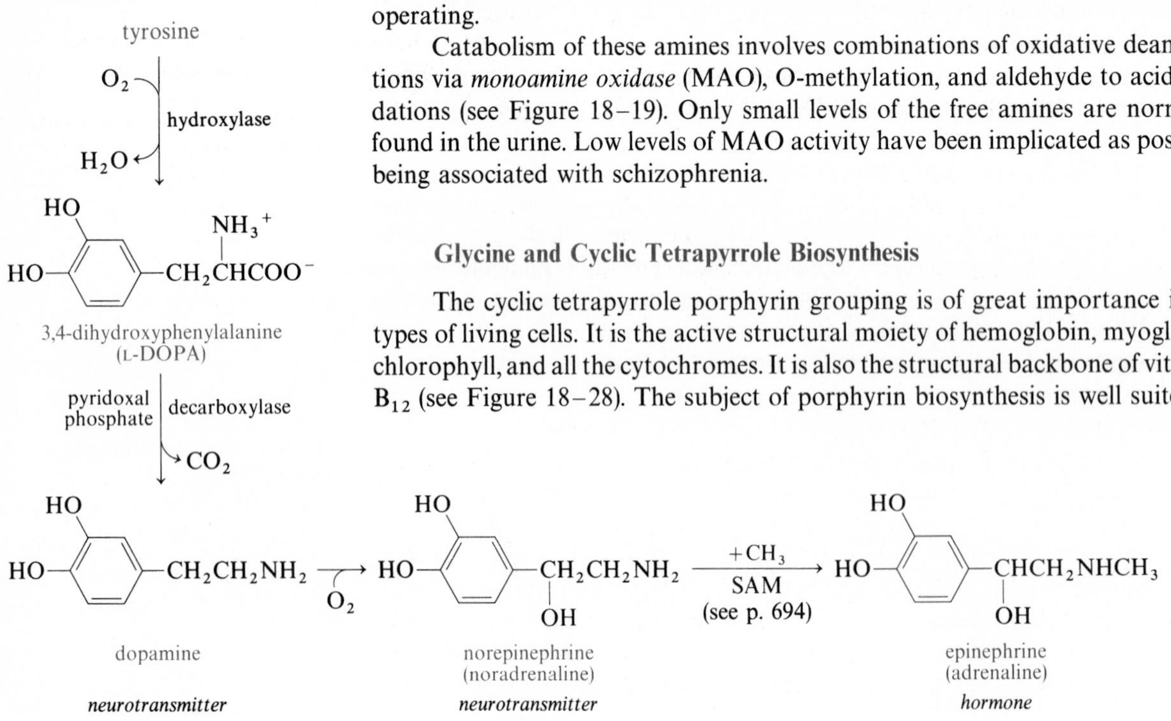

symptoms can be dramatically relieved by the administration of L-DOPA, the immediate metabolic precursor of dopamine.

In addition there has been and continues to be much research in evaluating whether various psychopathologies (such as schizophrenia) are associated with abnormally high levels of these catecholamines and other materials derived from them. No definitive results have yet been obtained, primarily because of bioethics—one obviously cannot perform biopsies on the brain of a living subject. The elevated concentrations may result from an overproduction of the primary catecholamines, from a defect in their catabolism, or from a defect in urinary excretion. Unknown reactions of other bioactive amines may also be operating.

Catabolism of these amines involves combinations of oxidative deaminations via *monoamine oxidase* (MAO), O-methylation, and aldehyde to acid oxidations (see Figure 18–19). Only small levels of the free amines are normally found in the urine. Low levels of MAO activity have been implicated as possibly being associated with schizophrenia.

Glycine and Cyclic Tetrapyrrole Biosynthesis

The cyclic tetrapyrrole porphyrin grouping is of great importance in all types of living cells. It is the active structural moiety of hemoglobin, myoglobin, chlorophyll, and all the cytochromes. It is also the structural backbone of vitamin B_{12} (see Figure 18–28). The subject of porphyrin biosynthesis is well suited to

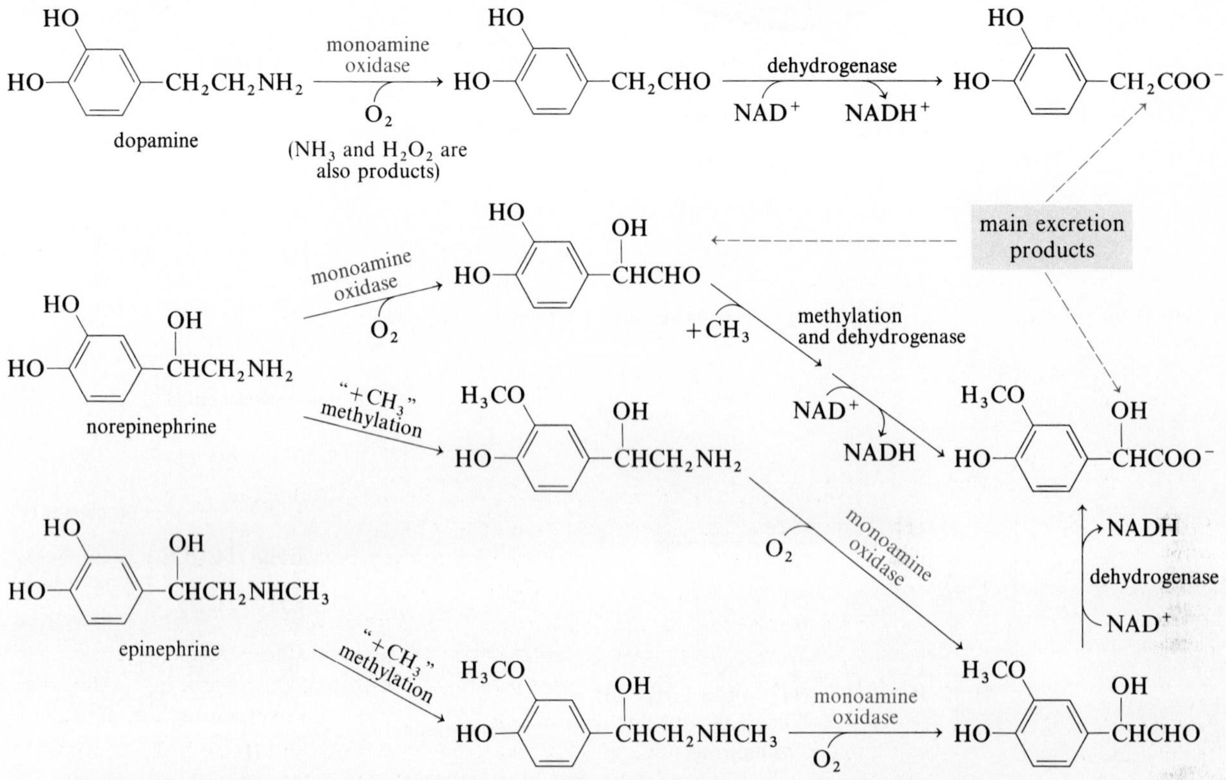

HO—⟨benzene⟩(HO)—CH₂CH₂NH₂ →[monoamine oxidase, O₂] HO—⟨benzene⟩(HO)—CH₂CHO →[dehydrogenase, NAD⁺ → NADH⁺] HO—⟨benzene⟩(HO)—CH₂COO⁻

dopamine

(NH₃ and H₂O₂ are also products)

HO—⟨benzene⟩(HO)—CHCH₂NH₂ (OH)

norepinephrine

→[monoamine oxidase, O₂] HO—⟨benzene⟩(HO)—CHCHO (OH)

→[" +CH₃ " methylation] H₃CO—⟨benzene⟩(HO)—CHCH₂NH₂ (OH)

main excretion products

→[methylation and dehydrogenase, NAD⁺ → NADH] H₃CO—⟨benzene⟩(HO)—CHCOO⁻ (OH)

→[monoamine oxidase, O₂]

HO—⟨benzene⟩(HO)—CHCH₂NHCH₃ (OH)

epinephrine

→[" +CH₃ " methylation] H₃CO—⟨benzene⟩(HO)—CHCH₂NHCH₃ (OH) →[monoamine oxidase, O₂] H₃CO—⟨benzene⟩(HO)—CHCHO (OH)

NADH ↕ dehydrogenase NAD⁺

FIGURE 18–19 Further metabolism of catecholamines. The participation of monoamine oxidase (MAO) is highlighted in color. Methylation reactions are described later in the chapter (see Figure 18–27).

this chapter, since the sole precursors are *glycine* and *succinyl-SCoA*. A few of the main steps are shown in Figure 18–20.

The process begins with a condensation of succinyl-SCoA and glycine to form *α-amino-β-ketoadipic acid*. This acid in turn is decarboxylated to yield *δ-aminolevulinic acid*. The next step involves an intermolecular condensation of two molecules of aminolevulinic acid to yield *porphobilinogen*, a substituted pyrrole. Four molecules of porphobilinogen are then condensed to form a *linear tetrapyrrole,* which then undergoes a ring closure to produce a *cyclic tetrapyrrole* species. Depending on the specificity of the enzymes that catalyze this latter step, the ring closing can either occur by a direct joining of the two ends of the linear tetrapyrrole or, as shown in Figure 18–20, be accompanied by a flipping of the pyrrole unit at the nonamino end into a different position before ring closure. The latter thus causes the pyrrole side chains to be arranged in a nonalternating fashion, that is, A-P, P-A, P-A, P-A, rather than P-A, P-A, P-A, P-A. In other words, a specific isomer is produced, called *uroporphyrinogen III*. From this point a specific combination of reactions involving modification of the side chains and reductions of the —CH₂— pyrrole links will yield the various porphyrins.

The structure of *protoporphyrin IX* is shown because it is the most common porphyrin system, being found in hemoglobin, myoglobin, and some cytochromes. The chelation of the porphyrin with the corresponding metal completes the formation of the active metalloporphyrin. Note that all four nitrogen atoms, the four methenyl links (—CH=), and four pyrrole carbons come from glycine. All of the other carbons originate from succinyl-SCoA. Since succinyl-SCoA is

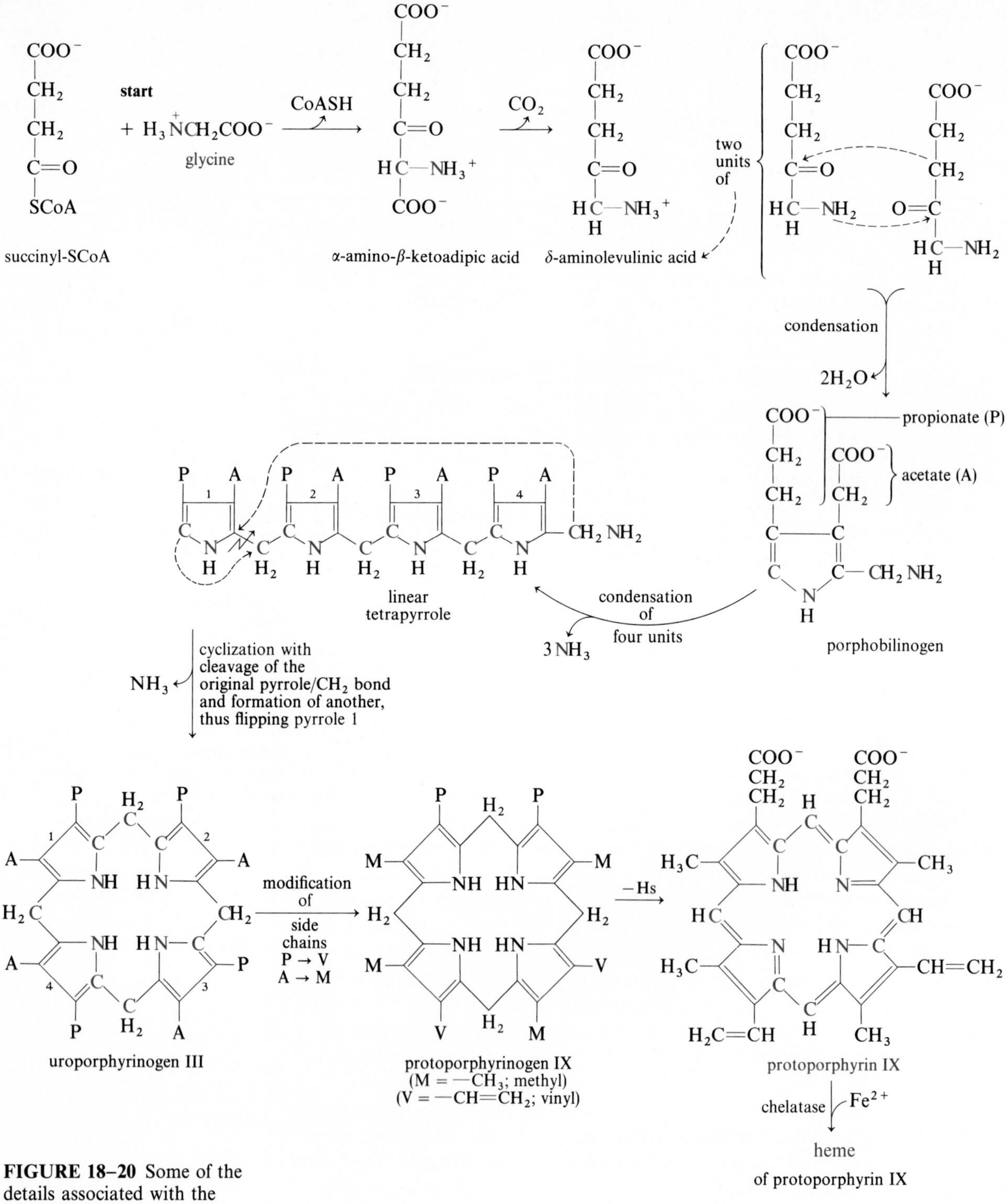

FIGURE 18–20 Some of the details associated with the biosynthesis of porphyrins and hemes.

an intermediate of the citric acid cycle, the ultimate source of these carbons can be carbohydrate, lipid, or any amino acid that can be degraded to an intermediate of the citric acid cycle.

Cysteine and Methionine: Sulfur Metabolism

Sulfur is one of the six major elements (C, H, O, N, P, S) of which most biomolecules are composed. So far, we have encountered sulfur in several substances of major importance to all living organisms. These substances include glutathione, thiamine pyrophosphate, biotin, lipoic acid, coenzyme A, chondroitin sulfates, sulfolipids, iron-sulfur proteins, and those proteins in which cysteine and methionine are present. There are several other sulfur-containing compounds that occur naturally.

ACTIVATION AND REDUCTION OF SULFATE   The ultimate source of sulfur for most organisms is inorganic sulfate, SO_4^{2-}. After entering a cell, sulfate is mobilized for metabolism by conversion first to *adenosine-5'-phosphosulfate* (APS) and then to *3'-phosphoadenosine-5'-phosphosulfate* (PAPS). Each step (see Figure 18–21) requires ATP. The chemical significance of the conversions is that inorganic sulfate in the free unreactive state SO_4^{2-} is converted to a more highly reactive sulfo-phospho anhydride species.

In humans and other animals PAPS serves as a donor of the sulfate group in the formation of organic sulfate esters such as heparin and the chondroitin

FIGURE 18–21 Activation of inorganic sulfate to PAPS and its subsequent reduction to inorganic sulfide. Also indicated is its role as a sulfo transfer agent.

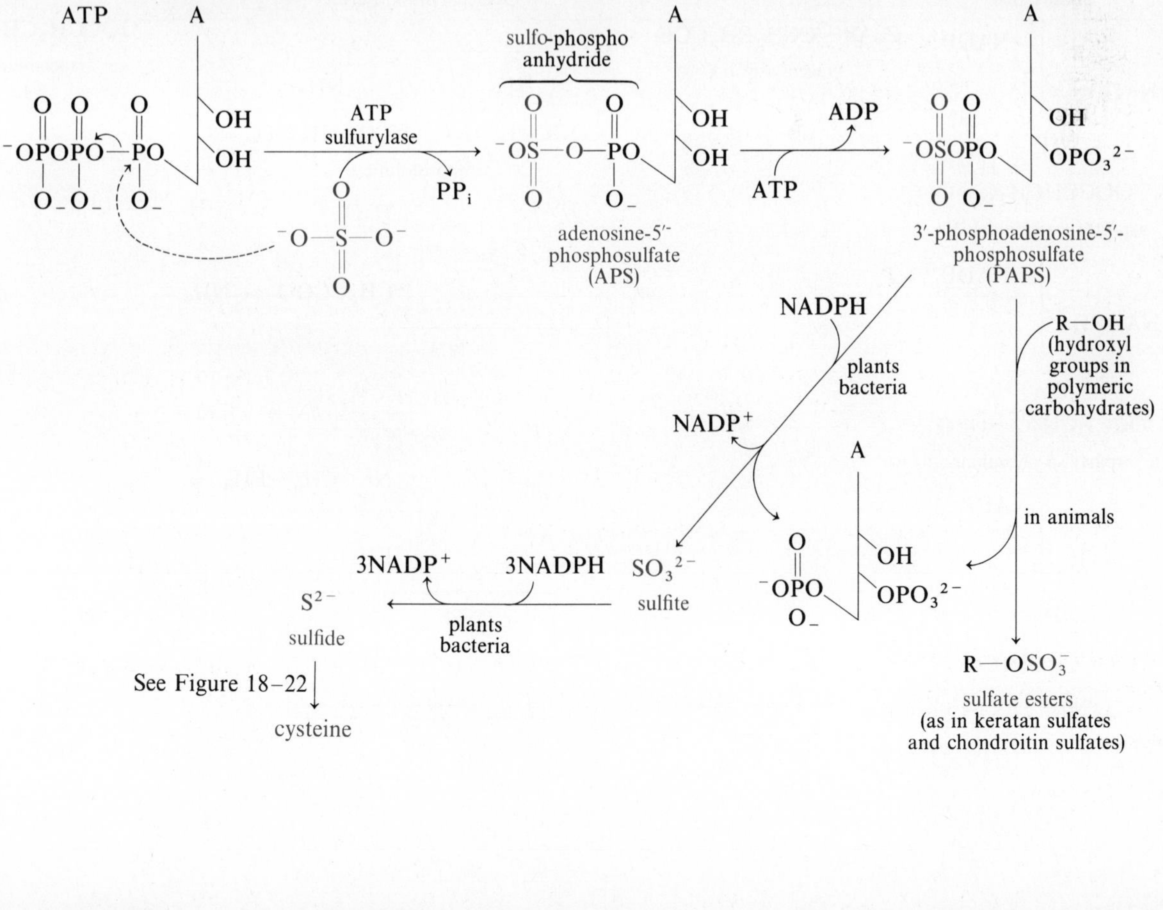

FIGURE 18–22 Cysteine biosynthesis. The assimilation of inorganic sulfide into organosulfur as cysteine is shown at the top. The precursor roles of cysteine (sulfur source) and aspartate (carbon source) for methionine biosynthesis are shown at the bottom. Except for the homocysteine → methionine conversion, humans and other animals lack the enzymes for these reactions.

sulfates (see Section 10–4). Animal organisms, however, are incapable of utilizing PAPS as a donor of sulfur to form other sulfur-containing compounds. This utilization requires that the sulfur of the sulfate grouping be reduced to sulfide S^{2-}, a conversion that does not occur in animals. Enzymes capable of the SO_4^{2-} to S^{2-} conversion do exist in plant and bacterial organisms. The proposed sequence is shown in Figure 18–21. It involves two separate reductase enzymes, each using NADPH as an electron donor. A total of 4NADPH is required to provide eight electrons for the S^{6+} (in sulfate) → S^{2-} conversion.

SULFIDE INTO ORGANOSULFUR In plants and bacteria the major pathway (see Figure 18–22) proposed for using inorganic sulfide to form an or-

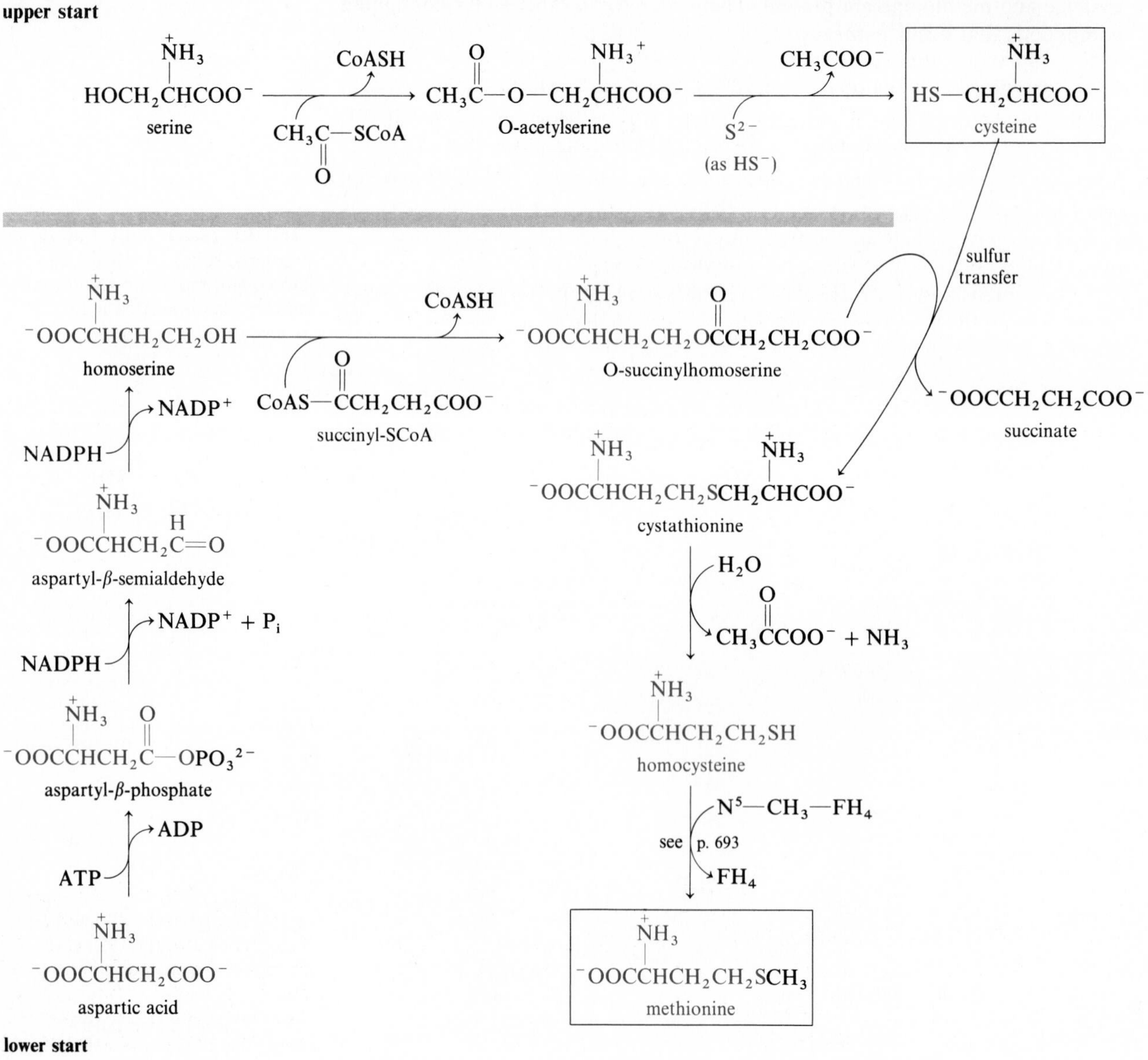

upper start

lower start

ganosulfur compound is *cysteine biosynthesis*. The carbon skeleton is provided by *serine* after an initial activation by conversion to an O-acetyl derivative. Cysteine is then the source of sulfur in the biosynthesis of *methionine,* with the carbon skeleton of methionine originating from *aspartic acid*. The sulfur transfer involves the condensation of cysteine with O-succinylhomoserine (which originates from aspartate and succinyl-SCoA) to yield cystathionine. A cleavage to homocysteine and a transmethylation reaction (see Figure 18–27) finally yield methionine.

Unable to convert cysteine to methionine, animals depend on a dietary supply of methionine. Although animals are also unable to form cysteine from SO_4^{2-}, cysteine is not an essential dietary amino acid because methionine and serine can be converted to cysteine. Methionine provides the S; serine provides the carbon skeleton. The pathway (see Figure 18–23) is an offshoot of the use of S-adenosylmethionine (SAM) as a methyl donor, a topic to be discussed shortly. After the reaction of transmethylation S-adenosylhomocysteine is formed; this can be cleaved to yield free homocysteine. A condensation with serine gives cystathionine, which in turn is degraded to yield cysteine.

Coenzyme A from Valine, Aspartic Acid, and Cysteine

The biosynthesis of coenzyme A is illustrated in Figure 18–24. As indicated, individual parts of the structure originate from ATP, *valine, aspartic acid,* and *cysteine*. Whereas plants and bacteria utilize the entire pathway, an-

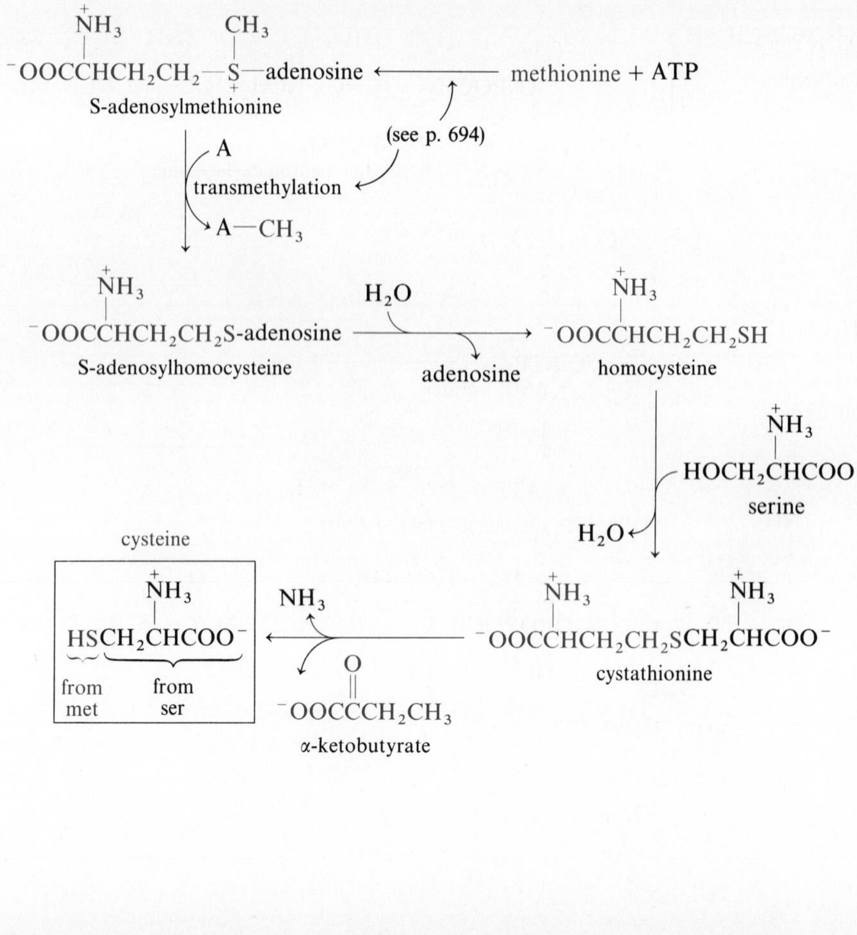

FIGURE 18–23 Pathway used by most organisms, including humans, to assemble cysteine from methionine (sulfur source) and serine (carbon source). Transmethylation reactions are described later in this chapter.

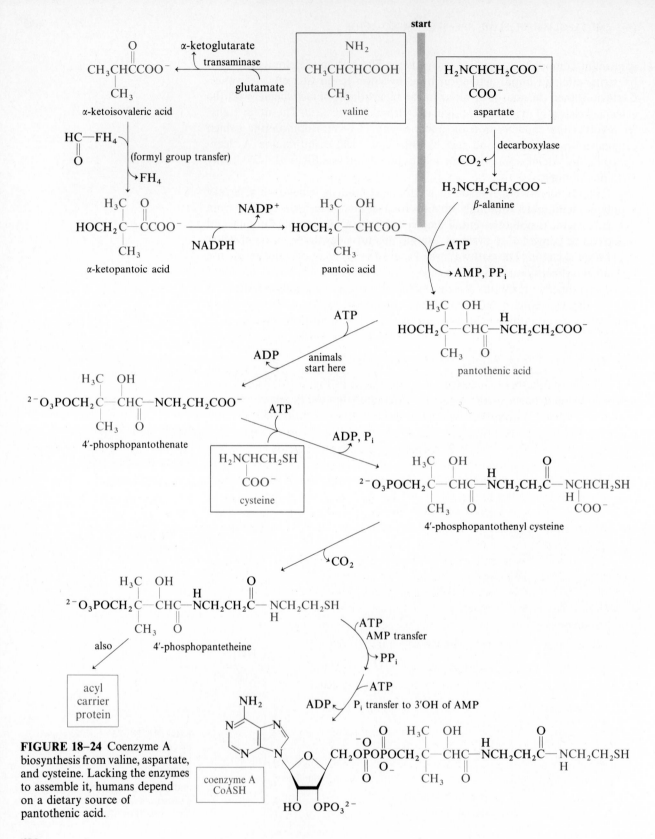

FIGURE 18–24 Coenzyme A biosynthesis from valine, aspartate, and cysteine. Lacking the enzymes to assemble it, humans depend on a dietary source of pantothenic acid.

690

imal organisms can perform only the conversions starting with pantothenic acid, explaining why pantothenic acid is an essential vitamin. Note that the phosphopantetheine intermediate will also be utilized in the formation of the acyl carrier protein (ACP–SH; see Section 17–2).

Serine and One-Carbon Folate Adducts

In addition to carboxylation reactions, another important group of transformations involving the transfer of a *one-carbon unit between two substrates* occurs in all types of cells. Included are reactions for the transfer of *methyl* ($-CH_3$), *hydroxymethyl* ($-CH_2OH$), *formyl* ($-CHO$), *methenyl* ($=CH-$), *methylene* ($-CH_2-$), and *formimino* ($-CH=NH$) groups. The most extensive type involves methyl group transfer, an area of metabolism involving participation of the amino acids serine and methionine.

The primary source (see Figure 18–25) from which these one-carbon units are obtained is *serine*. The biosynthesis of serine itself can be traced to 3-phosphoglycerate (an intermediate of glycolysis) via enzymes that by now you can readily identify. The entry of the $-CH_2OH$ carbon of serine into one-carbon metabolism involves the enzyme *serine hydroxymethyl transferase*. In this reaction the coenzyme *tetrahydrofolic acid* (FH_4) serves as a cosubstrate to which the $-CH_2OH$ is transferred, appearing in a folate adduct ultimately as a methylene (CH_2) group. Although our current focus is on the fate of the $-CH_2OH$ function, note that this reaction also explains the biosynthesis of *glycine,* which in turn has several metabolic functions (see Note 18–3).

Substance $N^5,N^{10}-CH_2-FH_4$ is only one example of one-carbon folate adducts. The others are identified in Figure 18–26. This display also identifies the enzyme-catalyzed steps responsible for the interconversion of these adducts and important metabolic functions of specific adducts in the *biosynthesis of methionine, purines,* and *pyrimidines.* Although serine is the major source of the one-carbon unit, a secondary source is formate, which is produced as a by-product in a small number of other reactions.

The connections to purine and pyrimidine biosynthesis are highlighted later in this chapter. Now we focus on the use of the $N^5-CH_3-FH_4$ adduct in methionine biosynthesis, from which the methyl group is disseminated into other structures.

Methionine: An Important Source of Methyl Groups

All of the following methyls have been encountered in previous chapters as specific examples of methylated structures in living cells:

- Each methyl group in the $(CH_3)_3$ of choline.
- Each methyl group in the $(CH_3)_3$ of carnitine.
- The methyl in epinephrine.
- The methyls in catecholamine degradation products.
- The methyl for modifying purine and pyrimidine bases in DNA and RNA.
- The methyl for the 5′ cap of eukaryotic messenger RNA.

NOTE 18–3

Glycine is used in the biosynthesis of the following:

Glutathione

Glycocholic acid

Creatine

Porphyrins

Purines

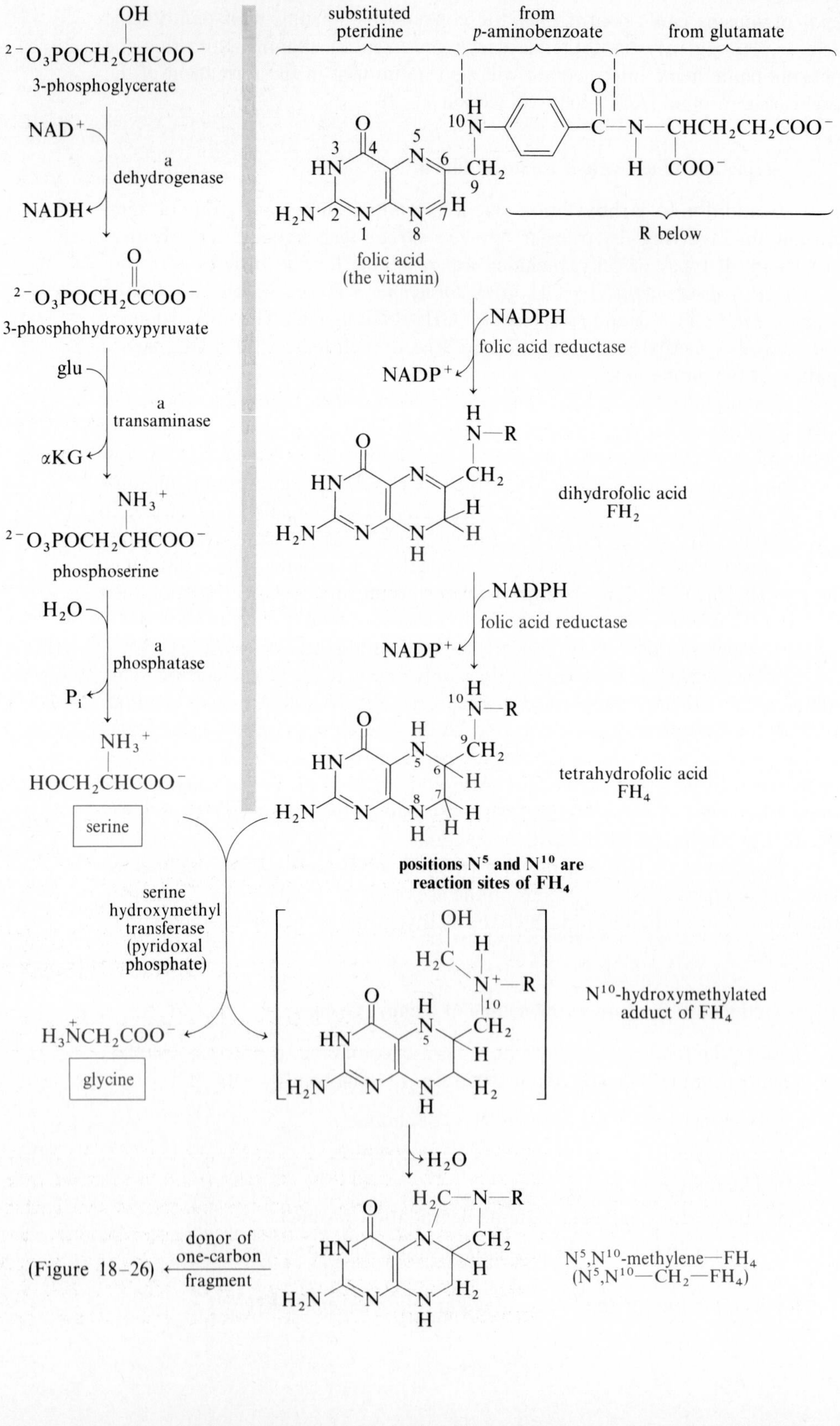

FIGURE 18–25 One-carbon transfer. The biosynthesis of serine is shown at the upper left. The structure of folic acid is given at the upper right. Mammalian cells use folate reductase for a two-step reduction to the tetrahydro FH₄ coenzyme species. The bottom of the figure shows the CH₂OH group of serine being transferred to FH₄. Further reactions involving one-carbon folate adducts are displayed in Figure 18–26.

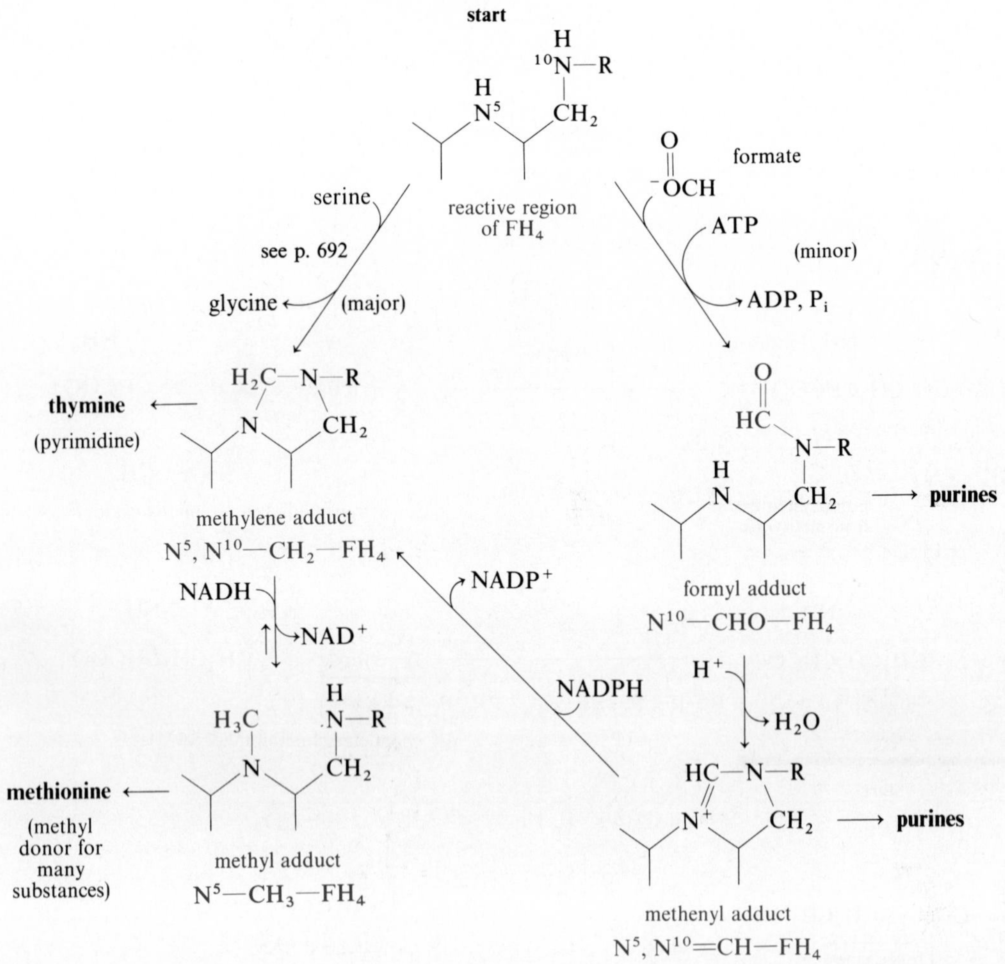

start

reactive region of FH$_4$

serine

see p. 692

glycine ← (major)

formate

ATP

(minor)

ADP, P$_i$

thymine ← (pyrimidine)

methylene adduct
N^5, N^{10}—CH$_2$—FH$_4$

NADH

NAD$^+$

NADP$^+$

purines

formyl adduct
N^{10}—CHO—FH$_4$

NADPH

H$^+$

H$_2$O

methionine ←
(methyl donor for many substances)

methyl adduct
N^5—CH$_3$—FH$_4$

purines

methenyl adduct
N^5, N^{10}=CH—FH$_4$

FIGURE 18–26 Summary of one-carbon folate adducts: formation, types of adducts, interconversions, and subsequent uses in the biosynthesis of purines, thymine, and methionine. Though not identified here by name, all reactions involve specific enzymes.

A common denominator for all these structures is that the *methyl groups are acquired from methionine.* The explanation begins with the reaction catalyzed by *homocysteine transmethylase* in which the N^5—CH$_3$—FH$_4$ adduct is used to methylate the SH function of homocysteine. (The participation of vitamin B$_{12}$ in this reaction is described in the next section.) The reaction is

$$\underset{\text{homocysteine}}{HSCH_2CH_2\overset{\overset{NH_3^+}{|}}{C}HCOO^-} + N^5-CH_3-FH_4 \xrightarrow[\text{(vitamin B}_{12}\text{)}]{\substack{\text{homocysteine} \\ \text{transmethylase}}} \underset{\text{methionine}}{H_3CSCH_2CH_2\overset{\overset{NH_3^+}{|}}{C}HCOO^-} + FH_4$$

To serve as an activated methyl donor, methionine first undergoes a reaction with ATP to convert its methyl sulfide function (H$_3$C—S—R) to a more reactive *methyl sulfonium* function

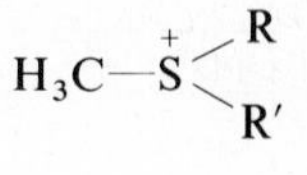

$$H_3C-\overset{+}{S}\begin{matrix} R \\ R' \end{matrix}$$

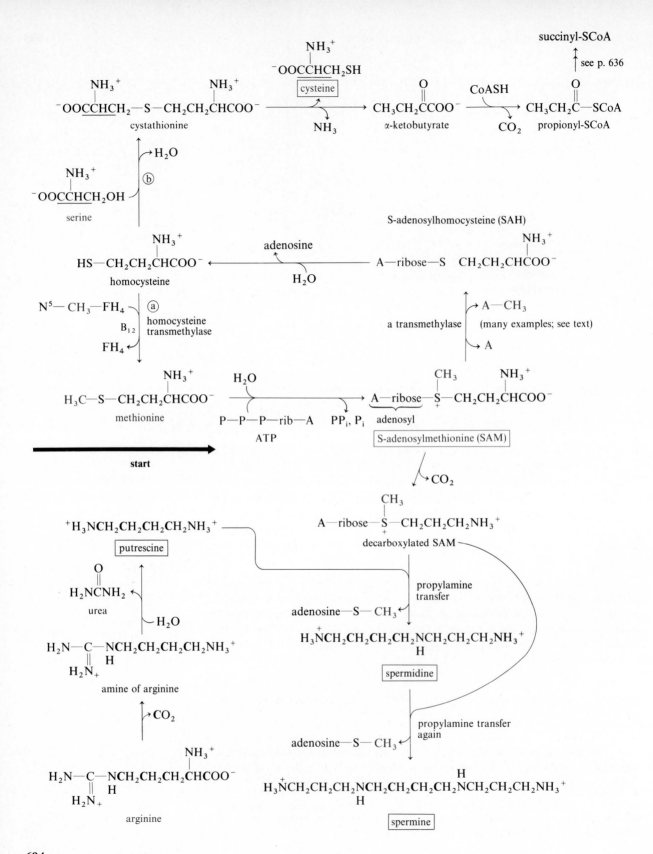

succinyl-SCoA

see p. 636

cysteine

cystathionine

α-ketobutyrate

propionyl-SCoA

CoASH

CO_2

serine

H_2O

ⓑ

S-adenosylhomocysteine (SAH)

adenosine

homocysteine

H_2O

A—ribose—S

N^5—CH_3—FH_4

B_{12}

ⓐ homocysteine transmethylase

FH_4

a transmethylase

A—CH_3 (many examples; see text)

A

methionine

H_2O

P—P—P—rib—A PP_i, P_i
ATP

adenosyl

S-adenosylmethionine (SAM)

start

CO_2

decarboxylated SAM

putrescine

urea

H_2O

amine of arginine

propylamine transfer

adenosine—S—CH_3

spermidine

CO_2

arginine

propylamine transfer again

adenosine—S—CH_3

spermine

The specific sulfonium product is *S-adenosylmethionine* (SAM) (see Figure 18–27), produced by the transfer of an adenosyl group from ATP to the S of methionine. With its active methyl SAM then serves as a universal methyl donor for several different transmethylase enzymes. For example, successive rounds of a SAM-dependent transmethylase operate to produce the $\overset{+}{R}N(CH_3)_3$ function in choline and carnitine.

The loss of the methyl group from SAM yields *S-adenosylhomocysteine* (SAH). Thereafter, SAH is acted on by a hydrolase to release free homocysteine, which can react again with N^5—CH_3—FH_4 to re-form methionine or act as a sulfur source for the biosynthesis of cysteine. For the cysteine route the carbon skeleton of cysteine is provided by serine, which condenses with homocysteine to form cystathionine. In turn, the cystathionine is cleaved to form cysteine and α-ketobutyrate. As in the pyruvate → acetyl-SCoA + CO_2 reaction, α-ketobutyrate can be oxidatively decarboxylated to *propionyl-SCoA*. Recall that propionyl-SCoA can then be converted to succinyl-SCoA via the intermediate formation of methylmalonyl-SCoA (see Figure 17–7).

Polyamines from Methionine and Arginine

Figure 18–27 also highlights another important metabolic role of methionine (via S-adenosylmethionine), linked in part to some aspects of *arginine metabolism,* namely, the biosynthesis of low–molecular weight *polyamines*. As for methionine, this phase of metabolism begins with a decarboxylation of SAM to yield a decarboxylated sulfonium species called S-adenosylmethamine. The arginine transformations include a decarboxylation and removal of the guanidine function as urea, with the latter reaction also accompanied by the formation of *putrescine* (1,4-diaminobutane). Methionine and arginine metabolism converge with the reaction of S-adenosylmethamine and putrescine, transferring a propylamine function from the former to the latter, to produce *spermidine*. A second propylamine transfer with spermidine and S-adenosylmethamine yields *spermine.*

At physiological pH 7 the three amines—putrescine, spermidine, and spermine—each carry positive charges, providing a basis for their binding interaction with DNA and possibly RNA. Polyamine binding will certainly contribute to the stabilization of the DNA double helix. Not yet certain is whether polyamine binding is also responsible in part for regulating the expression (transcription) of genes in DNA.

◄ **FIGURE 18–27** (facing page) Composite summary of some important aspects of methionine metabolism. Start your examination with the conversion of methionine to SAM, an active methyl sulfonium species. Upward from SAM are several transmethylation reactions involving SAM as a methyl donor for various substrates (A). The rest of the methionine structure is released as free homocysteine, which (a) can be remethylated to methionine or (b) be used as a source of sulfur in the assembly of cysteine from serine with the rest of the carbons channeled into succinyl-SCoA. Downward from SAM, in conjunction with two degradation reactions of arginine, there is a decarboxylated species of SAM serving as a donor of a propylamine grouping—all contributing to the formation of three polyamine species. Most of the individual enzymes are not identified.

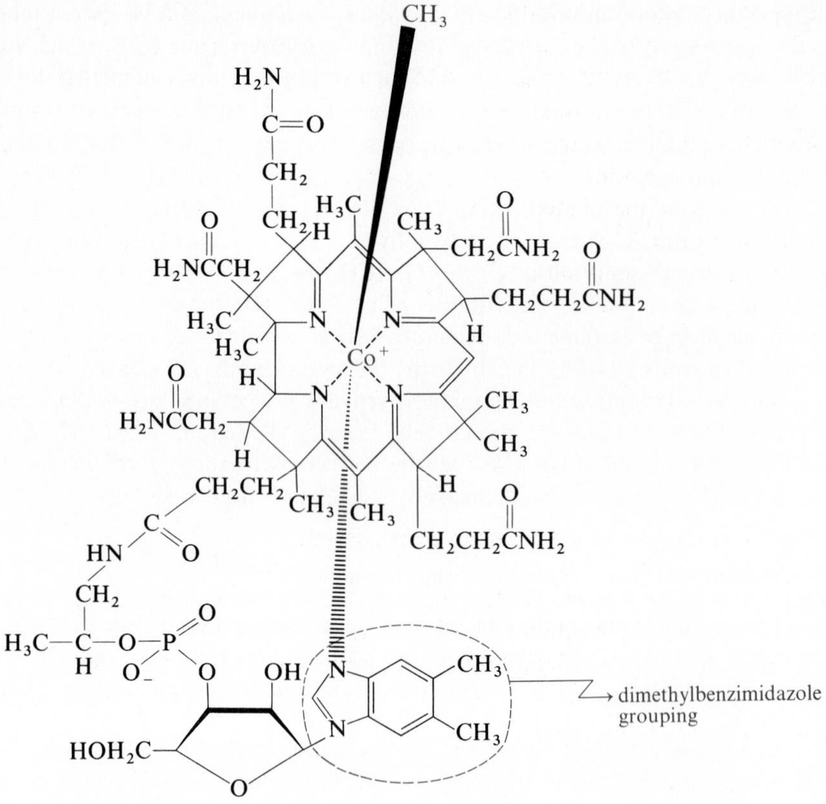

FIGURE 18–28 Methyl form of vitamin B_{12}. Other coenzyme forms of B_{12} have OH^-, H_2O, or deoxyadenosine (see Figure 18–29) in place of CH_3.

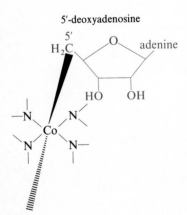

FIGURE 18–29 Deoxyadenosine form of vitamin B_{12}.

Vitamin B_{12} and B_{12}-Dependent Reactions

Since its discovery in 1948, vitamin B_{12} has attracted the interest and fascination of many researchers. It is a very complex substance in terms of both structure and biological function. Dorothy Hodgkin (Nobel Prize, 1964) spent seven years solving the structure, and knowledge of its participation in cellular processes is still developing. To date, about a dozen B_{12}-dependent reactions have been documented, most occurring in bacteria. Some of these reactions are described here. The daily intake requirement of B_{12} for humans is not known, but the condition of *pernicious anemia* brought on by a deficiency of the vitamin can be prevented by an intake of approximately 0.1 μg (0.0000001 g) per day. This low requirement in part reflects the small number of B_{12}-dependent reactions in humans but is also due to the presence of B_{12} supplied to the body by the intestinal bacteria capable of B_{12} synthesis.

The B_{12} structure is basically that of a tetrapyrrole compound, albeit a rather elaborate one (see Figure 18–28.) Compared with the more common tetrapyrrole grouping in heme, the distinctive features of B_{12} structures are as follows: (1) There is one fewer methenyl (—CH=) bridge between pyrroles, so two pyrroles are linked directly; to distinguish it from the prophyrin ring of heme, we call this system a *corrin ring;* (2) the corrin tetrapyrrole system is coordinated to *cobalt;* (3) one of the pyrrole side chains contains a *5,6-dimethylbenzimidazole* grouping that is also coordinated to the fifth position of the central cobalt; and (4) the pyrrole side chains consist of amide rather than acid groupings. Various coenzyme forms of B_{12} exist, differing in the sixth group coordinated to cobalt, which may be H_2O, OH^-, methyl, or deoxyadenosine (Figure 18–29). When B_{12} is isolated from fermentation broths, it is usually recovered with CN^- (cyanide) complexed to the sixth position of cobalt.

Two B_{12}-dependent reactions of importance in mammals are homocysteine → methionine and methylmalonyl-SCoA → succinyl-SCoA. In the first reaction B_{12} shuttles a methyl group from N^5—CH_3—FH_4 to homocysteine via the intermediate formation of methyl-B_{12} (see Figure 18–30).

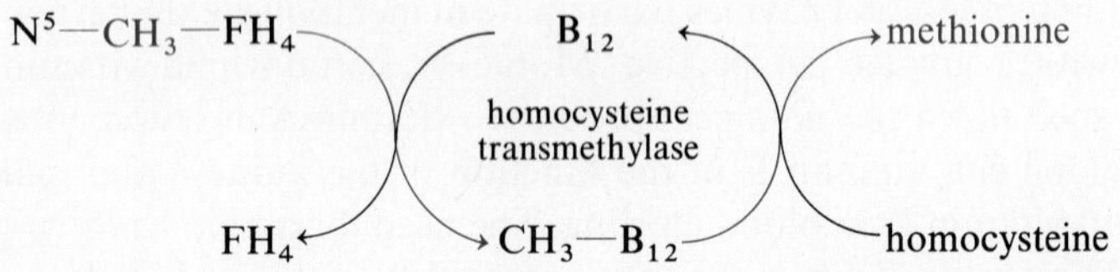

FIGURE 18–30 Methyl-B_{12} metabolism.

The isomerization of methylmalonyl-SCoA → succinyl-SCoA utilizes the deoxyadenosyl form of B_{12} (see Figure 18–31), which under the guidance of the enzyme participates in the chemistry of an intramolecular migration. (We will not examine the mechanism of this interesting chemistry.) Recall that this reaction is part of a set that accounts for the further metabolism of propionyl-SCoA, derived from the catabolism of some amino acids and fatty acids with an odd number of carbon atoms (see Section 17–1).

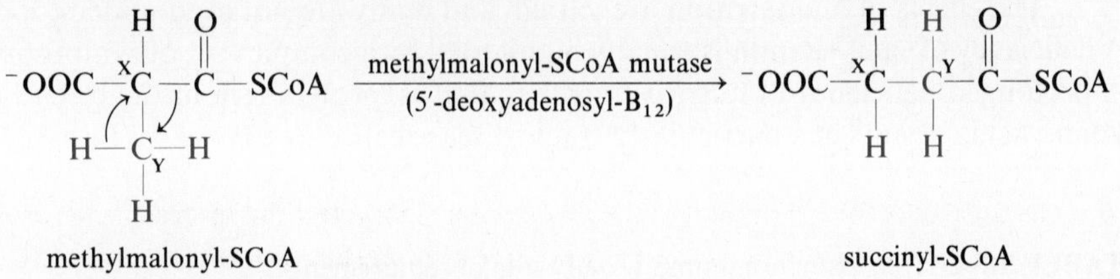

methylmalonyl-SCoA

succinyl-SCoA

FIGURE 18–31 Intramolecular rearrangement catalyzed by B_{12}-dependent mutase.

In some bacteria deoxyadenosyl-B_{12} is the coenzyme for ribonucleotide reductase, the enzyme that converts ribonucleotides to *deoxyribo*nucleotides, a reaction with obvious importance to DNA biosynthesis. We will discuss it later (see Section 18–7).

18–6 NUTRITIONAL CONSIDERATIONS

Obviously, every organism requires a supply of amino acids for normal growth and development. In many cases this requirement is satisfied by the organism itself owing to an enzymatic ability to synthesize each of the amino acids from other sources. Most microbes and the photosynthetic plants are prime examples. In fact, given supplies of carbon, energy, inorganic nitrogen, and varied inorganic nutrients, these organisms can produce not only all of the amino acids but also every other substance necessary for normal growth and development. Animal organisms—and humans in particular—are not quite so versatile. Many substances—the exact ones varying from one organism to another—must be supplied in the diet, since the organisms lack the necessary enzymes for their biosynthesis.

Two especially critical aspects of human nutrition concern an intake of the vitamins and several amino acids. As we have noted on several occasions throughout the last few chapters, the vitamins are required for many life processes in the area of general metabolism, where they function as coenzymes. At this point you may wish to return to our first treatment of coenzymes, where the relationship to vitamins was initially discussed (see Section 5–2). In view of your understanding of how coenzymes participate in metabolism, the review can now be made with a greater perspective. Moreover, certain lipid vitamins are involved in specialized physiological processes: vitamin A in vision, vitamin D in bone development, vitamin E in the function of the kidneys and male genital organs, and vitamin K in blood clotting. The need for amino acids is, of course, obvious. Thousands of different proteins must be synthesized, and as evidenced by the previous section, many amino acids have specialized metabolic functions. The major exogenous supply of amino acids is protein. The amino acids are made available during digestion by the action of the proteolytic enzymes of the gastrointestinal secretions: trypsin, chymotrypsin, pepsin, carboxypeptidase, and aminopeptidase. The free amino acids are then transported across the intestinal wall and carried throughout the body in the bloodstream.

An obvious question at this point is, Which of the 20 common amino acids are required by humans? The answer is provided in Table 18–2.

The effects of malnutrition are varied, and many are not even understood. A deficiency of each vitamin is usually manifested by a complex set of symptoms. A prolonged deficiency of nutritive protein, that is, protein rich in the essential amino acids, results in a particularly tragic disease called *kwashiorkor*. To satisfy

TABLE 18–2 Approximate minimum daily intake requirements[a] of essential amino acids[b]

L-AMINO ACID	INFANTS (mg/kg body weight)	MALE ADULT (g)	FEMALE ADULT[c] (g)
Histidine	30	0	0
Tryptophan	20	0.25	0.16
Phenylalanine[d]	90	1.1	0.22
Lysine	100	0.80	0.50
Threonine	90	0.50	0.31
Methionine	45	1.1	0.35
Leucine	150	1.1	0.62
Isoleucine	130	0.70	0.45
Valine	110	0.80	0.65

[a] Recommended levels are generally twice the minimal values.
[b] Small quantities of arginine are required under certain conditions.
[c] Greater intake of all acids is recommended during pregnancy and lactation.
[d] Since much of the physiological function of phenylalanine requires its conversion to tyrosine, roughly 75% of the phenylalanine requirement can be covered by tyrosine.

the need, the body tissues begin to degrade their own protein. The results are impaired development and function of all vital organs. An early death is quite common, particularly in infants feeding from a mother who herself is subsisting on a deficient diet. The hideous problem of malnutrition is widespread but is especially severe in developing countries. Recombinant DNA technology may soon provide some solutions: bacteria or algae able to convert waste material to edible and nutritious biomass, nutritive plants that grow in poor climates and without the need for expensive nitrogen fertilizers, and vaccines to protect livestock against killer viruses. More attention is also needed to harvesting the nutritive foods of the oceans.

18–7 NUCLEOTIDE METABOLISM

Biosynthesis of UMP and IMP

Nearly all organisms are capable of synthesizing pyrimidine and purine nucleotides in the same way from elementary substances such as CO_2, NH_3, aspartate, glycine, glutamine, and ribose. Two separate pathways are involved: one for pyrimidines, another for purines. The first nucleotide product of the pyrimidine pathway is *uridine-5'-monophosphate* (UMP), which then serves as the precursor of all other pyrimidine nucleotides. The first nucleotide product of the purine pathway is *inosine-5'-monophosphate* (IMP), which is then converted to all other purine nucleotides. See Figure 18–32.

Although the end product of each pathway is a monophosphonucleotide, it is the heterocyclic nitrogen base moiety that is actually being assembled: *uracil* in UMP and *hypoxanthine* (6-oxopurine) in IMP. The C and N ring atoms originate from various sources. However, in both processes the phosphoribose component is contributed intact by *5-phosphoribosyl-1-pyrophosphate* (PRPP) (see Figure 18–33), which in turn originates from ribose-5-phosphate and it in turn from glucose by such sources as the pentose pathway (see Section 13–9).

A summary of the assembly of the uracil and hypoxanthine ring structures follows. Obviously, the purine pathway is more involved, originating from more sources and requiring more steps. The sequences in which the reactions occur in each pathway are displayed in Figure 18–34 (for UMP) and Figure 18–35 (for IMP).

The net reactions of UMP and IMP biosynthesis are as follows:

$$CO_2 + \text{glutamine} + \text{aspartate} \xrightarrow[\text{PRPP} \quad PP_i \quad NAD^+ \quad NADH]{2ATP \qquad 2ADP, 2P_i} UMP + CO_2 + \text{glutamate}$$

$$CO_2 + 2(C_1—FH_4) + \text{glycine} + 2(\text{glutamine}) + \text{aspartate} \xrightarrow[\text{PRPP} \quad PP_i]{4ATP \quad 4ADP, 4P_i} IMP + 2(\text{glutamate}) + \text{fumarate}$$

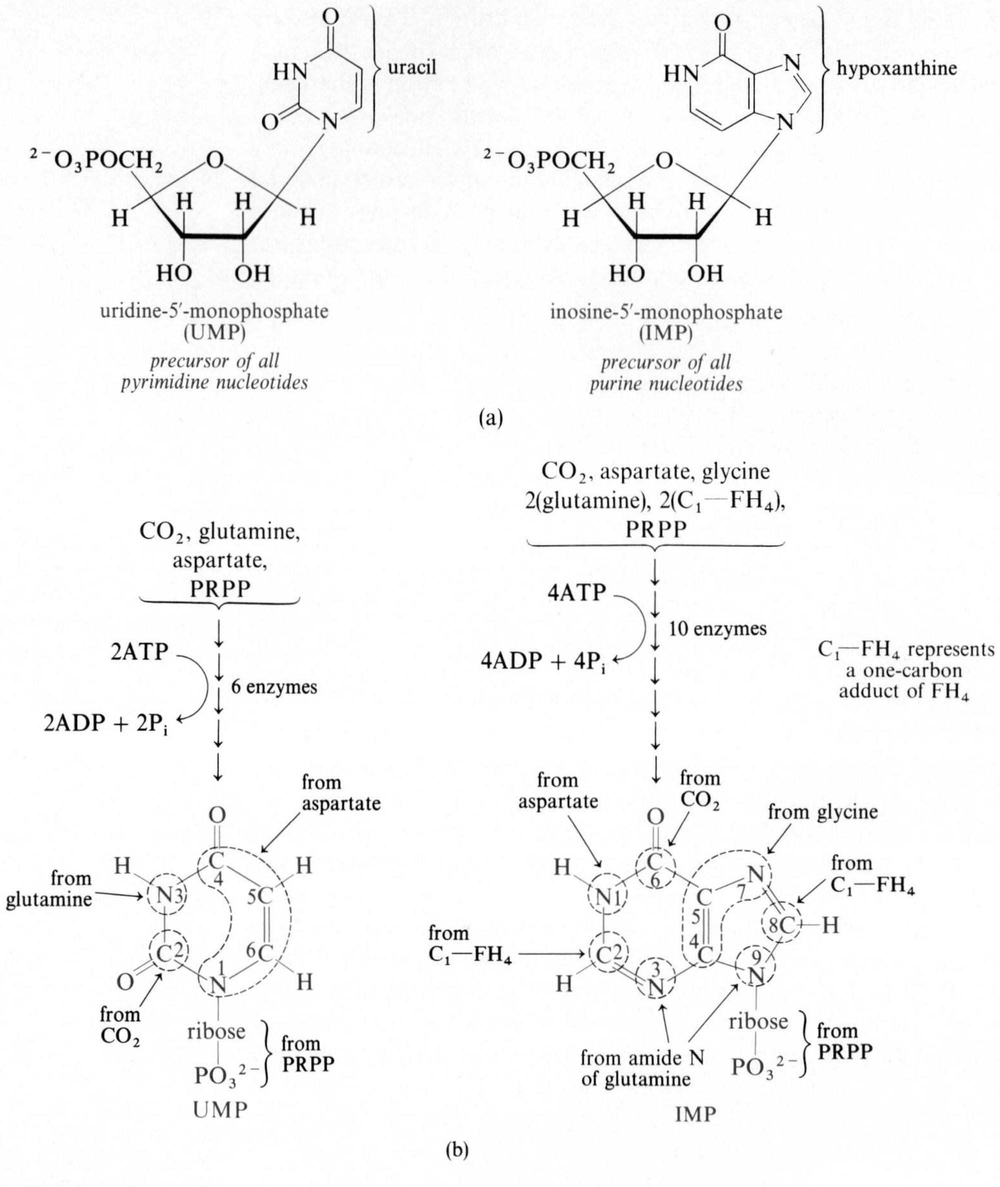

FIGURE 18–32 Biosynthesis of UMP and IMP. (a) Structures of UMP and IMP from which all pyrimidines and purine nucleotides are derived. (b) Biosynthetic origins of the uracil pyrimidine ring and the hypoxanthine purine ring. Details of biosynthesis are displayed in Figures 18–34 and 18–35.

The UMP biosynthesis (Figure 18–34) begins with the formation of *carbamyl phosphate* via transfer of an amino function from glutamine to CO_2 (as HCO_3^-). Located in the cytoplasm, the enzyme involved is *carbamyl phosphate synthetase II* (CPS II). Recall (Section 18–4) that a similar enzyme (CPS I) operates in mitochondria to start the urea cycle. Whereas CPS I uses NH_3 directly, the CPS II enzyme utilizes glutamine as a nitrogen source. Catalyzed by *aspartate transcarbamylase* (ATCase), the carbamyl phosphate then condenses with aspartate to form *carbamyl aspartate,* containing in noncyclic form all the C's and N's that will comprise the pyrimidine ring.

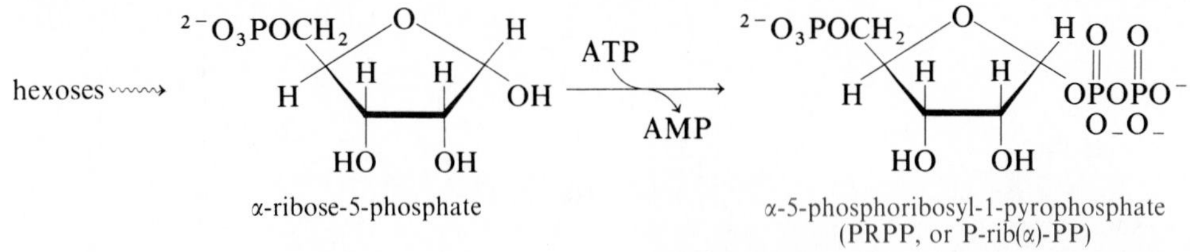

FIGURE 18–33 Formation of phosphoribosyl-1-pyrophosphate.

In the UMP pathway ATCase is the key regulatory enzyme, responding to strong allosteric inhibition by the pyrimidine nucleotide CTP and to allosteric activation by the purine nucleotide ATP. (Recall our previous consideration of ATCase as a model system, to introduce the principles of allosteric control of enzymes in Section 5–9.) An internal condensation then converts carbamyl aspartate into a cyclic product, *dihydroorotic acid,* which is then oxidized to give *orotic acid.* Formation of an intact nucleotide occurs with the condensation of orotic acid and PRPP, a condensation that occurs with an inversion of configuration at the anomeric carbon of ribose. A final decarboxylation yields UMP.

The IMP pathway (Figure 18–35) follows a different strategy, with the PRPP substrate being used immediately to anchor fragments of the purine ring as it is assembled. The first part positioned is an N atom from a glutamine

FIGURE 18–34 Complete pathway for the biosynthesis of the uracil pyrimidine ring and the parent pyrimidine nucleotide, UMP. Changes in the activity of aspartate transcarbamylase regulate the pathway in vivo.

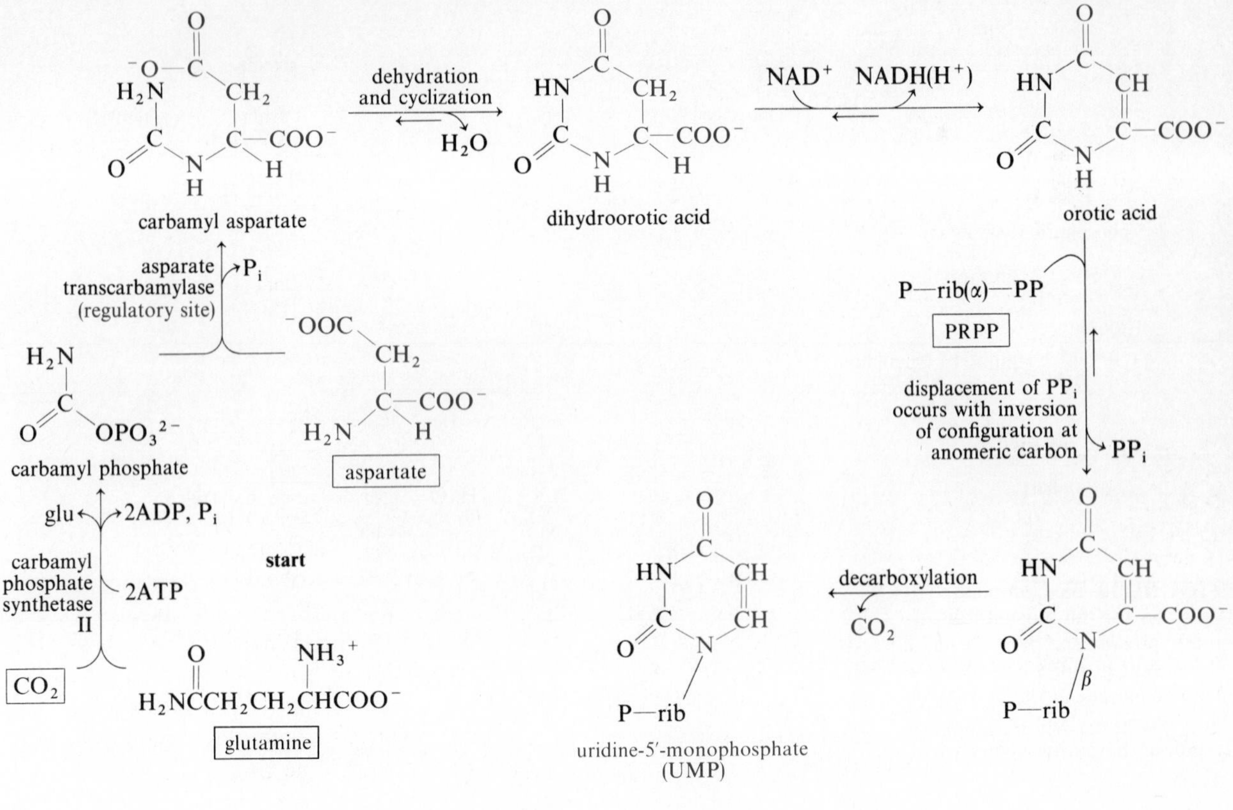

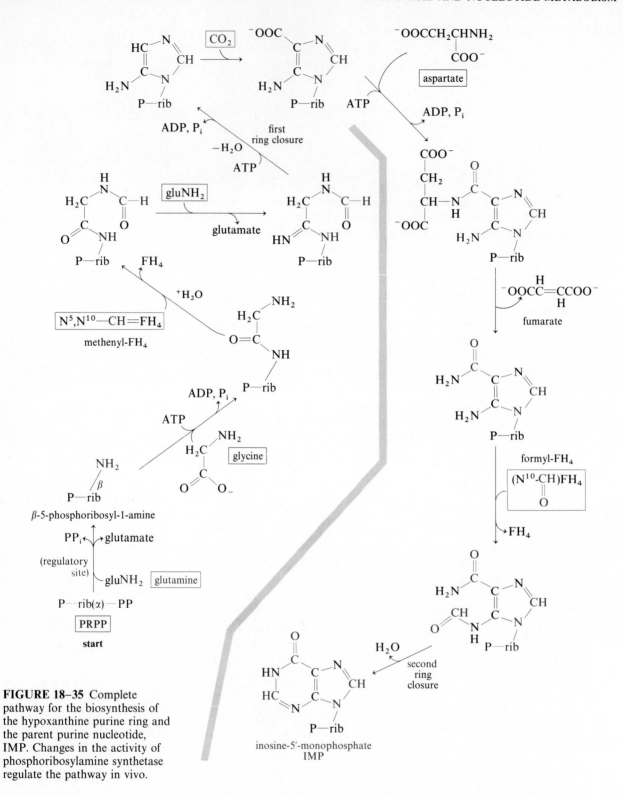

FIGURE 18–35 Complete pathway for the biosynthesis of the hypoxanthine purine ring and the parent purine nucleotide, IMP. Changes in the activity of phosphoribosylamine synthetase regulate the pathway in vivo.

donor, which eventually becomes N9 of the fully assembled purine ring. Thereafter, a progressive series of condensation reactions, including two ring-closing condensations, ultimately give rise to IMP. Note again that two of the purine C's originate from one-carbon folate adducts. A special role of a one-carbon folate adduct in pyrimidine biosynthesis is described later in the chapter.

Purine biosynthesis is also under metabolic control, with the key regulatory step being the first one responsible for forming phosphoribosylamine. Normal feedback inhibition signals include IMP, AMP, and GMP. Later in this chapter we will consider a disease condition that is related to purine overproduction caused in part by an absence of these normal feedback controls.

Biosynthesis of UTP, CTP, GTP, and ATP

The UMP and IMP biosynthesis represents only the first of three phases of nucleotide biosynthesis (see Figure 18–36). The second phase consists of the conversion of these monophosphonucleotide precursors to the *triphosphoribonucleotides:* UTP and CTP from UMP, and GTP and ATP from IMP. In addition to having specialized metabolic roles (such as UTP in carbohydrate metabolism and CTP in phospholipid metabolism), the four triphospho species are also the required substrates for the biosynthesis of RNA. The third phase involves the formation of the triphospho*deoxyribo*nucleotides required for DNA biosynthesis. The ribo → deoxy conversions and the formation of the third pyrimidine nucleotide, dTTP, will be discussed later.

We note here the difference between these reactions, particularly those involved with ATP formation, and reactions for the production of ATP by phosphorylation coupled to the electron transport process, or substrate-level phosphorylation reactions. They are distinct in two respects. First, the reactions to be discussed here are catalyzed by soluble enzymes in the cytoplasm rather than being associated with membranous systems such as mitochondria and chloroplasts. Second, the primary function of these reactions is to produce triphosphoribonucleotides, including ATP, for purposes other than supplying useful metabolic energy to the cell. Indeed, supplying energy would be rather foolish, since the eventual production of UTP, CTP, GTP, and ATP through UMP and IMP requires more energy than would be produced.

UTP AND CTP   The conversion of UMP to UTP proceeds via two successive ATP-dependent phosphorylations. *Nucleoside phosphate kinase* and *nucleoside diphosphate kinase* are the enzymes. We will refer to these enzymes as nucleotide kinases. Each enzyme displays a broad degree of specificity, acting equally well on any of the mono- or diphosphonucleotides, respectively:

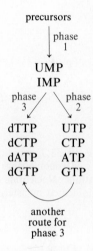

FIGURE 18–36 Brief outline of nucleotide biosynthesis.

$$\boxed{\text{UMP}} \xrightarrow[\text{ATP} \qquad \text{ADP}]{\substack{\text{nucleoside phosphate} \\ \text{kinase}}} \boxed{\text{UDP}} \xrightarrow[\text{ATP} \qquad \text{ADP}]{\substack{\text{nucleoside diphosphate} \\ \text{kinase}}} \boxed{\text{UTP}}$$

Some specific nucleotide kinases also exist.

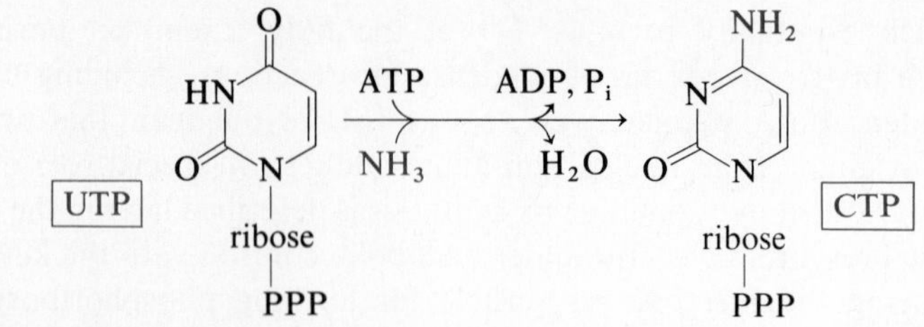

FIGURE 18–37 Amination of uracil to yield cytosine.

The CTP is then formed from UTP by amination at position 4 of the pyrimidine ring (see Figure 18–37), with ammonia as the nitrogen source. The enzyme catalyzing this endergonic reaction is likewise ATP-dependent and also requires GTP for optimum activity, suggesting an allosteric regulation. Indeed, this reaction is but one of several regulatory sites in nucleotide biosynthesis.

GTP AND ATP   The biosynthesis of GTP and ATP involves two different pathways from the same IMP precursor. Both are diagramed in Figure 18–38. In each case the hypoxanthine ring of IMP is first modified to yield the appropriate monophosphonucleotides, AMP and GMP, which are then phosphorylated (nucleotide kinase again) to ultimately yield ATP and GTP. Note that the C=O to C—NH_2 chemistry is different for adenine and guanine formation, and both are different from cytosine formation in UTP → CTP. In the formation of GMP the IMP precursor must first be oxidized (NAD^+) to produce C=O at position 2 to yield xanthosine-5′-monophosphate (XMP). Then glutamine serves as the N donor. In the AMP route aspartate is the N donor. For CTP (Figure 18–37) NH_3 is the N donor. However, each amination step requires an expenditure of energy: GTP in adenine formation and ATP in guanine formation.

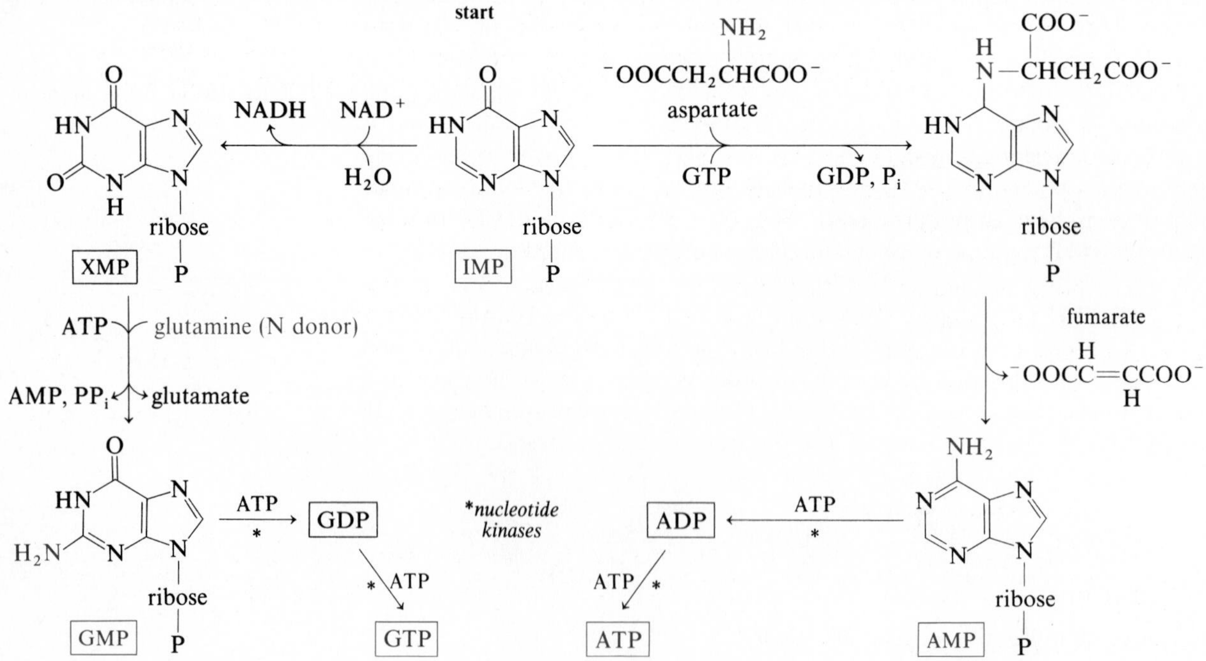

FIGURE 18–38 Formation of guanine and adenine rings from hypoxanthine. Nucleotide kinase activity indicated by asterisks.

Biosynthesis of dCTP, dGTP, dATP, and dTTP

The biosynthesis of DNA requires dCTP, dGTP, dATP, and dTTP as substrates. These deoxyribonucleotides are formed from their ribo counterparts by a process involving three proteins: *thioredoxin, thioredoxin reductase,* and *ribonucleotide reductase.*

Thioredoxin (TR) is a small protein (MW = 11,700 in *E. coli*) composed of a single polypeptide chain with one intrachain disulfide bond—hence TR-S$_2$ (see Figure 18–39). Thioredoxin reductase is an enzyme that catalyzes the NADPH-dependent reduction of —S—S— → —SH + HS— to yield reduced thioredoxin TR-(SH)$_2$. Ribonucleotide reductase is the enzyme catalyzing the reduction of ribo → 2′-deoxyribo, using TR-(SH)$_2$ as the reducing agent. Two different ribonucleotide reductases have been discovered. One contains non–heme iron and uses nucleoside diphosphates (NuDP) as substrates; this substance appears to be operating in mammalian tissues and in *E. coli*. The other reductase is a B$_{12}$-dependent enzyme and uses nucleoside triphosphates as substrates. However, both use TR-(SH)$_2$ as the electron donor.

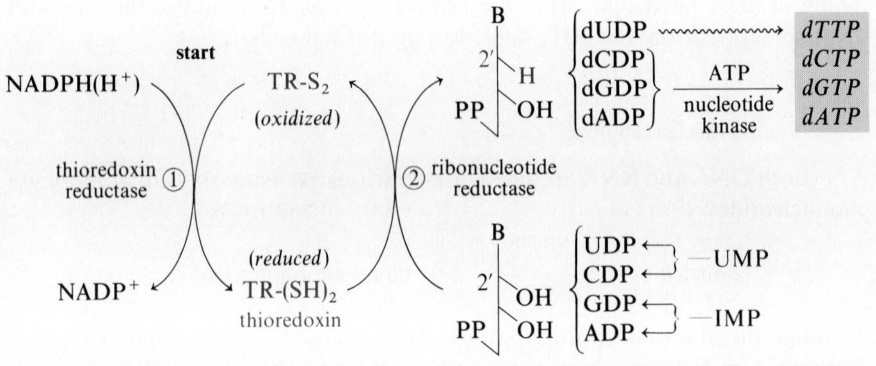

FIGURE 18–39 Reduction of ribonucleotides to 2′-deoxyribonucleotides. B represents nitrogen base.

When NuDPs are utilized, the products are dUDP, dCDP, dADP, and dGDP. The latter three are then phosphorylated (nucleotide kinase again) to yield dCTP, dATP, and dGTP. What happens to dUDP? As indicated in Figure 18–39, dUDP serves as the precursor for dTTP. The dUDP →→→ dTTP sequence is shown in Figure 18–40, highlighting the involvement of N^5,N^{10}—methylene—FH$_4$ as the methyl donor for the uracil → thymine conversion. This one-carbon transfer occurs after conversion of dUDP to dUMP and is followed by two phosphorylations.

FIGURE 18–40 Source of the methyl group in thymine. The anticancer drug, 5-fluorouracil, inhibits thymidylate synthetase.

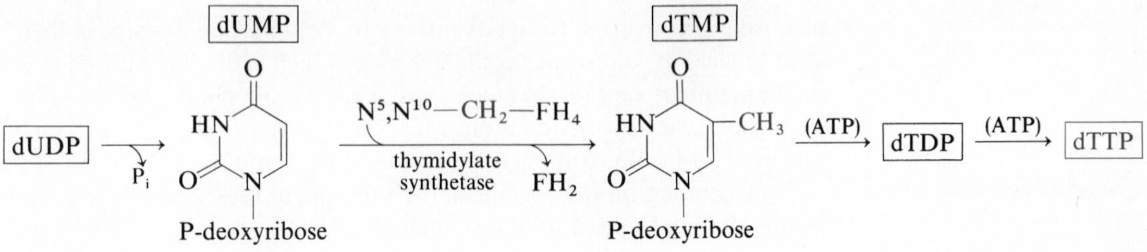

Anticancer Therapy

In the dUMP → dTMP conversion, note that dihydrofolic acid (FH$_2$) is produced as a consequence of using N^5,N^{10}-methylene—FH$_4$ as the source of the —CH$_3$ group. If normal metabolism is to be sustained, FH$_2$ (dihydrofolate) must be reduced back to FH$_4$ (tetrahydrofolate), a task performed (refer to Figure 18–25) by folate reductase and NADPH. Any inhibition of folate reductase would thus indirectly inhibit the dUMP → dTMP conversion and, of course, other FH$_4$-dependent processes. *Folate reductase inhibition* is in fact the basis of action for a group of antifolate drugs that have proven very effective as antitumor and anticancer agents. Two particularly effective drugs are *aminopterin* and *methotrexate*.

Thymidylate synthetase, the enzyme for dUMP → dTMP, is directly inhibited by *5-fluorouracil* and *5-fluoro-2'-deoxyuridine*. This specific inhibition of thymine production needed for DNA biosynthesis has also been exploited in the use of these substances as part of an anticancer chemotherapy treatment.

In rapidly growing and dividing cancer cells, the loss of thymine and FH$_4$ would arrest cell growth and even result in cell death. However, dosage and timing of these particular drugs are critical in order to minimize the effects of thymine deprivation and FH$_4$ depletion on normal cells.

Nucleotide Catabolism

Both DNA and RNA are acted on by various *nucleases* to yield monophosphonucleotides:

$$\text{nucleic acids} \xrightarrow[\text{and endonucleases}]{\substack{\text{combined action} \\ \text{of exonucleases}}} \text{monophosphonucleotides}$$

Although the monophosphonucleotides can be reused for RNA and DNA biosynthesis, our objective here is to consider their further degradation to other products. Because several types of reactions are associated with the degradation of deoxyribonucleotides, we will narrow our treatment even further by specifically examining the catabolism of ribonucleotides.

Purine Catabolism, Purine Salvage, and the Gouty Condition

PURINE CATABOLISM In most organisms the primary degradative pathways for GMP and AMP involve separate reactions that converge in the production of *xanthine* (see Figure 18–41). The conversion of guanine from GMP to xanthine is direct, whereas the formation of xanthine from adenine of AMP is indirect. After a dephosphorylation of AMP, the adenine moiety of adenosine is converted to hypoxanthine to yield *inosine*. Inosine is then hydrolyzed to yield *hypoxanthine* as the free base, which is finally oxidized to xanthine via the action of *xanthine oxidase*. This oxidase is interesting because it produces the highly toxic superoxide radical (O$_2^-$·) as a by-product. Here, then, is a definite need for the detoxifying role of superoxide dismutase (refer to Section 15–5).

Depending on the organism, the subsequent fate of xanthine is varied (see Figure 18–42). In most primates (including humans), birds, certain reptiles, and

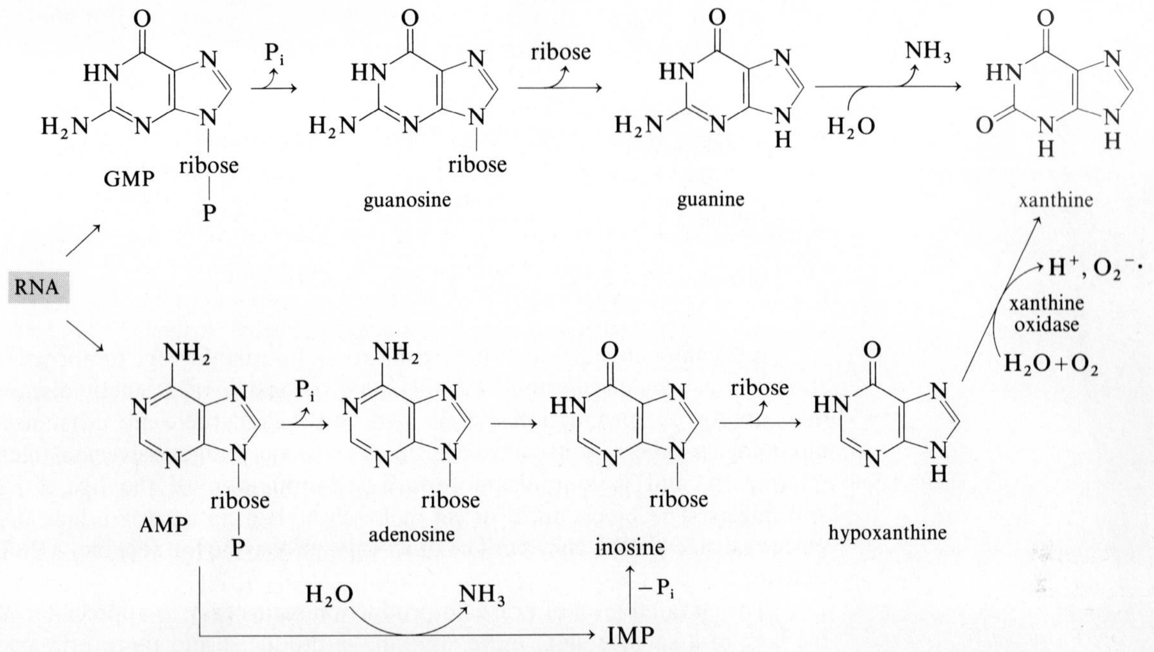

FIGURE 18–41 First stages of purine nucleotide degradation, which yield xanthine.

the majority of insects, xanthine is converted by the action (again) of xanthine oxidase to *uric acid,* which is excreted as the final end product of purine catabolism. In all other land animals *allantoin,* which is formed by the further oxidation of uric acid, is the final end product. In amphibians and fish allantoin is further degraded to *allantoic acid.* In many microorganisms allantoic acid is converted to *glyoxylate* and *urea.* All of these reactions are catalyzed by specific enzymes and clearly illustrate the themes of biochemical unity and diversity.

PURINE SALVAGE A large amount of ATP energy is expended to assemble the purine ring structure: 5ATP for the adenine of each AMP and 5ATP for the guanine of each GMP. Not surprising, then, is that cells contain enzymes that reclaim purine rings before their conversion to xanthine and then to uric acid for excretion. There are two such enzymes: *hypoxanthine-guanine phosphoribosyltransferase* (HPRT) and *adenine phosphoribosyltransferase* (APRT). These

FIGURE 18–42 Second stage of purine degradation.

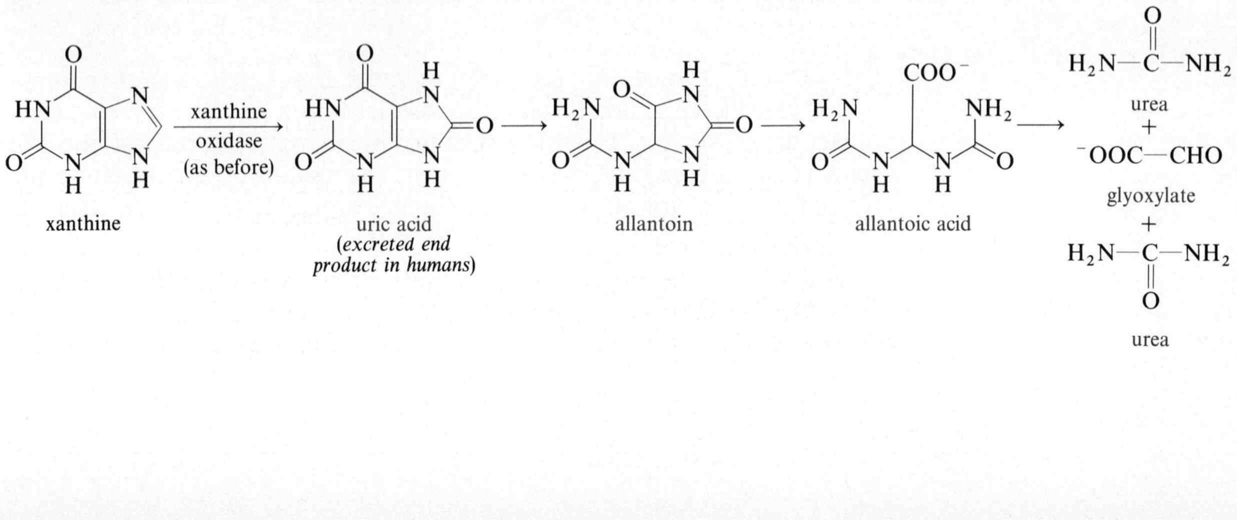

enzymes salvage free purines by converting them back to nucleotides, using PRPP as a donor of the phosphoribosyl grouping:

$$\text{guanine} + \text{PRPP} \xrightarrow[\text{phosphoribosyltransferase}]{\text{hypoxanthine-guanine}} \text{GMP} + \text{PP}_i$$

$$\text{hypoxanthine} + \text{PRPP} \xrightarrow{\text{same enzyme}} \text{IMP} + \text{PP}_i$$

$$\left.\right\} \text{not operative in Lesch-Nyhan disease}$$

$$\text{adenine} + \text{PRPP} \xrightarrow[\text{phosphoribosyltransferase}]{\text{adenine}} \text{AMP} + \text{PP}_i$$

A startling illustration of the significance of purine salvage to normal biochemistry is provided by the *Lesch-Nyhan syndrome,* a rare genetic disease in humans. An overproduction of uric acid occurs, and there are other neurophysiological aberrations such as excessive anxiety, aggressiveness, mental retardation, and a compulsion toward self-mutilation of the lips, tongue, and fingers. The biochemical defect in Lesch-Nyhan patients is a lack of the purine salvage HPRT enzyme. The other salvage enzyme for adenine, APRT, is present.

The elevated level of uric acid production seems easy to appreciate: With the lack of a salvage step, more xanthine is produced and more uric acid is formed. However, there is more to it than that. The failure to re-form GMP and IMP diminishes the normal regulatory control on the rate of production of purines. Earlier, remember we mentioned that the enzyme for PRPP + gln → P-rib-NH$_2$ + glu + PP$_i$ is a regulatory site of IMP biosynthesis. Substances GMP, IMP, and AMP are all feedback inhibitors of this enzyme. Normally, the HPRT enzyme, by re-forming GMP and IMP, would provide a metabolic signal to tone the activity of the regulatory enzyme (see Figure 18–43). In the Lesch-Nyhan individual the lack of HPRT eliminates this possibility and results in a greater rate of purine synthesis and an uninterrupted flow of the purines through xanthine to uric acid. What effects are felt in other areas of normal metabolism by this metabolic abnormality, and how are they related to the neurophysiological aberrations mentioned earlier? As yet, these questions remain unanswered.

The Lesch-Nyhan condition provides an obvious example of the complexity of metabolism and its control. In addition, we see an illustration of how one biochemical *regulatory defect* can severely disrupt normal body growth and development.

GOUT   Though the effect is not as severe as in Lesch-Nyhan syndrome, excess blood levels of uric acid are also associated with the more common hereditary disease of *gout.* The excess is caused by an overproduction of uric acid plus a failure of the kidneys to eliminate it. The result is the accumulation and crystallization of uric acid in the synovial fluid around bone joints (particularly in the big toe), resulting in intense pain. Many gout patients are treated with *allopurinol* (see Figure 18–44), which reduces uric acid production by acting as a *competitive inhibitor of xanthine oxidase.*

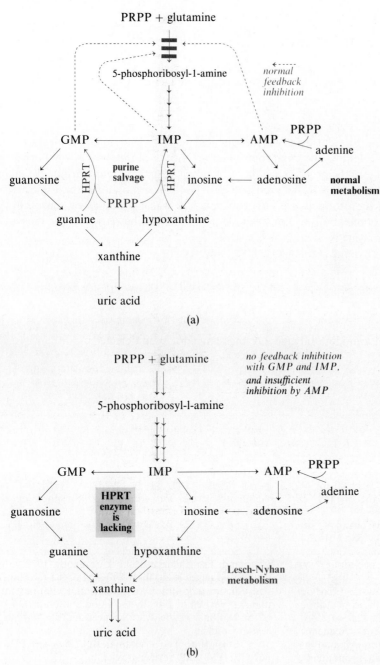

(a)

(b)

FIGURE 18–43 Purine metabolism. (a) Summary of the normal anabolic and catabolic pathways of purine metabolism. (b) Abnormal overproduction (color arrows) of uric acid in Lesch-Nyhan disease. The overproduction of uric acid stems from an inability to depress IMP biosynthesis. Enzyme HPRT is the salvage enzyme for guanine and hypoxanthine.

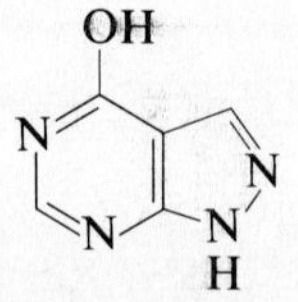

FIGURE 18–44 Allopurinol, an oral drug to treat gout.

Pyrimidine Catabolism

The catabolism of pyrimidine bases originating from pyrimidine-containing nucleotides apparently proceeds through one of several different pathways, depending on the organism. For example, one of the major pathways, occurring in humans and other animals, involves the reduction of uracil or

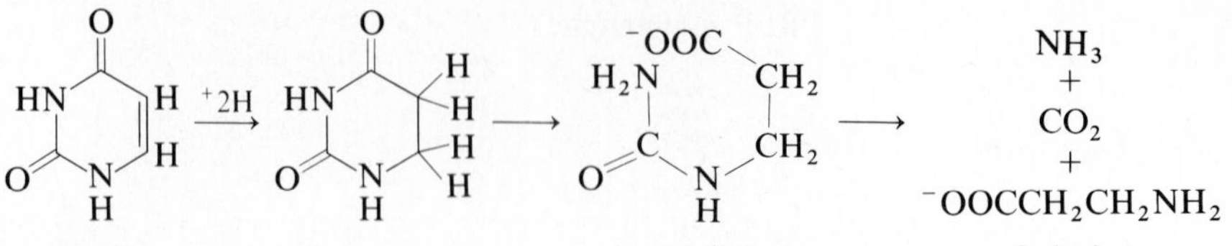

FIGURE 18–45 Catabolism of uracil.

thymine to yield a fully hydrogenated heterocyclic ring. The degradation of cytosine follows the same route, after an initial deamination to yield uracil. Ring cleavage of the product produced by uracil (cytosine) yields carbamyl-β-alanine, which is further hydrolyzed to CO_2, NH_3, and β-alanine (see Figure 18–45). All of the products are either excreted as waste products or reused in other areas of metabolism. For example, β-alanine can be reused in the biosynthesis of coenzyme A.

Thymine yields CO_2, NH_3, and

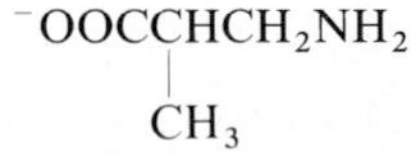

$$^-OOCCHCH_2NH_2$$
$$\quad | $$
$$\quad CH_3$$

Pyrimidine salvage enzymes have not been detected.

LITERATURE

BENKOVIC, S. J., and C. M. TATUM. "Mechanisms of Folate Cofactors." *Trends Biochem. Sci.,* **2**, 161–163 (1977). A short review article.

BLAKELY, R. L., and E. VITROLS. "The Control of Nucleotide Biosynthesis." *Annu. Rev. Biochem.,* **38**, 210–224 (1968). A review article summarizing the many regulatory enzymes associated with the pathways of nucleotide biosynthesis.

GREENBERG, D. M., and W. W. RODWELL. "Carbon Catabolism of Amino Acids" and "Biosynthesis of Amino Acids and Related Compounds." In *Metabolic Pathways,* 3rd ed., edited by D. Greenberg, vol. 3. New York: Academic Press, 1969. Two review articles giving a thorough analysis of catabolic and anabolic pathways of all the amino acids. Other articles in this volume are devoted to nitrogen metabolism of amino acids, sulfur metabolism, and metabolism of porphyrin compounds.

MEISTER, A. *Biochemistry of the Amino Acids.* 2nd ed. New York: Academic Press, 1965. A treatise in two volumes containing a comprehensive treatment of amino acid metabolism.

———. "On the Enzymology of Amino Acid Transport." *Science,* **180**, 33–39 (1973). A description of the γ-glutamyl cycle for membrane transport of amino acids.

SAFRANY, D. R. "Nitrogen Fixation." *Sci. Am.,* **231**, 64 (1974).

SOBER, H. A., ed. *Handbook of Biochemistry.* 2nd ed. Cleveland: The Chemical Rubber Company, 1970. A listing is given on pages K–50 through K–52 of the recommended daily dietary allowances of protein and vitamins for humans, as revised in 1968 by the Food and Nutrition Board of the National Academy of Sciences and the National Research Council.

STADTMAN, T. C. "Vitamin B_{12}." *Science,* **171**, 859–867 (1971). A good review article.

STREICHER, S. L., and R. C. VALENTINE. "Comparative Biochemistry of Nitrogen Fixation." *Annu. Rev. Biochem.,* **42**, 279–302 (1973).

UMBARGER, H. E. "Regulation of Amino Acid Metabolism." *Annu. Rev. Biochem.,* **38**, 323–370 (1969). A review article describing metabolite-mediated regulation of enzyme activity in multifunctional pathways, in general, and pathways of amino acid metabolism, in particular.

EXERCISES

18–1. Complete each of the following reactions:

(a) arginine + α-ketoglutarate $\xrightarrow{\text{transaminase}}$

(b) valine + α-ketoglutarate $\xrightarrow{\text{transaminase}}$

(c) L-lysine + FMN $\xrightarrow[\text{oxidase}]{\text{amino acid}}$

(d) cysteine + α-ketoglutarate $\xrightarrow{\text{transaminase}}$

(e) leucine $\xrightarrow[\text{decarboxylase}]{\text{leucine}}$

(g) pyruvate + NADH + H$^+$ + NH$_3$ $\xrightarrow{\text{dehydrogenase}}$

18–2. What relationship exists between the reactions given below, known to occur in liver, and the lack of a dietary requirement for arginine in adults?

18–3. Propose a series of reactions to explain how the carbon atoms of glucose could be converted to γ-aminobutyrate.

18–4. Propose a series of reactions to explain how a carbon atom of tryptophan could eventually be utilized in the biosynthesis of epinephrine. What specific carbon atom of tryptophan is involved?

18–5. In many organisms the immediate biosynthetic precursor of L-lysine is α, ε-diaminopimelic acid (structure below). What type of enzyme would catalyze this reaction; what coenzyme would be required; and what type of enzyme-substrate complex would be formed?

$$^-\text{OOCCHCH}_2\text{CH}_2\text{CH}_2\text{CHCOO}^- \xrightarrow{E} \text{L-lysine}$$
$$\quad\;\; |\qquad\qquad\qquad\quad |$$
$$\quad\;\; \text{NH}_3^+ \qquad\qquad\quad\;\; \text{NH}_3^+$$

α, ε-diaminopimelic acid

18–6. Write a net reaction for the operation of the urea cycle, linking it to ancillary reactions of the citric acid cycle and glutamate dehydrogenase.

18–7. Nothing would be gained if homocysteine were converted to methionine with S-adenosylmethionine functioning as the methyl donor. Explain.

18–8. The catabolism of valine involves (in the order given) (1) a transamination to give its alpha keto acid, A; (2) oxidative decarboxylation of A in the presence of CoASH to yield a four-carbon thioester, B; (3) dehydrogenation of B to yield an unsaturated derivative, C; (4) hydration of C to yield a β-hydroxy thioester, D; (5) hydrolysis of D to give the free acid, E; (6) an NAD$^+$-dependent oxidation of the —CH$_2$OH group in E to yield F, which contains —CHO; (7) conversion of F to methylmalonyl-SCoA; and (8) the last step, catalyzed by methylmalonyl-SCoA mutase. Illustrate the pathway, giving structures of all the intermediates.

18–9. Tryptophan is known to be converted into indole acetic acid, a potent growth hormone in plants, by the action of tryptophan decarboxylase, monamine oxidase, and an NAD-dependent dehydrogenase, in that order. Reproduce the complete pathway, showing the structures of all intermediate metabolites.

indole acetate

18–10. Propose a reaction sequence to account for the utilization of carbon from fatty acids in the biosynthesis of porphyrin compounds.

18–11. Mastocytosis is a physiological disorder resulting from the infiltration of mast cells into skin and such organs as the liver, spleen, and kidney. Clinically, a patient with this condition may possess an enlarged liver and spleen containing unusually high amounts of histamine. The latter is suggested to be caused by an excessive production of histamine as a result of the invading mast cells, rather than by a block in the degradation of histamine, which would likewise result in its accumulation. If this suggestion is so, what enzyme would you expect to find in large amounts in mast cells?

APPENDIXES

APPENDIX I   QUANTITATIVE MEASUREMENTS FROM LIGHT ABSORPTION

The pattern of energy absorption by a substance when light of varying wavelength passes through it is uniquely characteristic of the substance. The pattern is called an absorption spectrum and is useful in identification of structure. The spectrum may be made in the ultraviolet (UV) region (200–400 nm), in the visible (VIS) region (400–800 nm), or in the infrared (IR) region (800 nm–50 μm). The complexity of the spectrum depends on the nature of the substance and the light region examined. Individual peaks in the spectrum represent wavelengths of light at which absorption occurs, with the largest peak referred to as a wavelength of maximum absorption. Figure AI–1 displays the UV spectra of two different forms of the substance nicotinamide adenine dinucleotide (NAD). NAD^+ is an oxidized form; NADH is a reduced form. Despite great similarity in their structures (see Figure 12–18), each gives a distinct spectrum. Both absorb maximally at 260 nm, but the absorption by NAD^+ is greater. The most distinct difference is at 340 nm where NADH does absorb and NAD^+ does not.

The amount of light absorption is directly related to the concentration of the absorbing substance.

This correlation was first observed over one hundred years ago by Beer and Lambert. When light of wavelength λ and intensity I_0 passes through an absorbing sample of concentration c in a tube having a light path l, absorption will reduce the intensity of the transmitted light, I_T. See Figure AI–2.

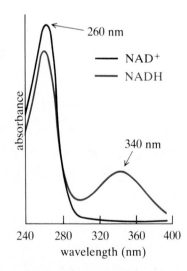

FIGURE AI–1 Ultraviolet absorption spectra of equimolar solutions of oxidized (NAD^+) and reduced (NADH) nicotinamide adenine dinucleotide.

712

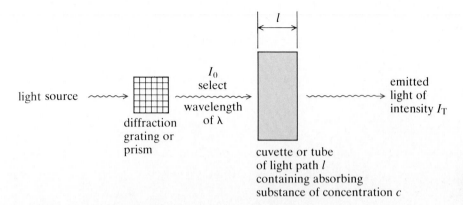

FIGURE AI–2 Absorption of monochromatic radiation.

The ratio of I_T/I_0 is called the *transmittance* T of the system. Beer and Lambert showed that

$$-\log \frac{I_0}{I_T} = -\log T \propto cl$$

Defining $-\log T$ as the *absorbance A,* and defining the proportionality constant as the extinction coefficient ε, we have the Beer-Lambert relationship:

$$A \equiv -\log T = \varepsilon cl$$

The relationship between absorbance and concentration is linear (see Figure AI–3):

$$y = mx$$

where $y = A$, $x = c$, and $m = \varepsilon l$.

Absorbance values are measured in an instrument called a spectrophotometer (also called a colorimeter). In practice, unknown concentrations of a substance are measured by comparing the absorbance of the unknown with the absorbance of known concentrations of the same substance measured under the exact same conditions: same l (usually 1 cm), same λ, same temperature, same solvent, same instrument, and so on. For obvious reasons absorption is usually measured at a wavelength of maximum absorption. A plot of absorbance values for known concentrations of the substance is called a *standard graph,* from which the unknown concentration can be obtained (see Figure AI–4). The unknown concentration can be read from the graph or be calculated from $c = A/\varepsilon l$ after evaluating the slope of the line, which is equal to εl.

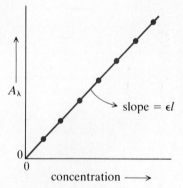

FIGURE AI–3 Linear relationship of the Beer-Lambert law. If one knows the value of l (usually 1 cm) and the slope, the value of the extinction coefficient ε can be calculated.

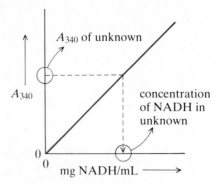

FIGURE AI–4 Graphical evaluation of unknown concentration of NADH from standard plot.

Absorption measurements in the visible region require that the sample be colored. If it is not naturally colored, it can be reacted with another reagent to produce a colored product. Table AI–1 summarizes some of the procedures used to produce color with substances routinely encountered in the biochemistry laboratory.

TABLE AI–1 Color-producing reactions in biochemical use

SUBSTANCE TO BE MEASURED	COLOR REAGENT	RESULT	COMMENT
Protein	Cu^{++} in strong base (Biuret reagent)	Blue-purple color, with maximum absorption at about 550 nm	Cu^{++} complexes to peptide bonds
Protein	Cu^{++} in strong base and phosphomolyb-dotungstic acid (Lowry reagent)	Blue-purple color, with maximum absorption at about 500 nm	Cu^{++} complexes to peptide bonds, and phosphomolybdo-tungstic acid reacts with tyrosine; more sensitive than Biuret reagent
Protein	Coomassie Brilliant Blue R; a dye material[a]	Color with maximum absorption at about 595 nm	Dye binds to protein; easier assay and more sensitive than Lowry reagent
RNA	Orcinol	Green color, with maximum absorption at 660 nm	
DNA	Diphenylamine	Blue color, with maximum absorption at 600 nm	
Inorganic phosphate	Molybdate/acid solution (Fiske-Subbarow reagent)	Blue color, with maximum absorption at 600 nm	

Note: Protein and nucleic acid can also be detected by measuring light absorption in the UV region. No chemical pretreatment is required. Proteins absorb at 280 nm (due to the presence of the phe, tyr, and trp), and nucleic acids absorb at 260 nm.

[a] The reagent is marketed under the name of BIO–RAD Protein Assay Reagent by BIO–RAD Laboratories.

APPENDIX II   RADIOISOTOPES

The use of radioactive isotopes (1) to detect extremely minute levels of activity with cells or purified substances, (2) to observe a particular process occurring simultaneously with many other processes, (3) to trace the interconversion of substances involving one or more reactions, and (4) to perform experiments of the types in 1, 2, and 3 with a high degree of quantitative precision and accuracy is indispensable to biochemical research.

Radioactive isotopes are unstable forms of elements that spontaneously disintegrate to more stable nuclear arrangements. Specific isotopes undergo this disintegration, also called radioactive decay, at very characteristic rates. The rate of disintegration is expressed as the *half-life* ($t_{1/2}$), the amount of time required for 50% of the material to decay. Disintegration is accompanied by the release of radiation: *alpha* (α) *radiation, beta* (β) *radiation, gamma* (γ) *radiation,* or a combination of these. The type of radiation emitted and its energy content are also characteristic of each radioactive isotope.

Radioisotopes that have extensive application in biochemistry are carbon 14, hydrogen 3 (tritium), sulfur 35, phosphorus 32, and phosphorus 33. All are pure β emitters (see Table AII–1). Radioactive forms of iodine are also valuable. Radioactive forms of oxygen and nitrogen are known, but their practical use is obviated because they decay very rapidly. Instead, the stable isotopic forms ^{18}O and ^{15}N are used, and detection and measurement are based on the heavier atomic masses. The same use is also made of deuterium (^2H), the stable isotope of hydrogen, and ^{13}C, a stable isotope of carbon.

TABLE AII–1 Comparison of radioisotopic forms of the primary biological elements

ISOTOPE	RADIATION EMITTED	HALF-LIFE	E (MeV)	NATURALLY ABUNDANT STABLE ISOTOPE (% OCCURRENCE)
^{14}C	β	5770 years	0.156	^{12}C (98.89%)
^3H	β	12.26 years	0.0186	^1H (99.985%)
^{35}S	β	86.7 days	0.167	^{32}S (95.0%)
^{32}P	β	14.3 days	1.71	^{31}P (100%)
^{33}P	β	25 days	0.25	
^{125}I	γ	57.4 days	0.035	^{127}I (100%)
^{131}I	β, γ	8.05 days	0.6 (β); 0.36 (γ)	
^{19}O	β, γ	29 s	3.25 (β); 0.20 (γ)	^{16}O (99.76%)
^{16}N	β, γ	7.35 s	4.3 (β); 6.13 (γ)	^{14}N (99.63%)

Note: Isotopes differ not only in relative instability but also in the energy value E of the emitted radiation. The energy content is directly correlated to hazardous biological effects as well as the efficiency with which the radiation can be detected (1 MeV = 3.83×10^{-14} cal).

Radioactivity can be detected and quantitatively measured by various procedures, such as Geiger-Müller (GM) detection, liquid scintillation, solid scintillation, and radioautography. The GM and scintillation systems convert the detection of radiation into pulses of current that are registered electronically. Each pulse of current is termed a *count,* with collection commonly expressed on the basis of 1 min—hence the term *count per minute* (cpm). If the instrument is capable of 100% efficiency, a single count corresponds to the disintegration (d) of a single radioactive atom, and cpm = dpm. Counting efficiency varies for different isotopes and for different counting procedures. The β counting is best done via liquid scintillation, which can achieve a 50%–60% efficiency even for the weak (low-energy) ^3H isotope. The ^{14}C efficiency is even higher, and ^{35}S and ^{32}P efficiencies are essentially 100%.

The formal expression for the amount of radioactivity is in terms of the unit *curie* (Ci), after Marie Curie who discovered the phenomenon of radioactivity. One curie is defined as 2.2×10^{12} disintegrations per minute. Sublevel amounts are expressed as millicuries (mCi) and microcuries (μCi). A more useful expression of radioactivity levels is the specific activity of a substance: the amount of radioactivity per unit amount of substance (such as Ci/mole, mCi/mmole, and μCi/mg).

An excellent treatment for students on the use of radioisotopes in biochemistry is contained in *Experimental Biochemistry* (2nd ed.), by J. M. Clark and R. L. Switzer (Freeman, 1977).

APPENDIX III   TABLE OF LOGARITHMS

N	0	1	2	3	4	5	6	7	8	9	N	0	1	2	3	4	5	6	7	8	9
10	0000	0043	0086	0128	0170	0212	0253	0294	0334	0374	55	7404	7412	7419	7427	7435	7443	7451	7459	7466	7474
11	0414	0453	0492	0531	0569	0607	0645	0682	0719	0755	56	7482	7490	7497	7505	7513	7520	7528	7536	7543	7551
12	0792	0828	0864	0899	0934	0969	1004	1038	1072	1106	57	7559	7566	7574	7582	7589	7597	7604	7612	7619	7627
13	1139	1173	1206	1239	1271	1303	1335	1367	1399	1430	58	7634	7642	7649	7657	7664	7672	7679	7686	7694	7701
14	1461	1492	1523	1553	1584	1614	1644	1673	1703	1732	59	7709	7716	7723	7731	7738	7745	7752	7760	7767	7774
15	1761	1790	1818	1847	1875	1903	1931	1959	1987	2014	60	7782	7789	7796	7803	7810	7818	7825	7832	7839	7846
16	2041	2068	2095	2122	2148	2175	2201	2227	2253	2279	61	7853	7860	7868	7875	7882	7889	7896	7903	7910	7917
17	2304	2330	2355	2380	2405	2430	2455	2480	2504	2529	62	7924	7931	7938	7945	7952	7959	7966	7973	7980	7987
18	2533	2577	2601	2625	2648	2672	2695	2718	2742	2765	63	7993	8000	8007	8014	8021	8028	8035	8041	8048	8055
19	2788	2810	2833	2856	2878	2900	2923	2945	2967	2989	64	8062	8069	8075	8082	8089	8096	8102	8109	8116	8122
20	3010	3032	3054	3075	3096	3118	3139	3160	3181	3201	65	8129	8136	8142	8149	8156	8162	8169	8176	8182	8189
21	3222	3243	3263	3284	3304	3324	3345	3365	3385	3404	66	8195	8202	8209	8215	8222	8228	8235	8241	8248	8254
22	3424	3444	3464	3483	3502	3522	3541	3560	3579	3598	67	8261	8267	8274	8280	8287	8293	8299	8306	8312	8319
23	3617	3636	3655	3674	3692	3711	3729	3747	3766	3784	68	8325	8331	8338	8344	8351	8357	8363	8370	8376	8382
24	3802	3820	3838	3856	3874	3892	3909	3927	3945	3962	69	8388	8395	8401	8407	8414	8420	8426	8432	8439	8445
25	3979	3997	4014	4031	4048	4065	4085	4099	4116	4133	70	8451	8457	8463	8470	8476	8482	8488	8494	8500	8506
26	4150	4166	4183	4200	4216	4232	4249	4265	4281	4298	71	8513	8519	8525	8531	8537	8543	8549	8555	8561	8567
27	4314	4330	4346	4362	4378	4393	4409	4425	4440	4456	72	8573	8579	8585	8591	8597	8603	8609	8615	8621	8627
28	4472	4487	4502	4518	4533	4548	4564	4579	4594	4609	73	8633	8639	8645	8651	8657	8663	8669	8675	8681	8686
29	4624	4639	4654	4669	4683	4698	4713	4728	4742	4757	74	8692	8698	8704	8710	8716	8722	8727	8733	8739	8745
30	4771	4786	4800	4814	4829	4843	4857	4871	4886	4900	75	8751	8756	8762	8768	8774	8779	8785	8791	8797	8802
31	4914	4928	4942	4955	4969	4983	4997	5011	5024	5038	76	8808	8814	8820	8825	8831	8837	8842	8848	8854	8859
32	5051	5065	5079	5092	5105	5119	5132	5145	5159	5172	77	8865	8871	8876	8882	8887	8893	8899	8904	8910	8915
33	5185	5198	5211	5224	5237	5250	5263	5276	5289	5302	78	8921	8927	8932	8938	8943	8949	8954	8960	8965	8971
34	5315	5328	5340	5353	5366	5378	5391	5403	5416	5428	79	8976	8982	8987	8993	8998	9004	9009	9015	9020	9025
35	5441	5453	5465	5478	5490	5502	5514	5527	5539	5551	80	9031	9036	9042	9047	9053	9058	9063	9069	9074	9079
36	5563	5575	5587	5599	5611	5623	5635	5647	5658	5670	81	9085	9090	9096	9101	9106	9112	9117	9122	9128	9133
37	5682	5694	5705	5717	5729	5740	5752	5763	5775	5786	82	9138	9143	9149	9154	9159	9165	9170	9175	9180	9186
38	5798	5809	5821	5832	5843	5855	5866	5877	5888	5899	83	9191	9196	9201	9206	9212	9217	9222	9227	9232	9238
39	5911	5922	5933	5944	5955	5966	5977	5988	5999	6010	84	9243	9248	9253	9258	9263	9269	9274	9279	9284	9289
40	6021	6031	6042	6053	6064	6075	6085	6096	6107	6117	85	9294	9299	9304	9309	9315	9320	9325	9330	9335	9340
41	6128	6138	6149	6160	6170	6180	6191	6201	6212	6222	86	9345	9350	9355	9360	9365	9370	9375	9380	9385	9390
42	6232	6243	6253	6263	6274	6284	6294	6304	6314	6325	87	9395	9400	9405	9410	9415	9420	9425	9430	9435	9440
43	6335	6345	6355	6365	6375	6385	6395	6405	6415	6425	88	9445	9450	9455	9460	9465	9469	9474	9479	9484	9489
44	6435	6444	6454	6464	6474	6484	6493	6503	6513	6522	89	9494	9499	9504	9509	9513	9518	9523	9528	9533	9538
45	6532	6542	6551	6561	6571	6580	6590	6599	6609	6618	90	9542	9547	9552	9557	9562	9566	9571	9576	9581	9586
46	6628	6637	6646	6656	6665	6675	6684	6693	6702	6712	91	9590	9595	9600	9605	9609	9614	9619	9624	9628	9633
47	6721	6730	6739	6749	6758	6767	6776	6785	6794	6803	92	9638	9643	9647	9652	9657	9661	9666	9671	9675	9680
48	6812	6821	6830	6839	6848	6857	6866	6875	6884	6893	93	9685	9689	9694	9699	9703	9708	9713	9717	9722	9727
49	6902	6911	6920	6928	6937	6946	6955	6964	6972	6981	94	9731	9736	9741	9745	9750	9754	9759	9763	9768	9773
50	6990	6998	7007	7016	7024	7033	7042	7050	7059	7067	95	9777	9782	9786	9791	9795	9800	9805	9809	9814	9818
51	7076	7084	7093	7101	7110	7118	7126	7135	7143	7152	96	9823	9827	9832	9836	9841	9845	9850	9854	9859	9863
52	7160	7168	7177	7185	7193	7202	7210	7218	7226	7235	97	9868	9872	9877	9881	9886	9890	9894	9899	9903	9908
53	7243	7251	7259	7267	7275	7284	7292	7300	7308	7316	98	9912	9917	9921	9926	9930	9934	9939	9943	9948	9952
54	7324	7332	7340	7348	7356	7364	7372	7380	7388	7396	99	9956	9961	9965	9969	9974	9978	9983	9987	9991	9996
N	0	1	2	3	4	5	6	7	8	9	N	0	1	2	3	4	5	6	7	8	9

INDEX